Lectures on
Quantum Field Theory

Lectures on
Quantum Field Theory

Ashok Das
University of Rochester, USA

World Scientific

NEW JERSEY · LONDON · SINGAPORE · BEIJING · SHANGHAI · HONG KONG · TAIPEI · CHENNAI

Published by

World Scientific Publishing Co. Pte. Ltd.
5 Toh Tuck Link, Singapore 596224
USA office: 27 Warren Street, Suite 401-402, Hackensack, NJ 07601
UK office: 57 Shelton Street, Covent Garden, London WC2H 9HE

British Library Cataloguing-in-Publication Data
A catalogue record for this book is available from the British Library.

LECTURES ON QUANTUM FIELD THEORY

Copyright © 2008 by World Scientific Publishing Co. Pte. Ltd.

All rights reserved. This book, or parts thereof, may not be reproduced in any form or by any means, electronic or mechanical, including photocopying, recording or any information storage and retrieval system now known or to be invented, without written permission from the Publisher.

For photocopying of material in this volume, please pay a copying fee through the Copyright Clearance Center, Inc., 222 Rosewood Drive, Danvers, MA 01923, USA. In this case permission to photocopy is not required from the publisher.

ISBN-13 978-981-283-285-6
ISBN-10 981-283-285-8
ISBN-13 978-981-283-286-3 (pbk)
ISBN-10 981-283-286-6 (pbk)

Printed in Singapore by World Scientific Printers

To
My friends and collaborators
Josif and Susumu
and
to
Ever caring and charming
Kiron and Momo

Preface

Over the past several years I have taught a two-semester graduate course on quantum field theory at the University of Rochester. In this course the ideas of quantum field theory are developed in a traditional manner through canonical quantization. This book consists of my lectures in this course. At Rochester, we also teach a separate course on quantum field theory based on the path integral approach and my lectures in that course have already been published by World Scientific in

A. Das, *Field Theory: A Path Integral Approach* (Second Edition), World Scientific, Singapore (2006).

The material in the present book should be thought of as complementary to this earlier book. In fact, in the present lectures, there is no attempt to develop the path integral methods, rather we use the results from path integrals with a brief discussion when needed.

The topics covered in the present book contain exactly the material discussed in the two-semester course except for Chapter 10 (Dirac quantization) and Chapter 11 (Discrete symmetries) which have been added for completeness and are normally discussed in another course. Quantum field theory is a vast subject and only selected topics, which I personally feel every graduate student in the subject should know, have been covered in these lectures. Needless to say, there are many other important topics which have not been discussed because of time constraints in the course (and space constraints in the book). However, all the material covered in this book has been presented in an informal (classroom like) setting with detailed derivations which should be helpful to students.

A book of this size is bound to have many possible sources of error. However, since my lectures have already been used by various people in different universities, I have been fortunate to have their feedback which I have incorporated into the book. In addition, several other people have read all the chapters carefully and I thank them all for their comments. In particular, it is a pleasure for me to thank Ms. Judy Mack and Professor Susumu Okubo for their tireless effort in going through the entire material. I am personally grateful to Dr. John Boersma for painstakingly and meticulously checking all the mathematical derivations. Of course, any remaining errors and typos are my own.

Like the subject itself, the list of references to topics in quantum field theory is enormous and it is simply impossible to do justice to everyone who has contributed to the growth of the subject. I have in no way attempted to give an exhaustive list of references to the subject. Instead I have listed only a few suggestive references at the end of each chapter in the hope that the readers can get to the other references from these sources.

The Feynman graphs in this book were drawn using Jaxodraw while most other figures were generated using PSTricks. I am grateful to the people who developed these extremely useful softwares. Finally, I would like to thank Dave Munson for helping out with various computer related problems.

Ashok Das
Rochester

Contents

Preface		vii
1	**Relativistic equations**	**1**
1.1	Introduction	1
1.2	Notations	2
1.3	Klein-Gordon equation	10
	1.3.1 Klein paradox	14
1.4	Dirac equation	19
1.5	References	26
2	**Solutions of the Dirac equation**	**27**
2.1	Plane wave solutions	27
2.2	Normalization of the wave function	34
2.3	Spin of the Dirac particle	40
2.4	Continuity equation	44
2.5	Dirac's hole theory	47
2.6	Properties of the Dirac matrices	49
	2.6.1 Fierz rearrangement	58
2.7	References	62
3	**Properties of the Dirac equation**	**65**
3.1	Lorentz transformations	65
3.2	Covariance of the Dirac equation	72
3.3	Transformation of bilinears	82
3.4	Projection operators, completeness relation	84
3.5	Helicity	92
3.6	Massless Dirac particle	94
3.7	Chirality	99
3.8	Non-relativistic limit of the Dirac equation	105
3.9	Electron in an external magnetic field	107
3.10	Foldy-Wouthuysen transformation	111

	3.11	Zitterbewegung	117
	3.12	References	122
4	**Representations of Lorentz and Poincaré groups**		125
	4.1	Symmetry algebras	125
		4.1.1 Rotation	125
		4.1.2 Translation	129
		4.1.3 Lorentz transformation	130
		4.1.4 Poincaré transformation	133
	4.2	Representations of the Lorentz group	135
		4.2.1 Similarity transformations and representations	140
	4.3	Unitary representations of the Poincaré group	147
		4.3.1 Massive representation	151
		4.3.2 Massless representation	155
	4.4	References	160
5	**Free Klein-Gordon field theory**		161
	5.1	Introduction	161
	5.2	Lagrangian density	163
	5.3	Quantization	167
	5.4	Field decomposition	171
	5.5	Creation and annihilation operators	175
	5.6	Energy eigenstates	186
	5.7	Physical meaning of energy eigenstates	190
	5.8	Green's functions	194
	5.9	Covariant commutation relations	205
	5.10	References	209
6	**Self-interacting scalar field theory**		211
	6.1	Nöther's theorem	211
		6.1.1 Space-time translation	215
	6.2	Self-interacting ϕ^4 theory	219
	6.3	Interaction picture and time evolution operator	223
	6.4	S-matrix	229
	6.5	Normal ordered product and Wick's theorem	233
	6.6	Time ordered products and Wick's theorem	241
	6.7	Spectral representation and dispersion relation	246
	6.8	References	254

7	Complex scalar field theory		257
	7.1	Quantization	257
	7.2	Field decomposition	260
	7.3	Charge operator	263
	7.4	Green's functions	268
	7.5	Spontaneous symmetry breaking and the Goldstone theorem	270
	7.6	Electromagnetic coupling	281
	7.7	References	283
8	Dirac field theory		285
	8.1	Pauli exclusion principle	285
	8.2	Quantization of the Dirac field	286
	8.3	Field decomposition	291
	8.4	Charge operator	297
	8.5	Green's functions	300
	8.6	Covariant anti-commutation relations	303
	8.7	Normal ordered and time ordered products	305
	8.8	Massless Dirac fields	308
	8.9	Yukawa interaction	312
	8.10	Feynman diagrams	318
	8.11	References	325
9	Maxwell field theory		327
	9.1	Maxwell's equations	327
	9.2	Canonical quantization	330
	9.3	Field decomposition	335
	9.4	Photon propagator	342
	9.5	Quantum electrodynamics	347
	9.6	Physical processes	350
	9.7	Ward-Takahashi identity in QED	355
	9.8	Covariant quantization of the Maxwell theory	360
	9.9	References	376
10	Dirac method for constrained systems		379
	10.1	Constrained systems	379
	10.2	Dirac method and Dirac bracket	384
	10.3	Particle moving on a sphere	390
	10.4	Relativistic particle	395
	10.5	Dirac field theory	401
	10.6	Maxwell field theory	407

	10.7	References	413
11	Discrete symmetries		415
	11.1	Parity.	415
		11.1.1 Parity in quantum mechanics	417
		11.1.2 Spin zero field	424
		11.1.3 Photon field	428
		11.1.4 Dirac field	429
	11.2	Charge conjugation	436
		11.2.1 Spin zero field	437
		11.2.2 Dirac field	441
		11.2.3 Majorana fermions	449
		11.2.4 Eigenstates of charge conjugation	453
	11.3	Time reversal	458
		11.3.1 Spin zero field and Maxwell's theory	464
		11.3.2 Dirac fields	467
		11.3.3 Consequences of $\mathcal{T}$ invariance	473
		11.3.4 Electric dipole moment of neutron	477
	11.4	$\mathcal{CPT}$ theorem	479
		11.4.1 Equality of mass for particles and antiparticles	479
		11.4.2 Electric charge for particles and antiparticles .	480
		11.4.3 Equality of lifetimes for particles and antiparticles	480
	11.5	References	482
12	Yang-Mills theory		485
	12.1	Non-Abelian gauge theories	485
	12.2	Canonical quantization of Yang-Mills theory	502
	12.3	Path integral quantization of gauge theories	512
	12.4	Path integral quantization of tensor fields	530
	12.5	References	542
13	BRST invariance and its consequences		545
	13.1	BRST symmetry	545
	13.2	Covariant quantization of Yang-Mills theory	550
	13.3	Unitarity	561
	13.4	Slavnov-Taylor identity	565
	13.5	Feynman rules	571
	13.6	Ghost free gauges	578
	13.7	References	581

14	Higgs phenomenon and the standard model	583
	14.1 Stückelberg formalism	583
	14.2 Higgs phenomenon	589
	14.3 The standard model	596
	14.3.1 Field content	599
	14.3.2 Lagrangian density	601
	14.3.3 Spontaneous symmetry breaking	605
	14.4 References	616
15	Regularization of Feynman diagrams	619
	15.1 Introduction	619
	15.2 Loop expansion	621
	15.3 Cut-off regularization	623
	15.3.1 Calculation in the Yukawa theory	631
	15.4 Pauli-Villars regularization	638
	15.5 Dimensional regularization	647
	15.5.1 Calculations in QED	656
	15.6 References	666
16	Renormalization theory	669
	16.1 Superficial degree of divergence	669
	16.2 A brief history of renormalization	679
	16.3 Schwinger-Dyson equation	690
	16.4 BPHZ renormalization	692
	16.5 Renormalization of gauge theories	721
	16.6 Anomalous Ward identity	724
	16.7 References	732
17	Renormalization group and equation	733
	17.1 Gell-Mann-Low equation	733
	17.2 Renormalization group	739
	17.3 Renormalization group equation	744
	17.4 Solving the renormalization group equation	748
	17.5 Callan-Symanzik equation	759
	17.6 References	766
Index		769

CHAPTER 1
Relativistic equations

1.1 Introduction

As we know, in single particle, non-relativistic quantum mechanics, we start with the Hamiltonian description of the corresponding classical, non-relativistic physical system and promote each of the observables to a Hermitian operator. The time evolution of the quantum mechanical system, in this case, is given by the time dependent Schrödinger equation which has the form

$$i\hbar \frac{\partial \psi}{\partial t} = H\psi. \tag{1.1}$$

Here $\psi(\mathbf{x}, t)$ represents the wave function of the system which corresponds to the probability amplitude for finding the particle at the coordinate $\mathbf{x}$ at a given time t and the Hamiltonian, H, has the generic form

$$H = \frac{\mathbf{p}^2}{2m} + V(\mathbf{x}), \tag{1.2}$$

with $\mathbf{p}$ denoting the momentum of the particle and $V(\mathbf{x})$ representing the potential through which the particle moves. (Throughout the book we will use a bold symbol to represent a three dimensional quantity.)

This formalism is clearly non-relativistic (non-covariant) which can be easily seen by noting that, even for a free particle, the dynamical equation (1.1) takes the form

$$i\hbar\frac{\partial\psi}{\partial t}=\frac{\mathbf{p}^2}{2m}\psi. \tag{1.3}$$

In the coordinate basis, the momentum operator has the form

$$\mathbf{p}\to-i\hbar\boldsymbol{\nabla}, \tag{1.4}$$

so that the time dependent Schrödinger equation, in this case, takes the form

$$i\hbar\frac{\partial\psi}{\partial t}=-\frac{\hbar^2}{2m}\boldsymbol{\nabla}^2\psi. \tag{1.5}$$

This equation is linear in the time derivative while it is quadratic in the space derivatives. Therefore, space and time are not treated on an equal footing in this case and, consequently, the equation cannot have the same form (covariant) in different Lorentz frames. A relativistic equation, on the other hand, must treat space and time coordinates on an equal footing and remain form invariant in all inertial frames (Lorentz frames). Let us also recall that, even for a simple fundamental system such as the Hydrogen atom, the ground state electron is fairly relativistic ($\frac{v}{c}$, for the ground state electron is of the order of the fine structure constant). Consequently, there is a need to generalize the quantum mechanical description to relativistic systems. In this chapter, we will study how we can systematically develop a quantum mechanical description of a single relativistic particle and the difficulties associated with such a description.

1.2 Notations

Before proceeding any further, let us fix our notations. We note that in the three dimensional Euclidean space, which we are all familiar with, a vector is labelled uniquely by its three components. (We denote three dimensional vectors in boldface.) Thus,

$$\begin{aligned}\mathbf{x}&=(x_1,x_2,x_3),\\ \mathbf{J}&=(J_1,J_2,J_3),\\ \mathbf{A}&=(A_1,A_2,A_3),\end{aligned} \tag{1.6}$$

where **x** and **J** represent respectively the position and the angular momentum vectors while **A** stands for any arbitrary vector. In such a space, as we know, the scalar product of any two arbitrary vectors is defined to be

$$\mathbf{A} \cdot \mathbf{B} = A_i B_i = \delta_{ij} A_i B_j = \delta^{ij} A_i B_j, \tag{1.7}$$

where repeated indices are assumed to be summed. The scalar product of two vectors is invariant under rotations of the three dimensional space which is the maximal symmetry group of the Euclidean space that leaves the origin invariant. This also allows us to define the length of a vector simply as

$$\mathbf{A}^2 = \mathbf{A} \cdot \mathbf{A} = A_i A_i = \delta_{ij} A_i A_j = \delta^{ij} A_i A_j. \tag{1.8}$$

The Kronecker delta, δ_{ij}, in this case, represents the metric of the Euclidean space and is trivial (in the sense that all the nonzero components are simply unity). Consequently, it does not matter whether we write the indices "up" or "down". Let us note from the definition of the length of a vector in Euclidean space that, for any vector, it is necessarily positive definite, namely,

$$\begin{aligned}\mathbf{A}^2 &\geq 0, \\ \mathbf{A}^2 &= 0, \quad \text{if and only if } \mathbf{A} = 0.\end{aligned} \tag{1.9}$$

When we treat space and time on an equal footing and enlarge our three dimensional Euclidean manifold to the four dimensional space-time manifold, we can again define vectors in this manifold. However, these would now consist of four components. Namely, any point in this manifold will be specified uniquely by four coordinates and, consequently, any vector would also have four components. However, unlike the case of the Euclidean space, there are now two distinct four vectors that we can define on this manifold, namely, ($\mu = 0, 1, 2, 3$ and we are being a little sloppy in representing the four vector by what may seem like its component)

$$\begin{aligned} x^\mu &= (ct, \mathbf{x}), \\ x_\mu &= (ct, -\mathbf{x}). \end{aligned} \quad (1.10)$$

Here c represents the speed of light (necessary to give the same dimension to all the components) and we note that the two four vectors simply represent the two distinct possible ways space and time components can be embedded into the four vector. On a more fundamental level, the two four vectors have distinct transformation properties under Lorentz transformations (in fact, one transforms inversely with respect to the other) and are known respectively as contravariant and covariant vectors.

The contravariant and the covariant vectors are related to each other through the metric tensor of the four dimensional manifold, commonly known as the Minkowski space, namely,

$$\begin{aligned} x_\mu &= \eta_{\mu\nu} x^\nu, \\ x^\mu &= \eta^{\mu\nu} x_\nu. \end{aligned} \quad (1.11)$$

From the forms of the contravariant and the covariant vectors in (1.10) as well as using (1.11), we can immediately read out the metric tensors for the four dimensional Minkowski space which are diagonal with the signature $(+,-,-,-)$. Namely, we can write them in the matrix form as

$$\eta^{\mu\nu} = \begin{pmatrix} 1 & 0 & 0 & 0 \\ 0 & -1 & 0 & 0 \\ 0 & 0 & -1 & 0 \\ 0 & 0 & 0 & -1 \end{pmatrix}, \quad (1.12)$$

$$\eta_{\mu\nu} = \begin{pmatrix} 1 & 0 & 0 & 0 \\ 0 & -1 & 0 & 0 \\ 0 & 0 & -1 & 0 \\ 0 & 0 & 0 & -1 \end{pmatrix}. \quad (1.13)$$

The contravariant metric tensor, $\eta^{\mu\nu}$, and the covariant metric tensor, $\eta_{\mu\nu}$, are inverses of each other, since they satisfy

$$\eta^{\mu\lambda}\eta_{\lambda\nu} = \delta^{\mu}{}_{\nu}. \tag{1.14}$$

Furthermore, each is symmetric as they are expected to be, namely,

$$\eta^{\mu\nu} = \eta^{\nu\mu}, \qquad \eta_{\mu\nu} = \eta_{\nu\mu}. \tag{1.15}$$

This particular choice of the metric is conventionally known as the Bjorken-Drell metric and this is what we will be using throughout these lectures. Different authors, however, use different metric conventions and you should be careful in reading the literature. (As is clear from the above discussion, the nonuniqueness in the choice of the metric tensors reflects the nonuniqueness of the embedding of space and time components into a four vector. Physical results, however, are independent of the choice of a metric.)

Given two arbitrary four vectors

$$\begin{aligned} A^{\mu} &= (A^0, \mathbf{A}), \\ B^{\mu} &= (B^0, \mathbf{B}), \end{aligned} \tag{1.16}$$

we can define an invariant scalar product of the two vectors as

$$\begin{aligned} A \cdot B &= A^{\mu}B_{\mu} = A_{\mu}B^{\mu} \\ &= \eta^{\mu\nu}A_{\mu}B_{\nu} = \eta_{\mu\nu}A^{\mu}B^{\nu} \\ &= A^0 B^0 - \mathbf{A} \cdot \mathbf{B}. \end{aligned} \tag{1.17}$$

Since the contravariant and the covariant vectors transform in an inverse manner, such a product is easily seen to be invariant under Lorentz transformations. This is the generalization of the scalar product of the three dimensional Euclidean space to the four dimensional Minkowski space and is invariant under Lorentz transformations which are the analogs of rotations in Minkowski space. In fact, any product of Lorentz tensors defines a scalar if all the Lorentz indices are contracted, namely, if there is no free Lorentz index. (Two Lorentz indices are said to be contracted if a contravariant and a covariant index are summed over all possible values.)

Given this, we note that the length of a (four) vector in Minkowski space can be determined to have the form

$$A^2 = A \cdot A = \eta^{\mu\nu} A_\mu A_\nu = \eta_{\mu\nu} A^\mu A^\nu = (A^0)^2 - \mathbf{A}^2. \tag{1.18}$$

Unlike the Euclidean space, however, here we see that the length of a vector need not always be positive semi-definite. In fact, if we look at the Minkowski space itself, we find that

$$x^2 = x^\mu x_\mu = \eta_{\mu\nu} x^\mu x^\nu = c^2 t^2 - \mathbf{x}^2. \tag{1.19}$$

This is the invariant length (of any point from the origin) in this space. The invariant length between two points infinitesimally close to each other follows from this to be

$$\mathrm{d}s^2 = c^2 \mathrm{d}\tau^2 = \eta_{\mu\nu} \mathrm{d}x^\mu \mathrm{d}x^\nu, \tag{1.20}$$

where τ is known as the proper time.

For coordinates which satisfy (see (1.19), we will set $c = 1$ from now on for simplicity)

$$x^2 = t^2 - \mathbf{x}^2 > 0, \tag{1.21}$$

we say that the region of space-time is time-like for obvious reasons. On the other hand, for coordinates which satisfy

$$x^2 = t^2 - \mathbf{x}^2 < 0, \tag{1.22}$$

the region of space-time is known as space-like. The boundary of the two regions, namely, the region for which

$$x^2 = t^2 - \mathbf{x}^2 = 0, \tag{1.23}$$

defines trajectories for light-like particles and is, consequently, known as the light-like region. (Light-like vectors, for which the invariant length vanishes, are nontrivial unlike the case of the Euclidean space.)

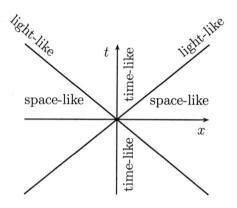

Figure 1.1: Different invariant regions of Minkowski space.

Thus, we see that, unlike the Euclidean space, the Minkowski space-time manifold separates into four invariant wedges (regions which do not mix under Lorentz transformations), which in a two dimensional projection has the form shown in Fig. 1.1. The different invariant wedges are known as

$$
\begin{aligned}
&t > 0, \quad x^2 \geq 0: \quad \text{future light cone}, \\
&t < 0, \quad x^2 \geq 0: \quad \text{past light cone}, \\
&x^2 < 0: \quad \text{space} - \text{like}.
\end{aligned}
\tag{1.24}
$$

All physical processes are assumed to take place in the future light cone or the forward light cone defined by

$$
t > 0 \quad \text{and} \quad x^2 \geq 0. \tag{1.25}
$$

Given the contravariant and the covariant coordinates, we can define the contragradient and the cogradient respectively as ($c = 1$)

$$\partial^\mu = \frac{\partial}{\partial x_\mu} = \left(\frac{\partial}{\partial t}, -\boldsymbol{\nabla}\right),$$

$$\partial_\mu = \frac{\partial}{\partial x^\mu} = \left(\frac{\partial}{\partial t}, \boldsymbol{\nabla}\right). \tag{1.26}$$

From these, we can construct the Lorentz invariant quadratic operator

$$\Box = \partial^2 = \partial^\mu \partial_\mu = \frac{\partial^2}{\partial t^2} - \boldsymbol{\nabla}^2, \tag{1.27}$$

which is known as the D'Alembertian. It is the generalization of the Laplacian to the four dimensional Minkowski space.

Let us note next that energy and momentum also define four vectors in this case. (Namely, they transform like four vectors under Lorentz transformations.) Thus, we can write (remember that $c = 1$, otherwise, we have to write $\frac{E}{c}$)

$$p^\mu = (E, \mathbf{p}),$$

$$p_\mu = (E, -\mathbf{p}). \tag{1.28}$$

Given the energy-momentum four vectors, we can construct the Lorentz scalar

$$p^2 = p^\mu p_\mu = E^2 - \mathbf{p}^2. \tag{1.29}$$

The Einstein relation for a free particle (remember $c = 1$)

$$E^2 = \mathbf{p}^2 + m^2, \tag{1.30}$$

where m represents the rest mass of the particle, can now be seen as the Lorentz invariant condition

$$p^2 = E^2 - \mathbf{p}^2 = m^2. \tag{1.31}$$

1.2 NOTATIONS

In other words, in this space, the energy and the momentum of a free particle must lie on a hyperbola satisfying the above relation.

We already know that the coordinate representation of the energy and the momentum operators takes the forms

$$\begin{aligned} E &\rightarrow i\hbar\frac{\partial}{\partial t}, \\ \mathbf{p} &\rightarrow -i\hbar\boldsymbol{\nabla}. \end{aligned} \tag{1.32}$$

We can combine these to write the coordinate representation for the energy-momentum four vector operator as

$$\begin{aligned} p^\mu &= i\hbar\partial^\mu = i\hbar\frac{\partial}{\partial x_\mu} = \left(i\hbar\frac{\partial}{\partial t}, -i\hbar\boldsymbol{\nabla}\right), \\ p_\mu &= i\hbar\partial_\mu = i\hbar\frac{\partial}{\partial x^\mu} = \left(i\hbar\frac{\partial}{\partial t}, i\hbar\boldsymbol{\nabla}\right). \end{aligned} \tag{1.33}$$

Finally, let us note that in the four dimensional space-time, we can construct two totally antisymmetric fourth rank tensors $\epsilon^{\mu\nu\lambda\rho}, \epsilon_{\mu\nu\lambda\rho}$, the four dimensional contravariant and covariant Levi-Civita tensors respectively. We will choose the normalization $\epsilon^{0123} = 1 = -\epsilon_{0123}$ so that

$$\epsilon^{0ijk} = \epsilon_{ijk} = -\epsilon_{0ijk}, \tag{1.34}$$

where ϵ_{ijk} denotes the three dimensional Levi-Civita tensor with $\epsilon_{123} = 1$. An anti-symmetric tensor such as $\epsilon^i{}_{jk}$ is then understood to denote

$$\epsilon^i{}_{jk} = \eta^{i\ell}\epsilon_{\ell jk}, \tag{1.35}$$

and so on. This completes the review of all the essential basic notation that we will be using in this book. We will introduce new notations as they arise in the context of our discussions.

1.3 Klein-Gordon equation

With all these basics, we are now ready to write down the simplest of the relativistic equations. Note that in the case of a non-relativistic particle, we start with the non-relativistic energy-momentum relation

$$E = \frac{\mathbf{p}^2}{2m} + V(\mathbf{x}), \tag{1.36}$$

and promote the dynamical variables (observables) to Hermitian operators to obtain the time-dependent Schrödinger equation (see (1.1))

$$i\hbar \frac{\partial \psi}{\partial t} = \left(-\frac{\hbar^2}{2m} \boldsymbol{\nabla}^2 + V(\mathbf{x}) \right) \psi. \tag{1.37}$$

Let us consider the simplest of relativistic systems, namely, a relativistic free particle of mass m. In this case, we have seen that the energy-momentum relation is none other than the Einstein relation (1.30), namely,

$$E^2 = \mathbf{p}^2 + m^2,$$
$$\text{or,} \quad E^2 - \mathbf{p}^2 = p^\mu p_\mu = m^2. \tag{1.38}$$

Thus, as before, promoting these to operators, we obtain the simplest relativistic quantum mechanical equation to be (see (1.33))

$$p^\mu p_\mu \phi = m^2 \phi,$$
$$\text{or,} \quad (i\hbar \partial^\mu)(i\hbar \partial_\mu) \phi = m^2 \phi,$$
$$\text{or,} \quad -\hbar^2 \Box \phi = m^2 \phi. \tag{1.39}$$

Setting $\hbar = 1$ from now on for simplicity, the equation above takes the form

$$(\Box + m^2)\phi = 0. \tag{1.40}$$

1.3 KLEIN-GORDON EQUATION

Since the operator in the parenthesis is a Lorentz scalar and since we assume the quantum mechanical wave function, $\phi(\mathbf{x}, t)$, to be a scalar function, this equation is invariant under Lorentz transformations.

This equation, (1.40), is known as the Klein-Gordon equation and, for $m = 0$, or when the rest mass vanishes, it reduces to the wave equation (recall Maxwell's equations). Like the wave equation, the Klein-Gordon equation also has plane wave solutions which are characteristic of free particle solutions. In fact, the functions

$$e^{\mp ik \cdot x} = e^{\mp ik_\mu x^\mu} = e^{\mp ik^\mu x_\mu} = e^{\mp i(k_0 t - \mathbf{k} \cdot \mathbf{x})}, \tag{1.41}$$

with $k^\mu = (k^0, \mathbf{k})$ are eigenfunctions of the energy-momentum operator, namely, using (1.33) (remember that $\hbar = 1$) we obtain

$$p^\mu e^{\mp ik \cdot x} = i\partial^\mu e^{\mp ik \cdot x} = i\frac{\partial}{\partial x_\mu} e^{\mp ik \cdot x} = \pm k^\mu e^{\mp ik \cdot x}, \tag{1.42}$$

so that $\pm k^\mu$ are the eigenvalues of the energy-momentum operator. (In fact, the eigenvalues should be $\pm \hbar k^\mu$, but we have set $\hbar = 1$.) This shows that the plane waves define a solution of the Klein-Gordon equation provided

$$k^2 - m^2 = (k^0)^2 - \mathbf{k}^2 - m^2 = 0,$$
$$\text{or,} \quad k^0 = \pm E = \pm \sqrt{\mathbf{k}^2 + m^2}. \tag{1.43}$$

Thus, we see the first peculiarity of the Klein-Gordon equation (which is a relativistic equation), namely, that it allows for both positive and negative energy solutions. This basically arises from the fact that, for a relativistic particle (even a free one), the energy-momentum relation is given by the Einstein relation which is a quadratic relation in E, as opposed to the case of a non-relativistic particle, where the energy-momentum relation is linear in E. If we accept the Klein-Gordon equation as describing a free, relativistic, quantum mechanical particle of mass m, then, we will see shortly that the presence of the negative energy solutions would render the theory inconsistent.

To proceed further, let us note that the Klein-Gordon equation and its complex conjugate (remember that a quantum mechanical wave function is, in general, complex), namely,

$$(\Box + m^2)\phi = 0,$$
$$(\Box + m^2)\phi^* = 0, \tag{1.44}$$

would imply

$$\phi^*\Box\phi - \phi\Box\phi^* = 0,$$
or, $\quad \partial_\mu (\phi^* \partial^\mu \phi - \phi \partial^\mu \phi^*) = 0,$

or, $\quad \dfrac{\partial}{\partial t}\left(\phi^* \dfrac{\partial \phi}{\partial t} - \phi \dfrac{\partial \phi^*}{\partial t}\right) - \boldsymbol{\nabla} \cdot (\phi^* \boldsymbol{\nabla}\phi - \phi \boldsymbol{\nabla}\phi^*) = 0. \tag{1.45}$

Defining the probability current density four vector as

$$J^\mu = (j^0, \mathbf{J}) = (\rho, \mathbf{J}), \tag{1.46}$$

where

$$\begin{aligned} \mathbf{J} &= \dfrac{1}{2im}(\phi^*\boldsymbol{\nabla}\phi - \phi\boldsymbol{\nabla}\phi^*), \\ \rho &= \dfrac{i}{2m}\left(\phi^*\dfrac{\partial \phi}{\partial t} - \phi\dfrac{\partial \phi^*}{\partial t}\right), \end{aligned} \tag{1.47}$$

we note that equation (1.45) can be written as a continuity equation for the probability current, namely,

$$\partial_\mu J^\mu = \dfrac{\partial \rho}{\partial t} + \boldsymbol{\nabla} \cdot \mathbf{J} = 0. \tag{1.48}$$

The probability current density,

$$\mathbf{J} = \dfrac{1}{2im}(\phi^*\boldsymbol{\nabla}\phi - \phi\boldsymbol{\nabla}\phi^*), \tag{1.49}$$

1.3 KLEIN-GORDON EQUATION

of course, has the same form as in non-relativistic quantum mechanics. However, we note that the form of the probability density (which results from the requirement of covariance)

$$\rho = \frac{i}{2m}\left(\phi^*\frac{\partial \phi}{\partial t} - \phi\frac{\partial \phi^*}{\partial t}\right), \tag{1.50}$$

is quite different from that in non-relativistic quantum mechanics and it is here that the problem of the negative energy states shows up. For example, even for the simplest of solutions, namely, plane waves of the form

$$\phi(x) = e^{-ik\cdot x}, \tag{1.51}$$

we obtain

$$\rho = \frac{i}{2m}(-ik^0 - ik^0) = \frac{k^0}{m} = \pm\frac{E}{m}. \tag{1.52}$$

Since energy can take both positive and negative values, it follows that ρ cannot truly represent the probability density which, by definition, has to be positive semi-definite. It is worth noting here that this problem really arises because the Klein-Gordon equation, unlike the time dependent Schrödinger equation, is second order in time derivatives. This has the consequence that the probability density involves a first order time derivative and that is how the problem of the negative energy states enters. (Note that if the equation is second order in the space derivatives, then covariance would require that it be second order in time derivative as well. This would, in turn, lead to the difficulty with the probability density being positive semi-definite.) One can, of course, ask whether we can restrict ourselves to positive energy solutions only in order to avoid the difficulty with the interpretation of ρ. Classically, we can do this. However, quantum mechanically, we cannot arbitrarily impose this for a variety of reasons. The simplest way to see this is to note that the positive energy solutions alone do not define a complete set of states (basis) in the Hilbert space and, consequently, even if we restrict the states to be of positive energy to begin with, negative energy states may be generated through quantum mechanical corrections. It is for these reasons that the Klein-Gordon equation was abandoned as a quantum mechanical equation for a single relativistic particle. However, as we will see later, this equation is quite meaningful as a relativistic field equation.

1.3.1 Klein paradox. Let us consider a charged scalar particle described by the Klein-Gordon equation (1.40) in an external electromagnetic field. We recall that the coupling of a charged particle to an electromagnetic field is given by the minimal coupling

$$p_\mu \to p_\mu - eA_\mu,$$
$$\text{or,} \quad \partial_\mu \to \partial_\mu + ieA_\mu, \tag{1.53}$$

where we have used the coordinate representation for the momentum as in (1.33) and A_μ denotes the vector potential associated with the electromagnetic field. In this case, therefore, the scalar particle will satisfy the minimally coupled Klein-Gordon equation

$$\left((\partial_\mu + ieA_\mu)(\partial^\mu + ieA^\mu) + m^2\right)\phi(x) = 0. \tag{1.54}$$

As a result, the probability current density in (1.46) can be determined to have the form

$$J^\mu = \frac{i}{2m}\left(\phi^*(x)\overleftrightarrow{\partial^\mu}\phi(x) + 2ieA^\mu\phi^*(x)\phi(x)\right), \tag{1.55}$$

where we have defined

$$A\overleftrightarrow{\partial^\mu}B = A(\partial^\mu B) - (\partial^\mu A)B. \tag{1.56}$$

With this general description, let us consider the scattering of a charged scalar (Klein-Gordon) particle with positive energy from a constant electrostatic potential. In this case, therefore, we have

$$\mathbf{A} = 0, \quad A^0 = \Phi = \text{constant}. \tag{1.57}$$

1.3 KLEIN-GORDON EQUATION

For simplicity, let us assume the constant electrostatic potential to be of the form

$$\Phi(z) = \begin{cases} 0, & z < 0, \\ \Phi_0, & z > 0, \end{cases} \tag{1.58}$$

and we assume that the particle is incident on the potential along the z-axis as shown in Fig. 1.2.

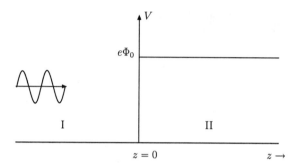

Figure 1.2: Klein-Gordon particle scattering from a constant electrostatic potential.

The dynamical equations will now be different in the two regions, $z < 0$ (region I) and $z > 0$ (region II), and have the forms

$$\left(\Box + m^2\right) \phi_\mathrm{I} = 0, \qquad z < 0,$$

$$\left(\Box + m^2 + 2ie\Phi_0 \frac{\partial}{\partial t} - e^2\Phi_0^2\right) \phi_\mathrm{II} = 0, \qquad z > 0. \tag{1.59}$$

In region I, there will be an incident as well as a reflected wave so that we can write

$$\phi_\mathrm{I}(t,z) = e^{-iEt}\left(e^{ipz} + A\,e^{-ipz}\right), \quad z < 0, \tag{1.60}$$

while in region II, we only expect a transmitted wave of the form

$$\phi_\mathrm{II}(t,z) = B\,e^{-iEt+ip'z}, \quad z > 0, \tag{1.61}$$

where A, B are related respectively to reflection and transmission coefficients. We note here that the continuity of the wave function at the boundary $z = 0$ requires that the energy be the same in the two regions.

For the wave functions in (1.60) and (1.61) to satisfy the respective equations in (1.59), we must have

$$\begin{aligned} E &= \sqrt{\mathbf{p}^2 + m^2}, \\ p' &= \pm\sqrt{(E - e\Phi_0)^2 - m^2} \\ &= \pm\sqrt{(E - e\Phi_0 + m)(E - e\Phi_0 - m)}. \end{aligned} \quad (1.62)$$

Here we have used the fact that the energy of the incident particle is positive and, therefore, the square root in the first equation in (1.62) is with a positive sign. However, the sign of the square root in the second relation remains to be fixed.

Let us note from the second relation in (1.62) that p' is real for both $E - e\Phi_0 > m$ (weak potential) and for $E - e\Phi_0 < -m$ (strong potential). However, for a potential of intermediate strength satisfying $-m < E - e\Phi_0 < m$, we note that p' is purely imaginary. Thus, the behavior of the transmitted wave depends on the strength of the potential. As a result, in this second case, we must have

$$p' = i|p'|, \quad -m < E - e\Phi_0 < m, \quad (1.63)$$

in order that the wave function is damped in region II. To determine the sign of the square root in the cases when p' is real, let us note from the second relation in (1.62) that the group velocity of the transmitted wave is given by

$$v_{\text{group}} = \frac{\partial E}{\partial p'} = \frac{p'}{E - e\Phi_0}. \quad (1.64)$$

Since we expect the transmitted wave to be travelling to the right, we determine from (1.64) that

$$\begin{aligned} p' &> 0, \quad \text{for } E - e\Phi_0 > 0, \\ p' &< 0 \quad \text{for } E - e\Phi_0 < 0. \end{aligned} \quad (1.65)$$

1.3 KLEIN-GORDON EQUATION

This, therefore, fixes the sign of the square root in the second relation in (1.62) for various cases.

Matching the wave functions in (1.60) and (1.61) and their first derivatives at the boundary $z = 0$, we determine

$$1 + A = B, \qquad 1 - A = \frac{p'}{p} B, \tag{1.66}$$

so that we determine

$$A = \frac{p - p'}{p + p'}, \qquad B = \frac{2p}{p + p'}. \tag{1.67}$$

Let us next determine the probability current densities associated with the different beams. From (1.55) as well as the form of the potential in (1.58) we obtain

$$\begin{aligned}
J_{\text{inc}} &= \hat{\mathbf{z}} \cdot \mathbf{J}_{\text{inc}} = \frac{p}{m}, \\
J_{\text{refl}} &= -\hat{\mathbf{z}} \cdot \mathbf{J}_{\text{refl}} = \frac{p}{m} \frac{(p-p')(p-(p')^*)}{(p+p')(p+(p')^*)}, \\
J_{\text{trans}} &= \hat{\mathbf{z}} \cdot \mathbf{J}_{\text{trans}} = \frac{(p'+(p')^*)}{2m} \frac{4p^2}{(p+p')(p+(p')^*)},
\end{aligned} \tag{1.68}$$

where we have used (1.67) as well as the fact that, while p is real and positive, p' can be positive or negative or even imaginary depending on the strength of the potential (see (1.63) and (1.65)). We can now determine the reflection and the transmission coefficients simply as

$$\begin{aligned}
R &= \frac{J_{\text{refl}}}{J_{\text{inc}}} = \frac{(p-p')(p-(p')^*)}{(p+p')(p+(p')^*)}, \\
T &= \frac{J_{\text{trans}}}{J_{\text{inc}}} = \frac{2p(p'+(p')^*)}{(p+p')(p+(p')^*)}.
\end{aligned} \tag{1.69}$$

We see from the reflection and the transmission coefficients that

$$R + T = \frac{(p-p')(p-(p')^*) + 2p(p'+(p')^*)}{(p+p')(p+(p')^*)} = 1, \tag{1.70}$$

so that the reflection and the transmission coefficients satisfy unitarity for all strengths of the potential.

However, let us now analyze the different cases of the potential strengths individually. First, for the case, $E - e\Phi_0 > m$ (weak potential), we see that p' is real and positive and we have

$$R = \left(\frac{p-p'}{p+p'}\right)^2 < 1, \quad T = \frac{4pp'}{(p+p')^2} > 0, \quad R + T = 1, \quad (1.71)$$

which corresponds to the normal scenario in scattering. For the case of an intermediate potential strength, $-m < E - e\Phi_0 < m$, we note from (1.63) that p' is purely imaginary in this case. As a result, it follows from (1.69) that

$$R = \frac{(p-p')(p+p')}{(p+p')(p-p')} = 1, \quad T = 0, \quad R + T = 1, \quad (1.72)$$

so that the incident beam is totally reflected and there is no transmission in this case. The third case of the strong potential, $E - e\Phi_0 < -m$, is the most interesting. In this case, we note from (1.65) that p' is real, but negative. As a result, from (1.69) we have

$$R = \left(\frac{p+|p'|}{p-|p'|}\right)^2 > 1, \quad T = -\frac{4p|p'|}{(p-|p'|)^2} < 0, \quad R + T = 1.$$
$$(1.73)$$

Namely, even though unitarity is not violated, in this case the transmission coefficient is negative and the reflection coefficient exceeds unity. This is known as the Klein paradox and it contradicts our intuition from the one particle scatterings studied in non-relativistic quantum mechanics. On the other hand, if we go beyond the one particle description and assume that a sufficiently strong enough electrostatic potential can produce particle-antiparticle pairs, there is no paradox. For example, the antiparticles are attracted by the barrier leading to a negative charged current moving to the right which explains the negative transmission coefficient. On the other hand, the particles are reflected from the barrier and add to the totally reflected incident particles (which is already seen for intermediate strength potentials) to give a reflection coefficient that exceeds unity.

1.4 Dirac equation

As we have seen, relativistic equations seem to imply the presence of both positive as well as negative energy solutions and that quantum mechanically, we need both these solutions to describe a physical system. Furthermore, as we have seen, the Klein-Gordon equation is second order in the time derivatives and this leads to the definition of the probability density which is first order in the time derivative. Together with the negative energy solutions, this implies that the probability density can become negative which is inconsistent with the definition of a probability density. It is clear, therefore, that even if we cannot avoid the negative energy solutions, we can still possibly obtain a consistent probability density provided we have a relativistic equation which is first order in the time derivative just like the time dependent Schrödinger equation. The difference, of course, is that Lorentz invariance would require space and time to be treated on an equal footing and, therefore, such an equation, if we can find it, must be first order in both space and time derivatives. Clearly, this can be done provided we have a linear relation between energy and momentum operators. Let us recall that the Einstein relation gives

$$E^2 = \mathbf{p}^2 + m^2. \tag{1.74}$$

The positive square root of this gives

$$E = \sqrt{\mathbf{p}^2 + m^2}, \tag{1.75}$$

which is far from a linear relation.

Although the naive square root of the Einstein relation does not lead to a linear relation between the energy and the momentum variables, a matrix square root may, in fact, lead to such a relation. This is exactly what Dirac proposed. Let us, for example, write the Einstein relation as

$$E^2 - \mathbf{p}^2 = m^2,$$
$$\text{or,} \quad p^2 = p^\mu p_\mu = m^2. \tag{1.76}$$

Let us consider this as a matrix relation (namely, an $n \times n$ identity matrix multiplying both sides). Let us further assume that there exist four linearly independent $n \times n$ matrices γ^μ, $\mu = 0,1,2,3$, which are space-time independent such that

$$\not{p} = \gamma^\mu p_\mu, \tag{1.77}$$

represents the matrix square root of p^2. If this is true, then, by definition, we have

$$\not{p}\not{p} = p^2 \mathbb{1},$$
$$\text{or,} \quad \gamma^\mu p_\mu \gamma^\nu p_\nu = p^2 \mathbb{1},$$
$$\text{or,} \quad \gamma^\mu \gamma^\nu p_\mu p_\nu = p^2 \mathbb{1},$$
$$\text{or,} \quad \frac{1}{2}(\gamma^\mu \gamma^\nu + \gamma^\nu \gamma^\mu) p_\mu p_\nu = p^2 \mathbb{1}. \tag{1.78}$$

Here $\mathbb{1}$ denotes the identity matrix (in the appropriate matrix space, in this case, n dimensional) and we have used the fact that the matrices, γ^μ, are constant to move them past the momentum operators. For the relation (1.78) to be true, it is clear that the matrices, γ^μ, have to satisfy the algebra ($\mu = 0,1,2,3$)

$$\gamma^\mu \gamma^\nu + \gamma^\nu \gamma^\mu = \left[\gamma^\mu, \gamma^\nu\right]_+ = 2\eta^{\mu\nu} \mathbb{1}. \tag{1.79}$$

Here the brackets with a subscript "+" stand for the anti-commutator of two quantities defined in (1.79) (sometimes it is also denoted by curly brackets which we will not use to avoid confusion with Poisson brackets) and this algebra is known as the Clifford algebra. We see that if we can find a set of four linearly independent constant matrices satisfying the Clifford algebra, then, we can obtain a matrix square root of p^2 which would be linear in energy and momentum.

Before going into an actual determination of such matrices, let us look at the consequences of such a possibility. In this case, the solutions of the equation (sign of the mass term is irrelevant and the mass term is multiplied by the identity matrix which we do not write explicitly)

1.4 DIRAC EQUATION

$$\not{p}\psi = m\psi, \tag{1.80}$$

would automatically satisfy the Einstein relation. Namely,

$$\not{p}(\not{p}\psi) = m\not{p}\psi,$$
$$\text{or,} \quad p^2\psi = m^2\psi. \tag{1.81}$$

Furthermore, since the new equation is linear in the energy and momentum variables, it will, consequently, be linear in the space and time derivatives. This is, of course, what we would like for a consistent definition of the probability density. The equation (1.80) (or its coordinate representation) is known as the Dirac equation.

To determine the matrices, γ^μ, and their dimensionality, let us note that the Clifford algebra in (1.79)

$$[\gamma^\mu, \gamma^\nu]_+ = 2\eta^{\mu\nu}\mathbb{1}, \qquad \mu = 0, 1, 2, 3, \tag{1.82}$$

can be written out explicitly as

$$(\gamma^0)^2 = \mathbb{1},$$
$$(\gamma^i)^2 = -\mathbb{1}, \qquad \text{for any fixed } i = 1, 2, 3,$$
$$\gamma^0\gamma^i + \gamma^i\gamma^0 = 0,$$
$$\gamma^i\gamma^j + \gamma^j\gamma^i = 0, \qquad i \neq j. \tag{1.83}$$

We can choose any one of the matrices to be diagonal and without loss of generality, let us choose

$$\gamma^0 = \begin{pmatrix} b_1 & 0 & \cdots & 0 \\ 0 & b_2 & \cdots & 0 \\ \vdots & & \ddots & \vdots \\ 0 & 0 & \cdots & b_n \end{pmatrix}. \tag{1.84}$$

From the fact that $(\gamma^0)^2 = \mathbb{1}$, we conclude that each of the diagonal elements in γ^0 must be ± 1, namely,

$$b_\alpha = \pm 1, \qquad \alpha = 1, 2, \cdots, n. \tag{1.85}$$

Let us note next that using the relations from the Clifford algebra in (1.83), for a fixed i, we obtain

$$\operatorname{Tr} \gamma^i \gamma^0 \gamma^i = \operatorname{Tr} \gamma^i (-\gamma^i \gamma^0) = -\operatorname{Tr} (\gamma^i)^2 \gamma^0 = \operatorname{Tr} \gamma^0, \tag{1.86}$$

where "Tr" denotes trace over the matrix indices. On the other hand, the cyclicity property of the trace, namely,

$$\operatorname{Tr} ABC = \operatorname{Tr} CAB, \tag{1.87}$$

leads to

$$\operatorname{Tr} \gamma^i \gamma^0 \gamma^i = \operatorname{Tr} (\gamma^i)^2 \gamma^0 = -\operatorname{Tr} \gamma^0. \tag{1.88}$$

Thus, comparing Eqs. (1.86) and (1.88), we obtain

$$\operatorname{Tr} \gamma^i \gamma^0 \gamma^i = \operatorname{Tr} \gamma^0 = -\operatorname{Tr} \gamma^0,$$
$$\text{or,} \quad \operatorname{Tr} \gamma^0 = 0. \tag{1.89}$$

For this to be true, we conclude that γ^0 must have as many diagonal elements with value $+1$ as with -1. Consequently, the γ^μ matrices must be even dimensional.

Let us assume that $n = 2N$. The simplest nontrivial matrix structure would arise for $N = 1$ when the matrices would be two dimensional (namely, 2×2 matrices). We know that the three Pauli matrices along with the identity matrix define a complete basis for 2×2 matrices. However, as we know, they do not satisfy the Clifford algebra. Namely, if we define $\sigma^\mu = (\mathbb{1}, \boldsymbol{\sigma})$, then,

$$\left[\sigma^\mu, \sigma^\nu\right]_+ \neq 2\eta^{\mu\nu} \mathbb{1}. \tag{1.90}$$

In fact, we know that in two dimensions, there cannot exist four anti-commuting matrices.

1.4 DIRAC EQUATION

The next choice is $N = 2$ for which the matrices will be four dimensional (4×4 matrices). In this case, we can find a set of four linearly independent, constant matrices which satisfy the Clifford algebra. A particular choice of these matrices, for example, has the form

$$\gamma^0 = \begin{pmatrix} \mathbb{1} & 0 \\ 0 & -\mathbb{1} \end{pmatrix},$$

$$\gamma^i = \begin{pmatrix} 0 & \sigma_i \\ -\sigma_i & 0 \end{pmatrix}, \qquad i = 1, 2, 3, \tag{1.91}$$

where each element of the 4×4 matrices represents a 2×2 matrix and the σ_i correspond to the three Pauli matrices. This particular choice of the Dirac matrices is commonly known as the Pauli-Dirac representation.

There are, of course, other representations for the γ^μ matrices. However, the physics of Dirac equation is independent of any particular representation for the γ^μ matrices. This can be easily seen by invoking Pauli's fundamental theorem which says that if there are two sets of matrices γ^μ and γ'^μ satisfying the Clifford algebra, then, they must be related by a similarity transformation. Namely, if

$$\left[\gamma^\mu, \gamma^\nu\right]_+ = 2\eta^{\mu\nu} \mathbb{1},$$

$$\left[\gamma'^\mu, \gamma'^\nu\right]_+ = 2\eta^{\mu\nu} \mathbb{1}, \tag{1.92}$$

then, there exists a constant, nonsingular matrix S such that (in fact, the similarity transformation is really a unitary transformation if we take the Hermiticity properties of the γ-matrices)

$$\gamma'^\mu = S\gamma^\mu S^{-1}. \tag{1.93}$$

Therefore, given the equation

$$(\gamma'^\mu p_\mu - m)\psi' = 0, \tag{1.94}$$

we obtain

$$(S\gamma^\mu S^{-1}p_\mu - m)\psi' = 0,$$
or, $$S(\gamma^\mu p_\mu - m)S^{-1}\psi' = 0,$$
or, $$(\gamma^\mu p_\mu - m)S^{-1}\psi' = 0,$$
or, $$(\gamma^\mu p_\mu - m)\psi = 0, \tag{1.95}$$

with $\psi = S^{-1}\psi'$. (The matrix S^{-1} can be moved past the momentum operator since it is assumed to be constant.) This shows that different representations of the γ^μ matrices are equivalent and merely correspond to a change in the basis of the wave function. As we know, a change of basis does not change physics.

To obtain the Hamiltonian for the Dirac equation, let us go to the coordinate representation where the Dirac equation (1.80) takes the form (remember $\hbar = 1$)

$$(i\slashed{\partial} - m)\psi = (i\gamma^\mu \partial_\mu - m)\psi = 0,$$
or, $$(i\gamma^0 \partial_0 + i\boldsymbol{\gamma} \cdot \boldsymbol{\nabla} - m)\psi = 0. \tag{1.96}$$

Multiplying with γ^0 from the left and using the fact that $(\gamma^0)^2 = \mathbb{1}$, we obtain

$$i\frac{\partial \psi}{\partial t} = (-i\gamma^0 \boldsymbol{\gamma} \cdot \boldsymbol{\nabla} + m\gamma^0)\psi. \tag{1.97}$$

Conventionally, one denotes

$$\beta = \gamma^0, \qquad \boldsymbol{\alpha} = \gamma^0 \boldsymbol{\gamma}. \tag{1.98}$$

In terms of these matrices, then, we can write (1.97) as

$$i\frac{\partial \psi}{\partial t} = (-i\boldsymbol{\alpha} \cdot \boldsymbol{\nabla} + m\beta)\psi = (\boldsymbol{\alpha} \cdot \mathbf{p} + \beta m)\psi. \tag{1.99}$$

This is a first order equation (in time derivative) like the Schrödinger equation and we can identify the Hamiltonian for the Dirac equation with

1.4 Dirac equation

$$H = \boldsymbol{\alpha} \cdot \mathbf{p} + \beta m. \tag{1.100}$$

In the particular representation of the γ^μ matrices in (1.91), we note that

$$\begin{aligned} \beta &= \gamma^0 = \begin{pmatrix} \mathbb{1} & 0 \\ 0 & -\mathbb{1} \end{pmatrix}, \\ \boldsymbol{\alpha} &= \gamma^0 \boldsymbol{\gamma} = \begin{pmatrix} \mathbb{1} & 0 \\ 0 & -\mathbb{1} \end{pmatrix} \begin{pmatrix} 0 & \boldsymbol{\sigma} \\ -\boldsymbol{\sigma} & 0 \end{pmatrix} = \begin{pmatrix} 0 & \boldsymbol{\sigma} \\ \boldsymbol{\sigma} & 0 \end{pmatrix}. \end{aligned} \tag{1.101}$$

We can now determine either from the definition in (1.98) and (1.79) or from the explicit representation in (1.101) that the matrices $\boldsymbol{\alpha}, \beta$ satisfy the anti-commutation relations

$$\begin{aligned} \left[\alpha^i, \alpha^j\right]_+ &= 2\delta^{ij}\,\mathbb{1}, \\ \left[\alpha^i, \beta\right]_+ &= 0, \end{aligned} \tag{1.102}$$

with $\beta^2 = \mathbb{1}$. We can, of course, directly check from this explicit representation that both β and $\boldsymbol{\alpha}$ are Hermitian matrices. But, independently, we also note from the form of the Hamiltonian in (1.100) that, in order for it to be Hermitian, we must have

$$\begin{aligned} \beta^\dagger &= \beta, \\ \boldsymbol{\alpha}^\dagger &= \boldsymbol{\alpha}. \end{aligned} \tag{1.103}$$

In terms of the γ^μ matrices, this translates to

$$\begin{aligned} \beta &= \gamma^0 = (\gamma^0)^\dagger = \beta^\dagger, \\ \boldsymbol{\alpha} &= \gamma^0 \boldsymbol{\gamma} = (\gamma^0 \boldsymbol{\gamma})^\dagger = \boldsymbol{\alpha}^\dagger. \end{aligned} \tag{1.104}$$

Equivalently, we can write

$$(\gamma^0)^\dagger = \gamma^0,$$
$$(\gamma^i)^\dagger = -\gamma^i. \qquad (1.105)$$

Namely, independent of the representation, the γ^μ matrices must satisfy the Hermiticity properties in (1.105). (With a little bit of more analysis, it can be seen that, in general, the Hermiticity properties of the γ^μ matrices are related to the choice of the metric tensor and this particular choice is associated with the Bjorken-Drell metric.) In the next chapter, we would study the plane wave solutions of the first order Dirac equation.

1.5 References

The material presented in this chapter is covered in many standard textbooks and we list below only a few of them.

1. L. I. Schiff, *Quantum Mechanics*, McGraw-Hill, New York, 1968.

2. J. D. Bjorken and S. Drell, *Relativistic Quantum Mechanics*, McGraw-Hill, New York, 1964.

3. C. Itzykson and J-B. Zuber, *Quantum Field Theory*, McGraw-Hill, New York, 1980.

4. A. Das, *Lectures on Quantum Mechanics*, Hindustan Publishing, New Delhi, India, 2003.

CHAPTER 2
Solutions of the Dirac equation

2.1 Plane wave solutions

The Dirac equation in the momentum representation (1.80)

$$(\not{p} - m)\psi = (\gamma^\mu p_\mu - m)\psi = 0, \tag{2.1}$$

or in the coordinate representation

$$(i\not{\partial} - m)\psi = (i\gamma^\mu \partial_\mu - m)\psi = 0, \tag{2.2}$$

defines a set of matrix equations. Since the Dirac matrices, γ^μ, are 4×4 matrices, the wave functions, in this case, are four component column matrices (column vectors). From the study of angular momentum, we know that multicomponent wave functions suggest a nontrivial spin angular momentum. (Other nontrivial internal symmetries can also lead to a multicomponent wavefunction, but here we are considering a simple system without any nontrivial internal symmetry.) Therefore, we expect the solutions of the Dirac equation to describe particles with spin. To understand what kind of particles are described by the Dirac equation, let us look at the plane wave solutions of the equation (which are supposed to describe free particles). Let us denote the four component wave function as (x stands for both space and time)

$$\psi(x) = \begin{pmatrix} \psi_1(x) \\ \psi_2(x) \\ \psi_3(x) \\ \psi_4(x) \end{pmatrix}, \tag{2.3}$$

with

$$\psi_\alpha(x) = e^{-ip\cdot x}\, u_\alpha(p), \qquad \alpha = 1,2,3,4. \tag{2.4}$$

Substituting this back into the Dirac equation, we obtain (we define $\not{A} = \gamma^\mu A_\mu$ for any four vector A_μ)

$$(i\gamma^\mu \partial_\mu - m)\psi(x) = 0,$$

or, $\quad (i\gamma^\mu(-ip_\mu) - m)u(p) = 0,$

or, $\quad (\not{p} - m)u(p) = 0, \tag{2.5}$

where the four component function, $u(p)$, has the form

$$u(p) = \begin{pmatrix} u_1(p) \\ u_2(p) \\ u_3(p) \\ u_4(p) \end{pmatrix}. \tag{2.6}$$

Let us simplify the calculation by restricting to motion along the z-axis. In other words, let us set

$$p_1 = p_2 = 0. \tag{2.7}$$

In this case, the equation takes the form

$$(\gamma^0 p_0 + \gamma^3 p_3 - m)u(p) = 0. \tag{2.8}$$

Taking the particular representation of the γ^μ matrices in (1.91), we can write this explicitly as

$$\begin{pmatrix} p_0 - m & 0 & p_3 & 0 \\ 0 & p_0 - m & 0 & -p_3 \\ -p_3 & 0 & -(p_0+m) & 0 \\ 0 & p_3 & 0 & -(p_0+m) \end{pmatrix} \begin{pmatrix} u_1(p) \\ u_2(p) \\ u_3(p) \\ u_4(p) \end{pmatrix} = 0. \tag{2.9}$$

2.1 Plane wave solutions

This is a set of four linear homogeneous equations and a nontrivial solution exists only if the determinant of the coefficient matrix vanishes. Thus, requiring,

$$\det \begin{pmatrix} p_0 - m & 0 & p_3 & 0 \\ 0 & p_0 - m & 0 & -p_3 \\ -p_3 & 0 & -(p_0 + m) & 0 \\ 0 & p_3 & 0 & -(p_0 + m) \end{pmatrix} = 0, \quad (2.10)$$

we obtain,

$$(p_0 - m)\left[(p_0 - m)(p_0 + m)^2 - p_3^2(p_0 + m)\right]$$
$$+ p_3\left[p_3^3 + (p_0 - m)(-p_3(p_0 + m))\right] = 0,$$
or, $\quad (p_0^2 - m^2)^2 - 2p_3^2(p_0^2 - m^2) + p_3^4 = 0,$
or, $\quad (p_0^2 - p_3^2 - m^2)^2 = 0,$
or, $\quad p_0^2 - p_3^2 - m^2 = 0. \quad (2.11)$

Thus, we see that a nontrivial plane wave solution of the Dirac equation exists only for the energy values

$$p_0 = \pm E = \pm \sqrt{p_3^2 + m^2}. \quad (2.12)$$

Furthermore, each of these energy values is doubly degenerate. Of course, we would expect the positive and the negative energy roots in (2.12) from Einstein's relation. However, the double degeneracy seems to be a reflection of the nontrivial spin structure of the wave function as we will see shortly.

The energy eigenvalues can also be obtained in a simpler fashion by noting that

$$\det(\gamma^\mu p_\mu - m) = 0,$$

or, $\quad \det \begin{pmatrix} (p_0 - m)\mathbb{1} & \sigma_3 p_3 \\ -\sigma_3 p_3 & -(p_0 + m)\mathbb{1} \end{pmatrix} = 0,$

or, $\quad \det\left(-(p_0^2 - m^2)\mathbb{1} + p_3^2 \mathbb{1}\right) = 0,$

or, $\quad \det\left(-(p_0^2 - p_3^2 - m^2)\mathbb{1}\right) = 0,$

or, $\quad p_0^2 - p_3^2 - m^2 = 0. \qquad (2.13)$

This is identical to (2.11) and the energy eigenvalues would then correspond to the roots of this equation given in (2.12). (Note that this method of evaluating a determinant is not valid for matrices involving submatrices that do not commute. In the present case, however, the submatrices $\mathbb{1}, \sigma_3$ are both diagonal and, therefore, commute which is why this simpler method works out.)

We can obtain the solutions of the Dirac equation by directly solving the set of four coupled equations in (2.9). Alternatively, we can introduce two component wave functions $\tilde{u}(p)$ and $\tilde{v}(p)$ and write

$$u(p) = \begin{pmatrix} \tilde{u}(p) \\ \tilde{v}(p) \end{pmatrix}, \qquad (2.14)$$

where

$$\tilde{u}(p) = \begin{pmatrix} u_1(p) \\ u_2(p) \end{pmatrix}, \qquad \tilde{v}(p) = \begin{pmatrix} u_3(p) \\ u_4(p) \end{pmatrix}. \qquad (2.15)$$

We note that for the positive energy solutions

$$p_0 = E_+ = E = \sqrt{p_3^2 + m^2}, \qquad (2.16)$$

the set of coupled equations takes the form

$$(\gamma^\mu p_\mu - m)u(p) = 0,$$

or, $\quad \begin{pmatrix} (E_+ - m)\mathbb{1} & \sigma_3 p_3 \\ -\sigma_3 p_3 & -(E_+ + m)\mathbb{1} \end{pmatrix} \begin{pmatrix} \tilde{u}(p) \\ \tilde{v}(p) \end{pmatrix} = 0. \qquad (2.17)$

2.1 Plane wave solutions

Writing out explicitly, (2.17) leads to

$$(E_+ - m)\tilde{u}(p) + \sigma_3 p_3 \tilde{v}(p) = 0,$$
$$\sigma_3 p_3 \tilde{u}(p) + (E_+ + m)\tilde{v}(p) = 0. \tag{2.18}$$

The two component function $\tilde{v}(p)$ can be solved in terms of $\tilde{u}(p)$ and we obtain from the second relation in (2.18)

$$\tilde{v}(p) = -\frac{\sigma_3 p_3}{E_+ + m}\tilde{u}(p). \tag{2.19}$$

Let us note here parenthetically that the first relation in (2.18) also leads to the same relation (as it should), namely,

$$\begin{aligned}
\tilde{v}(p) &= -\frac{(E_+ - m)}{p_3}\sigma_3 \tilde{u}(p) \\
&= -\frac{(E_+ - m)(E_+ + m)}{p_3(E_+ + m)}\sigma_3 \tilde{u}(p) \\
&= -\frac{(E_+^2 - m^2)}{p_3(E_+ + m)}\sigma_3 \tilde{u}(p) \\
&= -\frac{p_3^2}{p_3(E_+ + m)}\sigma_3 \tilde{u}(p) = -\frac{\sigma_3 p_3}{E_+ + m}\tilde{u}(p),
\end{aligned} \tag{2.20}$$

where we have used the property of the Pauli matrices, namely, $\sigma_3^2 = 1$. Note also that if the relation (2.19) obtained from the second equation in (2.18) is substituted into the first relation, it will hold identically. Therefore, the positive energy solution is completely given by the relation (2.19).

Choosing the two independent solutions for $\tilde{u}$ as

$$\tilde{u}(p) = \begin{pmatrix} 1 \\ 0 \end{pmatrix}, \quad \tilde{u}(p) = \begin{pmatrix} 0 \\ 1 \end{pmatrix}, \tag{2.21}$$

we obtain respectively

$$\tilde{v}(p) = -\frac{\sigma_3 p_3}{E_+ + m}\begin{pmatrix}1\\0\end{pmatrix} = \begin{pmatrix}-\frac{p_3}{E_+ + m}\\0\end{pmatrix}, \qquad (2.22)$$

and

$$\tilde{v}(p) = -\frac{\sigma_3 p_3}{E_+ + m}\begin{pmatrix}0\\1\end{pmatrix} = \begin{pmatrix}0\\\frac{p_3}{E_+ + m}\end{pmatrix}. \qquad (2.23)$$

This determines the two positive energy solutions of the Dirac equation (remember that the energy eigenvalues are doubly degenerate). (The question of which components can be chosen independently follows from an examination of the dynamical equations. Thus, for example, from the second of the two two-component Dirac equations in (2.18), we note that $\tilde{v}$ must vanish in the rest frame while $\tilde{u}$ remains arbitrary. Thus, $\tilde{u}$ can be thought of as the independent solution.)

Similarly, for the negative energy solutions we write

$$p_0 = E_- = -E = -\sqrt{p_3^2 + m^2}, \qquad (2.24)$$

and the set of equations (2.9) becomes

$$(E_- - m)\tilde{u}(p) + \sigma_3 p_3 \tilde{v}(p) = 0,$$
$$\sigma_3 p_3 \tilde{u}(p) + (E_- + m)\tilde{v}(p) = 0. \qquad (2.25)$$

We can solve these as

$$\tilde{u}(p) = -\frac{\sigma_3 p_3}{E_- - m}\tilde{v}(p). \qquad (2.26)$$

Choosing the independent solutions as

$$\tilde{v}(p) = \begin{pmatrix}1\\0\end{pmatrix}, \qquad \tilde{v}(p) = \begin{pmatrix}0\\1\end{pmatrix}, \qquad (2.27)$$

2.1 Plane wave solutions

we obtain respectively

$$\tilde{u}(p) = -\frac{\sigma_3 p_3}{E_- - m}\begin{pmatrix} 1 \\ 0 \end{pmatrix} = \begin{pmatrix} -\frac{p_3}{E_- - m} \\ 0 \end{pmatrix}, \tag{2.28}$$

and

$$\tilde{u}(p) = -\frac{\sigma_3 p_3}{E_- - m}\begin{pmatrix} 0 \\ 1 \end{pmatrix} = \begin{pmatrix} 0 \\ \frac{p_3}{E_- - m} \end{pmatrix}, \tag{2.29}$$

and these determine the two negative energy solutions of the Dirac equation.

The independent two component wave functions in (2.21) and (2.27) are reminiscent of the spin up and spin down states of a two component spinor. Thus, from the fact that we can write

$$u_+(p) = \begin{pmatrix} \tilde{u}(p) \\ -\frac{\sigma_3 p_3}{E_+ + m}\tilde{u}(p) \end{pmatrix}, \quad u_-(p) = \begin{pmatrix} -\frac{\sigma_3 p_3}{E_- - m}\tilde{v}(p) \\ \tilde{v}(p) \end{pmatrix}, \tag{2.30}$$

the positive and the negative energy solutions have the explicit forms

$$u_+^\uparrow(p) = \begin{pmatrix} 1 \\ 0 \\ -\frac{p_3}{E_+ + m} \\ 0 \end{pmatrix}, \quad u_+^\downarrow = \begin{pmatrix} 0 \\ 1 \\ 0 \\ \frac{p_3}{E_+ + m} \end{pmatrix}, \tag{2.31}$$

$$u_-^\uparrow(p) = \begin{pmatrix} -\frac{p_3}{E_- - m} \\ 0 \\ 1 \\ 0 \end{pmatrix}, \quad u_-^\downarrow = \begin{pmatrix} 0 \\ \frac{p_3}{E_- - m} \\ 0 \\ 1 \end{pmatrix}. \tag{2.32}$$

The notation is suggestive and implies that the wave function corresponds to that of a spin $\frac{1}{2}$ particle. (We will determine the spin of the Dirac particle shortly.) It is because of the presence of negative energy solutions that the wave function becomes a four component column matrix as opposed to the two component spinor we expect in non-relativistic systems. (The correct counting for the number of components of the wave function for a massive, relativistic particle of spin s in the presence of both positive and negative energies follows to be $2(2s+1)$, unlike the nonrelativistic counting $(2s+1)$.)

From the structure of the wave function, it is also clear that, for the case of general motion, where

$$p_1 \neq p_2 \neq 0, \tag{2.33}$$

the solutions take the forms (with $p_0 = E_\pm = \pm\sqrt{\mathbf{p}^2 + m^2}$)

$$u_+(p) = \begin{pmatrix} \tilde{u}(p) \\ \dfrac{\boldsymbol{\sigma}\cdot\mathbf{p}}{E_+ + m}\tilde{u}(p) \end{pmatrix}, \quad u_-(p) = \begin{pmatrix} \dfrac{\boldsymbol{\sigma}\cdot\mathbf{p}}{E_- - m}\tilde{v}(p) \\ \tilde{v}(p) \end{pmatrix}, \tag{2.34}$$

which can be explicitly verified. (The change in the sign in the dependent components comes from raising the index of the momentum, namely, $p_i = -p^i$.)

2.2 Normalization of the wave function

Let us note that if we define

$$E = E_+ = \sqrt{\mathbf{p}^2 + m^2} = -E_-, \tag{2.35}$$

then, we can write the solutions for motion along a general direction as

$$u_+(p) = \alpha \begin{pmatrix} \tilde{u}(p) \\ \dfrac{\boldsymbol{\sigma}\cdot\mathbf{p}}{E + m}\tilde{u}(p) \end{pmatrix}, \quad u_-(p) = \beta \begin{pmatrix} -\dfrac{\boldsymbol{\sigma}\cdot\mathbf{p}}{E + m}\tilde{v}(p) \\ \tilde{v}(p) \end{pmatrix}. \tag{2.36}$$

Here α and β are normalization constants to be determined. The two component spinors $\tilde{u}(p)$ and $\tilde{v}(p)$ in (2.21) and (2.27) respectively are normalized as (for the same spin components)

2.2 Normalization of the wave function

$$\tilde{u}^\dagger(p)\tilde{u}(p) = 1 = \tilde{v}^\dagger(p)\tilde{v}(p). \tag{2.37}$$

For different spin components, this product vanishes.

Given this, we can now calculate

$$\begin{aligned}
u_+^\dagger(p)u_+(p) &= \alpha^*\alpha \begin{pmatrix} \tilde{u}^\dagger(p) & \tilde{u}^\dagger(p)\dfrac{\boldsymbol{\sigma}\cdot\mathbf{p}}{E+m} \end{pmatrix} \begin{pmatrix} \tilde{u}(p) \\ \dfrac{\boldsymbol{\sigma}\cdot\mathbf{p}}{E+m}\tilde{u}(p) \end{pmatrix} \\
&= |\alpha|^2 \left(\tilde{u}^\dagger(p)\tilde{u}(p) + \tilde{u}^\dagger(p)\dfrac{(\boldsymbol{\sigma}\cdot\mathbf{p})(\boldsymbol{\sigma}\cdot\mathbf{p})}{(E+m)^2} \tilde{u}(p) \right) \\
&= |\alpha|^2 \left(1 + \dfrac{\mathbf{p}^2}{(E+m)^2} \right) \tilde{u}^\dagger(p)\tilde{u}(p) \\
&= |\alpha|^2 \left(\dfrac{E^2 + m^2 + 2Em + \mathbf{p}^2}{(E+m)^2} \right) \tilde{u}^\dagger(p)\tilde{u}(p) \\
&= |\alpha|^2 \dfrac{2E(E+m)}{(E+m)^2} \tilde{u}^\dagger(p)\tilde{u}(p) \\
&= \dfrac{2E}{E+m} |\alpha|^2 \, \tilde{u}^\dagger(p)\tilde{u}(p),
\end{aligned} \tag{2.38}$$

where we have used the familiar identity satisfied by the Pauli matrices

$$(\boldsymbol{\sigma}\cdot\mathbf{A})(\boldsymbol{\sigma}\cdot\mathbf{B}) = \mathbf{A}\cdot\mathbf{B} + i\boldsymbol{\sigma}\cdot(\mathbf{A}\times\mathbf{B}). \tag{2.39}$$

Similarly, for the negative energy solutions we have

$$\begin{aligned}
u_-^\dagger(p)u_-(p) &= \beta^*\beta \begin{pmatrix} -\tilde{v}^\dagger(p)\dfrac{\boldsymbol{\sigma}\cdot\mathbf{p}}{E+m} & \tilde{v}^\dagger(p) \end{pmatrix} \begin{pmatrix} -\dfrac{\boldsymbol{\sigma}\cdot\mathbf{p}}{E+m}\tilde{v}(p) \\ \tilde{v}(p) \end{pmatrix} \\
&= |\beta|^2 \left(\tilde{v}^\dagger(p)\dfrac{(\boldsymbol{\sigma}\cdot\mathbf{p})(\boldsymbol{\sigma}\cdot\mathbf{p})}{(E+m)^2} \tilde{v}(p) + \tilde{v}^\dagger(p)\tilde{v}(p) \right) \\
&= |\beta|^2 \left(\dfrac{\mathbf{p}^2}{(E+m)^2} + 1 \right) \tilde{v}^\dagger(p)\tilde{v}(p) \\
&= \dfrac{2E}{E+m} |\beta|^2 \, \tilde{v}^\dagger(p)\tilde{v}(p).
\end{aligned} \tag{2.40}$$

It is worth remarking here that although we have seen in (2.37) that, for the same spin components, $\tilde{u}^\dagger \tilde{u} = 1 = \tilde{v}^\dagger \tilde{v}$, we have carried along these factors in (2.38) and (2.40) simply because we have not specified their spin components.

In dealing with the Dirac equation, another wave function (known as the adjoint spinor) that plays an important role is defined to be

$$\overline{u}(p) = u^\dagger(p)\gamma^0. \tag{2.41}$$

Thus, for example,

$$\begin{aligned}
\overline{u}_+(p) &= \alpha^* \begin{pmatrix} \tilde{u}^\dagger(p) & \tilde{u}^\dagger(p)\dfrac{\boldsymbol{\sigma}\cdot\mathbf{p}}{E+m} \end{pmatrix} \begin{pmatrix} \mathbb{1} & 0 \\ 0 & -\mathbb{1} \end{pmatrix} \\
&= \alpha^* \begin{pmatrix} \tilde{u}^\dagger(p) & -\tilde{u}^\dagger(p)\dfrac{\boldsymbol{\sigma}\cdot\mathbf{p}}{E+m} \end{pmatrix}, \\
\overline{u}_-(p) &= \beta^* \begin{pmatrix} -\tilde{v}^\dagger(p)\dfrac{\boldsymbol{\sigma}\cdot\mathbf{p}}{E+m} & \tilde{v}^\dagger(p) \end{pmatrix} \begin{pmatrix} \mathbb{1} & 0 \\ 0 & -\mathbb{1} \end{pmatrix} \\
&= \beta^* \begin{pmatrix} -\tilde{v}^\dagger(p)\dfrac{\boldsymbol{\sigma}\cdot\mathbf{p}}{E+m} & -\tilde{v}^\dagger(p) \end{pmatrix}. \tag{2.42}
\end{aligned}$$

Thus, we see that the difference between the hermitian conjugate $u^\dagger$ and $\overline{u}$ is in the relative sign in the second of the two-component spinors.

We can also calculate the product

$$\begin{aligned}
\overline{u}_+(p) u_+(p) &= \alpha^* \alpha \begin{pmatrix} \tilde{u}^\dagger(p) & -\tilde{u}^\dagger(p)\dfrac{\boldsymbol{\sigma}\cdot\mathbf{p}}{E+m} \end{pmatrix} \begin{pmatrix} \tilde{u}(p) \\ \dfrac{\boldsymbol{\sigma}\cdot\mathbf{p}}{E+m}\tilde{u}(p) \end{pmatrix} \\
&= |\alpha|^2 \left(\tilde{u}^\dagger(p)\tilde{u}(p) - \tilde{u}^\dagger(p)\dfrac{(\boldsymbol{\sigma}\cdot\mathbf{p})(\boldsymbol{\sigma}\cdot\mathbf{k})}{(E+m)^2}\tilde{u}(p) \right) \\
&= |\alpha|^2 \left(1 - \dfrac{\mathbf{p}^2}{(E+m)^2} \right) \tilde{u}^\dagger(p)\tilde{u}(p) \\
&= |\alpha|^2 \left(\dfrac{E^2 + m^2 + 2Em - \mathbf{p}^2}{(E+m)^2} \right) \tilde{u}^\dagger(p)\tilde{u}(p)
\end{aligned}$$

2.2 NORMALIZATION OF THE WAVE FUNCTION

$$= |\alpha|^2 \frac{2m(E+m)}{(E+m)^2} \tilde{u}^\dagger(p)\tilde{u}(p)$$

$$= \frac{2m}{E+m} |\alpha|^2 \tilde{u}^\dagger(p)\tilde{u}(p). \tag{2.43}$$

Similarly, we can show that

$$\bar{u}_-(p)u_-(p) = -\frac{2m}{E+m} |\beta|^2 \tilde{v}^\dagger(p)\tilde{v}(p). \tag{2.44}$$

Our naive instinct will be to normalize the wave function, as in the non-relativistic case, by requiring (for the same spin components)

$$u_+^\dagger(p)u_+(p) = 1 = u_-^\dagger(p)u_-(p). \tag{2.45}$$

However, as we will see shortly, this is not a relativistic normalization. In fact, $u^\dagger u$, as we will see, is related to the probability density which transforms like the time component of a four vector. Thus, a relativistically covariant normalization would be to require (for the same spin components)

$$u_+^\dagger(p)u_+(p) = \frac{E}{m} = u_-^\dagger(p)u_-(p). \tag{2.46}$$

(Remember that this will correspond to the probability density and, therefore, must be positive. By the way, the motivation for such a normalization condition comes from the fact that, in the rest frame of the particle, this will reduce to $u_\pm^\dagger u_\pm = 1$ which corresponds to the non-relativistic normalization.) The independent wave functions for a free particle, $\psi_p(x) = e^{-ik\cdot x}u(p)$ with $p_0 = \pm E$, with this normalization condition, would give (for the same spin components)

$$\int d^3x \, \psi_p^\dagger(x)\psi_{p'}(x) = \frac{E}{m} (2\pi)^3 \delta^3(p-p'). \tag{2.47}$$

With the requirement (2.46), we determine from (2.38) and (2.40) (for the same spin components when (2.37) holds)

$$u_+^\dagger(p)u_+(p) = \frac{2E}{E+m}|\alpha|^2 = \frac{E}{m},$$

or, $\quad \alpha = \alpha^* = \sqrt{\dfrac{E+m}{2m}},$

$$u_-^\dagger(p)u_-(p) = \frac{2E}{E+m}|\beta|^2 = \frac{E}{m},$$

or, $\quad \beta = \beta^* = \sqrt{\dfrac{E+m}{2m}}.$ $\hspace{2cm}$ (2.48)

Therefore, with this normalization, we can write the normalized positive and negative energy solutions of the Dirac equation to be

$$\begin{aligned} u_+(p) &= \sqrt{\frac{E+m}{2m}} \begin{pmatrix} \tilde{u}(p) \\ \frac{\boldsymbol{\sigma}\cdot\mathbf{p}}{E+m}\tilde{u}(p) \end{pmatrix}, \\ u_-(p) &= \sqrt{\frac{E+m}{2m}} \begin{pmatrix} -\frac{\boldsymbol{\sigma}\cdot\mathbf{p}}{E+m}\tilde{v}(p) \\ \tilde{v}(p) \end{pmatrix}. \end{aligned} \qquad (2.49)$$

It is also clear that, with this normalization, we will obtain from (2.43) and (2.44) (for the same spin components)

$$\begin{aligned} \bar{u}_+(p)u_+(p) &= \frac{2m}{E+m}|\alpha|^2 = \frac{2m}{E+m}\frac{E+m}{2m} = 1, \\ \bar{u}_-(p)u_-(p) &= -\frac{2m}{E+m}|\beta|^2 = -\frac{2m}{E+m}\frac{E+m}{2m} = -1. \end{aligned} \qquad (2.50)$$

This particular product, therefore, appears to be a Lorentz invariant (scalar) and we will see later that this is indeed true.

Let us also note here that by construction the positive and the negative energy solutions are orthogonal. For example,

$$u_+^\dagger(p)u_-(p) = \alpha^*\beta \begin{pmatrix} \tilde{u}^\dagger(p) & \tilde{u}^\dagger(p)\dfrac{\boldsymbol{\sigma}\cdot\mathbf{p}}{E+m} \end{pmatrix} \begin{pmatrix} -\dfrac{\boldsymbol{\sigma}\cdot\mathbf{p}}{E+m}\tilde{v}(p) \\ \tilde{v}(p) \end{pmatrix}$$

$$= \alpha^*\beta \left(-\tilde{u}^\dagger(p)\,\dfrac{\boldsymbol{\sigma}\cdot\mathbf{p}}{E+m}\,\tilde{v}(p) + \tilde{u}^\dagger(p)\,\dfrac{\boldsymbol{\sigma}\cdot\mathbf{p}}{E+m}\,\tilde{v}(p) \right)$$

$$= 0. \tag{2.51}$$

Therefore, the solutions we have constructed correspond to four linearly independent, orthonormal solutions of the Dirac equation. Note, however, that

$$\begin{aligned}
\bar{u}_+(p)u_-(p) &= -2\alpha^*\beta\,\tilde{u}^\dagger(p)\,\dfrac{\boldsymbol{\sigma}\cdot\mathbf{p}}{E+m}\,\tilde{v}(p) \neq 0, \\
\bar{u}_-(p)u_+(p) &= -2\beta^*\alpha\,\tilde{v}^\dagger(p)\,\dfrac{\boldsymbol{\sigma}\cdot\mathbf{p}}{E+m}\,\tilde{u}(p) \neq 0.
\end{aligned} \tag{2.52}$$

While we will be using this particular normalization for massive particles, let us note that it becomes meaningless for massless particles. (There is no rest frame for a massless particle.) The probability density has to be well defined. Correspondingly, an alternative normalization which works well for both massive and massless particles is given by

$$u_+^\dagger(p)u_+(p) = E = u_-^\dagger(p)u_-(p). \tag{2.53}$$

This still behaves like the time component of a four vector (m is a Lorentz scalar). In this case, we will obtain from (2.38) and (2.40) (for the same spin components)

$$u_+^\dagger(p)u_+(p) = \dfrac{2E}{E+m}\,|\alpha|^2 = E,$$

or, $\quad \alpha = \alpha^* = \sqrt{\dfrac{E+m}{2}},$

$$u_-^\dagger(p)u_-(p) = \dfrac{2E}{E+m}\,|\beta|^2 = E,$$

or, $\quad \beta = \beta^* = \sqrt{\dfrac{E+m}{2}}. \tag{2.54}$

Correspondingly, in this case, we obtain

$$\begin{aligned}\bar{u}_+(p)u_+(p) &= \frac{2m}{E+m}\,|\alpha|^2 = m,\\ \bar{u}_-(p)u_-(p) &= -\frac{2m}{E+m}\,|\beta|^2 = -m,\end{aligned} \qquad (2.55)$$

which vanishes for a massless particle. This product continues to be a scalar. Let us note once again that this is a particularly convenient normalization for massless particles.

Let us note here parenthetically that, while the arbitrariness in the normalization of $u(p)$ may seem strange, it can be understood in light of what we have already pointed out earlier as follows. We can write the solution of the Dirac equation for a general motion (along an arbitrary direction) as

$$\psi(x) = \int d^4p\, a(p)\delta(p^2 - m^2)\, e^{-ip\cdot x}\, u(p), \qquad (2.56)$$

where $a(p)$ is a coefficient which depends on the normalization of $u(p)$ in such a way that the wave function would lead to a total probability normalized to unity,

$$\int d^3x\, \psi^\dagger(x)\psi(x) = 1. \qquad (2.57)$$

Namely, a particular choice of normalization for the $u(p)$ is compensated for by a specific choice of the coefficient function $a(p)$ so that the total probability integrates to unity. The true normalization is really contained in the total probability.

2.3 Spin of the Dirac particle

As we have argued repeatedly, the structure of the plane wave solutions of the Dirac equation is suggestive of the fact that the particle described by the Dirac equation has spin $\frac{1}{2}$. That this is indeed true can be seen explicitly as follows.

Let us define a four dimensional generalization of the Pauli matrices as (in this section, we will use the notations of three dimensional

2.3 SPIN OF THE DIRAC PARTICLE

Euclidean space since we will be dealing only with three dimensional vectors)

$$\tilde{\alpha}_i = \begin{pmatrix} \sigma_i & 0 \\ 0 & \sigma_i \end{pmatrix}, \quad i = 1, 2, 3. \tag{2.58}$$

It can, of course, be checked readily that this is related to the α_i matrices defined in (1.98) and (1.101) through the relation

$$\alpha_i = \begin{pmatrix} 0 & \sigma_i \\ \sigma_i & 0 \end{pmatrix} = \rho\tilde{\alpha}_i = \tilde{\alpha}_i\rho, \tag{2.59}$$

where

$$\rho = \begin{pmatrix} 0 & \mathbb{1} \\ \mathbb{1} & 0 \end{pmatrix}. \tag{2.60}$$

We note that $\rho^2 = \mathbb{1}$ so that we can invert the defining relation (2.59) and write

$$\tilde{\alpha}_i = \rho\alpha_i = \alpha_i\rho. \tag{2.61}$$

From the structures of the matrices α_i and $\tilde{\alpha}_i$ we conclude that

$$\begin{aligned}[] [\tilde{\alpha}_i, \tilde{\alpha}_j] &= \begin{pmatrix} [\sigma_i, \sigma_j] & 0 \\ 0 & [\sigma_i, \sigma_j] \end{pmatrix} \\ &= \begin{pmatrix} 2i\epsilon_{ijk}\sigma_k & 0 \\ 0 & 2i\epsilon_{ijk}\sigma_k \end{pmatrix} = 2i\epsilon_{ijk}\tilde{\alpha}_k. \end{aligned} \tag{2.62}$$

This shows that $\frac{1}{2}\tilde{\alpha}_i$ satisfies the angular momentum algebra (remember $\hbar = 1$) and this is why we call the matrices, $\tilde{\alpha}_i$, the generalized Pauli matrices. (Note, however, that this defines a reducible representation of spin generators since the matrices are block diagonal.) Furthermore, let us note that

$$\begin{aligned}[\tilde{\alpha}_i, \alpha_j] &= \begin{pmatrix} \sigma_i & 0 \\ 0 & \sigma_i \end{pmatrix} \begin{pmatrix} 0 & \sigma_j \\ \sigma_j & 0 \end{pmatrix} - \begin{pmatrix} 0 & \sigma_j \\ \sigma_j & 0 \end{pmatrix} \begin{pmatrix} \sigma_i & 0 \\ 0 & \sigma_i \end{pmatrix} \\ &= \begin{pmatrix} 0 & \sigma_i\sigma_j - \sigma_j\sigma_i \\ \sigma_i\sigma_j - \sigma_j\sigma_i & 0 \end{pmatrix} \\ &= \begin{pmatrix} 0 & [\sigma_i, \sigma_j] \\ [\sigma_i, \sigma_j] & 0 \end{pmatrix} \\ &= \begin{pmatrix} 0 & 2i\epsilon_{ijk}\sigma_k \\ 2i\epsilon_{ijk}\sigma_k & 0 \end{pmatrix} = 2i\epsilon_{ijk}\alpha_k, \\ [\tilde{\alpha}_i, \beta] &= \begin{pmatrix} [\sigma_i, \mathbb{1}] & 0 \\ 0 & -[\sigma_i, \mathbb{1}] \end{pmatrix} = 0. \end{aligned} \qquad (2.63)$$

With these relations at our disposal, let us look at the free Dirac Hamiltonian in (1.100) (remember that we are using three dimensional Euclidean notations in this section)

$$H = \boldsymbol{\alpha} \cdot \mathbf{p} + \beta m = \alpha_i p_i + \beta m. \qquad (2.64)$$

As we will see in the next chapter, the Dirac equation transforms covariantly under a Lorentz transformation. In other words, Lorentz transformations define a symmetry of the Dirac Hamiltonian and, therefore, rotations which correspond to a subset of the Lorentz transformations must also be a symmetry of the Dirac Hamiltonian. Consequently, the angular momentum operators which generate rotations should commute with the Dirac Hamiltonian. Let us recall that the orbital angular momentum operator is given by

$$L_i = \epsilon_{ijk} x_j p_k, \quad i, j, k = 1, 2, 3. \qquad (2.65)$$

Calculating the commutator of this operator with the Dirac Hamiltonian, we obtain

2.3 Spin of the Dirac particle

$$\begin{aligned}
[L_i, H] &= [\epsilon_{ijk} x_j p_k, \alpha_\ell p_\ell + \beta m] \\
&= [\epsilon_{ijk} x_j p_k, \alpha_\ell p_\ell] \\
&= \epsilon_{ijk} \alpha_\ell [x_j, p_\ell] p_k \\
&= \epsilon_{ijk} \alpha_\ell \left(i \delta_{j\ell}\right) p_k = i \epsilon_{ijk} \alpha_j p_k.
\end{aligned} \quad (2.66)$$

Here we have used the fact that since β is a constant matrix and m is a constant, the second term in the Hamiltonian drops out of the commutator. Thus, we note that the orbital angular momentum operator does not commute with the Dirac Hamiltonian. Consequently, the total angular momentum which should commute with the Hamiltonian must contain a spin part as well.

To determine the spin angular momentum, we note that

$$\begin{aligned}
[\tilde{\alpha}_i, H] &= [\tilde{\alpha}_i, \alpha_j p_j + \beta m] \\
&= [\tilde{\alpha}_i, \alpha_j] p_j + [\tilde{\alpha}_i, \beta] m \\
&= 2 i \epsilon_{ijk} \alpha_k p_j = -2 i \epsilon_{ijk} \alpha_j p_k,
\end{aligned} \quad (2.67)$$

so that combining this relation with (2.66) we obtain

$$\begin{aligned}
\left[L_i + \frac{1}{2} \tilde{\alpha}_i, H\right] &= [L_i, H] + \frac{1}{2} [\tilde{\alpha}_i, H] \\
&= i \epsilon_{ijk} \alpha_j p_k - i \epsilon_{ijk} \alpha_j p_k = 0.
\end{aligned} \quad (2.68)$$

In other words, the total angular momentum which should commute with the Hamiltonian, if rotations are a symmetry of the system, can be identified with

$$J_i = L_i + \frac{1}{2} \tilde{\alpha}_i. \quad (2.69)$$

In this case, therefore, we can identify the spin angular momentum operator with

$$S_i = \frac{1}{2}\tilde{\alpha}_i. \tag{2.70}$$

Note, in particular, that

$$S_3 = \frac{1}{2}\tilde{\alpha}_3 = \frac{1}{2}\begin{pmatrix} \sigma_3 & 0 \\ 0 & \sigma_3 \end{pmatrix}, \tag{2.71}$$

which has doubly degenerate eigenvalues $\pm\frac{1}{2}$. Therefore, we conclude that the particle described by the Dirac equation corresponds to a spin $\frac{1}{2}$ particle.

2.4 Continuity equation

The Dirac equation, written in the Hamiltonian form, is given by

$$i\frac{\partial\psi}{\partial t} = H\psi = \left(-i\boldsymbol{\alpha}\cdot\boldsymbol{\nabla} + \beta m\right)\psi. \tag{2.72}$$

Taking the Hermitian conjugate of this equation, we obtain

$$-i\frac{\partial\psi^\dagger}{\partial t} = \psi^\dagger\left(i\boldsymbol{\alpha}\cdot\overleftarrow{\boldsymbol{\nabla}} + \beta m\right), \tag{2.73}$$

where the gradient is assumed to act on $\psi^\dagger$. Multiplying (2.72) with $\psi^\dagger$ on the left and (2.73) with ψ on the right and subtracting the second from the first, we obtain

$$i\psi^\dagger\frac{\partial\psi}{\partial t} + i\frac{\partial\psi^\dagger}{\partial t}\psi = -i\left(\psi^\dagger\boldsymbol{\alpha}\cdot\boldsymbol{\nabla}\psi + (\boldsymbol{\nabla}\psi^\dagger)\cdot\boldsymbol{\alpha}\psi\right),$$

or, $$i\frac{\partial}{\partial t}(\psi^\dagger\psi) = -i\boldsymbol{\nabla}\cdot(\psi^\dagger\boldsymbol{\alpha}\psi),$$

or, $$\frac{\partial}{\partial t}(\psi^\dagger\psi) = -\boldsymbol{\nabla}\cdot(\psi^\dagger\boldsymbol{\alpha}\psi). \tag{2.74}$$

This is the continuity equation for the probability current density associated with the Dirac equation and we note that we can identify

2.4 CONTINUITY EQUATION

$$\rho = \psi^\dagger \psi = \text{probability density},$$
$$\mathbf{J} = \psi^\dagger \boldsymbol{\alpha} \psi = \text{probability current density}, \tag{2.75}$$

to write the continuity equation as

$$\frac{\partial \rho}{\partial t} = -\boldsymbol{\nabla} \cdot \mathbf{J}. \tag{2.76}$$

This suggests that we can write the current four vector as

$$J^\mu = (\rho, \mathbf{J}) = (\psi^\dagger \psi, \psi^\dagger \boldsymbol{\alpha} \psi), \tag{2.77}$$

so that the continuity equation can be written in the manifestly covariant form

$$\partial_\mu J^\mu = 0. \tag{2.78}$$

This, in fact, shows that the probability density, ρ, is the time component of j^μ (see (2.77)) and, therefore, must transform like the time coordinate under a Lorentz transformation. (We are, of course, yet to show that j^μ transforms like a four vector which we will do in the next chapter.) On the other hand, the total probability

$$P = \int d^3 x \, \rho = \int d^3 x \, \psi^\dagger \psi, \tag{2.79}$$

is a constant independent of any particular Lorentz frame. It is worth recalling that we have already used this Lorentz transformation property of ρ in defining the normalization of the wave function.

An alternative and more covariant way of deriving the continuity equation is to start with the covariant Dirac equation

$$(i\gamma^\mu \partial_\mu - m)\psi = 0, \tag{2.80}$$

and note that the Hermitian conjugate of ψ satisfies

$$\psi^\dagger\bigl(-i(\gamma^\mu)^\dagger \overleftarrow{\partial}_\mu - m\bigr) = 0. \tag{2.81}$$

Multiplying this equation with γ^0 on the right and using the fact that $(\gamma^0)^2 = \mathbb{1}$, we obtain

$$\overline{\psi}\bigl(-i\gamma^0(\gamma^\mu)^\dagger\gamma^0 \overleftarrow{\partial}_\mu - m\bigr) = 0,$$
$$\text{or,} \quad \overline{\psi}\bigl(-i\gamma^\mu \overleftarrow{\partial}_\mu - m\bigr) = 0, \tag{2.82}$$

where we have used the property of the gamma matrices that (for $\mu = 0, 1, 2, 3$)

$$\gamma^0 \gamma^\mu \gamma^0 = (\gamma^\mu)^\dagger,$$
$$\gamma^0 (\gamma^\mu)^\dagger \gamma^0 = \gamma^\mu. \tag{2.83}$$

Multiplying (2.80) with $\overline{\psi}$ on the left and (2.82) with ψ on the right and subtracting the second from the first, we obtain

$$i\bigl(\overline{\psi}\gamma^\mu \partial_\mu \psi + \overline{\psi}\gamma^\mu \overleftarrow{\partial}_\mu \psi\bigr) = 0,$$
$$\text{or,} \quad i\partial_\mu\bigl(\overline{\psi}\gamma^\mu \psi\bigr) = 0,$$
$$\text{or,} \quad \partial_\mu\bigl(\overline{\psi}\gamma^\mu \psi\bigr) = 0. \tag{2.84}$$

This is, in fact, the covariant continuity equation and we can identify

$$J^\mu = \overline{\psi}\gamma^\mu \psi. \tag{2.85}$$

Note from the definition in (2.85) that

$$\begin{aligned} J^0 &= \overline{\psi}\gamma^0 \psi = \psi^\dagger \gamma^0 \gamma^0 \psi = \psi^\dagger \psi = \rho, \\ \mathbf{J} &= \overline{\psi}\boldsymbol{\gamma}\psi = \psi^\dagger \gamma^0 \boldsymbol{\gamma}\psi = \psi^\dagger \boldsymbol{\alpha}\psi, \end{aligned} \tag{2.86}$$

which is what we had derived earlier in (2.77).

Let me conclude this discussion by noting that although the Dirac equation has both positive and negative energy solutions, because it is a first order equation (particularly in the time derivative), the probability density is independent of time derivative much like the Schrödinger equation. Consequently, the probability density, as we have seen explicitly in (2.38) and (2.40), can be defined to be positive definite even in the presence of negative energy solutions. This is rather different from the case of the Klein-Gordon equation that we have studied in chapter **1**.

2.5 Dirac's hole theory

We have seen that Dirac's equation leads to both positive and negative energy solutions. In the free particle case, for example, the energy eigenvalues are given by

$$p^0 = E_\pm = \pm E = \pm \sqrt{\mathbf{p}^2 + m^2}. \tag{2.87}$$

Thus, even for this simple case of a free particle the energy spectrum has the form shown in Fig. 2.1. We note from Fig. 2.1 (as well as from the equation above) that the positive and the negative energy solutions are separated by a gap of magnitude $2m$ (remember that we are using $c = 1$).

Figure 2.1: Energy spectrum for a free Dirac particle.

Even when the probability density is consistently defined, the presence of negative energy solutions leads to many conceptual difficulties. First of all, in such a case, we note that the energy spectrum is unbounded from below. Since physical systems have a tendency to go to the lowest energy state available, this implies that any such physical system (of Dirac particles) would make a transition to these unphysical energy states thereby leading to a collapse of all stable systems such as the Hydrogen atom. Classically, of course, we can restrict ourselves to the subspace of positive energy solutions. But as we have argued earlier within the context of the Klein-Gordon equation, quantum mechanically this is not acceptable. Namely, even if we start out with a positive energy solution, any perturbation would cause the energy to lower, destabilizing the physical system and leading to an ultimate collapse.

In the case of Dirac particles, however, there is a way out of this difficulty. Let us recall that the Dirac particles carry spin $\frac{1}{2}$ and are, therefore, fermions. To be specific, let us assume that the particles described by the Dirac equation are the spin $\frac{1}{2}$ electrons. Since fermions obey Pauli exclusion principle, any given energy state can at any time accommodate at the most two electrons with opposite spin projections. Taking advantage of this fact, Dirac postulated that the physical ground state (vacuum) in such a theory should be redefined for consistency. Namely, Dirac postulated that the ground state in such a theory is the state where all the negative energy states are filled with electrons. Thus, unlike the conventional picture of the ground state as being the state without any particle (quantum), here the ground state, in fact, contains an infinite number of negative energy particles. Furthermore, Dirac assumed that the electrons in the negative energy states are passive in the sense that they do not produce any observable effect such as charge, electromagnetic field etc. (Momentum and energy of these electrons are also assumed to be unobservable. This simply means that one redefines the values of all these observables with respect to this ground state.)

This redefinition of the vacuum automatically prevents the instability associated with matter. For example, a positive energy electron can no longer drop down to a negative energy state without violating the Pauli exclusion principle since the negative energy states are already filled. (Note that this would not work for a bosonic system such as particles described by the Klein-Gordon equation. It is only because fermions obey Pauli exclusion principle that this works for the Dirac equation.) On the other hand, it does predict some new physical phenomena which are experimentally observed. For example, if enough energy is provided to such a ground state, a negative energy electron can make a transition to a positive energy state and can appear as a positive energy electron. Furthermore, the absence of a negative energy electron can be thought of as a "hole" which would have exactly the same mass as the particle but otherwise opposite internal quantum numbers. This "hole" state is what we have come to recognize as the anti-particle – in this case, a positron – and the process under discussion is commonly referred to as pair creation (production). Thus, the Dirac theory predicts an anti-particle of equal mass for every Dirac particle. (The absence of a negative energy electron in the ground state can be thought of as the ground state plus a positive energy "hole" state with exactly opposite quantum numbers to neutralize its effects. The amount of energy necessary to excite a negative energy electron to a positive energy state is $E \geq 2m$.)

This is Dirac's theory of electrons and works quite well. However, we must recognize that it is inherently a many particle theory in the sense that the vacuum (ground state) of the theory is defined to contain infinitely many negative energy particles. (This unconventional definition of a vacuum state can be avoided in a second quantized field theory which we will study later.) In spite of this, the Dirac equation passes as a one particle equation primarily because of the Pauli exclusion principle. On the other hand, this is a general feature that combining quantum mechanics with relativity necessarily leads to a many particle theory.

2.6 Properties of the Dirac matrices

The Dirac matrices, γ^μ, were crucial in taking a matrix square root of the Einstein relation and, thereby, in defining a first order equation. In this section, we will study some of the useful properties of these matrices. As we have seen, the four Dirac matrices satisfy (in addition to the Clifford algebra)

$$\begin{aligned}(\gamma^0)^\dagger &= \gamma^0, \\ (\gamma^i)^\dagger &= -\gamma^i, \\ \text{Tr } \gamma^\mu &= 0, \qquad \mu = 0, 1, 2, 3, \qquad i = 1, 2, 3.\end{aligned} \qquad (2.88)$$

Since these are 4×4 matrices, a complete set of Dirac matrices must consist of 16 such matrices. Of course, the identity matrix will correspond to one of them.

To obtain the other basis matrices, let us define the following sets of matrices. Let

$$\gamma_5 = i\gamma^0\gamma^1\gamma^2\gamma^3 = -\frac{i}{4!}\,\epsilon_{\mu\nu\lambda\rho}\gamma^\mu\gamma^\nu\gamma^\lambda\gamma^\rho, \tag{2.89}$$

where

$$\epsilon^{0123} = 1 = -\epsilon_{0123}, \tag{2.90}$$

represents the four-dimensional generalization of the Levi-Civita tensor. Note that in our particular representation for the γ^μ matrices given in (1.91), we obtain

$$\begin{aligned}
\gamma_5 &= i\begin{pmatrix} \mathbb{1} & 0 \\ 0 & -\mathbb{1} \end{pmatrix}\begin{pmatrix} 0 & \sigma_1 \\ -\sigma_1 & 0 \end{pmatrix}\begin{pmatrix} 0 & \sigma_2 \\ -\sigma_2 & 0 \end{pmatrix}\begin{pmatrix} 0 & \sigma_3 \\ -\sigma_3 & 0 \end{pmatrix} \\
&= i\begin{pmatrix} 0 & \sigma_1 \\ \sigma_1 & 0 \end{pmatrix}\begin{pmatrix} -\sigma_2\sigma_3 & 0 \\ 0 & -\sigma_2\sigma_3 \end{pmatrix} \\
&= i\begin{pmatrix} 0 & -\sigma_1\sigma_2\sigma_3 \\ -\sigma_1\sigma_2\sigma_3 & 0 \end{pmatrix} \\
&= i\begin{pmatrix} 0 & -i\mathbb{1} \\ -i\mathbb{1} & 0 \end{pmatrix} = \begin{pmatrix} 0 & \mathbb{1} \\ \mathbb{1} & 0 \end{pmatrix},
\end{aligned} \tag{2.91}$$

where we have used the property of the Pauli matrices

$$\sigma_1\sigma_2\sigma_3 = i\mathbb{1}. \tag{2.92}$$

We recognize from (2.91) that we can identify this with the matrix ρ defined earlier in (2.60). Note that, by definition,

$$\gamma_5^2 = \mathbb{1}, \qquad \gamma_5^\dagger = \gamma_5, \tag{2.93}$$

2.6 Properties of the Dirac matrices

and that, since it is the product of all the four γ^μ matrices, it anticommutes with any one of them. Namely,

$$[\gamma_5, \gamma^\mu]_+ = 0. \tag{2.94}$$

Given the matrix γ_5, we can define four new matrices as

$$\gamma_5 \gamma^\mu, \qquad \mu = 0, 1, 2, 3. \tag{2.95}$$

Since we know the explicit forms of the matrices $\mathbb{1}$, γ^μ and γ_5 in our representation, let us write out the forms of $\gamma_5 \gamma^\mu$ also in this representation.

$$\begin{aligned}
\gamma_5 \gamma^0 &= \begin{pmatrix} 0 & \mathbb{1} \\ \mathbb{1} & 0 \end{pmatrix} \begin{pmatrix} \mathbb{1} & 0 \\ 0 & -\mathbb{1} \end{pmatrix} = \begin{pmatrix} 0 & -\mathbb{1} \\ \mathbb{1} & 0 \end{pmatrix}, \\
\gamma_5 \gamma^i &= \begin{pmatrix} 0 & \mathbb{1} \\ \mathbb{1} & 0 \end{pmatrix} \begin{pmatrix} 0 & \sigma_i \\ -\sigma_i & 0 \end{pmatrix} = \begin{pmatrix} -\sigma_i & 0 \\ 0 & \sigma_i \end{pmatrix}.
\end{aligned} \tag{2.96}$$

Finally, we can also define six anti-symmetric matrices, $\sigma^{\mu\nu}$, as ($\mu, \nu = 0, 1, 2, 3$)

$$\begin{aligned}
\sigma^{\mu\nu} &= -\sigma^{\nu\mu} = \frac{i}{2} [\gamma^\mu, \gamma^\nu] = \frac{i}{2} (\gamma^\mu \gamma^\nu - \gamma^\nu \gamma^\mu) \\
&= i(\eta^{\mu\nu} - \gamma^\nu \gamma^\mu) \\
&= -i(\eta^{\mu\nu} - \gamma^\mu \gamma^\nu),
\end{aligned} \tag{2.97}$$

whose explicit forms in our representation can be worked out to be ($i, j, k = 1, 2, 3$)

$$\sigma^{0i} = i\gamma^0\gamma^i = i\begin{pmatrix} \mathbb{1} & 0 \\ 0 & -\mathbb{1} \end{pmatrix}\begin{pmatrix} 0 & \sigma_i \\ -\sigma_i & 0 \end{pmatrix}$$

$$= \begin{pmatrix} 0 & i\sigma_i \\ i\sigma_i & 0 \end{pmatrix} = i\alpha_i,$$

$$\sigma^{ij} = i\gamma^i\gamma^j = i\begin{pmatrix} 0 & \sigma_i \\ -\sigma_i & 0 \end{pmatrix}\begin{pmatrix} 0 & \sigma_j \\ -\sigma_j & 0 \end{pmatrix}$$

$$= i\begin{pmatrix} -\sigma_i\sigma_j & 0 \\ 0 & -\sigma_i\sigma_j \end{pmatrix}$$

$$= i\begin{pmatrix} -i\epsilon_{ijk}\sigma_k & 0 \\ 0 & -i\epsilon_{ijk}\sigma_k \end{pmatrix} = \epsilon^{ijk}\tilde{\alpha}_k. \tag{2.98}$$

We have already seen in (2.70) that the matrices $\frac{1}{2}\tilde{\alpha}_i$ represent the spin operators for the Dirac particle. From (2.98) we conclude, therefore, that the matrices

$$\frac{1}{2}\tilde{\alpha}_i = \frac{1}{4}\epsilon_{ijk}\sigma^{jk}, \tag{2.99}$$

can be identified with the spin operators for the Dirac particle. (This relation can be obtained from (2.98) using the identity for products of Levi-Civita tensors, namely, $\epsilon_{ijk}\epsilon_{\ell jk} = 2\,\delta_{i\ell}$.)

We have thus constructed a set of sixteen Dirac matrices, namely,

$$\begin{aligned} \Gamma^{(S)} &= \mathbb{1}, & 1, \\ \Gamma^{(V)} &= \gamma^\mu, & 4, \\ \Gamma^{(T)} &= \sigma^{\mu\nu}, & 6, \\ \Gamma^{(A)} &= \gamma_5\gamma^\mu, & 4, \\ \Gamma^{(P)} &= \gamma_5, & 1, \end{aligned} \tag{2.100}$$

where the numbers on the right denote the number of matrices in each category and these, in fact, provide a basis for all the 4×4 matrices. Here, the notation is suggestive and stands for the fact that $\bar{\psi}\Gamma^{(S)}\psi$

2.6 Properties of the Dirac matrices

transforms like a scalar under Lorentz and parity transformations. Similarly, $\bar{\psi}\Gamma^{(V)}\psi$, $\bar{\psi}\Gamma^{(T)}\psi$, $\bar{\psi}\Gamma^{(A)}\psi$ and $\bar{\psi}\Gamma^{(P)}\psi$ behave respectively like a vector, tensor, axial-vector and a pseudo-scalar under a Lorentz and parity transformations as we will see in the next chapter.

Let us note here that each of the matrices, even within a given class, has its own hermiticity property. However, it can be checked that except for γ_5, which is defined to be Hermitian, all other matrices satisfy

$$\gamma^0 \Gamma^{(\alpha)} \gamma^0 = (\Gamma^{(\alpha)})^\dagger, \qquad \alpha = S, V, A, T. \tag{2.101}$$

In fact, it follows easily that

$$\begin{aligned}
\gamma^0 \mathbb{1} \gamma^0 &= (\gamma^0)^2 = \mathbb{1} = (\mathbb{1})^\dagger, \\
\gamma^0 \gamma^\mu \gamma^0 &= (\gamma^\mu)^\dagger, \\
\gamma^0 \gamma_5 \gamma^\mu \gamma^0 &= -\gamma_5 \gamma^0 \gamma^\mu \gamma^0 = -(\gamma_5)^\dagger (\gamma^\mu)^\dagger = (\gamma_5 \gamma^\mu)^\dagger,
\end{aligned} \tag{2.102}$$

where we have used the fact that γ_5 is Hermitian and it anti-commutes with γ^μ. Finally, from

$$\gamma^0 \gamma^\mu \gamma^\nu \gamma^0 = \gamma^0 \gamma^\mu \gamma^0 \gamma^0 \gamma^\nu \gamma^0 = (\gamma^\nu \gamma^\mu)^\dagger, \tag{2.103}$$

it follows that

$$\begin{aligned}
\gamma^0 \sigma^{\mu\nu} \gamma^0 &= \frac{i}{2} \gamma^0 (\gamma^\mu \gamma^\nu - \gamma^\nu \gamma^\mu) \gamma^0 \\
&= \frac{i}{2} \left((\gamma^\nu \gamma^\mu)^\dagger - (\gamma^\mu \gamma^\nu)^\dagger \right) \\
&= -\frac{i}{2} (\gamma^\mu \gamma^\nu - \gamma^\nu \gamma^\mu)^\dagger \\
&= (\sigma^{\mu\nu})^\dagger.
\end{aligned} \tag{2.104}$$

The Dirac matrices satisfy nontrivial (anti) commutation relations. We already know that

$$[\gamma^\mu, \gamma^\nu]_+ = 2\eta^{\mu\nu}\mathbb{1},$$
$$[\gamma_5, \gamma^\mu]_+ = 0. \qquad (2.105)$$

We can also calculate various other commutation relations in a straightforward and representation independent manner. For example,

$$\begin{aligned}[\gamma^\mu, \sigma^{\nu\lambda}] &= [\gamma^\mu, -i(\eta^{\nu\lambda} - \gamma^\nu\gamma^\lambda)] \\ &= i[\gamma^\mu, \gamma^\nu\gamma^\lambda] \\ &= i\left([\gamma^\mu, \gamma^\nu]_+ \gamma^\lambda - \gamma^\nu[\gamma^\mu, \gamma^\lambda]_+\right) \\ &= 2i\left(\eta^{\mu\nu}\gamma^\lambda - \eta^{\mu\lambda}\gamma^\nu\right). \qquad (2.106)\end{aligned}$$

In this derivation, we have used the fact that

$$\begin{aligned}[A, BC] &= ABC - BCA \\ &= (AB + BA)C - B(AC + CA) \\ &= [A, B]_+ C - B[A, C]_+. \qquad (2.107)\end{aligned}$$

We note here parenthetically that the commutator in (2.107) can also be expressed in terms of commutators (instead of anti-commutators) as

$$[A, BC] = [A, B]C + B[A, C]. \qquad (2.108)$$

However, since γ^μ matrices satisfy simple anti-commutation relations, the form in (2.107) is more useful for our purpose.

Similarly, for the commutator of two $\sigma^{\mu\nu}$ matrices, we obtain

$$\begin{aligned}
[\sigma^{\mu\nu}, \sigma^{\lambda\rho}] &= \left[-i(\eta^{\mu\nu} - \gamma^\mu\gamma^\nu), \sigma^{\lambda\rho}\right] \\
&= i[\gamma^\mu\gamma^\nu, \sigma^{\lambda\rho}] \\
&= i\gamma^\mu[\gamma^\nu, \sigma^{\lambda\rho}] + i[\gamma^\mu, \sigma^{\lambda\rho}]\gamma^\nu \\
&= i\gamma^\mu\left[2i\eta^{\nu\lambda}\gamma^\rho - 2i\eta^{\nu\rho}\gamma^\lambda\right] + i\left[2i\eta^{\mu\lambda}\gamma^\rho - 2i\eta^{\mu\rho}\gamma^\lambda\right]\gamma^\nu \\
&= -2i\left[\eta^{\mu\lambda}\bigl(i(\eta^{\nu\rho} - \gamma^\rho\gamma^\nu)\bigr) + \eta^{\nu\rho}\bigl(-i(\eta^{\mu\lambda} - \gamma^\mu\gamma^\lambda)\bigr)\right. \\
&\qquad \left. -\eta^{\mu\rho}\bigl(i(\eta^{\nu\lambda} - \gamma^\lambda\gamma^\nu)\bigr) - \eta^{\nu\lambda}(-i(\eta^{\mu\rho} - \gamma^\mu\gamma^\rho))\right] \\
&= -2i\left[\eta^{\mu\lambda}\sigma^{\nu\rho} + \eta^{\nu\rho}\sigma^{\mu\lambda} - \eta^{\mu\rho}\sigma^{\nu\lambda} - \eta^{\nu\lambda}\sigma^{\mu\rho}\right]. \quad (2.109)
\end{aligned}$$

Thus, we see that the $\sigma^{\mu\nu}$ matrices satisfy an algebra in the sense that the commutator of any two of them gives back a $\sigma^{\mu\nu}$ matrix. We will see in the next chapter that they provide a representation for the Lorentz algebra.

The various commutation and anti-commutation relations also lead to many algebraic simplifications in dealing with such matrices. This becomes particularly useful in calculating various amplitudes involving Dirac particles. Thus, for example, (these relations are true only in 4-dimensions)

$$\begin{aligned}
\gamma_\mu \gamma^\nu \gamma^\mu &= \gamma_\mu\left([\gamma^\nu, \gamma^\mu]_+ - \gamma^\mu\gamma^\nu\right) \\
&= 2\eta^{\nu\mu}\gamma_\mu - 4\gamma^\nu = 2\gamma^\nu - 4\gamma^\nu = -2\gamma^\nu, \quad (2.110)
\end{aligned}$$

where we have used $(\gamma_\mu = \eta_{\mu\nu}\gamma^\nu)$

$$\gamma_\mu \gamma^\mu = 4\,\mathbb{1}, \quad (2.111)$$

and it follows now that,

$$\gamma_\mu \slashed{A} \gamma^\mu = \gamma_\mu A_\nu \gamma^\nu \gamma^\mu = A_\nu \gamma_\mu \gamma^\nu \gamma^\mu = -2A_\nu \gamma^\nu = -2\slashed{A}. \quad (2.112)$$

Similarly,

$$\begin{aligned}
\gamma_\mu \gamma^\nu \gamma^\lambda \gamma^\mu &= \gamma_\mu \gamma^\nu \left([\gamma^\lambda, \gamma^\mu]_+ - \gamma^\mu \gamma^\lambda \right) \\
&= 2\eta^{\lambda\mu} \gamma_\mu \gamma^\nu + 2\gamma^\nu \gamma^\lambda \\
&= 2\gamma^\lambda \gamma^\nu + 2\gamma^\nu \gamma^\lambda = 2[\gamma^\lambda, \gamma^\nu]_+ = 4\eta^{\lambda\nu} \mathbb{1}, \quad (2.113)
\end{aligned}$$

and so on.

The commutation and anticommutation relations also come in handy when we are evaluating traces of products of such matrices. For example, we know from the cyclicity of traces that

$$\operatorname{Tr} \gamma^\mu \gamma^\nu = \operatorname{Tr} \gamma^\nu \gamma^\mu. \qquad (2.114)$$

Therefore, it follows (in 4-dimensions) that

$$\begin{aligned}
\operatorname{Tr} \gamma^\mu \gamma^\nu &= \frac{1}{2} \left(\operatorname{Tr} \gamma^\mu \gamma^\nu + \operatorname{Tr} \gamma^\nu \gamma^\mu \right) = \frac{1}{2} \operatorname{Tr} \left[\gamma^\mu, \gamma^\nu \right]_+ \\
&= \frac{1}{2} \operatorname{Tr} \left(2\eta^{\mu\nu} \mathbb{1} \right) = \eta^{\mu\nu} \operatorname{Tr} \mathbb{1} = 4\eta^{\mu\nu}, \\
\operatorname{Tr} \gamma_5 \gamma^\mu &= \operatorname{Tr} \gamma^\mu \gamma_5 = -\operatorname{Tr} \gamma_5 \gamma^\mu = 0. \qquad (2.115)
\end{aligned}$$

Even more complicated traces can be evaluated by using the basic relations we have developed so far. For example, we note that

$$\begin{aligned}
\operatorname{Tr} \gamma^\mu \gamma^\nu \gamma^\lambda \gamma^\rho &= \operatorname{Tr} \left[\left([\gamma^\mu, \gamma^\nu]_+ - \gamma^\nu \gamma^\mu \right) \gamma^\lambda \gamma^\rho \right] \\
&= \operatorname{Tr} \left(2\eta^{\mu\nu} \gamma^\lambda \gamma^\rho - \gamma^\nu \gamma^\mu \gamma^\lambda \gamma^\rho \right) \\
&= 8\eta^{\mu\nu} \eta^{\lambda\rho} - \operatorname{Tr} \gamma^\nu \left([\gamma^\mu, \gamma^\lambda]_+ - \gamma^\lambda \gamma^\mu \right) \gamma^\rho \\
&= 8\eta^{\mu\nu} \eta^{\lambda\rho} - 8\eta^{\mu\lambda} \eta^{\nu\rho} + \operatorname{Tr} \gamma^\nu \gamma^\lambda \left([\gamma^\mu, \gamma^\rho]_+ - \gamma^\rho \gamma^\mu \right) \\
&= 8\eta^{\mu\nu} \eta^{\lambda\rho} - 8\eta^{\mu\lambda} \eta^{\nu\rho} + 8\eta^{\nu\lambda} \eta^{\mu\rho} - \operatorname{Tr} \gamma^\nu \gamma^\lambda \gamma^\rho \gamma^\mu,
\end{aligned}$$

or, $\quad 2 \operatorname{Tr} \gamma^\mu \gamma^\nu \gamma^\lambda \gamma^\rho = 8\eta^{\mu\nu} \eta^{\lambda\rho} - 8\eta^{\mu\lambda} \eta^{\nu\rho} + 8\eta^{\nu\lambda} \eta^{\mu\rho},$

or, $\quad \operatorname{Tr} \gamma^\mu \gamma^\nu \gamma^\lambda \gamma^\rho = 4 \left(\eta^{\mu\nu} \eta^{\lambda\rho} - \eta^{\mu\lambda} \eta^{\nu\rho} + \eta^{\nu\lambda} \eta^{\mu\rho} \right), \quad (2.116)$

2.6 Properties of the Dirac matrices

and so on. We would use all these properties in the next chapter to study the covariance of the Dirac equation under a Lorentz transformation.

To conclude this section, let us note that we have constructed a particular representation for the Dirac matrices commonly known as the Pauli-Dirac representation. However, there are other equivalent representations possible which may be more useful for a particular system under study. For example, there exists a representation for the Dirac matrices where γ^μ are all purely imaginary. This is known as the Majorana representation and is quite useful in the study of Majorana fermions which are charge neutral fermions. Explicitly, the γ^μ_M matrices have the forms

$$\gamma^0_M = \begin{pmatrix} 0 & \sigma_2 \\ \sigma_2 & 0 \end{pmatrix}, \quad \gamma^1_M = \begin{pmatrix} i\sigma_3 & 0 \\ 0 & i\sigma_3 \end{pmatrix},$$

$$\gamma^2_M = \begin{pmatrix} 0 & -\sigma_2 \\ \sigma_2 & 0 \end{pmatrix}, \quad \gamma^3_M = \begin{pmatrix} -i\sigma_1 & 0 \\ 0 & -i\sigma_1 \end{pmatrix}. \quad (2.117)$$

It can be checked that the Dirac matrices in the Pauli-Dirac representation and the Majorana representation are related by the similarity (unitary) transformation

$$\gamma^\mu_M = S\gamma^\mu S^{-1}, \quad S = \frac{1}{\sqrt{2}}\gamma^0 \left(\mathbb{1} + \gamma^2\right). \quad (2.118)$$

Similarly, there exists yet another representation for the γ^μ matrices, namely,

$$\gamma^\mu_W = \begin{pmatrix} 0 & \sigma^\mu \\ \tilde{\sigma}^\mu & 0 \end{pmatrix}, \quad (2.119)$$

where

$$\sigma^\mu = (\mathbb{1}, \boldsymbol{\sigma}), \quad \tilde{\sigma}^\mu = (\mathbb{1}, -\boldsymbol{\sigma}). \quad (2.120)$$

This is known as the Weyl representation for the Dirac matrices and is quite useful in the study of massless fermions. It can be checked that the Weyl representation is related to the standard Pauli-Dirac representation through the similarity (unitary) transformation

$$\gamma_W^\mu = S\gamma^\mu S^{-1}, \quad S = \frac{1}{\sqrt{2}}\left(\mathbb{1} + \gamma_5\gamma^0\right). \tag{2.121}$$

2.6.1 Fierz rearrangement. As we have pointed out in (2.100), the sixteen Dirac matrices $\Gamma^{(a)}, a = S, V, T, A, P$ define a complete basis for 4×4 matrices. This is easily demonstrated by showing that they are linearly independent which is seen as follows.

We have explicitly constructed the sixteen matrices to correspond to the set

$$\Gamma^{(a)} = \{\mathbb{1}, \gamma^\mu, \sigma^{\mu\nu}, \gamma_5\gamma^\mu, \gamma_5\}. \tag{2.122}$$

From the properties of the γ^μ matrices, it can be easily checked that

$$\operatorname{Tr} \Gamma^{(a)}\Gamma^{(b)} = 0, \qquad a \neq b, \tag{2.123}$$

where "Tr" denotes trace over the matrix indices. As a result, given this set of matrices, we can construct the inverse set of matrices as

$$\Gamma_{(a)} = \frac{\Gamma^{(a)}}{\operatorname{Tr}\left(\Gamma^{(a)}\Gamma^{(a)}\right)}, \qquad a \text{ not summed}, \tag{2.124}$$

such that

$$\operatorname{Tr}\left(\Gamma_{(a)}\Gamma^{(b)}\right) = \delta_{(a)}^{(b)}. \tag{2.125}$$

Explicitly, we can write the inverse set of matrices as

$$\Gamma_{(a)} = \frac{1}{4}\{\mathbb{1}, \gamma_\mu, \sigma_{\mu\nu}, -\gamma_5\gamma_\mu, \gamma_5\}. \tag{2.126}$$

With this, the linear independence of the set of matrices in (2.122) is straightforward. For example, it follows now that if

2.6 Properties of the Dirac matrices

$$\sum_{(a)} C_{(a)} \Gamma^{(a)} = 0, \qquad (2.127)$$

then, multiplying (2.127) with $\Gamma_{(b)}$, where b is arbitrary, and taking trace over the matrix indices and using (2.125) we obtain

$$\operatorname{Tr} \Gamma_{(b)} \sum_{(a)} C_{(a)} \Gamma^{(a)} = 0,$$

or, $\quad \displaystyle\sum_{(a)} C_{(a)} \operatorname{Tr}\left(\Gamma_{(b)} \Gamma^{(a)}\right) = 0,$

or, $\quad \displaystyle\sum_{(a)} C_{(a)} \delta^{(a)}_{(b)} = 0,$

or, $\quad C_{(b)} = 0, \qquad (2.128)$

for any $b = S, V, T, A, P$. Therefore, (2.127) implies that all the coefficients of expansion must vanishing which shows that the set of sixteen matrices $\Gamma^{(a)}$ in (2.122) are linearly independent. As a result they constitute a basis for 4×4 matrices.

Since the set of matrices in (2.122) provide a basis for the 4×4 matrix space, any arbitrary 4×4 matrix M can be expanded as a linear superposition of these matrices, namely,

$$M = \sum_{(a)} C^{(M)}_{(a)} \Gamma^{(a)}. \qquad (2.129)$$

Multiplying this expression with $\Gamma_{(b)}$ and taking trace over the matrix indices, we obtain

$$\operatorname{Tr} \Gamma_{(b)} M = \sum_{(a)} C^{(M)}_{(a)} \operatorname{Tr}\left(\Gamma_{(b)} \Gamma^{(a)}\right) = \sum_{(a)} C^{(M)}_{(a)} \delta^{(a)}_{(b)},$$

or, $\quad C^{(M)}_{(b)} = \operatorname{Tr}\left(\Gamma_{(b)} M\right). \qquad (2.130)$

Substituting (2.130) into the expansion (2.129), we obtain

$$M = \sum_{(a)} \left(\operatorname{Tr} \Gamma_{(a)} M \right) \Gamma^{(a)}. \tag{2.131}$$

Introducing the matrix indices explicitly, this leads to

$$M_{\alpha\beta} = \sum_{(a)} \Gamma_{(a)\ \gamma\eta} M_{\eta\gamma} \Gamma^{(a)}_{\alpha\beta},$$

or, $\quad \sum_{(a)} \left(\Gamma_{(a)} \right)_{\gamma\eta} \left(\Gamma^{(a)} \right)_{\alpha\beta} = \delta_{\alpha\eta} \delta_{\beta\gamma}. \tag{2.132}$

Here $\alpha, \beta, \gamma, \eta = 1, 2, 3, 4$ and correspond to the matrix indices of the 4×4 matrices and we are assuming that the repeated indices are being summed.

Equation (2.132) describes a fundamental relation which expresses the completeness relation for the sixteen basis matrices. Just like any other completeness relation, it can be used effectively in many ways. For example, we note that if M and N denote two arbitrary 4×4 matrices, then using (2.132) we can derive (for simplicity, we will use the standard convention that the repeated index (a) as well as the matrix indices are being summed)

$$\left(\Gamma^{(a)} M \right)_{\gamma\beta} \left(\Gamma_{(a)} N \right)_{\alpha\delta}$$
$$= \left(\Gamma^{(a)} \right)_{\gamma\bar{\beta}} M_{\bar{\beta}\beta} \left(\Gamma_{(a)} \right)_{\alpha\bar{\delta}} N_{\bar{\delta}\delta} = M_{\bar{\beta}\beta} N_{\bar{\delta}\delta} \left(\Gamma^{(a)} \right)_{\gamma\bar{\beta}} \left(\Gamma_{(a)} \right)_{\alpha\bar{\delta}}$$
$$= M_{\bar{\beta}\beta} N_{\bar{\delta}\delta} \delta_{\gamma\bar{\delta}} \delta_{\bar{\beta}\alpha} = M_{\alpha\beta} N_{\gamma\delta},$$

$$\left(M \Gamma^{(a)} \right)_{\alpha\delta} \left(N \Gamma_{(a)} \right)_{\gamma\beta}$$
$$= M_{\alpha\bar{\delta}} \left(\Gamma^{(a)} \right)_{\bar{\delta}\delta} N_{\gamma\bar{\beta}} \left(\Gamma_{(a)} \right)_{\bar{\beta}\beta} = M_{\alpha\bar{\delta}} N_{\gamma\bar{\beta}} \left(\Gamma^{(a)} \right)_{\bar{\delta}\delta} \left(\Gamma_{(a)} \right)_{\bar{\beta}\beta}$$
$$= M_{\alpha\bar{\delta}} N_{\gamma\bar{\beta}} \delta_{\bar{\delta}\beta} \delta_{\delta\bar{\beta}} = M_{\alpha\beta} N_{\gamma\delta}. \tag{2.133}$$

Using the relations in (2.133), it is now straightforward to obtain

$$\left(\bar{\psi}_1 M \Gamma^{(a)} \psi_4\right) \left(\bar{\psi}_3 N \Gamma_{(a)} \psi_2\right)$$
$$= \bar{\psi}_{1\alpha} \left(M\Gamma^{(a)}\right)_{\alpha\delta} \psi_{4\delta} \bar{\psi}_{3\gamma} \left(N\Gamma_{(a)}\right)_{\gamma\beta} \psi_{2\beta}$$
$$= \bar{\psi}_{1\alpha} \psi_{4\delta} \bar{\psi}_{3\gamma} \psi_{2\beta} \left(M\Gamma^{(a)}\right)_{\alpha\delta} \left(N\Gamma_{(a)}\right)_{\gamma\beta}$$
$$= \bar{\psi}_{1\alpha} \psi_{4\delta} \bar{\psi}_{3\gamma} \psi_{2\beta} M_{\alpha\beta} N_{\gamma\delta} = \left(\bar{\psi}_1 M \psi_2\right)\left(\bar{\psi}_3 N \psi_4\right),$$

$$\left(\bar{\psi}_1 \Gamma^{(a)} M \psi_4\right) \left(\bar{\psi}_3 \Gamma_{(a)} N \psi_2\right)$$
$$= \bar{\psi}_{1\gamma} \left(\Gamma^{(a)} M\right)_{\gamma\beta} \psi_{4\beta} \bar{\psi}_{3\alpha} \left(\Gamma_{(a)} N\right)_{\alpha\delta} \psi_{2\delta}$$
$$= \bar{\psi}_{1\gamma} \psi_{4\beta} \bar{\psi}_{3\alpha} \psi_{2\delta} \left(\Gamma^{(a)} M\right)_{\gamma\beta} \left(\Gamma_{(a)} N\right)_{\alpha\delta}$$
$$= \bar{\psi}_{1\gamma} \psi_{4\beta} \bar{\psi}_{3\alpha} \psi_{2\delta} M_{\alpha\beta} N_{\gamma\delta} = \left(\bar{\psi}_3 M \psi_4\right)\left(\bar{\psi}_1 N \psi_2\right). \quad (2.134)$$

The two relations in (2.134) are known as the Fierz rearrangement identities which are very useful in calculating cross sections. In deriving these identities, we have assumed that the spinors are ordinary functions. On the other hand, if they correspond to anti-commuting fermion operators, the right-hand sides of the identities in (2.134) pick up a negative sign which arises from commuting the fermionic fields past one another.

Let us note that using the explicit forms for $\Gamma^{(a)}$ and $\Gamma_{(a)}$ in (2.122) and (2.124) respectively, we can write the first of the Fierz rearrangement identities in (2.134) as (assuming the spinors are ordinary functions)

$$\bar{\psi}_1 M \psi_2 \, \bar{\psi}_3 N \psi_4$$
$$= \frac{1}{4} \left[\bar{\psi}_1 M \psi_4 \, \bar{\psi}_3 N \psi_2 + \bar{\psi}_1 M \gamma^\mu \psi_4 \, \bar{\psi}_3 N \gamma_\mu \psi_2 \right.$$
$$+ \bar{\psi}_1 M \sigma^{\mu\nu} \psi_4 \, \bar{\psi}_3 N \sigma_{\mu\nu} \psi_2 - \bar{\psi}_1 M \gamma_5 \gamma^\mu \psi_4 \, \bar{\psi}_3 N \gamma_5 \gamma_\mu \psi_2$$
$$+ \left. \bar{\psi}_1 M \gamma_5 \psi_4 \, \bar{\psi}_3 N \gamma_5 \psi_2 \right]. \quad (2.135)$$

Since this is true for any matrices M, N and any spinors, we can define a new spinor $\tilde{\psi}_4 = N\psi_4$ to write the identity in (2.135) equivalently as

$$\bar{\psi}_1 M \psi_2 \, \bar{\psi}_3 N \psi_4 = \bar{\psi}_1 M \psi_2 \, \bar{\psi}_3 \tilde{\psi}_4$$
$$= \frac{1}{4} \left[\bar{\psi}_1 M \tilde{\psi}_4 \, \bar{\psi}_3 \psi_2 + \bar{\psi}_1 M \gamma^\mu \tilde{\psi}_4 \, \bar{\psi}_3 \gamma_\mu \psi_2 \right.$$
$$+ \bar{\psi}_1 M \sigma^{\mu\nu} \tilde{\psi}_4 \, \bar{\psi}_3 \sigma_{\mu\nu} \psi_2 - \bar{\psi}_1 M \gamma_5 \gamma^\mu \tilde{\psi}_4 \, \bar{\psi}_3 \gamma_5 \gamma_\mu \psi_2$$
$$\left. + \bar{\psi}_1 M \gamma_5 \tilde{\psi}_4 \, \bar{\psi}_3 \gamma_5 \psi_2 \right]$$
$$= \frac{1}{4} \left[\bar{\psi}_1 M N \psi_4 \, \bar{\psi}_3 \psi_2 + \bar{\psi}_1 M \gamma^\mu N \psi_4 \, \bar{\psi}_3 \gamma_\mu \psi_2 \right.$$
$$+ \bar{\psi}_1 M \sigma^{\mu\nu} N \psi_4 \, \bar{\psi}_3 \sigma_{\mu\nu} \psi_2 - \bar{\psi}_1 M \gamma_5 \gamma^\mu N \psi_4 \, \bar{\psi}_3 \gamma_5 \gamma_\mu \psi_2$$
$$\left. + \bar{\psi}_1 M \gamma_5 N \psi_4 \, \bar{\psi}_3 \gamma_5 \psi_2 \right], \tag{2.136}$$

which is often calculationally simpler. Thus, for example, if we choose

$$M = (\mathbb{1} - \gamma_5) \gamma^\mu, \quad N = (\mathbb{1} - \gamma_5) \gamma_\mu, \tag{2.137}$$

then using various properties of the gamma matrices derived earlier as well as (2.110) and (2.113), we obtain from (2.136)

$$\bar{\psi}_1 (\mathbb{1} - \gamma_5) \gamma^\mu \psi_2 \, \bar{\psi}_3 (\mathbb{1} - \gamma_5) \gamma_\mu \psi_4$$
$$= -\bar{\psi}_1 (\mathbb{1} - \gamma_5) \gamma^\mu \psi_4 \, \bar{\psi}_3 (\mathbb{1} - \gamma_5) \gamma_\mu \psi_2. \tag{2.138}$$

This is the well known fact from the weak interactions that the $V - A$ form of the weak interaction Hamiltonian proposed by Sudarshan and Marshak is form invariant under a Fierz rearrangement (the negative sign is there simply because we are considering spinor functions and will be absent for anti-commuting fermion fields).

2.7 References

1. L. I. Schiff, *Quantum Mechanics*, McGraw-Hill, New York, 1968.

2. J. D. Bjorken and S. Drell, *Relativistic Quantum Mechanics*, McGraw-Hill, New York, 1964.

3. C. Itzykson and J-B. Zuber, *Quantum Field Theory*, McGraw-Hill, New York, 1980.

4. A. Das, *Lectures on Quantum Mechanics*, Hindustan Publishing, New Delhi, India, 2003.

5. S. Okubo, *Real representations of finite Clifford algebras. I. Classification*, Journal of Mathematical Physics **32**, 1657 (1991).

6. E. C. G. Sudarshan and R. E. Marshak, *Proceedings of Padua-Venice conference on mesons and newly discovered particles*, (1957); Physical Review **109**, 1860 (1958).

CHAPTER 3
Properties of the Dirac equation

3.1 Lorentz transformations

In three dimensions, we are well acquainted with rotations. For example, we know that a rotation of coordinates around the z-axis by an angle θ can be represented as the transformation

$$\mathbf{x} \to \mathbf{x}' = R\mathbf{x}, \tag{3.1}$$

where R represents the rotation matrix such that

$$\begin{aligned} x'_1 &= \cos\theta\, x_1 - \sin\theta\, x_2, \\ x'_2 &= \sin\theta\, x_1 + \cos\theta\, x_2, \\ x'_3 &= x_3. \end{aligned} \tag{3.2}$$

Here we are using a three dimensional notation, but this can also be written in terms of the four vector notation we have developed. The rotation around the z-axis in (3.2) can also be written in matrix form as

$$\begin{pmatrix} x'_1 \\ x'_2 \\ x'_3 \end{pmatrix} = \begin{pmatrix} \cos\theta & -\sin\theta & 0 \\ \sin\theta & \cos\theta & 0 \\ 0 & 0 & 1 \end{pmatrix} \begin{pmatrix} x_1 \\ x_2 \\ x_3 \end{pmatrix}, \tag{3.3}$$

so that the coefficient matrix on the right hand side can be identified with the rotation matrix in (3.1). Thus, we see from (3.3) that a rotation around the 3-axis (z-axis) or in the 1-2 plane is denoted by an orthogonal matrix, R ($R^T R = \mathbb{1}$), with unit determinant. We also note from (3.2) that an infinitesimal rotation around the 3-axis (z-axis) takes the form

$$\begin{aligned}
x'_1 &= x_1 - \epsilon x_2, \\
x'_2 &= \epsilon x_1 + x_2, \\
x'_3 &= x_3,
\end{aligned} \quad (3.4)$$

where we have identified $\theta = \epsilon =$ infinitesimal. We observe here that the matrix representing the infinitesimal change under a rotation is anti-symmetric.

Under a Lorentz boost along the x-axis, we also know that the coordinates transform as (boost velocity $v = \beta$ since $c = 1$)

$$x^\mu \to x'^\mu, \quad (3.5)$$

such that

$$\begin{aligned}
x'^0 &= \gamma x^0 - \gamma \beta x^1, \\
x'^1 &= -\gamma \beta x^0 + \gamma x^1, \\
x'^2 &= x^2, \\
x'^3 &= x^3,
\end{aligned} \quad (3.6)$$

where the Lorentz factor γ is defined in terms of the boost velocity to be

$$\gamma = \frac{1}{\sqrt{1-\beta^2}}. \quad (3.7)$$

We recognize that (3.6) can also be written in the matrix form as

3.1 LORENTZ TRANSFORMATIONS

$$\begin{pmatrix} x'^0 \\ x'^1 \\ x'^2 \\ x'^3 \end{pmatrix} = \begin{pmatrix} \gamma & -\gamma\beta & 0 & 0 \\ -\gamma\beta & \gamma & 0 & 0 \\ 0 & 0 & 1 & 0 \\ 0 & 0 & 0 & 1 \end{pmatrix} \begin{pmatrix} x^0 \\ x^1 \\ x^2 \\ x^3 \end{pmatrix}$$

$$= \begin{pmatrix} \cosh\omega & -\sinh\omega & 0 & 0 \\ -\sinh\omega & \cosh\omega & 0 & 0 \\ 0 & 0 & 1 & 0 \\ 0 & 0 & 0 & 1 \end{pmatrix} \begin{pmatrix} x^0 \\ x^1 \\ x^2 \\ x^3 \end{pmatrix}, \quad (3.8)$$

where we have defined

$$\begin{aligned} \cosh\omega &= \gamma = \frac{1}{\sqrt{1-\beta^2}}, \\ \sinh\omega &= \gamma\beta = \frac{\beta}{\sqrt{1-\beta^2}}, \end{aligned} \quad (3.9)$$

so that

$$\cosh^2\omega - \sinh^2\omega = \frac{1}{1-\beta^2} - \frac{\beta^2}{1-\beta^2} = 1. \quad (3.10)$$

Since the range of the boost velocity is given by $-1 \leq \beta \leq 1$, we conclude from (3.9) that $-\infty \leq \omega \leq \infty$.

Thus, we note that a Lorentz boost along the x-direction can be written as

$$x'^\mu = \Lambda^\mu{}_\nu x^\nu, \quad (3.11)$$

where

$$\Lambda^\mu{}_\nu = \begin{pmatrix} \cosh\omega & -\sinh\omega & 0 & 0 \\ -\sinh\omega & \cosh\omega & 0 & 0 \\ 0 & 0 & 1 & 0 \\ 0 & 0 & 0 & 1 \end{pmatrix}. \quad (3.12)$$

From this, we can obtain,

$$\Lambda_\mu^{\ \nu} = \eta_{\mu\lambda}\,\eta^{\nu\rho}\,\Lambda^\lambda_{\ \rho} = \begin{pmatrix} \cosh\omega & \sinh\omega & 0 & 0 \\ \sinh\omega & \cosh\omega & 0 & 0 \\ 0 & 0 & 1 & 0 \\ 0 & 0 & 0 & 1 \end{pmatrix}, \qquad (3.13)$$

which would lead to the transformation of the covariant coordinate vector, namely,

$$x'_\mu = \Lambda_\mu^{\ \nu}\, x_\nu. \qquad (3.14)$$

The matrix representing the Lorentz transformation of the coordinates, $\Lambda^\mu_{\ \nu}$ (or $\Lambda_\mu^{\ \nu}$), is easily seen from (3.12) or (3.13) to be an orthogonal matrix in the sense that

$$(\Lambda^T)^\mu_{\ \nu}\,\Lambda^\nu_{\ \lambda} = \Lambda_\nu^{\ \mu}\,\Lambda^\nu_{\ \lambda} = \delta^\mu_\lambda, \qquad (3.15)$$

and also has a unit determinant, much like the rotation matrix R in (3.3). (Incidentally, (3.15) also shows that the covariant vector transforms in an inverse manner compared with the contravariant vector.) Therefore, we can think of the Lorentz boost along the 1-axis as a rotation in the 0-1 plane with an imaginary angle (so that we have hyperbolic functions instead of ordinary trigonometric functions). (That these rotations become complex is related to the fact that the metric has opposite signature for time and space components.) Furthermore, as we have seen, the "angle" of rotation, ω, (or the parameter of boost) can take any real value and, as a result, Lorentz boosts correspond to noncompact transformations unlike space rotations.

Let us finally note that if we are considering an infinitesimal Lorentz boost along the 1-axis (or a rotation in the 0-1 plane), then we can write ($\omega = \epsilon =$ infinitesimal)

$$\Lambda^\mu_{\ \nu} = \begin{pmatrix} 1 & -\epsilon & 0 & 0 \\ -\epsilon & 1 & 0 & 0 \\ 0 & 0 & 1 & 0 \\ 0 & 0 & 0 & 1 \end{pmatrix} = \delta^\mu_{\ \nu} + \epsilon^\mu_{\ \nu}, \qquad (3.16)$$

where,

$$\epsilon^\mu{}_\nu = \begin{pmatrix} 0 & -\epsilon & 0 & 0 \\ -\epsilon & 0 & 0 & 0 \\ 0 & 0 & 0 & 0 \\ 0 & 0 & 0 & 0 \end{pmatrix}. \tag{3.17}$$

It follows from this that

$$\epsilon^{\mu\nu} = \eta^{\nu\lambda} \epsilon^\mu{}_\lambda = \begin{pmatrix} 0 & \epsilon & 0 & 0 \\ -\epsilon & 0 & 0 & 0 \\ 0 & 0 & 0 & 0 \\ 0 & 0 & 0 & 0 \end{pmatrix} = -\epsilon^{\nu\mu}. \tag{3.18}$$

In other words, the matrix representing the change under an infinitesimal Lorentz boost is anti-symmetric just like the case of an infinitesimal rotation. In a general language, therefore, we note that we can combine rotations and Lorentz boosts into what are known as the homogeneous Lorentz transformations, which can be thought of as rotations in the four dimensional space-time.

General Lorentz transformations are defined as the transformations

$$x'^\mu = \Lambda^\mu{}_\nu x^\nu, \tag{3.19}$$

which leave the length of the vector invariant, namely,

$$\eta_{\mu\nu} x'^\mu x'^\nu = \eta_{\mu\nu} x^\mu x^\nu,$$

or, $\quad \eta_{\mu\nu} \Lambda^\mu{}_\rho x^\rho \Lambda^\nu{}_\sigma x^\sigma = \eta_{\mu\nu} x^\mu x^\nu,$

or, $\quad \eta_{\mu\nu} \Lambda^\mu{}_\rho \Lambda^\nu{}_\sigma x^\rho x^\sigma = \eta_{\rho\sigma} x^\rho x^\sigma,$

or, $\quad \Lambda_{\mu\rho} \Lambda^\mu{}_\sigma = \eta_{\rho\sigma},$

or, $\quad \Lambda_\mu{}^\rho \Lambda^\mu{}_\sigma = \delta^\rho_\sigma. \tag{3.20}$

This is, of course, what we have seen before in (3.15). Lorentz transformations define the maximal symmetry of the space-time manifold which leaves the origin invariant.

Choosing $\rho = \sigma = 0$, we can write out the relation (3.20) explicitly as

$$\Lambda_0{}^0 \Lambda^0{}_0 + \Lambda_i{}^0 \Lambda^i{}_0 = 1,$$
$$\text{or,} \quad \left(\Lambda^0{}_0\right)^2 - \left(\Lambda^i{}_0\right)^2 = 1,$$
$$\text{or,} \quad \left(\Lambda^0{}_0\right)^2 = 1 + \left(\Lambda^i{}_0\right)^2 \geq 1. \tag{3.21}$$

Therefore, we conclude that

$$\Lambda^0{}_0 \geq 1, \quad \text{or,} \quad \Lambda^0{}_0 \leq -1. \tag{3.22}$$

If $\Lambda^0{}_0 \geq 1$, then the transformation is called orthochronous. (The Greek prefix "ortho" means straight up. Thus, orthochronous means straight up in time. Namely, such a Lorentz transformation does not change the direction of time. Incidentally, "gonia" in Greek means an angle or a corner and, therefore, orthogonal means the corner that is straight up (perpendicular). In the same spirit, an orthodontist is someone who can make your teeth straight.) Note also that since

$$\Lambda^T \Lambda = \mathbb{1}, \tag{3.23}$$

we obtain (for clarity, we note that $\left(\Lambda^T\right)^\mu{}_\nu = \Lambda_\nu{}^\mu$ and $\left(\Lambda^T\right)_\mu{}^\nu = \Lambda^\nu{}_\mu$ as can be seen from (3.20))

$$(\det \Lambda)^2 = 1,$$
$$\text{or,} \quad \det \Lambda = \pm 1. \tag{3.24}$$

The set of homogeneous Lorentz transformations satisfying

$$\det \Lambda = 1, \quad \text{and} \quad \Lambda^0{}_0 \geq 1, \tag{3.25}$$

3.1 LORENTZ TRANSFORMATIONS

are known as the proper, orthochronous Lorentz transformations and constitute a set of continuous transformations that can be connected to the identity matrix. (Just to emphasize, we note that the set of transformations with $\det \Lambda = 1$ are known as proper transformations and the set for which $\Lambda^0{}_0 \geq 1$ are called orthochronous.) In general, however, there are four kinds of Lorentz transformations, namely,

$$\Lambda^0{}_0 \geq 1, \quad \det \Lambda = \pm 1,$$
$$\Lambda^0{}_0 \leq -1, \quad \det \Lambda = \pm 1. \tag{3.26}$$

Given the proper orthochronous Lorentz transformations, we can obtain the other Lorentz transformations by simply appending space reflection or time reflection or both (which are discrete transformations). Thus, if Λ_{prop} denotes a proper Lorentz transformation, then by adding space reflection, $\mathbf{x} \to -\mathbf{x}$, we obtain a Lorentz transformation

$$\Lambda = \Lambda_{\text{space}} \Lambda_{\text{prop}} = \begin{pmatrix} 1 & 0 & 0 & 0 \\ 0 & -1 & 0 & 0 \\ 0 & 0 & -1 & 0 \\ 0 & 0 & 0 & -1 \end{pmatrix} \Lambda_{\text{prop}}. \tag{3.27}$$

This would correspond to having $\Lambda^0{}_0 \geq 1$, $\det \Lambda = -1$ (which is orthochronous but no longer proper). If we add time reversal, $t \to -t$, to a proper orthochronous Lorentz transformation, then we obtain a Lorentz transformation

$$\Lambda = \Lambda_{\text{time}} \Lambda_{\text{prop}} = \begin{pmatrix} -1 & 0 & 0 & 0 \\ 0 & 1 & 0 & 0 \\ 0 & 0 & 1 & 0 \\ 0 & 0 & 0 & 1 \end{pmatrix} \Lambda_{\text{prop}}, \tag{3.28}$$

satisfying $\Lambda^0{}_0 \leq -1$ and $\det \Lambda = -1$ (which is neither proper nor orthochronous). Finally, if we add both space and time reflections, $x^\mu \to -x^\mu$, to a proper Lorentz tranformation, we obtain a Lorentz transformation

$$\Lambda = \Lambda_{\text{space-time}}\Lambda_{\text{prop}} = \begin{pmatrix} -1 & 0 & 0 & 0 \\ 0 & -1 & 0 & 0 \\ 0 & 0 & -1 & 0 \\ 0 & 0 & 0 & -1 \end{pmatrix}\Lambda_{\text{prop}}, \qquad (3.29)$$

with $\Lambda^0{}_0 \leq -1$ and $\det \Lambda = 1$ (which is proper but not orthochronous). These additional transformations, however, cannot be continuously connected to the identity matrix since they involve discrete reflections. In these lectures, we would refer to proper orthochronous Lorentz transformations as the Lorentz transformations.

3.2 Covariance of the Dirac equation

Given any dynamical equation of the form

$$L\psi = 0, \qquad (3.30)$$

where L is a linear operator, we say that it is covariant under a given transformation provided the transformed equation has the form

$$L'\psi' = 0, \qquad (3.31)$$

where ψ' represents the transformed wavefunction and L' stands for the transformed operator (namely, the operator L, with the transformed variables). In simple terms, covariance implies that a given equation is form invariant under a particular transformation (has the same form in different reference frames).

With this general definition, let us now consider the Dirac equation

$$(i\gamma^\mu \partial_\mu - m)\psi(x) = 0. \qquad (3.32)$$

Under a Lorentz transformation

$$x^\mu \to x'^\mu = \Lambda^\mu{}_\nu x^\nu, \qquad (3.33)$$

3.2 COVARIANCE OF THE DIRAC EQUATION

if the transformed equation has the form

$$(i\gamma^\mu \partial'_\mu - m)\psi'(x') = 0, \tag{3.34}$$

where $\psi'(x')$ is the Lorentz transformed wave function, then the Dirac equation would be covariant under a Lorentz transformation. Note that the Dirac matrices, γ^μ, are a set of four space-time independent matrices and, therefore, do not change under a Lorentz transformation.

Let us assume that, under a Lorentz transformation, the transformed wavefunction has the form

$$\psi'(x') = \psi'(\Lambda x) = S(\Lambda)\psi(x), \tag{3.35}$$

where $S(\Lambda)$ is a 4×4 matrix, since $\psi(x)$ is a four component spinor. Parenthetically, what this means is that we are finding a representation of the Lorentz transformation on the Hilbert space. In the notation of other symmetries that we know from studies in non-relativistic quantum mechanics, we can define an operator $L(\Lambda)$ to represent the Lorentz transformation on the coordinate states as (with indices suppressed)

$$|x\rangle \rightarrow |x'\rangle = |\Lambda x\rangle = L(\Lambda)|x\rangle. \tag{3.36}$$

However, since the Dirac wavefunction is a four component spinor, in addition to the change in the coordinates, the Lorentz transformation can also mix up the spinor components (much like angular momentum/rotation). Thus, we can define the Lorentz transformation acting on the Dirac Hilbert space (Hilbert space of states describing a Dirac particle) as, (with $S(\Lambda)$ representing the 4×4 matrix which rotates the matrix components of the wave function)

$$\begin{aligned} |\psi\rangle \rightarrow |\psi'\rangle &= L(\Lambda)S(\Lambda)|\psi\rangle \\ &= L(\Lambda)S(\Lambda) \int dx\, |x\rangle\langle x|\psi\rangle \\ &= \int dx\, L(\Lambda)|x\rangle S(\Lambda)\psi(x) \end{aligned}$$

$$= \int \mathrm{d}x \, |\Lambda x\rangle S(\Lambda)\psi(x), \tag{3.37}$$

where the wave function is recognized to be

$$\psi(x) = \langle x|\psi\rangle, \tag{3.38}$$

so that, from (3.37) we obtain (see (3.35))

$$\psi'(x') = \langle x'|\psi'\rangle = S(\Lambda)\psi(x). \tag{3.39}$$

Namely, the effect of the Lorentz transformation, on the wave function, can be represented by a matrix $S(\Lambda)$ which depends only on the parameter of transformation Λ and not on the space-time coordinates. A more physical way to understand this is to note that the Dirac wave function simply consists of four functions which do not change, but get rotated by the $S(\Lambda)$ matrix.

Since the Lorentz transformations are invertible, the matrix $S(\Lambda)$ must possess an inverse so that from (3.35) we can write

$$\psi(x) = S^{-1}(\Lambda)\psi'(x'). \tag{3.40}$$

Let us also note from (3.33) that

$$\frac{\partial x'^\mu}{\partial x^\nu} = \Lambda^\mu{}_\nu, \tag{3.41}$$

define a set of real quantities. Thus, we can write

$$(i\gamma^\mu \partial_\mu - m)\psi(x) = 0,$$

or, $\quad \left(i\gamma^\mu \dfrac{\partial x'^\nu}{\partial x^\mu} \dfrac{\partial}{\partial x'^\nu} - m\right) S^{-1}(\Lambda)\psi'(x') = 0,$

or, $\quad \left(i\gamma^\mu \Lambda^\nu{}_\mu \partial'_\nu - m\right) S^{-1}(\Lambda)\psi'(x') = 0,$

or, $\quad \left(i\Lambda^\mu{}_\nu \gamma^\nu \partial'_\mu - m\right) S^{-1}(\Lambda)\psi'(x') = 0,$

or, $\quad \left(i\Lambda^\mu{}_\nu S\gamma^\nu S^{-1} \partial'_\mu - m\right) \psi'(x') = 0, \tag{3.42}$

3.2 COVARIANCE OF THE DIRAC EQUATION

where we have used (3.40).

Therefore, we see from (3.42) that the Dirac equation will be form invariant (covariant) under a Lorentz transformation provided there exists a matrix $S(\Lambda)$, generating Lorentz transformations (for the Dirac wavefunction), such that

$$\Lambda^\mu_{\ \nu} S\gamma^\nu S^{-1} = \gamma^\mu,$$
$$\text{or,} \quad \Lambda^\mu_{\ \nu} \gamma^\nu = S^{-1}\gamma^\mu S. \tag{3.43}$$

Let us note that if we define

$$\gamma'^\mu = \Lambda^\mu_{\ \nu} \gamma^\nu, \tag{3.44}$$

then,

$$\begin{aligned}
\left[\gamma'^\mu, \gamma'^\nu\right]_+ &= \left[\Lambda^\mu_{\ \rho} \gamma^\rho, \Lambda^\nu_{\ \sigma} \gamma^\sigma\right]_+ \\
&= \Lambda^\mu_{\ \rho} \Lambda^\nu_{\ \sigma} \left[\gamma^\rho, \gamma^\sigma\right]_+ = \Lambda^\mu_{\ \rho} \Lambda^\nu_{\ \sigma} 2\eta^{\rho\sigma}\mathbb{1} \\
&= 2\Lambda^\mu_{\ \rho}\Lambda^{\nu\rho}\mathbb{1} = 2\eta^{\mu\nu}\mathbb{1},
\end{aligned} \tag{3.45}$$

where we have used the orthogonality of the Lorentz transformations (see (3.15)). Therefore, the matrices γ'^μ satisfy the Clifford algebra and, by Pauli's fundamental theorem, there must exist a matrix connecting the two representations, γ^μ and γ'^μ. It now follows from (3.43) that the matrix S exists and all we need to show is that it also generates Lorentz transformations in order to prove that the Dirac equation is covariant under a Lorentz transformation.

Next, let us note that since the parameters of Lorentz transformation are real (namely, $(\Lambda^*)^\mu_{\ \nu} = \Lambda^\mu_{\ \nu}$)

$$\gamma^0 (\Lambda^\mu_{\ \nu}\gamma^\nu)^\dagger \gamma^0 = \Lambda^\mu_{\ \nu}\gamma^0 (\gamma^\nu)^\dagger \gamma^0 = \Lambda^\mu_{\ \nu}\gamma^\nu,$$
$$\text{or,} \quad \gamma^0 (S^{-1}\gamma^\mu S)^\dagger \gamma^0 = S^{-1}\gamma^\mu S,$$
$$\text{or,} \quad S\gamma^0 S^\dagger \gamma^0 \gamma^0 \gamma^{\mu\dagger} \gamma^0 \gamma^0 S^{-1\dagger}\gamma^0 S^{-1} = \gamma^\mu,$$
$$\text{or,} \quad (S\gamma^0 S^\dagger \gamma^0)\gamma^\mu (S\gamma^0 S^\dagger \gamma^0)^{-1} = \gamma^\mu. \tag{3.46}$$

Here we have used (3.43) and the relations $(\gamma^0)^\dagger = \gamma^0 = (\gamma^0)^{-1}$ as well as $\gamma^0(\gamma^\mu)^\dagger\gamma^0 = \gamma^\mu$. It is clear from (3.46) that the matrix $S\gamma^0 S^\dagger \gamma^0$ commutes with all the γ^μ matrices and, therefore, must be proportional to the identity matrix (this can be easily checked by taking a linear combination of the sixteen basis matrices in (2.100) and calculating the commutator with γ^μ). As a result, we can denote

$$S\gamma^0 S^\dagger \gamma^0 = b\mathbb{1},$$
$$\text{or,} \quad S^\dagger \gamma^0 = b\gamma^0 S^{-1}. \qquad (3.47)$$

Taking the Hermitian conjugate of (3.47), we obtain

$$(S\gamma^0 S^\dagger \gamma^0)^\dagger = b^*\mathbb{1},$$
$$\text{or,} \quad \gamma^0 S \gamma^0 S^\dagger = b^*\mathbb{1},$$
$$\text{or,} \quad \gamma^0(\gamma^0 S\gamma^0 S^\dagger)\gamma^0 = b^*\mathbb{1},$$
$$\text{or,} \quad S\gamma^0 S^\dagger \gamma^0 = b^*\mathbb{1} = b\mathbb{1}, \qquad (3.48)$$

which, therefore, determines that the parameter b is real, namely,

$$b = b^*. \qquad (3.49)$$

We also note that $\det \gamma^0 = 1$ and since we are interested in proper Lorentz transformations, $\det S = 1$. Using these in (3.47), we determine

$$\det(S\gamma^0 S^\dagger \gamma^0) = \det(b\mathbb{1}),$$
$$\text{or,} \quad b^4 = 1. \qquad (3.50)$$

The real roots of this equation are

$$b = \pm 1. \qquad (3.51)$$

In fact, we can determine the unique value of b in the following way.

Let us note, using (3.43) and (3.47), that

$$\begin{aligned}
S^\dagger S &= S^\dagger \gamma^0 \gamma^0 S \\
&= b\gamma^0 S^{-1} \gamma^0 S \\
&= b\gamma^0 (\Lambda^0{}_\nu \gamma^\nu) \\
&= b\gamma^0 (\Lambda^0{}_0 \gamma^0 + \Lambda^0{}_i \gamma^i) \\
&= b(\Lambda^0{}_0 + \Lambda^0{}_i \gamma^0 \gamma^i),
\end{aligned}$$

$$\text{or,} \quad \operatorname{Tr} S^\dagger S = 4b\Lambda^0{}_0 \rangle 0, \tag{3.52}$$

which follows since $S^\dagger S$ represents a non-negative matrix. The two solutions of this equation are obvious

$$\begin{aligned}
&\Lambda^0{}_0 \geq 1, & b &> 0, \\
\text{or,} \quad &\Lambda^0{}_0 \leq -1, & b &< 0.
\end{aligned} \tag{3.53}$$

Since we are dealing with proper Lorentz transformations, we are assuming

$$\Lambda^0{}_0 \geq 1, \tag{3.54}$$

which implies (see (3.53)) that $b > 0$ and, therefore, it follows from (3.51) that

$$b = 1. \tag{3.55}$$

Thus, we conclude from (3.47) that

$$S\gamma^0 S^\dagger \gamma^0 = \mathbb{1},$$
$$\text{or,} \quad S^\dagger \gamma^0 = \gamma^0 S^{-1}. \tag{3.56}$$

These are some of the properties satisfied by the matrix S which will be useful in showing that it provides a representation for the Lorentz transformations.

Next, let us consider an infinitesimal Lorentz transformation of the form ($\epsilon^\mu_{\ \nu}$ infinitesimal)

$$x'^\mu = \Lambda^\mu_{\ \nu} x^\nu = (\delta^\mu_{\ \nu} + \epsilon^\mu_{\ \nu}) x^\nu = x^\mu + \epsilon^\mu_{\ \nu} x^\nu. \tag{3.57}$$

From our earlier discussion in (3.18), we recall that the infinitesimal transformation matrix is antisymmetric, namely,

$$\epsilon^{\mu\nu} = -\epsilon^{\nu\mu}. \tag{3.58}$$

For an infinitesimal transformation, therefore, we can expand the matrix $S(\Lambda)$ as

$$S(\Lambda) = S(\epsilon) = \mathbb{1} - \frac{i}{4} M_{\mu\nu} \epsilon^{\mu\nu}, \tag{3.59}$$

where the matrices $M_{\mu\nu}$ are assumed to be anti-symmetric in the Lorentz indices (for different values of the Lorentz indices, $M_{\mu\nu}$ denote matrices in the Dirac space),

$$M_{\mu\nu} = -M_{\nu\mu}, \tag{3.60}$$

since

$$\epsilon^{\mu\nu} = -\epsilon^{\nu\mu}. \tag{3.61}$$

We can also write

$$S^{-1}(\epsilon) = \mathbb{1} + \frac{i}{4} M_{\mu\nu} \epsilon^{\mu\nu}, \tag{3.62}$$

so that

$$\begin{aligned} S^{-1}(\epsilon) S(\epsilon) &= \left(\mathbb{1} + \frac{i}{4} M_{\mu\nu} \epsilon^{\mu\nu}\right) \left(\mathbb{1} - \frac{i}{4} M_{\sigma\tau} \epsilon^{\sigma\tau}\right) \\ &= \mathbb{1} + \frac{i}{4} M_{\mu\nu} \epsilon^{\mu\nu} - \frac{i}{4} M_{\mu\nu} \epsilon^{\mu\nu} + O(\epsilon^2) \\ &= \mathbb{1} + O(\epsilon^2). \end{aligned} \tag{3.63}$$

3.2 COVARIANCE OF THE DIRAC EQUATION

To the leading order, therefore, $S^{-1}(\epsilon)$ indeed represents the inverse of the matrix $S(\epsilon)$.

The defining relation for the matrix $S(\Lambda)$ in (3.43) now takes the form

$$S^{-1}(\epsilon)\gamma^\mu S(\epsilon) = \Lambda^\mu{}_\nu \gamma^\nu = (\delta^\mu{}_\nu + \epsilon^\mu{}_\nu)\gamma^\nu,$$

or, $\quad \left(\mathbb{1} + \dfrac{i}{4} M_{\lambda\rho}\epsilon^{\lambda\rho}\right)\gamma^\mu \left(\mathbb{1} - \dfrac{i}{4} M_{\sigma\tau}\epsilon^{\sigma\tau}\right) = \gamma^\mu + \epsilon^\mu{}_\nu \gamma^\nu,$

or, $\quad \gamma^\mu + \dfrac{i}{4} \epsilon^{\lambda\rho} M_{\lambda\rho}\gamma^\mu - \dfrac{i}{4} \epsilon^{\lambda\rho}\gamma^\mu M_{\lambda\rho} + 0(\epsilon^2) = \gamma^\mu + \epsilon^\mu{}_\nu \gamma^\nu,$

or, $\quad -\dfrac{i}{4} \epsilon_{\lambda\rho}[\gamma^\mu, M^{\lambda\rho}] = \epsilon^\mu{}_\nu \gamma^\nu. \qquad (3.64)$

At this point, let us recall the commutation relation (2.106)

$$[\gamma^\mu, \sigma^{\nu\lambda}] = 2i(\eta^{\mu\nu}\gamma^\lambda - \eta^{\mu\lambda}\gamma^\nu), \qquad (3.65)$$

and note from (3.64) that if we identify

$$M^{\lambda\rho} = \sigma^{\lambda\rho}, \qquad (3.66)$$

then,

$$\begin{aligned}
-\dfrac{i}{4} \epsilon_{\lambda\rho}[\gamma^\mu, M^{\lambda\rho}] &= -\dfrac{i}{4} \epsilon_{\lambda\rho}[\gamma^\mu, \sigma^{\lambda\rho}] \\
&= -\dfrac{i}{4} \epsilon_{\lambda\rho} \times 2i(\eta^{\mu\lambda}\gamma^\rho - \eta^{\mu\rho}\gamma^\lambda) \\
&= \dfrac{1}{2}(\epsilon^\mu{}_\rho \gamma^\rho + \epsilon^\mu{}_\lambda \gamma^\lambda) = \epsilon^\mu{}_\nu \gamma^\nu,
\end{aligned} \qquad (3.67)$$

which coincides with the right hand side of (3.64). Therefore, we see that for infinitesimal transformations, we have determined the form of $S(\epsilon)$ to be

$$S(\epsilon) = \mathbb{1} - \dfrac{i}{4} M_{\mu\nu}\epsilon^{\mu\nu} = \mathbb{1} - \dfrac{i}{4} \sigma_{\mu\nu}\epsilon^{\mu\nu}. \qquad (3.68)$$

Let us note here from the form of $S(\epsilon)$ that we can identify

$$\frac{1}{2}\sigma_{\mu\nu}, \tag{3.69}$$

with the generators of infinitesimal Lorentz transformations for the Dirac wave function. (The other factor of $\frac{1}{2}$ is there to avoid double counting.) We will see in the next chapter (when we study the representations of the Lorentz group) that the algebra (2.109) which the generators of the infinitesimal transformations, $\frac{1}{2}\sigma_{\mu\nu}$, satisfy can be identified with the Lorentz algebra (which also explains why they are closed under multiplication).

Thus, at least for infinitesimal Lorentz transformations, we have shown that there exists a $S(\Lambda)$ which satisies (3.43) and generates Lorentz transformations and as a result, the Dirac equation is form invariant (covariant) under such a Lorentz transformation. A finite transformation can, of course, be constructed out of a series of infinitesimal transformations and, consequently, the matrix $S(\Lambda)$ for a finite Lorentz transformation will be the product of a series of such infinitesimal matrices which leads to an exponentiation of the infinitesimal generators with the appropriate parameters of transformation.

For completeness, let us note that infinitesimal rotations around the 3-axis or in the 1-2 plane would correspond to choosing

$$\epsilon^{12} = \epsilon = -\epsilon^{21}, \tag{3.70}$$

with all other components of $\epsilon^{\mu\nu}$ vanishing. In such a case (see also (2.98)),

$$S(\epsilon) = \mathbb{1} - \frac{i}{2}\sigma_{12}\epsilon = \mathbb{1} + \frac{1}{2}\gamma_1\gamma_2\epsilon = \mathbb{1} - \frac{i}{2}\tilde{\alpha}_3\epsilon. \tag{3.71}$$

A finite rotation by angle θ in the 1-2 plane would, then, be obtained from an infinite sequence of infinitesimal transformations resulting in an exponentiation of the infinitesimal generators as

$$S(\theta) = e^{-\frac{i}{2}\tilde{\alpha}_3\theta}. \tag{3.72}$$

3.2 COVARIANCE OF THE DIRAC EQUATION

Note that since $\tilde{\alpha}_i^\dagger = \tilde{\alpha}_i$, we have $S^\dagger(\theta) = S^{-1}(\theta)$, namely, rotations define unitary transformations. Furthermore, recalling that

$$\tilde{\alpha}_3 = \begin{pmatrix} \sigma_3 & 0 \\ 0 & \sigma_3 \end{pmatrix}, \tag{3.73}$$

we have

$$(\tilde{\alpha}_3)^2 = \begin{pmatrix} \mathbb{1} & 0 \\ 0 & \mathbb{1} \end{pmatrix} = \mathbb{1}, \tag{3.74}$$

and, therefore, we can determine

$$S(\theta) = \cos\frac{\theta}{2} - i\tilde{\alpha}_3 \sin\frac{\theta}{2}. \tag{3.75}$$

This shows that

$$\begin{aligned} S(\theta + 2\pi) &= -\left(\cos\frac{\theta}{2} - i\tilde{\alpha}_3 \sin\frac{\theta}{2}\right) = -S(\theta), \\ S(\theta + 4\pi) &= S(\theta). \end{aligned} \tag{3.76}$$

That is, the rotation operator, in this case, is double valued and, therefore, corresponds to a spinor representation. This is, of course, consistent with the fact that the Dirac equation describes spin $\frac{1}{2}$ particles.

Let us next consider an infinitesimal rotation in the 0-1 plane, namely, we are considering an infinitesimal boost along the 1-axis. In this case, we can identify

$$\epsilon^{01} = \epsilon = -\epsilon^{10}, \tag{3.77}$$

with all other components of $\epsilon^{\mu\nu}$ vanishing, so that we can write (see also (2.98))

$$\begin{aligned} S(\epsilon) &= \mathbb{1} - \frac{i}{2}\sigma_{01}\epsilon = \mathbb{1} + \frac{1}{2}\gamma_0\gamma_1\epsilon \\ &= \mathbb{1} - \frac{1}{2}\alpha_1\epsilon. \end{aligned} \tag{3.78}$$

In this case, the matrix for a finite boost, ω, can be obtained through exponentiation as

$$S(\omega) = e^{-\frac{1}{2}\alpha_1\omega}. \tag{3.79}$$

Furthermore, recalling that

$$\alpha_1 = \begin{pmatrix} 0 & \sigma_1 \\ \sigma_1 & 0 \end{pmatrix}, \tag{3.80}$$

and, therefore,

$$\alpha_1^2 = \begin{pmatrix} \mathbb{1} & 0 \\ 0 & \mathbb{1} \end{pmatrix} = \mathbb{1}, \tag{3.81}$$

we can determine

$$S(\omega) = \cosh\frac{\omega}{2} - \alpha_1 \sinh\frac{\omega}{2}. \tag{3.82}$$

We note here that since $\alpha_1^\dagger = \alpha_1$,

$$S^\dagger(\omega) \neq S^{-1}(\omega). \tag{3.83}$$

That is, in this four dimensional space (namely, as 4×4 matrices), operators defining boosts are not unitary. This is related to the fact that Lorentz boosts are non-compact transformations and for such transformations, there does not exist any finite dimensional unitary representation. All the unitary representations are necessarily infinite dimensional.

3.3 Transformation of bilinears

In the last section, we have shown how to construct the matrix $S(\Lambda)$ for finite Lorentz transformations (for both rotations and boosts). Let us note next that, since under a Lorentz transformation

$$\psi'(x') = S(\Lambda)\psi(x), \tag{3.84}$$

it follows that

$$\begin{aligned}
\psi'^{\dagger}(x') &= \psi^{\dagger}(x)S^{\dagger}(\Lambda), \\
\overline{\psi}'(x') &= \psi'^{\dagger}(x')\gamma^0 = \psi^{\dagger}(x)S^{\dagger}(\Lambda)\gamma^0 \\
&= \psi^{\dagger}(x)\gamma^0 S^{-1}(\Lambda) = \overline{\psi}(x)S^{-1}(\Lambda),
\end{aligned} \tag{3.85}$$

where we have used the relation (3.56). In other words, we see that the adjoint wave function $\overline{\psi}(x)$ transforms inversely compared to the wave function $\psi(x)$ under a Lorentz transformation. This implies that a bilinear product such as $\overline{\psi}\psi$ would transform under a Lorentz transformation as

$$\begin{aligned}
\overline{\psi}(x)\psi(x) &\to \overline{\psi}'(x')\psi'(x') \\
&= \overline{\psi}(x)S^{-1}(\Lambda)S(\Lambda)\psi(x) = \overline{\psi}(x)\psi(x).
\end{aligned} \tag{3.86}$$

Namely, such a product will not change under a Lorentz transformation — would behave like a scalar — which is what we had discussed earlier in connection with the normalization of the Dirac wavefunction (see (2.50) and (2.55)).

Similarly, under a Lorentz transformation

$$\begin{aligned}
\overline{\psi}(x)\gamma^\mu\psi(x) &\to \overline{\psi}'(x')\gamma^\mu\psi'(x') \\
&= \overline{\psi}(x)S^{-1}(\Lambda)\gamma^\mu S(\Lambda)\psi(x) \\
&= \overline{\psi}(x)\Lambda^\mu{}_\nu\gamma^\nu\psi(x) = \Lambda^\mu{}_\nu\overline{\psi}(x)\gamma^\nu\psi(x),
\end{aligned} \tag{3.87}$$

where we have used (3.43). Thus, we see that if we define a current of the form $j^\mu(x) = \overline{\psi}(x)\gamma^\mu\psi(x)$, it would transform as a four vector under a proper Lorentz transformation, namely,

$$J^\mu(x) = \overline{\psi}(x)\gamma^\mu\psi(x) \to \Lambda^\mu{}_\nu J^\nu(x). \tag{3.88}$$

3.4 Projection operators, completeness relation

Let us note that the positive energy solutions of the Dirac equation satisfy

$$(\not{p} - m)u_+(p) = (\gamma^0 p^0 - \boldsymbol{\gamma}\cdot\mathbf{p} - m)u_+(p^0, \mathbf{p}) = 0, \quad (3.89)$$

where

$$p^0 = \omega = \sqrt{\mathbf{p}^2 + m^2}, \quad (3.90)$$

while the negative energy solutions satisfy

$$(-\gamma^0 p^0 - \boldsymbol{\gamma}\cdot\mathbf{p} - m)u_-(-p^0, \mathbf{p}) = 0, \quad (3.91)$$

with the same value of p^0 as in (3.90). It is customary to identify (see (2.49), the reason for this will become clear when we discuss the quantization of Dirac field theory later)

$$\begin{aligned} u(p) &= u_+(p) = \sqrt{\frac{E+m}{2m}}\begin{pmatrix} \tilde{u} \\ \frac{\boldsymbol{\sigma}\cdot\mathbf{p}}{E+m}\tilde{u} \end{pmatrix}, \\ v(p) &= u_-(-p^0, -\mathbf{p}) = u_-(-p) = \sqrt{\frac{E+m}{2m}}\begin{pmatrix} \frac{\boldsymbol{\sigma}\cdot\mathbf{p}}{E+m}\tilde{v} \\ \tilde{v} \end{pmatrix}, \end{aligned} \quad (3.92)$$

so that the equations satisfied by $u(p)$ and $v(p)$ (positive and negative energy solutions), (3.89) and (3.91), can be written as

$$(\not{p} - m)u(p) = 0, \quad (3.93)$$

3.4 PROJECTION OPERATORS, COMPLETENESS RELATION

and

$$(-\gamma^0 p^0 + \boldsymbol{\gamma}\cdot\mathbf{p} - m)v(p) = 0,$$

or, $(-\not{p} - m)v(p) = 0,$

or, $(\not{p} + m)v(p) = 0.$ (3.94)

Given these equations, the adjoint equations are easily obtained to be (taking the Hermitian conjugate and multiplying γ^0 on the right)

$$u^\dagger(p)\left((\not{p})^\dagger - m\right)\gamma^0 = 0,$$

or, $\bar{u}(p)(\not{p} - m) = 0,$

$$v^\dagger(p)\left((\not{p})^\dagger + m\right)\gamma^0 = 0,$$

or, $\bar{v}(p)(\not{p} + m) = 0,$ (3.95)

where we have used $(\gamma^\mu)^\dagger \gamma^0 = \gamma^0 \gamma^\mu$ (see (2.83)). As we have seen earlier there are two positive energy solutions and two negative energy solutions of the Dirac equation. Let us denote them by

$$u^r(p) \quad \text{and} \quad v^r(p), \qquad r = 1, 2, \qquad (3.96)$$

where r, as we had seen earlier, can represent the spin projection of the two component spinors (in terms of which the four component solutions were obtained). Let us also note that each of the four solutions really represents a four component spinor. Let us denote the spinor index by $\alpha = 1, 2, 3, 4$. With these notations, we can write down the Lorentz invariant conditions we had derived earlier from the normalization of a massive Dirac particle as (see (2.50))

$$\bar{u}^r(p)u^s(p) = \sum_{\alpha=1}^{4} \bar{u}^r_\alpha(p)u^s_\alpha(p) = \delta_{rs},$$

$$\bar{v}^r(p)v^s(p) = \sum_{\alpha=1}^{4} \bar{v}^r_\alpha(p)v^s_\alpha(p) = -\delta_{rs},$$

$$\bar{u}^r(p)v^s(p) = 0 = \bar{v}^r(p)u^s(p). \qquad (3.97)$$

Although we had noted earlier that $\bar{u}_+(p)u_-(p) \neq 0$, the last relation in (3.97) can be checked to be true simply because $v(p) = u_-(-p^0, -\mathbf{p})$, namely, because the direction of momentum changes for $v(p)$ (see the derivation in (2.52)). This also allows us to write

$$\sum_{r=1}^{2} [\bar{u}^r(p)u^r(p) - \bar{v}^r(p)v^r(p)] = 4. \tag{3.98}$$

For completeness we note here that it is easy to check

$$u^\dagger(p)v(-p) = 0 = v^\dagger(-p)u(p), \tag{3.99}$$

for any two spin components of the positive and the negative energy spinors.

From the form of the equations satisfied by the positive and the negative energy spinors, (3.93) and (3.94), it is clear that we can define projection operators for such solutions as

$$\Lambda_+(p) = \frac{\slashed{p}+m}{2m},$$
$$\Lambda_-(p) = \frac{-\slashed{p}+m}{2m}. \tag{3.100}$$

These are, of course, 4×4 matrices and their effect on the Dirac spinors is quite clear,

$$\begin{aligned}
\Lambda_+(p)u^r(p) &= \frac{\slashed{p}+m}{2m} u^r(p) \\
&= \frac{\slashed{p}-m+2m}{2m} u^r(p) = u^r(p), \\
\Lambda_+(p)v^r(p) &= \frac{\slashed{p}+m}{2m} v^r(p) = 0, \\
\Lambda_-(p)u^r(p) &= \frac{-\slashed{p}+m}{2m} u^r(p) = 0, \\
\Lambda_-(p)v^r(p) &= \frac{-\slashed{p}+m}{2m} v^r(p) \\
&= \frac{-\slashed{p}-m+2m}{2m} v^r(p) = v^r(p). \tag{3.101}
\end{aligned}$$

3.4 Projection operators, completeness relation

Similar relations also hold for the adjoint spinors and it is clear that $\Lambda_+(p)$ projects only on to the space of positive energy solutions, while $\Lambda_-(p)$ projects only on to the space of negative energy ones.

Let us note that

$$\begin{aligned}
\Lambda_+(p)\Lambda_+(p) &= \left(\frac{\not{p}+m}{2m}\right)^2 = \frac{p^2 + 2m\not{p} + m^2}{4m^2} \\
&= \frac{m^2 + 2m\not{p} + m^2}{4m^2} = \frac{2m(\not{p}+m)}{4m^2} \\
&= \frac{\not{p}+m}{2m} = \Lambda_+(p), \\
\Lambda_-(p)\Lambda_-(p) &= \left(\frac{-\not{p}+m}{2m}\right)^2 = \frac{p^2 - 2m\not{p} + m^2}{4m^2} \\
&= \frac{m^2 - 2m\not{p} + m^2}{4m^2} = \frac{2m(-\not{p}+m)}{4m^2} \\
&= \frac{-\not{p}+m}{2m} = \Lambda_-(p), \\
\Lambda_+(p)\Lambda_-(p) &= \frac{\not{p}+m}{2m} \times \frac{-\not{p}+m}{2m} \\
&= \frac{1}{4m^2}\left(-p^2 + m^2\right) = \frac{1}{4m^2}\left(-m^2 + m^2\right) \\
&= 0 = \Lambda_-(p)\Lambda_+(p),
\end{aligned} \tag{3.102}$$

where we have used $(\not{p})^2 = p^2$ and $p^2 = m^2$. Thus, we see that $\Lambda_\pm(p)$ are indeed projection operators and they are orthogonal to each other. Furthermore, let us also note that

$$\Lambda_+(p) + \Lambda_-(p) = \frac{\not{p}+m}{2m} + \frac{-\not{p}+m}{2m} = \mathbb{1}, \tag{3.103}$$

as it should be since all the solutions can be divided into either positive or negative energy ones.

Let us next consider the outer product of the spinor solutions. Let us define a 4×4 matrix P with elements

$$P_{\alpha\beta}(p) = \sum_{r=1}^{2} u_\alpha^r(p)\bar{u}_\beta^r(p), \qquad \alpha,\beta = 1,2,3,4. \qquad (3.104)$$

This matrix has the property that acting on a positive energy spinor it gives back the same spinor. Namely,

$$\begin{aligned}
(P(p)u^s(p))_\alpha &= \sum_{\beta=1}^{4} P_{\alpha\beta}(p) u_\beta^s(p) = \sum_{\beta=1}^{4}\sum_{r=1}^{2} u_\alpha^r(p)\bar{u}_\beta^r(p) u_\beta^s(p) \\
&= \sum_{r=1}^{2} u_\alpha^r(p)\delta_{rs} = u_\alpha^s(p), \\
(\bar{u}^s(p)P(p))_\alpha &= \sum_{\beta=1}^{4} \bar{u}_\beta^s(p) P_{\beta\alpha}(p) = \sum_{\beta=1}^{4}\sum_{r=1}^{2} \bar{u}_\beta^s(p) u_\beta^r(p) \bar{u}_\alpha^r(p) \\
&= \sum_{r=1}^{2} \delta_{rs} \bar{u}_\alpha^r(p) = \bar{u}_\alpha^s(p), \\
(P(p)v^s(p))_\alpha &= \sum_{\beta=1}^{4} P_{\alpha\beta}(p) v_\beta^s(p) \\
&= \sum_{\beta=1}^{4}\sum_{r=1}^{2} u_\alpha^r(p)\bar{u}_\beta^r(p) v_\beta^s(p) = 0. \qquad (3.105)
\end{aligned}$$

Thus, we see that the matrix P projects only on to the space of positive energy solutions and, therefore, we can identify

$$P_{\alpha\beta}(p) = (\Lambda_+(p))_{\alpha\beta},$$

or, $\displaystyle\sum_{r=1}^{2} u_\alpha^r(p)\bar{u}_\beta^r(p) = \left(\frac{\not{p}+m}{2m}\right)_{\alpha\beta}.$ $\qquad (3.106)$

Similarly, if we define

3.4 Projection operators, completeness relation

$$Q_{\alpha\beta}(p) = \sum_{r=1}^{2} v_\alpha^r(p)\bar{v}_\beta^r(p), \qquad (3.107)$$

then, it is straightforward to see that

$$\begin{aligned}
(Q(p)u^s(p))_\alpha &= \sum_{\beta=1}^{4} Q_{\alpha\beta}(p)u_\beta^s(p) \\
&= \sum_{\beta=1}^{4}\sum_{r=1}^{2} v_\alpha^r(p)\bar{v}_\beta^r(p)u_\beta^s(p) = 0, \\
(Q(p)v^s(p))_\alpha &= \sum_{\beta=1}^{4} Q_{\alpha\beta}(p)v_\beta^s(p) = \sum_{\beta=1}^{4}\sum_{r=1}^{2} v_\alpha^r(p)\bar{v}_\beta^r(p)v_\beta^s(p) \\
&= \sum_{r=1}^{2} v_\alpha^r(p)(-\delta_{rs}) = -v_\alpha^s(p), \\
(\bar{v}^s(p)Q(p))_\alpha &= \sum_{\beta=1}^{4} \bar{v}_\beta^s(p)Q_{\beta\alpha}(p) = \sum_{\beta=1}^{4}\sum_{r=1}^{2} \bar{v}_\beta^s(p)v_\beta^r(p)\bar{v}_\alpha^r(p) \\
&= \sum_{r=1}^{2}(-\delta_{rs})\bar{v}_\alpha^r = -\bar{v}_\alpha^s. \qquad (3.108)
\end{aligned}$$

Namely, the matrix Q projects only on to the space of negative energy solutions with a phase (a negative sign). Hence we can identify

$$Q_{\alpha\beta} = -(\Lambda_-(p))_{\alpha\beta},$$

or, $$\sum_{r=1}^{2} v_\alpha^r(p)\bar{v}_\beta^r(p) = \left(\frac{\not{p}-m}{2m}\right)_{\alpha\beta}. \qquad (3.109)$$

The completeness relation for the solutions of the Dirac equation now follows from the observation that (see (3.103))

$$P_{\alpha\beta} - Q_{\alpha\beta} = (\Lambda_+(p))_{\alpha\beta} + (\Lambda_-(p))_{\alpha\beta} = \delta_{\alpha\beta},$$

or, $$\sum_{r=1}^{2} [u_\alpha^r(p)\overline{u}_\beta^r(p) - v_\alpha^r(p)\overline{v}_\beta^r(p)] = \delta_{\alpha\beta}. \tag{3.110}$$

In a matrix notation, the completeness relation (3.110) can also be written as

$$\sum_{r=1}^{2} (u^r(p)\overline{u}^r(p) - v^r(p)\overline{v}^r(p)) = \mathbb{1}. \tag{3.111}$$

We note here that the relative negative sign between the two terms in (3.110) or in (3.111) can be understood as follows. As we have seen, $\overline{u}u$ and $\overline{v}v$ have opposite sign, the latter being negative while the former is positive. Hence, we can think of the space of solutions of the Dirac equation as an indefinite metric space. In such a space, the completeness relation does not involve a sum of terms with positive definite sign, rather it involves a sum with the metric structure of the space built in.

These relations are particularly useful in simplifying the evaluations of transition amplitudes and probabilities. For example, let us suppose that we are interested in a transition amplitude which has the form

$$\overline{u}^r(p) M u^s(p'), \tag{3.112}$$

where M stands for a 4×4 matrix. If the initial and the final states are the same, this may represent the expectation value of a given operator in a given electron state and will have the form (r not summed)

$$\langle M \rangle = \overline{u}^r(p) M u^r(p). \tag{3.113}$$

If we are not interested in the expectation value in a particular electron state, but rather wish to obtain an average over the two possible

3.4 PROJECTION OPERATORS, COMPLETENESS RELATION 91

electron states (in experiments we may want to average over the spin polarization states), then we will have

$$\begin{aligned}
\overline{\langle M \rangle} &= \frac{1}{2} \sum_{r=1}^{2} \overline{u}^r(p) M u^r(p) \\
&= \frac{1}{2} \sum_{r=1}^{2} \sum_{\alpha,\beta=1}^{4} \overline{u}_\alpha^r(p) M_{\alpha\beta} u_\beta^r(p) \\
&= \frac{1}{2} \sum_{\alpha,\beta=1}^{4} \sum_{r=1}^{2} M_{\alpha\beta} u_\beta^r(p) \overline{u}_\alpha^r(p) \\
&= \frac{1}{2} \sum_{\alpha,\beta=1}^{4} M_{\alpha\beta} (\Lambda_+(p))_{\beta\alpha} = \frac{1}{2} \operatorname{Tr} M \Lambda_+(p). \quad (3.114)
\end{aligned}$$

Similarly, if we have a transition from a given electron state to another and if we are interested in a process where we average over the initial electron states and sum over the final electron states (for example, think of an experiment with unpolarized initial electron states where the final spin polarization is not measured), the probability for such a transition will be determined from

$$\begin{aligned}
&\frac{1}{2} \sum_{r,s=1}^{2} \left(\overline{u}^r(p) M u^s(p') \right) \left(\overline{u}^r(p) M u^s(p') \right)^\dagger \\
&= \frac{1}{2} \sum_{r,s=1}^{2} \left(\overline{u}^r(p) M u^s(p') \right) \left(u^{\dagger s}(p') M^\dagger (\overline{u}^r)^\dagger(p) \right) \\
&= \frac{1}{2} \sum_{r,s=1}^{2} \overline{u}^r(p) M u^s(p') \overline{u}^s(p') \gamma^0 M^\dagger \gamma^0 u^r(p) \\
&= \frac{1}{2} \sum_{r,s=1}^{2} \sum_{\alpha,\beta,\sigma,\lambda=1}^{4} \overline{u}_\alpha^r(p) M_{\alpha\beta} u_\beta^s(p') \overline{u}_\sigma^s(p') \left(\gamma^0 M^\dagger \gamma^0 \right)_{\sigma\lambda} u_\lambda^r(p) \\
&= \frac{1}{2} \sum_{\alpha,\beta,\sigma,\lambda=1}^{4} \sum_{r,s=1}^{2} M_{\alpha\beta} u_\beta^s(p') \overline{u}_\sigma^s(p') \left(\gamma^0 M^\dagger \gamma^0 \right)_{\sigma\lambda} u_\lambda^r(p) \overline{u}_\alpha^r(p)
\end{aligned}$$

$$= \frac{1}{2} \sum_{\alpha,\beta,\sigma,\lambda=1}^{4} M_{\alpha\beta} \left(\Lambda_+(p')\right)_{\beta\sigma} \left(\gamma^0 M^\dagger \gamma^0\right)_{\sigma\lambda} \left(\Lambda_+(p)\right)_{\lambda\alpha}$$

$$= \frac{1}{2} \operatorname{Tr} \left[M\Lambda_+(p')\gamma^0 M^\dagger \gamma^0 \Lambda_+(p)\right]. \tag{3.115}$$

The trace is over the 4×4 matrix indices and can be easily performed using the properties of the Dirac matrices that we have discussed earlier in section **2.6**.

3.5 Helicity

As we have seen, the Dirac Hamiltonian

$$H = \boldsymbol{\alpha} \cdot \mathbf{p} + \beta m, \tag{3.116}$$

does not commute either with the orbital angular momentum or with spin (rather, it commutes with the total angular momentum). Thus, unlike the case of non-relativistic systems where we specify a given energy state by the projection of spin along the z-axis (namely, by the eigenvalue of S_z), in the relativistic case this is not useful since spin is not a constant of motion. In fact, we have already seen that the spin operator

$$\mathbf{S} = \frac{1}{2}\tilde{\boldsymbol{\alpha}}, \qquad \tilde{\alpha}_i = \begin{pmatrix} \sigma_i & 0 \\ 0 & \sigma_i \end{pmatrix}, \tag{3.117}$$

satisfies the commutation relation (see (2.67))

$$[S_i, H] = \left[\frac{1}{2}\tilde{\alpha}_i, H\right] = -i\epsilon_{ijk}\alpha_j p_k. \tag{3.118}$$

As a consequence, it can be easily checked that the plane wave solutions which we had derived earlier are not eigenstates of the spin operator. Note, however, that for a particle at rest, spin commutes with the Hamiltonian (since in this frame $\mathbf{p} = 0$) and such solutions can be labelled by the spin projection.

On the other hand, we note that since momentum commutes with the Dirac Hamiltonian, namely,

3.5 HELICITY

$$[p_i, H] = [p_i, \boldsymbol{\alpha} \cdot \mathbf{p} + \beta m] = 0, \qquad (3.119)$$

the operator $\mathbf{S} \cdot \mathbf{p}$ does also (momentum and spin commute and, therefore, the order of these operators in the product is not relevant). Namely,

$$[S_i p_i, H] = [S_i, H] p_i = -i\epsilon_{ijk}\alpha_j p_k p_i = 0. \qquad (3.120)$$

Therefore, this operator is a constant of motion. The normalized operator

$$h = \frac{\mathbf{S} \cdot \mathbf{p}}{|\mathbf{p}|}, \qquad (3.121)$$

measures the longitudinal component of the spin of the particle or the projection of the spin along the direction of motion. This is known as the helicity operator and we note that since the Hamiltonian commutes with helicity, the eigenstates of energy can also be labelled by the helicity eigenvalues. Note that

$$h^2 = \left(\frac{\mathbf{S} \cdot \mathbf{p}}{|\mathbf{p}|}\right)^2 = \frac{1}{4}\mathbb{1}, \qquad (3.122)$$

where we have used (this is the generalization of the identity satisfied by the Pauli matrices)

$$(\mathbf{S} \cdot \mathbf{p})(\mathbf{S} \cdot \mathbf{p}) = \frac{1}{4}(\tilde{\boldsymbol{\alpha}} \cdot \mathbf{p})(\tilde{\boldsymbol{\alpha}} \cdot \mathbf{p}) = \frac{1}{4}\mathbf{p}^2 \mathbb{1}. \qquad (3.123)$$

Therefore, the eigenvalues of the helicity operator, for a Dirac particle, can only be $\pm\frac{1}{2}$ and we can label the positive and the negative energy solutions also as $u(p, h)$, $v(p, h)$ with $h = \pm\frac{1}{2}$ (the two helicity eigenvalues). The normalization relations in this case will take the forms

$$\begin{aligned}
\overline{u}(p, h) u(p, h') &= \delta_{hh'} = -\overline{v}(p, h) v(p, h'), \\
\overline{u}(p, h) v(p, h') &= 0 = \overline{v}(p, h) u(p, h').
\end{aligned} \qquad (3.124)$$

Furthermore, the completeness relation, (3.110) or (3.111), can now be written as

$$\sum_{h=\pm\frac{1}{2}} [u(p,h)\bar{u}(p,h) - v(p,h)\bar{v}(p,h)] = \mathbb{1}. \tag{3.125}$$

3.6 Massless Dirac particle

Let us consider the free Dirac equation for a massive spin $\frac{1}{2}$ particle,

$$(\gamma^\mu p_\mu - m)\, u(p) = 0,$$
$$\text{or,} \quad (\gamma^0 p^0 - \boldsymbol{\gamma}\cdot\mathbf{p} - m)\, u(p) = 0, \tag{3.126}$$

where we are not assuming any relation between p^0 and $\mathbf{p}$ as yet. Let us represent the four component spinor (as before) as

$$u(p) = \begin{pmatrix} u_1(p) \\ u_2(p) \end{pmatrix}, \tag{3.127}$$

where $u_1(p)$ and $u_2(p)$ are two component spinors. In terms of $u_1(p)$ and $u_2(p)$, the Dirac equation takes the form

$$\begin{pmatrix} (p^0 - m)\mathbb{1} & -\boldsymbol{\sigma}\cdot\mathbf{p} \\ \boldsymbol{\sigma}\cdot\mathbf{p} & -(p^0 + m)\mathbb{1} \end{pmatrix} \begin{pmatrix} u_1 \\ u_2 \end{pmatrix} = 0. \tag{3.128}$$

Explicitly, this leads to the two (2-component) coupled equations

$$\begin{aligned} (p^0 - m)\, u_1(p) - \boldsymbol{\sigma}\cdot\mathbf{p}\, u_2(p) &= 0, \\ \boldsymbol{\sigma}\cdot\mathbf{p}\, u_1(p) - (p^0 + m)\, u_2(p) &= 0, \end{aligned} \tag{3.129}$$

which can also be written as

$$\begin{aligned} p^0 u_1(p) - \boldsymbol{\sigma}\cdot\mathbf{p}\, u_2(p) &= m u_1(p), \\ p^0 u_2(p) - \boldsymbol{\sigma}\cdot\mathbf{p}\, u_1(p) &= -m u_2(p). \end{aligned} \tag{3.130}$$

3.6 MASSLESS DIRAC PARTICLE

Taking the sum and the difference of the two equations in (3.130), we obtain

$$\left(p^0 - \boldsymbol{\sigma} \cdot \mathbf{p}\right)(u_1(p) + u_2(p)) = m\left(u_1(p) - u_2(p)\right),$$
$$\left(p^0 + \boldsymbol{\sigma} \cdot \mathbf{p}\right)(u_1(p) - u_2(p)) = m\left(u_1(p) + u_2(p)\right). \tag{3.131}$$

We note that if we define two new (2-component) spinors as

$$u_L(p) = \frac{1}{2}(u_1(p) - u_2(p)),$$
$$u_R(p) = \frac{1}{2}(u_1(p) + u_2(p)), \tag{3.132}$$

then, the equations in (3.131) can be rewritten as a set of two coupled (2-component) spinor equations of the form

$$p^0 u_R(p) - \boldsymbol{\sigma} \cdot \mathbf{p}\, u_R(p) = m u_L(p),$$
$$p^0 u_L(p) + \boldsymbol{\sigma} \cdot \mathbf{p}\, u_L(p) = m u_R(p). \tag{3.133}$$

We note that it is the mass term which couples the two equations.

Let us note that in the limit $m \to 0$, the two equations in (3.133) reduce to two (2-component) spinor equations which are decoupled and have the simpler forms

$$p^0 u_R(p) = \boldsymbol{\sigma} \cdot \mathbf{p}\, u_R(p),$$
$$p^0 u_L(p) = -\boldsymbol{\sigma} \cdot \mathbf{p}\, u_L(p). \tag{3.134}$$

These two equations, like the Dirac equation, can be shown to be covariant under proper Lorentz transformations (as they should be, since vanishing of the mass which is a Lorentz scalar should not change the behavior of the equation under proper Lorentz transformations). These equations, however, are not invariant under parity or space reflection and are known as the Weyl equations. The corresponding two component spinors u_L and u_R are also known as Weyl spinors.

Let us note that

$$p^0 u_R(p) = \boldsymbol{\sigma} \cdot \mathbf{p}\, u_R(p),$$

or, $\quad (p^0)^2 u_R(p) = \boldsymbol{\sigma} \cdot \mathbf{p}\, p^0 u_R(p)$

$$= (\boldsymbol{\sigma} \cdot \mathbf{p})(\boldsymbol{\sigma} \cdot \mathbf{p}) u_R(p) = \mathbf{p}^2 u_R(p),$$

or, $\quad \left((p^0)^2 - \mathbf{p}^2\right) u_R(p) = 0. \qquad (3.135)$

Similarly, we can show that $u_L(p)$ also satisfies

$$\left((p^0)^2 - \mathbf{p}^2\right) u_L(p) = 0. \qquad (3.136)$$

Thus, for a nontrivial solution of these equations to exist, we must have

$$(p^0)^2 - \mathbf{p}^2 = 0, \qquad (3.137)$$

which is the Einstein relation for a massless particle. It is clear, therefore, that for such solutions, we must have

$$p^0 = \pm |\mathbf{p}|. \qquad (3.138)$$

For $p^0 = |\mathbf{p}|$, namely, for the positive energy solutions, we note that

$$p^0 u_R(p) = \boldsymbol{\sigma} \cdot \mathbf{p}\, u_R(p),$$

or, $\quad \dfrac{\boldsymbol{\sigma} \cdot \mathbf{p}}{|\mathbf{p}|} u_R(p) = u_R(p), \qquad (3.139)$

while

$$p^0 u_L(p) = -\boldsymbol{\sigma} \cdot \mathbf{p}\, u_L(p),$$

or, $\quad \dfrac{\boldsymbol{\sigma} \cdot \mathbf{p}}{|\mathbf{p}|} u_L(p) = -u_L(p). \qquad (3.140)$

In other words, the two different Weyl equations really describe particles with opposite helicity. Recalling that $\frac{1}{2}\boldsymbol{\sigma}$ denotes the spin operator for a two component spinor, we note that $u_L(p)$ describes a particle with helicity $-\frac{1}{2}$ or a particle with spin anti-parallel to its direction of motion. If we think of spin as arising from a circular motion, then we conclude that for such a particle, the circular motion would correspond to that of a left-handed screw. Correspondingly, such a particle is called a left-handed particle (which is the reason for the subscript L). On the other hand, $u_R(p)$ describes a particle with helicity $+\frac{1}{2}$ or a particle with spin parallel to its direction of motion. Such a particle is known as a right-handed particle since its spin motion would correspond to that of a right-handed screw. This is shown in Fig. 3.1 and we note here that this nomenclature is opposite of what is commonly used in optics. (Handedness is also referred to as chirality and these spinors can be shown to be eigenstates of the γ_5 matrix which can also be understood more easily from the chiral symmetry associated with massless Dirac systems.)

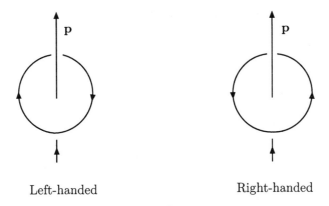

Figure 3.1: Right-handed and left-handed particles with spins parallel and anti-parallel to the direction of motion.

As we know, the electron neutrino emitted in a beta decay

$$^A X^{Z+1} \rightarrow {^A Y^Z} + e^+ + \nu_e, \tag{3.141}$$

is massless (present experiments suggest they are almost massless) and, therefore, can be described by a two component Weyl equation. We also know, experimentally, that ν_e is left-handed, namely, its helicity is $-\frac{1}{2}$. In the hole theoretic language, then, the absence of a negative energy neutrino would appear as a "hole" with the momentum reversed. Therefore, the anti-neutrino, in this description, will have opposite helicity or will be right-handed. Alternatively, the neutrino is left-handed and hence satisfies the equation

$$p^0 u_L = -\boldsymbol{\sigma} \cdot \mathbf{p}\, u_L, \tag{3.142}$$

and has negative helicity. It is helicity which is the conserved quantum number and, hence, the absence of a negative energy neutrino would appear as a "hole" with opposite helicity. That the anti-neutrino is right-handed is, of course, observed in experiments such as

$$n \to p + e^- + \bar{\nu}_e. \tag{3.143}$$

A very heuristic way to conclude that parity is violated in processes involving neutrinos is as follows. The neutrino is described by the equation

$$\frac{\boldsymbol{\sigma} \cdot \mathbf{p}}{|\mathbf{p}|} u_L(p) = -u_L(p). \tag{3.144}$$

Under parity or space reflection,

$$\mathbf{x} \to -\mathbf{x},$$
$$\mathbf{p} \to -\mathbf{p},$$
$$|\mathbf{p}| \to |\mathbf{p}|,$$
$$\mathbf{L} = \mathbf{x} \times \mathbf{p} \to (-\mathbf{x}) \times (-\mathbf{p}) = \mathbf{x} \times \mathbf{p} = \mathbf{L}. \tag{3.145}$$

Since $\boldsymbol{\sigma}$ represents an angular momentum, we conclude that it must transform under parity like $\mathbf{L}$, so that under a space reflection

$$\frac{\boldsymbol{\sigma} \cdot \mathbf{p}}{|\mathbf{p}|} \to \frac{\boldsymbol{\sigma} \cdot (-\mathbf{p})}{|\mathbf{p}|} = -\frac{\boldsymbol{\sigma} \cdot \mathbf{p}}{|\mathbf{p}|}. \tag{3.146}$$

Consequently, the neutrino equation is not invariant under parity, and processes involving neutrinos, therefore, would violate parity. This has been experimentally verified in a number of processes.

3.7 Chirality

With the normalization for massless spinors discussed in (2.53) and (2.54), the solutions of the massless Dirac equation ($m = 0$)

$$\not{p}u(p) = 0 = \not{p}v(p), \tag{3.147}$$

can be written as (see also (3.92))

$$u(p) = \sqrt{\frac{E}{2}} \begin{pmatrix} \tilde{u}(p) \\ \frac{\sigma \cdot \mathbf{p}}{E} \tilde{u}(p) \end{pmatrix} = \sqrt{\frac{|\mathbf{p}|}{2}} \begin{pmatrix} \tilde{u}(p) \\ \frac{\sigma \cdot \mathbf{p}}{|\mathbf{p}|} \tilde{u}(p) \end{pmatrix},$$

$$v(p) = \sqrt{\frac{E}{2}} \begin{pmatrix} \frac{\sigma \cdot \mathbf{p}}{E} \tilde{v}(p) \\ \tilde{v}(p) \end{pmatrix} = \sqrt{\frac{|\mathbf{p}|}{2}} \begin{pmatrix} \frac{\sigma \cdot \mathbf{p}}{|\mathbf{p}|} \tilde{v}(p) \\ \tilde{v}(p) \end{pmatrix}. \tag{3.148}$$

From the structure of the massless Dirac equation (3.147), we note that if $u(p)$ (or $v(p)$) is a solution, then $\gamma_5 u(p)$ (or $\gamma_5 v(p)$) is also a solution. Therefore, the solutions of the massless Dirac equation can be classified according to the eigenvalues of γ_5 also known as the chirality or the handedness.

This can also be seen from the fact that the Hamiltonian for a massless Dirac fermion (see (1.100))

$$H = \boldsymbol{\alpha} \cdot \mathbf{p}, \tag{3.149}$$

commutes with γ_5 (in fact, in the Pauli-Dirac representation $\gamma_5 = \rho$ defined in (2.60) and ρ commutes with $\boldsymbol{\alpha}$, see, for example, (2.61)). Since

$$\gamma_5^2 = \mathbb{1}, \tag{3.150}$$

it follows that the eigenvalues of γ_5 are ± 1 and spinors with the eigenvalue $+1$, namely,

$$\gamma_5 u_R(p) = u_R(p), \quad \gamma_5 v_R(p) = v_R(p), \tag{3.151}$$

are known as right-handed (positive chirality) spinors while those with the eigenvalue -1, namely,

$$\gamma_5 u_L(p) = -u_L(p), \quad \gamma_5 v_L(p) = -v_L(p), \tag{3.152}$$

are called left-handed (negative chirality) spinors. We note that if the fermion is massive ($m \neq 0$), then the Dirac Hamiltonian (1.100) would no longer commute with γ_5 and in this case chirality would not be a good quantum number to label the states with.

Given a general spinor, the right-handed and the left-handed components can be obtained through the projection operators ($\mathbb{1}$ denotes the identity matrix in the appropriate space)

$$\begin{aligned}
u_R(p) &= P_R u(p) = \frac{1}{2}(\mathbb{1} + \gamma_5) u(p) \\
&= \sqrt{\frac{|\mathbf{p}|}{2}} \begin{pmatrix} \frac{1}{2}(\mathbb{1} + \frac{\boldsymbol{\sigma} \cdot \mathbf{p}}{|\mathbf{p}|}) \tilde{u}(p) \\ \frac{1}{2}(\mathbb{1} + \frac{\boldsymbol{\sigma} \cdot \mathbf{p}}{|\mathbf{p}|}) \tilde{u}(p) \end{pmatrix}, \\
u_L(p) &= P_L u(p) = \frac{1}{2}(\mathbb{1} - \gamma_5) u(p) \\
&= \sqrt{\frac{|\mathbf{p}|}{2}} \begin{pmatrix} \frac{1}{2}(\mathbb{1} - \frac{\boldsymbol{\sigma} \cdot \mathbf{p}}{|\mathbf{p}|}) \tilde{u}(p) \\ -\frac{1}{2}(\mathbb{1} - \frac{\boldsymbol{\sigma} \cdot \mathbf{p}}{|\mathbf{p}|}) \tilde{u}(p) \end{pmatrix}, \\
v_R(p) &= P_R v(p) = \frac{1}{2}(\mathbb{1} + \gamma_5) v(p) \\
&= \sqrt{\frac{|\mathbf{p}|}{2}} \begin{pmatrix} \frac{1}{2}(\mathbb{1} + \frac{\boldsymbol{\sigma} \cdot \mathbf{p}}{|\mathbf{p}|}) \tilde{v}(p) \\ \frac{1}{2}(\mathbb{1} + \frac{\boldsymbol{\sigma} \cdot \mathbf{p}}{|\mathbf{p}|}) \tilde{v}(p) \end{pmatrix}, \\
v_L(p) &= P_L v(p) = \frac{1}{2}(\mathbb{1} - \gamma_5) v(p) \\
&= \sqrt{\frac{|\mathbf{p}|}{2}} \begin{pmatrix} -\frac{1}{2}(\mathbb{1} - \frac{\boldsymbol{\sigma} \cdot \mathbf{p}}{|\mathbf{p}|}) \tilde{v}(p) \\ \frac{1}{2}(\mathbb{1} - \frac{\boldsymbol{\sigma} \cdot \mathbf{p}}{|\mathbf{p}|}) \tilde{v}(p) \end{pmatrix},
\end{aligned} \tag{3.153}$$

where we have defined

3.7 CHIRALITY

$$P_R = \frac{1}{2}(\mathbb{1} + \gamma_5), \quad P_L = \frac{1}{2}(\mathbb{1} - \gamma_5). \tag{3.154}$$

We note that by definition these projection operators satisfy

$$\begin{aligned}(P_R)^2 &= P_R, \quad (P_L)^2 = P_L, \\ P_R P_L &= 0 = P_L P_R, \quad P_R + P_L = 1,\end{aligned} \tag{3.155}$$

which implies that any four component spinor can be uniquely decomposed into a right-handed and a left-handed component. (In the Pauli-Dirac representation, these projection operators have the explicit forms (see (2.91))

$$P_R = \frac{1}{2}\begin{pmatrix} \mathbb{1} & \mathbb{1} \\ \mathbb{1} & \mathbb{1} \end{pmatrix}, \quad P_L = \frac{1}{2}\begin{pmatrix} \mathbb{1} & -\mathbb{1} \\ -\mathbb{1} & \mathbb{1} \end{pmatrix}. \tag{3.156}$$

We note from (3.153) that in the massless Dirac theory, the four component spinors can be effectively described by two component spinors. This is connected with our earlier observation (see section **3.6**) that in the massless limit, the Dirac equation reduces to two decoupled two component Weyl equations (recall that it is the mass term which generally couples these two spinors). The reducibility of the spinors is best seen in the Weyl representation for the Dirac matrices discussed in (2.119). However, we will continue our discussion in the Pauli-Dirac representation of the Dirac matrices which we have used throughout. From the definition of the helicity operator in (3.121) (for the two component spinors $\mathbf{S} = \frac{1}{2}\boldsymbol{\sigma}$), we note that spinors of the form

$$\chi^{(\pm)}(\mathbf{p}) = \frac{1}{2}\left(1 \pm \frac{\boldsymbol{\sigma} \cdot \mathbf{p}}{|\mathbf{p}|}\right)\tilde{\chi}, \tag{3.157}$$

correspond to states with definite helicity, namely,

$$\begin{aligned}h\chi^{(\pm)}(\mathbf{p}) &= \frac{\boldsymbol{\sigma} \cdot \mathbf{p}}{2|\mathbf{p}|}\frac{1}{2}\left(1 \pm \frac{\boldsymbol{\sigma} \cdot \mathbf{p}}{|\mathbf{p}|}\right)\tilde{\chi} \\ &= \pm\frac{1}{2} \times \frac{1}{2}\left(1 \pm \frac{\boldsymbol{\sigma} \cdot \mathbf{p}}{|\mathbf{p}|}\right)\tilde{\chi} = \pm\frac{1}{2}\chi^{(\pm)}(\mathbf{p}),\end{aligned} \tag{3.158}$$

so that the right-handed (four component) spinors in (3.153) are described by two component spinors with positive helicity while the left-handed (four component) spinors are described in terms of two component spinors of negative helicity. Explicitly, we see from (3.153) and (3.157) that we can identify

$$u_R(\mathbf{p}) = \sqrt{\frac{|\mathbf{p}|}{2}} \begin{pmatrix} u^{(+)}(\mathbf{p}) \\ u^{(+)}(\mathbf{p}) \end{pmatrix}, \quad v_R(\mathbf{p}) = \sqrt{\frac{|\mathbf{p}|}{2}} \begin{pmatrix} v^{(+)}(\mathbf{p}) \\ v^{(+)}(\mathbf{p}) \end{pmatrix},$$

$$u_L(\mathbf{p}) = \sqrt{\frac{|\mathbf{p}|}{2}} \begin{pmatrix} u^{(-)}(\mathbf{p}) \\ -u^{(-)}(\mathbf{p}) \end{pmatrix}, \quad v_L(\mathbf{p}) = \sqrt{\frac{|\mathbf{p}|}{2}} \begin{pmatrix} -v^{(-)}(\mathbf{p}) \\ v^{(-)}(\mathbf{p}) \end{pmatrix}. \quad (3.159)$$

We note here that the operators

$$P^{(\pm)} = \frac{1}{2}\left(\mathbb{1} \pm \frac{\boldsymbol{\sigma} \cdot \mathbf{p}}{|\mathbf{p}|}\right) = \frac{1}{2}(\mathbb{1} \pm \boldsymbol{\sigma} \cdot \hat{\mathbf{p}}), \quad (3.160)$$

can also be written in a covariant notation as

$$P^{(+)} = \frac{1}{2}\tilde{\sigma}^\mu \hat{p}_\mu, \quad P^{(-)} = \frac{1}{2}\sigma^\mu \hat{p}_\mu, \quad (3.161)$$

with $\sigma^\mu, \tilde{\sigma}^\mu$ defined in (2.120) and $\hat{p}_\mu = (1, -\hat{\mathbf{p}})$. It is straightforward to check that they satisfy the relations

$$\begin{aligned}(P^{(+)})^2 &= P^{(+)}, \quad (P^{(-)})^2 = P^{(-)}, \\ P^{(+)}P^{(-)} &= 0 = P^{(-)}P^{(+)}, \quad P^{(+)} + P^{(-)} = \mathbb{1}, \end{aligned} \quad (3.162)$$

and, therefore, define projection operators into the space of positive and negative helicity two component spinors. They can be easily generalized to a reducible representation of operators acting on the four component spinors and have the form (see (2.70) or (3.117))

$$\begin{aligned} P^{(\pm)}_{4\times 4} &= \begin{pmatrix} P^{(\pm)} & 0 \\ 0 & P^{(\pm)} \end{pmatrix} \\ &= \begin{pmatrix} \frac{1}{2}(\mathbb{1} \pm \boldsymbol{\sigma} \cdot \hat{\mathbf{p}}) & 0 \\ 0 & \frac{1}{2}(\mathbb{1} \pm \boldsymbol{\sigma} \cdot \hat{\mathbf{p}}) \end{pmatrix}, \end{aligned} \quad (3.163)$$

3.7 CHIRALITY

and it is straightforward to check from (3.156) and (3.163) that

$$[P_{L,R}, P^{(\pm)}_{4\times 4}] = 0, \tag{3.164}$$

which is the reason the spinors can be simultaneous eigenstates of chirality and helicity (when mass vanishes). In fact, from (3.159) as well as (3.163) we see that the right-handed spinors with chirality $+1$ are characterized by helicity $+1$ while the left-handed spinors with chirality -1 have helicity -1.

For completeness as well as for later use, let us derive some properties of these spinors. We note from (3.157) that we can write the positive and the negative energy solutions as (we will do this in detail for the right-handed spinors and only quote the results for the left-handed spinors)

$$u^{(+)}(\mathbf{p}) = \frac{1}{2}(\mathbb{1} + \boldsymbol{\sigma} \cdot \hat{\mathbf{p}})\tilde{u}, \quad v^{(+)}(\mathbf{p}) = \frac{1}{2}(\mathbb{1} + \boldsymbol{\sigma} \cdot \hat{\mathbf{p}})\tilde{v}. \tag{3.165}$$

Each of these spinors is one dimensional and together they span the two dimensional spinor space. We can choose $\tilde{u}$ and $\tilde{v}$ to be normalized so that we have

$$\tilde{u}^\dagger \tilde{u} = 1 = \tilde{v}^\dagger \tilde{v}, \quad \tilde{u}\tilde{u}^\dagger + \tilde{v}\tilde{v}^\dagger = \mathbb{1}. \tag{3.166}$$

For example, we can choose

$$\tilde{u} = \begin{pmatrix} 1 \\ 0 \end{pmatrix}, \quad \tilde{v} = \begin{pmatrix} 0 \\ 1 \end{pmatrix}, \tag{3.167}$$

such that when $p_1 = p_2 = 0$, the helicity spinors simply reduce to eigenstates of σ_3. Furthermore, we can also define normalized spinors $u^{(+)}$ and $v^{(+)}$. For example, with the choice of the basis in (3.167), the normalized spinors take the forms (here we are using the three dimensional notation so that $p_i = (\mathbf{p})_i$)

$$\begin{aligned} u^{(+)}(\mathbf{p}) &= \frac{1}{\sqrt{2|\mathbf{p}|(|\mathbf{p}| + p_3)}} \begin{pmatrix} |\mathbf{p}| + p_3 \\ p_1 + ip_2 \end{pmatrix}, \\ v^{(+)}(\mathbf{p}) &= \frac{1}{\sqrt{2|\mathbf{p}|(|\mathbf{p}| - p_3)}} \begin{pmatrix} p_1 - ip_2 \\ |\mathbf{p}| - p_3 \end{pmatrix}. \end{aligned} \tag{3.168}$$

However, we do not need to use any particular representation for our discussions. In general, the positive helicity spinors satisfy

$$u^{(+)\dagger}(\mathbf{p})u^{(+)}(\mathbf{p}) = v^{(+)\dagger}(\mathbf{p})v^{(+)}(\mathbf{p}) = 1,$$
$$u^{(+)\dagger}(\mathbf{p})v^{(+)}(-\mathbf{p}) = 0 = v^{(+)\dagger}(-\mathbf{p})u^{(+)}(\mathbf{p}),$$
$$u^{(+)}(\mathbf{p})u^{(+)\dagger}(\mathbf{p}) = v^{(+)}(\mathbf{p})v^{(+)\dagger}(\mathbf{p}) = \frac{1}{2}(\mathbb{1} + \boldsymbol{\sigma} \cdot \hat{\mathbf{p}}), \quad (3.169)$$

which can be checked from the explicit forms of the spinors in (3.168). Here we note that the second relation follows from the fact that a positive helicity spinor changes into an orthogonal negative helicity spinor when the direction of the momentum is reversed (which is also manifest in the projection operator).

Given the form of the right-handed spinors in (3.159), as well as (3.169) it now follows in a straightforward manner that

$$u_R^\dagger(\mathbf{p})u_R(\mathbf{p}) = |\mathbf{p}| = v_R^\dagger(\mathbf{p})v_R(\mathbf{p}),$$
$$u_R^\dagger(\mathbf{p})v_R(-\mathbf{p}) = v_R^\dagger(-\mathbf{p})u_R(\mathbf{p}) = 0,$$
$$u_R(\mathbf{p})u_R^\dagger(\mathbf{p}) = v_R(\mathbf{p})v_R^\dagger(\mathbf{p}) = \frac{|\mathbf{p}|}{2}\begin{pmatrix} P^{(+)} & P^{(+)} \\ P^{(+)} & P^{(+)} \end{pmatrix}. \quad (3.170)$$

The completeness relation in (3.170) can be simplified by noting the following identity. We note that with $p_0 = |\mathbf{p}|$, we can write

$$\frac{1}{4}\not{p}\gamma^0(\mathbb{1} + \gamma_5)$$
$$= \frac{1}{4}\begin{pmatrix} |\mathbf{p}| & -\boldsymbol{\sigma} \cdot \mathbf{p} \\ \boldsymbol{\sigma} \cdot \mathbf{p} & -|\mathbf{p}| \end{pmatrix}\begin{pmatrix} \mathbb{1} & 0 \\ 0 & -\mathbb{1} \end{pmatrix}\begin{pmatrix} \mathbb{1} & \mathbb{1} \\ \mathbb{1} & \mathbb{1} \end{pmatrix}$$
$$= \frac{|\mathbf{p}|}{4}\begin{pmatrix} \mathbb{1} & \boldsymbol{\sigma} \cdot \hat{\mathbf{p}} \\ \boldsymbol{\sigma} \cdot \hat{\mathbf{p}} & \mathbb{1} \end{pmatrix}\begin{pmatrix} \mathbb{1} & \mathbb{1} \\ \mathbb{1} & \mathbb{1} \end{pmatrix}$$
$$= \frac{|\mathbf{p}|}{2}\begin{pmatrix} P^{(+)} & P^{(+)} \\ P^{(+)} & P^{(+)} \end{pmatrix}, \quad (3.171)$$

so that we can write the completeness relation in (3.170) as

$$u_R(\mathbf{p})u_R^\dagger(\mathbf{p}) = v_R(\mathbf{p})v_R^\dagger(\mathbf{p}) = \frac{1}{4}\not{p}\gamma^0(\mathbb{1}+\gamma_5), \qquad (3.172)$$

which can also be derived using the methods in section **3.4**. We conclude this section by noting (without going into details) that similar relations can be derived for the left-handed spinors and take the forms

$$u_L^\dagger(\mathbf{p})u_L(\mathbf{p}) = |\mathbf{p}| = v_L^\dagger(\mathbf{p})v_L(\mathbf{p}),$$
$$u_L^\dagger(\mathbf{p})v_L(-\mathbf{p}) = 0 = v_L^\dagger(-\mathbf{p})u_L(\mathbf{p}),$$
$$u_L(\mathbf{p})u_L^\dagger(\mathbf{p}) = v_L(\mathbf{p})v_L^\dagger(\mathbf{p}) = \frac{1}{4}\not{p}\gamma^0(\mathbb{1}-\gamma_5), \qquad (3.173)$$

3.8 Non-relativistic limit of the Dirac equation

Let us recall that the positive energy solutions of the Dirac equation have the form (see (2.49))

$$u_+(p) = \sqrt{\frac{E+m}{2m}}\begin{pmatrix}\tilde{u} \\ \frac{\boldsymbol{\sigma}\cdot\mathbf{p}}{E+m}\tilde{u}\end{pmatrix} = \begin{pmatrix}u_L(p) \\ u_S(p)\end{pmatrix}, \qquad (3.174)$$

while the negative energy solutions have the form

$$u_-(p) = \sqrt{\frac{E+m}{2m}}\begin{pmatrix}-\frac{\boldsymbol{\sigma}\cdot\mathbf{p}}{E+m}\tilde{v} \\ \tilde{v}\end{pmatrix} = \begin{pmatrix}v_S(p) \\ v_L(p)\end{pmatrix}. \qquad (3.175)$$

In (3.174) we have defined

$$u_L(p) = \sqrt{\frac{E+m}{2m}}\,\tilde{u},$$
$$u_S(p) = \sqrt{\frac{E+m}{2m}}\,\frac{\boldsymbol{\sigma}\cdot\mathbf{p}}{E+m}\tilde{u} = \frac{\boldsymbol{\sigma}\cdot\mathbf{p}}{E+m}u_L(p), \qquad (3.176)$$

and we emphasize that the subscript "L" here does not stand for the left-handed particles introduced in the last section. Similarly, in (3.175) we have denoted

$$v_L(p) = \sqrt{\frac{E+m}{2m}}\,\tilde{v},$$

$$v_S(p) = -\sqrt{\frac{E+m}{2m}}\,\frac{\boldsymbol{\sigma}\cdot\mathbf{p}}{E+m}\,\tilde{v} = -\frac{\boldsymbol{\sigma}\cdot\mathbf{p}}{E+m}\,v_L(p). \qquad (3.177)$$

It is clear that in the non-relativistic limit, when $|\mathbf{p}| \ll m$, the component $u_S(p)$ is much smaller than (of the order of $\frac{v}{c}$) $u_L(p)$ and correspondingly, $u_L(p)$ and $u_S(p)$ are known as the large and the small components of the positive energy Dirac solution. Similarly, $v_L(p)$ and $v_S(p)$ are also known as the large and the small components of the negative energy solution. In the non-relativistic limit, we expect the large components to give the dominant contribution to the wave function.

Let us next look at the positive energy solutions in (3.174), which satisfy the equation

$$Hu_+(p) = Eu_+(p),$$

or, $\quad (\boldsymbol{\alpha}\cdot\mathbf{p} + \beta m)\,u_+(p) = Eu_+(p),$

or, $\quad \begin{pmatrix} m\mathbb{1} & \boldsymbol{\sigma}\cdot\mathbf{p} \\ \boldsymbol{\sigma}\cdot\mathbf{p} & -m\mathbb{1} \end{pmatrix} \begin{pmatrix} u_L(p) \\ u_S(p) \end{pmatrix} = E \begin{pmatrix} u_L(p) \\ u_S(p) \end{pmatrix}. \qquad (3.178)$

This would lead to the two (2-component) equations

$$\boldsymbol{\sigma}\cdot\mathbf{p}\,u_S(p) = (E-m)u_L(p),$$
$$\boldsymbol{\sigma}\cdot\mathbf{p}\,u_L(p) = (E+m)u_S(p). \qquad (3.179)$$

We note that the second equation in (3.179) gives the relation

$$u_S(p) = \frac{\boldsymbol{\sigma}\cdot\mathbf{p}}{E+m}\,u_L(p), \qquad (3.180)$$

while, with the substitution of this, the first equation in (3.179) takes the form

$$(\boldsymbol{\sigma} \cdot \mathbf{p}) \frac{\boldsymbol{\sigma} \cdot \mathbf{p}}{E+m} u_L(p) = (E-m) u_L(p),$$

or, $\quad \dfrac{\mathbf{p}^2}{2m} u_L(p) \simeq (E-m) u_L(p),$ \hfill (3.181)

where we have used the fact that for a non-relativistic system, $|\mathbf{p}| \ll m$, and, therefore, $E \approx m$ (recall that we have set $c = 1$). Furthermore, if we identify the non-relativistic energy (without the rest mass term) as

$$E_{\text{NR}} = E - m, \tag{3.182}$$

then, equation (3.181) has the form

$$\frac{\mathbf{p}^2}{2m} u_L(p) = E_{\text{NR}}\, u_L(p). \tag{3.183}$$

Namely, the Dirac equation in this case reduces to the Schrödinger equation for a two component spinor which we are familiar with. This is, of course, what we know for a free non-relativistic electron (spin $\frac{1}{2}$ particle).

3.9 Electron in an external magnetic field

The coupling of a charged particle to an external electromagnetic field can be achieved through what is conventionally known as the minimal coupling. This preserves the gauge invariance associated with the Maxwell's equations and corresponds to defining

$$p_\mu \to p_\mu - e A_\mu, \tag{3.184}$$

where e denotes the charge of the particle and A_μ represents the four vector potential of the associated electromagnetic field. Since the coordinate representation of p_μ is given by (see (1.33) and remember that we are choosing $\hbar = 1$)

$$p_\mu \to i\partial_\mu, \tag{3.185}$$

the minimal coupling prescription also corresponds to defining (in the coordinate representation)

$$\partial_\mu \to \partial_\mu + ieA_\mu. \tag{3.186}$$

Let us next consider an electron interacting with a time independent external magnetic field. In this case, we have

$$\begin{aligned} A_0 &= 0 = \phi, \\ \mathbf{B} &= (\nabla \times \mathbf{A}), \end{aligned} \tag{3.187}$$

where we are assuming that $\mathbf{A} = \mathbf{A}(\mathbf{x})$. The Dirac equation for the positive energy electrons, in this case, takes the form

$$(\boldsymbol{\alpha} \cdot (\mathbf{p} - e\mathbf{A}) + \beta m) u(p) = E u(p),$$

or, $\begin{pmatrix} m\mathbb{1} & \boldsymbol{\sigma} \cdot (\mathbf{p} - e\mathbf{A}) \\ \boldsymbol{\sigma} \cdot (\mathbf{p} - e\mathbf{A}) & -m\mathbb{1} \end{pmatrix} \begin{pmatrix} u_L(p) \\ u_S(p) \end{pmatrix} = E \begin{pmatrix} u_L(p) \\ u_S(p) \end{pmatrix}.$
$$\tag{3.188}$$

Explicitly, we can write the two (2-component) equations as

$$\begin{aligned} \boldsymbol{\sigma} \cdot (\mathbf{p} - e\mathbf{A}) u_S(p) &= (E - m) u_L(p), \\ \boldsymbol{\sigma} \cdot (\mathbf{p} - e\mathbf{A}) u_L(p) &= (E + m) u_S(p). \end{aligned} \tag{3.189}$$

In this case, the second equation in (3.189) leads to

$$u_S(p) = \frac{\boldsymbol{\sigma} \cdot (\mathbf{p} - e\mathbf{A})}{E + m} u_L(p) \simeq \frac{\boldsymbol{\sigma} \cdot (\mathbf{p} - e\mathbf{A})}{2m} u_L(p), \tag{3.190}$$

where in the last relation, we have used $|\mathbf{p}| \ll m$ in the non-relativistic limit. Substituting this back into the first equation in (3.189), we obtain

3.9 ELECTRON IN AN EXTERNAL MAGNETIC FIELD

$$(\boldsymbol{\sigma} \cdot (\mathbf{p} - e\mathbf{A})) \frac{(\boldsymbol{\sigma} \cdot (\mathbf{p} - e\mathbf{A}))}{2m} u_L(p) \simeq (E - m)u_L(p). \quad (3.191)$$

Let us simplify the expression on the left hand side of (3.191) using the following identity for the Pauli matrices

$$\begin{aligned}
(\boldsymbol{\sigma} \cdot (\mathbf{p} - e\mathbf{A}))\,(\boldsymbol{\sigma} \cdot (\mathbf{p} - e\mathbf{A})) \\
= (\mathbf{p} - e\mathbf{A}) \cdot (\mathbf{p} - e\mathbf{A}) + i\boldsymbol{\sigma} \cdot ((\mathbf{p} - e\mathbf{A}) \times (\mathbf{p} - e\mathbf{A})) \\
= (\mathbf{p} - e\mathbf{A})^2 - ie\boldsymbol{\sigma} \cdot (\mathbf{p} \times \mathbf{A} + \mathbf{A} \times \mathbf{p}).
\end{aligned} \quad (3.192)$$

Note that (here, we are going to use purely three dimensional notation for simplicity)

$$\begin{aligned}
(\mathbf{p} \times \mathbf{A} + \mathbf{A} \times \mathbf{p})_i &= \epsilon_{ijk}(p_j A_k + A_j p_k) = \epsilon_{ijk}(p_j A_k - A_k p_j) \\
&= \epsilon_{ijk}[p_j, A_k] = -i\epsilon_{ijk}[\nabla_j, A_k] \\
&= -i(\boldsymbol{\nabla} \times \mathbf{A})_i = -i(\mathbf{B})_i.
\end{aligned} \quad (3.193)$$

We can use this in (3.192) to write

$$\begin{aligned}
(\boldsymbol{\sigma} \cdot (\mathbf{p} - e\mathbf{A}))\,(\boldsymbol{\sigma} \cdot (\mathbf{p} - e\mathbf{A})) \\
= (\mathbf{p} - e\mathbf{A})^2 - ie\boldsymbol{\sigma} \cdot (-i\mathbf{B}) \\
= (\mathbf{p} - e\mathbf{A})^2 - e\boldsymbol{\sigma} \cdot \mathbf{B}.
\end{aligned} \quad (3.194)$$

Consequently, in the non-relativistic limit, when we can approximate the Dirac equation by that satisfied by the two component spinor $u_L(p)$, equation (3.191) takes the form

$$\frac{1}{2m}\left((\mathbf{p} - e\mathbf{A})^2 - e\boldsymbol{\sigma} \cdot \mathbf{B}\right) u_L(p) = (E - m)u_L(p),$$

$$\text{or,} \quad \left(\frac{1}{2m}(\mathbf{p} - e\mathbf{A})^2 - \frac{e}{2m}\boldsymbol{\sigma} \cdot \mathbf{B}\right) u_L(p) \simeq E_{\text{NR}} u_L(p), \quad (3.195)$$

where we have identified (as before)

$$E_{\text{NR}} = E - m. \tag{3.196}$$

We recognize (3.195) to be the Schrödinger equation for a charged electron with a minimal coupling to an external vector field along with a magnetic dipole interaction with the external magnetic field. Namely, a minimally coupled Dirac particle automatically leads, in the non-relativistic limit, to a magnetic dipole interaction (recall that in the non-relativistic theory, we have to add such an interaction by hand) and we can identify the magnetic moment operator associated with the electron to correspond to

$$\boldsymbol{\mu} = \frac{e}{2m}\,\boldsymbol{\sigma}. \tag{3.197}$$

Of course, this shows that a point Dirac particle has a magnetic moment corresponding to a gyro-magnetic ratio

$$g = 2. \tag{3.198}$$

Let us recall that the magnetic moment of a particle is defined to be ($c = 1$)

$$\boldsymbol{\mu} = g\,\frac{e}{2m}\,\mathbf{S}. \tag{3.199}$$

Since $\mathbf{S} = \frac{1}{2}\boldsymbol{\sigma}$ for a two component electron, comparing with (3.197) we obtain $g = 2$. Quantum mechanical corrections (higher order corrections) in an interacting theory such as quantum electrodynamics, however, change this value slightly and the experimental deviation of g from the value of 2 ($g - 2$ experiment) for the electron agrees exceptionally well with the theoretical predictions of quantum electrodynamics. Particles with a nontrivial structure (that is particles which are not point like and have extended structures), however, can have g-factors quite different from 2. In this case, one says that there is an anomalous contribution to the magnetic moment. Thus, for example, for the proton and the neutron, we know that the magnetic moments are given by

$$\mu_N = -1.91 \, \mu_{nm},$$
$$\mu_P = 2.79 \, \mu_{nm}, \qquad (3.200)$$

where the nuclear magneton is defined to be

$$\mu_{nm} = \frac{|e|}{2m_P}, \qquad (3.201)$$

with m_P denoting the mass of the proton.

Anomalous magnetic moments can be accommodated through an additional interaction Hamiltonian (in the Dirac system) of the form (this is known as a non-minimal coupling)

$$H_I = \frac{e\kappa}{2m} \sigma^{\mu\nu} F_{\mu\nu}, \qquad (3.202)$$

where

$$F_{\mu\nu} = \partial_\mu A_\nu - \partial_\nu A_\mu, \qquad (3.203)$$

denotes the electromagnetic field strength tensor and κ represents the anomalous magnetic moment of the particle. This is commonly known as the Pauli coupling or the Pauli interaction.

3.10 Foldy-Wouthuysen transformation

In the last two sections, we have described how the non-relativistic limit of a Dirac theory can be taken in a simple manner. In the non-relativistic limit, the relevant expansion parameter is $\frac{|\mathbf{p}|}{m}$ and the method works quite well in the lowest order of expansion, as we have seen explicitly. However, at higher orders, this method runs into difficulty. For example, if we were to calculate the electric dipole interaction of an electron in a background electromagnetic field using the method described in the earlier sections, the electric dipole moment becomes imaginary at order $\frac{1}{m^2}$ (namely, the Hamiltonian becomes non-Hermitian). This puzzling feature can be understood in a simple manner as follows. The process of eliminating the "small" components from the Dirac equation described in the earlier sections can be understood mathematically as

$$u(p) = \begin{pmatrix} u_L(p) \\ u_S(p) \end{pmatrix} = \begin{pmatrix} u_L(p) \\ Au_L(p) \end{pmatrix} \xrightarrow{T} \begin{pmatrix} u_L(p) \\ 0 \end{pmatrix}, \quad (3.204)$$

where the matrix A, in the case of the free Dirac equation, for example, has the form (see (3.180))

$$A = \frac{\boldsymbol{\sigma} \cdot \mathbf{p}}{E + m}, \quad (3.205)$$

for the positive energy spinors. The matrix T that takes us to the two component "large" spinors in (3.204) has the form

$$T = \begin{pmatrix} \mathbb{1} & 0 \\ -A & \mathbb{1} \end{pmatrix}. \quad (3.206)$$

It is clear from the form of the matrix in (3.206) that it is not unitary and this is the reason that the Hamiltonian becomes non-Hermitian at higher orders in the inverse mass expansion (non-relativistic expansion). This difficulty in taking a consistent non-relativistic limit to any order in the expansion in $\frac{1}{m}$ was successfully solved by Foldy and Wouthuysen and also independently by Tani which we describe below.

Since the lack of unitarity in (3.206) is the source of the problem in taking the non-relativistic limit consistently, the main idea in the works of Foldy-Wouthuysen as well as Tani is to ensure that the relevant transformation used in going to the non-relativistic limit is manifestly unitary. Thus, for example, let us look at the free Dirac theory where we know that the Hamiltonian has the form (see (1.100) as well as (1.101))

$$H = \boldsymbol{\alpha} \cdot \mathbf{p} + \beta m = \gamma^0 \left(\boldsymbol{\gamma} \cdot \mathbf{p} + m \right). \quad (3.207)$$

Let us next look for a unitary transformation that will diagonalize the Hamiltonian in (3.207). In this case, such a transformation would

3.10 FOLDY-WOUTHUYSEN TRANSFORMATION

also transform the spinor into two 2-component spinors that will be decoupled and we do not have to eliminate one in favor of the other (namely, avoid the problem with "large" and "small" spinors). Let us consider a transformation of the type

$$U(\theta) = e^{\frac{1}{2m} \gamma \cdot \mathbf{p} \theta}, \tag{3.208}$$

where the real scalar parameter of the transformation is a function of $\mathbf{p}$ and m,

$$\theta = \theta(|\mathbf{p}|, m). \tag{3.209}$$

From the properties of the gamma matrices in (1.83) or (1.91), we note that

$$(\gamma)^\dagger = -\gamma, \quad (\gamma \cdot \mathbf{p})^2 = -\mathbf{p}^2 = -|\mathbf{p}|^2, \tag{3.210}$$

and using this we can simplify and write

$$\begin{aligned} U(\theta) &= \sum_{n=0}^{\infty} \left[\frac{1}{(2n)!} \left(\frac{\gamma \cdot \mathbf{p}\theta}{2m} \right)^{2n} + \frac{1}{(2n+1)!} \left(\frac{\gamma \cdot \mathbf{p}\theta}{2m} \right)^{2n+1} \right] \\ &= \cos\left(\frac{|\mathbf{p}|\theta}{2m} \right) + \frac{\gamma \cdot \mathbf{p}}{|\mathbf{p}|} \sin\left(\frac{|\mathbf{p}|\theta}{2m} \right). \end{aligned} \tag{3.211}$$

It follows now that

$$U^\dagger(\theta) = \cos\left(\frac{|\mathbf{p}|\theta}{2m} \right) - \frac{\gamma \cdot \mathbf{p}}{|\mathbf{p}|} \sin\left(\frac{|\mathbf{p}|\theta}{2m} \right), \tag{3.212}$$

which leads to

$$\begin{aligned}
U(\theta)U^\dagger(\theta) &= \left(\cos\left(\frac{|\mathbf{p}|\theta}{2m}\right) + \frac{\boldsymbol{\gamma}\cdot\mathbf{p}}{|\mathbf{p}|}\sin\left(\frac{|\mathbf{p}|\theta}{2m}\right)\right) \\
&\quad \times \left(\cos\left(\frac{|\mathbf{p}|\theta}{2m}\right) - \frac{\boldsymbol{\gamma}\cdot\mathbf{p}}{|\mathbf{p}|}\sin\left(\frac{|\mathbf{p}|\theta}{2m}\right)\right) \\
&= \cos^2\left(\frac{|\mathbf{p}|\theta}{2m}\right) - \left(\frac{\boldsymbol{\gamma}\cdot\mathbf{p}}{|\mathbf{p}|}\right)^2\sin^2\left(\frac{|\mathbf{p}|\theta}{2m}\right) \\
&= \cos^2\left(\frac{|\mathbf{p}|\theta}{2m}\right) + \sin^2\left(\frac{|\mathbf{p}|\theta}{2m}\right) = \mathbb{1}. \quad (3.213)
\end{aligned}$$

Namely, the transformation (3.208) is indeed unitary.

Under the unitary transformation (3.208), the free Dirac Hamiltonian (3.207) would transform as

$$\begin{aligned}
H \to H' &= U(\theta)HU^\dagger(\theta) \\
&= \left(\cos\left(\frac{|\mathbf{p}|\theta}{2m}\right) + \frac{\boldsymbol{\gamma}\cdot\mathbf{p}}{|\mathbf{p}|}\sin\left(\frac{|\mathbf{p}|\theta}{2m}\right)\right)\gamma^0(\boldsymbol{\gamma}\cdot\mathbf{p}+m) \\
&\quad \times \left(\cos\left(\frac{|\mathbf{p}|\theta}{2m}\right) - \frac{\boldsymbol{\gamma}\cdot\mathbf{p}}{|\mathbf{p}|}\sin\left(\frac{|\mathbf{p}|\theta}{2m}\right)\right) \\
&= \gamma^0\left(\cos\left(\frac{|\mathbf{p}|\theta}{2m}\right) - \frac{\boldsymbol{\gamma}\cdot\mathbf{p}}{|\mathbf{p}|}\sin\left(\frac{|\mathbf{p}|\theta}{2m}\right)\right)(\boldsymbol{\gamma}\cdot\mathbf{p}+m) \\
&\quad \times \left(\cos\left(\frac{|\mathbf{p}|\theta}{2m}\right) - \frac{\boldsymbol{\gamma}\cdot\mathbf{p}}{|\mathbf{p}|}\sin\left(\frac{|\mathbf{p}|\theta}{2m}\right)\right) \\
&= \gamma^0(\boldsymbol{\gamma}\cdot\mathbf{p}+m)\left(\cos\left(\frac{|\mathbf{p}|\theta}{2m}\right) - \frac{\boldsymbol{\gamma}\cdot\mathbf{p}}{|\mathbf{p}|}\sin\left(\frac{|\mathbf{p}|\theta}{2m}\right)\right)^2 \\
&= \gamma^0(\boldsymbol{\gamma}\cdot\mathbf{p}+m)\left(\cos\left(\frac{|\mathbf{p}|\theta}{m}\right) - \frac{\boldsymbol{\gamma}\cdot\mathbf{p}}{|\mathbf{p}|}\sin\left(\frac{|\mathbf{p}|\theta}{m}\right)\right) \\
&= \gamma^0\left[\left(m\cos\left(\frac{|\mathbf{p}|\theta}{m}\right) + |\mathbf{p}|\sin\left(\frac{|\mathbf{p}|\theta}{m}\right)\right)\right. \\
&\quad \left. + \left(|\mathbf{p}|\cos\left(\frac{|\mathbf{p}|\theta}{m}\right) - m\sin\left(\frac{|\mathbf{p}|\theta}{m}\right)\right)\frac{\boldsymbol{\gamma}\cdot\mathbf{p}}{|\mathbf{p}|}\right]. \quad (3.214)
\end{aligned}$$

3.10 FOLDY-WOUTHUYSEN TRANSFORMATION

So far, our discussion has been quite general and the parameter of the transformation, θ, has been arbitrary. However, if we want the transformation to diagonalize the Hamiltonian, it is clear from (3.214) that we can choose the parameter of transformation to satisfy

$$|\mathbf{p}| \cos\left(\frac{|\mathbf{p}|\theta}{m}\right) - m \sin\left(\frac{|\mathbf{p}|\theta}{m}\right) = 0,$$

or, $\quad \tan\left(\dfrac{|\mathbf{p}|\theta}{m}\right) = \dfrac{|\mathbf{p}|}{m},$

or, $\quad \theta = \dfrac{m}{|\mathbf{p}|} \arctan\left(\dfrac{|\mathbf{p}|}{m}\right).$ \hfill (3.215)

In this case, we have

$$\cos\left(\frac{|\mathbf{p}|\theta}{m}\right) = \frac{m}{\sqrt{\mathbf{p}^2+m^2}}, \quad \sin\left(\frac{|\mathbf{p}|\theta}{m}\right) = \frac{|\mathbf{p}|}{\sqrt{\mathbf{p}^2+m^2}}, \quad (3.216)$$

which, from (3.214), leads to the diagonalized Hamiltonian

$$H' = \gamma^0 \left(\frac{m^2}{\sqrt{\mathbf{p}^2+m^2}} + \frac{\mathbf{p}^2}{\sqrt{\mathbf{p}^2+m^2}}\right) = \gamma^0 \sqrt{\mathbf{p}^2+m^2}. \quad (3.217)$$

We see from (3.217) that the Hamiltonian is now diagonalized in the positive and the negative energy spaces. As a result, the two components of the transformed spinor

$$u'(p) = U(\theta) u(p) = \begin{pmatrix} u'_1(p) \\ u'_2(p) \end{pmatrix}, \quad (3.218)$$

would be decoupled in the energy eigenvalue equation and we can without any difficulty restrict ourselves to the positive energy sector where the energy eigenvalue equation takes the form

$$\sqrt{\mathbf{p}^2+m^2}\, u'_1(p) = E u'_1(p). \quad (3.219)$$

For $|\mathbf{p}| \ll m$, this leads to the non-relativistic equation in (3.183) to the lowest order and it can be expanded to any order in $\frac{1}{m}$ without any problem. We also note that with the parameter θ determined in (3.215), the unitary transformation in (3.211) takes the form

$$U_{\text{FW}}(\theta) = \cos\left(\frac{1}{2}\arctan\left(\frac{|\mathbf{p}|}{m}\right)\right) + \frac{\gamma \cdot \mathbf{p}}{|\mathbf{p}|}\sin\left(\frac{1}{2}\arctan\left(\frac{|\mathbf{p}|}{m}\right)\right), \tag{3.220}$$

which has a natural non-relativistic expansion in powers of $\frac{|\mathbf{p}|}{m}$. This analysis can be generalized even in the presence of interactions and the higher order terms in the interaction Hamiltonian are all well behaved without any problem of non-hermiticity.

There is a second limit of the Dirac equation, namely, the ultrarelativistic limit $|\mathbf{p}| \gg m$, for which the generalized Foldy-Wouthuysen transformation (3.211) is also quite useful. In this case, the transformation is known as the Cini-Touschek transformation and is obtained as follows. Let us note from (3.214) that if we choose the parameter of transformation to satisfy

$$m\cos\left(\frac{|\mathbf{p}|\theta}{m}\right) + |\mathbf{p}|\sin\left(\frac{|\mathbf{p}|\theta}{m}\right) = 0,$$

$$\text{or,} \quad \tan\left(\frac{|\mathbf{p}|\theta}{m}\right) = -\frac{m}{|\mathbf{p}|},$$

$$\text{or,} \quad \theta = -\frac{m}{|\mathbf{p}|}\arctan\left(\frac{m}{|\mathbf{p}|}\right), \tag{3.221}$$

this would lead to

$$\cos\left(\frac{|\mathbf{p}|\theta}{m}\right) = \frac{|\mathbf{p}|}{\sqrt{\mathbf{p}^2 + m^2}},$$

$$\sin\left(\frac{|\mathbf{p}|\theta}{m}\right) = -\frac{m}{\sqrt{\mathbf{p}^2 + m^2}}. \tag{3.222}$$

As a result, in this case, the transformed Hamiltonian (3.214) will have the form

$$H' = \gamma^0 \left(\frac{\mathbf{p}^2}{\sqrt{\mathbf{p}^2 + m^2}} + \frac{m^2}{\sqrt{\mathbf{p}^2 + m^2}} \right) \frac{\boldsymbol{\gamma} \cdot \mathbf{p}}{|\mathbf{p}|}$$

$$= \gamma^0 \frac{\sqrt{\mathbf{p}^2 + m^2}}{|\mathbf{p}|} \boldsymbol{\gamma} \cdot \mathbf{p} = \frac{\sqrt{\mathbf{p}^2 + m^2}}{|\mathbf{p}|} \boldsymbol{\alpha} \cdot \mathbf{p}, \quad (3.223)$$

which has a natural expansion in powers of $\frac{m}{|\mathbf{p}|}$. In fact, in this case, the unitary transformation (3.211) has the form

$$U_{\mathrm{CT}}(\theta) = \cos\left(\frac{1}{2}\arctan\left(\frac{m}{|\mathbf{p}|}\right)\right) - \frac{\boldsymbol{\gamma} \cdot \mathbf{p}}{|\mathbf{p}|} \sin\left(\frac{1}{2}\arctan\left(\frac{m}{|\mathbf{p}|}\right)\right), \quad (3.224)$$

which clearly has a natural expansion in powers of $\frac{m}{|\mathbf{p}|}$ (ultrarelativistic expansion). Therefore, we can think of the Foldy-Wouthuysen transformation (3.220) as transforming away the $\boldsymbol{\alpha} \cdot \mathbf{p}$ term in the Hamiltonian (3.207) while the Cini-Touschek transformation rotates away the mass term βm from the Hamiltonian (3.207).

3.11 Zitterbewegung

The presence of negative energy solutions for the Dirac equation leads to various interesting consequences. For example, let us consider the free Dirac Hamiltonian (1.100)

$$H = \boldsymbol{\alpha} \cdot \mathbf{p} + \beta m. \quad (3.225)$$

In the Heisenberg picture, where operators carry time dependence and states are time independent, the Heisenberg equations of motion take the forms ($\hbar = 1$)

$$\dot{x}^i = \frac{1}{i}[x^i, H] = \frac{1}{i}[x^i, \alpha^j p^j + \beta m]$$

$$= \frac{1}{i}[x^i, \alpha^j p^j] = \frac{1}{i}\alpha^j [x^i, p^j]$$

$$= \frac{1}{i}\alpha^j (i\delta^{ij}) = \alpha^i,$$

or, $\dot{\mathbf{x}} = \boldsymbol{\alpha},$

$$\dot{p}^i = \frac{1}{i}[p^i, H] = \frac{1}{i}[p^i, \alpha^j p^j + \beta m] = 0,$$

or, $\dot{\mathbf{p}} = 0.$ (3.226)

Here a dot denotes differentiation with respect to time.

The second equation in (3.226) shows that the momentum is a constant of motion as it should be for a free particle. The first equation, on the other hand, identifies $\boldsymbol{\alpha}(t)$ with the velocity operator. Let us recall that, by definition,

$$\boldsymbol{\alpha}(t) = e^{iHt}\boldsymbol{\alpha}\, e^{-iHt}, \qquad (3.227)$$

where we have denoted the operator in the Schrödinger picture by

$$\boldsymbol{\alpha}(0) = \boldsymbol{\alpha}. \qquad (3.228)$$

Furthermore, using (1.101) we conclude that

$$[\boldsymbol{\alpha}, H] = [\boldsymbol{\alpha}, \boldsymbol{\alpha}\cdot\mathbf{p} + \beta m] \neq 0. \qquad (3.229)$$

As a result, it follows that

$$[\boldsymbol{\alpha}(t), H] \neq 0. \qquad (3.230)$$

In other words, even though the momentum of a free particle is a constant of motion, the velocity is not. Secondly, since the eigenvalues of $\boldsymbol{\alpha}$ are ± 1 (see, for example, (1.101)), it follows that the

3.11 ZITTERBEWEGUNG

eigenvalues of $\boldsymbol{\alpha}(t)$ are ± 1 as well. This is easily understood from the fact that the eigenvalues of an operator do not change under a unitary transformation. More explicitly, we note that if

$$\boldsymbol{\alpha}|\psi\rangle = \boldsymbol{\lambda}|\psi\rangle, \tag{3.231}$$

where $\boldsymbol{\lambda}$ denotes the eigenvalue of the velocity operator $\boldsymbol{\alpha}$, then, it follows that

$$e^{iHt}\boldsymbol{\alpha} e^{-iHt} e^{iHt}|\psi\rangle = \boldsymbol{\lambda} e^{iHt}|\psi\rangle,$$
or, $\quad \boldsymbol{\alpha}(t)\left(e^{iHt}|\psi\rangle\right) = \boldsymbol{\lambda}\left(e^{iHt}|\psi\rangle\right),$
or, $\quad \boldsymbol{\alpha}(t)|\psi'\rangle = \boldsymbol{\lambda}|\psi'\rangle, \tag{3.232}$

where we have identified

$$|\psi'\rangle = e^{iHt}|\psi\rangle. \tag{3.233}$$

Equation (3.232) shows that the eigenvalues of $\boldsymbol{\alpha}(t)$ are the same as those of $\boldsymbol{\alpha}$ (only the eigenfunctions are transformed) and, therefore, are ± 1. This would seem to imply that the velocity of an electron is equal to the speed of light which is unacceptable even classically, since the electron is a massive particle.

These peculiarities of the relativistic theory can be understood as follows. We note from Heisenberg's equations of motion that the time derivative of the velocity operator is given by

$$\begin{aligned}
\dot{\boldsymbol{\alpha}}(t) &= \frac{1}{i}\left[\boldsymbol{\alpha}(t), H\right] \\
&= -i\left(2\boldsymbol{\alpha}(t)H - \left[\boldsymbol{\alpha}(t), H\right]_+\right) \\
&= -2i\boldsymbol{\alpha}(t)H + 2i\mathbf{p}.
\end{aligned} \tag{3.234}$$

Here we have used the relations (see (1.102))

$$\begin{aligned}
\left[\alpha^i, \alpha^j\right]_+ &= 2\delta^{ij}\mathbb{1}, \\
\left[\alpha^i, \beta\right]_+ &= 0,
\end{aligned} \tag{3.235}$$

as well as the fact that momentum commutes with the Hamiltonian. Let us note next that both **p** and H are constants of motion. Therefore, differentiating (3.234) with respect to time, we obtain

$$\ddot{\boldsymbol{\alpha}}(t) = -2i\dot{\boldsymbol{\alpha}}(t)H,$$
$$\text{or,} \quad \dot{\boldsymbol{\alpha}}(t) = \dot{\boldsymbol{\alpha}}(0)e^{-2iHt}. \tag{3.236}$$

On the other hand, from (3.234) we have

$$\dot{\boldsymbol{\alpha}}(0) = -2i\boldsymbol{\alpha}(0)H + 2i\mathbf{p} = -2i(\boldsymbol{\alpha}(0) - \mathbf{p}H^{-1})H. \tag{3.237}$$

Substituting this back into (3.236), we obtain

$$\dot{\boldsymbol{\alpha}}(t) = -2i(\boldsymbol{\alpha}(0) - \mathbf{p}\,H^{-1})He^{-2iHt}. \tag{3.238}$$

Furthermore, using this relation in (3.234), we finally determine

$$-2i(\boldsymbol{\alpha}(0) - \mathbf{p}\,H^{-1})He^{-2iHt} = -2i\boldsymbol{\alpha}(t)H + 2i\mathbf{p},$$
$$\text{or,} \quad \boldsymbol{\alpha}(t) = \mathbf{p}H^{-1} + (\boldsymbol{\alpha}(0) - \mathbf{p}H^{-1})e^{-2iHt},$$
$$\text{or,} \quad \boldsymbol{\alpha}(t) = \frac{\mathbf{p}}{H} + \left(\boldsymbol{\alpha}(0) - \frac{\mathbf{p}}{H}\right)e^{-2iHt}. \tag{3.239}$$

The first term in (3.239) is quite expected. For example, in an eigenstate of momentum it would have the form $\frac{\mathbf{p}}{E}$ which is the true relativistic expression for velocity. We note that, for a relativistic particle, ($c = 1$)

$$E = m\gamma, \qquad \mathbf{p} = m\gamma\mathbf{v}, \tag{3.240}$$

so that

$$\frac{\mathbf{p}}{E} = \frac{m\gamma\mathbf{v}}{m\gamma} = \mathbf{v}, \tag{3.241}$$

which is the first term in (3.239). It is the second term, however, which is unexpected. It represents an additional component to the

3.11 ZITTERBEWEGUNG

velocity which is oscillating at a very high frequency (for an electron at rest, for example, the energy is $\approx .5\text{MeV}$ corresponding to a frequency of the order of $10^{21}/\text{sec.}$) and gives a time dependence to $\boldsymbol{\alpha}(t)$. Let us also note from (3.226) that since

$$\dot{\mathbf{x}}(t) = \boldsymbol{\alpha}(t) = \frac{\mathbf{p}}{H} + \left(\boldsymbol{\alpha}(0) - \frac{\mathbf{p}}{H}\right)e^{-2iHt}, \qquad (3.242)$$

integrating this over time, we obtain

$$\mathbf{x}(t) = \mathbf{a} + \frac{\mathbf{p}}{H}t + \frac{i}{2}\left(\boldsymbol{\alpha}(0) - \frac{\mathbf{p}}{H}\right)H^{-1}e^{-2iHt}, \qquad (3.243)$$

where $\mathbf{a}$ is a constant. The first two terms in (3.243) are again what we will expect classically for uniform motion. However, the third term represents an additional contribution to the electron trajectory which is oscillatory with a very high frequency. Its occurrence is quite surprising, since there is no potential whatsoever in the problem. This quivering motion of the electron was first studied by Schrödinger and is known as Zitterbewegung ("jittery motion").

The unconventional operator relations in (3.239) and (3.243) can be shown in the Schrödinger picture to arise from the presence of negative energy solutions. In fact, it is easy to check that for a positive energy electron state

$$u(p) = \sqrt{\frac{E+m}{2m}}\begin{pmatrix} \tilde{u} \\ \frac{\boldsymbol{\sigma}\cdot\mathbf{p}}{E+m}\tilde{u} \end{pmatrix}, \qquad (3.244)$$

we have

$$u^\dagger(p)\left(\boldsymbol{\alpha}(0) - \frac{\mathbf{p}}{H}\right)u(p)$$

$$= \frac{E+m}{2m}\begin{pmatrix} \tilde{u}^\dagger & \tilde{u}^\dagger\frac{\boldsymbol{\sigma}\cdot\mathbf{p}}{E+m}\end{pmatrix}\begin{pmatrix} -\frac{\mathbf{p}}{E} & \boldsymbol{\sigma} \\ \boldsymbol{\sigma} & -\frac{\mathbf{p}}{E}\end{pmatrix}\begin{pmatrix} \tilde{u} \\ \frac{\boldsymbol{\sigma}\cdot\mathbf{p}}{E+m}\tilde{u}\end{pmatrix}$$

$$= \frac{E+m}{2m}\begin{pmatrix} \tilde{u}^\dagger & \tilde{u}^\dagger\frac{\boldsymbol{\sigma}\cdot\mathbf{p}}{E+m}\end{pmatrix}\begin{pmatrix} -\frac{\mathbf{p}}{E}\tilde{u} + \boldsymbol{\sigma}\frac{\boldsymbol{\sigma}\cdot\mathbf{p}}{E+m}\tilde{u} \\ \boldsymbol{\sigma}\tilde{u} - \frac{\mathbf{p}}{E}\frac{\boldsymbol{\sigma}\cdot\mathbf{p}}{E+m}\tilde{u}\end{pmatrix}$$

$$= \frac{E+m}{2m}\left(-\frac{\mathbf{p}}{E}\tilde{u}^\dagger\tilde{u} + \tilde{u}^\dagger\boldsymbol{\sigma}\frac{\boldsymbol{\sigma}\cdot\mathbf{p}}{E+m}\tilde{u}\right.$$

$$\left.+\tilde{u}^\dagger\frac{\boldsymbol{\sigma}\cdot\mathbf{p}}{E+m}\boldsymbol{\sigma}\tilde{u} - \frac{\mathbf{p}}{E}\tilde{u}^\dagger\frac{(\boldsymbol{\sigma}\cdot\mathbf{p})^2}{(E+m)^2}\tilde{u}\right)$$

$$= \frac{E+m}{2m}\left(-\frac{\mathbf{p}}{E}\left(1+\frac{\mathbf{p}^2}{(E+m)^2}\right) + \frac{2\mathbf{p}}{E+m}\right)\tilde{u}^\dagger\tilde{u}$$

$$= \frac{E+m}{2m}\left(-\frac{\mathbf{p}}{E}\times\frac{2E}{E+m} + \frac{2\mathbf{p}}{E+m}\right)\tilde{u}^\dagger\tilde{u} = 0. \qquad (3.245)$$

This shows that even though the operator relations are unconventional, in a positive energy electron state

$$\langle\boldsymbol{\alpha}(t)\rangle_+ = \left\langle\frac{\mathbf{p}}{H}\right\rangle_+, \qquad (3.246)$$

as we should expect from the Ehrenfest theorem. This shows that even though the eigenvalues of the operator $\boldsymbol{\alpha}(t)$ are ± 1 corresponding to motion with the speed of light, the physical velocity of the electron (observed velocity which is the expectation value of the operator in the positive energy electron state) is what we would expect. This also shows that the eigenstates of the velocity operator, $\boldsymbol{\alpha}(t)$, which are not simultaneous eigenstates of the Hamiltonian must necessarily contain both positive and negative energy solutions as superposition and that the extra terms have non-zero value only in the transition between a positive energy and a negative energy state. (This makes clear that neglecting the negative energy solutions of the Dirac equation would lead to inconsistencies.)

3.12 References

1. L. I. Schiff, *Quantum Mechanics*, McGraw-Hill, New York, 1968.

2. J. D. Bjorken and S. Drell, *Relativistic Quantum Mechanics*, McGraw-Hill, New York, 1964.

3. C. Itzykson and J-B. Zuber, *Quantum Field Theory*, McGraw-Hill, New York, 1980.

4. M. Cini and B. Touschek, Nuovo Cimento **7**, 422 (1958).

5. L. L. Foldy and S. A. Wouthuysen, Physical Review **78**, 29 (1950).

6. S. Okubo, Progress of Theoretical Physics **12**, 102 (1954); *ibid.* **12**, 603 (1954).

7. S. Tani, Progress of Theoretical Physics **6**, 267 (1951).

CHAPTER 4
Representations of Lorentz and Poincaré groups

4.1 Symmetry algebras

Relativistic theories, as we have discussed, should be invariant under Lorentz transformations. In addition, experimentally we know that space-time translations also define a symmetry of physical theories. In this chapter, therefore, we will study the symmetry algebras of the Lorentz and the Poincaré groups as well as their representations which are essential in constructing physical theories. But, let us start with rotations which we have already discussed briefly in the last chapter. In studying the symmetry algebras of continuous symmetry transformations, it is sufficient to study the behavior of infinitesimal transformations since any finite transformation can be built out of infinitesimal transformations. Furthermore, the symmetry algebra associated with a continuous symmetry group is given by the algebra of the generators of infinitesimal transformations. It is worth noting here that, for space-time symmetries, the symmetry algebras can be easily obtained from the coordinate representation of the symmetry generators and that is the approach we will follow in our discussions.

4.1.1 Rotation. Let us consider an arbitrary, infinitesimal three dimensional rotation of the form

$$\delta x^i = \epsilon^i{}_{jk} x^j \alpha^k, \qquad i,j,k = 1,2,3, \tag{4.1}$$

where α^k represents the infinitesimal constant parameter of rotation around the k-th axis (there are three of them). (Let us recall our notation from (1.34) and (1.35) here for clarity. ϵ_{ijk} denotes the three dimensional Levi-Civita tensor with $\epsilon_{123} = 1$. $\epsilon^i{}_{jk} = \eta^{i\ell}\epsilon_{\ell jk}$, etc.) If we now identify (In the last chapter we had denoted the infinitesimal transformation matrices by $\epsilon^i{}_j, \epsilon^\mu{}_\nu$, but here we denote them by $\omega^i{}_j, \omega^\mu{}_\nu$ in order to avoid confusion.)

$$\omega^i{}_j = \epsilon^i{}_{jk}\alpha^k, \tag{4.2}$$

then, we note that

$$\omega^{ij} = -\omega^{ji}, \tag{4.3}$$

and that the infinitesimal rotation around the k-th axis can also be represented in the form (taking place in the i-j plane)

$$\delta x^i = \omega^i{}_j x^j. \tag{4.4}$$

This is, of course, the form of the rotation that we had discussed in the last chapter.

Let us next define an infinitesimal vector operator (also known as the tangent vector field operator) for rotations (an operator in the coordinate basis) of the form

$$\begin{aligned}\widehat{R}(\omega) &= \frac{1}{2}\omega^{ij}M_{ij} = \frac{1}{2}\omega^{ij}\left(x_i\partial_j - x_j\partial_i\right) \\ &= \omega^{ij}x_i\partial_j = -\omega^{ij}x_j\partial_i.\end{aligned} \tag{4.5}$$

It follows now that

$$\begin{aligned}\widehat{R}(\omega)x^i &= -\omega^{kj}x_j\partial_k x^i = -\omega^{kj}x_j\delta^i_k \\ &= -\omega^{ij}x_j = -\omega^i{}_j x^j.\end{aligned} \tag{4.6}$$

In other words, we can write the infinitesimal rotations also as

$$\delta x^i = -\widehat{R}(\omega)x^i = \omega^i{}_j x^j. \tag{4.7}$$

4.1 SYMMETRY ALGEBRAS

Namely, the vector operator, $\widehat{R}(\omega)$ in (4.5) generates infinitesimal rotations and the operators, M_{ij}, are known as the generators of infinitesimal rotations.

The Lie algebra of the group of rotations can be obtained from the algebra of the vector operators themselves. Thus, we note that

$$\begin{aligned}
\left[\widehat{R}(\omega), \widehat{R}(\overline{\omega})\right] &= \left[\omega^{ij} x_i \partial_j, \overline{\omega}^{k\ell} x_k \partial_\ell\right] \\
&= \omega^{ij} \overline{\omega}^{k\ell} [x_i \partial_j, x_k \partial_\ell] \\
&= \omega^{ij} \overline{\omega}^{k\ell} \left(x_i [\partial_j, x_k] \partial_\ell + x_k [x_i, \partial_\ell] \partial_j\right) \\
&= \omega^{ij} \overline{\omega}^{k\ell} \left(\eta_{jk} x_i \partial_\ell - \eta_{i\ell} x_k \partial_j\right) \\
&= \omega^{ij} \overline{\omega}_j{}^\ell x_i \partial_\ell - \omega^{ij} \overline{\omega}{}^k{}_i x_k \partial_j \\
&= \left(\omega^{ij} \overline{\omega}_j{}^k - \overline{\omega}^{ij} \omega_j{}^k\right) x_i \partial_k \\
&= \widehat{R}(\widetilde{\omega}), \quad (4.8)
\end{aligned}$$

where we have identified

$$\widetilde{\omega}^{ij} = \left(\omega^{ik} \overline{\omega}_k{}^j - \overline{\omega}^{ik} \omega_k{}^j\right) = -\widetilde{\omega}^{ji} = [\omega, \overline{\omega}]^{ij}. \quad (4.9)$$

Namely, two rotations do not commute, rather, they give back a rotation. Such an algebra is called a non-Abelian (non-commutative) algebra. Using the form of $\widehat{R}(\omega)$ in (4.5), namely,

$$\widehat{R}(\omega) = \frac{1}{2} \omega^{ij} M_{ij}, \quad (4.10)$$

we can obtain the algebra satisfied by the generators of infinitesimal rotations, M_{ij}, from the algebra of the vector operators in (4.8). Alternatively, we can calculate them directly as

$$[M_{ij}, M_{k\ell}] = [x_i\partial_j - x_j\partial_i, x_k\partial_\ell - x_\ell\partial_k]$$
$$= [x_i\partial_j, x_k\partial_\ell] - [x_i\partial_j, x_\ell\partial_k] - [x_j\partial_i, x_k\partial_\ell] + [x_j\partial_i, x_\ell\partial_k]$$
$$= (\eta_{jk}x_i\partial_\ell - \eta_{i\ell}x_k\partial_j) - (\eta_{j\ell}x_i\partial_k - \eta_{ik}x_\ell\partial_j)$$
$$\quad - (\eta_{ik}x_j\partial_\ell - \eta_{j\ell}x_k\partial_i) + (\eta_{i\ell}x_j\partial_k - \eta_{jk}x_\ell\partial_i)$$
$$= -\eta_{ik}(x_j\partial_\ell - x_\ell\partial_j) - \eta_{j\ell}(x_i\partial_k - x_k\partial_i)$$
$$\quad + \eta_{i\ell}(x_j\partial_k - x_k\partial_j) + \eta_{jk}(x_i\partial_\ell - x_\ell\partial_i)$$
$$= -\eta_{ik}M_{j\ell} - \eta_{j\ell}M_{ik} + \eta_{i\ell}M_{jk} + \eta_{jk}M_{i\ell}. \qquad (4.11)$$

This is the Lie algebra for the group of rotations. If we would like the generators to be Hermitian quantum mechanical operators corresponding to a unitary representation, then we may define the operators, M_{ij}, with a factor of "i". But up to a rescaling, (4.11) represents the Lie algebra of the group $SO(3)$ or equivalently $SU(2)$. To obtain the familiar algebra of the angular momentum operators, we note that we can define (recall that in the four vector notation $J_i = -(\mathbf{J})_i$)

$$M_{ij} = (x_i\partial_j - x_j\partial_i) = \epsilon_{ij}{}^k J_k,$$
$$\text{or,} \quad J_i = -\frac{1}{2}\epsilon_i{}^{jk}M_{jk} = -\frac{1}{2}\epsilon_i{}^{jk}(x_j\partial_k - x_k\partial_j), \qquad (4.12)$$

which gives the familiar orbital angular momentum operators. Using this, then, we obtain ($p,q,r,s = 1,2,3$)

$$[J_i, J_j] = \left[\frac{1}{2}\epsilon_i{}^{pq}M_{pq}, \frac{1}{2}\epsilon_j{}^{rs}M_{rs}\right]$$
$$= \frac{1}{4}\epsilon_i{}^{pq}\epsilon_j{}^{rs}[M_{pq}, M_{rs}]$$
$$= \frac{1}{4}\epsilon_i{}^{pq}\epsilon_j{}^{rs}(-\eta_{pr}M_{qs} - \eta_{qs}M_{pr} + \eta_{ps}M_{qr} + \eta_{qr}M_{ps})$$
$$= -\frac{1}{4}\epsilon_{ir}{}^q\epsilon_j{}^{rs}M_{qs} - \frac{1}{4}\epsilon_i{}^p{}_s\epsilon_j{}^{rs}M_{pr}$$

$$+\frac{1}{4}\epsilon_{is}{}^q\epsilon_j{}^{rs}M_{qr} + \frac{1}{4}\epsilon_i{}^p{}_r\epsilon_j{}^{rs}M_{ps}$$

$$= -\frac{1}{4}\epsilon_{ri}{}^q\epsilon_j{}^{sr}M_{qs} - \frac{1}{4}\epsilon_i{}^q{}_r\epsilon_j{}^{sr}M_{qs}$$

$$-\frac{1}{4}\epsilon_{ri}{}^q\epsilon_j{}^{sr}M_{qs} - \frac{1}{4}\epsilon_i{}^q{}_r\epsilon_j{}^{sr}M_{qs}$$

$$= -\epsilon_{ri}{}^q\epsilon_j{}^{sr}M_{qs},$$

or, $\quad [J_i, J_j] = \epsilon_{ij}{}^r J_r,$ \hfill (4.13)

where in the last step we have used the Jacobi identity for the structure constants of $SO(3)$ or $SU(2)$ (or the identity satisfied by the Levi-Civita tensors), namely,

$$\epsilon_{ri}{}^q\epsilon_j{}^{sr} + \epsilon_r{}^{sq}\epsilon_{ij}{}^r + \epsilon_{rj}{}^q\epsilon^s{}_i{}^r = 0, \tag{4.14}$$

which, then, leads to (see (4.12))

$$\epsilon_{ri}{}^q\epsilon_j{}^{sr}M_{qs} = -\frac{1}{2}\epsilon_{ij}{}^r\epsilon_r{}^{sq}M_{qs} = -\epsilon_{ij}{}^r J_r. \tag{4.15}$$

The algebra of the generators in (4.11) or (4.13) is, of course, the Lie algebra of $SO(3)$ or $SU(2)$ (or the familiar algebra of angular momentum operators) up to a rescaling.

4.1.2 Translation. In the same spirit, let us note that a constant infinitesimal space-time translation of the form

$$\delta x^\mu = \epsilon^\mu, \tag{4.16}$$

can be generated by the infinitesimal vector operator

$$\widehat{R}(\epsilon) = -\epsilon^\mu P_\mu = -\epsilon^\mu \partial_\mu, \tag{4.17}$$

so that

$$\widehat{R}(\epsilon)x^\mu = -\epsilon^\nu \partial_\nu x^\mu = -\epsilon^\nu \delta^\mu_\nu = -\epsilon^\mu, \tag{4.18}$$

and we can write

$$\delta x^\mu = -\widehat{R}(\epsilon)x^\mu. \tag{4.19}$$

The Lie algebra associated with translations is then obtained from

$$\left[\widehat{R}(\epsilon), \widehat{R}(\bar{\epsilon})\right] = [\epsilon^\mu \partial_\mu, \bar{\epsilon}^\nu \partial_\nu] = \epsilon^\mu \bar{\epsilon}^\nu [\partial_\mu, \partial_\nu] = 0. \tag{4.20}$$

In other words, two translations commute and the corresponding relation for the generators is

$$[P_\mu, P_\nu] = [\partial_\mu, \partial_\nu] = 0. \tag{4.21}$$

Namely, translations form an Abelian (commuting) group while the three dimensional rotations form a non-Abelian group.

4.1.3 Lorentz transformation. As we have seen in the last chapter, a proper Lorentz transformation can be thought of as a rotation in the four dimensional Minkowski space-time and has the infinitesimal form

$$\delta x^\mu = \omega^\mu{}_\nu x^\nu, \tag{4.22}$$

where, as we have seen in (3.18), the infinitesimal, constant parameters of transformation satisfy

$$\omega^{\mu\nu} = -\omega^{\nu\mu}, \qquad \mu, \nu = 0, 1, 2, 3. \tag{4.23}$$

As in the case of rotations, let us note that if we define an infinitesimal vector operator

$$\begin{aligned}\widehat{R}(\omega) &= \frac{1}{2}\omega^{\mu\nu} M_{\mu\nu} = \frac{1}{2}\omega^{\mu\nu}\left(x_\mu \partial_\nu - x_\nu \partial_\mu\right) \\ &= \omega^{\mu\nu} x_\mu \partial_\nu = -\omega^{\mu\nu} x_\nu \partial_\mu,\end{aligned} \tag{4.24}$$

then, we obtain

4.1 SYMMETRY ALGEBRAS 131

$$\begin{aligned}\widehat{R}(\omega)x^\mu &= -\omega^{\lambda\nu}x_\nu \partial_\lambda x^\mu = -\omega^{\lambda\nu}x_\nu \delta^\mu_\lambda = -\omega^{\mu\nu}x_\nu \\ &= -\omega^\mu{}_\nu x^\nu = -\delta x^\mu.\end{aligned} \quad (4.25)$$

Therefore, we can think of $\widehat{R}(\omega)$ as the vector operator generating infinitesimal proper Lorentz transformations and the operators, $M_{\mu\nu} = -M_{\nu\mu}$, as the generators of the infinitesimal transformations. We also note that we can identify the infinitesimal generators of spatial rotations with (see (4.12))

$$M_{ij} = -M_{ji} = x_i \partial_j - x_j \partial_i = \epsilon_{ij}{}^k J_k, \quad (4.26)$$

and the generators of infinitesimal boosts with

$$M_{0i} = -M_{i0} = x_0 \partial_i - x_i \partial_0 = K_i. \quad (4.27)$$

As before, we can determine the group properties of the Lorentz transformations from the algebra of the vector operators generating the transformations. Thus,

$$\begin{aligned}\left[\widehat{R}(\omega), \widehat{R}(\overline{\omega})\right] &= \left[\frac{1}{2}\omega^{\mu\nu}M_{\mu\nu}, \frac{1}{2}\overline{\omega}^{\lambda\rho}M_{\lambda\rho}\right] \\ &= \omega^{\mu\nu}\overline{\omega}^{\lambda\rho}[x_\mu \partial_\nu, x_\lambda \partial_\rho] \\ &= \omega^{\mu\nu}\overline{\omega}^{\lambda\rho}(\eta_{\nu\lambda}x_\mu \partial_\rho - \eta_{\mu\rho}x_\lambda \partial_\nu) \\ &= \omega^\mu{}_\lambda \overline{\omega}^{\lambda\rho} x_\mu \partial_\rho - \omega^{\mu\nu}\overline{\omega}^\lambda{}_\mu x_\lambda \partial_\nu \\ &= \left(\omega^\mu{}_\lambda \overline{\omega}^{\lambda\nu} - \overline{\omega}^\mu{}_\lambda \omega^{\lambda\nu}\right) x_\mu \partial_\nu \\ &= \widetilde{\omega}^{\mu\nu} x_\mu \partial_\nu = \widehat{R}(\widetilde{\omega}),\end{aligned} \quad (4.28)$$

where, as in the case of rotations, we have

$$\widetilde{\omega}^{\mu\nu} = -\widetilde{\omega}^{\nu\mu} = \omega^\mu{}_\lambda \overline{\omega}^{\lambda\nu} - \overline{\omega}^\mu{}_\lambda \omega^{\lambda\nu} = [\omega, \overline{\omega}]^{\mu\nu}. \quad (4.29)$$

This shows that the algebra of the vector operators is closed and that Lorentz transformations define a non-Abelian group.

The algebra of the generators can also be calculated directly and has the form

$$[M_{\mu\nu}, M_{\lambda\rho}] = [x_\mu \partial_\nu - x_\nu \partial_\mu, x_\lambda \partial_\rho - x_\rho \partial_\lambda]$$
$$= [x_\mu \partial_\nu, x_\lambda \partial_\rho] - [x_\mu \partial_\nu, x_\rho \partial_\lambda]$$
$$- [x_\nu \partial_\mu, x_\lambda \partial_\rho] + [x_\nu \partial_\mu, x_\rho \partial_\lambda]$$
$$= (\eta_{\nu\lambda} x_\mu \partial_\rho - \eta_{\mu\rho} x_\lambda \partial_\nu) - (\eta_{\nu\rho} x_\mu \partial_\lambda - \eta_{\mu\lambda} x_\rho \partial_\nu)$$
$$- (\eta_{\mu\lambda} x_\nu \partial_\rho - \eta_{\nu\rho} x_\lambda \partial_\mu) + (\eta_{\mu\rho} x_\nu \partial_\lambda - \eta_{\nu\lambda} x_\rho \partial_\mu)$$
$$= -\eta_{\mu\lambda} (x_\nu \partial_\rho - x_\rho \partial_\nu) - \eta_{\nu\rho} (x_\mu \partial_\lambda - x_\lambda \partial_\mu)$$
$$+ \eta_{\mu\rho} (x_\nu \partial_\lambda - x_\lambda \partial_\nu) + \eta_{\nu\lambda} (x_\mu \partial_\rho - x_\rho \partial_\mu)$$
$$= -\eta_{\mu\lambda} M_{\nu\rho} - \eta_{\nu\rho} M_{\mu\lambda} + \eta_{\mu\rho} M_{\nu\lambda} + \eta_{\nu\lambda} M_{\mu\rho}. \tag{4.30}$$

This, therefore, gives the Lie algebra associated with Lorentz transformations. As we have seen these transformations correspond to rotations, in this case, in four dimensions and, therefore, the Lie algebra of the generators is isomorphic to that of the group $SO(4)$. In fact, we note that the number of generators for $SO(4)$ which is (for $SO(n)$, it is $\frac{n(n-1)}{2}$)

$$\frac{1}{2} \times 4 \times (4 - 1) = 2 \times 3 = 6, \tag{4.31}$$

coincides exactly with the six generators we have (namely, three rotations and three boosts). However, since the rotations are in Minkowski space-time whose metric is not Euclidean it is more appropriate to identify the Lie algebra as that of the group $SO(3,1)$. (Namely, Lorentz transformations (boosts) are non-compact unlike rotations in Euclidean space.)

We end this section by pointing out that the algebra in (2.109) coincides with (4.30) (up to a scaling). This implies that, up to a scaling, the matrices $\sigma_{\mu\nu}$ provide a representation for the generators of the Lorentz group. This is what we had seen explicitly in (3.69) in connection with the discussion of covariance of the Dirac equation.

4.1 SYMMETRY ALGEBRAS

4.1.4 Poincaré transformation. If, in addition to infinitesimal Lorentz transformations, we also consider infinitesimal translations, the general transformation of the coordinates takes the form

$$\delta x^\mu = \epsilon^\mu + \omega^\mu{}_\nu x^\nu, \tag{4.32}$$

where $\epsilon^\mu, \omega^\mu{}_\nu$ denote respectively the parameters of infinitesimal translation and Lorentz transformation. The transformations in (4.32) are known as the (infinitesimal) Poincaré transformations or the inhomogeneous Lorentz transformations. Clearly, in this case, if we define an infinitesimal vector operator as

$$\begin{aligned}\widehat{R}(\epsilon,\omega) &= -\epsilon^\mu P_\mu + \frac{1}{2}\omega^{\mu\nu} M_{\mu\nu} \\ &= -\epsilon^\mu \partial_\mu + \omega^{\mu\nu} x_\mu \partial_\nu = -\epsilon^\mu \partial_\mu - \omega^{\mu\nu} x_\nu \partial_\mu \\ &= -\left(\epsilon^\mu \partial_\mu + \omega^{\mu\nu} x_\nu \partial_\mu\right),\end{aligned} \tag{4.33}$$

then, acting on the coordinates, it generates infinitesimal Poincaré transformations. Namely,

$$\begin{aligned}\widehat{R}(\epsilon,\omega)x^\mu &= -\left(\epsilon^\nu \partial_\nu + \omega^{\lambda\nu} x_\nu \partial_\lambda\right)x^\mu = -(\epsilon^\nu \delta^\mu_\nu + \omega^{\lambda\nu} x_\nu \delta^\mu_\lambda) \\ &= -(\epsilon^\mu + \omega^\mu{}_\nu x^\nu) = -\delta x^\mu.\end{aligned} \tag{4.34}$$

The algebra of the vector operators for the Poincaré transformations can also be easily calculated as

$$\begin{aligned}\left[\widehat{R}(\epsilon,\omega), \widehat{R}(\overline{\epsilon},\overline{\omega})\right] &= \left[\epsilon^\mu \partial_\mu + \omega^{\mu\nu} x_\nu \partial_\mu, \overline{\epsilon}^\lambda \partial_\lambda + \overline{\omega}^{\lambda\rho} x_\rho \partial_\lambda\right] \\ &= \epsilon^\mu \overline{\omega}^{\lambda\rho}\left[\partial_\mu, x_\rho \partial_\lambda\right] + \omega^{\mu\nu} \overline{\epsilon}^\lambda \left[x_\nu \partial_\mu, \partial_\lambda\right] \\ &\quad + \omega^{\mu\nu} \overline{\omega}^{\lambda\rho}\left[x_\nu \partial_\mu, x_\rho \partial_\lambda\right] \\ &= \epsilon^\mu \overline{\omega}^{\lambda\rho}\eta_{\mu\rho}\partial_\lambda + \omega^{\mu\nu} \overline{\epsilon}^\lambda\left(-\eta_{\nu\lambda}\right)\partial_\mu \\ &\quad + \omega^{\mu\nu} \overline{\omega}^{\lambda\rho}\left(\eta_{\mu\rho} x_\nu \partial_\lambda - \eta_{\nu\lambda} x_\rho \partial_\mu\right) \\ &= -\left(\omega^\mu{}_\nu \overline{\epsilon}^\nu - \overline{\omega}^\mu{}_\nu \epsilon^\nu\right)\partial_\mu + \left(\omega^{\mu\lambda}\overline{\omega}_\lambda{}^\nu - \overline{\omega}^{\mu\lambda}\omega_\lambda{}^\nu\right)x_\mu \partial_\nu \\ &= -\left(\widetilde{\epsilon}^\mu \partial_\mu + \widetilde{\omega}^{\mu\nu} x_\nu \partial_\mu\right) = \widehat{R}\left(\widetilde{\epsilon}, \widetilde{\omega}\right),\end{aligned} \tag{4.35}$$

where we have identified

$$\tilde{\epsilon}^\mu = (\omega^\mu{}_\nu \bar{\epsilon}^\nu - \bar{\omega}^\mu{}_\nu \epsilon^\nu),$$
$$\tilde{\omega}^{\mu\nu} = \left(\omega^{\mu\lambda}\bar{\omega}_\lambda{}^\nu - \bar{\omega}^{\mu\lambda}\omega_\lambda{}^\nu\right) = [\omega, \bar{\omega}]^{\mu\nu}. \qquad (4.36)$$

We can also calculate the algebra of the generators of Poincaré group. We already know the commutation relations $[M_{\mu\nu}, M_{\lambda\rho}]$ as well as $[P_\mu, P_\nu]$ (see (4.30) and (4.21)). Therefore, the only relation that needs to be calculated is the commutator between the generators of translation and Lorentz transformations. Note that

$$\begin{aligned}[] [P_\mu, M_{\nu\lambda}] &= [\partial_\mu, x_\nu \partial_\lambda - x_\lambda \partial_\nu] = \eta_{\mu\nu}\partial_\lambda - \eta_{\mu\lambda}\partial_\nu \\ &= \eta_{\mu\nu}P_\lambda - \eta_{\mu\lambda}P_\nu, \end{aligned} \qquad (4.37)$$

which simply shows that under a Lorentz transformation, P_μ behaves like a covariant four vector. (This is seen by recalling that $\frac{1}{2}\omega^{\mu\nu}M_{\mu\nu}$ generates infinitesimal Lorentz transformations. The commutator of a generator (multiplied by the appropriate transformation parameter) with any operator gives the infinitesimal change in that operator under the transformation generated by that particular generator. For change in the coordinate four vector under an infinitesimal Lorentz transformation, see, for example, (4.22) and (3.57).)

Thus, combining with our earlier results on the algebra of the translation group, (4.21), as well as the homogeneous Lorentz group, (4.30), we conclude that the Lie algebra associated with the Poincaré transformations (inhomogeneous Lorentz group) is given by

$$\begin{aligned}[] [P_\mu, P_\nu] &= 0, \\ [P_\mu, M_{\nu\lambda}] &= \eta_{\mu\nu}P_\lambda - \eta_{\mu\lambda}P_\nu, \\ [M_{\mu\nu}, M_{\lambda\rho}] &= -\eta_{\mu\lambda}M_{\nu\rho} - \eta_{\nu\rho}M_{\mu\lambda} + \eta_{\mu\rho}M_{\nu\lambda} + \eta_{\nu\lambda}M_{\mu\rho}. \end{aligned} \qquad (4.38)$$

We note that the algebra of translations defines an Abelian subalgebra of the Poincaré algebra (4.38). However, since the generators of translations do not commute with the generators of Lorentz transformations, Poincaré algebra cannot be written as a direct sum of those for translations and Lorentz transformations. Namely,

$$\text{Poincaré algebra} \neq t^4 \oplus so(3,1). \tag{4.39}$$

Rather, it is what is known as a semi-direct sum of the two algebras. (The general convention is to denote groups by capital letters while the algebras are represented by lower case letters.)

4.2 Representations of the Lorentz group

Let us next come back to the homogeneous Lorentz group and note that the Lie algebra in this case is given by (4.30)

$$[M_{\mu\nu}, M_{\lambda\rho}] = (-\eta_{\mu\lambda}M_{\nu\rho} - \eta_{\nu\rho}M_{\mu\lambda} + \eta_{\mu\rho}M_{\nu\lambda} + \eta_{\nu\lambda}M_{\mu\rho}). \tag{4.40}$$

We recall from (4.12), (4.26) and (4.27) that we can identify the angular momentum and boost operators as

$$\begin{aligned} J_i &= -\frac{1}{2}\epsilon_i{}^{jk}M_{jk}, \\ K_i &= M_{0i}, \qquad i = 1,2,3. \end{aligned} \tag{4.41}$$

Written out in terms of these generators, the Lorentz algebra takes the form

$$\begin{aligned}
[K_i, K_j] &= [M_{0i}, M_{0j}] = -M_{ij} = -\epsilon_{ij}{}^k J_k, \\
[K_i, J_j] &= \left[M_{0i}, -\frac{1}{2}\epsilon_j{}^{k\ell}M_{k\ell}\right] \\
&= -\frac{1}{2}\epsilon_j{}^{k\ell}\left(-\eta_{i\ell}M_{0k} + \eta_{ik}M_{0\ell}\right) \\
&= \frac{1}{2}\epsilon_j{}^k{}_i K_k - \frac{1}{2}\epsilon_{ji}{}^\ell K_\ell \\
&= \epsilon_{ij}{}^k K_k = [J_i, K_j], \\
[J_i, J_j] &= \epsilon_{ij}{}^k J_k, \tag{4.42}
\end{aligned}$$

where we have used (4.13) in the last relation.

This is a set of coupled commutation relations. Let us define a set of new generators as linear superpositions of J_i and K_i as (this is also known as changing the basis of the algebra)

$$A_i = \tfrac{1}{2}(J_i + iK_i),$$
$$B_i = \tfrac{1}{2}(J_i - iK_i), \qquad (4.43)$$

which also leads to the inverse relations

$$J_i = (A_i + B_i),$$
$$K_i = -i(A_i - B_i). \qquad (4.44)$$

Parenthetically, let us note from the form of the algebra in (4.42) that we can assign the following hermiticity properties to the generators, namely,

$$J_i^\dagger = -J_i, \qquad K_i^\dagger = K_i. \qquad (4.45)$$

This unconventional hermiticity for J_i arises because, in choosing the coordinate representation for the generators, we have not been particularly careful about choosing Hermitian operators. As a consequence of (4.45), we have

$$A_i^\dagger = -A_i, \qquad B_i^\dagger = -B_i. \qquad (4.46)$$

Namely, the generators in the new basis are all anti-Hermitian. The opposite hermiticity property of the generators of boosts, K_i, (compared to J_i) is connected with the fact that such transformations are non-compact and, consequently, the finite dimensional representations of boosts are non-unitary (hence the opposite Hermiticity of K_i). However, infinite dimensional representations are unitary, as can be seen from the hermiticity of the generators in the coordinate basis, namely, if we define the generators with a factor of "i",

$$M_{\mu\nu} = i(x_\mu \partial_\nu - x_\nu \partial_\mu) = M_{\mu\nu}^\dagger. \qquad (4.47)$$

4.2 Representations of the Lorentz group

In the new basis (4.43), the Lorentz algebra (4.42) takes the form

$$
\begin{aligned}
[A_i, A_j] &= \left[\frac{1}{2}(J_i + iK_i), \frac{1}{2}(J_j + iK_j)\right] \\
&= \frac{1}{4}\left([J_i, J_j] + i[J_i, K_j] + i[K_i, J_j] - [K_i, K_j]\right) \\
&= \frac{1}{4}\left(\epsilon_{ij}{}^k J_k + i\epsilon_{ij}{}^k K_k + i\epsilon_{ij}{}^k K_k + \epsilon_{ij}{}^k J_k\right) \\
&= \frac{1}{2}\epsilon_{ij}{}^k (J_k + iK_k) = \epsilon_{ij}{}^k A_k, \\
[B_i, B_j] &= \left[\frac{1}{2}(J_i - iK_i), \frac{1}{2}(J_j - iK_j)\right] \\
&= \frac{1}{4}\left([J_i, J_j] - i[J_i, K_j] - i[K_i, J_j] - [K_i, K_j]\right) \\
&= \frac{1}{4}\left(\epsilon_{ij}{}^k J_k - i\epsilon_{ij}{}^k K_k - i\epsilon_{ij}{}^k K_k + \epsilon_{ij}{}^k J_k\right) \\
&= \frac{1}{2}\epsilon_{ij}{}^k (J_k - iK_k) = \epsilon_{ij}{}^k B_k, \\
[A_i, B_j] &= \left[\frac{1}{2}(J_i + iK_i), \frac{1}{2}(J_j - iK_j)\right] \\
&= \frac{1}{4}\left([J_i, J_j] - i[J_i, K_j] + i[K_i, J_j] + [K_i, K_j]\right) \\
&= \frac{1}{4}\left(\epsilon_{ij}{}^k J_k - i\epsilon_{ij}{}^k K_k + i\epsilon_{ij}{}^k K_k - \epsilon_{ij}{}^k J_k\right) \\
&= 0. \tag{4.48}
\end{aligned}
$$

In other words, in this new basis, the algebra separates into two angular momentum algebras which are decoupled. Mathematically, one says that the Lorentz algebra is isomorphic to the direct sum of two angular momentum algebras,

$$so(3,1) \simeq so(3) \oplus so(3) \simeq su(2) \oplus su(2). \tag{4.49}$$

Incidentally, as we have already seen in the last chapter, the Lorentz group is double valued (doubly connected). Therefore, it is more meaningful to consider the simply connected universal covering group of $SO(3,1)$ which is $SL(2,\mathbb{C})$ to describe the Lorentz transformations, much the same way we consider the universal covering group $SU(2)$ of $SO(3)$ to describe rotations.

The finite dimensional unitary representations of each of the angular momentum algebras is well known. Denoting by j_A and j_B the eigenvalues of the Casimir operators $\mathbf{A}^2$ and $\mathbf{B}^2$ respectively for the two algebras, we have

$$\begin{aligned} j_A &= 0, \frac{1}{2}, 1, \frac{3}{2}, \ldots, \\ j_B &= 0, \frac{1}{2}, 1, \frac{3}{2}, \ldots. \end{aligned} \qquad (4.50)$$

An irreducible nonunitary representation of the homogeneous Lorentz group, therefore, can be specified uniquely once we know the values of j_A and j_B and is labelled as $D^{(j_A,j_B)}$ (just as the representation of the rotation group is denoted by $D^{(j)}$). Namely, this represents the operator implementing finite transformations on the Hilbert space of states or wave functions as

$$\begin{aligned} \psi'^{(j_A,j_B)}(x') &= D^{(j_A,j_B)}(\Lambda)\psi^{(j_A,j_B)}(x) \\ &= D^{(j_A)}(\Lambda)D^{(j_B)}(\Lambda)\psi^{(j_A,j_B)}(x), \end{aligned} \qquad (4.51)$$

where Λ represents the finite Lorentz transformation parameter. Explicitly, we can write (This is the generalization of the $S(\Lambda)$ matrix that we studied in (3.35) in connection with the covariance of the Dirac equation.)

$$\begin{aligned} D^{(j_A,j_B)}(\Lambda) &= e^{-i\left(\theta_A^i A_i^{(j_A)} + \theta_B^i B_i^{(j_B)}\right)} \\ &= e^{-i\left(\theta^i J_i^{(j_A,j_B)} + \delta^i K_i^{(j_A,j_B)}\right)}, \end{aligned} \qquad (4.52)$$

where the finite parameters of rotation and boost can be identified with

$$\theta^i = \frac{1}{2}\left(\theta_A^i + \theta_B^i\right), \quad \delta^i = \frac{i}{2}\left(\theta_A^i - \theta_B^i\right). \qquad (4.53)$$

4.2 REPRESENTATIONS OF THE LORENTZ GROUP

Such a representation labelled by (j_A, j_B) will have the dimensionality

$$\text{dimensionality}\left(D^{(j_A, j_B)}\right) = (2j_A + 1)(2j_B + 1), \tag{4.54}$$

and its spin content follows from the fact that (see (4.44))

$$\mathbf{J} = \mathbf{A} + \mathbf{B}. \tag{4.55}$$

Consequently, from our knowledge of the addition of angular momenta, we conclude that the values of the spin in a given representation characterized by (j_A, j_B) can lie between

$$j = |j_A - j_B|, |j_A - j_B| + 1, \ldots, j_A + j_B. \tag{4.56}$$

The first few low lying representations of the Lorentz group are as follows. For $j_A = j_B = 0$,

$$D^{(0,0)}, \text{ dimensionality} = 1, \ j = 0, \tag{4.57}$$

corresponds to a scalar representation with zero spin (which acts on the wave function of a Klein-Gordon particle). For $j_A = \frac{1}{2}, j_B = 0$,

$$D^{(\frac{1}{2},0)}, \text{ dimensionality} = 2, \ j = \frac{1}{2}, \tag{4.58}$$

corresponds to a two component spinor representation with spin $\frac{1}{2}$. Similarly, for $j_A = 0, j_B = \frac{1}{2}$,

$$D^{(0,\frac{1}{2})}, \text{ dimensionality} = 2, \ j = \frac{1}{2}, \tag{4.59}$$

also corresponds to a two component spinor representation with spin $\frac{1}{2}$. These two representations are inequivalent and, in fact, are complex conjugates of each other and can be identified to act on the wave functions of the two kinds of massless Dirac particles (Weyl fermions) we had discussed in the last chapter. For $j_A = j_B = \frac{1}{2}$,

$$D^{(\frac{1}{2},\frac{1}{2})}, \text{ dimensionality} = 4, \; j = 0, 1, \tag{4.60}$$

is known as a four component vector representation and can be identified with a spin content of 0 and 1 for the components. (Note that a four vector such as x^μ has a spin zero component, namely, t and a spin 1 component $\mathbf{x}$ (under rotations) and the same is true for any other four vector.) It may be puzzling as to where the four component Dirac spinor fits into this description. It actually corresponds to a reducible representation of the Lorentz group of the form

$$D^{(0,\frac{1}{2})} \otimes D^{(\frac{1}{2},0)}, \text{ dimensionality} = 2 \times 2 = 4, \; j = \frac{1}{2}. \tag{4.61}$$

This discussion can similarly be carried over to higher dimensional representations.

4.2.1 Similarity transformations and representations. Let us now construct explicitly a few of the low order representations for the generators of the Lorentz group. To compare with the results that we had derived earlier, we now consider Hermitian generators by letting $M_{\mu\nu} \to iM_{\mu\nu}$ as in (4.47). (Namely, we scale all the generators J_i, K_i, A_i, B_i by a factor of i.)

From (4.50), we note that for the first few low order representations, we have (We note here that the negative sign in the spin $\frac{1}{2}$ representation in (4.62) arises because $\epsilon_{ij}^{\;k} = -\epsilon_{ijk}$ in (4.48).)

$$A_i^{(0)} = 0, \quad A_i^{(\frac{1}{2})} = -\frac{1}{2}\sigma_i, \quad \cdots,$$

$$B_i^{(0)} = 0, \quad B_i^{(\frac{1}{2})} = -\frac{1}{2}\sigma_i, \quad \cdots. \tag{4.62}$$

Using (4.44), this leads to the first two nontrivial representations for the angular momentum and boost operators of the forms

$$\begin{aligned} J_i^{(\frac{1}{2},0)} &= A_i^{(\frac{1}{2})} + B_i^{(0)} = -\frac{1}{2}\sigma_i, \\ K_i^{(\frac{1}{2},0)} &= -i\left(A_i^{(\frac{1}{2})} - B_i^{(0)}\right) = \frac{i}{2}\sigma_i, \end{aligned} \tag{4.63}$$

4.2 REPRESENTATIONS OF THE LORENTZ GROUP

and

$$J_i^{(0,\frac{1}{2})} = A_i^{(0)} + B_i^{(\frac{1}{2})} = -\frac{1}{2}\sigma_i,$$
$$K_i^{(0,\frac{1}{2})} = -i\left(A_i^{(0)} - B_i^{(\frac{1}{2})}\right) = -\frac{i}{2}\sigma_i. \tag{4.64}$$

Equations (4.63) and (4.64) give the two inequivalent representations of dimensionality 2 as we have noted earlier. Two representations are said to be equivalent, if there exists a similarity transformation relating the two. For example, if we can find a similarity transformation S leading to

$$J_i^{(0,\frac{1}{2})} = S^{-1} J_i^{(\frac{1}{2},0)} S, \quad K_i^{(0,\frac{1}{2})} = S^{-1} K_i^{(\frac{1}{2},0)} S, \tag{4.65}$$

then, we would say that the two representations $\left(\frac{1}{2}, 0\right)$ and $\left(0, \frac{1}{2}\right)$ are equivalent. In fact, from (4.63) and (4.64) we see that the condition (4.65) would imply the existence of an invertible matrix S such that

$$S^{-1}\sigma_i S = \sigma_i, \quad S^{-1}\sigma_i S = -\sigma_i, \tag{4.66}$$

which is clearly impossible. Therefore, the two representations labelled by $\left(\frac{1}{2}, 0\right)$ and $\left(0, \frac{1}{2}\right)$ are inequivalent representations. They provide the representations of angular momentum and boost for the left-handed and the right-handed Weyl particles.

From (4.63) and (4.64), we can obtain the representation of the Lorentz generators for the reducible four component Dirac spinors as

$$J_i^{(\text{Dirac})} = J_i^{(\frac{1}{2},0)} \oplus J_i^{(0,\frac{1}{2})} = \begin{pmatrix} -\frac{1}{2}\sigma_i & 0 \\ 0 & -\frac{1}{2}\sigma_i \end{pmatrix},$$
$$K_i^{(\text{Dirac})} = K_i^{(\frac{1}{2},0)} \oplus K_i^{(0,\frac{1}{2})} = \begin{pmatrix} \frac{i}{2}\sigma_i & 0 \\ 0 & -\frac{i}{2}\sigma_i \end{pmatrix}. \tag{4.67}$$

However, we note that these do not resemble the generators of the Lorentz algebra defined in (3.69) and (2.98) (or (3.71) and (3.78)).

This puzzle can be understood as follows. We note that in the Weyl representation for the gamma matrices defined in (2.119),

$$\sigma_W^{0i} = \frac{i}{2}[\gamma_W^0, \gamma_W^i] = \begin{pmatrix} -i\sigma_i & 0 \\ 0 & i\sigma_i \end{pmatrix},$$

$$\sigma_W^{ij} = \frac{i}{2}[\gamma_W^i, \gamma_W^j] = \epsilon_{ijk}\begin{pmatrix} \sigma_k & 0 \\ 0 & \sigma_k \end{pmatrix}. \quad (4.68)$$

As a result, we note that the angular momentum and boost operators in (4.67) are obtained from

$$M_{\mu\nu} = \frac{1}{2}\sigma_{\mu\nu}^W, \quad (4.69)$$

and, consequently, give a representation of the Lorentz generators in the Weyl representation. On the other hand, if we would like the generators in the standard Pauli-Dirac representation (which is what we had used in our earlier discussions), we can apply the inverse similarity transformation in (2.121) to obtain

$$\begin{aligned}
J_{i\ PD}^{(\text{Dirac})} &= S^{-1} J_i^{(\text{Dirac})} S \\
&= \frac{1}{2}\begin{pmatrix} 1 & 1 \\ -1 & 1 \end{pmatrix}\begin{pmatrix} -\frac{1}{2}\sigma_i & 0 \\ 0 & -\frac{1}{2}\sigma_i \end{pmatrix}\begin{pmatrix} 1 & -1 \\ 1 & 1 \end{pmatrix} \\
&= \begin{pmatrix} -\frac{1}{2}\sigma_i & 0 \\ 0 & -\frac{1}{2}\sigma_i \end{pmatrix} = -\frac{1}{2}\epsilon_i^{\ jk}\sigma_{jk}^{PD}, \\
K_{i\ PD}^{(\text{Dirac})} &= S^{-1} K_i^{(\text{Dirac})} S \\
&= \frac{1}{2}\begin{pmatrix} 1 & 1 \\ -1 & 1 \end{pmatrix}\begin{pmatrix} \frac{i}{2}\sigma_i & 0 \\ 0 & -\frac{i}{2}\sigma_i \end{pmatrix}\begin{pmatrix} 1 & -1 \\ 1 & 1 \end{pmatrix} \\
&= \begin{pmatrix} 0 & -\frac{i}{2}\sigma_i \\ -\frac{i}{2}\sigma_i & 0 \end{pmatrix} = \frac{1}{2}\sigma_{0i}^{PD}. \quad (4.70)
\end{aligned}$$

Therefore, we note that the generators in (4.67) and in our earlier discussion in (3.69) and (2.98) (see also (3.71) and (3.78)) are equivalent since they are connected by a similarity transformation that relates the Weyl representation of the Dirac matrices in (2.119) to the standard Pauli-Dirac representation.

There is yet another interesting example which sheds light on similarity transformations between representation. For example, from the infinitesimal change in the coordinates under a Lorentz transformation (see, for example, (3.18)), we can determine a representation for the generators of the Lorentz transformations belonging to the representation for the four vectors. On the other hand, as we discussed earlier, from the Lie algebra point of view the four vector representation corresponds to $j_A = j_B = \frac{1}{2}$ (see (4.60)) and we can construct the representations for $\mathbf{J}$ and $\mathbf{K}$ in this case as well from a knowledge of the addition of angular momenta. Surprisingly, the two representations for the generators constructed from two different perspectives (for the same four vector representation) appear rather different and, therefore, there must be a similarity transformation relating the two representations. Let us illustrate this for the simpler case of rotations. The case for Lorentz transformations follows in a parallel manner.

Let us consider a three dimensional infinitesimal rotation of coordinates around the z-axis as described in (3.4). (Here we will use 3-dimensional Euclidean notation without worrying about raising and lowering of the indices.) Representing the infinitesimal change in the coordinates as

$$\delta x_i = i\epsilon \left(J_3\right)_{ij} x_j, \tag{4.71}$$

we can immediately read out from (3.4) the matrix structure of the generator J_3 to be

$$J_3 = \begin{pmatrix} 0 & i & 0 \\ -i & 0 & 0 \\ 0 & 0 & 0 \end{pmatrix}. \tag{4.72}$$

Similarly, considering infinitesimal rotations of the coordinates around the x-axis and the y-axis respectively, we can deduce the matrix form of the corresponding generators to be

$$J_1 = \begin{pmatrix} 0 & 0 & 0 \\ 0 & 0 & i \\ 0 & -i & 0 \end{pmatrix}, \quad J_2 = \begin{pmatrix} 0 & 0 & i \\ 0 & 0 & 0 \\ -i & 0 & 0 \end{pmatrix}. \tag{4.73}$$

It can be directly checked from the matrix structures in (4.72) and (4.73) that they satisfy

$$[J_i, J_j] = i\epsilon_{ijk} J_k, \quad i, j, k = 1, 2, 3, \tag{4.74}$$

and, therefore provide a representation for the generators of rotations. This is, in fact, the representation in the space of three vectors which would correspond to $j = 1$.

On the other hand, it is well known from the study of the representations of the angular momentum algebra that the generators in the representation $j = 1$ have the forms[1]

$$J_1^{(LA)} = \frac{1}{\sqrt{2}} \begin{pmatrix} 0 & 1 & 0 \\ 1 & 0 & 1 \\ 0 & 1 & 0 \end{pmatrix}, \quad J_2^{(LA)} = \frac{i}{\sqrt{2}} \begin{pmatrix} 0 & -1 & 0 \\ 1 & 0 & -1 \\ 0 & 1 & 0 \end{pmatrix},$$

$$J_3^{(LA)} = \begin{pmatrix} 1 & 0 & 0 \\ 0 & 0 & 0 \\ 0 & 0 & -1 \end{pmatrix}, \tag{4.75}$$

which look really different from the generators in (4.72) and (4.73) in spite of the fact that they belong to the same representation for $j = 1$. This puzzle can be resolved by noting that there is a similarity transformation that connects the two representations and, therefore, they are equivalent.

To construct the similarity transformation (which actually is a unitary transformation), let us note that the generators obtained from the Lie algebra are constructed by choosing the generator $J_3^{(LA)}$

---

[1] See, for example, *Quantum Mechanics: A Modern Introduction*, A. Das and A. C. Melissinos (Gordon and Breach), page 289 or *Lectures on Quantum Mechanics*, A. Das (Hindustan Book Agency, New Delhi), page 182 (note there is a typo in the sign of the 23 element for L_2 in this reference)

4.2 REPRESENTATIONS OF THE LORENTZ GROUP

to be diagonal. Let us note from (4.72) that the three normalized eigenstates of J_3 have the forms

$$\psi^{(j_3=1)} = \begin{pmatrix} \frac{1}{\sqrt{2}} \\ -\frac{i}{\sqrt{2}} \\ 0 \end{pmatrix}, \quad \psi^{(j_3=0)} = \begin{pmatrix} 0 \\ 0 \\ -1 \end{pmatrix},$$

$$\psi^{(j_3=-1)} = \begin{pmatrix} -\frac{1}{\sqrt{2}} \\ -\frac{i}{\sqrt{2}} \\ 0 \end{pmatrix}. \tag{4.76}$$

Let us construct a unitary matrix from the three eigenstates in (4.76) which will diagonalize the matrix J_3,

$$U = \begin{pmatrix} \frac{1}{\sqrt{2}} & 0 & -\frac{1}{\sqrt{2}} \\ -\frac{i}{\sqrt{2}} & 0 & -\frac{i}{\sqrt{2}} \\ 0 & -1 & 0 \end{pmatrix}, \quad U^\dagger = \begin{pmatrix} \frac{1}{\sqrt{2}} & \frac{i}{\sqrt{2}} & 0 \\ 0 & 0 & -1 \\ -\frac{1}{\sqrt{2}} & \frac{i}{\sqrt{2}} & 0 \end{pmatrix}. \tag{4.77}$$

If we now define a similarity transformation (unitary)

$$S = U, \quad S^{-1} = U^\dagger, \tag{4.78}$$

then, it is straightforward to check

$$S^{-1} J_1 S$$

$$= \begin{pmatrix} \frac{1}{\sqrt{2}} & \frac{i}{\sqrt{2}} & 0 \\ 0 & 0 & -1 \\ -\frac{1}{\sqrt{2}} & \frac{i}{\sqrt{2}} & 0 \end{pmatrix} \begin{pmatrix} 0 & 0 & 0 \\ 0 & 0 & i \\ 0 & -i & 0 \end{pmatrix} \begin{pmatrix} \frac{1}{\sqrt{2}} & 0 & -\frac{1}{\sqrt{2}} \\ -\frac{i}{\sqrt{2}} & 0 & -\frac{i}{\sqrt{2}} \\ 0 & -1 & 0 \end{pmatrix}$$

$$= \begin{pmatrix} \frac{1}{\sqrt{2}} & \frac{i}{\sqrt{2}} & 0 \\ 0 & 0 & -1 \\ -\frac{1}{\sqrt{2}} & \frac{i}{\sqrt{2}} & 0 \end{pmatrix} \begin{pmatrix} 0 & 0 & 0 \\ 0 & -i & 0 \\ -\frac{1}{\sqrt{2}} & 0 & -\frac{1}{\sqrt{2}} \end{pmatrix}$$

$$= \frac{1}{\sqrt{2}} \begin{pmatrix} 0 & 1 & 0 \\ 1 & 0 & 1 \\ 0 & 1 & 0 \end{pmatrix} = J_1^{(LA)},$$

$S^{-1} J_2 S$

$$= \begin{pmatrix} \frac{1}{\sqrt{2}} & \frac{i}{\sqrt{2}} & 0 \\ 0 & 0 & -1 \\ -\frac{1}{\sqrt{2}} & \frac{i}{\sqrt{2}} & 0 \end{pmatrix} \begin{pmatrix} 0 & 0 & i \\ 0 & 0 & 0 \\ -i & 0 & 0 \end{pmatrix} \begin{pmatrix} \frac{1}{\sqrt{2}} & 0 & -\frac{1}{\sqrt{2}} \\ -\frac{i}{\sqrt{2}} & 0 & -\frac{i}{\sqrt{2}} \\ 0 & -1 & 0 \end{pmatrix}$$

$$= \begin{pmatrix} \frac{1}{\sqrt{2}} & \frac{i}{\sqrt{2}} & 0 \\ 0 & 0 & -1 \\ -\frac{1}{\sqrt{2}} & \frac{i}{\sqrt{2}} & 0 \end{pmatrix} \begin{pmatrix} 0 & -i & 0 \\ 0 & 0 & 0 \\ -\frac{i}{\sqrt{2}} & 0 & \frac{i}{\sqrt{2}} \end{pmatrix}$$

$$= \frac{i}{\sqrt{2}} \begin{pmatrix} 0 & -1 & 0 \\ 1 & 0 & -1 \\ 0 & 1 & 0 \end{pmatrix} = J_2^{(LA)},$$

$S^{-1} J_3 S$

$$= \begin{pmatrix} \frac{1}{\sqrt{2}} & \frac{i}{\sqrt{2}} & 0 \\ 0 & 0 & -1 \\ -\frac{1}{\sqrt{2}} & \frac{i}{\sqrt{2}} & 0 \end{pmatrix} \begin{pmatrix} 0 & i & 0 \\ -i & 0 & 0 \\ 0 & 0 & 0 \end{pmatrix} \begin{pmatrix} \frac{1}{\sqrt{2}} & 0 & -\frac{1}{\sqrt{2}} \\ -\frac{i}{\sqrt{2}} & 0 & -\frac{i}{\sqrt{2}} \\ 0 & -1 & 0 \end{pmatrix}$$

$$= \begin{pmatrix} \frac{1}{\sqrt{2}} & \frac{i}{\sqrt{2}} & 0 \\ 0 & 0 & -1 \\ -\frac{1}{\sqrt{2}} & \frac{i}{\sqrt{2}} & 0 \end{pmatrix} \begin{pmatrix} \frac{1}{\sqrt{2}} & 0 & \frac{1}{\sqrt{2}} \\ -\frac{i}{\sqrt{2}} & 0 & \frac{i}{\sqrt{2}} \\ 0 & 0 & 0 \end{pmatrix}$$

$$= \begin{pmatrix} 1 & 0 & 0 \\ 0 & 0 & 0 \\ 0 & 0 & -1 \end{pmatrix} = J_3^{(LA)}. \tag{4.79}$$

This shows explicitly that the two representations for **J** corresponding to $j = 1$ in (4.72), (4.73) and (4.75) which look rather different are, in fact, related by a similarity transformation and, therefore, are equivalent.

4.3 Unitary representations of the Poincaré group

Since we are interested in physical theories which are invariant under translations as well as homogeneous Lorentz transformations, it is useful to study the representations of the Poincaré group. This would help us in understanding the kinds of theories we can consider and the nature of the states they can have. Since Poincaré group is non-compact (like the Lorentz group), it is known that it has only infinite dimensional unitary representations except for the trivial representation that is one dimensional. Therefore, we seek to find unitary representations in some infinite dimensional Hilbert space where the generators $P_\mu, M_{\mu\nu}$ act as Hermitian operators.

In order to determine the unitary representations, let us note that the operator

$$P^2 = \eta^{\mu\nu} P_\mu P_\nu = P^\mu P_\mu, \tag{4.80}$$

defines a quadratic Casimir operator of the Poincaré algebra (4.38) since it commutes with all the ten generators, namely,

$$\begin{aligned}
\left[P^2, P_\mu\right] &= \left[P^\lambda P_\lambda, P_\mu\right] = 0, \\
\left[P^2, M_{\mu\nu}\right] &= \left[P^\lambda P_\lambda, M_{\mu\nu}\right] \\
&= P^\lambda \left[P_\lambda, M_{\mu\nu}\right] + \left[P_\lambda, M_{\mu\nu}\right] P^\lambda \\
&= P^\lambda \left(\eta_{\lambda\mu} P_\nu - \eta_{\lambda\nu} P_\mu\right) + \left(\eta_{\lambda\mu} P_\nu - \eta_{\lambda\nu} P_\mu\right) P^\lambda \\
&= P_\mu P_\nu - P_\nu P_\mu + P_\nu P_\mu - P_\mu P_\nu = 0. \tag{4.81}
\end{aligned}$$

The last relation in (4.81) can be intuitively understood as follows. The operators $M_{\mu\nu}$ generate infinitesimal Lorentz transformations through commutation relations and the relation above, which is supposed to characterize the infinitesimal transformation of P^2, simply implies that P^2 does not change under a Lorentz transformation (it is a Lorentz scalar) which is to be expected since it does not have any free Lorentz index.

Let us define a new vector operator, known as the Pauli-Lubanski operator, from the generators of the Poincaré group as

$$W^\mu = \frac{1}{2}\epsilon^{\mu\nu\lambda\rho}P_\nu M_{\lambda\rho} = \frac{1}{2}\epsilon^{\mu\nu\lambda\rho}M_{\lambda\rho}P_\nu. \tag{4.82}$$

The commutator between P_μ and $M_{\nu\lambda}$ introduces metric tensors (see (4.38)) which vanish when contracted with the anti-symmetric Levi-Civita tensor. As a result, the order of P_μ and $M_{\nu\lambda}$ are irrelevant in the definition of the Pauli-Lubanski operator. Furthermore, we note that

$$P_\mu W^\mu = \frac{1}{2}\epsilon^{\mu\nu\lambda\rho}P_\mu P_\nu M_{\lambda\rho} = 0, \tag{4.83}$$

which follows from the fact that the generators of translation commute. It follows from (4.83) that, in general, the vector W_μ is orthogonal to P_μ. (However, this is not true for massless theories as we will see shortly.) In general, therefore, (4.83) implies that the Pauli-Lubanski operator has only three independent components (both in the massive and massless cases). Let us define the dual of the generators of Lorentz transformation as

$$\widetilde{M}^{\mu\nu} = \frac{1}{2}\epsilon^{\mu\nu\lambda\rho}M_{\lambda\rho},$$
$$M_{\mu\nu} = -\frac{1}{2}\epsilon_{\mu\nu\lambda\rho}\widetilde{M}^{\lambda\rho}. \tag{4.84}$$

With this, we can write (4.82) also as

$$W^\mu = P_\nu \widetilde{M}^{\mu\nu} = \widetilde{M}^{\mu\nu}P_\nu, \tag{4.85}$$

where the order of the operators is once again not important.

Let us next calculate the commutators between W^μ and the ten generators of the Poincaré group. First, we have

$$
\begin{aligned}
[W^\mu, P_\sigma] &= \left[\frac{1}{2}\epsilon^{\mu\nu\lambda\rho}P_\nu M_{\lambda\rho}, P_\sigma\right] \\
&= \frac{1}{2}\epsilon^{\mu\nu\lambda\rho}P_\nu [M_{\lambda\rho}, P_\sigma] \\
&= \frac{1}{2}\epsilon^{\mu\nu\lambda\rho}P_\nu (-\eta_{\lambda\sigma}P_\rho + \eta_{\rho\sigma}P_\lambda) \\
&= 0, \quad (4.86)
\end{aligned}
$$

which follows from the the fact that momenta commute. Consequently, any function of W_μ and, in particular $W_\mu W^\mu$, will also commute with the generators of translation. We also note that

$$
\begin{aligned}
\left[\widetilde{M}_{\mu\nu}, M_{\sigma\tau}\right] &= \left[\frac{1}{2}\epsilon_{\mu\nu}{}^{\lambda\rho}M_{\lambda\rho}, M_{\sigma\tau}\right] \\
&= \frac{1}{2}\epsilon_{\mu\nu}{}^{\lambda\rho}[M_{\lambda\rho}, M_{\sigma\tau}] \\
&= \frac{1}{2}\epsilon_{\mu\nu}{}^{\lambda\rho}(-\eta_{\lambda\sigma}M_{\rho\tau} - \eta_{\rho\tau}M_{\lambda\sigma} + \eta_{\lambda\tau}M_{\rho\sigma} + \eta_{\rho\sigma}M_{\lambda\tau}) \\
&= -\frac{1}{2}\epsilon_{\mu\nu\sigma}{}^\rho M_{\rho\tau} + \frac{1}{2}\epsilon_{\mu\nu\tau}{}^\rho M_{\rho\sigma} + \frac{1}{2}\epsilon_{\mu\nu\tau}{}^\rho M_{\rho\sigma} - \frac{1}{2}\epsilon_{\mu\nu\sigma}{}^\rho M_{\rho\tau} \\
&= -\epsilon_{\mu\nu\sigma}{}^\rho M_{\rho\tau} + \epsilon_{\mu\nu\tau}{}^\rho M_{\rho\sigma} \\
&= -\epsilon_{\mu\nu\sigma}{}^\rho \left(-\frac{1}{2}\epsilon_{\rho\tau}{}^{\delta\varsigma}\widetilde{M}_{\delta\varsigma}\right) + \epsilon_{\mu\nu\tau}{}^\rho \left(-\frac{1}{2}\epsilon_{\rho\sigma}{}^{\delta\varsigma}\widetilde{M}_{\delta\varsigma}\right) \\
&= -\frac{1}{2}\epsilon_{\mu\nu\sigma}{}^\rho \epsilon_\tau{}^{\delta\varsigma}{}_\rho \widetilde{M}_{\delta\varsigma} + \frac{1}{2}\epsilon_{\mu\nu\tau}{}^\rho \epsilon_\sigma{}^{\delta\varsigma}{}_\rho \widetilde{M}_{\delta\varsigma} \\
&= \frac{1}{2}\left[\eta_{\mu\tau}\left(\delta^\delta_\nu \delta^\varsigma_\sigma - \delta^\delta_\sigma \delta^\varsigma_\nu\right) + \delta^\delta_\mu \left(\delta^\varsigma_\nu \eta_{\tau\sigma} - \delta^\varsigma_\sigma \eta_{\tau\nu}\right)\right. \\
&\quad \left. + \delta^\varsigma_\mu \left(\eta_{\nu\tau}\delta^\delta_\sigma - \eta_{\sigma\tau}\delta^\delta_\nu\right) - \sigma \leftrightarrow \tau\right]\widetilde{M}_{\delta\varsigma} \\
&= \frac{1}{2}\left[\eta_{\mu\tau}\left(\widetilde{M}_{\nu\sigma} - \widetilde{M}_{\sigma\nu}\right) + \widetilde{M}_{\mu\nu}\eta_{\tau\sigma} - \widetilde{M}_{\mu\sigma}\eta_{\tau\nu}\right. \\
&\quad \left. + \eta_{\nu\tau}\widetilde{M}_{\sigma\mu} - \eta_{\sigma\tau}\widetilde{M}_{\nu\mu} - \sigma \leftrightarrow \tau\right] \\
&= \frac{1}{2}\left(2\eta_{\mu\tau}\widetilde{M}_{\nu\sigma} - 2\eta_{\nu\tau}\widetilde{M}_{\mu\sigma} - 2\eta_{\mu\sigma}\widetilde{M}_{\nu\tau} + 2\eta_{\nu\sigma}\widetilde{M}_{\mu\tau}\right)
\end{aligned}
$$

$$= -\eta_{\mu\sigma}\widetilde{M}_{\nu\tau} - \eta_{\nu\tau}\widetilde{M}_{\mu\sigma} + \eta_{\mu\tau}\widetilde{M}_{\nu\sigma} + \eta_{\nu\sigma}\widetilde{M}_{\mu\tau}. \tag{4.87}$$

Here we have used the identity satisfied by the four dimensional Levi-Civita tensors,

$$\begin{aligned}\epsilon_{\mu\nu\lambda}{}^{\rho}\epsilon_{\sigma\tau\zeta\rho} &= -\eta_{\mu\sigma}\left(\eta_{\nu\tau}\eta_{\lambda\zeta} - \eta_{\nu\zeta}\eta_{\lambda\tau}\right) - \eta_{\mu\tau}\left(\eta_{\nu\zeta}\eta_{\lambda\sigma} - \eta_{\nu\sigma}\eta_{\lambda\zeta}\right)\\ &\quad -\eta_{\mu\zeta}\left(\eta_{\nu\sigma}\eta_{\lambda\tau} - \eta_{\nu\tau}\eta_{\lambda\sigma}\right). \end{aligned} \tag{4.88}$$

Equation (4.87) simply says that under a Lorentz transformation, the operator $\widetilde{M}_{\mu\nu}$ behaves exactly like the generators of Lorentz transformation (see (4.30)). Namely, it behaves like a second rank antisymmetric tensor under a Lorentz transformation. Using this, then, we can now evaluate

$$\begin{aligned}[W^{\mu}, M_{\sigma\tau}] &= \left[P_{\nu}\widetilde{M}^{\mu\nu}, M_{\sigma\tau}\right]\\ &= [P_{\nu}, M_{\sigma\tau}]\widetilde{M}^{\mu\nu} + P_{\nu}\left[\widetilde{M}^{\mu\nu}, M_{\sigma\tau}\right]\\ &= (\eta_{\nu\sigma}P_{\tau} - \eta_{\nu\tau}P_{\sigma})\widetilde{M}^{\mu\nu}\\ &\quad + P_{\nu}\left(-\delta^{\mu}_{\sigma}\widetilde{M}^{\nu}{}_{\tau} - \delta^{\nu}_{\tau}\widetilde{M}^{\mu}{}_{\sigma} + \delta^{\mu}_{\tau}\widetilde{M}^{\nu}{}_{\sigma} + \delta^{\nu}_{\sigma}\widetilde{M}^{\mu}{}_{\tau}\right)\\ &= P_{\tau}\widetilde{M}^{\mu}{}_{\sigma} - P_{\sigma}\widetilde{M}^{\mu}{}_{\tau} - \delta^{\mu}_{\sigma}P_{\nu}\widetilde{M}^{\nu}{}_{\tau} - P_{\tau}\widetilde{M}^{\mu}{}_{\sigma}\\ &\quad + \delta^{\mu}_{\tau}P_{\nu}\widetilde{M}^{\nu}{}_{\sigma} + P_{\sigma}\widetilde{M}^{\mu}{}_{\tau}\\ &= \delta^{\mu}_{\sigma}P_{\nu}\widetilde{M}_{\tau}{}^{\nu} - \delta^{\mu}_{\tau}P_{\nu}\widetilde{M}_{\sigma}{}^{\nu} = \delta^{\mu}_{\sigma}W_{\tau} - \delta^{\mu}_{\tau}W_{\sigma}. \end{aligned}\tag{4.89}$$

In other words, we see that the operator W^{μ} transforms precisely the same way as does the generator of translation or the P^{μ} operator under a Lorentz transformation. Namely, it transforms like a vector which we should expect since it has a free Lorentz index. Let us note here, for completeness as well as for later use, that

$$\begin{aligned}[W_{\mu}, W_{\nu}] &= \left[W_{\mu}, \frac{1}{2}\epsilon_{\nu}{}^{\lambda\rho\sigma}M_{\rho\sigma}P_{\lambda}\right] = \frac{1}{2}\epsilon_{\nu}{}^{\lambda\rho\sigma}\left[W_{\mu}, M_{\rho\sigma}\right]P_{\lambda}\\ &= \frac{1}{2}\epsilon_{\nu}{}^{\lambda\rho\sigma}\left(\eta_{\mu\rho}W_{\sigma} - \eta_{\mu\sigma}W_{\rho}\right)P_{\lambda}\\ &= -\epsilon_{\mu\nu}{}^{\lambda\rho}W_{\lambda}P_{\rho}. \end{aligned}\tag{4.90}$$

It follows now from (4.89) that

$$[W^\mu W_\mu, M_{\sigma\tau}] = W^\mu [W_\mu, M_{\sigma\tau}] + [W_\mu, M_{\sigma\tau}] W^\mu$$
$$= W^\mu (\eta_{\mu\sigma} W_\tau - \eta_{\mu\tau} W_\sigma) + (\eta_{\mu\sigma} W_\tau - \eta_{\mu\tau} W_\sigma) W^\mu$$
$$= W_\sigma W_\tau - W_\tau W_\sigma + W_\tau W_\sigma - W_\sigma W_\tau = 0, \tag{4.91}$$

which is to be expected since $W_\mu W^\mu$ is a Lorentz scalar. Therefore, we conclude that if we define an operator

$$W^2 = W^\mu W_\mu, \tag{4.92}$$

then, this would also represent a Casimir operator of the Poincaré algebra since W_μ commutes with the generators of translation (see (4.86)). It can be shown that P^2 and W^2 represent the only Casimir operators of the algebra and, consequently, the representations can be labelled by the eigenvalues of these operators. In fact, let us note from this analysis that a Casimir operator for the Poincaré algebra must necessarily be a Lorentz scalar (since it has to commute with $M_{\mu\nu}$). There are other Lorentz scalars that can be constructed from P_μ and $M_{\mu\nu}$ such as

$$M_{\mu\nu} M^{\mu\nu}, \quad M_{\mu\nu} \widetilde{M}^{\mu\nu}, \quad L^2 = L_\mu L^\mu \text{ with } L_\mu = P^\nu M_{\nu\mu}. \tag{4.93}$$

However, it is easy to check that these do not commute with the generators of translation and, therefore, cannot represent Casimir operators of the algebra.

The irreducible representations of the Poincaré group can be classified into two distinct categories, which we treat separately.

4.3.1 Massive representation. To find unitary irreducible representations of the Poincaré algebra, we choose the basis vectors of the representation to be eigenstates of the momentum operators. Namely, without loss of generality, we can choose the momentum operators, P_μ, to be diagonal (they satisfy an Abelian subalgebra). The eigenstates of the momentum operators $|p\rangle$ are, of course, labelled by the momentum eigenvalues, p_μ, satisfying

$$P_\mu|p\rangle = p_\mu|p\rangle, \qquad (4.94)$$

and in this basis, the eigenvalues of the operator $P^2 = P^\mu P_\mu$ are obvious, namely,

$$P^2|p\rangle = p^\mu p_\mu|p\rangle, \qquad (4.95)$$

where

$$p^2 = p^\mu p_\mu = m^2. \qquad (4.96)$$

Here m denotes the rest mass of the single particle state and we assume the rest mass to be non-zero. However, the eigenvalues of W^2 are not so obvious. Therefore, let us study this operator in some detail. We recall that

$$W^\mu = \frac{1}{2}\epsilon^{\mu\nu\lambda\rho}P_\nu M_{\lambda\rho} = \frac{1}{2}\epsilon^{\mu\nu\lambda\rho}M_{\lambda\rho}P_\nu. \qquad (4.97)$$

Therefore, using (4.88), we have

$$\begin{aligned}
W^2 = W^\mu W_\mu &= \frac{1}{4}\epsilon^{\mu\nu\lambda\rho}M_{\lambda\rho}P_\nu \epsilon_\mu{}^{\sigma\tau\varsigma}M_{\tau\varsigma}P_\sigma \\
&= \frac{1}{4}\epsilon^{\mu\nu\lambda\rho}\epsilon_\mu{}^{\sigma\tau\varsigma}M_{\lambda\rho}P_\nu M_{\tau\varsigma}P_\sigma \\
&= \frac{1}{4}\left[-\eta^{\nu\sigma}\left(\eta^{\lambda\tau}\eta^{\rho\varsigma} - \eta^{\lambda\varsigma}\eta^{\rho\tau}\right) - \eta^{\nu\tau}\left(\eta^{\lambda\varsigma}\eta^{\rho\sigma} - \eta^{\lambda\sigma}\eta^{\rho\varsigma}\right) \right. \\
&\quad \left. - \eta^{\nu\varsigma}\left(\eta^{\lambda\sigma}\eta^{\rho\tau} - \eta^{\lambda\tau}\eta^{\rho\sigma}\right)\right]M_{\lambda\rho}P_\nu M_{\tau\varsigma}P_\sigma \\
&= -\frac{1}{2}M_{\lambda\rho}P_\nu M^{\lambda\rho}P^\nu - \frac{1}{4}M_{\lambda\rho}P_\nu M^{\nu\lambda}P^\rho + \frac{1}{4}M_{\lambda\rho}P_\nu M^{\nu\rho}P^\lambda \\
&\quad - \frac{1}{4}M_{\lambda\rho}P_\nu M^{\rho\nu}P^\lambda + \frac{1}{4}M_{\lambda\rho}P_\nu M^{\lambda\nu}P^\rho \\
&= -\frac{1}{2}M_{\lambda\rho}P_\nu M^{\lambda\rho}P^\nu - M_{\lambda\rho}P_\nu M^{\nu\lambda}P^\rho
\end{aligned}$$

$$= -\frac{1}{2} M_{\lambda\rho} \left(M^{\lambda\rho} P^2 + \delta^\lambda_\nu P^\rho P^\nu - \delta^\rho_\nu P^\lambda P^\nu \right)$$
$$- M_{\lambda\rho} \left(M^{\nu\lambda} P_\nu P^\rho + 4 P^\lambda P^\rho - \delta^\lambda_\nu P^\nu P^\rho \right)$$
$$= -\frac{1}{2} M^{\lambda\rho} M_{\lambda\rho} P^2 - M_{\lambda\rho} M^{\nu\lambda} P_\nu P^\rho, \tag{4.98}$$

where we have simplified terms in the intermediate steps using the anti-symmetry of the Lorentz generators.

To understand the meaning of this operator, let us go to the rest frame of the massive particle. In this frame,

$$p_\mu = (m, 0, 0, 0), \qquad p^2 = p^\mu p_\mu = m^2, \tag{4.99}$$

and the operator W^2 acting on such a state, takes the form

$$W^2 = -\frac{1}{2} m^2 M^{\lambda\rho} M_{\lambda\rho} - m^2 M_{\lambda 0} M^{0\lambda}$$
$$= -\frac{1}{2} m^2 \left(2 M^{0\lambda} M_{0\lambda} + M^{ij} M_{ij} \right) + m^2 M^{0\lambda} M_{0\lambda}$$
$$= -\frac{1}{2} m^2 M^{ij} M_{ij}. \tag{4.100}$$

Recalling that (see (4.26))

$$M_{ij} = \epsilon_{ij}{}^k J_k, \tag{4.101}$$

where J_k represents the total angular momentum of the particle, we obtain

$$W^2 = -\frac{1}{2} m^2 \epsilon^{ij}{}_k J^k \, \epsilon_{ij}{}^\ell J_\ell$$
$$= -\frac{1}{2} m^2 \left(-2 \delta^\ell_k \right) J^k J_\ell$$
$$= m^2 J^k J_k = -m^2 \mathbf{J}^2. \tag{4.102}$$

The result in (4.102) can also be derived in an alternative manner which is simpler and quite instructive. Let us note that in the rest frame (4.99), the Pauli-Lubanski operator (4.82) has the form

$$W^0 = \frac{1}{2}\epsilon^{0ijk}M_{jk}p_i = 0,$$

$$W^i = \frac{1}{2}\epsilon^{i\nu\lambda\rho}M_{\lambda\rho}p_\nu = -\frac{m}{2}\epsilon_{ijk}M_{jk}$$

$$= -\frac{m}{2}\epsilon_i{}^{jk}M_{jk} = mJ_i, \qquad (4.103)$$

where we have used (1.34) as well as (4.12). It follows now that

$$W^2 = W^\mu W_\mu = W^i W_i = m^2 J^i J_i = -m^2 \mathbf{J}^2, \qquad (4.104)$$

which is the result obtained in (4.102). Therefore, for a massive particle, we can think of W^2 as being proportional to $\mathbf{J}^2$ and in the rest frame of the particle, this simply measures the spin of the particle. That is, for a massive particle at rest, we find

$$\text{eigenvalues of } W^2 \ : \ w^2 = -m^2 s(s+1). \qquad (4.105)$$

Thus, we see that the representations with $p^2 \neq 0$ can be labelled by the eigenvalues of the Casimir operators, (m, s), namely the mass and the spin of a particle and the dimensionality of such a representation will be $(2s + 1)$ (for both positive as well as negative energy).

The dimensionality of the representation can also be understood in an alternative manner as follows. For a state at rest with momentum of the form $p_\mu = (m, 0, 0, 0)$, we can ask what Lorentz transformations would leave such a vector invariant. Clearly, these would define an invariant subgroup of the Lorentz group and will lead to the degeneracy of states. It is not hard to see that all possible 3-dimensional rotations would leave such a vector invariant. Namely, rotations around the x or the y or the z axis will not change the time component of a four vector (recall that it is the spin 0 component of a four vector) and, therefore, would define the stability group of such vectors. Technically, one says that the 3-dimensional rotations define the "little" group of a time-like vector and this method of determining the representation is known as the "induced" representation. Therefore, all the degenerate states can be labelled not just by the eigenvalue of the momentum, but also by the eigenvalues of three dimensional rotations, namely, $s = 0, \frac{1}{2}, 1, \cdots$ and $m_s = s, s-1, \cdots, -s+1, -s$. This defines the $2s+1$ dimensional representation for a massive particle of spin s.

This can also be seen algebraically. Namely, a state at rest is an eigenstate of the P_0 operator. From the Lorentz algebra, we note that (see (4.30))

$$[P_0, M_{ij}] = 0. \tag{4.106}$$

Namely, the operators M_{ij}, which generate 3-dimensional rotations and are related to the angular momentum operators, commute with P_0. Consequently, the eigenstates of P_0 are invariant under three dimensional rotations and are simultaneous eigenstates of the angular momentum operators as well and such spaces are $(2s+1)$ dimensional. In closing, let us note from (4.103) that, up to a normalization factor, the three nontrivial Pauli-Lubanski operators correspond to the generators of symmetry of the "little group" in the rest frame.

4.3.2 Massless representation. In contrast to the massive representations of the Poincaré group, the representations for a massless particle are slightly more involved. The basic reason behind this is that the "little" group of a light-like vector is not so obvious. In this case, we note that (see (4.95))

$$p^\mu p_\mu = 0, \qquad p_\mu = (p, 0, 0, -p), \qquad p \neq 0. \tag{4.107}$$

Consequently, acting on states in such a vector space, we would have (see (4.83))

$$P^2 = 0. \tag{4.108}$$

However, from (4.83) we see that our states in the representation should also satisfy

$$P_\mu W^\mu = 0. \tag{4.109}$$

There now appear two distinct possibilities for the action of the Casimir W^2 on the states of the representation, namely,

$$W^2 = W^\mu W_\mu \neq 0, \quad \text{or,} \quad W^2 = W^\mu W_\mu = 0. \tag{4.110}$$

In the first case, namely, for a massless particle if $W^2 \neq 0$, then it can be shown (we will see this at the end of this section) that the representations are infinite dimensional with an infinity of spin values. Such representations do not correspond to physical particles and, consequently, we will not consider such representations.

On the other hand, in the second case where $W^2 = 0$ on the states of the representation, we can easily show that the action of W^μ in such a space is proportional to that of the momentum vector, namely, acting on states in such a space, W^μ has the form

$$W^\mu = -hp^\mu, \tag{4.111}$$

where h represents a proportionality factor (operator). To determine h, let us recall that

$$W^\mu = \frac{1}{2} \epsilon^{\mu\nu\lambda\rho} P_\nu M_{\lambda\rho}, \tag{4.112}$$

from which it follows that acting on a general momentum basis state $|p\rangle$ (not necessarily restricting to massless states), it would lead to (see (4.88))

$$W^0 = \frac{1}{2} \epsilon^{0ijk} p_i M_{jk} = \frac{1}{2} \epsilon_{ijk} p_i \left(\epsilon_{jk}{}^\ell J_\ell \right) = -\mathbf{p} \cdot \mathbf{J} = -\mathbf{J} \cdot \mathbf{p}. \tag{4.113}$$

Comparing with (4.111) we conclude that in this space

$$h = \frac{\mathbf{J} \cdot \mathbf{p}}{p^0}. \tag{4.114}$$

This is nothing other than the helicity operator (since $\mathbf{L} \cdot \mathbf{p} = 0$) and, therefore, the simultaneous eigenstates of P^2 and W^2 would correspond to the eigenstates of momentum and helicity. For completeness, let us note here that in the light-like frame (4.107), the Pauli-Lubanski operator (4.82) takes the form

$$W^0 = \frac{1}{2}\epsilon^{0ijk}M_{jk}P_i = \frac{1}{2}\epsilon_{ijk}M_{jk}p_i = -pM_{12},$$

$$W^1 = \frac{1}{2}\epsilon^{1\nu\lambda\rho}M_{\lambda\rho}p_\nu = p(M_{02} - M_{23}),$$

$$W^2 = \frac{1}{2}\epsilon^{2\nu\lambda\rho}M_{\lambda\rho}p_\nu = -p(M_{01} - M_{13}),$$

$$W^3 = \frac{1}{2}\epsilon^{3\nu\lambda\rho}M_{\lambda\rho}p_\nu = -pM_{12} = W^0. \tag{4.115}$$

We see from both (4.103) and (4.115) that the Pauli-Lubanski operator indeed has only three independent components because of the transversality condition (4.83), as we had pointed out earlier. We also note from (4.115) that, in the massless case, W^0, indeed represents the helicity operator up to a normalization as we had noted in (4.113). It follows now from (4.115) that (the contributions from W^0 and W^3 cancel out)

$$W^2 = W^\mu W_\mu = W^1 W_1 + W^2 W_2$$
$$= -p^2\left((M_{01} - M_{13})^2 + (M_{02} - M_{23})^2\right). \tag{4.116}$$

Let us now determine the dimensionality of the massless representations algebraically. Let us recall that we are considering a massless state with momentum of the form $p_\mu = (p, 0, 0, -p)$ and we would like to determine the "little" group of symmetries associated with such a vector. We recognize that in this case, the set of Lorentz transformations which would leave this four vector invariant must include rotations around the z-axis. This can be seen intuitively from the fact that the motion of the particle is along the z axis, but also algebraically by recognizing that a light-like vector of the form being considered is an eigenstate of the operator $P_0 - P_3$, namely,

$$(P_0 - P_3)|p\rangle = 2p|p\rangle. \tag{4.117}$$

Furthermore, from the Poincaré algebra in (4.38), we see that

$$[P_0 - P_3, M_{12}] = 0, \tag{4.118}$$

so that rotations around the z-axis define a symmetry of the light-like vector (state) that we are considering. To determine the other symmetries of a light-like vector, let us define two new operators as

$$\Pi_1 = M_{01} - M_{13} = \frac{W_2}{p}, \qquad \Pi_2 = M_{02} - M_{23} = -\frac{W_1}{p}. \tag{4.119}$$

It follows now that these operators commute with $P_0 - P_3$ in the space of light-like states, namely,

$$\begin{aligned}
{[P_0 - P_3, \Pi_1]}\,|p\rangle &= [P_0 - P_3, M_{01} - M_{13}]\,|p\rangle = 2P_1|p\rangle = 0, \\
{[P_0 - P_3, \Pi_2]}\,|p\rangle &= [P_0 - P_3, M_{02} - M_{23}]\,|p\rangle = 2P_2|p\rangle = 0,
\end{aligned} \tag{4.120}$$

and, therefore, also define symmetries of light-like states. These represent all the symmetries of the light-like vector (state). We note that the algebra of the symmetry generators takes the form

$$\begin{aligned}
{[M_{12}, \Pi_1]} &= [M_{12}, M_{01} - M_{13}] = \Pi_2, \\
{[M_{12}, \Pi_2]} &= [M_{12}, M_{02} - M_{23}] = -\Pi_1, \\
{[\Pi_1, \Pi_2]} &= [M_{01} - M_{13}, M_{02} - M_{23}] = 0.
\end{aligned} \tag{4.121}$$

Namely, it is isomorphic to the algebra of the Euclidean group in two dimensions, $\mathcal{E}_2$ (which consists of translations and rotation). Thus, we say that the stability group or the "little" group of a light-like vector is $\mathcal{E}_2$. Clearly, M_{12} is the generator of rotations around the z axis or in the two dimensional plane and Π_1, Π_2 have the same commutation relations as those of translations in this two dimensional space. Furthermore, comparing with $W^i, i = 1, 2, 3$ in (4.115), we see that up to a normalization, the three independent Pauli-Lubanski operators are, in fact, the generators of symmetry of the "little" group, as we had also seen in the massive case. This may seem puzzling, but can be easily understood as follows. We note from (4.90) that in the momentum basis states (where p_ρ is a number), the Pauli-Lubanski operators satisfy an algebra and, therefore, can be thought of as generators of some transformations. The meaning of the transformations, then, follows from (4.86) as the transformations that leave p_μ invariant. Namely, they generate transformations which will leave the momentum basis states invariant. This is, of course, what we have been investigating within the context of "little" groups.

Let us note from (4.121) that $\Pi_1 \mp i\Pi_2$ correspond respectively to raising and lowering operators for M_{12}, namely,

$$[M_{12}, \Pi_1 \mp i\Pi_2] = \pm i \left(\Pi_1 \mp i\Pi_2\right). \tag{4.122}$$

Let us also note for completeness that the Casimir of the $\mathcal{E}_2$ algebra is given by

$$\Pi^2 = \Pi_1^2 + \Pi_2^2 = (\Pi_1 \mp i\Pi_2)(\Pi_1 \pm i\Pi_2), \tag{4.123}$$

and comparing with (4.116), we see that in the space of light-like momentum states $W^2 \propto \Pi^2$. Since Π_1, Π_2 correspond to generators of "translation", their eigenvalues can take any value. As a result, if $W^2 \neq 0$ in this space, we note from (4.122) that spin can take an infinite number of values which, as we have already pointed out, does not correspond to any physical system. On the other hand, if $W^2 = 0$ in this space of states, then it follows from (4.123) that (h corresponds to the helicity quantum number)

$$(\Pi_1 \pm i\Pi_2)\left|p, h\right\rangle = 0. \tag{4.124}$$

(Alternatively, we can say that $\Pi_1|p, h\rangle = 0 = \Pi_2|p, h\rangle$ and this is the reason for the assertion in (4.111).) This corresponds to the one dimensional representation of $\mathcal{E}_2$ known as the "degenerate" representation. Clearly, such a state would correspond to the highest or the lowest helicity state. Furthermore, if our theory is also invariant under parity (or space reflection), the space of physical states would also include the state with the opposite helicity (recall that helicity changes sign under space reflection, see (3.146)). As a result, massless theories with nontrivial spin that are parity invariant would have two dimensional representations corresponding to the highest and the lowest helicity states, independent of the spin of the particle. On the other hand, if the theory is not parity invariant, the dimensionality of the representation will be one dimensional, as we have seen explicitly in the case of massless fermion theories describing neutrinos.

Incidentally, the fact that the massless representations have to be one dimensional, in general, can be seen in a heuristic way as follows. Let us consider spin as arising from a circular motion. Then, it is clear that since a massless particle moves at the speed of light, the only consistent circular motion a massless particle can have, is in a plane perpendicular to the direction of motion (otherwise, some component of the velocity would exceed the speed of light). In other words, in such a case, spin can only be either parallel or anti-parallel to the direction of motion leading to the one dimensional nature of the representation. However, if parity (space reflection) is a symmetry of the system, then we must have states corresponding to both the circular motions leading to the two dimensional representation.

4.4 References

1. E. Wigner, *On unitary representations of the inhomogeneous Lorentz group*, Annals of Mathematics **40**, 149 (1939).

2. V. Bargmann, *Irreducible unitary representations of the Lorentz group*, Annals of Mathematics **48**, 568 (1947).

3. E. Wigner, *Group Theory and its Applications to the Quantum Mechanics of Atomic Spectra*, Academic Press, New York (1959).

CHAPTER 5
Free Klein-Gordon field theory

5.1 Introduction

It is clear by now that the quantum mechanical description of a single relativistic particle runs into difficulties. Combining relativity with quantum mechanics necessarily seems to lead to a many particle theory. Intuitively, what this means is that a relativistic particle has a large enough energy that it can create particles and, therefore, a consistent theory describing such a particle cannot truly be a single particle theory. While the Dirac theory inherently describes a many particle theory, it manages to manifest itself as a single particle theory only because of the Pauli principle. Even so, it is not adequate to describe various decay processes such as

$$n \to p + e^- + \overline{\nu}_e, \tag{5.1}$$

where distinct fermions are involved.

If, on the other hand, we want to have a theory which describes infinitely many degrees of freedom, then the natural basic object for such a theory is a field variable which is a continuous function of space and time. It is clear from the study of classical electromagnetic fields or even from that of the displacement field for the oscillations of a string that such quantities naturally lead to infinitely many modes of oscillation each of which, upon quantization, can lead to a particle-like behavior and, therefore, can describe many particles. The main question in dealing with fields, however, is how do we choose dynamical equations of motion and how do we quantize such equations. The answer to the first question is quite easy from all of our discussions

Figure 5.1: Oscillating string fixed at both ends.

so far. We would like to have a relativistic, covariant theory for fundamental physical processes and, therefore, each field must belong to a specific representation of the Lorentz group. Secondly, the dynamical equation involving the field variables must be covariant under Poincaré transformations. Given a relativistic dynamical equation, the quantization of the fields as operators can then be carried out in the Hamiltonian formalism. However, since the Hamiltonian is not a manifestly covariant concept, it is generally preferrable to work in the Lagrangian formulation which is manifestly covariant. On the other hand, whenever any ambiguity arises in the quantization, one reverts back to the Hamiltonian formalism for its resolution.

With this in mind, let us look at the simplest of field theories, namely, the classical free Klein-Gordon theory where we assume that the field variable $\phi(x) \equiv \phi(\mathbf{x}, t)$ is real and that it behaves like a scalar under a Lorentz transformation,

$$\phi(x) = \phi^*(x), \qquad \phi'(x') = \phi(x), \tag{5.2}$$

where $\phi'(x')$ denotes the Lorentz transformed field. We also assume that the field falls off rapidly at spatial infinity (this is necessary for the theory to be well defined), namely,

$$\lim_{|\mathbf{x}| \to \infty} \phi(x) \to 0. \tag{5.3}$$

The derivatives of the field are also assumed to fall off at infinity. The classical, free Klein-Gordon equation with mass m, as we have seen (see (1.40)), has the form

$$\left(\Box + m^2\right) \phi(x) = \left(\partial_\mu \partial^\mu + m^2\right) \phi(x) = 0. \tag{5.4}$$

This equation is clearly invariant under Lorentz transformations. (Note that the operator in the parenthesis in the momentum space corresponds to $(P^2 - m^2)$. As we have seen, P^2 denotes one of the Casimirs of the Poincaré algebra and m is a constant.) It is a second order (hyperbolic) equation and can be classically solved uniquely if we are given the initial values (at $t = 0$) $\phi(\mathbf{x}, 0)$ and $\dot{\phi}(\mathbf{x}, 0)$ (a dot denotes a derivative with respect to t). The specification of the initial values would appear to single out a preferred hyper-surface which can possibly destroy covariance under a Lorentz transformation. However, it can be easily shown that the initial values can be prescribed on any arbitrary space-like surface and the equation can still be uniquely solved.

5.2 Lagrangian density

The next question that we would like to ask is whether there exists a Lagrangian or an action which would lead to the Klein-Gordon equation under a minimum action principle. To that end, we follow exactly what we normally do in point particle mechanics. Namely, let us assume that there exists an action of the form

$$S = \int_{t_i}^{t_f} dt\, L = \int_{t_i}^{t_f} dt\, d^3x\, \mathcal{L} = \int_{t_i}^{t_f} d^4x\, \mathcal{L}, \tag{5.5}$$

where we have written the Lagrangian in terms of a Lagrangian density as

$$L = \int d^3x\, \mathcal{L}. \tag{5.6}$$

Furthermore, we assume that the Lagrangian density depends only on the field variables and their first derivatives, namely,

$$\mathcal{L} = \mathcal{L}\left(\phi(x), \partial_\mu \phi(x)\right). \tag{5.7}$$

In general, of course, a Lagrangian density can depend on higher order derivatives. However, for equations which are at most second order in the derivatives, the Lagrangian density can depend at the most on the first order derivatives of the field variables. These are the kinds of equations we will be interested in and correspondingly we will assume this dependence of the Lagrangian density on the field variables through out.

We can now generalize the variational principle of point particle mechanics and ask under what conditions will the action be stationary if we change the fields arbitrarily and infinitesimally as

$$\phi(x) \to \phi'(x) = \phi(x) + \delta\phi(x), \tag{5.8}$$

subject to the boundary conditions

$$\delta\phi(\mathbf{x}, t_i) = \delta\phi(\mathbf{x}, t_f) = 0. \tag{5.9}$$

Note that under an infinitesimal change

$$\begin{aligned}
\delta S &= \int_{t_i}^{t_f} \mathrm{d}^4x \, \delta\mathcal{L} \\
&= \int_{t_i}^{t_f} \mathrm{d}^4x \left(\frac{\partial \mathcal{L}}{\partial \phi(x)} \delta\phi(x) + \frac{\partial \mathcal{L}}{\partial \partial_\mu \phi(x)} \delta(\partial_\mu \phi(x)) \right) \\
&= \int_{t_i}^{t_f} \mathrm{d}^4x \left(\frac{\partial \mathcal{L}}{\partial \phi(x)} \delta\phi(x) + \frac{\partial \mathcal{L}}{\partial \partial_\mu \phi(x)} \partial_\mu \delta\phi(x) \right) \\
&= \int_{t_i}^{t_f} \mathrm{d}^4x \left[\frac{\partial \mathcal{L}}{\partial \phi(x)} \delta\phi(x) + \partial_\mu \left(\frac{\partial \mathcal{L}}{\partial \partial_\mu \phi(x)} \delta\phi(x) \right) \right. \\
&\qquad \left. - \partial_\mu \frac{\partial \mathcal{L}}{\partial \partial_\mu \phi(x)} \delta\phi(x) \right] \\
&= \int_{t_i}^{t_f} \mathrm{d}^4x \left(\frac{\partial \mathcal{L}}{\partial \phi(x)} - \partial_\mu \frac{\partial \mathcal{L}}{\partial \partial_\mu \phi(x)} \right) \delta\phi(x) \\
&\quad + \int \mathrm{d}^3x \left. \frac{\partial \mathcal{L}}{\partial \dot\phi(x)} \delta\phi(x) \right|_{t_i}^{t_f}.
\end{aligned} \tag{5.10}$$

Here we have used Gauss' theorem (as well as the vanishing of field variables asymptotically (5.3)) to simplify the surface term. We have

5.2 LAGRANGIAN DENSITY

also used the fact that $\delta(\partial_\mu \phi) = \partial_\mu(\delta\phi)$ which is easily seen from (consider one dimension for simplicity)

$$\begin{aligned}
\delta(\partial_x \phi) &= \lim_{\epsilon \to 0} \frac{1}{\epsilon} \left(\phi'(x+\epsilon) - \phi'(x) - (\phi(x+\epsilon) - \phi(x)) \right) \\
&= \lim_{\epsilon \to 0} \frac{1}{\epsilon} \left((\phi'(x+\epsilon) - \phi(x+\epsilon)) - (\phi'(x) - \phi(x)) \right) \\
&= \lim_{\epsilon \to 0} \frac{1}{\epsilon} (\delta\phi(x+\epsilon) - \delta\phi(x)) \\
&= \partial_x(\delta\phi).
\end{aligned} \qquad (5.11)$$

We note that the "surface term" – the last term in the expression (5.10) – vanishes because of the boundary conditions (5.9), namely,

$$\delta\phi(\mathbf{x}, t_i) = 0 = \delta\phi(\mathbf{x}, t_f). \qquad (5.12)$$

Thus, we conclude that under an arbitrary, infinitesimal change in the field variable, the change in the action is given by

$$\delta S = \int_{t_i}^{t_f} \mathrm{d}^4 x \left(\frac{\partial \mathcal{L}}{\partial \phi(x)} - \partial_\mu \frac{\partial \mathcal{L}}{\partial \partial_\mu \phi(x)} \right) \delta\phi(x). \qquad (5.13)$$

Therefore, the action will be stationary under an arbitrary change in the field variable subject to the boundary conditions provided (namely, $\delta S = 0$ only if),

$$\frac{\partial \mathcal{L}}{\partial \phi(x)} - \partial_\mu \frac{\partial \mathcal{L}}{\partial \partial_\mu \phi(x)} = 0. \qquad (5.14)$$

This is known as the Euler-Lagrange equation associated with the action S or the Lagrangian density $\mathcal{L}(\phi, \partial_\mu \phi(x))$.

As in point particle mechanics, we can think of the dynamical equations as arising from the minimum principle associated with a given action. Thus, for the Klein-Gordon equation,

$$\partial_\mu \partial^\mu \phi + m^2 \phi = 0, \qquad (5.15)$$

we can ask whether there exists a Lagrangian density whose Euler-Lagrange equation will lead to this dynamical equation. Note that if we choose

$$\mathcal{L} = \frac{1}{2} \partial_\mu \phi \partial^\mu \phi - \frac{m^2}{2} \phi^2, \qquad (5.16)$$

then

$$\frac{\partial \mathcal{L}}{\partial \phi(x)} = -m^2 \phi(x),$$

$$\frac{\partial \mathcal{L}}{\partial \partial_\mu \phi(x)} = \partial^\mu \phi(x). \qquad (5.17)$$

Therefore, the Euler-Lagrange equation, in this case, gives

$$\frac{\partial \mathcal{L}}{\partial \phi(x)} - \partial_\mu \frac{\partial \mathcal{L}}{\partial \partial_\mu \phi(x)} = -m^2 \phi(x) - \partial_\mu \partial^\mu \phi(x) = 0,$$

or, $\quad \left(\partial_\mu \partial^\mu + m^2 \right) \phi(x) = 0. \qquad (5.18)$

In other words, we see that the Lagrangian density

$$\mathcal{L} = \frac{1}{2} \partial_\mu \phi \partial^\mu \phi - \frac{m^2}{2} \phi^2, \qquad (5.19)$$

gives the Klein-Gordon equation for a free, real scalar field as its Euler Lagrange equation. Note that the Lagrangian density $\mathcal{L}$ in (5.19) is manifestly Lorentz invariant. It is worth noting here that one can add a constant to $\mathcal{L}$ without changing the equation of motion. This would simply correspond to adding a zero point energy. Similarly, adding a total divergence to $\mathcal{L}$ also does not change the equations of motion. Of course, in this case, the action does not change either, if we assume the fields to fall off rapidly at infinite separation (see (5.3)).

5.3 Quantization

Since we have a Lagrangian description for the classical Klein-Gordon field theory, we can try to develop a Hamiltonian description for it (which we need for quantization) in complete parallel to what we do for point particle mechanics. In the case of a point particle, the Lagrangian is a function of the coordinate and the velocity of the particle (a dot denotes a derivative with respect to t),

$$L = L(x, \dot{x}). \tag{5.20}$$

The conjugate momentum associated with the dynamical variable x is defined as

$$p = \frac{\partial L}{\partial \dot{x}}, \tag{5.21}$$

and this leads to the definition of the Hamiltonian in the form (this is an example of a Legendre transformation)

$$H(x, p) = p\dot{x} - L(x, \dot{x}). \tag{5.22}$$

The classical canonical Poisson bracket relations between the coordinate and the conjugate momentum have the familiar form

$$\begin{aligned} \{x, x\} &= 0 = \{p, p\}, \\ \{x, p\} &= 1. \end{aligned} \tag{5.23}$$

These, in turn, lead to the dynamical equations in the Hamiltonian form as

$$\begin{aligned} \dot{x} &= \{x, H\}, \\ \dot{p} &= \{p, H\}. \end{aligned} \tag{5.24}$$

In the case of the classical scalar (Klein-Gordon) field theory, on the other hand, the basic variables are $\phi(x)$ and $\partial_\mu \phi(x)$ and we have from (5.19) (x^μ in this case merely labels space-time coordinates)

$$\mathcal{L} = \mathcal{L}\left(\phi(x), \partial_\mu \phi(x)\right). \tag{5.25}$$

In an analogous manner to the point particle mechanics in (5.21), we define a momentum canonically conjugate to the field variable $\phi(x)$ as

$$\Pi(x) = \frac{\partial \mathcal{L}}{\partial \dot{\phi}(x)}, \tag{5.26}$$

which will, in turn, allow us to define a Hamiltonian density as

$$\mathcal{H} = \Pi(x)\dot{\phi}(x) - \mathcal{L}, \tag{5.27}$$

and a Hamiltonian

$$H = \int d^3 x\, \mathcal{H}. \tag{5.28}$$

If we further assume the equal time canonical Poisson bracket relations between the field variable $\phi(x)$ and the conjugate momentum $\Pi(x)$ to be (see (5.23))

$$\begin{aligned}
\{\phi(x), \phi(y)\}_{x^0=y^0} &= 0 = \{\Pi(x), \Pi(y)\}_{x^0=y^0}, \\
\{\phi(x), \Pi(y)\}_{x^0=y^0} &= \delta^3(x - y),
\end{aligned} \tag{5.29}$$

then, we can show that the dynamical equation (namely, the Klein-Gordon equation in this case) can be written in the Hamiltonian form as

$$\begin{aligned}
\dot{\phi}(x) &= \{\phi(x), H\}, \\
\dot{\Pi}(x) &= \{\Pi(x), H\}.
\end{aligned} \tag{5.30}$$

Let us examine all of this in detail for the case of the free Klein-Gordon theory. Note that the Lagrangian density in this case is given by

5.3 QUANTIZATION

$$\begin{aligned}\mathcal{L} &= \frac{1}{2}\partial_\mu\phi\partial^\mu\phi - \frac{m^2}{2}\phi^2 \\ &= \frac{1}{2}\dot{\phi}^2 - \frac{1}{2}\boldsymbol{\nabla}\phi\cdot\boldsymbol{\nabla}\phi - \frac{m^2}{2}\phi^2.\end{aligned} \quad (5.31)$$

Therefore, we can identify the conjugate momentum with

$$\Pi(x) = \frac{\partial \mathcal{L}}{\partial \dot{\phi}(x)} = \dot{\phi}(x). \quad (5.32)$$

Consequently, this leads to the Hamiltonian density

$$\begin{aligned}\mathcal{H} &= \Pi(x)\dot{\phi}(x) - \mathcal{L} \\ &= \Pi(x)\dot{\phi}(x) - \frac{1}{2}\dot{\phi}^2 + \frac{1}{2}\boldsymbol{\nabla}\phi\cdot\boldsymbol{\nabla}\phi + \frac{m^2}{2}\phi^2(x) \\ &= \Pi(x)\Pi(x) - \frac{1}{2}\Pi^2(x) + \frac{1}{2}\boldsymbol{\nabla}\phi\cdot\boldsymbol{\nabla}\phi + \frac{m^2}{2}\phi^2(x) \\ &= \frac{1}{2}\Pi^2(x) + \frac{1}{2}\boldsymbol{\nabla}\phi\cdot\boldsymbol{\nabla}\phi + \frac{m^2}{2}\phi^2(x),\end{aligned} \quad (5.33)$$

which gives the Hamiltonian for the theory

$$\begin{aligned}H &= \int \mathrm{d}^3 x\, \mathcal{H} \\ &= \int \mathrm{d}^3 x \left(\frac{1}{2}\Pi^2(x) + \frac{1}{2}\boldsymbol{\nabla}\phi\cdot\boldsymbol{\nabla}\phi + \frac{m^2}{2}\phi^2(x)\right).\end{aligned} \quad (5.34)$$

With the equal time canonical Poisson bracket relations defined to be (see (5.29))

$$\begin{aligned}\{\phi(x),\phi(y)\}_{x^0=y^0} &= 0 = \{\Pi(x),\Pi(y)\}_{x^0=y^0}, \\ \{\phi(x),\Pi(y)\}_{x^0=y^0} &= \delta^3(x-y),\end{aligned} \quad (5.35)$$

we obtain

$$\dot\phi(x) = \{\phi(x), H\}$$
$$= \left\{\phi(x), \int d^3y\left(\frac{1}{2}\Pi^2(y) + \frac{1}{2}\nabla_y\phi\cdot\nabla_y\phi + \frac{m^2}{2}\phi^2(y)\right)\right\}_{x^0=y^0}$$
$$= \frac{1}{2}\int d^3y\,\{\phi(x), \Pi^2(y)\}_{x^0=y^0}$$
$$= \int d^3y\,\Pi(y)\,\{\phi(x), \Pi(y)\}\bigg|_{x^0=y^0}$$
$$= \int d^3y\,\Pi(\mathbf{y}, x^0)\,\delta^3(x-y) = \Pi(x), \tag{5.36}$$

where we have used the fact that, since the Hamiltonian is time independent (this can be checked easily using the equations of motion and will also be shown in (5.73)), we can choose the time coordinate for the fields in its definition to coincide with x^0 in order that we can use the equal time Poisson bracket relations (5.29). Furthermore, we also have

$$\dot\Pi(x) = \{\Pi(x), H\}$$
$$= \left\{\Pi(x), \int d^3y\left(\frac{1}{2}\Pi^2(y) + \frac{1}{2}\nabla_y\phi\cdot\nabla_y\phi + \frac{m^2}{2}\phi^2(y)\right)\right\}_{x^0=y^0}$$
$$= \int d^3y\,[\nabla_y\phi(y)\cdot\{\Pi(x), \nabla_y\phi(y)\} + m^2\phi(y)\,\{\Pi(x), \phi(y)\}]_{x^0=y^0}$$
$$= \int d^3y\,[\nabla_y\phi(\mathbf{y}, x^0)\cdot\nabla_y(-\delta^3(x-y))$$
$$+ m^2\phi(\mathbf{y}, x^0)(-\delta^3(x-y))]$$
$$= \nabla\cdot\nabla\phi(x) - m^2\phi(x). \tag{5.37}$$

Thus, the Hamiltonian equations (5.36) and (5.37), in this case, lead to

$$\dot\phi(x) = \Pi(x),$$
$$\dot\Pi(x) = \nabla\cdot\nabla\phi(x) - m^2\phi(x) = \nabla^2\phi(x) - m^2\phi(x). \tag{5.38}$$

These are first order equations as Hamiltonian equations should be and from these, we obtain the second order equation

$$\ddot{\phi}(x) = \dot{\Pi}(x) = \boldsymbol{\nabla}^2 \phi(x) - m^2 \phi(x),$$
or, $\quad \ddot{\phi} - \boldsymbol{\nabla}^2 \phi(x) + m^2 \phi(x) = 0,$
or, $\quad \left(\partial_\mu \partial^\mu + m^2\right) \phi(x) = 0.$ (5.39)

This is, of course, the free Klein-Gordon equation (1.40) (which is a second order Euler-Lagrange equation) and this discussion, therefore, brings out the classical Hamiltonian description of the classical Klein-Gordon field theory.

Once we have the classical Hamiltonian description of a physical system, the quantization of such a system is quite straightforward. In the quantum theory, we are supposed to treat $\phi(x)$ and $\Pi(x)$ as Hermitian operators (because we are dealing with real fields, see (5.2)) satisfying the equal time commutation relations (remember $\hbar = 1$)

$$\begin{aligned}
\left[\phi(x), \phi(y)\right]_{x^0=y^0} &= 0 = \left[\Pi(x), \Pi(y)\right]_{x^0=y^0}, \\
\left[\phi(x), \Pi(y)\right]_{x^0=y^0} &= i\delta^3(x - y).
\end{aligned}$$ (5.40)

Note that this is the same Klein-Gordon equation as we had studied earlier in chapter **1**. However, here we are treating $\phi(x)$ as an operator which is quantized and not as a wavefunction as we had done earlier. This is conventionally referred to as second quantization. (Namely, in the first quantization we quantize x's and p's, but not the wave function. In the second quantization, we quantize the "wavefunction".)

5.4 Field decomposition

Given any complete set of solutions of the classical Klein-Gordon equation, we can expand the field operator $\phi(x)$ in such a basis. We already know that the plane wave solutions (1.41) of the Klein-Gordon equation define a complete basis and, therefore, we can expand the field operator in this basis as

$$\phi(x) = \frac{1}{(2\pi)^{\frac{3}{2}}} \int d^4k\, e^{-ik\cdot x} \widetilde{\phi}(k), \tag{5.41}$$

where we recognize that the operators $\phi(x)$ and $\widetilde{\phi}(k)$ are Fourier transforms of each other (up to a multiplicative factor). (Recall that $k \cdot x = k^0 x^0 - \mathbf{k} \cdot \mathbf{x}$ and the factor $\frac{1}{(2\pi)^{3/2}}$ is introduced for later convenience. We also emphasize here that the hermiticity properties of a given field and its Fourier transform are, in general, different.) Since $\phi(x)$ satisfies the Klein-Gordon equation

$$\left(\partial_\mu \partial^\mu + m^2\right) \phi(x) = 0, \tag{5.42}$$

using the field expansion (5.41) in the Klein-Gordon equation, we obtain

$$\frac{1}{(2\pi)^{\frac{3}{2}}} \left(\partial_\mu \partial^\mu + m^2\right) \int d^4k\, e^{-ik\cdot x} \widetilde{\phi}(k) = 0,$$

or, $\quad \dfrac{1}{(2\pi)^{\frac{3}{2}}} \displaystyle\int d^4k \left(-k^2 + m^2\right) \widetilde{\phi}(k) e^{-ik\cdot x} = 0. \tag{5.43}$

This shows that unless $k^2 = m^2$, $\widetilde{\phi}(k) = 0$. Namely, the Fourier transform has non-vanishing contribution (support) only on the mass-shell of the particle defined by

$$k^2 = (k^0)^2 - \mathbf{k}^2 = m^2. \tag{5.44}$$

As a result, we can denote

$$\widetilde{\phi}(k) = \delta\left(k^2 - m^2\right) a(k), \tag{5.45}$$

where the operator $a(k)$ is no longer constrained by the equation of motion. Substituting this back into the expansion (5.41), we obtain

$$\phi(x) = \frac{1}{(2\pi)^{\frac{3}{2}}} \int d^4k\, \delta\left(k^2 - m^2\right) e^{-ik\cdot x} a(k). \tag{5.46}$$

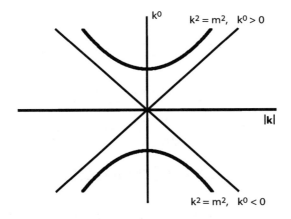

Figure 5.2: Mass shell on which $\widetilde{\phi}(k)$ has support.

We note that the argument of the delta function vanishes for

$$k^2 = m^2,$$
or, $$(k^0)^2 = \mathbf{k}^2 + m^2,$$
or, $$k^0 = \pm\sqrt{\mathbf{k}^2 + m^2} \equiv \pm E_k. \tag{5.47}$$

Correspondingly, we can write

$$\begin{aligned} \delta\left(k^2 - m^2\right) &= \delta\left((k^0)^2 - E_k^2\right) \\ &= \frac{1}{2|k^0|}\left[\delta\left(k^0 - E_k\right) + \delta\left(k^0 + E_k\right)\right] \\ &= \frac{1}{2E_k}\left[\delta\left(k^0 - E_k\right) + \delta\left(k^0 + E_k\right)\right]. \end{aligned} \tag{5.48}$$

Using this relation in (5.46), the field expansion becomes

$$\phi(x) = \frac{1}{(2\pi)^{\frac{3}{2}}} \int dk^0 d^3k \frac{1}{2E_k} \left[\delta\left(k^0 - E_k\right) + \delta\left(k^0 + E_k\right)\right]$$
$$\times e^{-ik^0 x^0 + i\mathbf{k}\cdot\mathbf{x}} a(k^0, \mathbf{k})$$
$$= \frac{1}{(2\pi)^{\frac{3}{2}}} \int d^3k \frac{1}{2E_k} \left(e^{-iE_k x^0 + i\mathbf{k}\cdot\mathbf{x}} a(E_k, \mathbf{k})\right.$$
$$\left. + e^{iE_k x^0 + i\mathbf{k}\cdot\mathbf{x}} a(-E_k, \mathbf{k})\right). \tag{5.49}$$

Changing $\mathbf{k} \to -\mathbf{k}$ in the second term in (5.49) and identifying $k^0 = E_k > 0$, we obtain

$$\phi(x) = \frac{1}{(2\pi)^{\frac{3}{2}}} \int \frac{d^3k}{2k^0} \left(e^{-ik\cdot x} a(k) + e^{ik\cdot x} a(-k)\right). \tag{5.50}$$

Let us note that we are dealing with a Hermitian field (see (5.2)). Therefore,

$$\phi^\dagger(x) = \phi(x),$$

or, $\quad \dfrac{1}{(2\pi)^{\frac{3}{2}}} \int \dfrac{d^3k}{2k^0} \left(e^{ik\cdot x} a^\dagger(k) + e^{-ik\cdot x} a^\dagger(-k)\right)$

$$= \frac{1}{(2\pi)^{\frac{3}{2}}} \int \frac{d^3k}{2k^0} \left(e^{-ik\cdot x} a(k) + e^{ik\cdot x} a(-k)\right). \tag{5.51}$$

Comparing the left-hand side and the right-hand side of (5.51), we obtain

$$a^\dagger(k) = a(-k), \qquad a^\dagger(-k) = a(k), \tag{5.52}$$

so that we can write

$$\phi(x) = \frac{1}{(2\pi)^{\frac{3}{2}}} \int \frac{d^3k}{2k^0} \left(e^{-ik\cdot x} a(k) + e^{ik\cdot x} a^\dagger(k)\right). \tag{5.53}$$

5.5 CREATION AND ANNIHILATION OPERATORS

Here we are supposed to understand that

$$k^0 = E_k = \sqrt{\mathbf{k}^2 + m^2} > 0. \tag{5.54}$$

It is clear from (5.54) that $a(k)$ and $a^\dagger(k)$ are really functions of the three momentum $\mathbf{k}$ alone. Thus, conventionally one defines

$$a(\mathbf{k}) = \frac{a(k)}{\sqrt{2k^0}}, \qquad a^\dagger(\mathbf{k}) = \frac{a^\dagger(k)}{\sqrt{2k^0}}, \tag{5.55}$$

and in terms of these operators we can write the field operator (5.53) as

$$\phi(x) = \int \frac{d^3k}{\sqrt{(2\pi)^3 2k^0}} \left(e^{-ik\cdot x} a(\mathbf{k}) + e^{ik\cdot x} a^\dagger(\mathbf{k}) \right). \tag{5.56}$$

This unique decomposition of the field operator $\phi(x)$ into positive and negative energy (frequency) parts is quite significant as we will see in the course of our discussions. In fact, we often denote the positive and the negative energy parts of the field operator as

$$\phi^{(+)}(x) = \int \frac{d^3k}{\sqrt{(2\pi)^3 2k^0}} e^{-ik\cdot x} a(\mathbf{k}),$$

$$\phi^{(-)}(x) = \int \frac{d^3k}{\sqrt{(2\pi)^3 2k^0}} e^{ik\cdot x} a^\dagger(\mathbf{k}), \tag{5.57}$$

so that we can write the field operator as a sum of its positive and negative energy parts,

$$\phi(x) = \phi^{(+)}(x) + \phi^{(-)}(x). \tag{5.58}$$

5.5 Creation and annihilation operators

Given the field expansion (5.56)

$$\phi(x) = \int \frac{d^3k}{\sqrt{(2\pi)^3 2k^0}} \left(e^{-ik\cdot x} a(\mathbf{k}) + e^{ik\cdot x} a^\dagger(\mathbf{k}) \right), \tag{5.59}$$

we obtain the expansion for the conjugate momentum in (5.32) to be

$$\begin{aligned}\Pi(x) &= \dot\phi(x) \\ &= -i \int d^3k \sqrt{\frac{k^0}{2(2\pi)^3}} \left(e^{-ik\cdot x} a(\mathbf{k}) - e^{ik\cdot x} a^\dagger(\mathbf{k}) \right).\end{aligned} \tag{5.60}$$

The two defining relations in (5.59) and (5.60) are invertible. Namely, we note from these relations that we can write the operators $a(\mathbf{k})$ and $a^\dagger(\mathbf{k})$ in terms of $\phi(x)$ and $\Pi(x)$ as (recall that $\Pi(x) = \dot\phi(x)$)

$$\begin{aligned}a(\mathbf{k}) &= \frac{1}{\sqrt{(2\pi)^3 2k^0}} \int d^3x\, e^{ik\cdot x} \left(k^0 \phi(x) + i\Pi(x) \right) \\ &= \frac{i}{\sqrt{(2\pi)^3 2k^0}} \int d^3x\, e^{ik\cdot x} \overleftrightarrow{\partial_t} \phi(x), \\ a^\dagger(\mathbf{k}) &= \frac{1}{\sqrt{(2\pi)^3 2k^0}} \int d^3x\, e^{-ik\cdot x} \left(k^0 \phi(x) - i\Pi(x) \right) \\ &= -\frac{i}{\sqrt{(2\pi)^3 2k^0}} \int d^3x\, e^{-ik\cdot x} \overleftrightarrow{\partial_t} \phi(x),\end{aligned} \tag{5.61}$$

where $\overleftrightarrow{\partial_t}$ is defined in (1.56). It is important to emphasize here that in the expressions in (5.61), it is assumed that $k^0 = E_k$ (namely, k^0 is not an independent variable). It is clear from (5.61) that since the left-hand sides are independent of time, the expressions on the right-hand side must also be time independent. This can also be checked explicitly. For example, taking the time derivative of the first relation in (5.61) we obtain (neglecting the overall multiplicative factors)

$$\partial_t \left(\int d^3x \, e^{ik\cdot x} \overleftrightarrow{\partial_t} \phi(x) \right)$$

$$= \int d^3x \left(e^{ik\cdot x} (\partial_t^2 \phi(x)) - (\partial_t^2 e^{ik\cdot x}) \phi(x) \right)$$

$$= \int d^3x \left(e^{ik\cdot x} (\nabla^2 - m^2) \phi(x) - (\partial_t^2 e^{ik\cdot x}) \phi(x) \right)$$

$$= \int d^3x \, (-\partial_t^2 + \nabla^2 - m^2) \, e^{ik\cdot x} \, \phi(x)$$

$$= \int d^3x \, ((k^0)^2 - \mathbf{k}^2 - m^2) \, e^{ik\cdot x} \phi(x) = 0. \tag{5.62}$$

Here, in the intermediate steps, we have used the fact that $\phi(x)$ satisfies the Klein-Gordon equation, integrated the space derivatives by parts assuming that the surface terms vanish (see (5.3)) and have made the identification $k^0 = E_k = \sqrt{\mathbf{k}^2 + m^2}$. Similarly, the time independence of the second expression in (5.61) can also be derived in a simple manner (or note that the second relation is the Hermitian conjugate of the first and, therefore, must be time independent). This shows explicitly that the integrals in (5.61) are indeed time independent.

Imposing now the quantization relations between $\phi(x)$ and $\Pi(x)$ (see (5.40)) we obtain

$$[\phi(x), \phi(y)]_{x^0=y^0} = \int \frac{d^3k}{\sqrt{(2\pi)^3 2k^0}} \frac{d^3k'}{\sqrt{(2\pi)^3 2k'^0}}$$

$$\times \left(e^{-ik\cdot x - ik'\cdot y} [a(\mathbf{k}), a(\mathbf{k}')] + e^{-ik\cdot x + ik'\cdot y} [a(\mathbf{k}), a^\dagger(\mathbf{k}')] \right.$$

$$\left. + e^{ik\cdot x - ik'\cdot y} [a^\dagger(\mathbf{k}), a(\mathbf{k}')] + e^{ik\cdot x + ik'\cdot y} [a^\dagger(\mathbf{k}), a^\dagger(\mathbf{k}')] \right)_{x^0=y^0}$$

$$= 0, \tag{5.63}$$

$$[\Pi(x), \Pi(y)]_{x^0=y^0} = -\int d^3k \, d^3k' \, \frac{\sqrt{k^0 k'^0}}{2(2\pi)^3}$$

$$\times \left(e^{-ik\cdot x - ik'\cdot y} [a(\mathbf{k}), a(\mathbf{k}')] - e^{-ik\cdot x + ik'\cdot y} [a(\mathbf{k}), a^\dagger(\mathbf{k}')] \right.$$

$$- e^{ik\cdot x - ik'\cdot y} \left[a^\dagger(\mathbf{k}), a(\mathbf{k}')\right] + e^{ik\cdot x + ik'\cdot y} \left[a^\dagger(\mathbf{k}), a^\dagger(\mathbf{k}')\right]\Big)_{x^0=y^0}$$

$$= 0, \tag{5.64}$$

$$[\phi(x), \Pi(y)]_{x^0=y^0} = -\frac{i}{(2\pi)^3} \int d^3k\, d^3k' \sqrt{\frac{k'^0}{4k^0}}$$

$$\times \left(e^{-ik\cdot x - ik'\cdot y} \left[a(\mathbf{k}), a(\mathbf{k}')\right] - e^{-ik\cdot x + ik'\cdot y} \left[a(\mathbf{k}), a^\dagger(\mathbf{k}')\right] \right.$$

$$\left. + e^{ik\cdot x - ik'\cdot y} \left[a^\dagger(\mathbf{k}), a(\mathbf{k}')\right] - e^{ik\cdot x + ik'\cdot y} \left[a^\dagger(\mathbf{k}), a^\dagger(\mathbf{k}')\right] \right)_{x^0=y^0}$$

$$= i\delta^3(x-y). \tag{5.65}$$

From these, we can deduce the fundamental commutation relations between the coefficients of expansion to be (This nice form arises because of the redefinition (5.55). Any other redefinition will introduce a multiplicative factor into the commutation relations.)

$$\left[a(\mathbf{k}), a(\mathbf{k}')\right] = 0 = \left[a^\dagger(\mathbf{k}), a^\dagger(\mathbf{k}')\right],$$

$$\left[a(\mathbf{k}), a^\dagger(\mathbf{k}')\right] = \delta^3\left(k - k'\right). \tag{5.66}$$

This shows that the operators $a(\mathbf{k})$ and $a^\dagger(\mathbf{k})$, which are the coefficients of expansion of the field operator in a plane wave basis, have commutation relations analogous to the annihilation and creation operators of a harmonic oscillator. However, it appears that there is an infinite number of such operators in the present case – one for every value of the momentum $\mathbf{k}$.

The commutation relations in (5.66) can also be obtained more directly from the inversion formulae in (5.61). Thus, for example, recalling that the integrals are ime independent, we have

$$\left[a(\mathbf{k}), a^\dagger(\mathbf{k}')\right]$$

$$= \int \frac{d^3x\, d^3y}{(2\pi)^3 \sqrt{4k^0 k'^0}} e^{ik\cdot x - ik'\cdot y}$$

$$\times \left[k^0 \phi(x) + i\Pi(x), k'^0 \phi(y) - i\Pi(y)\right]\Big|_{x^0=y^0}$$

5.5 Creation and annihilation operators

$$
\begin{aligned}
&= \int \frac{\mathrm{d}^3x \mathrm{d}^3y}{(2\pi)^3\sqrt{4k^0 k'^0}} e^{i k \cdot x - i k' \cdot y} \\
&\quad \times \left(-ik^0 \left[\phi(x), \Pi(y)\right] + ik'^0 \left[\Pi(x), \phi(y)\right] \right)_{x^0 = y^0} \\
&= \frac{(k^0 + k'^0)}{\sqrt{4k^0 k'^0}} \int \frac{\mathrm{d}^3x \mathrm{d}^3y}{(2\pi)^3} e^{i k \cdot x - i k' \cdot y} \delta^3(x - y) \bigg|_{x^0 = y^0} \\
&= \frac{(k^0 + k'^0)}{\sqrt{4k^0 k'^0}} \int \frac{\mathrm{d}^3x}{(2\pi)^3} e^{i(k - k') \cdot x} \\
&= \frac{(k^0 + k'^0)}{\sqrt{4k^0 k'^0}} e^{i(k^0 - k'^0)x^0} \delta^3(k - k') = \delta^3(k - k'), \quad (5.67)
\end{aligned}
$$

where we have used the fact that for $\mathbf{k}' = \mathbf{k}$, we have $k'^0 = k^0$. The other two commutation relations in (5.66) can also be derived in the same manner

$$
\begin{aligned}
&\left[a(\mathbf{k}), a(\mathbf{k}') \right] \\
&= \int \frac{\mathrm{d}^3x \mathrm{d}^3y}{(2\pi)^3\sqrt{4k^0 k'^0}} e^{i k \cdot x + i k' \cdot y} \\
&\quad \times \left[k^0 \phi(x) + i\Pi(x), k'^0 \phi(y) + i\Pi(y) \right] \bigg|_{x^0 = y^0} \\
&= \int \frac{\mathrm{d}^3x \mathrm{d}^3y}{(2\pi)^3\sqrt{4k^0 k'^0}} e^{i k \cdot x + i k' \cdot y} \\
&\quad \times \left(ik^0 \left[\phi(x), \Pi(y)\right] + ik'^0 \left[\Pi(x), \phi(y)\right] \right)_{x^0 = y^0} \\
&= -\frac{(k^0 - k'^0)}{\sqrt{4k^0 k'^0}} \int \frac{\mathrm{d}^3x \mathrm{d}^3y}{(2\pi)^3} e^{i k \cdot x + i k' \cdot y} \delta^3(x - y) \bigg|_{x^0 = y^0} \\
&= -\frac{(k^0 - k'^0)}{\sqrt{4k^0 k'^0}} \int \frac{\mathrm{d}^3x}{(2\pi)^3} e^{i(k + k') \cdot x} \\
&= -\frac{(k^0 - k'^0)}{\sqrt{4k^0 k'^0}} e^{i(k^0 + k'^0)x^0} \delta^3(k + k') = 0,
\end{aligned}
$$

$$
\left[a^\dagger(\mathbf{k}), a^\dagger(\mathbf{k}') \right]
$$

$$
= \int \frac{\mathrm{d}^3x \mathrm{d}^3y}{(2\pi)^3\sqrt{4k^0 k'^0}} e^{-i k \cdot x - i k' \cdot y}
$$

$$\times \left[k^0 \phi(x) - i\Pi(x), k'^0 \phi(y) - i\Pi(y) \right] \Big|_{x^0=y^0}$$

$$= \int \frac{d^3x \, d^3y}{(2\pi)^3 \sqrt{4k^0 k'^0}} e^{-ik\cdot x - ik'\cdot y}$$

$$\times \left(-ik^0 \left[\phi(x), \Pi(y) \right] - ik'^0 \left[\Pi(x), \phi(y) \right] \right)_{x^0=y^0}$$

$$= \frac{(k^0 - k'^0)}{\sqrt{4k^0 k'^0}} \int \frac{d^3x \, d^3y}{(2\pi)^3} e^{-ik\cdot x - ik'\cdot y} \delta^3(x-y) \Big|_{x^0=y^0}$$

$$= \frac{(k^0 - k'^0)}{\sqrt{4k^0 k'^0}} \int \frac{d^3x}{(2\pi)^3} e^{-i(k+k')\cdot x}$$

$$= \frac{(k^0 - k'^0)}{\sqrt{4k^0 k'^0}} e^{-i(k^0 + k'^0)x^0} \delta^3(k+k') = 0, \tag{5.68}$$

where we have used the fact that $k'^0 = k^0$ for $\mathbf{k}' = -\mathbf{k}$.

To understand the meaning of these operators further, let us look at the Hamiltonian of the system. We note that

$$\int d^3x \, \Pi^2(x) = -\int d^3x \int \frac{d^3k \, d^3k'}{2(2\pi)^3} \sqrt{k^0 k'^0}$$

$$\times \left(e^{-i(k+k')\cdot x} a(\mathbf{k}) a(\mathbf{k}') - e^{-i(k-k')\cdot x} a(\mathbf{k}) a^\dagger(\mathbf{k}') \right.$$

$$\left. - e^{i(k-k')\cdot x} a^\dagger(\mathbf{k}) a(\mathbf{k}') + e^{i(k+k')\cdot x} a^\dagger(\mathbf{k}) a^\dagger(\mathbf{k}') \right)$$

$$= -\int \frac{d^3k \, d^3k'}{2} \sqrt{k^0 k'^0} \left(e^{-i(k^0+k'^0)x^0} \delta^3(k+k') \, a(\mathbf{k}) a(\mathbf{k}') \right.$$

$$- e^{-i(k^0-k'^0)x^0} \delta^3(k-k') \, a(\mathbf{k}) a^\dagger(\mathbf{k}')$$

$$- e^{i(k^0-k'^0)x^0} \delta^3(k-k') \, a^\dagger(\mathbf{k}) a(\mathbf{k}')$$

$$\left. + e^{i(k^0+k'^0)x^0} \delta^3(k+k') \, a^\dagger(\mathbf{k}) a^\dagger(\mathbf{k}') \right)$$

$$= -\frac{1}{2} \int d^3k \, k^0 \left(e^{-2ik^0 x^0} a(\mathbf{k}) a(-\mathbf{k}) - a(\mathbf{k}) a^\dagger(\mathbf{k}) \right.$$

$$\left. - a^\dagger(\mathbf{k}) a(\mathbf{k}) + e^{2ik^0 x^0} a^\dagger(\mathbf{k}) a^\dagger(-\mathbf{k}) \right). \tag{5.69}$$

Here, we have used the fact that, independent of whether $\mathbf{k}' = \pm \mathbf{k}$,

5.5 CREATION AND ANNIHILATION OPERATORS 181

$$k'^0 = \sqrt{(\mathbf{k}')^2 + m^2} = \sqrt{\mathbf{k}^2 + m^2} = k^0. \tag{5.70}$$

Similarly, we can calculate

$$\int d^3x \, \boldsymbol{\nabla}\phi(x) \cdot \boldsymbol{\nabla}\phi(x) = -\frac{1}{(2\pi)^3} \int d^3x \int \frac{d^3k}{\sqrt{2k^0}} \frac{d^3k'}{\sqrt{2k'^0}} \mathbf{k} \cdot \mathbf{k}'$$
$$\times \left(e^{-i(k+k')\cdot x} a(\mathbf{k})a(\mathbf{k}') - e^{-i(k-k')\cdot x} a(\mathbf{k})a^\dagger(\mathbf{k}') \right.$$
$$\left. - e^{i(k-k')\cdot x} a^\dagger(\mathbf{k})a(\mathbf{k}') + e^{i(k+k')\cdot x} a^\dagger(\mathbf{k})a^\dagger(\mathbf{k}') \right)$$
$$= -\int \frac{d^3k}{\sqrt{2k^0}} \frac{d^3k'}{\sqrt{2k'^0}} \mathbf{k} \cdot \mathbf{k}'$$
$$\times \left(e^{-i(k^0+k'^0)x^0} \delta^3\left(k+k'\right) a(\mathbf{k})a(\mathbf{k}') \right.$$
$$- e^{-i(k^0-k'^0)x^0} \delta^3\left(k-k'\right) a(\mathbf{k})a^\dagger(\mathbf{k}')$$
$$- e^{i(k^0-k'^0)x^0} \delta^3\left(k-k'\right) a^\dagger(\mathbf{k})a(\mathbf{k}')$$
$$\left. + e^{i(k^0+k'^0)x^0} \delta^3\left(k+k'\right) a^\dagger(\mathbf{k})a^\dagger(\mathbf{k}') \right)$$
$$= -\frac{1}{2}\int d^3k \frac{\mathbf{k}^2}{k^0} \left(-e^{-2ik^0x^0} a(\mathbf{k})a(-\mathbf{k}) - a(\mathbf{k})a^\dagger(\mathbf{k}) \right.$$
$$\left. - a^\dagger(\mathbf{k})a(\mathbf{k}) - e^{2ik^0x^0} a^\dagger(\mathbf{k})a^\dagger(-\mathbf{k}) \right), \tag{5.71}$$

$$\int d^3x \, \phi^2(x) = \frac{1}{(2\pi)^3} \int d^3x \int \frac{d^3k}{\sqrt{2k^0}} \frac{d^3k'}{\sqrt{2k'^0}}$$
$$\times \left(e^{-i(k+k')\cdot x} a(\mathbf{k})a(\mathbf{k}') + e^{-i(k-k')\cdot x} a(\mathbf{k})a^\dagger(\mathbf{k}') \right.$$
$$\left. + e^{i(k-k')\cdot x} a^\dagger(\mathbf{k})a(\mathbf{k}') + e^{i(k+k')\cdot x} a^\dagger(\mathbf{k})a^\dagger(\mathbf{k}') \right)$$
$$= \int \frac{d^3k}{\sqrt{2k^0}} \frac{d^3k'}{\sqrt{2k'^0}}$$
$$\times \left(e^{-i(k^0+k'^0)x^0} \delta^3\left(k+k'\right) a(\mathbf{k})a(\mathbf{k}') \right.$$
$$\left. + e^{-i(k^0-k'^0)x^0} \delta^3\left(k-k'\right) a(\mathbf{k})a^\dagger(\mathbf{k}') \right.$$

$$+ e^{i(k^0 - k'^0)x^0} \delta^3\left(k - k'\right) a^\dagger(\mathbf{k}) a(\mathbf{k}')$$
$$+ \left. e^{i(k^0 + k'^0)x^0} \delta^3\left(k + k'\right) a^\dagger(\mathbf{k}) a^\dagger(\mathbf{k}')\right)$$
$$= \frac{1}{2} \int d^3k \, \frac{1}{k^0} \left(e^{-2ik^0 x^0} a(\mathbf{k}) a(-\mathbf{k}) + a(\mathbf{k}) a^\dagger(\mathbf{k}) \right.$$
$$\left. + a^\dagger(\mathbf{k}) a(\mathbf{k}) + e^{2ik^0 x^0} a^\dagger(\mathbf{k}) a^\dagger(-\mathbf{k}) \right). \tag{5.72}$$

Substituting (5.69), (5.71) and (5.72) into the expression for the Hamiltonian (5.34), we obtain

$$H = \int d^3x \, \mathcal{H}$$
$$= \frac{1}{2} \int d^3x \left(\Pi^2(x) + \boldsymbol{\nabla}\phi(x) \cdot \boldsymbol{\nabla}\phi(x) + m^2 \phi^2(x) \right)$$
$$= -\frac{1}{4} \int d^3k \left[k^0 \left(e^{-2ik^0 x^0} a(\mathbf{k}) a(-\mathbf{k}) - a(\mathbf{k}) a^\dagger(\mathbf{k}) \right.\right.$$
$$\left. - a^\dagger(\mathbf{k}) a(\mathbf{k}) + e^{2ik^0 x^0} a^\dagger(\mathbf{k}) a^\dagger(-\mathbf{k}) \right)$$
$$+ \frac{\mathbf{k}^2}{k^0} \left(-e^{-2ik^0 x^0} a(\mathbf{k}) a(-\mathbf{k}) - a(\mathbf{k}) a^\dagger(\mathbf{k}) \right.$$
$$\left. - a^\dagger(\mathbf{k}) a(\mathbf{k}) - e^{2ik^0 x^0} a^\dagger(\mathbf{k}) a^\dagger(-\mathbf{k}) \right)$$
$$- \frac{m^2}{k^0} \left(e^{-2ik^0 x^0} a(\mathbf{k}) a(-\mathbf{k}) + a(\mathbf{k}) a^\dagger(\mathbf{k}) \right.$$
$$\left.\left. + a^\dagger(\mathbf{k}) a(\mathbf{k}) + e^{2ik^0 x^0} a^\dagger(\mathbf{k}) a^\dagger(-\mathbf{k}) \right) \right]$$
$$= -\frac{1}{4} \int \frac{d^3k}{k^0} \left[\left((k^0)^2 - \mathbf{k}^2 - m^2 \right) \right.$$
$$\times \left(e^{-2ik^0 x^0} a(\mathbf{k}) a(-\mathbf{k}) + e^{2ik^0 x^0} a^\dagger(\mathbf{k}) a^\dagger(-\mathbf{k}) \right)$$
$$\left. - \left((k^0)^2 + \mathbf{k}^2 + m^2 \right) \left(a(\mathbf{k}) a^\dagger(\mathbf{k}) + a^\dagger(\mathbf{k}) a(\mathbf{k}) \right) \right]$$
$$= \frac{1}{2} \int d^3k \, k^0 \left(a(\mathbf{k}) a^\dagger(\mathbf{k}) + a^\dagger(\mathbf{k}) a(\mathbf{k}) \right)$$

5.5 CREATION AND ANNIHILATION OPERATORS

$$= \int d^3k \frac{E_k}{2} \left(a(\mathbf{k}) a^\dagger(\mathbf{k}) + a^\dagger(\mathbf{k}) a(\mathbf{k}) \right), \qquad (5.73)$$

where we have used the relation (5.54)

$$k^0 = \sqrt{\mathbf{k}^2 + m^2} = E_k. \qquad (5.74)$$

Thus, we see from (5.73) that the Hamiltonian for the free Klein-Gordon field theory is indeed time independent and is the sum of the Hamiltonians for an infinite number of harmonic oscillators of frequency labelled by E_k. It follows now that

$$[a(\mathbf{k}), H] = \left[a(\mathbf{k}), \int d^3k' \frac{E_{k'}}{2} \left(a(\mathbf{k}') a^\dagger(\mathbf{k}') + a^\dagger(\mathbf{k}') a(\mathbf{k}') \right) \right]$$

$$= \int d^3k' \frac{E_{k'}}{2} \left(a(\mathbf{k}') \left[a(\mathbf{k}), a^\dagger(\mathbf{k}') \right] + \left[a(\mathbf{k}), a^\dagger(\mathbf{k}') \right] a(\mathbf{k}') \right)$$

$$= \int d^3k' \frac{E_{k'}}{2} \left(a(\mathbf{k}') \delta^3 \left(k - k' \right) + \delta^3 \left(k - k' \right) a(\mathbf{k}') \right)$$

$$= E_k \, a(\mathbf{k}),$$

$$\left[a^\dagger(\mathbf{k}), H \right] = \left[a^\dagger(\mathbf{k}), \int d^3k' \frac{E_{k'}}{2} \left(a(\mathbf{k}') a^\dagger(\mathbf{k}') + a^\dagger(\mathbf{k}') a(\mathbf{k}') \right) \right]$$

$$= \int d^3k' \frac{E_{k'}}{2} \left(\left[a^\dagger(\mathbf{k}), a(\mathbf{k}') \right] a^\dagger(\mathbf{k}') + a^\dagger(\mathbf{k}') \left[a^\dagger(\mathbf{k}), a(\mathbf{k}') \right] \right)$$

$$= \int d^3k' \frac{E_{k'}}{2} \left(-\delta^3 \left(k - k' \right) a^\dagger(\mathbf{k}') - a^\dagger(\mathbf{k}') \delta^3 \left(k - k' \right) \right)$$

$$= -E_k \, a^\dagger(\mathbf{k}). \qquad (5.75)$$

The two relations in (5.75) show that the operators $a(\mathbf{k})$ and $a^\dagger(\mathbf{k})$ annihilate and create respectively a quantum of energy E_k. Thus, we can indeed think of them as the annihilation and creation operators of a harmonic oscillator system. From the form of the Hamiltonian for the Klein-Gordon field in (5.73), we then conclude that we can think of this as the Hamiltonian for an infinite collection of decoupled harmonic oscillators. Clearly, the zero point energy of such a system would be infinite. But since this is an additive constant we can always rescale (shift) our energy to measure from zero which we take to be the ground state energy. This is conveniently done by normal ordering of the operators. Let us recall that in the passage from classical to a quantum theory, the ordering of the operators is ambiguous. Taking advantage of this, we define normal ordering as the ordering in which the creation operators stand to the left of the annihilation operators. Thus, (for bosons)

$$\begin{aligned}
\left(a(\mathbf{k})a^\dagger(\mathbf{k}')\right)^{\text{N.O.}} &= \; :a(\mathbf{k})a^\dagger(\mathbf{k}'): = a^\dagger(\mathbf{k}')a(\mathbf{k}), \\
\left(a^\dagger(\mathbf{k})a(\mathbf{k}')\right)^{\text{N.O.}} &= \; :a^\dagger(\mathbf{k})a(\mathbf{k}'): = a^\dagger(\mathbf{k})a(\mathbf{k}'), \\
\left(a(\mathbf{k})a(\mathbf{k}')\right)^{\text{N.O.}} &= \; :a(\mathbf{k})a(\mathbf{k}'): = a(\mathbf{k})a(\mathbf{k}') = a(\mathbf{k}')a(\mathbf{k}), \\
\left(a^\dagger(\mathbf{k})a^\dagger(\mathbf{k}')\right)^{\text{N.O.}} &= \; :a^\dagger(\mathbf{k})a^\dagger(\mathbf{k}'): \\
&= a^\dagger(\mathbf{k})a^\dagger(\mathbf{k}') = a^\dagger(\mathbf{k}')a^\dagger(\mathbf{k}). \quad (5.76)
\end{aligned}$$

If we normal order the Hamiltonian using (5.76), then, we have from (5.73) (namely, we assume that the quantum theory is defined by the normal ordered Hamiltonian)

$$\begin{aligned}
H^{\text{N.O.}} &= \int d^3k \, \frac{E_k}{2} \; :\left(a(\mathbf{k})a^\dagger(\mathbf{k}) + a^\dagger(\mathbf{k})a(\mathbf{k})\right): \\
&= \int d^3k \, \frac{E_k}{2} \left(a^\dagger(\mathbf{k})a(\mathbf{k}) + a^\dagger(\mathbf{k})a(\mathbf{k})\right) \\
&= \int d^3k \, E_k \, a^\dagger(\mathbf{k})a(\mathbf{k}) = \int d^3k \, E_k \, N(\mathbf{k}), \quad (5.77)
\end{aligned}$$

where we have defined the number operator for an oscillator with momentum $\mathbf{k}$ as in the case of a simple harmonic oscillator as

$$N(\mathbf{k}) = a^\dagger(\mathbf{k})a(\mathbf{k}), \quad (5.78)$$

and we can define the total number operator for all the oscillators (infinite number of them) in the system as

$$N = \int d^3k\, N(\mathbf{k}) = \int d^3k\, a^\dagger(\mathbf{k})a(\mathbf{k}). \tag{5.79}$$

It is worth noting from the derivation in (5.77) how the normal ordering has redefined away an infinite zero point energy (which basically would arise from commuting $a(\mathbf{k})a^\dagger(\mathbf{k})$ to bring it to the normal ordered form). It now follows from the definition of the number operators in (5.78) and (5.79) that

$$\begin{aligned}
\left[a(\mathbf{k}), N(\mathbf{k}')\right] &= \left[a(\mathbf{k}), a^\dagger(\mathbf{k}')a(\mathbf{k}')\right] \\
&= \left[a(\mathbf{k}), a^\dagger(\mathbf{k}')\right] a(\mathbf{k}') \\
&= a(\mathbf{k}')\delta^3(k-k'), \\
\left[a^\dagger(\mathbf{k}), N(\mathbf{k}')\right] &= \left[a^\dagger(\mathbf{k}), a^\dagger(\mathbf{k}')a(\mathbf{k}')\right] \\
&= a^\dagger(\mathbf{k}')\left[a^\dagger(\mathbf{k}), a(\mathbf{k}')\right] \\
&= -a^\dagger(\mathbf{k}')\delta^3(k-k'). \end{aligned} \tag{5.80}$$

Equation (5.80) then leads to

$$\begin{aligned}
\left[a(\mathbf{k}), N\right] &= \left[a(\mathbf{k}), \int d^3k'\, N(\mathbf{k}')\right] \\
&= \int d^3k'\, a(\mathbf{k}')\delta^3(k-k') = a(\mathbf{k}), \\
\left[a^\dagger(\mathbf{k}), N\right] &= \left[a^\dagger(\mathbf{k}), \int d^3k'\, N(\mathbf{k}')\right] \\
&= \int d^3k'\, \left(-a^\dagger(\mathbf{k}')\delta^3(k-k')\right) = -a^\dagger(\mathbf{k}), \end{aligned} \tag{5.81}$$

which reflects again the fact that $a(\mathbf{k})$ and $a^\dagger(\mathbf{k})$ merely lower or raise the number of quanta of momentum $\mathbf{k}$ by one unit.

5.6 Energy eigenstates

From now on, let us consider the normal ordered Hamiltonian (5.77) (as describing the quantum free Klein-Gordon theory) of the form

$$H = \int d^3k\, E_k\, a^\dagger(\mathbf{k})a(\mathbf{k}). \tag{5.82}$$

Let us consider an energy eigenstate $|E\rangle$ of this Hamiltonian satisfying

$$H|E\rangle = E|E\rangle, \tag{5.83}$$

where we are assuming that the state $|E\rangle$ is normalized and that the eigenvalue E is discrete for simplicity, which can be achieved by quantizing the system in a box. It is clear from (5.77) and (5.83) that

$$\begin{aligned}
E &= \langle E|H|E\rangle \\
&= \langle E|\int d^3k\, E_k a^\dagger(\mathbf{k})a(\mathbf{k})|E\rangle \\
&= \int d^3k\, E_k \langle E|a^\dagger(\mathbf{k})a(\mathbf{k})|E\rangle \\
&\geq 0.
\end{aligned} \tag{5.84}$$

This is because both E_k as well as the inner product of states is positive definite (namely, the inner product represents the norm of the state $a(\mathbf{k})|E\rangle$) and, consequently, the integrand is positive semidefinite. This shows that the eigenvalues of the Hamiltonian have to be positive semidefinite and in the second quantized field theory, we do not have the problem of negative energy states that we had in the single particle theory.

Let us next recall that the operators $a(\mathbf{k})$ and $a^\dagger(\mathbf{k})$ satisfy the commutation relations (see (5.75))

$$\begin{aligned}
[a(\mathbf{k}), H] &= E_k a(\mathbf{k}), \\
\left[a^\dagger(\mathbf{k}), H\right] &= -E_k a^\dagger(\mathbf{k}).
\end{aligned} \tag{5.85}$$

5.6 ENERGY EIGENSTATES

As a result, we have

$$[a(\mathbf{k}), H]|E\rangle = E_k a(\mathbf{k})|E\rangle,$$
or, $\quad a(\mathbf{k})H|E\rangle - Ha(\mathbf{k})|E\rangle = E_k a(\mathbf{k})|E\rangle,$
or, $\quad H\{a(\mathbf{k})|E\rangle\} = (E - E_k)\{a(\mathbf{k})|E\rangle\}. \qquad (5.86)$

Namely, if $|E\rangle$ represents an energy eigenstate with eigenvalue E, then $\{a(\mathbf{k})|E\rangle\}$ is also an eigenstate of energy with the lower energy value $E - E_k$. The annihilation operator $a(\mathbf{k})$ lowers the energy eigenvalue as it should. Similarly, we note that

$$\left[a^\dagger(\mathbf{k}), H\right]|E\rangle = -E_k a^\dagger(\mathbf{k})|E\rangle,$$
or, $\quad a^\dagger(\mathbf{k})H|E\rangle - Ha^\dagger(\mathbf{k})|E\rangle = -E_k a^\dagger(\mathbf{k})|E\rangle,$
or, $\quad H\left\{a^\dagger(\mathbf{k})|E\rangle\right\} = (E + E_k)\left\{a^\dagger(\mathbf{k})|E\rangle\right\}. \qquad (5.87)$

In other words, if $|E\rangle$ is an energy eigenstate with energy eigenvalue E, then $a^\dagger(\mathbf{k})$ acting on it gives another energy eigenstate with the higher value of energy $E+E_k$. The creation operator $a^\dagger(\mathbf{k})$, therefore, raises the energy value.

Let us next note that since $a(\mathbf{k})$ acting on an energy eigenstate lowers the energy, it would appear that by applying the annihilation operator successively, we can lower the energy eigenvalue arbitrarily. But as we have seen in (5.84), the energy eigenvalue is bounded from below ($E \geq 0$). Therefore, there must exist a state (energy eigenstate) with a minimum energy, $|E_{\min}\rangle$, such that

$$a(\mathbf{k})|E_{\min}\rangle = 0, \qquad (5.88)$$

and we cannot lower the energy any further. In such a state, clearly,

$$E_{\min} = \int d^3k \; E_k \langle E_{\min}|a^\dagger(\mathbf{k})a(\mathbf{k})|E_{\min}\rangle = 0. \qquad (5.89)$$

This is, therefore, the ground state of the system or the vacuum state (recall that by normal ordering the Hamiltonian, we had redefined the energy of the ground state to be zero). This ground state or the vacuum state is denoted by $|0\rangle$ and satisfies (we assume that the state is normalized)

$$\begin{aligned}
|E_{\min}\rangle &\equiv |0\rangle, \\
a(\mathbf{k})|0\rangle &= 0 = \langle 0|a^\dagger(\mathbf{k}), \\
\langle 0|0\rangle &= 1, \\
N|0\rangle &= 0, \\
H|0\rangle &= 0.
\end{aligned} \quad (5.90)$$

In other words, the vacuum state contains no quantum of energy or no particle if energy is related to that of particles. (Incidentally, it is clear now that the expectation value of any normal ordered operator in the vacuum state, conventionally called the vacuum expectation value or vev, would vanish.)

Given the vacuum state, one can build up states of higher energy by simply applying the creation operator. A general energy eigenstate with higher energy will have the form (up to normalizations)

$$\left(a^\dagger(\mathbf{k}_1)\right)^{n_1} \left(a^\dagger(\mathbf{k}_2)\right)^{n_2} \cdots \left(a^\dagger(\mathbf{k}_\ell)\right)^{n_\ell} |0\rangle. \quad (5.91)$$

From our study of the harmonic oscillator we know that such states will be eigenstates of the number operator and we can denote such a state (normalized) as

$$\begin{aligned}
&|n_1, k_1; n_2, k_2; \cdots ; n_\ell, k_\ell\rangle \\
&= \frac{\left(a^\dagger(\mathbf{k}_1)\right)^{n_1}}{\sqrt{n_1!}} \frac{\left(a^\dagger(\mathbf{k}_2)\right)^{n_2}}{\sqrt{n_2!}} \cdots \frac{\left(a^\dagger(\mathbf{k}_\ell)\right)^{n_\ell}}{\sqrt{n_\ell!}} |0\rangle.
\end{aligned} \quad (5.92)$$

This will be an eigenstate of the total number operator satisfying

$$\begin{aligned}
&N|n_1, k_1; n_2, k_2; \cdots ; n_\ell, k_\ell\rangle \\
&= \int d^3k\, N(\mathbf{k})|n_1, k_1; n_2, k_2; \cdots ; n_\ell, k_\ell\rangle \\
&= (n_1 + n_2 + \cdots + n_\ell)\, |n_1, k_1; n_2, k_2; \cdots ; n_\ell, k_\ell\rangle.
\end{aligned} \quad (5.93)$$

5.6 ENERGY EIGENSTATES

This relation can be easily shown using the identity (see also (5.80))

$$\left[N(\mathbf{k}), (a^\dagger(\mathbf{k}_1))^{n_1}\right] = a^\dagger(\mathbf{k}) \left[a(\mathbf{k}), \left(a^\dagger(\mathbf{k}_1)\right)^{n_1}\right]$$
$$= n_1 \left(a^\dagger(\mathbf{k}_1)\right)^{n_1} \delta^3(k - k_1), \quad (5.94)$$

which also leads to

$$N(\mathbf{k})|n_1, k_1; \cdots; n_\ell, k_\ell\rangle \quad (5.95)$$
$$= \left(n_1 \delta^3(k - k_1) + \cdots + n_\ell \delta^3(k - k_\ell)\right)|n_1, k_1; \cdots; n_\ell, k_\ell\rangle.$$

Namely, the state $|n_1, k_1; \cdots; n_\ell, k_\ell\rangle$ contains n_1 quanta with four momentum k_1^μ, n_2 quanta with four momentum k_2^μ and so on.

Let us further note that since

$$H = \int d^3k \, E_k \, a^\dagger(\mathbf{k}) a(\mathbf{k}) = \int d^3k \, E_k \, N(\mathbf{k}), \quad (5.96)$$

if we define the momentum operator as (This is, of course, the natural definition from considerations of covariance. However, the momentum operator can also be derived from Nöether's theorem and coincides exactly with this in the normal ordered form as we will see later.)

$$\mathbf{P} = \int d^3k \, \mathbf{k} \, a^\dagger(\mathbf{k}) a(\mathbf{k}) = \int d^3k \, \mathbf{k} \, N(\mathbf{k}), \quad (5.97)$$

then

$$H|n_1, k_1; n_2, k_2; \cdots; n_\ell, k_\ell\rangle$$
$$= \int d^3k \, E_k \, N(\mathbf{k})|n_1, k_1; n_2, k_2; \cdots; n_\ell, k_\ell\rangle \quad (5.98)$$
$$= (n_1 E_{k_1} + n_2 E_{k_2} + \cdots + n_\ell E_{k_\ell})|n_1, k_1; n_2, k_2; \cdots; n_\ell, k_\ell\rangle,$$

and

$$\begin{aligned}
\mathbf{P}|n_1,k_1;n_2,k_2;\cdots;n_\ell,k_\ell\rangle &\\
= \int d^3k\, \mathbf{k}\, N(\mathbf{k})&|n_1,k_1;n_2,k_2;\cdots;n_\ell,k_\ell\rangle \quad (5.99)\\
= (n_1\mathbf{k}_1 + n_2\mathbf{k}_2 + \cdots &+ n_\ell\mathbf{k}_\ell)|n_1,k_1;n_2,k_2;\cdots;n_\ell,k_\ell\rangle.
\end{aligned}$$

Namely, these states are eigenstates of H and $\mathbf{P}$ with total energy and momentum given by

$$\begin{aligned}
E &= n_1 E_{k_1} + n_2 E_{k_2} + \cdots + n_\ell E_{k_\ell},\\
\mathbf{p} &= n_1 \mathbf{k}_1 + n_2 \mathbf{k}_2 + \cdots + n_\ell \mathbf{k}_\ell.
\end{aligned} \quad (5.100)$$

We can show in a straightforward manner (using the fundamental commutation relations in (5.66) as well as the properties (5.90)) that these states are orthonormal and define a complete basis of states for the Hilbert space of the Klein-Gordon theory.

5.7 Physical meaning of energy eigenstates

To obtain the physical meaning of these energy eigenstates, let us analyze the state obtained from the vacuum by applying a single creation operator,

$$|k\rangle = a^\dagger(\mathbf{k})|0\rangle. \quad (5.101)$$

This state satisfies

$$\begin{aligned}
N|k\rangle &= |k\rangle,\\
H|k\rangle &= E_k|k\rangle,\\
\mathbf{P}|k\rangle &= \mathbf{k}|k\rangle.
\end{aligned} \quad (5.102)$$

It follows from (5.102) that

$$(H^2 - \mathbf{P}^2)|k\rangle = (E_k^2 - \mathbf{k}^2)|k\rangle = m^2|k\rangle. \quad (5.103)$$

5.7 Physical meaning of energy eigenstates

This state, therefore, satisfies the single particle Klein-Gordon equation with positive energy (it provides a representation for the Poincaré group) and, therefore, we can think of this as the one particle state with four momentum $k^\mu = (E_k, \mathbf{k})$. Similarly, we can show that the state

$$|n_1, k_1; n_2, k_2; \cdots ; n_\ell, k_\ell\rangle, \tag{5.104}$$

can be thought of as the state with n_1 particles with $k_1^\mu = (E_{k_1}, \mathbf{k}_1)$, n_2 particles with $k_2^\mu = (E_{k_2}, \mathbf{k}_2)$ and so on. Note that the particles described by such states are necessarily identical (although with different energy and momentum) and, as a result, field theories naturally describe systems of many identical particles.

Let us next consider the state that is produced by the field operator acting on the vacuum,

$$|\phi(x)\rangle = \phi(x)|0\rangle = \phi^{(-)}(x)|0\rangle. \tag{5.105}$$

Since $\phi(x)$ is linear in the creation and the annihilation operators, it is clear that

$$\langle 0|\phi(x)|0\rangle = 0. \tag{5.106}$$

In fact, the only nonzero matrix element of the field operator involving the vacuum state is given by

$$\langle k|\phi(x)|0\rangle = \langle k|\phi(x)\rangle. \tag{5.107}$$

This is, of course, an ordinary function representing the projection of the state $|\phi(x)\rangle$ on to the one particle state $|k\rangle$. This function satisfies

$$\left(\partial_\mu \partial^\mu + m^2\right) \langle k|\phi(x)\rangle = \langle k| \left(\partial_\mu \partial^\mu + m^2\right) \phi(x)|0\rangle = 0, \tag{5.108}$$

where we have used (5.105) as well as the fact that only the field operator $\phi(x)$ depends on the space-time coordinates on which the

Klein-Gordon operator $(\Box + m^2)$ can act. Therefore, $\langle k|\phi(x)\rangle$ defines a solution of the classical Klein-Gordon equation and we can relate this function to the single particle Klein-Gordon wave function with positive energy. In the present case, we note explicitly that

$$\begin{aligned}
\langle \phi(x)|k\rangle &= \langle k|\phi(x)\rangle^* = \langle 0|a(\mathbf{k})\phi^{(-)}(x)|0\rangle^* \\
&= \left\langle 0 \left| a(\mathbf{k}) \int \frac{\mathrm{d}^3 k'}{\sqrt{(2\pi)^3 2k'^0}} a^\dagger(\mathbf{k}') e^{ik' \cdot x} \right| 0 \right\rangle^* \\
&= \left[\int \frac{\mathrm{d}^3 k'}{\sqrt{(2\pi)^3 2k'^0}} e^{ik' \cdot x} \langle 0|a(\mathbf{k}) a^\dagger(\mathbf{k}')|0\rangle \right]^* \\
&= \int \frac{\mathrm{d}^3 k'}{\sqrt{(2\pi)^3 2k'^0}} \\
&\quad \times \left[e^{ik' \cdot x} \left\langle 0 \left| \left[a(\mathbf{k}), a^\dagger(\mathbf{k}')\right] + a^\dagger(\mathbf{k}')a(\mathbf{k}) \right| 0 \right\rangle \right]^* \\
&= \int \frac{\mathrm{d}^3 k'}{\sqrt{(2\pi)^3 2k'^0}} e^{-ik' \cdot x} \delta^3(\mathbf{k} - \mathbf{k}') \\
&= \frac{1}{\sqrt{(2\pi)^3 2k^0}} e^{-ik \cdot x}, \quad (5.109)
\end{aligned}$$

with $k^0 = E_k = \sqrt{\mathbf{k}^2 + m^2} > 0$. We recognize the function in (5.109) to be the positive energy plane wave solutions of the single particle Klein-Gordon equation. This brings out the connection between the second quantized theory and the first quantized theory.

Let us note here parenthetically that the wave function for a single particle is really identified with $\langle \phi(x)|k\rangle$. This can be understood by noting that $|\phi(x)\rangle$ is like the coordinate basis state $|x\rangle$ (which is the eigenstate of the operator representing the dynamical variable) of the first quantized description. If $|\psi\rangle$ denotes a state of the system, then, in the first quantized description, the wave function is given by

$$\psi(x) = \langle x|\psi\rangle. \quad (5.110)$$

In the second quantized description, $|k\rangle$ represents an energy eigenstate of the system describing a single particle (namely, $|\psi\rangle = |k\rangle$).

5.7 Physical meaning of energy eigenstates

Thus, the wave functional for a single particle with a definite energy-momentum has the form

$$\psi(\phi(x)) = \langle \phi(x)|\psi\rangle = \langle \phi(x)|k\rangle, \tag{5.111}$$

which is what we have seen explicitly in (5.109).

The state

$$|\phi(x)\rangle = \phi^{(-)}(x)|0\rangle, \tag{5.112}$$

can be thought of as the one particle state in the configuration space – describing the quantum mechanical state of the single particle at the coordinate x. As we have seen, such a state contains a superposition of all possible momentum states (which follows from the field decomposition in (5.56)). We can similarly construct multi-particle states in the configuration space of the form

$$|\phi(x_1), \phi(x_2), \cdots, \phi(x_n)\rangle = \phi^{(-)}(x_1)\phi^{(-)}(x_2)\cdots\phi^{(-)}(x_n)|0\rangle, \tag{5.113}$$

which will describe a state with n particles at coordinates $x_1, \cdots, x_n$. Such states are physically meaningful only if the time coordinates are equal (namely, $x_1^0 = x_2^0 = \cdots = x_n^0$). Furthermore, such configuration states are automatically symmetric as we would expect for a system of identical Bose particles. They can also be shown to define a complete basis and describe what is known as the Fock space for the system. These states naturally lead to quantum mechanical probabilities which are non-negative. Thus, for example, the absolute square of the amplitude (with $x_1^0 = x_2^0$)

$$\langle k, k'|\phi(x_1), \phi(x_2)\rangle = \langle k, k'|\phi^{(-)}(x_1)\phi^{(-)}(x_2)|0\rangle, \tag{5.114}$$

would give the probability for finding two Klein-Gordon particles with four momenta $k^\mu = (E_k, \mathbf{k})$ and $k'^\mu = (E_{k'}, \mathbf{k}')$ at the coordinates $\mathbf{x}_1$ and $\mathbf{x}_2$ at a given time. Such probabilities are, by construction, positive semi-definite and there is no problem of negative probabilities in the second quantized description. Thus, the second quantized noninteracting Klein-Gordon theory is free from the problems of the first quantized theory that we had discussed earlier, namely, all the physical states have positive semidefinite energy and that the probabilities are non-negative as they should be.

5.8 Green's functions

Even though we have been examining the free Klein-Gordon field theory so far, eventually we would like to study interactions of the system. When there are interactions present, the Klein-Gordon equation will modify. For example, for the simple case of the Klein-Gordon field interacting with an external source $J(x)$ (which is a c-number function or a classical function), the equation will take the form

$$\left(\partial_\mu \partial^\mu + m^2\right)\phi(x) = J(x), \tag{5.115}$$

which can be obtained from the Lagrangian density

$$\mathcal{L} = \frac{1}{2}\partial_\mu\phi\partial^\mu\phi - \frac{m^2}{2}\phi^2 + J\phi, \tag{5.116}$$

as the Euler-Lagrange equation (5.14). In a realistic theory the field can have self-interactions or can interact with other dynamical fields of the system, which we will study in the following chapters. Even classically we know that we can solve an inhomogeneous equation of the kind (5.115) provided we know the Green's function of the system. The Green's function for a given inhomogeneous equation is defined as the solution of the equation with a delta source, namely, for the Klein-Gordon equation, the Green's function $G(x-y)$ satisfies

$$\left(\partial_\mu \partial^\mu + m^2\right) G(x-y) = -\delta^4(x-y). \tag{5.117}$$

Here, the translation invariance of the Green's function arises because the right-hand side of the equation is invariant under translations. Furthermore, this equation is covariant (in fact, it is invariant under a Lorentz transformation) leading to the fact that the Green's function is Poincaré invariant. If we know the Green's function in (5.117),

5.8 GREEN'S FUNCTIONS

then the particular solution of the inhomogeneous equation (5.115) can be written as

$$\phi(x) = -\int d^4y\, G(x-y)J(y), \tag{5.118}$$

so that

$$\begin{aligned}\left(\partial_\mu\partial^\mu + m^2\right)\phi(x) &= -\int d^4y\left(\partial_{x\mu}\partial_x^\mu + m^2\right)G(x-y)J(y) \\ &= -\int d^4y\left(-\delta^4(x-y)\right)J(y) \\ &= J(x).\end{aligned} \tag{5.119}$$

We note here that a homogeneous solution can always be added to the particular solution depending on the system under study (and sometimes in order to implement appropriate boundary conditions).

The Green's function is, therefore, an important concept in studying the solutions of a system when interactions are present and can be easily determined by going over to the momentum space. Thus, defining the Fourier transforms

$$\begin{aligned}\delta^4(x-y) &= \frac{1}{(2\pi)^4}\int d^4k\, e^{-ik\cdot(x-y)}, \\ G(x-y) &= \int \frac{d^4k}{(2\pi)^4} e^{-ik\cdot(x-y)} \widetilde{G}(k),\end{aligned} \tag{5.120}$$

and substituting them back into (5.117), we obtain

$$\left(\partial_\mu\partial^\mu + m^2\right)G(x-y) = -\delta^4(x-y),$$

or, $\quad\dfrac{1}{(2\pi)^4}\left(-k^2 + m^2\right)\widetilde{G}(k) = -\dfrac{1}{(2\pi)^4},$

or, $\quad \widetilde{G}(k) = \dfrac{1}{k^2 - m^2}.$ \hfill (5.121)

In other words, the Fourier transformation turns the partial differential equation into an algebraic equation which is trivial to solve.

Thus, substituting (5.121) back into (5.120), we determine the Green's function for the Klein-Gordon equation to be

$$G(x - y) = \int \frac{d^4k}{(2\pi)^4} e^{-ik \cdot (x-y)} \widetilde{G}(k)$$

$$= \int \frac{d^4k}{(2\pi)^4} \frac{e^{-ik \cdot (x-y)}}{k^2 - m^2}. \tag{5.122}$$

This explicitly shows that the Green's function is Poincaré invariant.

Let us note that the integrand of the Green's function in (5.122) has poles at

$$k^2 - m^2 = 0,$$
$$\text{or,} \quad (k^0)^2 = \mathbf{k}^2 + m^2,$$
$$\text{or,} \quad k^0 = \pm E_k, \tag{5.123}$$

which lie on the real axis in the complex k^0-plane as shown in Fig. 5.3. Therefore, the Green's function is not uniquely defined until we specify a contour of integration in the complex k^0-plane and specifying a contour simply corresponds to specifying a boundary condition for the Green's function.

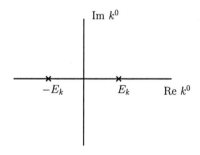

Figure 5.3: Poles of the integrand in the complex k^0-plane.

If we choose a contour of the form shown in Fig. 5.4, then, this is equivalent to moving the poles infinitesimally into the upper half plane. Mathematically, this is expressed by writing

5.8 GREEN'S FUNCTIONS

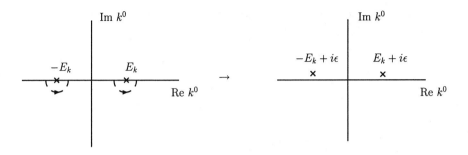

Figure 5.4: Choice of the contour in the complex k^0-plane for the advanced Green's function.

$$G^A(x-y) = \lim_{\eta \to 0^+} \int \frac{d^4k}{(2\pi)^4} \frac{e^{-ik\cdot(x-y)}}{k^2 - m^2 - ik^0\eta}. \tag{5.124}$$

In this case, the poles of the integrand occur at

$$k^2 - m^2 - ik^0\eta = (k^0)^2 - \mathbf{k}^2 - m^2 - ik^0\eta = 0,$$

or, $\quad \left(k^0 - \dfrac{i\eta}{2}\right)^2 \simeq \mathbf{k}^2 + m^2,$

or, $\quad k^0 - i\epsilon = \pm E_k,$

or, $\quad k^0 = \pm E_k + i\epsilon, \tag{5.125}$

where we have identified $\epsilon = \frac{\eta}{2}$.

Since in this case the poles in (5.124) are in the upper half plane, if we enclose the contour in the lower half plane, the integral will vanish. We note that if $x^0 - y^0 > 0$, then we must close the contour in the lower half plane for the damping of the exponential in (5.124). Therefore, we conclude that

$$G^A(x-y) = 0, \qquad \text{for } x^0 - y^0 > 0. \tag{5.126}$$

For $x^0 - y^0 < 0$, on the other hand, we have to close the contour in the upper half plane (for the damping of the exponential) in which case we can evaluate the integral by the method of residues. For $x^0 - y^0 < 0$, then, we have

$$\begin{aligned}
G^A(x-y) &= \lim_{\epsilon \to 0^+} \int \frac{d^3k}{(2\pi)^4} \int dk^0 \, \frac{e^{-ik\cdot(x-y)}}{(k^0 - i\epsilon + E_k)(k^0 - i\epsilon - E_k)} \\
&= \lim_{\epsilon \to 0^+} 2\pi i \int \frac{d^3k}{(2\pi)^4} \left[\frac{e^{-i(E_k+i\epsilon)(x^0-y^0)+i\mathbf{k}\cdot(\mathbf{x}-\mathbf{y})}}{2E_k} \right. \\
&\qquad\qquad\qquad \left. + \frac{e^{-i(-E_k+i\epsilon)(x^0-y^0)+i\mathbf{k}\cdot(\mathbf{x}-\mathbf{y})}}{-2E_k} \right] \\
&= \frac{i}{2} \int \frac{d^3k}{(2\pi)^3} \frac{e^{i\mathbf{k}\cdot(\mathbf{x}-\mathbf{y})}}{E_k} \left(e^{-iE_k(x^0-y^0)} - e^{iE_k(x^0-y^0)} \right) \\
&= \int \frac{d^3k}{(2\pi)^3} \frac{e^{i\mathbf{k}\cdot(\mathbf{x}-\mathbf{y})}}{E_k} \sin E_k (x^0 - y^0). \qquad (5.127)
\end{aligned}$$

This is known as the advanced Green's function and has support only in the past light cone. Explicitly, we can write (we denote the argument of the Green's function by x for simplicity)

$$G^A(x) = \begin{cases} 0, & \text{for } x^0 > 0, \\ \int \frac{d^3k}{(2\pi)^3} \frac{e^{i\mathbf{k}\cdot\mathbf{x}}}{E_k} \sin E_k x^0, & \text{for } x^0 < 0, \end{cases}$$

or, $\quad G^A(x) = \theta(-x^0) \int \frac{d^3k}{(2\pi)^3} \frac{e^{i\mathbf{k}\cdot\mathbf{x}}}{E_k} \sin E_k x^0.$ \qquad (5.128)

Similarly, if we choose the contour in the complex k^0-plane as shown in Fig. 5.5 or equivalently if we push both the poles to the lower half plane, we can express this mathematically by defining

$$G^R(x) = \lim_{\eta \to 0^+} \int \frac{d^4k}{(2\pi)^4} \frac{e^{-ik\cdot x}}{k^2 - m^2 + ik^0\eta}. \qquad (5.129)$$

In this case, the poles of the integrand in (5.129) would appear at

5.8 GREEN'S FUNCTIONS

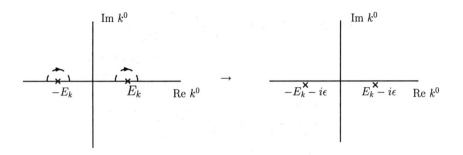

Figure 5.5: Choice of the contour in the complex k^0-plane for the retarded Green's function.

$$k^2 - m^2 + ik^0\eta = (k^0)^2 - \mathbf{k}^2 - m^2 + ik^0\eta = 0,$$

or, $\left(k^0 + \dfrac{i\eta}{2}\right)^2 \simeq \mathbf{k}^2 + m^2 = E_k^2,$

or, $k^0 + i\epsilon \simeq \pm E_k,$

or, $k^0 = \pm E_k - i\epsilon,$ \hfill (5.130)

where we have again identified $\epsilon = \frac{\eta}{2}$. In this case, since both the poles are in the lower half plane, if we close the contour in the upper half plane, then the integral will vanish. Note that if $x^0 < 0$, we have to close the contour in the upper half plane for the damping of the exponential in (5.129) and, therefore, we conclude that

$$G^R(x) = 0, \qquad x^0 < 0. \tag{5.131}$$

For $x^0 > 0$, on the other hand, we have to close the contour in the lower half plane (for the damping of the exponential) and the method of residues gives

$$G^R(x) = \lim_{\epsilon \to 0^+} \int \frac{d^3k}{(2\pi)^4} \int dk^0 \frac{e^{-ik\cdot x}}{(k^0 + i\epsilon + E_k)(k^0 + i\epsilon - E_k)}$$

$$= \lim_{\epsilon \to 0^+} (-2\pi i) \int \frac{d^3k}{(2\pi)^4} \left[\frac{e^{-i(E_k - i\epsilon)x^0 + i\mathbf{k}\cdot\mathbf{x}}}{2E_k} \right.$$

$$\left. + \frac{e^{-i(-E_k - i\epsilon)x^0 + i\mathbf{k}\cdot\mathbf{x}}}{-2E_k} \right]$$

$$= -\frac{i}{2} \int \frac{d^3k}{(2\pi)^3} \frac{e^{i\mathbf{k}\cdot\mathbf{x}}}{E_k} \left(e^{-iE_k x^0} - e^{iE_k x^0} \right)$$

$$= -\int \frac{d^3k}{(2\pi)^3} \frac{e^{i\mathbf{k}\cdot\mathbf{x}}}{E_k} \sin E_k x^0. \qquad (5.132)$$

The overall negative sign in (5.132) arises because the contour is clockwise in the lower half plane. Thus, we see that we can write

$$G^R(x) = \begin{cases} 0, & \text{for } x^0 < 0, \\ -\int \frac{d^3k}{(2\pi)^3} \frac{e^{i\mathbf{k}\cdot\mathbf{x}}}{E_k} \sin E_k x^0, & \text{for } x^0 > 0, \end{cases}$$

or, $\quad G^R(x) = -\theta(x^0) \int \frac{d^3k}{(2\pi)^3} \frac{e^{i\mathbf{k}\cdot\mathbf{x}}}{E_k} \sin E_k x^0. \qquad (5.133)$

Note that this Green's function has support only in the future light cone and is known as the retarded Green's function. (The advanced and the retarded Green's functions have similar form except for their support and the overall sign.)

It is clear from (5.128) and (5.133) that

$$G^R(x) = G^A(-x). \qquad (5.134)$$

Both the retarded and the advanced Green's functions satisfy the inhomogeneous equation (5.117) with a delta function source. Therefore, the average of the two, namely,

$$\frac{1}{2} \left(G^A(x) + G^R(x) \right), \qquad (5.135)$$

would also define a Green's function corresponding to the contour picking up the principal values at the poles. On the other hand, the difference between $G^A(x)$ and $G^R(x)$ would define a function which would satisfy the homogeneous (Klein-Gordon) equation and would not strictly correspond to a Green's function of the theory. Conventionally such a function is known as the Schwinger function and is defined to be

$$\begin{aligned} G(x) &= G^A(x) - G^R(x) \\ &= \theta\left(-x^0\right) \int \frac{d^3k}{(2\pi)^3} \frac{e^{i\mathbf{k}\cdot\mathbf{x}}}{E_k} \sin E_k x^0 \\ &\quad + \theta\left(x^0\right) \int \frac{d^3k}{(2\pi)^3} \frac{e^{i\mathbf{k}\cdot\mathbf{x}}}{E_k} \sin E_k x^0 \\ &= \left(\theta\left(x^0\right) + \theta\left(-x^0\right)\right) \int \frac{d^3k}{(2\pi)^3} \frac{e^{i\mathbf{k}\cdot\mathbf{x}}}{E_k} \sin E_k x^0 \\ &= \int \frac{d^3k}{(2\pi)^3} \frac{e^{i\mathbf{k}\cdot\mathbf{x}}}{E_k} \sin E_k x^0. \end{aligned} \qquad (5.136)$$

This clearly satisfies the homogeneous Klein-Gordon equation

$$\left(\partial_\mu \partial^\mu + m^2\right) G(x) = 0, \qquad (5.137)$$

and is antisymmetric, namely,

$$G(x) = -G(-x). \qquad (5.138)$$

Furthermore, it is easily seen from (5.136) that the Schwinger function is real

$$(G(x))^* = G(x). \qquad (5.139)$$

The retarded and the advanced Green's functions play an important role in many classical calculations as well as in many calculations in statistical mechanics. Quantum mechanically, however, there is another Green's function, known as the Feynman Green's function, which is more useful in the calculation of scattering matrix elements. This is defined with the contour in the complex k^0-plane as shown in Fig. 5.6, which corresponds to pushing one of the poles (on the left) to the upper half plane while moving the other (on the right)

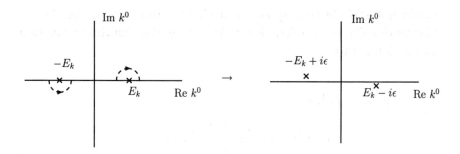

Figure 5.6: Choice of the contour in the complex k^0-plane for the Feynman Green's function.

to the lower half plane. Mathematically, this can be implemented by defining

$$G_F(x) = \lim_{\eta \to 0^+} \int \frac{d^4 k}{(2\pi)^4} \frac{e^{-ik \cdot x}}{k^2 - m^2 + i\eta}. \tag{5.140}$$

In this case, the poles of the integrand in (5.140) occur at

$$k^2 - m^2 + i\eta = 0,$$
$$\text{or,} \quad (k^0)^2 - E_k^2 + i\eta = 0,$$
$$\text{or,} \quad (k^0)^2 - \left(E_k - \frac{i\eta}{2E_k}\right)^2 \simeq 0,$$
$$\text{or,} \quad k^0 = \pm \left(E_k - \frac{i\eta}{2E_k}\right) = \pm (E_k - i\epsilon), \tag{5.141}$$

where we have identified $\epsilon = \frac{\eta}{2E_k}$ with $\eta \to 0^+$. Since there are poles in both halves of the complex k^0-plane, in this case, there will be a nontrivial contribution independent of whether we close the contour in the upper half or in the lower half plane. For $x^0 < 0$ we have to close the contour in the upper half plane for the damping of the exponential and this will pick up the residue of the pole at

$$k^0 = -E_k + i\epsilon. \tag{5.142}$$

5.8 GREEN'S FUNCTIONS

Therefore, for $x^0 < 0$, we have

$$
\begin{aligned}
G_F(x) &= \lim_{\epsilon \to 0^+} \int \frac{d^3k}{(2\pi)^4} \int dk^0 \, \frac{e^{-ik\cdot x}}{(k^0 - E_k + i\epsilon)(k^0 + E_k - i\epsilon)} \\
&= \lim_{\epsilon \to 0^+} 2\pi i \int \frac{d^3k}{(2\pi)^4} \frac{e^{-i(-E_k+i\epsilon)x^0+i\mathbf{k}\cdot\mathbf{x}}}{2(-E_k + i\epsilon)} \\
&= -\frac{i}{2} \int \frac{d^3k}{(2\pi)^3} \frac{e^{iE_k x^0 + i\mathbf{k}\cdot\mathbf{x}}}{E_k} \\
&= -\frac{i}{2} \int \frac{d^3k}{(2\pi)^3} \frac{e^{ik\cdot x}}{k^0} \\
&= G^{(-)}(x),
\end{aligned}
\qquad (5.143)
$$

where, we have used $\mathbf{k} \leftrightarrow -\mathbf{k}$ in the last step (and we understand that $k^0 = E_k$). On the other hand, if $x^0 > 0$, then we have to close the contour in the lower half plane which will pick up the residue of the pole at

$$k^0 = E_k - i\epsilon. \qquad (5.144)$$

Therefore, for $x^0 > 0$, we have

$$
\begin{aligned}
G_F(x) &= \lim_{\epsilon \to 0^+} \int \frac{d^3k}{(2\pi)^4} \int dk^0 \, \frac{e^{-ik\cdot x}}{(k^0 - E_k + i\epsilon)(k^0 + E_k - i\epsilon)} \\
&= \lim_{\epsilon \to 0^+} (-2\pi i) \int \frac{d^3k}{(2\pi)^4} \frac{e^{-i(E_k-i\epsilon)x^0+i\mathbf{k}\cdot\mathbf{x}}}{2(E_k - i\epsilon)} \\
&= -\frac{i}{2} \int \frac{d^3k}{(2\pi)^3} \frac{e^{-iE_k x^0 + i\mathbf{k}\cdot\mathbf{x}}}{E_k} \\
&= -\frac{i}{2} \int \frac{d^3k}{(2\pi)^3} \frac{e^{-ik\cdot x}}{k^0} \\
&= -G^{(+)}(x).
\end{aligned}
\qquad (5.145)
$$

Here the contour is clockwise leading to an overall negative sign. The Feynman Green's function clearly has support in both the future as well as the past light cones and can be written as

$$G_F(x) = -\theta\left(x^0\right) G^{(+)}(x) + \theta\left(-x^0\right) G^{(-)}(x). \tag{5.146}$$

Note from (5.143) and (5.145) that

$$G^{(+)}(-x) = \frac{i}{2} \int \frac{d^3 k}{(2\pi)^3} \frac{e^{ik \cdot x}}{k^0} = -G^{(-)}(x). \tag{5.147}$$

Therefore, the Feynman Green's function is an even function, namely,

$$G_F(-x) = G_F(x). \tag{5.148}$$

There is yet another choice of the contour and, therefore, another Green's function that is quite useful in studies of finite temperature field theory (as well as in the study of unitarity relations such as the cutting rules). It is defined as

$$\overline{G}(x) = \lim_{\eta \to 0^+} \int \frac{d^4 k}{(2\pi)^4} \frac{e^{-ik \cdot x}}{k^2 - m^2 - i\eta}, \tag{5.149}$$

for which the pole at $k^0 = -E_k$ is pushed down to the lower half plane while the pole at $k^0 = E_k$ is moved up to the upper half plane. However, we will not go into this in more detail here. We note from the structure of $G^{(-)}(x)$ in (5.143) and $G^{(+)}(x)$ in (5.145) that we can write the Schwinger function (5.136) as

$$G(x) = \int \frac{d^3 k}{(2\pi)^3} \frac{e^{i\mathbf{k}\cdot\mathbf{x}}}{E_k} \sin E_k x^0 = G^{(+)}(x) + G^{(-)}(x). \tag{5.150}$$

The functions $G^{(+)}(x)$ and $G^{(-)}(x)$ (see (5.143) and (5.145)) can be written in the manifestly covariant forms

$$\begin{aligned}
G^{(+)}(x) &= i \int \frac{d^4 k}{(2\pi)^3} \theta(k^0) \delta(k^2 - m^2) e^{-ik \cdot x}, \\
G^{(-)}(x) &= -i \int \frac{d^4 k}{(2\pi)^3} \theta(-k^0) \delta(k^2 - m^2) e^{-ik \cdot x},
\end{aligned} \tag{5.151}$$

5.9 COVARIANT COMMUTATION RELATIONS

and are correspondingly known as the positive and the negative energy (frequency) Green's functions. They satisfy (see (5.151))

$$\left(G^{(+)}(x)\right)^* = G^{(-)}(x), \tag{5.152}$$

and all the Green's functions can be expressed as linear combinations of these two fundamental Green's functions. We note from (5.151) that the Schwinger function has the manifestly covariant representation

$$\begin{aligned} G(x) &= G^{(+)}(x) + G^{(-)}(x) \\ &= i \int \frac{\mathrm{d}^4 k}{(2\pi)^3} \, \epsilon(k^0) \delta(k^2 - m^2) \, e^{-ik \cdot x}, \end{aligned} \tag{5.153}$$

where $\epsilon(x) = (\theta(x) - \theta(-x))$ is known as the sign function (or the alternating step function). For completeness, we note from (5.128), (5.133) and (5.150) that we can write

$$\begin{aligned} G^A(x) &= \theta(-x^0) G(x) = \theta(-x^0) \left(G^{(+)}(x) + G^{(-)}(x) \right), \\ G^R(x) &= -\theta(x^0) G(x) = -\theta(x^0) \left(G^{(+)}(x) + G^{(-)}(x) \right), \end{aligned} \tag{5.154}$$

which will be useful in our later discussions.

5.9 Covariant commutation relations

To make contact between the quantum field theory and the various Green's functions that we have constructed, let us calculate some of the covariant commutation relations satisfied by the field operators. Let us recall from (5.58) that

$$\phi(x) = \phi^{(+)}(x) + \phi^{(-)}(x), \tag{5.155}$$

where we note from (5.57) that (we recall that $k^0 = E_k = \sqrt{\mathbf{k}^2 + m^2} > 0$)

$$\phi^{(+)}(x) = \int \frac{d^3k}{\sqrt{(2\pi)^3 2k^0}} e^{-ik\cdot x} a(\mathbf{k}),$$

$$\phi^{(-)}(x) = \int \frac{d^3k}{\sqrt{(2\pi)^3 2k^0}} e^{ik\cdot x} a^\dagger(\mathbf{k}). \tag{5.156}$$

Since we know the commutation relations (5.66) between the creation and the annihilation operators,

$$\left[a(\mathbf{k}), a(\mathbf{k}')\right] = 0 = \left[a^\dagger(\mathbf{k}), a^\dagger(\mathbf{k}')\right],$$

$$\left[a(\mathbf{k}), a^\dagger(\mathbf{k}')\right] = \delta^3\left(k - k'\right), \tag{5.157}$$

it is straightforward to evaluate

$$\left[\phi^{(+)}(x), \phi^{(+)}(y)\right]$$

$$= \left[\int \frac{d^3k}{\sqrt{(2\pi)^3 2k^0}} e^{-ik\cdot x} a(\mathbf{k}), \int \frac{d^3k'}{\sqrt{(2\pi)^3 2k'^0}} e^{-ik'\cdot y} a(\mathbf{k}')\right]$$

$$= \frac{1}{(2\pi)^3} \int \frac{d^3k}{\sqrt{2k^0}} \frac{d^3k'}{\sqrt{2k'^0}} e^{-ik\cdot x} e^{-ik'\cdot y} \left[a(\mathbf{k}), a(\mathbf{k}')\right]$$

$$= 0,$$

$$\left[\phi^{(-)}(x), \phi^{(-)}(y)\right]$$

$$= \left[\int \frac{d^3k}{\sqrt{(2\pi)^3 2k^0}} e^{ik\cdot x} a^\dagger(\mathbf{k}), \int \frac{d^3k'}{\sqrt{(2\pi)^3 2k'^0}} e^{ik'\cdot y} a^\dagger(\mathbf{k}')\right]$$

$$= \frac{1}{(2\pi)^3} \int \frac{d^3k}{\sqrt{2k^0}} \frac{d^3k'}{\sqrt{2k'^0}} e^{ik\cdot x} e^{ik'\cdot y} \left[a^\dagger(\mathbf{k}), a^\dagger(\mathbf{k}')\right]$$

$$= 0,$$

$$\left[\phi^{(+)}(x), \phi^{(-)}(y)\right]$$

$$= \left[\int \frac{d^3k}{\sqrt{(2\pi)^3 2k^0}} e^{-ik\cdot x} a(\mathbf{k}), \int \frac{d^3k'}{\sqrt{(2\pi)^3 2k'^0}} e^{ik'\cdot y} a^\dagger(\mathbf{k}')\right]$$

$$= \frac{1}{(2\pi)^3} \int \frac{d^3k}{\sqrt{2k^0}} \frac{d^3k'}{\sqrt{2k'^0}} e^{-ik\cdot x + ik'\cdot y} \left[a(\mathbf{k}), a^\dagger(\mathbf{k}')\right]$$

$$= \frac{1}{(2\pi)^3} \int \frac{d^3k}{\sqrt{2k^0}} \frac{d^3k'}{\sqrt{2k'^0}} e^{-ik\cdot x + ik'\cdot y} \delta^3(k-k')$$

$$= \frac{1}{(2\pi)^3} \int \frac{d^3k}{2k^0} e^{-ik\cdot(x-y)}$$

$$= -iG^{(+)}(x-y). \tag{5.158}$$

It follows now that

$$\left[\phi^{(-)}(x), \phi^{(+)}(y)\right] = iG^{(+)}(y-x) = -iG^{(-)}(x-y). \tag{5.159}$$

Note that the field operators in the commutators (5.158) and (5.159) are not restricted to have equal time arguments any more (unlike in (5.40)) and, furthermore, using the results in (5.158) and (5.159), we can evaluate

$$[\phi(x), \phi(y)] = \left[\phi^{(+)}(x) + \phi^{(-)}(x), \phi^{(+)}(y) + \phi^{(-)}(y)\right]$$

$$= \left[\phi^{(+)}(x), \phi^{(-)}(y)\right] + \left[\phi^{(-)}(x), \phi^{(+)}(y)\right]$$

$$= -iG^{(+)}(x-y) - iG^{(-)}(x-y)$$

$$= -iG(x-y)$$

$$= -i \int \frac{d^3k}{(2\pi)^3} \frac{e^{i\mathbf{k}\cdot(\mathbf{x}-\mathbf{y})}}{E_k} \sin E_k \left(x^0 - y^0\right). \tag{5.160}$$

Namely, the commutator of two field operators, at unequal times, is proportional to the Schwinger Green's function. It is now clear that the Schwinger function satisfies the homogeneous equation

$$\left(\partial_{x\,\mu}\partial_x^\mu + m^2\right) G(x-y) = 0, \tag{5.161}$$

simply because the field operator $\phi(x)$ satisfies the Klein-Gordon equation,

$$\left(\partial_\mu \partial^\mu + m^2\right) \phi(x) = 0. \tag{5.162}$$

These commutation relations are known as covariant commutation relations simply because the (scalar) Green's functions are invariant under Lorentz transformations, namely,

$$G(\Lambda x) = G(x), \tag{5.163}$$

where $\Lambda^\mu_{\ \nu}$ represents a Lorentz transformation. Equation (5.160) also shows that two field operators $\phi(x)$ and $\phi(y)$ do not commute for arbitrary values of the coordinates. (The non-commutativity is a reflection of the Hamiltonian dynamics, as can be checked even in the simple classical Poisson bracket relations for a one dimensional free particle or a harmonic oscillator.) On the other hand, for $x^0 = y^0$

$$\begin{aligned} G(x-y)|_{x^0=y^0} &= \int \frac{\mathrm{d}^3 k}{(2\pi)^3} \frac{e^{i\mathbf{k}\cdot(\mathbf{x}-\mathbf{y})}}{E_k} \sin E_k \left(x^0 - y^0\right)\Big|_{x^0=y^0} \\ &= 0, \end{aligned} \tag{5.164}$$

which is expected from the anti-symmetry of the Schwinger function (5.138) and (5.160) leads to the familiar relation (see (5.40))

$$[\phi(x), \phi(y)]_{x^0=y^0} = -iG(x-y)\big|_{x^0=y^0} = 0. \tag{5.165}$$

We note that by a Lorentz transformation, this relation can also be seen to imply that

$$[\phi(x), \phi(y)] = 0, \quad \text{for} \quad (x-y)^2 < 0. \tag{5.166}$$

In other words, for space-like separations, the commutator of two field operators vanishes. This is consistent with our intuitive expectation. Namely, a light signal cannot connect two space-like points and, therefore, two measurements at space-like separations should not influence each other. This is known as the principle of microscopic causality. We also note that

$$\left.\frac{\partial G(x-y)}{\partial y^0}\right|_{x^0=y^0} = -\int \frac{d^3k}{(2\pi)^3} \frac{e^{i\mathbf{k}\cdot(\mathbf{x}-\mathbf{y})}}{E_k} E_k \cos E_k \left(x^0-y^0\right)\big|_{x^0=y^0}$$

$$= -\int \frac{d^3k}{(2\pi)^3} e^{i\mathbf{k}\cdot(\mathbf{x}-\mathbf{y})} = -\delta^3(\mathbf{x}-\mathbf{y}). \tag{5.167}$$

This is consistent with our earlier field quantization rule (5.40), namely,

$$[\phi(x),\Pi(y)]_{x^0=y^0} = \left[\phi(x),\dot{\phi}(y)\right]_{x^0=y^0}$$

$$= -i \left.\frac{\partial G(x-y)}{\partial y^0}\right|_{x^0=y^0} = i\delta^3(\mathbf{x}-\mathbf{y}). \tag{5.168}$$

To conclude this section, let us note from (5.160) as well as the relations in (5.136), (5.150) and (5.154) that the retarded and the advanced Green's function can also be expressed in terms of covariant commutation relations as

$$\theta(x^0-y^0)\left[\phi(x),\phi(y)\right]$$
$$= -i\theta(x^0-y^0)\left(G^{(+)}(x-y)+G^{(-)}(x-y)\right)$$
$$= -i\theta(x^0-y^0)G(x-y) = iG^R(x-y),$$
$$\theta(y^0-x^0)\left[\phi(x),\phi(y)\right]$$
$$= -i\theta(y^0-x^0)\left(G^{(+)}(x-y)+G^{(-)}(x-y)\right)$$
$$= -i\theta(y^0-x^0)G(x-y) = -iG^A(x-y), \tag{5.169}$$

which will be useful later.

5.10 References

1. J. D. Bjorken and S. Drell, *Relativistic Quantum Fields*, McGraw-Hill, New York, 1964.

2. S. Schweber, *Introduction to Relativistic Quantum Field Theory*, Row, Peterson, Evanston (1961).

3. P. Roman, *Introduction to Quantum Field Theory*, John Wliley, New York (1969).

4. C. Itzykson and J-B. Zuber, *Quantum Field Theory*, McGraw-Hill, New York, 1980.

5. A. Das, *Lectures on Quantum Mechanics*, Hindustan Publishing, New Delhi, India, 2003.

Chapter 6
Self-interacting scalar field theory

6.1 Nöther's theorem

In trying to extend the free Klein-Gordon field theory to include interactions, the first question that we face is how we can choose one interaction term over another. Obviously, we have to be guided by some symmetry principles and the question we have to answer is how we can incorporate the concept of symmetry into the field theoretic framework.

Let us suppose that we have a dynamical system described by the action

$$S = \int d^4x \, \mathcal{L}, \tag{6.1}$$

where, in the present case, we assume

$$\mathcal{L} = \mathcal{L}(\phi(x), \partial_\mu \phi(x)). \tag{6.2}$$

Under a general transformation of the form

$$\begin{aligned} x^\mu &\to x'^\mu, \\ \phi(x) &\to \phi'(x'), \\ \partial_\mu \phi(x) &\to \partial'_\mu \phi'(x'), \end{aligned} \tag{6.3}$$

we say that the dynamics described by the action (6.1) is invariant under the transformations (6.3) if the action does not change under these transformations. Namely, if

$$S = \int \mathrm{d}^4x\, \mathcal{L}\left(\phi(x), \partial_\mu \phi(x)\right) = \int \mathrm{d}^4x'\, \mathcal{L}(\phi'(x'), \partial'_\mu \phi'(x')), \quad (6.4)$$

then, the transformations in (6.3) define a symmetry of the system. It is clear that in such a case, the Euler-Lagrange equations for the primed and the unprimed systems would remain form invariant. It is worth emphasizing here that (6.3) includes a very interesting class of transformations where the space-time coordinates do not change, namely,

$$x'^\mu = x^\mu, \quad (6.5)$$

and only the dynamical variables of the theory transform. Such transformations are known as internal symmetry transformations to be contrasted with space-time transformations where space-time coordinates transform along with the dynamical variables, as indicated in (6.3). Our discussion of symmetries applies to both space-timetransformations where space-time coordinates are transformed as well as internal symmetry transformations where space-time coordinates are unaffected by the transformation.

Symmetries have interesting consequences for continuous transformations. Thus, let us consider an infinitesimal transformation with

$$\left|\frac{\partial x'}{\partial x}\right| = 1, \quad (6.6)$$

which holds for most global space-time symmetries as well as internal symmetries. In such a case, invariance of the action under the infinitesimal forms of the transformations in (6.3) would imply

$$\delta S = \int \mathrm{d}^4x'\, \mathcal{L}\left(\phi'(x'), \partial'_\mu \phi'(x')\right) - \int \mathrm{d}^4x\, \mathcal{L}\left(\phi(x), \partial_\mu \phi(x)\right) = 0,$$

or, $\quad \int \mathrm{d}^4x\, \mathcal{L}\left(\phi'(x), \partial_\mu \phi'(x)\right) - \int \mathrm{d}^4x\, \mathcal{L}\left(\phi(x), \partial_\mu \phi(x)\right) = 0,$

or, $\quad \int \mathrm{d}^4x\, \left(\mathcal{L}\left(\phi'(x), \partial_\mu \phi'(x)\right) - \mathcal{L}\left(\phi(x), \partial_\mu \phi(x)\right)\right) = 0, \quad (6.7)$

6.1 NÖTHER'S THEOREM

where we have identified the integration variable in the first term on the right-hand side to be x (instead of x') in the intermediate step. Clearly, (6.7) will hold if

$$\mathcal{L}\left(\phi'(x), \partial_\mu \phi'(x)\right) - \mathcal{L}\left(\phi(x), \partial_\mu \phi(x)\right) = \partial_\mu K^\mu, \tag{6.8}$$

which must hold independent of the use of equations of motion for the system under study. This is quite general and it is possible to have symmetry transformations for which $K^\mu = 0$. In fact, as we will see later, this is indeed the case for internal symmetry transformations.

On the other hand, defining the infinitesimal change in the field variable as (this is known as the Lie derivative of the field variable up to a sign)

$$\phi'(x) - \phi(x) = \delta\phi(x), \tag{6.9}$$

so that

$$\delta(\partial_\mu \phi(x)) = \partial_\mu \phi'(x) - \partial_\mu \phi(x) = \partial_\mu \delta\phi(x), \tag{6.10}$$

we can calculate explicitly (we remember that $\delta\phi(x)$ is infinitesimal and, therefore, keep only linear terms in $\delta\phi(x)$)

$$\begin{aligned}
\mathcal{L}&\left(\phi'(x), \partial_\mu \phi'(x)\right) - \mathcal{L}\left(\phi(x), \partial_\mu \phi(x)\right) \\
&= \mathcal{L}(\phi(x), \partial_\mu \phi(x)) + \delta\phi(x) \frac{\partial \mathcal{L}}{\partial \phi(x)} + \delta(\partial_\mu \phi(x)) \frac{\partial \mathcal{L}}{\partial \partial_\mu \phi(x)} \\
&\quad - \mathcal{L}(\phi(x), \partial_\mu \phi(x)) \\
&= \delta\phi(x) \frac{\partial \mathcal{L}}{\partial \phi(x)} + (\partial_\mu \delta\phi(x)) \frac{\partial \mathcal{L}}{\partial \partial_\mu \phi(x)} \\
&= \delta\phi(x) \partial_\mu \frac{\partial \mathcal{L}}{\partial \partial_\mu \phi(x)} + (\partial_\mu \delta\phi(x)) \frac{\partial \mathcal{L}}{\partial \partial_\mu \phi(x)} \\
&= \partial_\mu \left(\delta\phi(x) \frac{\partial \mathcal{L}}{\partial \partial_\mu \phi(x)}\right).
\end{aligned} \tag{6.11}$$

Here we have used the Euler-Lagrange equation in the intermediate steps. (Note that, for any infinitesimal variation, this is always true when the equations of motion are used. This is just the principle of least action. However, not all such variations will define a symmetry. Namely, in general, there is no restriction on $\delta\phi$ and, therefore, on ϕ'. For a symmetry, on the other hand, ϕ' must also satisfy the same equation of motion in the transformed frame as the original field.)

Comparing (6.8) and (6.11), we obtain

$$\partial_\mu \left(\delta\phi(x) \frac{\partial \mathcal{L}}{\partial \partial_\mu \phi(x)} \right) = \partial_\mu K^\mu$$

or, $\quad \partial_\mu \left(\delta\phi(x) \dfrac{\partial \mathcal{L}}{\partial \partial_\mu \phi(x)} - K^\mu \right) = 0.$ \hfill (6.12)

This shows that whenever there is a continuous symmetry associated with a system, we can define a current

$$J^\mu(x) = \delta\phi(x) \frac{\partial \mathcal{L}}{\partial \partial_\mu \phi(x)} - K^\mu, \tag{6.13}$$

which is conserved, namely,

$$\partial_\mu J^\mu(x) = 0. \tag{6.14}$$

Several comments are in order here. First, the conserved current independent of the parameter of transformation is not always a vector. Its tensor structure is determined by the tensor structure of the infinitesimal parameter of transformation. Second, given a conserved current, we can define a charge

$$Q = \int d^3x \, J^0(t, \mathbf{x}), \tag{6.15}$$

which will be a constant (in time) and will generate the symmetry transformations. In fact, we note that

$$\begin{aligned}
\frac{dQ}{dt} &= \int d^3x\, \partial_0 J^0 \\
&= \int d^3x\, \left(\partial_0 J^0 + \partial_i J^i\right) \\
&= \int d^3x\, \partial_\mu J^\mu = 0.
\end{aligned} \quad (6.16)$$

Here we have used the fact that, with the usual assumptions on the asymptotic fall off of the field operators (see (5.3)), the integral of a total divergence ($\nabla \cdot \mathbf{J}$) vanishes. In the operator language, the time independence of Q also means that it commutes with the Hamiltonian of the theory

$$[Q, H] = 0, \quad (6.17)$$

and, consequently, Q and H can have simultaneous eigenstates. In this case, Q corresponds to the generator of the infinitesimal symmetry transformations of the theory. Thus, if we have a continuous symmetry in the theory (transformations under which the action is invariant), there exists a conserved charge which is the generator of these infinitesimal symmetry transformations. The converse is also true. Namely, if there exists a conserved charge in a given theory, then it generates infinitesimal transformations (through commutation relations) which define a symmetry of the theory. This is known as Nöther's theorem and is quite important in the study of symmetries in dynamical systems.

6.1.1 Space-time translation. As an example of the consequences of Nöther's theorem, let us consider the simple case of an infinitesimal global space-time translation defined by

$$x^\mu \to x'^\mu = x^\mu + \epsilon^\mu,$$
$$\text{or,} \quad \delta x^\mu = x'^\mu - x^\mu = \epsilon^\mu, \quad (6.18)$$

where the parameter of translation ϵ^μ is assumed to be infinitesimal and constant (global). Since, we are dealing with a scalar field,

$$\phi'(x') = \phi(x), \tag{6.19}$$

and in this case we obtain the change in the field to correspond to (see (6.9) and this is how it corresponds to the negative of the Lie derivative)

$$\begin{aligned}
\delta\phi(x) &= \phi'(x) - \phi(x) = \phi'(x) - \phi'(x') \\
&= -\left(\phi'(x') - \phi'(x)\right) = -\epsilon^\mu \partial_\mu \phi'(x) \\
&= -\epsilon^\mu \partial_\mu \phi(x),
\end{aligned} \tag{6.20}$$

where in the last step, we have identified $\phi'(x) = \phi(x)$ simply because the parameter ϵ^μ multiplying on the right-hand side is already infinitesimal and any further correction coming from $\phi'(x)$ will only be of higher order. With this, then, we can now calculate explicitly the change in the Lagrangian density (see (6.7) and (6.8))

$$\begin{aligned}
&\mathcal{L}\left(\phi'(x), \partial_\mu \phi'(x)\right) - \mathcal{L}\left(\phi(x), \partial_\mu \phi(x)\right) \\
&= \delta\phi(x) \frac{\partial \mathcal{L}}{\partial \phi(x)} + (\partial_\nu \delta\phi(x)) \frac{\partial \mathcal{L}}{\partial \partial_\nu \phi(x)} \\
&= -\epsilon^\mu \partial_\mu \phi(x) \frac{\partial \mathcal{L}}{\partial \phi(x)} - \epsilon^\mu (\partial_\mu \partial_\nu \phi(x)) \frac{\partial \mathcal{L}}{\partial \partial_\nu \phi(x)} \\
&= -\epsilon^\mu \partial_\mu \mathcal{L} = \partial_\mu K^\mu,
\end{aligned} \tag{6.21}$$

where we have used (5.11) and (6.10). We see that since the change in the Lagrangian density is a total divergence, the action is invariant under infinitesimal translations which define a symmetry of the system. We can now identify from (6.21) that (ϵ^μ is a constant parameter)

$$K^\mu = -\epsilon^\mu \mathcal{L}. \tag{6.22}$$

Next, we note that under the transformation (6.20),

$$\delta\phi(x)\,\frac{\partial \mathcal{L}}{\partial \partial_\mu \phi(x)} = -\epsilon^\nu (\partial_\nu \phi(x))\,\frac{\partial \mathcal{L}}{\partial \partial_\mu \phi(x)}, \qquad (6.23)$$

so that we obtain the current associated with the symmetry transformation, depending on the parameter of transformation, to be (see (6.13))

$$\begin{aligned}
J_\epsilon^\mu(x) &= \delta\phi\,\frac{\partial \mathcal{L}}{\partial \partial_\mu \phi(x)} - K^\mu \\
&= -\epsilon^\nu (\partial_\nu \phi(x))\,\frac{\partial \mathcal{L}}{\partial \partial_\mu \phi(x)} + \epsilon^\mu \mathcal{L} \\
&= -\epsilon_\nu \left((\partial^\nu \phi(x))\,\frac{\partial \mathcal{L}}{\partial \partial_\mu \phi(x)} - \eta^{\mu\nu} \mathcal{L}\right) = -\epsilon_\nu T^{\mu\nu}. \quad (6.24)
\end{aligned}$$

It is clear, therefore, that the conserved current independent of the parameter can be identified with

$$T^{\mu\nu} = (\partial^\nu \phi(x))\,\frac{\partial \mathcal{L}}{\partial \partial_\mu \phi(x)} - \eta^{\mu\nu} \mathcal{L}, \qquad (6.25)$$

which can be easily checked to satisfy (using the equations of motion)

$$\partial_\mu T^{\mu\nu} = 0. \qquad (6.26)$$

$T^{\mu\nu}$ is known as the stress tensor of the theory and can always be defined to be symmetric (It is the source for the gravitational field $g_{\mu\nu}$ just as the electromagnetic current j^μ is the source for the electromagnetic potential A_μ. Therefore, even when the naive Nöther procedure does not lead to a symmetric stress tensor, we can always define an improved symmetric stress tensor by coupling the theory to a gravitational background and taking variation with respect to the gravitational background.), namely,

$$T^{\mu\nu} = T^{\nu\mu}. \qquad (6.27)$$

The conserved current, in this case, is a second rank tensor simply because the parameter of transformation is a four vector. (In general, the current is a tensor one rank higher than the parameter of transformation.) As a result, the conserved charges associated with the symmetry transformation (see (6.15)), in this case, would correspond to the components of a four vector which we identify with the energy-momentum operator as

$$P^\mu = \int d^3x\, T^{0\mu}. \tag{6.28}$$

This is consistent with our physical intuition that the energy-momentum, P^μ, should generate infinitesimal space-time translations.

For the free Klein-Gordon theory, let us recall that the Lagrangian density has the form

$$\mathcal{L} = \frac{1}{2} \partial_\mu \phi \partial^\mu \phi - \frac{m^2}{2} \phi^2. \tag{6.29}$$

Therefore, we obtain

$$\frac{\partial \mathcal{L}}{\partial \partial_\mu \phi(x)} = \partial^\mu \phi(x), \tag{6.30}$$

and this leads to the explicit form for the stress tensor (6.25)

$$\begin{aligned} T^{\mu\nu} &= \partial^\nu \phi(x) \frac{\partial \mathcal{L}}{\partial \partial_\mu \phi(x)} - \eta^{\mu\nu} \mathcal{L} \\ &= \partial^\nu \phi(x) \partial^\mu \phi(x) - \eta^{\mu\nu} \mathcal{L} = T^{\nu\mu}, \end{aligned} \tag{6.31}$$

which, in the present case, is manifestly symmetric. We note that, for the free theory,

$$P^0 = \int d^3x\, T^{00}$$
$$= \int d^3x \left[\left(\dot\phi(x)\right)^2 - \frac{1}{2}\left(\dot\phi(x)\right)^2 + \frac{1}{2}\boldsymbol{\nabla}\phi\cdot\boldsymbol{\nabla}\phi + \frac{m^2}{2}\phi^2\right]$$
$$= \int d^3x \left[\frac{1}{2}\left(\dot\phi(x)\right)^2 + \frac{1}{2}\boldsymbol{\nabla}\phi\cdot\boldsymbol{\nabla}\phi + \frac{m^2}{2}\phi^2\right]$$
$$= H,$$
$$P^i = \int d^3x\, T^{0i}$$
$$= \int d^3x\, \partial^i\phi(x)\dot\phi(x),$$
$$\mathbf{P} = -\int d^3x\, \boldsymbol{\nabla}\phi(x)\dot\phi(x). \tag{6.32}$$

We see that $P^0 = H$ coincides with the form of the Hamiltonian we had derived earlier in (5.34). We had also given an expression for the momentum operator in terms of creation and annihilation operators in (5.97) and it can be checked that $\mathbf{P}$ in (6.32) corresponds exactly to the former expression when normal ordered and expressed in terms of creation and annihilation operators.

6.2 Self-interacting ϕ^4 theory

Let us denote the free part of the Klein-Gordon Lagrangian density as

$$\mathcal{L}_0 = \frac{1}{2}\partial_\mu\phi\partial^\mu\phi - \frac{m^2}{2}\phi^2. \tag{6.33}$$

To include interactions, we note that the Lagrangian density for the interaction must be invariant under Lorentz transformations as well as translations (Poincaré invariant). Furthermore, since the free Lagrangian density is invariant under the discrete transformation

$$\phi(x) \leftrightarrow -\phi(x), \tag{6.34}$$

we would like to preserve this symmetry in the interactions as well. With these conditions, then, the simplest interaction Lagrangian density involving only the scalar fields which we can think of has the form

$$\mathcal{L}_I = -\frac{\lambda}{4!}\phi^4(x), \qquad \lambda > 0. \tag{6.35}$$

The restriction on the coupling constant, λ, is there so that the Hamiltonian will be positive definite, which in turn allows us to define a vacuum state of the theory. (We will see this shortly. Let us also note here that an interaction Lagrangian density of the form $-\frac{g}{3!}\phi^3$, which does not respect the discrete symmetry (6.34), would lead to a potential (and, therefore, a Hamiltonian) which is unbounded from below for any value of the coupling constant g.) As a result, the fully self-interacting theory of a real Klein-Gordon field is described by a Lagrangian density

$$\mathcal{L} = \mathcal{L}_0 + \mathcal{L}_I. \tag{6.36}$$

While there are other interaction Lagrangian densities that we can construct consistent with our symmetry requirements, for various other reasons we can show that (6.35) represents the only meaningful interaction term for such a quantum field theory. To explain this briefly, let us introduce the concept of canonical dimensions. Let us recall that the action for a quantum mechanical particle has the generic form

$$\begin{aligned} S &= \int dt\,(p\dot{q} + \cdots) \\ &= \int (p\,dq + \cdots), \end{aligned} \tag{6.37}$$

and has, in fact, the same canonical dimension as $\hbar$. Consequently, in units of $\hbar = c = 1$ (which we have been using), the action is dimensionless. In these units, we can show that the canonical dimension of any variable can be expressed in powers of an arbitrary mass dimension $[M]$. Thus,

6.2 SELF-INTERACTING ϕ^4 THEORY

$$[L] = [T] = [x^\mu] = [M]^{-1},$$
$$[\partial_\mu] = [M]. \tag{6.38}$$

With this we can now study the canonical dimension of the free part of the Klein-Gordon action and this leads to

$$[S_0] = \left[\int d^4x \left(\frac{1}{2}\partial_\mu\phi\partial^\mu\phi - \frac{m^2}{2}\phi^2\right)\right] = [M]^0,$$

or, $[d^4x][\partial_\mu][\phi][\partial^\mu][\phi] = [M]^0,$

or, $[M]^{-4}[M][\phi][M][\phi] = [M]^0,$

or, $([\phi])^2 = [M]^2,$

or, $[\phi] = [M]. \tag{6.39}$

In other words, in these units, the Klein-Gordon field variable in four dimensions has a canonical dimension 1. (The dimensionality of the field variable depends on the number of space-time dimensions which we will see later.) The mass parameter, of course, has a canonical dimension 1 and, therefore, the mass term in the Lagrangian density (the second term) automatically leads to a dimensionless action in these units with the canonical dimension of the field already determined. From the canonical dimension of the interaction term in the Lagrangian density in (6.35), we obtain

$$[S_I] = \left[-\int d^4x \frac{\lambda}{4!}\phi^4\right] = [M]^0,$$

or, $[d^4x][\lambda]([\phi])^4 = [M]^0,$

or, $[M]^{-4}[\lambda][M]^4 = [M]^0,$

or, $[\lambda] = [M]^0. \tag{6.40}$

Thus, we conclude that the coupling constant or the interaction strength for the ϕ^4-self-interaction in (6.35) is dimensionless. In general, let us note that for a monomial interaction action of the form

$$\tilde{S}_I = -\frac{g}{n!} \int d^4x \, \phi^n,$$

$$\left[\tilde{S}_I\right] = [g]\,[d^4x]\,[\phi^n] = [M]^0,$$

or, $\quad [g]\,[M]^{-4}\,[M]^n = [M]^0,$

or, $\quad [g] = [M]^{4-n}.$ \hfill (6.41)

Therefore, for $n > 4$, the coupling constant of the interaction Lagrangian density will have inverse dimensions of mass. In such a case, we can show that the transition amplitudes (scattering amplitudes) in the quantum theory will become divergent in such a way that meaningful physical results cannot be extracted from such theories. Such theories are known as non-renormalizable theories. If we want to restrict to renormalizable theories which can give rise to meaningful physical predictions, the coupling constants in the theory cannot have dimensions of inverse mass. (We will see this in a later chapter when we discuss renormalization of quantum field theories.) This, therefore, restricts $n \leq 4$ and, consequently, the ϕ^4 interaction is the only physically allowed interaction in this case, consistent with our symmetry requirements (see, for example, (6.34)).

Since the interaction Lagrangian density does not involve derivatives of fields, the canonical momentum conjugate to the field variable of the theory continues to be (see (5.32))

$$\Pi(x) = \frac{\partial \mathcal{L}}{\partial \dot{\phi}(x)} = \dot{\phi}(x), \hfill (6.42)$$

so that from (6.32) we obtain the Hamiltonian density for the interacting theory to be

$$\begin{aligned}
\mathcal{H} &= \Pi \dot{\phi} - \mathcal{L} \\
&= \Pi^2 - \left(\frac{1}{2}\dot{\phi}^2 - \frac{1}{2}\boldsymbol{\nabla}\phi \cdot \boldsymbol{\nabla}\phi - \frac{m^2}{2}\phi^2 - \frac{\lambda}{4!}\phi^4\right) \\
&= \Pi^2 - \frac{1}{2}\Pi^2 + \frac{1}{2}\boldsymbol{\nabla}\phi \cdot \boldsymbol{\nabla}\phi + \frac{m^2}{2}\phi^2 + \frac{\lambda}{4!}\phi^4
\end{aligned}$$

$$\begin{aligned}
&= \frac{1}{2}\Pi^2 + \frac{1}{2}\nabla\phi\cdot\nabla\phi + \frac{m^2}{2}\phi^2 + \frac{\lambda}{4!}\phi^4 \\
&= \mathcal{H}_0 + \mathcal{H}_I,
\end{aligned} \tag{6.43}$$

which leads to

$$\begin{aligned}
H &= \int d^3x\, \mathcal{H} \\
&= \int d^3x \left(\frac{1}{2}\Pi^2 + \frac{1}{2}\nabla\phi\cdot\nabla\phi + \frac{m^2}{2}\phi^2 + \frac{\lambda}{4!}\phi^4 \right) \\
&= H_0 + H_I.
\end{aligned} \tag{6.44}$$

It is now clear that for $\lambda > 0$, each term in the integrand in (6.44) is positive definite and, therefore, the Hamiltonian will be bounded from below leading to a meaningful vacuum state. For $\lambda < 0$, on the other hand, the Hamiltonian is indefinite because of the interaction term. Consequently, the ground state of the free Hamiltonian will not be stable under perturbations. This is why we restrict to $\lambda > 0$ in the interacting theory. The other point that should be emphasized here is that our quantum Hamiltonian should be normal ordered (even though we are not indicating the normal ordering explicitly).

6.3 Interaction picture and time evolution operator

In quantum mechanics as well as in quantum field theory, neither the states in the Hilbert space nor the operators acting on state vectors are observables. Rather, the observables correspond to expectation values of Hermitian operators in quantum states. As a result, the quantum description allows for a unitary change in the states as well as the operators without changing the expectation values which correspond to physical quantities. This leads to different possible pictures for describing the same quantum mechanical system through distinct time evolutions.

Our discussion so far has been within the context of the Heisenberg picture where the field operator $\phi(\mathbf{x}, t)$ carries time dependence. In this picture, as we know, the operators carry time dependence while the state vectors do not and the dynamical equations are given by the Heisenberg equations of motion. There is, of course, also the

Schrödinger picture where the operators are time independent but the state vectors carry all the time dependence and the time evolution of states is given by the Schrödinger equation governed by the total Hamiltonian of the system. The two pictures are related by the unitary transformation

$$\begin{aligned}|\psi(t)\rangle^{(S)} &= e^{-iHt}|\psi\rangle^{(H)},\\ O^{(S)} &= e^{-iHt}O^{(H)}(t)e^{iHt},\end{aligned} \qquad (6.45)$$

where H denotes the total Hamiltonian of the system in the Heisenberg picture (we do not put a superscript denoting this to avoid possible confusion) which is time independent. It follows from (6.45) that

$$^{(H)}\langle\psi|O^{(H)}(t)|\psi\rangle^{(H)} = {}^{(S)}\langle\psi(t)|O^{(S)}|\psi(t)\rangle^{(S)}. \qquad (6.46)$$

In writing (6.45), we have assumed that at $t = 0$ both the pictures coincide, namely,

$$\begin{aligned}|\psi(0)\rangle^{(S)} &= |\psi\rangle^{(H)},\\ O^{(S)} &= O^{(H)}(0).\end{aligned} \qquad (6.47)$$

Thus, for example, in the Schrödinger picture, the Klein-Gordon field operator will be given by

$$\phi^{(S)}(\mathbf{x}) = e^{-iHt}\phi^{(H)}(\mathbf{x},t)e^{iHt} = \phi^{(H)}(\mathbf{x},0). \qquad (6.48)$$

The total Hamiltonian operator in the Schrödinger picture coincides with that in the Heisenberg picture (see, for example, (6.45)) and is, of course, time independent. In the Schrödinger picture, the derivation of the time evolution operator is exactly analogous to the discussion in non-relativistic quantum mechanics and has the form

$$U^{(S)}(t,t_0) = e^{-iH(t-t_0)}. \qquad (6.49)$$

6.3 INTERACTION PICTURE AND TIME EVOLUTION OPERATOR

However, the Schrödinger picture is not very desirable for relativistic theories since the field operators in the Schrödinger picture are not manifestly Lorentz covariant. (However, it is worth emphasizing here that for certain applications, this picture is useful.) Although the Heisenberg picture has a manifestly covariant description, in relativistic interacting field theories, the more convenient description goes under the name of interaction picture which is, in some sense, intermediate between the Schrödinger and the Heisenberg pictures. Here, we define

$$\begin{aligned}
|\psi(t)\rangle^{(IP)} &= e^{iH_0^{(S)}t}|\psi(t)\rangle^{(S)} = e^{iH_0^{(S)}t}e^{-iHt}|\psi\rangle^{(H)}, \\
O^{(IP)}(t) &= e^{iH_0^{(S)}t} O^{(S)} e^{-iH_0^{(S)}t} \\
&= e^{iH_0^{(S)}t} e^{-iHt} O^{(H)}(t) e^{iHt} e^{-iH_0^{(S)}t}. \qquad (6.50)
\end{aligned}$$

Note, once again, that we can identify

$$\begin{aligned}
|\psi(0)\rangle^{(IP)} &= |\psi(0)\rangle^{(S)} = |\psi\rangle^{(H)}, \\
O^{(IP)}(0) &= O^{(S)} = O^{(H)}(0), \\
H_0^{(IP)}(t) &= e^{iH_0^{(S)}t} H_0^{(S)} e^{-iH_0^{(S)}t} = H_0^{(S)}. \qquad (6.51)
\end{aligned}$$

In the interaction picture, therefore, both the operators as well as the state vectors carry time dependence. In this case, we can determine that (remember that $\hbar = 1$)

$$\begin{aligned}
i\frac{\partial |\psi(t)\rangle^{(IP)}}{\partial t} &= i\frac{\partial}{\partial t}\left(e^{iH_0^{(S)}t} e^{-iHt}|\psi\rangle^{(H)}\right) \\
&= -H_0^{(S)} e^{iH_0^{(S)}t} e^{-iHt}|\psi\rangle^{(H)} + e^{iH_0^{(S)}t}\left(He^{-iHt}\right)|\psi\rangle^{(H)} \\
&= -H_0^{(IP)}|\psi(t)\rangle^{(IP)} + e^{iH_0^{(S)}t} H e^{-iH_0^{(S)}t} e^{iH_0^{(S)}t} e^{-iHt}|\psi\rangle^{(H)} \\
&= -H_0^{(IP)}|\psi(t)\rangle^{(IP)} + H^{(IP)}|\psi(t)\rangle^{(IP)} \\
&= H_I^{(IP)}(t)|\psi(t)\rangle^{(IP)}, \qquad (6.52)
\end{aligned}$$

where we have used the fact that the total Hamiltonian H is the same in both the Schrödinger as well as the Heisenberg pictures and that, by definition,

$$H_I^{(IP)}(t) = e^{iH_0^{(S)}t} H_I^{(S)} e^{-iH_0^{(S)}t}. \tag{6.53}$$

Similarly, we can derive the time evolution for the operators to be

$$\begin{aligned}
\frac{\partial O^{(IP)}}{\partial t} &= \frac{\partial}{\partial t} \left(e^{iH_0^{(S)}t} O^{(S)} e^{-iH_0^{(S)}t} \right) \\
&= iH_0^{(S)} O^{(IP)}(t) - iO^{(IP)}(t) H_0^{(S)} \\
&= -i\left[O^{(IP)}(t), H_0^{(S)}\right] = \frac{1}{i}\left[O^{(IP)}(t), H_0^{(IP)}\right]. \tag{6.54}
\end{aligned}$$

In other words, the time evolution of the state vectors, in this picture, is governed by the Schrödinger equation with the interaction Hamiltonian playing the role of the Hamiltonian, while the dynamical evolution of the operators is governed by the free Hamiltonian through the Heisenberg equations of motion. We note from (6.51) that the free Hamiltonian in the interaction picture is time independent and, therefore, the field operators can have a plane wave expansion. This shows that once we define commutation relations for the operators in the free theory, in the interaction picture they continue to hold even in the presence of interactions.

From now on, let us drop the superscript (IP) with the understanding that we are nonetheless working in the interaction picture. In this picture, if we define the time evolution operator through the relation

$$|\psi(t)\rangle = U(t, t_0) |\psi(t_0)\rangle, \tag{6.55}$$

where the time evolution operator in the interaction picture in (6.55) can be seen from (6.49) and (6.50) to have the explicit form

$$U(t, t_0) = e^{iH_0^{(S)}t} e^{-iH(t-t_0)} e^{-iH_0^{(S)}t_0}, \tag{6.56}$$

then, it is easy to show that the time evolution operator satisfies the properties,

6.3 INTERACTION PICTURE AND TIME EVOLUTION OPERATOR

$$U(t,t) = \mathbb{1},$$
$$U(t_2,t_1)U(t_1,t_0) = U(t_2,t_0),$$
$$U^\dagger(t,t_0) = U^{-1}(t,t_0) = U(t_0,t). \tag{6.57}$$

Furthermore, from (6.52) we see that the time evolution operator in (6.55) satisfies the equation (this also follows from (6.56) with the use of (6.50))

$$i\frac{\partial |\psi(t)\rangle}{\partial t} = H_I(t)|\psi(t)\rangle,$$
or, $\quad i\dfrac{\partial U(t,t_0)}{\partial t} = H_I(t)U(t,t_0).$ \hfill (6.58)

We note from (6.53) that the interaction Hamiltonian in the interaction picture is time dependent. (Since H_0 is time independent in the interaction picture (see (6.51)), it follows that the total Hamiltonian is time independent in the interaction picture.)

We can now solve (6.58) iteratively subject to the initial condition in (6.57), to obtain

$$U(t,t_0) = \mathbb{1} - i\int_{t_0}^t dt_1\, H_I(t_1)\, U(t_1,t_0)$$
$$= \mathbb{1} - i\int_{t_0}^t dt_1\, H_I(t_1)\left(\mathbb{1} - i\int_{t_0}^{t_1} dt_2\, H_I(t_2)\, U(t_2,t_0)\right)$$
$$= \mathbb{1} - i\int_{t_0}^t dt_1\, H_I(t_1) + (-i)^2 \int_{t_0}^t dt_1 \int_{t_0}^{t_1} dt_2\, H_I(t_1) H_I(t_2)$$
$$+ \cdots + \cdots$$
$$+ (-i)^n \int_{t_0}^t dt_1 \int_{t_0}^{t_1} dt_2 \cdots \int_{t_0}^{t_{n-1}} dt_n\, H_I(t_1) H_I(t_2)\cdots H_I(t_n)$$
$$+ \cdots . \tag{6.59}$$

To bring this to a more convenient form, let us look at the second order term in (6.59) and note that

$$(-i)^2 \int_{t_0}^{t} dt_1 \int_{t_0}^{t_1} dt_2\, H_I(t_1) H_I(t_2)$$

$$= (-i)^2 \int_{t_0}^{t} dt_2 \int_{t_0}^{t_2} dt_1\, H_I(t_2) H_I(t_1)$$

$$= \frac{(-i)^2}{2} \left(\int_{t_0}^{t} dt_1 \int_{t_0}^{t_1} dt_2\, H_I(t_1) H_I(t_2) \right.$$
$$\left. + \int_{t_0}^{t} dt_2 \int_{t_0}^{t_2} dt_1\, H_I(t_2) H_I(t_1) \right)$$

$$= \frac{(-i)^2}{2} \left(\int_{t_0}^{t} dt_1 \int_{t_0}^{t_1} dt_2\, \theta(t_1 - t_2) H_I(t_1) H_I(t_2) \right.$$
$$\left. + \int_{t_0}^{t} dt_2 \int_{t_0}^{t_2} dt_1\, \theta(t_2 - t_1) H_I(t_2) H_I(t_1) \right)$$

$$= \frac{(-i)^2}{2} \left(\int_{t_0}^{t} dt_1 \int_{t_0}^{t} dt_2\, \theta(t_1 - t_2) H_I(t_1) H_I(t_2) \right.$$
$$\left. + \int_{t_0}^{t} dt_2 \int_{t_0}^{t} dt_1\, \theta(t_2 - t_1) H_I(t_2) H_I(t_1) \right)$$

$$= \frac{(-i)^2}{2} \iint_{t_0}^{t} dt_1 dt_2 \big(\theta(t_1 - t_2) H_I(t_1) H_I(t_2)$$
$$+ \theta(t_2 - t_1) H_I(t_2) H_I(t_1) \big)$$

$$= \frac{(-i)^2}{2!} \int_{t_0}^{t} \int_{t_0}^{t} dt_1 dt_2\, T\left(H_I(t_1) H_I(t_2)\right), \tag{6.60}$$

where T denotes the time ordering operator and is conventionally defined for two bosonic operators as

$$T(A(t_1)B(t_2)) = \theta(t_1 - t_2)A(t_1)B(t_2) + \theta(t_2 - t_1)B(t_2)A(t_1), \tag{6.61}$$

with the operator at later time standing to the left of the operator at earlier time. In a similar manner, we can show that the n-th term in the series in (6.59) can be written as

$$\frac{(-i)^n}{n!} \int_{t_0}^{t} \cdots \int_{t_0}^{t} dt_1 \cdots dt_n \, T\left(H_I(t_1) \cdots H_I(t_n)\right). \tag{6.62}$$

Therefore, the iterative solution for the time evolution operator in (6.59) takes the form

$$U(t,t_0) = \mathbb{1} - i \int_{t_0}^{t} dt_1 \, H_I(t_1)$$

$$+ \frac{(-i)^2}{2!} T \int_{t_0}^{t} \int_{t_0}^{t} dt_1 dt_2 \, H_I(t_1) H_I(t_2)$$

$$+ \cdots + \frac{(-i)^n}{n!} T \int_{t_0}^{t} \cdots \int_{t_0}^{t} dt_1 \cdots dt_n \, H_I(t_1) \cdots H_I(t_n)$$

$$+ \cdots$$

or, $\quad U(t,t_0) = T\left(e^{-i \int_{t_0}^{t} dt' H_I(t')}\right). \tag{6.63}$

In other words, in the interaction picture the time evolution operator is given by the time ordered exponential involving the integral of only the interaction Hamiltonian. (This is a formal definition which is to be understood in the sense of the expansion described in (6.63).)

6.4 S-matrix

In the non-relativistic quantum mechanical scattering problems, we assume that the initial and the final states are plane wave states corresponding to free particles. Similarly, in the scattering of particles in relativistic quantum field theory, we also assume that the incoming particles at $t = -\infty$ as well as the outgoing particles at $t = \infty$ are described by free particle states. This can clearly be implemented by assuming that the interaction switches off adiabatically at $t = \pm\infty$ (adibatic hypothesis). For example, we can implement this by modifying the interaction Hamiltonian as

$$H_I(t) \to \lim_{\eta \to 0^+} H_I^{(\eta)}(t) = \lim_{\eta \to 0^+} e^{-\eta|t|} H_I(t), \tag{6.64}$$

so that in the infinite past as well as in the infinite future, the interaction vanishes. As a result, we can take the initial and the final states to be the eigenstates of the free Hamiltonian which we know to be free particle energy-momentum states. The condition (6.64) can be thought of as the relevant boundary condition for the scattering problem under study. We are essentially assuming here that any smooth function describing the adiabatic switching off of the interaction (and not just the specific form in (6.64)) leads to the same result for the rate of transition in the physical scattering of particles.

Let us denote the initial state at infinite past as the free particle state

$$|\psi_i(-\infty)\rangle = |i\rangle. \tag{6.65}$$

Then, the state into which this will evolve at $t = \infty$ is defined from (6.55) to be

$$|\psi_i(\infty)\rangle = U(\infty, -\infty)|\psi_i(-\infty)\rangle = U(\infty, -\infty)|i\rangle = S|i\rangle, \tag{6.66}$$

where we have identified

$$S = U(\infty, -\infty). \tag{6.67}$$

Therefore, the probability amplitude for an initial state $|i\rangle$ to be in the final free particle state $|f\rangle$ at $t = \infty$, which is the definition of the scattering amplitude, is obtained from (6.66) to be

$$S_{fi} = \langle f|\psi_i(\infty)\rangle = \langle f|U(\infty, -\infty)|i\rangle = \langle f|S|i\rangle. \tag{6.68}$$

Consequently, the S-matrix (or the scattering matrix) of the theory can be identified with the time evolution operator (6.67) which has the explicit perturbative expansion of the form (see (6.63))

$$\begin{aligned}
S &= U(\infty, -\infty) = \lim_{\eta \to 0^+} T \left(e^{-i \int_{-\infty}^{\infty} dt\, H_I^{(\eta)}(t)} \right) \\
&= \mathbb{1} - \lim_{\eta \to 0^+} i \int_{-\infty}^{\infty} dt\, H_I^{(\eta)}(t) \\
&\quad + \lim_{\eta \to 0^+} \frac{(-i)^2}{2!} \int\!\!\int_{-\infty}^{\infty} dt\, dt'\, T\left(H_I^{(\eta)}(t) H_I^{(\eta)}(t') \right) \\
&\quad + \cdots .
\end{aligned} \quad (6.69)$$

Furthermore, the S-matrix (the scattering matrix) is unitary since the time evolution operator is (see (6.57)).

It is clear from the expansion above that the adiabatic switching of the interaction in (6.64) provides a regularized meaning to the oscillatory terms in (6.69) through the appropriate boundary conditions. This also leads to the notion of "in" and "out" states which are quite important in a formal description of scattering theory. These are asymptotic free states as $t \to -\infty$ and $t \to \infty$ respectively. These states can be constructed by noting that the three pictures (Heisenberg, Schrödinger and interaction) coincide at $t = 0$ (see (6.51)) and since the Heisenberg states are time independent, they are uniquely given by the states (in any picture) at $t = 0$. Let us suppose that at $t \to -\infty$ we have a free incoming state denoted as $|\Psi(-\infty)\rangle^{(\text{in})} = |i\rangle$. Then, this state would be related to the Heisenberg state as

$$|\Psi\rangle^{(H)} = |\Psi(0)\rangle^{(\text{in})} = U(0, -\infty)|\Psi(-\infty)\rangle^{(\text{in})}, \quad (6.70)$$

where the time evolution operator $U(0, -\infty)$ is, in general, not well defined if we do not use an adiabatic interaction of the form, say, in (6.64). Let us see this explicitly in the case of the linear term in the expansion of the time evolution operator in (6.63). In this case, the linear contribution to the right-hand side of (6.70) will have the form

$$\lim_{\eta \to 0^+} -i \int_{-\infty}^{0} dt\, H_I^{(\eta)}(t) |\Psi(-\infty)\rangle^{(\text{in})}$$

$$= \lim_{\eta \to 0^+} -i \int_{-\infty}^{0} dt\, e^{\eta t} e^{iH_0^{(S)}t} H_I^{(S)} e^{-iH_0^{(S)}t} |i\rangle$$

$$= \lim_{\eta \to 0^+} -i \int_{-\infty}^{0} dt\, e^{-i(E_i - H_0^{(S)} + i\eta)t} H_I^{(S)} |i\rangle$$

$$= \lim_{\eta \to 0^+} \frac{1}{E_i - H_0^{(S)} + i\eta} H_I^{(S)} |\Psi(-\infty)\rangle^{(\text{in})}, \tag{6.71}$$

where in the intermediate steps, we have used the fact that $H_0^{(IP)} = H_0^{(S)}$ and that $|\Psi(-\infty)\rangle^{(\text{in})} = |i\rangle$ is a free state with E_i the energy eigenvalue of $H_0^{(IP)}$. We also note that the integral in (6.71) is not defined at the lower limit in the absence of the adiabatic factor $e^{\eta t}$. This clarifies how the boundary condition (6.64) naturally provides a regularization for the formal definition of the time evolution operator $U(0, -\infty)$.

In a parallel manner, if we assume that as $t \to \infty$, we have a free outgoing state $|\Psi(\infty)\rangle^{(\text{out})}$, then this would be related to the Heisenberg state as

$$|\Psi\rangle^{(H)} = |\Psi(0)\rangle^{(\text{out})} = U(0, \infty) |\Psi(\infty)\rangle^{(\text{out})}, \tag{6.72}$$

and an analysis as in (6.71) would show that in this case, the regularizing factor would have to be $e^{-\eta t}$ without which $U(0, \infty)$ would not be well defined at the upper limit. This shows that the asymptotically free "in" and "out" states can be defined in a unique manner by relating them to the Heisenberg state. From (6.70) and (6.72) we see that even though both the "in" and the "out" states are related to the same Heisenberg state, they are not identical because their definitions use different regularizing factors. Rather, the "in" and the "out" states define a complete space of states at $t \to -\infty$ and $t \to \infty$ respectively. Furthermore, from their definitions in (6.70) and (6.72) we see that they are related to each other as

$$|\Psi(\infty)\rangle^{(\text{out})} = U^\dagger(0,\infty)|\Psi\rangle^{(H)} = U(\infty,0)|\Psi\rangle^{(H)}$$
$$= U(\infty,0)U(0,-\infty)|\Psi(-\infty)\rangle^{(\text{in})}$$
$$= U(\infty,-\infty)|\Psi(-\infty)\rangle^{(\text{in})} = S|\Psi(-\infty)\rangle^{(\text{in})}, \qquad (6.73)$$

where we have used the properties of the time evolution operator given in (6.57). This shows that the "in" and the "out" states are related by the S-matrix in (6.67) (and, therefore, by a phase since the S-matrix is unitary).

The first term in the expansion of the S-matrix in (6.69) corresponds to no scattering at all. Correspondingly, we can define

$$S - \mathbb{1} = \text{T}, \qquad (6.74)$$

where the T-matrix represents the true nontrivial effects of scattering. It is clear from this discussion that in calculating the S-matrix elements, S_{fi}, we need to evaluate the matrix elements of time ordered products of operators between free particle states. Furthermore, since our (interaction) Hamiltonian is assumed to be normal ordered, such a calculation will involve time ordered products of normal ordered products and this is where the Wick's theorem comes in handy. The calculation can be carried out in two steps. First, we will derive a simple relation for the product of factors each of which is normal ordered and then using this relation, we will simplify the time ordered product of factors of normal ordered terms. Later we will see that the Feynman rules are a wonderful and simple way of systematizing these results.

6.5 Normal ordered product and Wick's theorem

The normal ordered product, as we have defined earlier in (5.76), simply corresponds to arranging factors in the product so that the creation operators stand to the left of the annihilation operators. In terms of field operators, this is equivalent to saying that the negative energy (frequency) parts of the field operator stand to the left of the positive energy (frequency) parts (recall that the positive energy part of the field operator, $\phi^{(+)}$, contains the annihilation operator while the negative energy part, $\phi^{(-)}$, has the creation operator so that this prescription is equivalent to the definition of normal ordering in (5.76)). Thus, for the bosonic scalar field, we have

$$\begin{aligned}
N\left(\phi(x)\right) &\equiv\, :\phi(x): = \phi(x) = \phi^{(+)}(x) + \phi^{(-)}(x), \\
N\left(\phi^{(-)}(x)\phi^{(+)}(y)\right) &\equiv\, :\phi^{(-)}(x)\phi^{(+)}(y): = \phi^{(-)}(x)\phi^{(+)}(y), \\
N\left(\phi^{(+)}(x)\phi^{(-)}(y)\right) &\equiv\, :\phi^{(+)}(x)\phi^{(-)}(y): = \phi^{(-)}(y)\phi^{(+)}(x), \\
N\left(\phi^{(+)}(x)\phi^{(+)}(y)\right) &\equiv\, :\phi^{(+)}(x)\phi^{(+)}(y): = \phi^{(+)}(x)\phi^{(+)}(y), \\
N\left(\phi^{(-)}(x)\phi^{(-)}(y)\right) &\equiv\, :\phi^{(-)}(x)\phi^{(-)}(y): = \phi^{(-)}(x)\phi^{(-)}(y).
\end{aligned} \tag{6.75}$$

We note that, as far as bosonic operators are concerned, the order of the factors inside a normal ordered product is not important. Furthermore, using the relations in (6.75), we can write

$$\begin{aligned}
N\left(\phi(x)\phi(y)\right) &\equiv\, :\phi(x)\phi(y): \\
&=\, :\left(\phi^{(+)}(x) + \phi^{(-)}(x)\right)\left(\phi^{(+)}(y) + \phi^{(-)}(y)\right): \\
&=\, :\left(\phi^{(+)}(x)\phi^{(+)}(y) + \phi^{(+)}(x)\phi^{(-)}(y)\right. \\
&\quad \left. + \phi^{(-)}(x)\phi^{(+)}(y) + \phi^{(-)}(x)\phi^{(-)}(y)\right): \\
&= \phi^{(+)}(x)\phi^{(+)}(y) + \phi^{(-)}(y)\phi^{(+)}(x) \\
&\quad + \phi^{(-)}(x)\phi^{(+)}(y) + \phi^{(-)}(x)\phi^{(-)}(y).
\end{aligned} \tag{6.76}$$

Since the vacuum state is, by definition, annihilated by the annihilation operator, we can equivalently say that

$$\phi^{(+)}(x)|0\rangle = 0 = \langle 0|\phi^{(-)}(x). \tag{6.77}$$

It is clear, therefore, that the vacuum expectation value of any normal ordered product vanishes. For example, we note from (6.76) that

$$\langle 0|N\left(\phi(x)\phi(y)\right)|0\rangle$$
$$= \langle 0|\phi^{(+)}(x)\phi^{(+)}(y) + \phi^{(-)}(y)\phi^{(+)}(x) + \phi^{(-)}(x)\phi^{(+)}(x)$$
$$+\phi^{(-)}(x)\phi^{(-)}(y)|0\rangle = 0. \tag{6.78}$$

Therefore, normal ordering is quite useful since we would like the vacuum expectation value of any observable to vanish simply because the vacuum contains no particles – it is empty and, therefore, does not carry any nontrivial quantum number. This gives another reason to choose the prescription of normal ordering for physical observables. Furthermore, this also shows that if we can rewrite any product of operators (including normal ordered factors) in terms of normal ordered terms, the calculation of S-matrix elements will simplify enormously.

To see how we can write a product of operators in normal ordered form, let us note that, in the simple example of the product of two field operators, we have

$$\phi(x)\phi(y) = \left(\phi^{(+)}(x) + \phi^{(-)}(x)\right)\left(\phi^{(+)}(y) + \phi^{(-)}(y)\right)$$
$$= \phi^{(+)}(x)\phi^{(+)}(y) + \phi^{(+)}(x)\phi^{(-)}(y) + \phi^{(-)}(x)\phi^{(+)}(y)$$
$$+\phi^{(-)}(x)\phi^{(-)}(y)$$
$$= \phi^{(+)}(x)\phi^{(+)}(y) + \phi^{(-)}(y)\phi^{(+)}(x) + \left[\phi^{(+)}(x), \phi^{(-)}(y)\right]$$
$$+\phi^{(-)}(x)\phi^{(+)}(y) + \phi^{(-)}(x)\phi^{(-)}(y)$$
$$= :\phi(x)\phi(y): -iG^{(+)}(x-y), \tag{6.79}$$

where we have used (5.158) and (6.76). Let us denote (6.79) as

$$\phi(x)\phi(y) = :\phi(x)\phi(y): + \underbrace{\phi(x)\phi(y)}, \tag{6.80}$$

with the pairing of the two field operators defined to be related to the invariant positive energy Green's function in (5.151) as

$$\underbrace{\phi(x)\phi(y)} = -iG^{(+)}(x-y). \tag{6.81}$$

We note that $G^{(+)}(x-y)$ is not a symmetric function (so that the pairing should be read carefully as from left to right), but since it is a c-number function, we note from (6.80) that we can also identify

$$\underbrace{\phi(x)\phi(y)} = \langle 0|\phi(x)\phi(y)|0\rangle = -iG^{(+)}(x-y), \qquad (6.82)$$

because the vacuum expectation value of the normal ordered product in (6.80) is zero. This brings out yet another connection between Green's functions and the quantum field theory.

Since we choose the Hamiltonian to be normal ordered, let us next see how we can combine a product of normal ordered factors into normal ordered terms. Clearly, this will be necessary in developing a perturbation expansion for the quantum field theory where the interaction Hamiltonian is normal ordered. We know that by definition

$$:\phi(x): = \phi(x) = \phi^{(+)}(x) + \phi^{(-)}(x). \qquad (6.83)$$

Therefore,

$$:\phi(x)::\phi(y): =\, :\phi(x):\phi^{(+)}(y) + \,:\phi(x):\phi^{(-)}(y). \qquad (6.84)$$

We recognize that the first term in (6.84) simply corresponds to a normal ordered product

$$:\phi(x):\phi^{(+)}(y) = \phi(x)\phi^{(+)}(y) = \,:\phi(x)\phi^{(+)}(y):. \qquad (6.85)$$

On the other hand, the second term in (6.84) gives

$$\begin{aligned}
:\phi(x):\phi^{(-)}(y) &= \phi(x)\phi^{(-)}(y) \\
&= \phi^{(-)}(y)\phi(x) + [\phi(x),\phi^{(-)}(y)] \\
&= \,:\phi(x)\phi^{(-)}(y): + [\phi^{(+)}(x),\phi^{(-)}(y)] \\
&= \,:\phi(x)\phi^{(-)}(y): - iG^{(+)}(x-y).
\end{aligned} \qquad (6.86)$$

Thus adding the two terms in (6.85) and (6.86) we obtain

$$\begin{aligned}
:\phi(x)::\phi(y): &=\, :\phi(x):\phi^{(+)}(y) + :\phi(x):\phi^{(-)}(y) \\
&=\, :\phi(x)\phi^{(+)}(y): + :\phi(x)\phi^{(-)}(y): - iG^{(+)}(x-y) \\
&=\, :\phi(x)\phi(y): + \underbrace{\phi(x)\phi(y)}.
\end{aligned} \qquad (6.87)$$

This is, of course, what we had seen earlier in (6.79) and (6.80), namely,

$$\begin{aligned}
\phi(x)\phi(y) &=\, :\phi(x)\phi(y): - iG^{(+)}(x-y) \\
&=\, :\phi(x)\phi(y): + \underbrace{\phi(x)\phi(y)},
\end{aligned} \qquad (6.88)$$

but (6.87) shows how a product of two simple normal ordered factors can be expressed as a normal ordered term and a pairing.

Next, let us look at a product of the form

$$\begin{aligned}
&:\phi(x)\phi(y)::\phi(z): \\
&=\, :\phi(x)\phi(y):\phi^{(+)}(z) + :\phi(x)\phi(y):\phi^{(-)}(z) \\
&=\, :\phi(x)\phi(y)\phi^{(+)}(z): + \phi^{(-)}(z):\phi(x)\phi(y): \\
&\quad + \big[\,:\phi(x)\phi(y):,\phi^{(-)}(z)\big] \\
&=\, :\phi(x)\phi(y)\phi^{(+)}(z): + :\phi(x)\phi(y)\phi^{(-)}(z): \\
&\quad + \big[\phi(x)\phi(y) + iG^{(+)}(x-y),\phi^{(-)}(z)\big] \\
&=\, :\phi(x)\phi(y)\phi(z): + \big[\phi^{(+)}(x),\phi^{(-)}(z)\big]\phi(y) \\
&\quad + \phi(x)\big[\phi^{(+)}(y),\phi^{(-)}(z)\big] \\
&=\, :\phi(x)\phi(y)\phi(z): - iG^{(+)}(x-z)\phi(y) - iG^{(+)}(y-z)\phi(x) \\
&=\, :\phi(x)\phi(y)\phi(z): + \underbrace{\phi(x)\phi(z):\phi(y):} + \underbrace{\phi(y)\phi(z):\phi(x):}.
\end{aligned} \qquad (6.89)$$

Finally, let us note that

$$: \phi(x)\phi(y) :: \phi(z)\phi(w) :$$
$$= \; : \phi(x)\phi(y) : \phi^{(+)}(z)\phi^{(+)}(w) + \; : \phi(x)\phi(y) : \phi^{(-)}(z)\phi^{(+)}(w)$$
$$+ \; : \phi(x)\phi(y) : \phi^{(-)}(w)\phi^{(+)}(z) + \; : \phi(x)\phi(y) : \phi^{(-)}(z)\phi^{(-)}(w)$$
$$= \; : \phi(x)\phi(y)\phi^{(+)}(z)\phi^{(+)}(w) : + \phi^{(-)}(z) : \phi(x)\phi(y) : \phi^{(+)}(w)$$
$$+ [: \phi(x)\phi(y) :, \phi^{(-)}(z)]\phi^{(+)}(w) + \phi^{(-)}(w) : \phi(x)\phi(y) : \phi^{(+)}(z)$$
$$+ [: \phi(x)\phi(y) :, \phi^{(-)}(w)]\phi^{(+)}(z) + \phi^{(-)}(z)\phi^{(-)}(w) : \phi(x)\phi(y) :$$
$$+ [: \phi(x)\phi(y) :, \phi^{(-)}(z)\phi^{(-)}(w)]$$
$$= \; : \phi(x)\phi(y)\phi^{(+)}(z)\phi^{(+)}(w) : + \; : \phi(x)\phi(y)\phi^{(-)}(z)\phi^{(+)}(w) :$$
$$+ \; : \phi(x)\phi(y)\phi^{(+)}(z)\phi^{(-)}(w) : + \; : \phi(x)\phi(y)\phi^{(-)}(z)\phi^{(-)}(w) :$$
$$+ [: \phi(x)\phi(y) :, \phi^{(-)}(z)]\phi(w) + [: \phi(x)\phi(y) :, \phi^{(-)}(w)]\phi^{(+)}(z)$$
$$+ \phi^{(-)}(z)[: \phi(x)\phi(y) :, \phi^{(-)}(w)]$$
$$= \; : \phi(x)\phi(y)\phi(z)\phi(w) : -iG^{(+)}(x-z)\phi(y)\phi(w)$$
$$-iG^{(+)}(y-z)\phi(x)\phi(w) - iG^{(+)}(x-w)\phi(y)\phi^{(+)}(z)$$
$$-iG^{(+)}(y-w)\phi(x)\phi^{(+)}(z) - iG^{(+)}(x-w)\phi^{(-)}(z)\phi(y)$$
$$-iG^{(+)}(y-w)\phi^{(-)}(z)\phi(x)$$
$$= \; : \phi(x)\phi(y)\phi(z)\phi(w) : + \underbrace{\phi(x)\phi(z)}\big(: \phi(y)\phi(w) : + \underbrace{\phi(y)\phi(w)}\big)$$
$$+ \underbrace{\phi(y)\phi(z)}\big(: \phi(x)\phi(w) : + \underbrace{\phi(x)\phi(w)}\big)$$
$$+ \underbrace{\phi(x)\phi(w)} : \phi(y)\phi(z) : + \underbrace{\phi(y)\phi(w)} : \phi(x)\phi(z) :$$
$$= \; : \phi(x)\phi(y)\phi(z)\phi(w) : + \underbrace{\phi(x)\phi(z)} : \phi(y)\phi(w) :$$
$$+ \underbrace{\phi(x)\phi(w)} : \phi(y)\phi(z) : + \underbrace{\phi(y)\phi(z)} : \phi(x)\phi(w) :$$
$$+ \underbrace{\phi(y)\phi(w)} : \phi(x)\phi(z) : + \underbrace{\phi(x)\phi(z)}\,\underbrace{\phi(y)\phi(w)}$$
$$+ \underbrace{\phi(x)\phi(w)}\,\underbrace{\phi(y)\phi(z)}. \tag{6.90}$$

These simple examples demonstrate how a product of normal ordered

6.5 Normal ordered product and Wick's theorem

factors can be written in terms of normal ordered products with all possible pairings between fields in different factors (but no pairing between two fields within the same factor). This now leads us to an important theorem known as the Wick's theorem (this is one form of it) which says that a product of factors of normal ordered field operators can be written as a sum of terms with normal ordered terms and all possible pairings between field operators except for pairings between operators within the same normal ordered factor.

Let us note that any product of operators (and not just a product of normal ordered factors) can be written in terms of normal ordered products. For example, we already know from (6.80) that

$$\phi(x)\phi(y) = :\phi(x)\phi(y): + \underbrace{\phi(x)\phi(y)}. \tag{6.91}$$

It follows, therefore, that

$$\begin{aligned}
\phi(x)\phi(y)\phi(z) &= :\phi(x)\phi(y):\phi(z) + \underbrace{\phi(x)\phi(y)}\phi(z) \\
&= :\phi(x)\phi(y)::\phi(z): + \underbrace{\phi(x)\phi(y)}:\phi(z): \\
&= :\phi(x)\phi(y)\phi(z): + \underbrace{\phi(x)\phi(z)}:\phi(y): \\
&\quad + \underbrace{\phi(y)\phi(z)}:\phi(x): + \underbrace{\phi(x)\phi(y)}:\phi(z): . \tag{6.92}
\end{aligned}$$

Similarly,

$$\begin{aligned}
&\phi(x)\phi(y)\phi(z)\phi(w) \\
&= \left(:\phi(x)\phi(y): + \underbrace{\phi(x)\phi(y)}\right)\left(:\phi(z)\phi(w): + \underbrace{\phi(z)\phi(w)}\right) \\
&= :\phi(x)\phi(y)::\phi(z)\phi(w): + \underbrace{\phi(x)\phi(y)}:\phi(z)\phi(w): \\
&\quad + \underbrace{\phi(z)\phi(w)}:\phi(x)\phi(y): + \underbrace{\phi(x)\phi(y)}\underbrace{\phi(z)\phi(w)} \\
&= :\phi(x)\phi(y)\phi(z)\phi(w): + \underbrace{\phi(x)\phi(z)}:\phi(y)\phi(w): \\
&\quad + \underbrace{\phi(x)\phi(w)}:\phi(y)\phi(z): + \underbrace{\phi(y)\phi(z)}:\phi(x)\phi(w):
\end{aligned}$$

$$+ \underbrace{\phi(y)\phi(w)} : \phi(x)\phi(z) : + \underbrace{\phi(x)\phi(y)} : \phi(z)\phi(w) :$$

$$+ \underbrace{\phi(z)\phi(w)} : \phi(x)\phi(y) : + \underbrace{\phi(x)\phi(y)} \underbrace{\phi(z)\phi(w)}$$

$$+ \underbrace{\phi(x)\phi(z)} \underbrace{\phi(y)\phi(w)} + \underbrace{\phi(x)\phi(w)} \underbrace{\phi(y)\phi(z)}. \tag{6.93}$$

In other words, the product of any number of field operators can be expressed as a sum of terms involving normal ordered products with all possible pairings (pairings are from left to right since the positive energy Green's function $G^{(+)}(x-y)$ is not symmetric). This has to be contrasted with the products of normal ordered terms which also had a similar expansion except that there were no pairings between terms within the same factor. In general, Wick's theorem for a product of field operators says that

$$\phi(x_1)\phi(x_2)\cdots\phi(x_n) =: \phi(x_1)\cdots\phi(x_n) :$$
$$+ \underbrace{\phi(x_1)\phi(x_2)} : \phi(x_3)\cdots\phi(x_n) :$$
$$+ \underbrace{\phi(x_1)\phi(x_3)} : \phi(x_2)\phi(x_4)\cdots\phi(x_n) :$$
$$+ \cdots + \underbrace{\phi(x_{n-1})\phi(x_n)} : \phi(x_1)\cdots\phi(x_{n-2}) :$$
$$+ \cdots$$
$$+ \underbrace{\phi(x_1)\phi(x_2)}\underbrace{\phi(x_3)\phi(x_4)}\underbrace{\phi(x_5)\phi(x_6)} : \phi(x_7)\cdots\phi(x_n) :$$
$$+ \cdots . \tag{6.94}$$

This is quite significant since it says that the vacuum expectation value of the product of any number of factors involving field operators (whether the factors are normal ordered or not) is given by the terms where all the fields are completely paired. (We are interested in vacuum expectation values because any matrix element can be written as a vacuum expectation value.) An immediate consequence of this result is that the vacuum expectation value of an odd number of field operators must vanish (since not all field operators can be paired).

This construction can also be extended to the case where only some of the factors are normal ordered. For example, we have already noted in (6.89) that

$$: \phi(x)\phi(y) : \phi(z) =: \phi(x)\phi(y) :: \phi(z) :$$
$$=: \phi(x)\phi(y)\phi(z): + \underbrace{\phi(x)\phi(z)} :\phi(y): + \underbrace{\phi(y)\phi(z)} :\phi(x):. \quad (6.95)$$

We can now extend this result as (using (6.79) as well as (6.90))

$$: \phi(x)\phi(y) : \phi(z)\phi(w) =: \phi(x)\phi(y) : \left(: \phi(z)\phi(w) : + \underbrace{\phi(z)\phi(w)} \right)$$
$$= \; : \phi(x)\phi(y)\phi(z)\phi(w) : + \underbrace{\phi(x)\phi(z)} : \phi(y)\phi(w) :$$
$$+ \underbrace{\phi(x)\phi(w)} : \phi(y)\phi(z) : + \underbrace{\phi(y)\phi(z)} : \phi(x)\phi(w) :$$
$$+ \underbrace{\phi(y)\phi(w)} : \phi(x)\phi(z) : + \underbrace{\phi(z)\phi(w)} : \phi(x)\phi(y) :$$
$$+ \underbrace{\phi(x)\phi(z)}\,\underbrace{\phi(y)\phi(w)} + \underbrace{\phi(x)\phi(w)}\,\underbrace{\phi(y)\phi(z)}, \quad (6.96)$$

and so on.

6.6 Time ordered products and Wick's theorem

There is a second kind of operator product, known as the time ordered product, which plays a fundamental role in quantum field theories. We have already come across time ordering in the definition of the time evolution operator in (6.63) as well as in the definition of the S-matrix. The time ordered product of two bosonic operators is defined to be (see (6.61))

$$T(\phi(x)\phi(y)) = T(\phi(y)\phi(x)) = \begin{cases} \phi(x)\phi(y), & x^0 > y^0, \\ \phi(y)\phi(x), & y^0 > x^0. \end{cases} \quad (6.97)$$

Note that the order of the factors, inside time ordering, is not important for bosons. More compactly, we can write the time ordered product of two field operators in (6.97) as

$$T(\phi(x)\phi(y)) = \theta\left(x^0 - y^0\right)\phi(x)\phi(y) + \theta\left(y^0 - x^0\right)\phi(y)\phi(x), \quad (6.98)$$

where $\theta(x)$ represents the step function, defined to be

$$\theta(x) = \begin{cases} 1, & \text{for } x > 0, \\ 0, & \text{for } x < 0. \end{cases} \qquad (6.99)$$

We note here that, if the two time arguments are equal, there is no ambiguity in the definition of the time ordered product in (6.97) or (6.98) since the field operators commute at equal times. However, for operators which do not commute at equal times, the time ordered product has to be defined more carefully. Fortunately, in most of our discussions, this would not be a problem.

We note, in this simple example, that the equal time limit can be approached by assuming that $\theta(0) = \frac{1}{2}$. However, in more complicated situations, the equal time limit has to be taken in a consistent limiting manner. Thus, for example, we may choose the equal time limit as $x^0 \to y^0 + 0^+$ or $y^0 \to x^0 + 0^+$ with the appropriate limiting behavior of the step function. But, whatever limiting procedure we choose must be consistent through out. Let us note here that an integral representation for the step function which has the limiting behavior built in is given by

$$\theta(x) = \lim_{\epsilon \to 0^+} -\int \frac{dk}{2\pi i} \frac{e^{-ikx}}{k + i\epsilon}. \qquad (6.100)$$

Of course, by definition

$$T(\phi(x)) = \phi(x) =: \phi(x) :. \qquad (6.101)$$

If we have a product of three fields or more, we can generalize the definition of time ordered product in a straightforward manner. Thus, for the product of three field operators, we have

$$T\left(\phi(x)\phi(y)\phi(z)\right)$$
$$= \theta\left(x^0 - y^0\right)\theta\left(y^0 - z^0\right)\phi(x)\phi(y)\phi(z)$$
$$+ \theta\left(x^0 - z^0\right)\theta\left(z^0 - y^0\right)\phi(x)\phi(z)\phi(y)$$
$$+ \theta\left(y^0 - z^0\right)\theta\left(z^0 - x^0\right)\phi(y)\phi(z)\phi(x)$$
$$+ \theta\left(y^0 - x^0\right)\theta\left(x^0 - z^0\right)\phi(y)\phi(x)\phi(z)$$
$$+ \theta\left(z^0 - x^0\right)\theta\left(x^0 - y^0\right)\phi(z)\phi(x)\phi(y)$$
$$+ \theta\left(z^0 - y^0\right)\theta\left(y^0 - x^0\right)\phi(z)\phi(y)\phi(x), \quad (6.102)$$

and so on.

Given the definition of time ordered products, we can now relate them to normal ordered products as follows.

$$T\left(\phi(x)\right) = \phi(x) = :\phi(x):,$$
$$T\left(\phi(x)\phi(y)\right) = \theta\left(x^0 - y^0\right)\phi(x)\phi(y) + \theta\left(y^0 - x^0\right)\phi(y)\phi(x)$$
$$= \theta\left(x^0 - y^0\right)\left(:\phi(x)\phi(y): -i\, G^{(+)}(x-y)\right)$$
$$+ \theta\left(y^0 - x^0\right)\left(:\phi(y)\phi(x): -iG^{(+)}(y-x)\right). \quad (6.103)$$

Remembering that the order of the terms inside normal ordering does not matter for bosons, we obtain from (6.103)

$$T\left(\phi(x)\phi(y)\right)$$
$$= \left(\theta\left(x^0 - y^0\right) + \theta\left(y^0 - x^0\right)\right):\phi(x)\phi(y):$$
$$\quad -i\theta\left(x^0 - y^0\right)G^{(+)}(x-y) + i\theta\left(y^0 - x^0\right)G^{(-)}(x-y)$$
$$= :\phi(x)\phi(y): + i\Big(-\theta\left(x^0 - y^0\right)G^{(+)}(x-y)$$
$$\quad + \theta\left(y^0 - x^0\right)G^{(-)}(x-y)\Big)$$
$$= :\phi(x)\phi(y): + iG_F(x-y), \quad (6.104)$$

where we have used (5.147) as well as the definition of the Feynman Green's function in (5.146) in the intermediate step.

Let us write (6.104) as

$$T\left(\phi(x)\phi(y)\right) =: \phi(x)\phi(y) : + \overline{\phi(x)\phi(y)}, \tag{6.105}$$

where the contraction of two fields is defined as the c-number Feynman Green's function, namely,

$$\overline{\phi(x)\phi(y)} = iG_F(x-y). \tag{6.106}$$

Note that, since the Feynman Green's function is symmetric (it is an even function, see (5.148)), the order of the contraction is not important unlike the pairing of fields defined in (6.81). Furthermore, because the vacuum expectation value of a normal ordered product vanishes, we immediately identify

$$\langle 0|T\left(\phi(x)\phi(y)\right)|0\rangle = \overline{\phi(x)\phi(y)} = iG_F(x-y). \tag{6.107}$$

Namely, the Feynman Green's function can be identified with the vacuum expectation value of the time ordered product of two field operators which brings out yet another connection between the Green's functions and the quantum field theory. As we have seen earlier, the Feynman Green's function, unlike the Schwinger function, satisfies

$$\left(\partial_{x\mu}\partial^{x\mu} + m^2\right)G_F(x-y) = -\delta^4(x-y). \tag{6.108}$$

Since we know how to express the product of any number of field operators in terms of normal ordered products and pairings, we can carry through this construction to a time ordered product of any number of field operators as well. Let us simply note the results here.

6.6 Time ordered products and Wick's theorem

$$T\left(\phi(x)\phi(y)\phi(z)\right)$$
$$= \;:\phi(x)\phi(y)\phi(z): + \overline{\phi(x)\phi}(y):\phi(z):$$
$$+ \overline{\phi(y)\phi}(z):\phi(x): + \overline{\phi(x)\phi}(z):\phi(y):,$$

$$T\left(\phi(x)\phi(y)\phi(z)\phi(w)\right)$$
$$= \;:\phi(x)\phi(y)\phi(z)\phi(w): + \overline{\phi(x)\phi}(y):\phi(z)\phi(w):$$
$$+ \overline{\phi(x)\phi}(z):\phi(y)\phi(w): + \overline{\phi(x)\phi}(w):\phi(y)\phi(z):$$
$$+ \overline{\phi(y)\phi}(z):\phi(x)\phi(w): + \overline{\phi(y)\phi}(w):\phi(x)\phi(z):$$
$$+ \overline{\phi(z)\phi}(w):\phi(x)\phi(y): + \overline{\phi(x)\phi}(y)\,\overline{\phi(z)\phi}(w)$$
$$+ \overline{\phi(x)\phi}(z)\,\overline{\phi(y)\phi}(w) + \overline{\phi(x)\phi}(w)\,\overline{\phi(y)\phi}(z). \quad (6.109)$$

Here, we have to remember that

$$\overline{\phi(x)\phi}(y) = iG_F(x-y) = \overline{\phi(y)\phi}(x), \quad (6.110)$$

namely, this is an even function and the two contractions are, therefore, not distinct. This again shows that the time ordered product of any number of fields can be written as a sum of normal ordered terms with all possible distinct contractions. Namely,

$$T\left(\phi(x_1)\phi(x_2)\cdots\phi(x_n)\right) =: \phi(x_1)\phi(x_2)\cdots\phi(x_n):$$
$$+ \;\overline{\phi(x_1)\phi}(x_2):\phi(x_3)\cdots\phi(x_n):$$
$$+ \;\cdots$$
$$+ \;\overline{\phi(x_1)\phi}(x_2)\,\overline{\phi(x_3)\phi}(x_4):\phi(x_5)\cdots\phi(x_n):$$
$$+ \;\cdots. \quad (6.111)$$

This is another form of Wick's theorem and without going into derivations, we note here that a time ordered product of normal ordered products can be written as a sum of normal ordered terms with all possible contractions except for the contractions between field operators within a given normal ordered factor. It follows now that since the vacuum expectation value of a normal ordered product vanishes, the vacuum expectation value of a time ordered product of field operators (where the factors may or may not be normal ordered) is given only by the terms where all the field operators have been pairwise contracted – namely, such a vacuum expectation value, if it is nonzero, will involve products of Feynman Green's functions. Let us note here that the vacuum expectation value of a product of operators is a fundamental quantity in a quantum field theory since any matrix element can be written as a vacuum expectation value. As a result, the scattering matrix elements (see (6.68)) can be written as vacuum expectation values of time ordered products of fields (coming from normal ordered interaction terms if the Hamiltonian is normal ordered as well as field operators coming from the initial and the final states if they do not correspond to vacuum states). Therefore, this clarifies why it is the Feynman Green's function that is so important in the calculation of scattering matrix elements in a relativistic quantum field theory.

6.7 Spectral representation and dispersion relation

It is very rarely that we can solve an interacting field theory exactly. When we cannot solve a theory, it is useful to derive as much information as we can about the theory from its invariance and symmetry properties. In this section, we will discuss briefly two such interesting topics that often play an important role in studies of field theory.

We have already seen that vacuum expectation values of field operators are quite important and, in fact, in the case of free fields we have already seen how the vacuum expectation values of two free field operators can be related to various Green's functions. For example, let us note from (5.160) and (5.153) that, for free fields we can write the Schwinger function as (the subsequent arguments hold for any other vacuum expectation value of two fields)

6.7 SPECTRAL REPRESENTATION AND DISPERSION RELATION

$$\langle 0|\,[\phi(x),\phi(y)]\,|0\rangle = -iG(x-y)$$
$$= \int \frac{d^4k}{(2\pi)^3}\,\epsilon(k_0)\delta(k^2-m^2)\,e^{-ik\cdot(x-y)}. \quad (6.112)$$

Let us now consider two arbitrary operators $A(x)$ and $B(x)$ that do not necessarily obey the free field equations (these could, for example, represent the field operators in the fully interacting theory) and define

$$\langle 0|\,[A(x),B(y)]\,|0\rangle = -iG_{AB}(x-y), \quad (6.113)$$

and we would like to derive as much information as is possible about this function from the known facts about the quantum field theory.

Let us recall that the vacuum and the one particle states of the theory (see (5.90), (5.101) and (5.102)) satisfy

$$P_\mu|0\rangle = 0,$$
$$P_\mu|k\rangle = k_\mu|k\rangle, \quad k^0 = E_k = \sqrt{\mathbf{k}^2+m^2} > 0. \quad (6.114)$$

The true physical vacuum $|0\rangle$ is a Lorentz invariant state. Furthermore, we note that since P_μ generates translations of coordinates, we can write

$$A(x) = e^{iP\cdot x}A(0)e^{-iP\cdot x}, \quad B(y) = e^{iP\cdot y}B(0)e^{-iP\cdot y}. \quad (6.115)$$

As a result, inserting a complete set of energy-momentum basis states into (6.113) we obtain

$$G_{AB}(x-y) = i\langle 0|\,[A(x),B(y)]\,|0\rangle$$
$$= i\sum_n [\langle 0|A(x)|n\rangle\langle n|B(y)|0\rangle - \langle 0|B(y)|n\rangle\langle n|A(x)|0\rangle]$$
$$= i\sum_n \Big[e^{-ik_n\cdot(x-y)}\,\langle 0|A(0)|n\rangle\langle n|B(0)|0\rangle$$
$$\qquad - e^{ik_n\cdot(x-y)}\,\langle 0|B(0)|n\rangle\langle n|A(0)|0\rangle\Big], \quad (6.116)$$

where we have used (6.115) as well as the properties of the state vectors (6.114), in particular,

$$P_\mu |n\rangle = k_{n\mu} |n\rangle. \tag{6.117}$$

Note that since the operators A and B are not free field operators, the complete intermediate states $|n\rangle$ need not involve only single particle states.

In particular, if we identify $A(x) = B(x)$, and recall that the energy eigenstates are defined for positive energy ($k_{n0} > 0$), then (6.116) leads to

$$\begin{aligned}
G_{AA}(x-y) &= i\langle 0| \left[A(x), A(y)\right] |0\rangle \\
&= i \sum_n |\langle 0|A(0)|n\rangle|^2 \theta(k_{n0}) \left(e^{-ik_n \cdot (x-y)} - e^{ik_n \cdot (x-y)}\right) \\
&= \frac{i}{(2\pi)^3} \sum_n \int d^4q\, (2\pi)^3 \theta(q_0) \delta^4(q-k_n) |\langle 0|A(0)|n\rangle|^2 \\
&\quad \times \left(e^{-iq \cdot (x-y)} - e^{iq \cdot (x-y)}\right) \\
&= \frac{i}{(2\pi)^3} \int d^4q\, \rho(q)\, \theta(q_0) \left(e^{-iq \cdot (x-y)} - e^{iq \cdot (x-y)}\right), \tag{6.118}
\end{aligned}$$

where

$$\rho(q) = (2\pi)^3 \sum_n \delta^4(q - k_n) |\langle 0|A(0)|0\rangle|^2, \tag{6.119}$$

is known as the spectral function associated with the vacuum expectation value of the commutator and measures all the contribution coming from the matrix elements for a given q. Because of the Lorentz invariance of the theory, it is straightforward to show that the spectral function is really a function of q^2, namely,

$$\rho(q) = \rho(q^2), \tag{6.120}$$

so that we can write from (6.118)

6.7 SPECTRAL REPRESENTATION AND DISPERSION RELATION

$$\begin{aligned}G_{AA}(x-y) &= \frac{i}{(2\pi)^3}\int d^4q\, \rho(q^2)\theta(q_0)\left(e^{-iq\cdot(x-y)} - e^{iq\cdot(x-y)}\right)\\
&= \int d^4q\, \rho(q^2)\, \frac{i}{(2\pi)^3}\, \epsilon(q_0)\, e^{-iq\cdot(x-y)}\\
&= \int_0^\infty d\sigma^2\, \rho(\sigma^2)\, i\!\int \frac{d^4q}{(2\pi)^3}\, \epsilon(q_0)\delta(q^2-\sigma^2)\, e^{-iq\cdot(x-y)}\\
&= \int_0^\infty d\sigma^2\, \rho(\sigma^2)\, G(x-y, \sigma^2), \quad (6.121)\end{aligned}$$

where we have used (5.153) (or (6.112)) in the last step. Relation (6.121) is known as the spectral representation of the vacuum expectation value of the commutator and expresses it in terms of the vacuum expectation value of the free field commutator in (6.112). Here σ^2 can be thought of as the spectral parameter (eigenvalue of P^2) of the free Klein-Gordon equation and (6.121) represents the vacuum expectation value in terms of contributions coming from different mass values (which is the reason for the name spectral representation or spectral decomposition).

The second topic that we will discuss in this section goes under the name of dispersion relations and can be profitably used in the study of scattering theory. This can be very simply understood as follows. Let us suppose that we have a retarded function of time which we denote by $f(t)$ (see, for example, the retarded Green's function in (5.133)). Since the function is a retarded function (namely, it can be written as $f(t) = \theta(t)g(t)$ and since $\theta(t)\theta(t) = \theta(t)$), we can write

$$f(t) = \theta(t)f(t). \quad (6.122)$$

We are suppressing here any dependence of the function $f(t)$ on other arguments that are not relevant to our discussion. Let us assume that the Fourier transform $f(k_0)$ of the function exists and is given by

$$f(t) = \int_{-\infty}^{\infty} \frac{dk_0}{2\pi}\, e^{-ik_0 t}\, f(k_0). \quad (6.123)$$

Using the integral representation of the step function

$$\theta(t) = \lim_{\epsilon \to 0^+} -\frac{1}{2\pi i} \int_{-\infty}^{\infty} dk_0 \, \frac{e^{-ik_0 t}}{k_0 + i\epsilon}, \tag{6.124}$$

from (6.122) we obtain

$$\begin{aligned} f(t) &= \lim_{\epsilon \to 0^+} \left(-\frac{1}{2\pi i} \int_{-\infty}^{\infty} dk_0 \, \frac{e^{-ik_0 t}}{k_0 + i\epsilon} \right) \int_{-\infty}^{\infty} \frac{dk_0'}{2\pi} e^{-ik_0' t} f(k_0') \\ &= \lim_{\epsilon \to 0^+} -\iint_{-\infty}^{\infty} \frac{dk_0 dk_0'}{(2\pi)^2 i} \frac{e^{-i(k_0 + k_0')t}}{k_0 + i\epsilon} f(k_0') \\ &= \lim_{\epsilon \to 0^+} -\iint_{-\infty}^{\infty} \frac{dk_0 dk_0'}{(2\pi)^2 i} \frac{e^{-ik_0 t}}{k_0 - k_0' + i\epsilon} f(k_0') \\ &= \lim_{\epsilon \to 0^+} \int_{-\infty}^{\infty} \frac{dk_0}{2\pi} e^{-ik_0 t} \int_{-\infty}^{\infty} \frac{dk_0'}{2\pi i} \frac{f(k_0')}{k_0' - k_0 - i\epsilon}. \end{aligned} \tag{6.125}$$

Comparing (6.125) with (6.123), we conclude that the Fourier transform of a retarded function must satisfy a relation

$$f(k_0) = \lim_{\epsilon \to 0^+} \frac{1}{2\pi i} \int_{-\infty}^{\infty} dk_0' \, \frac{f(k_0')}{k_0' - k_0 - i\epsilon}, \tag{6.126}$$

which is the basic dispersion relation for the Fourier transform of a the retarded function. (We note here that in the conventional discussion of dispersion relations, the conjugate variable k_0 is identified with ω.)

Decomposing the denominator in (6.126) in the standard manner

$$\lim_{\epsilon \to 0^+} \frac{1}{k_0' - k_0 - i\epsilon} = \frac{1}{k_0' - k_0} + i\pi \delta(k_0' - k_0), \tag{6.127}$$

where the first term on the right-hand side represents the principal part, the dispersion relation (6.126) can also be written as

$$f(k_0) = \frac{1}{\pi i} \int_{-\infty}^{\infty} dk_0' \, \frac{f(k_0')}{k_0' - k_0}, \tag{6.128}$$

6.7 Spectral representation and dispersion relation

where the principal value for the pole is understood. Comparing the real and the imaginary parts in (6.128) we obtain

$$\begin{aligned}
\operatorname{Re} f(k_0) &= \frac{1}{\pi} \int_{-\infty}^{\infty} dk_0' \frac{\operatorname{Im} f(k_0')}{k_0' - k_0}, \\
\operatorname{Im} f(k_0) &= -\frac{1}{\pi} \int_{-\infty}^{\infty} dk_0' \frac{\operatorname{Re} f(k_0')}{k_0' - k_0}.
\end{aligned} \qquad (6.129)$$

This is quite interesting because it says that the real part of the Fourier transform of the retarded function can be obtained from a knowledge of its imaginary part in the entire complex k_0 plane and *vice versa*. This form of the dispersion relation, where the real part of an amplitude is related to the imaginary part (or the other way around), is known as the Kramers-Kronig relation. Let us recall from the optical theorem in scattering in quantum mechanics that the imaginary part of the forward scattering amplitude is related to the total scattering cross section. Therefore, the first of the relations in (6.129) can also be seen to imply that the real part of a forward scattering amplitude at a fixed energy (frequency) can be given as the integral over all energies of the total cross section. In optics, the first of the relations in (6.129) can be used to determine the real part of the refractive index of a medium in terms of its imaginary parts (which is how Kramers and Kronig had used such a dispersion relation). Using (6.129) we can also write

$$\begin{aligned}
f(k_0) &= \operatorname{Re} f(k_0) + i \operatorname{Im} f(k_0) \\
&= \frac{1}{\pi} \int_{-\infty}^{\infty} dk_0' \left[\frac{\operatorname{Im} f(k_0')}{k_0' - k_0} + i\pi \delta(k_0' - k_0) \operatorname{Im} f(k_0') \right] \\
&= \frac{1}{\pi} \int_{-\infty}^{\infty} dk_0' \frac{\operatorname{Im} f(k_0')}{k_0' - k_0 - i\epsilon},
\end{aligned} \qquad (6.130)$$

where $\epsilon \to 0^+$ is assumed. This relation says that we can determine the complete function $f(k_0)$ (which is the Fourier transform of a retarded function $f(t)$) from a knowledge of its imaginary part in the entire complex k_0 plane. There is a similar relation where the function can be written in terms of its real part alone. Without going into details we note here that dispersion relations inherently imply causality in a theory.

Our derivation for the dispersion relations has been quite formal and it is clear that (6.130) will hold (and is meaningful) only if

$$\lim_{|k_0|\to\infty} \operatorname{Im} f(k_0) \to 0. \tag{6.131}$$

This is quite interesting for it says that, as long as only the imaginary part (or the real part in an alternate formulation) of the Fourier transform of a retarded function vanishes for large values of the argument, the dispersion relation (6.130) would hold. This can be contrasted with the vanishing of an analytic function for large values of its argument (which is what we use in complex analysis) which would imply that both the real and the imaginary parts of the function have to vanish in this regime. If $\operatorname{Im} f(k_0)$ does not fall off fast enough, then we can define a subtracted dispersion relation that will be true and this is done as follows. Let $\bar{k}_0$ denote a fixed point in the complex k_0 plane and let us define a new function

$$F(k_0) = \frac{f(k_0) - f(\bar{k}_0)}{k_0 - \bar{k}_0}, \tag{6.132}$$

Clearly, the imaginary part of this function will vanish for large values of its argument provided $\operatorname{Im} f(k_0)$ does not grow for large values of its argument. Therefore, for this function we can write a dispersion relation of the form (6.130)

$$\begin{aligned}
F(k_0) &= \lim_{\epsilon\to 0^+} \frac{1}{\pi} \int_{-\infty}^{\infty} dk_0' \frac{\operatorname{Im} F(k_0')}{k_0' - k_0 - i\epsilon} \\
&= \lim_{\epsilon\to 0^+} \frac{1}{\pi} \int_{-\infty}^{\infty} dk_0' \frac{\operatorname{Im}\left(f(k_0') - f(\bar{k}_0)\right)}{(k_0' - \bar{k}_0)(k_0' - k_0 - i\epsilon)} \\
&= \lim_{\epsilon\to 0^+} \frac{1}{\pi} \int_{-\infty}^{\infty} dk_0' \frac{\operatorname{Im}\left(f(k_0') - f(\bar{k}_0)\right)}{(k_0' - \bar{k}_0 - i\epsilon)(k_0' - k_0 - i\epsilon)} \\
&= \lim_{\epsilon\to 0^+} \frac{1}{\pi(k_0 - \bar{k}_0)} \int_{-\infty}^{\infty} dk_0' \operatorname{Im}\left(f(k_0') - f(\bar{k}_0)\right) \\
&\quad \times \left[\frac{1}{k_0' - k_0 - i\epsilon} - \frac{1}{k_0' - \bar{k}_0 - i\epsilon}\right]. \tag{6.133}
\end{aligned}$$

6.7 Spectral representation and dispersion relation

Using the definition in (6.132), we can also write this relation as

$$f(k_0) - f(\bar{k}_0) = \lim_{\epsilon \to 0^+} \frac{1}{\pi} \int_{-\infty}^{\infty} dk_0' \operatorname{Im}\left(f(k_0') - f(\bar{k}_0)\right)$$
$$\times \left[\frac{1}{k_0' - k_0 - i\epsilon} - \frac{1}{k_0' - \bar{k}_0 - i\epsilon} \right]. \quad (6.134)$$

In many cases of physical interest, the imaginary part of a retarded function is odd, namely,

$$\operatorname{Im} f(k_0) = -\operatorname{Im}(-k_0). \quad (6.135)$$

If this is true and if we choose the point of subtraction $\bar{k}_0 = 0$, then the relation (6.134) takes the form

$$f(k_0) - f(0) = \lim_{\epsilon \to 0^+} \frac{1}{\pi} \int_{-\infty}^{\infty} dk_0' \left[\frac{\operatorname{Im} f(k_0')}{k_0' - k_0 - i\epsilon} - \frac{\operatorname{Im} f(k_0')}{k_0' - i\epsilon} \right]. \quad (6.136)$$

In this case, we truly have a subtracted dispersion relation. However, if (6.135) does not hold, then the appropriate dispersion relation is given by (6.134). (Higher subtractions are needed if the function has a more singular behaviour at infinity.)

As a simple example of the discussions on the dispersion relation (6.130), let us look at the momentum space retarded propagator defined in (5.129) (in our metric $k^0 = k_0$)

$$G^R(k_0, \mathbf{k}) = \lim_{\eta \to 0^+} \frac{1}{k_0^2 - E_k^2 + ik_0\eta}. \quad (6.137)$$

This has the imaginary part given by

$$\operatorname{Im} G^R(k_0, \mathbf{k}) = -\pi \epsilon(k_0)\delta(k_0^2 - E_k^2)$$
$$= -\frac{\pi}{2E_k}\left[\delta(k_0 - E_k) - \delta(k_0 + E_k)\right]. \quad (6.138)$$

We note that the imaginary part, in this case, satisfies (6.135). Furthermore, it vanishes for large values of k_0 and, therefore, we expect (6.130) to hold in this case. Explicitly we note that

$$\lim_{\epsilon \to 0^+} \frac{1}{\pi} \int_{-\infty}^{\infty} dk_0' \, \frac{\operatorname{Im} G^R(k_0', \mathbf{k})}{k_0' - k_0 - i\epsilon}$$

$$= \lim_{\epsilon \to 0^+} -\frac{\pi}{2\pi E_k} \int dk_0' \, \frac{1}{k_0' - k_0 - i\epsilon} \left[\delta(k_0' - E_k) - \delta(k_0' + E_k) \right]$$

$$= \lim_{\epsilon \to 0^+} -\frac{1}{2E_k} \left[\frac{1}{E_k - k_0 - i\epsilon} - \frac{1}{-E_k - k_0 - i\epsilon} \right]$$

$$= \lim_{\epsilon \to 0^+} \frac{1}{(k_0 + i\epsilon)^2 - E_k^2}$$

$$= \lim_{\eta \to 0^+} \frac{1}{k_0^2 - E_k^2 + i k_0 \eta}, \tag{6.139}$$

where we have used the fact that ϵ is infinitesimal and have identified, as in (5.130), $\eta = 2\epsilon$. Equation (6.139) can be compared with the dispersion relation in (6.130).

6.8 References

1. J. Schwinger, Physical Review **74**, 1439 (1948); *ibid.* **75**, 651 (1949); *ibid.* **76**, 790 (1949); S. Tomonaga, Progress of Theoretical Physics **1**, 27 (1946).

2. F. J. Dyson, Physical Review **75**, 486 (1949); em ibid. **82**, 428 (1951).

3. G. C. Wick, Physical Review **80**, 268 (1950).

4. E. L. Hill, Reviews of Modern Physics **23**, 253 (1957).

5. S. Schweber, *Introduction to Relativistic Quantum Field Theory*, Row, Peterson, Evanston (1961).

6. J. D. Bjorken and S. Drell, *Relativistic Quantum Fields*, McGraw-Hill, New York, 1964.

7. P. Roman, *Introduction to Quantum Field Theory*, John Wliley, New York (1969).

8. C. Itzykson and J-B. Zuber, *Quantum Field Theory*, McGraw-Hill, New York, 1980.

9. N. N. Bogoliubov and D. V. Shirkov, *Introduction to the theory of Quantized Fields*, Nauka, Moscow (1984).

10. A. Das, *Lectures on Quantum Mechanics*, Hindustan Publishing, New Delhi, India, 2003.

CHAPTER 7
Complex scalar field theory

7.1 Quantization

The natural field theoretic example to consider next is that of a free Klein-Gordon field theory where the field operator is not Hermitian. In this case, the dynamical equations take the forms

$$\left(\partial_\mu \partial^\mu + m^2\right) \phi(x) = 0,$$
$$\left(\partial_\mu \partial^\mu + m^2\right) \phi^\dagger(x) = 0, \qquad \phi^\dagger(x) \neq \phi(x). \qquad (7.1)$$

Therefore, there are now two distinct equations. The physical meaning of this system can be understood by noting that we can express the complex field operator in terms of two real (Hermitian) field operators as

$$\phi(x) = \frac{1}{\sqrt{2}} \left(\phi_1(x) + i\phi_2(x)\right),$$
$$\phi^\dagger(x) = \frac{1}{\sqrt{2}} \left(\phi_1(x) - i\phi_2(x)\right), \qquad (7.2)$$

where $\phi_1(x)$ and $\phi_2(x)$ are two distinct, spin zero (scalar) fields which are Hermitian. The relations in (7.2) can be inverted to give

$$\phi_1(x) = \frac{1}{\sqrt{2}} \left(\phi(x) + \phi^\dagger(x)\right),$$
$$\phi_2(x) = -\frac{i}{\sqrt{2}} \left(\phi(x) - \phi^\dagger(x)\right). \qquad (7.3)$$

It follows now that we can express the dynamical equations in (7.1) in terms of the real field variables as

$$\left(\partial_\mu \partial^\mu + m^2\right) \phi_1(x) = 0,$$
$$\left(\partial_\mu \partial^\mu + m^2\right) \phi_2(x) = 0. \qquad (7.4)$$

Namely, such a system equivalently describes two free, independent spin zero Klein-Gordon fields (and, therefore, two distinct particles) with degenerate mass.

Therefore, we can think of the complex Klein-Gordon field theory as describing two independent field degrees of freedom – either in terms of $\phi_1(x)$ and $\phi_2(x)$ or through $\phi(x)$ and $\phi^\dagger(x)$ – each of which is treated as an independent variable. The Lagrangian density which will give the dynamical equations in (7.1) or (7.4) as the Euler-Lagrange equations is easily determined to be

$$\begin{aligned} \mathcal{L} &= \frac{1}{2} \partial_\mu \phi_1 \partial^\mu \phi_1 - \frac{m^2}{2} \phi_1^2 + \frac{1}{2} \partial_\mu \phi_2 \partial^\mu \phi_2 - \frac{m^2}{2} \phi_2^2 \\ &= \frac{1}{2} \partial_\mu \left(\phi_1 - i\phi_2\right) \partial^\mu \left(\phi_1 + i\phi_2\right) - \frac{m^2}{2} \left(\phi_1 - i\phi_2\right)\left(\phi_1 + i\phi_2\right) \\ &= \partial_\mu \phi^\dagger \partial^\mu \phi - m^2 \phi^\dagger \phi, \qquad (7.5) \end{aligned}$$

and is manifestly Hermitian. (Note that we have not been careful about the order of the factors in rearranging terms in (7.5) because distinct degrees of freedom are expected to commute.)

We can quantize this theory in one of the two equivalent ways. If we treat $\phi_1(x)$ and $\phi_2(x)$ as the independent variables, then we can define the momenta conjugate to the two field operators as

$$\Pi_i(x) = \frac{\partial \mathcal{L}}{\partial \dot{\phi}_i(x)} = \dot{\phi}_i(x), \qquad i = 1, 2. \qquad (7.6)$$

Consequently, the equal time canonical commutation relations follow to be

$$\begin{aligned} \left[\phi_i(x), \phi_j(y)\right]_{x^0 = y^0} &= 0 = \left[\Pi_i(x), \Pi_j(y)\right]_{x^0 = y^0}, \\ \left[\phi_i(x), \Pi_j(y)\right]_{x^0 = y^0} &= i\delta_{ij} \delta^3(x - y). \end{aligned} \qquad (7.7)$$

7.1 QUANTIZATION

In this case, the Hamiltonian density is obtained in a straightforward manner to be of the form

$$\begin{aligned}
\mathcal{H} &= \sum_{i=1}^{2} \left(\Pi_i \dot{\phi}_i\right) - \mathcal{L} \\
&= \Pi_1^2 + \Pi_2^2 - \frac{1}{2}\Pi_1^2 + \frac{1}{2}\nabla\phi_1 \cdot \nabla\phi_1 + \frac{m^2}{2}\phi_1^2 \\
&\quad - \frac{1}{2}\Pi_2^2 + \frac{1}{2}\nabla\phi_2 \cdot \nabla\phi_2 + \frac{m^2}{2}\phi_2^2 \\
&= \frac{1}{2}\Pi_1^2 + \frac{1}{2}\nabla\phi_1 \cdot \nabla\phi_1 + \frac{m^2}{2}\phi_1^2 \\
&\quad + \frac{1}{2}\Pi_2^2 + \frac{1}{2}\nabla\phi_2 \cdot \nabla\phi_2 + \frac{m^2}{2}\phi_2^2 \\
&= \sum_{i=1}^{2} \left(\frac{1}{2}\Pi_i^2 + \frac{1}{2}\nabla\phi_i \cdot \nabla\phi_i + \frac{m^2}{2}\phi_i^2\right),
\end{aligned} \tag{7.8}$$

where we have used the identification in (7.6). We see that the Hamiltonian density of the free complex scalar field theory is simply the sum of the Hamiltonian densities for two non-interacting real Klein-Gordon field theories with degenerate mass and

$$H = \int d^3x\, \mathcal{H}. \tag{7.9}$$

Alternatively, if we treat $\phi(x)$ and $\phi^\dagger(x)$ as the independent dynamical variables, then we can define the momenta conjugate to the field operators from (7.5) to be

$$\begin{aligned}
\Pi(x) &= \frac{\partial \mathcal{L}}{\partial \dot{\phi}^\dagger(x)} = \dot{\phi}(x) \\
&= \frac{1}{\sqrt{2}}\left(\dot{\phi}_1(x) + i\dot{\phi}_2(x)\right) = \frac{1}{\sqrt{2}}\left(\Pi_1(x) + i\Pi_2(x)\right),
\end{aligned}$$

$$\Pi^\dagger(x) = \frac{\partial \mathcal{L}}{\partial \dot{\phi}(x)} = \dot{\phi}^\dagger(x)$$
$$= \frac{1}{\sqrt{2}}\left(\dot{\phi}_1(x) - i\dot{\phi}_2(x)\right) = \frac{1}{\sqrt{2}}(\Pi_1(x) - i\Pi_2(x)). \quad (7.10)$$

As a result, the equal time canonical commutation relations, in this case, take the forms (we do not write the equal time condition explicitly for simplicity, although it is to be understood that $x^0 = y^0$ in all of the relations in (7.11))

$$[\phi(x), \phi(y)] = \left[\phi(x), \phi^\dagger(y)\right] = \left[\phi^\dagger(x), \phi^\dagger(y)\right] = 0,$$
$$[\Pi(x), \Pi(y)] = \left[\Pi(x), \Pi^\dagger(y)\right] = \left[\Pi^\dagger(x), \Pi^\dagger(y)\right] = 0,$$
$$\left[\phi(x), \Pi^\dagger(y)\right] = \left[\phi^\dagger(x), \Pi(y)\right] = i\delta^3(x-y). \quad (7.11)$$

Furthermore, using (7.5) and (7.10), we can derive the Hamiltonian density in this formulation to have the form

$$\begin{aligned}\mathcal{H} &= \Pi\dot{\phi}^\dagger + \Pi^\dagger\dot{\phi} - \mathcal{L} \\ &= \Pi\Pi^\dagger + \Pi^\dagger\Pi - \dot{\phi}^\dagger\dot{\phi} + \boldsymbol{\nabla}\phi^\dagger \cdot \boldsymbol{\nabla}\phi + m^2\phi^\dagger\phi \\ &= 2\Pi^\dagger\Pi - \Pi^\dagger\Pi + \boldsymbol{\nabla}\phi^\dagger \cdot \boldsymbol{\nabla}\phi + m^2\phi^\dagger\phi \\ &= \Pi^\dagger\Pi + \boldsymbol{\nabla}\phi^\dagger \cdot \boldsymbol{\nabla}\phi + m^2\phi^\dagger\phi, \quad (7.12)\end{aligned}$$

which can be compared with (7.8) and

$$H = \int d^3x\, \mathcal{H}. \quad (7.13)$$

Note that the Hamiltonian, in this case, is manifestly Hermitian as it should be even though the dynamical variables are not.

7.2 Field decomposition

As in the case of the Hermitian (real) Klein-Gordon field, we can expand the field $\phi(x)$ in the basis of plane wave solutions of the Klein-Gordon equation (see (5.56)), in the present case, of the form

7.2 FIELD DECOMPOSITION

$$\phi_i(x) = \int \frac{d^3k}{\sqrt{(2\pi)^3 2k^0}} \left(e^{-ik\cdot x} a_i(\mathbf{k}) + e^{ik\cdot x} a_i^\dagger(\mathbf{k})\right), \quad i = 1, 2, \tag{7.14}$$

with $k^0 = E_k = \sqrt{\mathbf{k}^2 + m^2} > 0$. As before, $a_i(\mathbf{k})$ and $a_i^\dagger(\mathbf{k})$, with $i = 1, 2$, can then be interpreted as the annihilation and the creation operators for the two kinds of quanta associated with the fields $\phi_i(x)$. The commutation relations between these creation and annihilation operators can then be obtained from the canonical quantization relations to be (see (5.66))

$$\left[a_i(\mathbf{k}), a_j(\mathbf{k}')\right] = 0 = \left[a_i^\dagger(\mathbf{k}), a_j^\dagger(\mathbf{k}')\right],$$

$$\left[a_i(\mathbf{k}), a_j^\dagger(\mathbf{k}')\right] = \delta_{ij} \delta^3(k - k'). \tag{7.15}$$

Given the decomposition in (7.14), it is clear that the field operators $\phi(x)$ and $\phi^\dagger(x)$ can also be expanded in the plane wave basis as (see (7.2))

$$\begin{aligned}
\phi(x) &= \frac{1}{\sqrt{2}} (\phi_1(x) + i\phi_2(x)) \\
&= \frac{1}{\sqrt{2}} \int \frac{d^3k}{\sqrt{(2\pi)^3 2k^0}} \left[e^{-ik\cdot x} (a_1(\mathbf{k}) + ia_2(\mathbf{k})) \right. \\
&\quad \left. + e^{ik\cdot x} \left(a_1^\dagger(\mathbf{k}) + ia_2^\dagger(\mathbf{k})\right)\right] \\
&= \int \frac{d^3k}{\sqrt{(2\pi)^3 2k^0}} \left(e^{-ik\cdot x} a(\mathbf{k}) + e^{ik\cdot x} b^\dagger(\mathbf{k})\right), \\
\phi^\dagger(x) &= \int \frac{d^3k}{\sqrt{(2\pi)^3 2k^0}} \left(e^{-ik\cdot x} b(\mathbf{k}) + e^{ik\cdot x} a^\dagger(\mathbf{k})\right), \tag{7.16}
\end{aligned}$$

where we have defined

$$\begin{aligned}
a(\mathbf{k}) &= \frac{1}{\sqrt{2}} (a_1(\mathbf{k}) + ia_2(\mathbf{k})), \\
b(\mathbf{k}) &= \frac{1}{\sqrt{2}} (a_1(\mathbf{k}) - ia_2(\mathbf{k})), \tag{7.17}
\end{aligned}$$

which correspond to the annihilation operators associated with the quanta for the two fields $\phi(x)$ and $\phi^\dagger(x)$ respectively.

The commutation relations between the $a(\mathbf{k})$, $b(\mathbf{k})$, $a^\dagger(\mathbf{k})$ and $b^\dagger(\mathbf{k})$ can now be obtained from (7.15) and the definition in (7.17) (or directly from (7.11) and (7.16)) and take the form

$$\left[a(\mathbf{k}), a(\mathbf{k}')\right] = \left[b(\mathbf{k}), b(\mathbf{k}')\right] = \left[a(\mathbf{k}), b(\mathbf{k}')\right] = 0,$$

$$\left[a^\dagger(\mathbf{k}), a^\dagger(\mathbf{k}')\right] = \left[a^\dagger(\mathbf{k}), b^\dagger(\mathbf{k}')\right] = \left[b^\dagger(\mathbf{k}), b^\dagger(\mathbf{k}')\right] = 0,$$

$$\left[a(\mathbf{k}), b^\dagger(\mathbf{k}')\right] = \left[b(\mathbf{k}), a^\dagger(\mathbf{k}')\right] = 0,$$

$$\left[a(\mathbf{k}), a^\dagger(\mathbf{k}')\right] = \left[b(\mathbf{k}), b^\dagger(\mathbf{k}')\right] = \delta^3(k - k'). \tag{7.18}$$

The Hamiltonian for the system (see (7.8) or (7.12)) can be written, in terms of the creation and the annihilation operators, in two equivalent ways as

$$\begin{aligned} H &= \int \mathrm{d}^3 x\, \mathcal{H} \\ &= \int \mathrm{d}^3 k\, \frac{E_k}{2} \left(a_1^\dagger(\mathbf{k}) a_1(\mathbf{k}) + a_1(\mathbf{k}) a_1^\dagger(\mathbf{k}) \right. \\ &\qquad\qquad\qquad \left. + a_2^\dagger(\mathbf{k}) a_2(\mathbf{k}) + a_2(\mathbf{k}) a_2^\dagger(\mathbf{k}) \right) \\ &= \int \mathrm{d}^3 k\, \frac{E_k}{2} \left(a^\dagger(\mathbf{k}) a(\mathbf{k}) + a(\mathbf{k}) a^\dagger(\mathbf{k}) \right. \\ &\qquad\qquad\qquad \left. + b^\dagger(\mathbf{k}) b(\mathbf{k}) + b(\mathbf{k}) b^\dagger(\mathbf{k}) \right). \end{aligned} \tag{7.19}$$

If we normal order the Hamiltonian in (7.19) with respect to the $a(\mathbf{k})$ and $b(\mathbf{k})$ operators, we can write

$$:H: = \int \mathrm{d}^3 k\, E_k \left(a^\dagger(\mathbf{k}) a(\mathbf{k}) + b^\dagger(\mathbf{k}) b(\mathbf{k}) \right). \tag{7.20}$$

(We can equivalently normal order the Hamiltonian (7.19) with respect to the $a_i(\mathbf{k})$ operators and in that case the Hamiltonian would be a sum of two Klein-Gordon Hamiltonians that we have already studied in chapter **5**.)

The vacuum state in such a case can be defined to be the lowest energy state annihilated by both $a(\mathbf{k})$ and $b(\mathbf{k})$. Namely,

$$a(\mathbf{k})|0\rangle = 0 = b(\mathbf{k})|0\rangle = \langle 0|a^\dagger(\mathbf{k}) = \langle 0|b^\dagger(\mathbf{k}), \quad H|0\rangle = 0. \tag{7.21}$$

However, there will now be two distinct one particle states degenerate in mass (and, therefore, energy) given by

$$\begin{aligned} |k\rangle &= a^\dagger(\mathbf{k})|0\rangle, \\ |\tilde{k}\rangle &= b^\dagger(\mathbf{k})|0\rangle. \end{aligned} \tag{7.22}$$

Similarly, we can also construct higher order states which will also show a degeneracy that was not present in the case of the real Klein-Gordon theory. The degeneracy can be understood as follows.

7.3 Charge operator

To understand the physical meaning of the two distinct and degenerate one particle states, let us analyze the symmetry properties of this system. After all, as we know from our studies in non-relativistic quantum mechanics, degeneracies arise when there is a symmetry present in the system under study. The theory is, of course, manifestly Lorentz invariant and translation invariant, as was also the case for a real Klein-Gordon theory. However, in the present case, there is in addition an internal symmetry associated with the system. Namely, under the transformation

$$\begin{aligned} \phi(x) &\to \phi'(x) = e^{-i\theta}\phi(x), \\ \phi^\dagger(x) &\to \phi'^\dagger(x) = \phi^\dagger(x)e^{i\theta}, \end{aligned} \tag{7.23}$$

where θ is a real constant parameter of transformation (a global transformation), as we will see shortly, the theory is invariant. Such a transformation is not a space-time symmetry transformation since the space-time coordinates are not changed by this transformation. Rather, such a transformation is known as an internal symmetry transformation. Infinitesimally, the transformation in (7.23) takes the form ($\alpha = \epsilon =$ infinitesimal)

$$\begin{aligned}\delta\phi(x) &= \phi'(x) - \phi(x) = -i\epsilon\phi(x),\\ \delta\phi^\dagger(x) &= \phi'^\dagger(x) - \phi^\dagger(x) = i\epsilon\phi^\dagger(x),\end{aligned} \qquad (7.24)$$

where ϵ denotes the infinitesimal constant parameter of transformation.

We note that under the global phase transformation (7.23) (or the infinitesimal form of it in (7.24)), the change in the Lagrangian density (see (6.8)) is given by

$$\begin{aligned}&\mathcal{L}\left(\phi'(x), \partial_\mu\phi'(x), \phi'^\dagger(x), \partial_\mu\phi'^\dagger(x)\right)\\ &\quad -\mathcal{L}\left(\phi(x), \partial_\mu\phi(x), \phi^\dagger(x), \partial_\mu\phi^\dagger(x)\right)\\ &= \partial_\mu\phi'^\dagger\partial^\mu\phi' - m^2\phi'^\dagger\phi' - \partial_\mu\phi^\dagger\partial^\mu\phi + m^2\phi^\dagger\phi\\ &= \partial_\mu\phi^\dagger e^{i\theta}e^{-i\theta}\partial^\mu\phi - m^2\phi^\dagger e^{i\theta}e^{-i\theta}\phi - \partial_\mu\phi^\dagger\partial^\mu\phi + m^2\phi^\dagger\phi\\ &= 0.\end{aligned} \qquad (7.25)$$

The derivatives do not act on the exponentials since θ is a constant parameter (independent of space-time coordinates). Thus, in this case, the Lagrangian density (7.5) is invariant under the global phase transformation in (7.23) (or the infinitesimal form (7.24)) showing that it is a symmetry of the system and in this case we have

$$K^\mu = 0. \qquad (7.26)$$

On the other hand, for infinitesimal transformations in (7.24), we have (see (6.13))

7.3 Charge operator

$$\delta\phi \frac{\partial \mathcal{L}}{\partial \partial_\mu \phi} + \delta\phi^\dagger \frac{\partial \mathcal{L}}{\partial \partial_\mu \phi^\dagger}$$
$$= -i\epsilon\phi(x)\partial^\mu \phi^\dagger(x) + i\epsilon\phi^\dagger(x)\partial^\mu \phi(x)$$
$$= i\epsilon \left(\phi^\dagger(x)\partial^\mu \phi(x) - \phi(x)\partial^\mu \phi^\dagger(x) \right)$$
$$= i\epsilon \left(\phi^\dagger(x)\partial^\mu \phi(x) - (\partial^\mu \phi^\dagger(x))\phi(x) \right)$$
$$= i\epsilon \phi^\dagger(x)\overleftrightarrow{\partial^\mu}\phi(x). \tag{7.27}$$

(Note here that classically the order of the factors is not important. Even in a quantum theory, as we have seen, under normal ordering the order of factors does not matter (for a bosonic theory). Of course, we normal order all observables such as currents and charges and, therefore, we have freely rearranged the order of factors in (7.27).)

Therefore, we can define the Nöther current associated with this symmetry to be (see (6.13))

$$J^\mu_\epsilon = \delta\phi \frac{\partial \mathcal{L}}{\partial \partial_\mu \phi} + \delta\phi^\dagger \frac{\partial \mathcal{L}}{\partial \partial_\mu \phi^\dagger} - K^\mu = i\epsilon\phi^\dagger(x)\overleftrightarrow{\partial^\mu}\phi(x), \tag{7.28}$$

and the parameter independent current follows to be

$$J^\mu = i\phi^\dagger(x)\overleftrightarrow{\partial^\mu}\phi(x). \tag{7.29}$$

This is a vector current since the parameter of transformation is a scalar. (In fact, this is the probability current density (up to a normalization) which we had studied in the first quantized theory. But here it is an operator.) Furthermore, the current can be easily checked to be conserved using the equations of motion in (7.1)

$$\partial_\mu J^\mu(x) = 0. \tag{7.30}$$

Using the field decomposition (7.16), we can show that the associated conserved charge of the theory can be written in terms of the creation and annihilation operators as

$$\begin{aligned} Q &= \int \mathrm{d}^3x\, J^0(x) = i \int \mathrm{d}^3x\, \phi^\dagger(x) \overleftrightarrow{\partial^0} \phi(x) \\ &= i \int \mathrm{d}^3x \left(\phi^\dagger(x)\dot\phi(x) - \dot\phi^\dagger(x)\phi(x) \right) \\ &= \int \mathrm{d}^3k \left(a^\dagger(\mathbf{k})a(\mathbf{k}) - b(\mathbf{k})b^\dagger(\mathbf{k}) \right). \end{aligned} \qquad (7.31)$$

If we normal order, the charge operator takes the form

$$:Q: = \int \mathrm{d}^3k \left(a^\dagger(\mathbf{k})a(\mathbf{k}) - b^\dagger(\mathbf{k})b(\mathbf{k}) \right). \qquad (7.32)$$

Since this charge operator is associated with a global phase transformation of the kind associated with electromagnetic interactions (namely, a $U(1)$ phase transformation) we can identify this operator as the electric charge operator. We note that the normal ordered charge operator (7.32) annihilates the vacuum,

$$Q|0\rangle = \int \mathrm{d}^3k \left(a^\dagger(\mathbf{k})a(\mathbf{k}) - b^\dagger(\mathbf{k})b(\mathbf{k}) \right) |0\rangle = 0, \qquad (7.33)$$

so that the vacuum does not carry any charge. Here, we are suppressing the normal ordering symbol for simplicity with the understanding that the charge operator has the normal ordered form in (7.32). Acting on the first of the two one particle states in (7.22), the charge operator gives

$$\begin{aligned} Q|k\rangle &= \int \mathrm{d}^3k' \left(a^\dagger(\mathbf{k'})a(\mathbf{k'}) - b^\dagger(\mathbf{k'})b(\mathbf{k'}) \right) a^\dagger(\mathbf{k})|0\rangle \\ &= \int \mathrm{d}^3k'\, a^\dagger(\mathbf{k'})a(\mathbf{k'})a^\dagger(\mathbf{k})|0\rangle \\ &= \int \mathrm{d}^3k'\, a^\dagger(\mathbf{k'}) \left(\left[a(\mathbf{k'}), a^\dagger(\mathbf{k}) \right] + a^\dagger(\mathbf{k})a(\mathbf{k'}) \right) |0\rangle \\ &= \int \mathrm{d}^3k'\, a^\dagger(\mathbf{k'}) \delta^3(\mathbf{k'}-\mathbf{k})|0\rangle = a^\dagger(\mathbf{k})|0\rangle = |k\rangle. \quad (7.34) \end{aligned}$$

7.3 CHARGE OPERATOR

In other words, the one particle state $|k\rangle$ is an eigenstate of the charge operator with eigenvalue $(+1)$. In contrast, acting on the second one particle state in (7.22), the charge operator leads to

$$\begin{aligned} Q|\tilde{k}\rangle &= \int d^3k' \left(a^\dagger(\mathbf{k}')a(\mathbf{k}') - b^\dagger(\mathbf{k}')b(\mathbf{k}') \right) b^\dagger(\mathbf{k})|0\rangle \\ &= -\int d^3k'\, b^\dagger(\mathbf{k}')b(\mathbf{k}')b^\dagger(\mathbf{k})|0\rangle \\ &= -\int d^3k'\, b^\dagger(\mathbf{k}') \left(\left[b(\mathbf{k}'), b^\dagger(\mathbf{k})\right] + b^\dagger(\mathbf{k})b(\mathbf{k}') \right) |0\rangle \\ &= -\int d^3k'\, b^\dagger(\mathbf{k}')\delta^3(k'-k)|0\rangle = -b^\dagger(\mathbf{k})|0\rangle \\ &= -|\tilde{k}\rangle. \end{aligned} \qquad (7.35)$$

Namely, the second one particle state of the theory $|\tilde{k}\rangle$ is also an eigenstate of the charge operator but with the eigenvalue (-1).

The meaning of the two one particle states is now clear. They describe two distinct particles with degenerate mass but with exactly opposite electric charge. These are, of course, the particle and the anti-particle states – but both of them now correspond to positive energy states. Therefore, the second quantized theory leads to a much more satisfactory description of particles and anti-particles. The vacuum in this bosonic theory, in particular, contains no particle and this should be contrasted with the definition of the Dirac vacuum discussed in section **2.5**. Furthermore, from the form of the Hamiltonian (7.20)

$$H = \int d^3k\, E_k \left(a^\dagger(\mathbf{k})a(\mathbf{k}) + b^\dagger(\mathbf{k})b(\mathbf{k}) \right), \qquad (7.36)$$

we note that it is invariant under the discrete transformation

$$a(\mathbf{k}) \leftrightarrow b(\mathbf{k}). \qquad (7.37)$$

That is, the second quantized description is symmetric under the interchange of particles and anti-particles. This transformation is otherwise known as the charge conjugation symmetry which we will study in a later chapter. Thus, to conclude this section, we note that the complex Klein-Gordon field theory describes spin zero particles carrying electric charge while the real Klein-Gordon field theory describes charge neutral spin zero particles.

7.4 Green's functions

In chapter **6** we have already seen that the Feynman Green's functions play the most important role in the calculations of the S-matrix in a quantum field theory. Therefore, in this section, we will construct only the Feynman Green's function for the complex Klein-Gordon field theory. Since we have already constructed the Feynman Green's function for the real Klein-Gordon field theory in (5.146), let us discuss the Feynman Green's function in the basis of the complex $\phi(x)$ and $\phi^\dagger(x)$ fields.

We note from (6.107) that the Feynman Green's function can be related to the vacuum expectation value of the time ordered product of two field operators. Using the field decomposition in (7.16) as well as the commutation relations in (7.18) we now obtain

$$\langle 0|\phi(x)\phi(y)|0\rangle$$
$$= \iint \frac{\mathrm{d}^3k\mathrm{d}^3k'}{2(2\pi)^3\sqrt{k^0 k'^0}} \langle 0|\big(e^{-ik\cdot x}a(\mathbf{k}) + e^{ik\cdot x}b^\dagger(\mathbf{k})\big)$$
$$\times \big(e^{-ik'\cdot y}a(\mathbf{k'}) + e^{ik'\cdot y}b^\dagger(\mathbf{k'})\big)|0\rangle$$
$$= 0,$$

$$\langle 0|\phi^\dagger(x)\phi^\dagger(y)|0\rangle$$
$$= \iint \frac{\mathrm{d}^3k\mathrm{d}^3k'}{2(2\pi)^3\sqrt{k^0 k'^0}} \langle 0|\big(e^{-ik\cdot x}b(\mathbf{k}) + e^{ik\cdot x}a^\dagger(\mathbf{k})\big)$$
$$\times \big(e^{-ik'\cdot y}b(\mathbf{k'}) + e^{ik'\cdot y}a^\dagger(\mathbf{k'})\big)|0\rangle$$
$$= 0,$$

$$\langle 0|\phi(x)\phi^\dagger(y)|0\rangle$$
$$= \iint \frac{\mathrm{d}^3k\mathrm{d}^3k'}{2(2\pi)^3\sqrt{k^0 k'^0}} \langle 0|\big(e^{-ik\cdot x}a(\mathbf{k}) + e^{ik\cdot x}b^\dagger(\mathbf{k})\big)$$

$$\times \left(e^{-ik'\cdot y}b(\mathbf{k}') + e^{ik'\cdot y}a^{\dagger}(\mathbf{k}')\right)|0\rangle$$

$$= \iint \frac{\mathrm{d}^3k\,\mathrm{d}^3k'}{2(2\pi)^3\sqrt{k^0k'^0}} e^{-ik\cdot x + ik'\cdot y}\langle 0|a(\mathbf{k})a^{\dagger}(\mathbf{k}')|0\rangle$$

$$= \iint \frac{\mathrm{d}^3k\,\mathrm{d}^3k'}{2(2\pi)^3\sqrt{k^0k'^0}} e^{-ik\cdot x + ik'\cdot y}\delta^3(k-k')$$

$$= \int \frac{\mathrm{d}^3k}{(2\pi)^3 2k^0} e^{-ik\cdot(x-y)}$$

$$= -iG^{(+)}(x-y),$$

$$\langle 0|\phi^{\dagger}(y)\phi(x)|0\rangle$$

$$= \iint \frac{\mathrm{d}^3k\,\mathrm{d}^3k'}{2(2\pi)^3\sqrt{k^0k'^0}} \langle 0|\left(e^{-ik\cdot y}b(\mathbf{k}) + e^{ik\cdot y}a^{\dagger}(\mathbf{k})\right)$$

$$\times \left(e^{-ik'\cdot x}a(\mathbf{k}') + e^{ik'\cdot x}b^{\dagger}(\mathbf{k}')\right)|0\rangle$$

$$= \iint \frac{\mathrm{d}^3k\,\mathrm{d}^3k'}{2(2\pi)^3\sqrt{k^0k'^0}} e^{-ik\cdot y + ik'\cdot x}\langle 0|b(\mathbf{k})b^{\dagger}(\mathbf{k}')|0\rangle$$

$$= \iint \frac{\mathrm{d}^3k\,\mathrm{d}^3k'}{2(2\pi)^3\sqrt{k^0k'^0}} e^{ik'\cdot x - ik\cdot y}\delta^3(k-k')$$

$$= \int \frac{\mathrm{d}^3k}{(2\pi)^3 2k^0} e^{ik\cdot(x-y)}$$

$$= iG^{(-)}(x-y). \tag{7.38}$$

Here we have used (7.21) as well as the definitions of the positive and negative energy Green's functions in (5.143) and (5.145).

Given the relations in (7.38), it now follows that

$$\langle 0|T(\phi(x)\phi(y))|0\rangle$$
$$= \theta(x^0 - y^0)\langle 0|\phi(x)\phi(y)|0\rangle + \theta(y^0 - x^0)\langle 0|\phi(y)\phi(x)|0\rangle$$
$$= 0,$$

$$\langle 0|T(\phi^{\dagger}(x)\phi^{\dagger}(y))|0\rangle$$
$$= \theta(x^0 - y^0)\langle 0|\phi^{\dagger}(x)\phi^{\dagger}(y)|0\rangle + \theta(y^0 - x^0)\langle 0|\phi^{\dagger}(y)\phi^{\dagger}(x)|0\rangle$$
$$= 0. \tag{7.39}$$

On the other hand, we have

$$\begin{aligned}
&\langle 0|T(\phi(x)\phi^\dagger(y))|0\rangle \\
&= \theta(x^0-y^0)\langle 0|\phi(x)\phi^\dagger(y)|0\rangle + \theta(y^0-x^0)\langle 0|\phi^\dagger(y)\phi(x)|0\rangle \\
&= -i\theta(x^0-y^0)G^{(+)}(x-y) + i\theta(y^0-x^0)G^{(-)}(x-y) \\
&= iG_F(x-y),
\end{aligned}$$

$$\begin{aligned}
\langle 0|T(\phi^\dagger(x)\phi(y))|0\rangle &= \langle 0|T(\phi(y)\phi^\dagger(x))|0\rangle \\
&= iG_F(y-x) = iG_F(x-y).
\end{aligned} \qquad (7.40)$$

Here we have used the definition of the Feynman Green's function in (5.146). Thus, we see that in this case, there is only one independent Feynman Green's function that is related to the vacuum expectation value of the time ordered product of $\phi(x)\phi^\dagger(y)$.

7.5 Spontaneous symmetry breaking and the Goldstone theorem

We can introduce interactions into the complex Klein-Gordon field theory much the same way we did for the real Klein-Gordon field theory. However, in this case we should look for interactions that would preserve the internal global symmetry in (7.23). It is easy to check that the Lagrangian density with the quartic interaction

$$\mathcal{L} = \partial_\mu \phi^\dagger \partial^\mu \phi - m^2 \phi^\dagger \phi - \frac{\lambda}{4}(\phi^\dagger \phi)^2, \quad \lambda > 0, \qquad (7.41)$$

describes the most general (renormalizable) self-interacting theory for the complex Klein-Gordon field. Denoting $\phi_1 = \sigma$ and $\phi_2 = \chi$ in the definition in (7.2) (in order to be consistent with the standard notation used in the discussion of spontaneous symmetry breaking in the literature), namely, defining

$$\phi(x) = \frac{1}{\sqrt{2}}\left(\sigma(x) + i\chi(x)\right), \qquad (7.42)$$

7.5 SPONTANEOUS SYMMETRY BREAKING AND THE GOLDSTONE THEOREM

we note that the Lagrangian density (7.41) can also be written as

$$\mathcal{L} = \frac{1}{2}(\partial_\mu \sigma \partial^\mu \sigma + \partial_\mu \chi \partial^\mu \chi) - \frac{m^2}{2}(\sigma^2 + \chi^2)$$
$$- \frac{\lambda}{16}(\sigma^2 + \chi^2)^2. \tag{7.43}$$

Namely, this describes the theory of two real Klein-Gordon fields with quartic self-interactions and which are also interacting with each other. The Lagrangian density (7.41) or (7.43) is invariant under the global phase transformation (7.23) or the infinitesimal form in (7.24). In the real basis of (7.42), the infinitesimal transformations (7.24) can be shown to have the form

$$\delta\sigma(x) = \epsilon\chi(x), \quad \delta\chi(x) = -\epsilon\sigma(x). \tag{7.44}$$

The Hamiltonian for the interacting theory (7.41) has the form

$$H = \int d^3x \left[\Pi^\dagger \Pi + \boldsymbol{\nabla}\phi^\dagger \cdot \boldsymbol{\nabla}\phi + m^2 \phi^\dagger \phi + \frac{\lambda}{4}(\phi^\dagger \phi)^2 \right], \tag{7.45}$$

and we note that, for constant field configurations, the minimum of energy coincides with the minimum of the potential (We note that if the field ϕ is not constant, the other (kinetic energy) terms would only add positively to the energy.)

$$V = m^2 \phi^\dagger \phi + \frac{\lambda}{4}(\phi^\dagger \phi)^2$$
$$= \frac{m^2}{2}(\sigma^2 + \chi^2) + \frac{\lambda}{16}(\sigma^2 + \chi^2)^2. \tag{7.46}$$

In general, the minimum of the potential is obtained by requiring

$$\frac{\partial V}{\partial \phi} = 0 = \frac{\partial V}{\partial \phi^\dagger},$$

or, $\quad \dfrac{\partial V}{\partial \sigma} = 0 = \dfrac{\partial V}{\partial \chi}. \tag{7.47}$

In the present case, since the potential in (7.46) is a sum of positive terms, it is clear that the minimum occurs at (the subscript c is supposed to denote a classical field, a concept that we do not get into here)

$$\phi_c = \phi_c^\dagger = 0, \quad \text{or, equivalently } \sigma_c = \chi_c = 0. \tag{7.48}$$

As we know from studies in classical mechanics, perturbation around the minimum of the potential is stable. In quantum theory, this translates to the fact that we can define the ground state of the theory such that

$$\phi_c = \langle 0|\phi(x)|0\rangle = 0, \quad \phi_c^\dagger = \langle 0|\phi^\dagger(x)|0\rangle = 0,$$
$$\text{or,} \quad \sigma_c = \langle 0|\sigma(x)|0\rangle = 0, \quad \chi_c = \langle 0|\chi(x)|0\rangle = 0, \tag{7.49}$$

and perturbation theory developed around such a vacuum would be stable. We note from our discussion of the real Klein-Gordon theory that the plane wave expansion for the field operator in (5.56) has exactly this property (see (5.106)).

Let us next consider the same Lagrangian density for the complex scalar field as in (7.41) except that we change the sign of the mass term

$$\mathcal{L} = \partial_\mu \phi^\dagger \partial^\mu \phi + m^2 \phi^\dagger \phi - \frac{\lambda}{4}(\phi^\dagger \phi)^2, \quad \lambda > 0. \tag{7.50}$$

In terms of the real fields σ and χ defined in (7.42), this Lagrangian density takes the form

$$\mathcal{L} = \frac{1}{2}(\partial_\mu \sigma \partial^\mu \sigma + \partial_\mu \chi \partial^\mu \chi) + \frac{m^2}{2}(\sigma^2 + \chi^2)$$
$$- \frac{\lambda}{16}(\sigma^2 + \chi^2)^2. \tag{7.51}$$

Both the Lagrangian densities in (7.50) and (7.51) are clearly invariant under the global phase transformation in (7.23) (or (7.24) and (7.44)). Changing the sign of the mass term does not change the symmetry behaviour, but it does change the Hilbert space structure of the theory in a very interesting way. To begin with, let us note that in the absence of interaction ($\lambda = 0$), the theory (7.41) would lead to the dynamical equation (compare with (1.40) or (5.4))

$$(\Box - m^2)\phi(x) = 0, \tag{7.52}$$

corresponding to the Einstein relation

$$k^2 = -m^2,$$
$$\text{or,} \quad k^0 = \sqrt{\mathbf{k}^2 - m^2}. \tag{7.53}$$

This corresponds to a particle with an imaginary mass and has the immediate consequence that the group velocity for the particle motion (this is also the same as the phase velocity in the present case) exceeds unity

$$\frac{\partial k^0}{\partial |\mathbf{k}|} = \frac{|\mathbf{k}|}{\sqrt{\mathbf{k}^2 - m^2}} > 1. \tag{7.54}$$

Namely, in this case, the particle will travel faster than the speed of light (recall that in our units, $c = 1$) leading to an acausal propagation (and, therefore, a violation of causality). Such particles are known as tachyons and a theory with tachyons, in general, has many undesirable features.

Let us note that, for constant field configurations the potential of the theory (7.50) or (7.51) has the form

$$\begin{aligned} V &= -m^2\phi^\dagger\phi + \frac{\lambda}{4}(\phi^\dagger\phi)^2 \\ &= -\frac{m^2}{2}(\sigma^2 + \chi^2) + \frac{\lambda}{16}(\sigma^2 + \chi^2)^2. \end{aligned} \tag{7.55}$$

The minimum of this potential is obtained from (7.48)

$$\begin{aligned}
\frac{\partial V}{\partial \sigma} &= -m^2 \sigma + \frac{\lambda}{4} \sigma \left(\sigma^2 + \chi^2\right) \\
&= \sigma \left(-m^2 + \frac{\lambda}{4}\left(\sigma^2 + \chi^2\right)\right) = 0, \\
\frac{\partial V}{\partial \chi} &= -m^2 \chi + \frac{\lambda}{4} \chi \left(\sigma^2 + \chi^2\right) \\
&= \chi \left(-m^2 + \frac{\lambda}{4}\left(\sigma^2 + \chi^2\right)\right) = 0.
\end{aligned} \qquad (7.56)$$

There now appear to be two sets of solutions to the equations in (7.56), namely,

$$\sigma_c = 0 = \chi_c, \quad \sigma_c^2 + \chi_c^2 = \frac{4m^2}{\lambda}. \qquad (7.57)$$

For the second of these solutions, for example, we can choose

$$\sigma_c = \pm \frac{2m}{\sqrt{\lambda}}, \quad \chi_c = 0. \qquad (7.58)$$

To determine the true minimum, let us analyze the second derivatives of the potential at the two extrema in (7.57) which lead to

$$\begin{aligned}
\left.\frac{\partial^2 V}{\partial \sigma^2}\right|_{\sigma_c=0=\chi_c} &= \left.\frac{\partial^2 V}{\partial \chi^2}\right|_{\sigma_c=0=\chi_c} = -m^2, \\
\left.\frac{\partial^2 V}{\partial \sigma \partial \chi}\right|_{\sigma_c=0=\chi_c} &= 0,
\end{aligned} \qquad (7.59)$$

while

$$\begin{aligned}
\left.\frac{\partial^2 V}{\partial \sigma^2}\right|_{\sigma_c=\pm\frac{2m}{\sqrt{\lambda}},\chi_c=0} &= -m^2 + \frac{3\lambda}{4}\left(\frac{4m^2}{\lambda}\right) = 2m^2, \\
\left.\frac{\partial^2 V}{\partial \chi^2}\right|_{\sigma_c=\pm\frac{2m}{\sqrt{\lambda}},\chi_c=0} &= -m^2 + \frac{\lambda}{4}\left(\frac{4m^2}{\lambda}\right) = 0, \\
\left.\frac{\partial^2 V}{\partial \sigma \partial \chi}\right|_{\sigma_c=\pm\frac{2m}{\sqrt{\lambda}},\chi_c=0} &= 0.
\end{aligned} \qquad (7.60)$$

In other words we note that out of the two solutions in (7.57), the first corresponds to a local maximum. On the other hand, the second solution in (7.57) is a true minimum and we conclude that the true minimum of the potential occurs for

$$2\phi_c^\dagger \phi_c = \sigma_c^2 + \chi_c^2 = \frac{4m^2}{\lambda}. \tag{7.61}$$

In fact, we note explicitly from the form of the potential in (7.55) that

$$V(\sigma_c = \chi_c = 0) = 0,$$

$$V((\sigma_c^2 + \chi_c^2) = \frac{4m^2}{\lambda}) = -\frac{m^2}{2}\left(\frac{4m^2}{\lambda}\right) + \frac{\lambda}{16}\left(\frac{4m^2}{\lambda}\right)^2$$

$$= -\frac{2m^4}{\lambda} + \frac{m^4}{\lambda} = -\frac{m^4}{\lambda}, \tag{7.62}$$

so that the second solution, indeed, leads to a lower energy. All of this is more easily seen from the graph of the potential in Fig. 7.1.

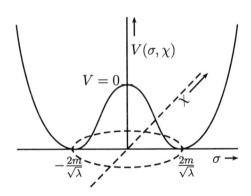

Figure 7.1: Potential leading to spontaneous breaking of symmetry.

There are several interesting things to note from this analysis. First, we note that there is an infinity of minima (in the σ-χ field space) of the potential given by the circle (7.61). Each point on this circle is related by the symmetry transformation (7.44). However, the choice of any particular point on this circle as the minimum of the potential to develop perturbation theory (or to build the quantum theory) breaks the symmetry (7.44) spontaneously. In this case, the action or the Hamiltonian of the theory is invariant under the symmetry transformation, but the ground state is not. This has to be contrasted with the case where an interaction term in the Hamiltonian can break the symmetry of a theory explicitly. In the quantum theory, such a breaking manifests in the following manner. We recall that any choice for the minimum of the potential would translate into the properties of the vacuum state of the quantum theory. Thus, for example, if we choose the minimum in (7.61) to correspond to the point (in the σ-χ field space)

$$\sigma_c = \frac{2m}{\sqrt{\lambda}}, \quad \chi_c = 0, \tag{7.63}$$

then, in the quantum theory this would imply (see (7.49))

$$\langle 0|\sigma(x)|0\rangle = \frac{2m}{\sqrt{\lambda}}, \quad \langle 0|\chi(x)|0\rangle = 0. \tag{7.64}$$

We recall that conserved charges are the generators of infinitesimal symmetry transformations of the theory and generate the transformations through commutators. In this case, the infinitesimal transformations in (7.44) can be written in terms of the commutators with the charge operator as

$$\begin{aligned} \delta\sigma(x) &= -i\epsilon\,[Q,\sigma(x)] = \epsilon\chi(x), \\ \delta\chi(x) &= -i\epsilon\,[Q,\chi(x)] = -\epsilon\sigma(x). \end{aligned} \tag{7.65}$$

From (7.64), it now follows that

$$\begin{aligned} \langle 0|\delta\chi(x)|0\rangle &= -i\epsilon\langle 0|\,[Q,\chi(x)]\,|0\rangle \\ &= -\epsilon\langle 0|\sigma(x)|0\rangle = -\frac{2m}{\sqrt{\lambda}}\epsilon, \end{aligned} \tag{7.66}$$

which implies that

7.5 SPONTANEOUS SYMMETRY BREAKING AND THE GOLDSTONE THEOREM

$$Q|0\rangle \neq 0. \tag{7.67}$$

This explicitly shows that the vacuum state of the theory is not invariant under the symmetry transformation in this case, although the Hamiltonian of the theory is, namely,

$$[Q, H] = 0. \tag{7.68}$$

Since Q does not annihilate the vacuum of the present theory, let us denote

$$Q|0\rangle = |\chi\rangle \tag{7.69}$$

It follows now from (7.68) and (7.69) that

$$[Q, H]|0\rangle = 0,$$
or, $\quad (QH - HQ)|0\rangle = 0,$
or, $\quad HQ|0\rangle = E_0 Q|0\rangle,$
or, $\quad H|\chi\rangle = E_0|\chi\rangle, \tag{7.70}$

where we have denoted the energy of the vacuum $|0\rangle$ as E_0 and this derivation shows that the state $|\chi\rangle$ defined in (7.69) would appear to be degenerate with the vacuum state $|0\rangle$ in energy. Therefore, it is suggestive that we can think of this state as another vacuum state. The problem with this interpretation is that the state is not normalizable. This can be easily seen from (7.31) and (7.69) as follows. (note that Q is Hermitian)

$$\begin{aligned}
\langle \chi | \chi \rangle &= \langle 0 | QQ | 0 \rangle \\
&= \langle 0 | \int d^3x\, J^0(\mathbf{x}, t) Q | 0 \rangle \\
&= \int d^3x\, \langle 0 | e^{iP\cdot x} J^0(0) e^{-iP\cdot x} Q | 0 \rangle.
\end{aligned} \tag{7.71}$$

We have already seen that the Hamiltonian commutes with Q expressing the fact that it is independent of time. Since Q does not depend on spatial coordinates, it follows that the momentum operator also commutes with Q and, therefore, we have

$$[P_\mu, Q] = 0, \tag{7.72}$$

and as a consequence, (7.72) leads to

$$\begin{aligned} \langle \chi | \chi \rangle &= \int d^3x \, \langle 0| e^{iP\cdot x} J^0(0) Q e^{-iP\cdot x} |0 \rangle \\ &= \int d^3x \, \langle 0| J^0(0) Q |0 \rangle \\ &= \langle 0| J^0(0) Q |0 \rangle \int d^3x \longrightarrow \infty, \end{aligned} \tag{7.73}$$

where we have used the property of the ground state

$$P_\mu |0\rangle = 0. \tag{7.74}$$

In other words, the state $|\chi\rangle$ is not normalizeable and hence cannot be thought of as another vacuum. This analysis also shows that the finite transformation operator

$$U(\theta) = e^{-i\theta Q}, \tag{7.75}$$

does not act unitarily on the Hilbert space. In fact, it is straightforward to show that the charge Q does not exist when there is spontaneous breakdown of the symmetry and the formally unitary operator in (7.75) takes a state out of the Hilbert space. The field whose variation (under the infinitesimal symmetry transformation) develops a nonzero vacuum expectation value is conventionally known as the Goldstone field (Nambu-Goldstone boson in the case of spontaneous breakdown of a bosonic symmetry) and in this case we see from (7.66) that $\chi(x)$ corresponds to the Nambu-Goldstone boson of the theory.

The second interesting observation that follows from the analysis in (7.60) is as follows. We note that the second derivatives of the potential (with respect to field variables) give the mass (squared) parameters for the field operators (particles) in a theory and (7.60) shows that in this case of spontaneous breakdown of the symmetry, one of the fields has become massive with a mass $\sqrt{2}m$ while the other is massless (in spite of the fact that the free theory had tachyons (see (7.53) and (7.54))). We note from Fig. 7.1 that the massless mode would correspond to oscillations along the valley of minima while the massive mode would represent the orthogonal oscillations. We can see this quantitatively in the quantum theory as follows. With the choice of the solutions for the minima as in (7.64)

$$\langle \sigma \rangle = \langle 0|\sigma(x)|0\rangle = \frac{2m}{\sqrt{\lambda}}, \quad \langle \chi \rangle = \langle 0|\chi(x)|0\rangle = 0, \tag{7.76}$$

we can shift the field variables as (in classical mechanics this corresponds to analyzing small oscillations around the minimum of a potential)

$$\sigma(x) \to \sigma(x) + \langle \sigma \rangle = \sigma(x) + \frac{2m}{\sqrt{\lambda}},$$
$$\chi(x) \to \chi(x) + \langle \chi \rangle = \chi(x), \tag{7.77}$$

so that the shifted fields have vanishing vacuum expectation values (vevs)

$$\langle 0|\sigma(x)|0\rangle = 0 = \langle 0|\chi(x)|0\rangle. \tag{7.78}$$

(Explicitly, the shift in (7.77), say for the first term, is given by $\sigma(x) = \sigma'(x) + \langle \sigma \rangle$ with the later identification $\sigma'(x) = \sigma(x)$ as the dynamical variable.)

Under the shift (7.77), the Lagrangian density in (7.51) becomes

$$\begin{aligned}
\mathcal{L} &= \frac{1}{2}\partial_\mu\sigma\partial^\mu\sigma + \frac{1}{2}\partial_\mu\chi\partial^\mu\chi + \frac{m^2}{2}\left(\left(\sigma+\frac{2m}{\sqrt{\lambda}}\right)^2 + \chi^2\right) \\
&\quad - \frac{\lambda}{16}\left(\left(\sigma+\frac{2m}{\sqrt{\lambda}}\right)^2 + \chi^2\right)^2 \\
&= \frac{1}{2}\partial_\mu\sigma\partial^\mu\sigma + \frac{1}{2}\partial_\mu\chi\partial^\mu\chi + \frac{m^2}{2}\left(\sigma^2 + \chi^2 + \frac{4m}{\sqrt{\lambda}}\sigma + \frac{4m^2}{\lambda}\right) \\
&\quad - \frac{\lambda}{16}\left(\sigma^2 + \chi^2 + \frac{4m}{\sqrt{\lambda}}\sigma + \frac{4m^2}{\lambda}\right)^2 \\
&= \frac{1}{2}\partial_\mu\sigma\partial^\mu\sigma + \frac{1}{2}\partial_\mu\chi\partial^\mu\chi + \frac{m^2}{2}\left(\sigma^2 + \chi^2 + \frac{4m}{\sqrt{\lambda}}\sigma + \frac{4m^2}{\lambda}\right) \\
&\quad - \frac{\lambda}{16}\left((\sigma^2+\chi^2)^2 + \frac{16m^2}{\lambda}\sigma^2 + \frac{16m^4}{\lambda^2} + \frac{8m}{\sqrt{\lambda}}\sigma(\sigma^2+\chi^2)\right. \\
&\quad \left. + \frac{8m^2}{\lambda}(\sigma^2+\chi^2) + \frac{32m^3}{\sqrt{\lambda^3}}\sigma\right) \\
&= \frac{1}{2}\partial_\mu\sigma\partial^\mu\sigma + \frac{1}{2}\partial_\mu\chi\partial^\mu\chi - m^2\sigma^2 + \frac{m^4}{\lambda} \\
&\quad - \frac{m\sqrt{\lambda}}{2}\sigma(\sigma^2+\chi^2) - \frac{\lambda}{16}(\sigma^2+\chi^2)^2. \quad (7.79)
\end{aligned}$$

This shows explicitly that perturbed around the true minimum, one of the scalar fields (in this case $\chi(x)$) becomes massless while the other scalar field (in this case $\sigma(x)$) is massive with a nontachyonic mass $\sqrt{2}m$. The massless boson is known as the Nambu-Goldstone boson (this is the one whose field variation picks up a nonvanishing vacuum expectation value as shown in (7.66)). This leads to a general theorem known as the Goldstone theorem which says that if in a manifestly Lorentz invariant theory with a positive definite metric Hilbert space, a continuous symmetry is spontaneously broken, then there must appear massless excitations.

We are, of course, aware of this phenomenon (known as the spontaneous symmetry breaking phenomenon) from studies in other areas of physics. For example, if we look at the Hamiltonian describing a magnet (for example, in the case of the Ising model), the Hamiltonian is invariant under rotations,

$$H = -g \sum_i \mathbf{s}_i \cdot \mathbf{s}_{i+1}, \tag{7.80}$$

where the negative sign in the Hamiltonian ($g > 0$) is for ferromagnets. The ground state of the theory is clearly the state (spin configuration) where all the spins are aligned. (For example, there are two such vacuum configurations allowed in one dimension, namely, when the spins are all pointing up or down.) However, it is clear that once we have chosen a particular ground state (vacuum) where all the spins are pointing in a given direction, we have effectively chosen a definite orientation in the physical space. Consequently the rotational symmetry is spontaneously broken. The analogue of the massless Nambu-Goldstone bosons in this case correspond to the long range correlations associated with the spin waves.

The Goldstone theorem is interesting in the sense that it can provide a possible reason for the existence of massless spin zero bosons if they are found in nature. Unfortunately, we do not observe any massless spin 0 bosons in nature. (The π meson which is the lightest among the spin zero bosons has a mass $m_\pi \sim 140$ MeV.)

7.6 Electromagnetic coupling

It is clear from our discussions so far that the complex Klein-Gordon field theory describes spin zero charged particles whereas the real Klein-Gordon field theory describes spin zero charge neutral particles. Given the complex Klein-Gordon field theory, we can ask how such a system can be coupled to a background electromagnetic field. The prescription is again through the minimal coupling discussed in (1.53),

$$\partial_\mu \to \partial_\mu + ieA_\mu, \tag{7.81}$$

where "e" denotes the charge carried by the particle. Thus, the Lagrangian density describing the interaction of charged spin zero fields with a background electromagnetic field (without the quartic self-interaction term) is given by

$$\begin{aligned}
\mathcal{L} &= \partial_\mu \phi^\dagger \partial^\mu \phi - m^2 \phi^\dagger \phi \\
&\to ((\partial_\mu + ieA_\mu)\phi)^\dagger (\partial^\mu + ieA^\mu)\phi - m^2 \phi^\dagger \phi \\
&= (\partial_\mu - ieA_\mu)\phi^\dagger (\partial^\mu + ieA^\mu)\phi - m^2 \phi^\dagger \phi \\
&= (D_\mu \phi)^\dagger (D^\mu \phi) - m^2 \phi^\dagger \phi.
\end{aligned} \qquad (7.82)$$

The modified derivative D_μ introduced in (7.82)

$$D_\mu \phi = \partial_\mu \phi + ieA_\mu \phi, \qquad (7.83)$$

is known as the covariant derivative since it transforms covariantly under a local phase transformation (or equivalently under a gauge transformation) as we will see shortly.

As we can verify readily, the Lagrangian density (7.82) is invariant under a local $(U(1))$ phase transformation of the form

$$\begin{aligned}
\phi(x) &\to \phi'(x) = e^{-i\theta(x)} \phi(x), \\
\phi^\dagger(x) &\to \phi'^\dagger(x) = \phi^\dagger(x) e^{i\theta(x)}, \\
A_\mu(x) &\to A'_\mu(x) = A_\mu(x) + \frac{1}{e} \partial_\mu \theta(x),
\end{aligned} \qquad (7.84)$$

where the parameter of transformation $\theta(x)$ is now a local function of space-time coordinates (unlike in (7.23)). In fact, note that under this transformation,

$$\begin{aligned}
D_\mu \phi(x) &\to D'_\mu \phi'(x) \\
&= \left(\partial_\mu + ieA'_\mu(x)\right) \phi'(x) \\
&= \left(\partial_\mu + ie\left(A_\mu(x) + \frac{1}{e}(\partial_\mu \theta(x))\right)\right) e^{-i\theta(x)} \phi(x) \\
&= e^{-i\theta(x)} \left(-i(\partial_\mu \theta(x)) + \partial_\mu + ieA_\mu(x) + i(\partial_\mu \theta(x))\right) \phi(x) \\
&= e^{-i\theta(x)} (\partial_\mu + ieA_\mu(x)) \phi(x) = e^{-i\theta(x)} D_\mu \phi(x).
\end{aligned} \qquad (7.85)$$

In other words, the covariant derivative acting on a field transforms exactly like the field itself (covariantly) under such a transformation (see (7.84)). (In contrast, note that $\partial_\mu \phi(x)$ does not transform covariantly.) It follows from this observation that under the local transformation (7.84)

$$\begin{aligned}\mathcal{L}(\phi(x), \phi^\dagger(x), A_\mu(x)) &\to \mathcal{L}(\phi'(x), \phi'^\dagger(x), A'_\mu(x)) \\ &= \left(D'_\mu \phi'(x)\right)^\dagger \left(D'^\mu \phi'(x)\right) - m^2 \phi'^\dagger(x) \phi'(x) \\ &= (D_\mu \phi(x))^\dagger e^{i\theta(x)} e^{-i\theta(x)} D^\mu \phi(x) - m^2 \phi^\dagger(x) e^{i\theta(x)} e^{-i\theta(x)} \phi(x) \\ &= (D_\mu \phi(x))^\dagger (D^\mu \phi(x)) - m^2 \phi^\dagger(x) \phi(x) \\ &= \mathcal{L}\left(\phi, \phi^\dagger, A_\mu\right).\end{aligned} \quad (7.86)$$

Consequently, the theory described by (7.82), which is obtained by minimally coupling the complex scalar field to the electromagnetic field, is invariant under the local phase transformation (7.84). Local phase transformations are commonly known as gauge transformations and we note that, unlike the global phase invariance in (7.23), the local phase invariance (7.84) requires an additional field (A_μ). This is a very general feature in a quantum field theory, namely, invariance under a local transformation necessarily requires additional fields known as gauge fields. Of course, the electromagnetic field in the present discussion has no dynamics (it is a background field and not a dynamical field) which we will develop in chapter **9**.

7.7 References

1. J. Goldstone, Nuovo Cimento **19**, 154 (1961).

2. Y. Nambu and G. Jona-Lasinio, Physical Review **122**, 345 (1961).

3. J. Goldstone, A. Salam and S. Weinberg, Physical Review **127**, 965 (1962).

4. S. Schweber, *Introduction to Relativistic Quantum Field Theory*, Row, Peterson, Evanston (1961).

5. J. D. Bjorken and S. Drell, *Relativistic Quantum Fields*, McGraw-Hill, New York, 1964.

6. J. Bernstein, Reviews of Modern Physics **46**, 7 (1974).

7. P. Roman, *Introduction to Quantum Field Theory*, John Wliley, New York (1969).

8. C. Itzykson and J-B. Zuber, *Quantum Field Theory*, McGraw-Hill, New York, 1980.

CHAPTER 8
Dirac field theory

8.1 Pauli exclusion principle

So far we have discussed field theories which describe spin zero bosonic particles. In trying to go beyond and study the theory of spin $\frac{1}{2}$ fields, we note that spin $\frac{1}{2}$ particles such as electrons are fermions. Unlike Bose particles, fermions obey the Pauli exclusion principle which simply says that there can at the most be one fermion in a given state. (There can be more fermions only if they are non-identical.) Thus, in dealing with such systems, we have to find a mechanism for incorporating the Pauli principle into our theory.

Let us consider an oscillator described by the annihilation and the creation operators a_F and $a_F^\dagger$ respectively. The number operator for the system is given as usual by

$$N_F = a_F^\dagger a_F. \tag{8.1}$$

It is easy to show that we can assign Fermi-Dirac statistics to such an oscillator (and, therefore, incorporate the Pauli principle) by requiring that the creation and the annihilation operators satisfy anti-commutation relations as opposed to the conventional commutation relations for the bosonic oscillator. For example, if we require

$$\begin{aligned}
{[a_F, a_F]}_+ &= a_F^2 + a_F^2 = 0, \\
{[a_F^\dagger, a_F^\dagger]}_+ &= \left(a_F^\dagger\right)^2 + \left(a_F^\dagger\right)^2 = 0, \\
{[a_F, a_F^\dagger]}_+ &= a_F a_F^\dagger + a_F^\dagger a_F = 1,
\end{aligned} \tag{8.2}$$

then, we obtain

$$\begin{aligned} N_F^2 &= a_F^\dagger a_F a_F^\dagger a_F \\ &= a_F^\dagger \left([a_F, a_F^\dagger]_+ - a_F^\dagger a_F \right) a_F \\ &= a_F^\dagger (1 - a_F^\dagger a_F) a_F = a_F^\dagger a_F = N_F. \end{aligned} \tag{8.3}$$

In other words, the anti-commutation relations in (8.2) automatically lead to

$$N_F (N_F - 1) = 0, \tag{8.4}$$

which shows that the eigenvalues of the number operator, in such a theory, can only be

$$n_F = 0, 1. \tag{8.5}$$

This is, of course, what the Pauli exclusion principle would say. Namely, there can be at the most one quantum in a given state. In fact, if this system is analyzed in more detail, it leads to the fact that with the anti-commutation relations, the wave function of the system will have the necessary antisymmetry property characteristic of fermions. It is clear, therefore, that in dealing with fermionic field operators, we will be naturally dealing with anti-commuting variables (these variables are commonly known as Grassmann variables) and anti-commutation relations.

8.2 Quantization of the Dirac field

As we have seen in (1.96), the Dirac equation for a massive spin $\frac{1}{2}$ particle, has the form

$$(i\gamma^\mu \partial_\mu - m)\psi = 0, \tag{8.6}$$

where $\psi_\alpha(x)$, $\alpha = 1, 2, 3, 4$, is a four component complex spinor function. This equation is manifestly Lorentz covariant and the adjoint equation has the form

8.2 Quantization of the Dirac field

$$\overline{\psi}(i\gamma^\mu \overleftarrow{\partial}_\mu + m) = 0, \tag{8.7}$$

where the adjoint spinor is defined to be (see also (2.41))

$$\overline{\psi}(x) = \psi^\dagger(x)\gamma^0. \tag{8.8}$$

In trying to second quantize the Dirac equation (theory), we note that we can treat $\psi(x)$ and $\overline{\psi}(x)$ as independent field operators as in the case of the complex Klein-Gordon field theory. However, as we have seen in chapter **3** as well as in chapter **4**, in this case, the Dirac field operators are expected to belong to the (reducible) spin $\frac{1}{2}$ representation of the Lorentz group (see (4.61)) describing fermionic particles and, consequently, need to be treated as anti-commuting operators.

In dealing with the Dirac field theory, we would like to decide on a Lorentz invariant Lagrangian density which would give rise to the two Dirac equations (8.6) and (8.7) as the Euler-Lagrange equations. Clearly, the Lagrangian density of the form

$$\mathcal{L} = \overline{\psi}\left(i\gamma^\mu \partial_\mu - m\right)\psi = i\overline{\psi}\slashed{\partial}\psi - m\overline{\psi}\psi, \tag{8.9}$$

is manifestly Lorentz invariant (recall the transformation of the Dirac bilinears under a Lorentz transformation discussed in section **3.3**). In deriving the Euler-Lagrange equations, however, let us note that since ψ and $\overline{\psi}$ are anti-commuting (Grassmann) variables, we have to be careful about signs that may arise in taking derivative operators (with respect to these variables) past other Grassmann variables. In general, we will always use the convention of taking fermionic (Grassmann) derivatives from the left. With this, the Euler-Lagrange equations following from the Lagrangian density (8.9) give

$$\frac{\partial \mathcal{L}}{\partial \overline{\psi}} = (i\gamma^\mu \partial_\mu - m)\psi = 0,$$

$$\frac{\partial \mathcal{L}}{\partial \psi} - \partial_\mu \frac{\partial \mathcal{L}}{\partial \partial_\mu \psi} = m\overline{\psi} - \partial_\mu(-i\overline{\psi}\gamma^\mu) = \overline{\psi}(i\gamma^\mu \overleftarrow{\partial}_\mu + m) = 0, \tag{8.10}$$

which are exactly the two Dirac equations (8.6) and (8.7) that we are interested in.

Even though the Lagrangian density (8.9) is Lorentz invariant and gives the two Dirac equations as the Euler-Lagrange equations, it is not Hermitian. In fact, note that

$$\begin{aligned}\mathcal{L}^\dagger &= \left(\overline{\psi}\left(i\gamma^\mu\partial_\mu - m\right)\psi\right)^\dagger \\ &= \psi^\dagger\left(-i\gamma^{\mu\dagger}\overleftarrow{\partial_\mu} - m\right)\overline{\psi}^\dagger \\ &= \psi^\dagger\left(-i\gamma^{\mu\dagger}\overleftarrow{\partial_\mu} - m\right)\gamma^0\psi \\ &= \overline{\psi}\left(-i\gamma^\mu\overleftarrow{\partial_\mu} - m\right)\psi \\ &= \overline{\psi}\left(i\gamma^\mu\partial_\mu - m\right)\psi - i\partial_\mu\left(\overline{\psi}\gamma^\mu\psi\right) \\ &\neq \mathcal{L},\end{aligned} \qquad (8.11)$$

where we have used (2.83), (8.8) as well as the fact that γ^0 is Hermitian. It is worth noting here that even though the Lagrangian density (8.9) is not Hermitian, the action associated with it is since a total divergence does not contribute to the action. An alternate Lagrangian density which is Lorentz invariant, Hermitian and which can also be checked to give the two Dirac equations as the Euler-Lagrange equations has the form

$$\mathcal{L}' = \frac{1}{2}\overline{\psi}\left(i\gamma^\mu\partial_\mu - m\right)\psi - \frac{1}{2}\overline{\psi}\left(i\gamma^\mu\overleftarrow{\partial_\mu} + m\right)\psi. \qquad (8.12)$$

However, we note that the Lagrangian density in (8.12) differs from $\mathcal{L}$ in (8.9) only by a total divergence. Namely,

$$\begin{aligned}\mathcal{L}' &= \frac{1}{2}\overline{\psi}\left(i\gamma^\mu\partial_\mu - m\right)\psi - \frac{1}{2}\partial_\mu\left(i\overline{\psi}\gamma^\mu\psi\right) + \frac{1}{2}\overline{\psi}\left(i\gamma^\mu\partial_\mu - m\right)\psi \\ &= \overline{\psi}\left(i\gamma^\mu\partial_\mu - m\right)\psi - \frac{i}{2}\partial_\mu\left(\overline{\psi}\gamma^\mu\psi\right) \\ &= \mathcal{L} - \frac{i}{2}\partial_\mu\left(\overline{\psi}\gamma^\mu\psi\right).\end{aligned} \qquad (8.13)$$

8.2 QUANTIZATION OF THE DIRAC FIELD

For this reason as well as for simplicity, we use $\mathcal{L}$ in (8.9) to describe the dynamics of the system. However, in the presence of gravitation, it is more appropriate to use the Lagrangian density in (8.12) which naturally leads to a symmetric stress tensor (see (6.27)) for the theory.

If we work with the Lagrangian density (8.9) to describe the Dirac theory (equations), we can define momenta conjugate to the field variables ψ and $\overline{\psi}$ as (we would be careful with taking derivatives from the left which is our convention for fermionic derivatives)

$$\Pi_\psi = \frac{\partial \mathcal{L}}{\partial \dot\psi} = -i\overline{\psi}\gamma^0 = -i\psi^\dagger,$$

$$\Pi_{\overline\psi} = \frac{\partial \mathcal{L}}{\partial \dot{\overline\psi}} = 0. \qquad (8.14)$$

Consequently, the equal time canonical anti-commutation relations for the field variables and their conjugate momenta take the forms (we do not write the equal time condition explicitly for simplicity, but we should understand that $x^0 = y^0$ in all the relations in (8.15) and (8.16))

$$[\psi_\alpha(x), \psi_\beta(y)]_+ = 0 = [(\Pi_\psi)_\alpha(x), (\Pi_\psi)_\beta(y)]_+,$$
$$[\psi_\alpha(x), (\Pi_\psi)_\beta(y)]_+ = -i\delta_{\alpha\beta}\delta^3(x-y). \qquad (8.15)$$

The negative sign in the last relation in (8.15) is a reflection of our choice of left derivatives. Using the first relation in (8.14), the second and the third relations in (8.15) can also be written as

$$[(\Pi_\psi)_\alpha(x), (\Pi_\psi)_\beta(y)]_+ = 0,$$
$$\text{or,}\ [\psi^\dagger_\alpha(x), \psi^\dagger_\beta(y)]_+ = 0,$$
$$[\psi_\alpha(x), (\Pi_\psi)_\beta(y)]_+ = [\psi_\alpha(x), -i\psi^\dagger_\beta(y)]_+ = -i\delta_{\alpha\beta}\delta^3(x-y),$$
$$\text{or,}\ [\psi_\alpha(x), \psi^\dagger_\beta(y)]_+ = \delta_{\alpha\beta}\delta^3(x-y). \qquad (8.16)$$

(We note that we have not written any anti-commutation relation for $\overline{\psi}_\alpha(x)$ and $(\Pi_{\overline{\psi}})_\alpha(x)$ simply because the relations in (8.14) can be thought of as constraints on the dynamics of the system. Such systems can be systematically studied through the method of Dirac brackets and such an analysis leads to the conclusion that the relations (8.15) and (8.16) are all that we need for quantizing the Dirac theory. We will discuss the method of Dirac in more detail in chapter **10**.)

The Hamiltonian density for the Dirac field theory (8.9) has the form (The negative sign in the first term arises from our convention of choosing left derivatives for fermions.)

$$\begin{aligned}\mathcal{H} &= -\Pi_\psi \dot{\psi} - \mathcal{L} \\ &= -\left(-i\psi^\dagger\right)\dot{\psi} - i\psi^\dagger\dot{\psi} - i\overline{\psi}\boldsymbol{\gamma}\cdot\boldsymbol{\nabla}\psi + m\overline{\psi}\psi \\ &= -i\overline{\psi}\boldsymbol{\gamma}\cdot\boldsymbol{\nabla}\psi + m\overline{\psi}\psi,\end{aligned} \qquad (8.17)$$

leading to the Hamiltonian

$$H = \int d^3x\, \mathcal{H} = \int d^3x \left(-i\overline{\psi}\boldsymbol{\gamma}\cdot\boldsymbol{\nabla}\psi + m\overline{\psi}\psi\right). \qquad (8.18)$$

Using the field anti-commutation relations (8.15) and (8.16), it is now easy to check that this Hamiltonian leads to the two Dirac equations as the Hamiltonian equations. For example,

$$\begin{aligned}i\dot{\psi}_\alpha(x) &= [\psi_\alpha(x), H] \\ &= \int d^3y \Big([\psi_\alpha(x),\psi^\dagger_\beta(y)]_+ \left(-i\gamma^0\boldsymbol{\gamma}\cdot\boldsymbol{\nabla}_y\psi(y)\right)_\beta \\ &\qquad\quad + m[\psi_\alpha(x),\psi^\dagger_\beta(y)]_+ \left(\gamma^0\psi(y)\right)_\beta\Big)_{x^0=y^0} \\ &= \int d^3y \Big(\delta_{\alpha\beta}\delta^3(x-y)\left(-i\gamma^0\boldsymbol{\gamma}\cdot\boldsymbol{\nabla}_y\psi(y)\right)_\beta \\ &\qquad\quad + m\delta_{\alpha\beta}\delta^3(x-y)\left(\gamma^0\psi(y)\right)_\beta\Big)_{x^0=y^0} \\ &= -i\left(\gamma^0\boldsymbol{\gamma}\cdot\boldsymbol{\nabla}\psi(x)\right)_\alpha + m\left(\gamma^0\psi(x)\right)_\alpha,\end{aligned} \qquad (8.19)$$

8.2 QUANTIZATION OF THE DIRAC FIELD

where we have used the fact that the Hamiltonian is independent of time to identify $x^0 = y^0$ and this equation can be written in the matrix form as

$$i\dot\psi(x) = -i\gamma^0\boldsymbol{\gamma}\cdot\boldsymbol{\nabla}\psi(x) + m\gamma^0\psi(x),$$
$$\text{or,}\quad i\gamma^0\dot\psi(x) = -i\boldsymbol{\gamma}\cdot\boldsymbol{\nabla}\psi(x) + m\psi(x),$$
$$\text{or,}\quad (i\gamma^\mu\partial_\mu - m)\psi(x) = 0. \tag{8.20}$$

Similarly, we obtain

$$\begin{aligned}
i\dot\psi^\dagger_\alpha(x) &= [\psi^\dagger_\alpha(x), H] \\
&= \int d^3y \Big(i\,(\overline\psi(y)\boldsymbol{\gamma}\cdot\boldsymbol{\nabla}_y)_\beta\, [\psi^\dagger_\alpha(x),\psi_\beta(y)]_+ \\
&\qquad\qquad - m\overline\psi_\beta(y)[\psi^\dagger_\alpha(x),\psi_\beta(y)]_+ \Big)_{x^0=y^0} \\
&= \int d^3y \Big(i\,(\overline\psi(y)\boldsymbol{\gamma}\cdot\boldsymbol{\nabla}_y)_\beta\, (\delta_{\alpha\beta}\delta^3(x-y)) \\
&\qquad\qquad - m\overline\psi_\beta(y)\delta_{\alpha\beta}\delta^3(x-y) \Big)_{x^0=y^0} \\
&= -i(\overline\psi(x)\boldsymbol{\gamma}\cdot\overleftarrow{\boldsymbol{\nabla}})_\alpha - m\overline\psi_\alpha(x),
\end{aligned} \tag{8.21}$$

and this can be written in the matrix form as

$$i\dot{\overline\psi}{}^\dagger = -i\overline\psi\boldsymbol{\gamma}\cdot\overleftarrow{\boldsymbol{\nabla}} - m\overline\psi,$$
$$\text{or,}\quad i\dot{\overline\psi}\gamma^0 = -i\overline\psi\boldsymbol{\gamma}\cdot\overleftarrow{\boldsymbol{\nabla}} - m\overline\psi,$$
$$\text{or,}\quad \overline\psi(i\gamma^\mu\overleftarrow{\partial_\mu} + m) = 0. \tag{8.22}$$

Namely, the anti-commutation relations (8.15) and (8.16) as well as the Hamiltonian (8.18) indeed reproduce the two Dirac equations (8.6) and (8.7) as the Hamiltonian (Heisenberg) equations.

8.3 Field decomposition

We have already seen that the positive and the negative energy plane wave spinor solutions of the Dirac equation define a complete basis for the space of four component spinor functions (see the discussion in section **3.4**). Therefore, we can expand the Dirac field operator in the basis of these plane wave solutions. Taking advantage of our experience with the complex Klein-Gordon field theory, we write ($k^0 = E_k = \sqrt{\mathbf{k}^2 + m^2} > 0$)

$$\psi_\alpha(x) = \sum_{s=\pm\frac{1}{2}} \int d^3k \sqrt{\frac{m}{(2\pi)^3 k^0}}$$
$$\times \left(e^{-ik\cdot x} c(\mathbf{k}, s) u_\alpha(k, s) + e^{ik\cdot x} d^\dagger(\mathbf{k}, s) v_\alpha(k, s)\right), \quad (8.23)$$

where $c(\mathbf{k}, s)$ and $d^\dagger(\mathbf{k}, s)$ are anti-commuting operators, $u_\alpha(k, s)$ and $v_\alpha(k, s)$ denote the positive and the negative energy spinors defined in (3.93) and (3.94) and the field (8.23) automatically satisfies the Dirac equation in this basis

$$(i\slashed{\partial} - m)\psi(x) = 0. \quad (8.24)$$

The specific form of the prefactor in (8.23) has been chosen keeping in mind the massive normalization of the plane wave solutions in (2.46).

As in the case of the Klein-Gordon field (see section **5.5**), let us note that the field decomposition in (8.23) can be inverted to give ($k^0 = E_k = \sqrt{\mathbf{k}^2 + m^2} > 0$)

$$\int d^3x \sqrt{\frac{m}{(2\pi)^3 k^0}} e^{ik\cdot x} u_\alpha^\dagger(k, s) \psi_\alpha(x)$$
$$= \sum_{s'} \iint \frac{d^3x\, d^3k'}{(2\pi)^3} \frac{m}{\sqrt{k^0 k'^0}} \left(e^{i(k-k')\cdot x} u_\alpha^\dagger(k, s) u_\alpha(k', s') c(\mathbf{k}', s')\right.$$
$$\left. + e^{i(k+k')\cdot x} u_\alpha^\dagger(k, s) v_\alpha(k', s') d^\dagger(\mathbf{k}', s')\right)$$

8.3 FIELD DECOMPOSITION

$$\begin{aligned}
&= \sum_{s'} \int d^3k' \frac{m}{\sqrt{k^0 k'^0}} \left(\delta^3(k-k') u_\alpha^\dagger(k,s) u_\alpha(k',s') c(k',s') \right. \\
&\quad \left. + \delta^3(k+k') e^{2ik^0 x^0} u_\alpha^\dagger(k,s) v_\alpha(k',s') d^\dagger(k',s') \right) \\
&= \sum_{s'} \frac{m}{k^0} \frac{k^0}{m} \delta_{ss'} c(\mathbf{k},s') = c(\mathbf{k},s),
\end{aligned} \qquad (8.25)$$

where we have used the massive normalization for the spinors discussed in (2.46) (see also (3.92)). In a similar manner, it can be shown that

$$\int d^3x \sqrt{\frac{m}{(2\pi)^3 k^0}} e^{-ik \cdot x} v_\alpha^\dagger(k,s) \psi_\alpha(x) = d^\dagger(\mathbf{k},s). \qquad (8.26)$$

It follows from (8.25) and (8.26) that

$$\begin{aligned}
c^\dagger(\mathbf{k},s) &= \int d^3x \sqrt{\frac{m}{(2\pi)^3 k^0}} e^{-ik \cdot x} \psi_\alpha^\dagger(x) u_\alpha(k,s), \\
d(\mathbf{k},s) &= \int d^3x \sqrt{\frac{m}{(2\pi)^3 k^0}} e^{ik \cdot x} \psi_\alpha^\dagger(x) v_\alpha(k,s).
\end{aligned} \qquad (8.27)$$

Furthermore, we know the equal time quantization conditions for the field operators from (8.15) and (8.16) to be

$$\begin{aligned}
\left[\psi_\alpha(x), \psi_\beta(y)\right]_{+, x^0 = y^0} &= 0, \\
\left[\psi_\alpha^\dagger(x), \psi_\beta^\dagger(y)\right]_{+, x^0 = y^0} &= 0, \\
\left[\psi_\alpha(x), \psi_\beta^\dagger(y)\right]_{+, x^0 = y^0} &= \delta_{\alpha\beta} \delta^3(x-y).
\end{aligned} \qquad (8.28)$$

Therefore, using the inversion relations (8.25)-(8.27), we can now obtain the anti-commutation relations satisfied by the operators $c(\mathbf{k},s)$, $c^\dagger(\mathbf{k},s)$ and $d(\mathbf{k},s), d^\dagger(\mathbf{k},s)$ in a straightforward manner from the quantization conditions (8.28) and they correspond to

$$\left[c(\mathbf{k},s), c^\dagger(\mathbf{k}',s')\right]_+ = \delta_{ss'} \delta^3(k-k') = \left[d(\mathbf{k},s), d^\dagger(\mathbf{k}',s')\right]_+, \quad (8.29)$$

with all other anti-commutators vanishing. This shows that we can think of $c(\mathbf{k}, s)$, $c^\dagger(\mathbf{k}, s)$ as well as $d(\mathbf{k}, s)$, $d^\dagger(\mathbf{k}, s)$ as annihilation and creation operators for the two kinds of fermionic quanta in this theory (much like the complex Klein-Gordon field theory).

The Hamiltonian (8.18) of the theory, in this case, can be written out in terms of creation and annihilation operators as

$$H = \int d^3x \left(-i\overline{\psi}\gamma \cdot \nabla\psi + m\overline{\psi}\psi\right) = \int d^3x\, \overline{\psi}\left(-i\gamma \cdot \nabla + m\right)\psi$$

$$= \int \frac{d^3x}{(2\pi)^3} \sum_{s,s'=\pm\frac{1}{2}} \int d^3k' \sqrt{\frac{m}{k'^0}}\, d^3k\, \sqrt{\frac{m}{k^0}}$$

$$\times \left(\overline{u}(k',s')c^\dagger(\mathbf{k}',s')e^{ik'\cdot x} + \overline{v}(k',s')d(\mathbf{k}',s')e^{-ik'\cdot x}\right)$$

$$\times (-i\gamma \cdot \nabla + m)$$

$$\times \left(e^{-ik\cdot x}c(\mathbf{k},s)u(k,s) + e^{ik\cdot x}d^\dagger(\mathbf{k},s)v(k,s)\right)$$

$$= \sum_{s,s'=\pm\frac{1}{2}} \int \frac{d^3x}{(2\pi)^3} d^3k'\, d^3k\, \sqrt{\frac{m}{k'^0}}\sqrt{\frac{m}{k^0}}$$

$$\times \left(\overline{u}\left(k',s'\right)c^\dagger(\mathbf{k}'s')e^{ik'\cdot x} + \overline{v}\left(k',s'\right)d(\mathbf{k}',s')e^{-ik'\cdot x}\right)$$

$$\times \Big(e^{-ik\cdot x}c(\mathbf{k},s)(\gamma\cdot k + m)u(k,s)$$

$$+ e^{ik\cdot x}d^\dagger(\mathbf{k},s)(-\gamma\cdot k + m)v(k,s)\Big). \tag{8.30}$$

If we now use the relations (3.93) and (3.94)

$$(\slashed{k} - m)\, u(k,s) = (\gamma^0 k^0 - \gamma\cdot\mathbf{k} - m)u(k,s) = 0,$$
$$(\slashed{k} + m)\, v(k,s) = (\gamma^0 k^0 - \gamma\cdot\mathbf{k} + m)v(k,s) = 0, \tag{8.31}$$

we can simplify the Hamiltonian in (8.30) and write

8.3 Field decomposition

$$\begin{aligned}
H = & \sum_{s,s'=\pm\frac{1}{2}} \int \frac{\mathrm{d}^3 x}{(2\pi)^3} \mathrm{d}^3 k' \mathrm{d}^3 k \sqrt{\frac{m}{k'^0}} \sqrt{\frac{m}{k^0}} k^0 \\
& \times \left(\overline{u}\left(k',s'\right) c^\dagger(\mathbf{k}',s') e^{ik'\cdot x} + \overline{v}\left(k',s'\right) d(\mathbf{k}',s') e^{-ik'\cdot x} \right) \\
& \times \gamma^0 \left(e^{-ik\cdot x} c(\mathbf{k},s) u(k,s) - e^{ik\cdot x} d^\dagger(\mathbf{k},s) v(k,s) \right) \\
= & \sum_{s,s'=\pm\frac{1}{2}} \int \mathrm{d}^3 k \mathrm{d}^3 k' \sqrt{\frac{m}{k'^0}} \sqrt{\frac{m}{k^0}} k^0 \\
& \left[\delta^3(k-k') \left(\overline{u}\left(k',s'\right) \gamma^0 u(k,s) c^\dagger(\mathbf{k}',s') c(\mathbf{k},s) \right. \right. \\
& \left. - \overline{v}\left(k',s'\right) \gamma^0 v(k,s) d(\mathbf{k}',s') d^\dagger(\mathbf{k},s) \right) \\
& - \delta^3(k+k') \left(e^{i(k'^0+k^0)x^0} \overline{u}\left(k',s'\right) \gamma^0 v(k,s) c^\dagger(\mathbf{k}',s') d^\dagger(\mathbf{k},s) \right. \\
& \left. \left. - e^{-i(k'^0+k^0)x^0} \overline{v}\left(k',s'\right) \gamma^0 u(k,s) d(\mathbf{k}',s') c(\mathbf{k},s) \right) \right] \\
= & \sum_{s,s'=\pm\frac{1}{2}} m \int \mathrm{d}^3 k \left[u^\dagger\left(k,s'\right) u(k,s) c^\dagger(\mathbf{k},s') c(\mathbf{k},s) \right. \\
& - v^\dagger\left(k,s'\right) v(k,s) d(\mathbf{k},s') d^\dagger(\mathbf{k},s) \\
& - e^{2ik^0 x^0} u^\dagger\left(k^0, -\mathbf{k}, s'\right) v(k,s) c^\dagger(-\mathbf{k},s') d^\dagger(\mathbf{k},s) \\
& \left. + e^{-2ik^0 x^0} v^\dagger\left(k^0, -\mathbf{k}, s'\right) u(k,s) d(-\mathbf{k},s') c(\mathbf{k},s) \right] \\
= & \sum_{s,s'=\pm\frac{1}{2}} m \int \mathrm{d}^3 k \left(\frac{E_k}{m} \right) \delta_{ss'} \left(c^\dagger(\mathbf{k},s') c(\mathbf{k},s) - d(\mathbf{k},s') d^\dagger(\mathbf{k},s) \right) \\
= & \sum_{s=\pm\frac{1}{2}} \int \mathrm{d}^3 k \, E_k \left(c^\dagger(\mathbf{k},s) c(\mathbf{k},s) - d(\mathbf{k},s) d^\dagger(\mathbf{k},s) \right), \quad (8.32)
\end{aligned}$$

where we have used the orthogonality relations for the plane wave solutions following from (2.46) (see also (3.92)).

We are yet to normal order the Hamiltonian. The normal ordering for products of fermionic operators is defined exactly the same way as for the bosonic operators, namely, we write products with the creation operators standing to the left of the annihilation operators. However, every time we move a fermionic operator past another in order to bring the product into its normal ordered form, we pick up a negative sign since they anti-commute (as opposed to bosonic variables). For example, we now have

$$\begin{aligned} : d^\dagger(\mathbf{k},s)d(\mathbf{k}',s') : &= d^\dagger(\mathbf{k},s)d(\mathbf{k}',s'), \\ : d(\mathbf{k}',s')d^\dagger(\mathbf{k},s) : &= -d^\dagger(\mathbf{k},s)d(\mathbf{k}',s'), \end{aligned} \tag{8.33}$$

so that the order of factors inside a product that is normal ordered becomes important. Using (8.33), the normal ordered Hamiltonian for the free Dirac field theory in (8.32) is obtained to have the form

$$: H := \sum_{s=\pm\frac{1}{2}} \int d^3k\, E_k \left(c^\dagger(\mathbf{k},s)c(\mathbf{k},s) + d^\dagger(\mathbf{k},s)d(\mathbf{k},s) \right). \tag{8.34}$$

As in the case of the complex (Klein-Gordon) scalar field theory (see (7.21)), we can define the vacuum state of the theory to correspond to the state which is annihilated by both $c(\mathbf{k},s)$ and $d(\mathbf{k},s)$ (The Hamiltonian is understood to be normal ordered as in (8.34) even though we do not put the normal ordering symbol explicitly.)

$$\begin{aligned} c(\mathbf{k},s)|0\rangle &= 0 = \langle 0|c^\dagger(\mathbf{k},s), \\ d(\mathbf{k},s)|0\rangle &= 0 = \langle 0|d^\dagger(\mathbf{k},s), \\ H|0\rangle &= 0. \end{aligned} \tag{8.35}$$

The one particle states of the theory can now be defined as (see (7.22))

$$\begin{aligned} |k,s\rangle &= c^\dagger(\mathbf{k},s)|0\rangle, \\ \widetilde{|k,s\rangle} &= d^\dagger(\mathbf{k},s)|0\rangle. \end{aligned} \tag{8.36}$$

8.4 Charge operator

Both these states can be checked to be eigenstates of the Hamiltonian (8.34) with energy eigenvalue $k^0 = E_k = \sqrt{\mathbf{k}^2 + m^2} > 0$. Therefore, as in the case of the complex scalar field theory, there are two distinct one particle states with degenerate mass (energy) and to understand the meaning of these states, we need to analyze the symmetry properties of the theory.

8.4 Charge operator

The Dirac Lagrangian density (8.9) (or (8.12)), like the Lagrangian density for the complex Klein-Gordon field theory can be seen to be invariant under a global $U(1)$ phase transformation. We note that under the global phase transformation (α is the constant parameter of transformation)

$$\begin{aligned} \psi(x) &\rightarrow \psi'(x) = e^{-i\theta}\psi(x), \\ \overline{\psi}(x) &\rightarrow \overline{\psi}'(x) = \overline{\psi}(x)e^{i\theta}, \end{aligned} \quad (8.37)$$

or the infinitesimal form of it (ϵ is the constant infinitesimal parameter of transformation)

$$\begin{aligned} \delta_\epsilon \psi &= \psi'(x) - \psi(x) = -i\epsilon\psi(x), \\ \delta_\epsilon \overline{\psi} &= \overline{\psi}'(x) - \overline{\psi}(x) = i\epsilon\overline{\psi}(x), \end{aligned} \quad (8.38)$$

the Lagrangian density (8.9)

$$\mathcal{L} = \overline{\psi}\left(i\gamma^\mu \partial_\mu - m\right)\psi, \quad (8.39)$$

transforms as

$$\begin{aligned} \mathcal{L}(\psi, \overline{\psi}) &\rightarrow \mathcal{L}(\psi', \overline{\psi}') \\ &= \overline{\psi}'\left(i\gamma^\mu \partial_\mu - m\right)\psi' \\ &= \overline{\psi}e^{i\theta}\left(i\gamma^\mu \partial_\mu - m\right)e^{-i\theta}\psi \\ &= \overline{\psi}\left(i\gamma^\mu \partial_\mu - m\right)\psi = \mathcal{L}(\psi, \overline{\psi}). \end{aligned} \quad (8.40)$$

Namely, the Lagrangian density (8.9) is invariant under the global phase transformation (8.37) (or (8.38)). In this case, as in the case of the complex scalar field theory in (7.26), we note that

$$K^\mu = 0. \tag{8.41}$$

On the other hand, using the infinitesimal transformation (8.38), we obtain (see (6.13) and (7.27) and we note that the form of the expression in (8.42) is consistent with the convention of taking left derivatives for the fermion fields)

$$\delta_\epsilon \overline{\psi}_\alpha \frac{\partial \mathcal{L}}{\partial \partial_\mu \overline{\psi}_\alpha} + \delta_\epsilon \psi_\alpha \frac{\partial \mathcal{L}}{\partial \partial_\mu \psi_\alpha}$$
$$= (-i\epsilon\psi)_\alpha (-i\overline{\psi}\gamma^\mu)_\alpha = \epsilon\overline{\psi}(x)\gamma^\mu \psi(x), \tag{8.42}$$

where we have rearranged the fermionic operators keeping in mind their anti-commuting nature.

Thus, the Nöther current associated with the internal symmetry transformation (8.37) (or (8.38)) is obtained from (6.13) to be

$$J^\mu_\epsilon(x) = \epsilon\overline{\psi}(x)\gamma^\mu \psi(x) = \epsilon J^\mu(x), \tag{8.43}$$

where

$$J^\mu(x) = \overline{\psi}(x)\gamma^\mu \psi(x), \tag{8.44}$$

represents the current independent of the parameter of transformation. It is a vector current as in the case of the complex Klein-Gordon field theory (see (7.29)) and it can be checked using the equations of motion (8.6) and (8.7) that this current is conserved,

$$\partial_\mu J^\mu(x) = 0. \tag{8.45}$$

The conserved charge associated with the symmetry transformation (8.37) of the system is given by (note that in the first quantized theory, this would correspond to the total probability)

8.4 CHARGE OPERATOR

$$Q = \int d^3x \, J^0(x) = \int d^3x \, \psi^\dagger(x)\psi(x). \tag{8.46}$$

We note here that the charge (8.46) is associated with an Abelian symmetry transformation (8.37) (or (8.38)). Therefore, as in the case of the complex Klein-Gordon field we can associate the charge operator with that of the electric charge. But, since we are dealing with a fermion system, it is also possible to associate the charge operator with a fermion number (such as baryon number, lepton number, etc) which are associated with $U(1)$ symmetry transformations as well. Using the field decomposition in (8.23), we can express (8.46) in terms of creation and annihilation operators as

$$: Q := \sum_{s=\pm\frac{1}{2}} \int d^3k \left(c^\dagger(\mathbf{k},s) c(\mathbf{k},s) - d^\dagger(\mathbf{k},s) d(\mathbf{k},s) \right). \tag{8.47}$$

As in the case of the complex Klein-Gordon field, we can now show that the charge operator (normal ordered) satisfies (see (7.33) and (7.34))

$$\begin{aligned}
Q|0\rangle &= 0, \\
Q|k,s\rangle &= \sum_{s'=\pm\frac{1}{2}} \int d^3k' \Big(c^\dagger(\mathbf{k}',s') c(\mathbf{k}',s') \\
&\qquad - d^\dagger(\mathbf{k}',s') d(\mathbf{k}',s') \Big) c^\dagger(\mathbf{k},s)|0\rangle \\
&= \sum_{s'=\pm\frac{1}{2}} \int d^3k' \, c^\dagger(\mathbf{k}',s') \Big([c(\mathbf{k}',s'), c^\dagger(\mathbf{k},s)]_+ \\
&\qquad - c^\dagger(\mathbf{k},s) c(\mathbf{k}',s') \Big) |0\rangle \\
&= \sum_{s'=\pm\frac{1}{2}} \int d^3k' c^\dagger(\mathbf{k}',s') \delta_{ss'} \delta^3(k'-k) |0\rangle \\
&= c^\dagger(\mathbf{k},s)|0\rangle = |k,s\rangle,
\end{aligned}$$

$$\begin{aligned}Q|\widetilde{k,s}\rangle &= \sum_{s'=\pm\frac{1}{2}}\int d^3k'\Big(c^\dagger(\mathbf{k}',s')c(\mathbf{k}',s') \\ &\quad - d^\dagger(\mathbf{k}',s')d(\mathbf{k}',s')\Big)d^\dagger(\mathbf{k},s)|0\rangle \\ &= -\sum_{s'=\pm\frac{1}{2}}\int d^3k'\, d^\dagger(\mathbf{k}',s')\Big([d(\mathbf{k}',s'),d^\dagger(\mathbf{k},s)]_+ \\ &\quad - d^\dagger(\mathbf{k},s)d(\mathbf{k}',s')\Big)|0\rangle \\ &= -\sum_{s'=\pm\frac{1}{2}}\int d^3k'\, d^\dagger(\mathbf{k}',s')\delta_{ss'}\delta^3(k-k')|0\rangle \\ &= -d^\dagger(\mathbf{k},s)|0\rangle = -|\widetilde{k,s}\rangle. \end{aligned} \qquad (8.48)$$

Namely, the vacuum state of the theory is charge neutral while the two distinct one particle states, which are degenerate in mass and are created by the operators $c^\dagger(\mathbf{k},s)$ and $d^\dagger(\mathbf{k},s)$ respectively, carry opposite charge. Thus, we can identify them with the particle and the anti-particle states of the theory with positive energy.

This shows that, in the second quantized description of the Dirac theory, the vacuum does not contain any particle as we would intuitively expect (compare this with the first quantized description of the vacuum in section **2.5**) and that the (normal ordered) Hamiltonian in (8.34) is symmetric under the discrete transformation (see also (7.37))

$$c(\mathbf{k},s) \leftrightarrow d(\mathbf{k},s), \qquad (8.49)$$

namely, the second quantized description of the Dirac theory is symmetric under the interchange of particles and anti-particles, as we have also seen in the case of the complex Klein-Gordon field theory.

8.5 Green's functions

As we have discussed within the context of the Klein-Gordon theory, we can define the Green's function for the Dirac field theory as satisfying the Dirac equation with a delta function potential (source) as

8.5 Green's functions

$$(i\partial\!\!\!/ - m) S(x - y) = \delta^4(x - y). \tag{8.50}$$

Here, $S(x - y)$ is a 4×4 matrix function in the spinor space. The right-hand side is also a matrix, but it is the trivial 4×4 identity matrix (which we do not write explicitly). Let us recall the identity

$$\begin{aligned}
(i\partial\!\!\!/ - m)(i\partial\!\!\!/ + m) &= (i\partial\!\!\!/)^2 - m^2 \\
&= -\gamma^\mu \partial_\mu \gamma^\nu \partial_\nu - m^2 = -\gamma^\mu \gamma^\nu \partial_\mu \partial_\nu - m^2 \\
&= -\frac{1}{2}[\gamma^\mu, \gamma^\nu]_+ \partial_\mu \partial_\nu - m^2 = -\eta^{\mu\nu} \partial_\mu \partial_\nu - m^2 \\
&= -\left(\partial_\mu \partial^\mu + m^2\right).
\end{aligned} \tag{8.51}$$

Furthermore, from (5.117) we know that the Green's function for the Klein-Gordon equation satisfies

$$\left(\partial_\mu \partial^\mu + m^2\right) G(x - y) = -\delta^4(x - y). \tag{8.52}$$

From these then, we can identify

$$S(x - y) = (i\partial\!\!\!/ + m) G(x - y), \tag{8.53}$$

so that

$$\begin{aligned}
(i\partial\!\!\!/ - m) S(x - y) &= (i\partial\!\!\!/ - m)(i\partial\!\!\!/ + m) G(x - y) \\
&= -\left(\partial_\mu \partial^\mu + m^2\right) G(x - y) \\
&= \delta^4(x - y).
\end{aligned} \tag{8.54}$$

Therefore, we do not have to calculate the Green's function for the Dirac field theory separately.

Although the relation (8.53) holds for any Green's function of the theory (retarded, advanced, Feynman, etc.), we would concentrate mainly on the Feynman Green's function since that is the most useful in the calculation of S-matrix elements. We note that the relation (8.53) can be written in the momentum space as

$$S(k) = (\slashed{k} + m) G(k), \tag{8.55}$$

so that we can write, for example, (see (5.140))

$$S_F(k) = (\slashed{k} + m) G_F(k) = \lim_{\epsilon \to 0^+} \frac{\slashed{k} + m}{k^2 - m^2 + i\epsilon}, \tag{8.56}$$

which can also be written as

$$S_F(k) = \lim_{\epsilon \to 0^+} \frac{1}{\slashed{k} - m + i\epsilon}. \tag{8.57}$$

From (8.53) we can also obtain the positive and the negative energy Green's functions for the Dirac field theory as (see (5.143) and (5.145) where $k^0 = E_k = \sqrt{\mathbf{k}^2 + m^2} > 0$)

$$\begin{aligned}
S^{(+)}(x) &= (i\slashed{\partial} + m) G^{(+)}(x) \\
&= (i\slashed{\partial} + m) \frac{i}{2} \int \frac{d^3k}{(2\pi)^3} \frac{1}{k^0} e^{-ik\cdot x} \\
&= \frac{i}{2} \int \frac{d^3k}{(2\pi)^3} \frac{\slashed{k} + m}{k^0} e^{-ik\cdot x}, \\
S^{(-)}(x) &= (i\slashed{\partial} + m) G^{(-)}(x) \\
&= (i\slashed{\partial} + m) \left(-\frac{i}{2}\right) \int \frac{d^3k}{(2\pi)^3} \frac{1}{k^0} e^{ik\cdot x} \\
&= -\frac{i}{2} \int \frac{d^3k}{(2\pi)^3} \frac{(-\slashed{k} + m)}{k^0} e^{ik\cdot x} \\
&= \frac{i}{2} \int \frac{d^3k}{(2\pi)^3} \frac{(\slashed{k} - m)}{k^0} e^{ik\cdot x}. \tag{8.58}
\end{aligned}$$

We can express the Feynman Green's function in terms of the positive and the negative energy Green's functions as well (see (5.146))

$$S_F(x) = (i\partial\!\!\!/ + m)\, G_F(x)$$
$$= (i\partial\!\!\!/ + m)\left[-\theta(x^0)\, G^{(+)}(x) + \theta(-x^0)\, G^{(-)}(x)\right]$$
$$= -\theta(x^0)\, S^{(+)}(x) + \theta(-x^0)\, S^{(-)}(x), \tag{8.59}$$

where we have used the fact that the terms involving derivatives of the step function cancel out (namely, $\delta(x^0)(G^{(+)}(x) + G^{(-)}(x)) = 0$). Similarly, other Green's functions for the Dirac field theory can also be easily obtained from the corresponding functions for the Klein-Gordon field theory.

8.6 Covariant anti-commutation relations

Let us decompose the field operator into its positive and negative energy (frequency) parts as

$$\psi_\alpha(x) = \psi_\alpha^{(+)}(x) + \psi_\alpha^{(-)}(x), \tag{8.60}$$

where

$$\psi_\alpha^{(+)}(x) = \sum_{s=\pm\frac{1}{2}} \int d^3k\, \sqrt{\frac{m}{(2\pi)^3 k^0}}\, e^{-ik\cdot x}\, c(\mathbf{k},s) u_\alpha(k,s),$$

$$\psi_\alpha^{(-)}(x) = \sum_{s=\pm\frac{1}{2}} \int d^3k\, \sqrt{\frac{m}{(2\pi)^3 k^0}}\, e^{ik\cdot x}\, d^\dagger(\mathbf{k},s) v_\alpha(k,s). \tag{8.61}$$

Similarly, we can decompose the adjoint field operator (8.8) also as

$$\overline{\psi}_\alpha(x) = \overline{\psi}_\alpha^{(+)} + \overline{\psi}_\alpha^{(-)}(x), \tag{8.62}$$

where

$$\overline{\psi}_\alpha^{(+)}(x) = \sum_{s=\pm\frac{1}{2}} \int d^3k\, \sqrt{\frac{m}{(2\pi)^3 k^0}}\, e^{-ik\cdot x}\, d(\mathbf{k},s) \overline{v}_\alpha(k,s),$$

$$\overline{\psi}_\alpha^{(-)}(x) = \sum_{s=\pm\frac{1}{2}} \int d^3k\, \sqrt{\frac{m}{(2\pi)^3 k^0}}\, e^{ik\cdot x}\, c^\dagger(\mathbf{k},s) \overline{u}_\alpha(k,s). \tag{8.63}$$

It is clear from the basic anti-commutation relations for the creation and the annihilation operators $c(\mathbf{k}, s)$, $c^\dagger(\mathbf{k}, s)$, $d(\mathbf{k}, s)$ and $d^\dagger(\mathbf{k}, s)$ in (8.29) that the only nontrivial anti-commutation relations among these four field components will have the form (these are not at equal times)

$$
\begin{aligned}
\left[\psi_\alpha^{(+)}(x), \overline{\psi}_\beta^{(-)}(y)\right]_+ &= \sum_{s,s'=\pm\frac{1}{2}} \iint \frac{d^3k\, d^3k'}{(2\pi)^3} \sqrt{\frac{m^2}{k^0 k'^0}} \\
&\quad \times e^{-ik\cdot x + ik'\cdot y} u_\alpha(k,s) \overline{u}_\beta(k',s') \left[c(\mathbf{k},s), c^\dagger(\mathbf{k}',s')\right]_+ \\
&= \sum_{s,s'=\pm\frac{1}{2}} \iint \frac{d^3k\, d^3k'}{(2\pi)^3} \sqrt{\frac{m^2}{k^0 k'^0}} \\
&\quad \times e^{-ik\cdot x + ik'\cdot y} u_\alpha(k,s) \overline{u}_\beta(k',s') \delta_{ss'} \delta^3(k-k') \\
&= \sum_{s=\pm\frac{1}{2}} \int \frac{d^3k}{(2\pi)^3} \frac{m}{k^0} e^{-ik\cdot(x-y)} u_\alpha(k,s) \overline{u}_\beta(k,s) \\
&= \int \frac{d^3k}{(2\pi)^3} \frac{m}{k^0} \frac{(\not{k}+m)_{\alpha\beta}}{2m} e^{-ik\cdot(x-y)} \\
&= \frac{1}{2} \int \frac{d^3k}{(2\pi)^3} \frac{(\not{k}+m)_{\alpha\beta}}{k^0} e^{-ik\cdot(x-y)} \\
&= -i S_{\alpha\beta}^{(+)}(x-y),
\end{aligned}
\tag{8.64}
$$

where we have used the completeness relation in (3.106) as well as the identification in (8.58). Similarly, we have

$$
\begin{aligned}
\left[\psi_\alpha^{(-)}(x), \overline{\psi}_\beta^{(+)}(y)\right]_+ &= \sum_{s,s'=\pm\frac{1}{2}} \iint \frac{d^3k\, d^3k'}{(2\pi)^3} \sqrt{\frac{m^2}{k^0 k'^0}} \\
&\quad \times e^{ik\cdot x - ik'\cdot y} v_\alpha(k,s) \overline{v}_\beta(k',s') \left[d^\dagger(\mathbf{k},s), d(\mathbf{k}',s')\right]_+ \\
&= \sum_{s,s'=\pm\frac{1}{2}} \int \frac{d^3k\, d^3k'}{(2\pi)^3} \sqrt{\frac{m^2}{k^0 k'^0}}
\end{aligned}
$$

$$\times e^{ik\cdot x - ik'\cdot y} v_\alpha(k,s) \overline{v}_\beta(k',s') \delta_{ss'} \delta^3(k-k')$$

$$= \sum_{s=\pm\frac{1}{2}} \int \frac{\mathrm{d}^3 k}{(2\pi)^3} \frac{m}{k^0} e^{ik\cdot(x-y)} v_\alpha(k,s) \overline{v}_\beta(k,s)$$

$$= \int \frac{\mathrm{d}^3 k}{(2\pi)^3} \frac{m}{k^0} \frac{(\not{k}-m)_{\alpha\beta}}{2m} e^{ik\cdot(x-y)}$$

$$= \frac{1}{2} \int \frac{\mathrm{d}^3 k}{(2\pi)^3} \frac{(\not{k}-m)_{\alpha\beta}}{k^0} e^{ik\cdot(x-y)}$$

$$= -iS^{(-)}_{\alpha\beta}(x-y), \tag{8.65}$$

where we have used (3.109). As a result, we can write

$$\begin{aligned}
\left[\psi_\alpha(x), \overline{\psi}_\beta(y)\right]_+ &= \left[\psi^{(+)}_\alpha(x), \overline{\psi}^{(-)}_\beta(y)\right]_+ + \left[\psi^{(-)}_\alpha(x), \overline{\psi}^{(+)}_\beta(y)\right]_+ \\
&= -iS^{(+)}_{\alpha\beta}(x-y) - iS^{(-)}_{\alpha\beta}(x-y) \\
&= -i\left(S^{(+)}(x-y) + S^{(-)}(x-y)\right)_{\alpha\beta} \\
&= -iS_{\alpha\beta}(x-y), \tag{8.66}
\end{aligned}$$

which is the analogue of the Schwinger function (5.136) (see also (5.160)) for the Dirac field theory and these relations are analogous to the covariant commutation relations for the Klein-Gordon field operators discussed in section **5.9**.

8.7 Normal ordered and time ordered products

The definitions of normal ordered and time ordered products for Dirac field operators are still the same as discussed in sections **6.5** and **6.6**. However, since fermionic field operators anti-commute (Grassmann variables), in rearranging terms to bring them to a particular form, we need to be careful about negative signs that can arise from changing the order of any two such operators. Thus, by definition the normal ordered product of two field operators yields (note from (8.61) and (8.63) that the positive and the negative energy parts of the field operators contain annihilation and creation operators respectively)

$$\begin{aligned}
: \psi_\alpha^{(-)}(x)\psi_\beta^{(+)}(y) : &= \psi_\alpha^{(-)}(x)\psi_\beta^{(+)}(y) \\
: \psi_\beta^{(+)}(y)\psi_\alpha^{(-)}(x) : &= -\psi_\alpha^{(-)}(x)\psi_\beta^{(+)}(y) \\
&= -:\psi_\alpha^{(-)}(x)\overline{\psi}_\beta^{(+)}(y): ,
\end{aligned} \qquad (8.67)$$

and we note that, unlike the bosonic case, here the order of factors inside normal ordering can lead to additional signs.

From the definition in (8.67), it is clear that

$$\begin{aligned}
: \psi_\alpha(x)\overline{\psi}_\beta(y) : &= : \left(\psi_\alpha^{(+)}(x) + \psi_\alpha^{(-)}(x)\right)\left(\overline{\psi}_\beta^{(+)}(y) + \overline{\psi}_\beta^{(-)}(y)\right) : \\
&= : \left(\psi_\alpha^{(+)}(x)\overline{\psi}_\beta^{(+)}(y) + \psi_\alpha^{(+)}(x)\overline{\psi}_\beta^{(-)}(y) \right. \\
&\quad \left. + \psi_\alpha^{(-)}(x)\overline{\psi}_\beta^{(+)}(y) + \psi_\alpha^{(-)}(x)\overline{\psi}_\beta^{(-)}(y)\right) : \\
&= \psi_\alpha^{(+)}(x)\overline{\psi}_\beta^{(+)}(y) - \overline{\psi}_\beta^{(-)}(y)\psi_\alpha^{(+)}(x) \\
&\quad + \psi_\alpha^{(-)}(x)\overline{\psi}_\beta^{(+)}(y) + \psi_\alpha^{(-)}(x)\overline{\psi}_\beta^{(-)}(y) \\
&= \psi_\alpha(x)\overline{\psi}_\beta(y) - \left[\psi_\alpha^{(+)}(x), \overline{\psi}_\beta^{(-)}(y)\right]_+ \\
&= \psi_\alpha(x)\overline{\psi}_\beta(y) + iS_{\alpha\beta}^{(+)}(x-y),
\end{aligned} \qquad (8.68)$$

where in the last step we have made the identification in (8.64). In other words, we can write (8.68) as

$$\psi_\alpha(x)\overline{\psi}_\beta(y) = : \psi_\alpha(x)\overline{\psi}_\beta(y) : -iS_{\alpha\beta}^{(+)}(x-y). \qquad (8.69)$$

Similarly, using (8.65), it is easy to show that

$$\overline{\psi}_\beta(y)\psi_\alpha(x) = : \overline{\psi}_\beta(y)\psi_\alpha(x) : -iS_{\alpha\beta}^{(-)}(x-y). \qquad (8.70)$$

As a result, we see that we can denote

$$\langle 0|\psi_\alpha(x)\overline{\psi}_\beta(y)|0\rangle = -iS_{\alpha\beta}^{(+)}(x-y) = \underbrace{\psi_\alpha(x)\overline{\psi}_\beta(y)}, \qquad (8.71)$$

8.7 Normal ordered and time ordered products

which can be compared with (6.81). The Wick's theorem for normal ordered products of Dirac fields can now be developed in a way completely analogous to the discussions in section **6.5** which we will not repeat here.

The time ordered product of Dirac field operators is again defined with time ordering of the operators from left to right (largest time on the left) keeping in mind the appropriate negative signs associated with rearranging fermionic operators as

$$T(\psi_\alpha(x)\overline{\psi}_\beta(y))$$
$$= \theta(x^0 - y^0)\psi_\alpha(x)\overline{\psi}_\beta(y) - \theta(y^0 - x^0)\overline{\psi}_\beta(y)\psi_\alpha(x). \quad (8.72)$$

This can be expressed in terms of the normal ordered products as

$$\begin{aligned}
T\left(\psi_\alpha(x)\overline{\psi}_\beta(y)\right) &= \theta(x^0 - y^0)\left(:\psi_\alpha(x)\overline{\psi}_\beta(y): -iS^{(+)}_{\alpha\beta}(x-y)\right) \\
&\quad - \theta(y^0 - x^0)\left(:\overline{\psi}_\beta(y)\psi_\alpha(x): -iS^{(-)}_{\alpha\beta}(x-y)\right) \\
&= \theta(x^0 - y^0):\psi_\alpha(x)\overline{\psi}_\beta(y): + \theta(y^0 - x^0):\psi_\alpha(x)\overline{\psi}_\beta(y): \\
&\quad + i\left(-\theta(x^0 - y^0)S^{(+)}_{\alpha\beta}(x-y) + \theta(y^0 - x^0)S^{(-)}_{\alpha\beta}(x-y)\right) \\
&= \left(\theta(x^0 - y^0) + \theta(y^0 - x^0)\right):\psi_\alpha(x)\overline{\psi}_\beta(y): + iS_{F,\alpha\beta}(x-y) \\
&= :\psi_\alpha(x)\overline{\psi}_\beta(y): + iS_{F,\alpha\beta}(x-y), \quad (8.73)
\end{aligned}$$

where we have used the definition in (8.59) and which can be compared with (6.104). It follows now from (8.73) that the Feynman Green's function is related to the vacuum expectation value of the time ordered product of $\psi_\alpha(x)$ and $\overline{\psi}_\beta(y)$,

$$\langle 0|T\left(\psi_\alpha(x)\overline{\psi}_\beta(y)\right)|0\rangle = iS_{F,\alpha\beta}(x-y) = \overline{\psi_\alpha(x)\overline{\psi}_\beta(y)}. \quad (8.74)$$

Once again, we can develop Wick's theorem for time ordered products of Dirac fields in a straightforward manner as discussed in section **6.6**.

8.8 Massless Dirac fields

As we have seen in section **3.7**, when the Dirac particle is massless, it is most natural to decompose it into its chirality states. Correspondingly, we can also decompose a Dirac field into a left-handed and a right-handed component as

$$\psi_R(x) = \frac{1}{2}(\mathbb{1} + \gamma_5)\psi(x), \quad \psi_L(x) = \frac{1}{2}(\mathbb{1} - \gamma_5). \tag{8.75}$$

As a result, the free Dirac Lagrangian density for a massless spinor field naturally decomposes as

$$\mathcal{L} = i\overline{\psi}\slashed{\partial}\psi = i\overline{\psi}_R\slashed{\partial}\psi_R + i\overline{\psi}_L\slashed{\partial}\psi_L. \tag{8.76}$$

Such massless left-handed and right-handed Dirac fields (Weyl fields) arise naturally in physical theories (as we will see, for example, in the case of the standard model in **14.3**). Therefore, in this section we will discuss briefly the quantization of such fields.

Let us consider the field quantization of a right-handed Dirac field (Weyl field). The discussion for the left-handed field can be carried out in an analogous manner. Using the definition of the chiral spinors in section **3.7**, we can expand the field as (see (8.23))

$$\psi_R(x) = \int \frac{\mathrm{d}^3 k}{\sqrt{(2\pi)^3 |\mathbf{k}|}} \left(e^{-ik\cdot x} c(\mathbf{k}) u_R(\mathbf{k}) + e^{ik\cdot x} d^\dagger(\mathbf{k}) v_R(\mathbf{k}) \right), \tag{8.77}$$

where, as before, we have identified $k^0 = E_k = |\mathbf{k}|$. Unlike in (8.23), we note here that there is no sum over spin polarization because the chiral spinors are charaterized by their unique chirality or helicity (see, for example, the discussion following (3.164)). Furthermore, the multiplicative factors in (8.77) is different from those in (8.23) because we are using here the normalization (2.53) and (2.54) which is appropriate for massless spinors.

Using the orthonormality relations for the right-handed spinors described in (3.170) we can invert the decomposition in (8.77) to obtain (here $k^0 = |\mathbf{k}|$)

8.8 Massless Dirac fields

$$\int \frac{\mathrm{d}^3 x}{\sqrt{(2\pi)^3 |\mathbf{k}|}} e^{ik\cdot x} u_R^\dagger(\mathbf{k}) \psi_R(x)$$

$$= \int \frac{\mathrm{d}^3 k' \mathrm{d}^3 x}{(2\pi)^3} \frac{1}{\sqrt{|\mathbf{k}||\mathbf{k}'|}} \Big[e^{i(k-k')\cdot x} c(\mathbf{k}') u_R^\dagger(\mathbf{k}) u_R(\mathbf{k}')$$

$$+ e^{i(k+k')\cdot x} d^\dagger(\mathbf{k}') u_R^\dagger(\mathbf{k}) v_R(\mathbf{k}') \Big]$$

$$= \int \frac{\mathrm{d}^3 k'}{\sqrt{|\mathbf{k}||\mathbf{k}'|}} \Big[\delta^3(k-k') c(\mathbf{k}') u_R^\dagger(\mathbf{k}) u_R(\mathbf{k}')$$

$$+ \delta^3(k+k') e^{2i|\mathbf{k}|x^0} d^\dagger(\mathbf{k}') u_R^\dagger(\mathbf{k}) v_R(\mathbf{k}') \Big]$$

$$= c(\mathbf{k}). \tag{8.78}$$

Similarly, it can also be shown that (it is assumed that $k^0 = |\mathbf{k}|$)

$$d^\dagger(\mathbf{k}) = \int \frac{\mathrm{d}^3 x}{\sqrt{(2\pi)^3 |\mathbf{k}|}} e^{-ik\cdot x} v_R^\dagger(\mathbf{k}) \psi_R(x). \tag{8.79}$$

As a result, from the equal-time anti-commutation relations for the field variables following from the Lagrangian density for the right-handed spinor field (see, for example, (8.76)), it can be shown that the nontrivial anti-commutation relations between the creation and the annihilation operators take the form

$$\left[c(\mathbf{k}), c^\dagger(\mathbf{k}') \right]_+ = \delta^3(k - k') = \left[d(\mathbf{k}), d^\dagger(\mathbf{k}') \right]_+. \tag{8.80}$$

The Hilbert space of the theory can now be built with the vacuum defined as satisfying

$$c(\mathbf{k})|0\rangle = 0 = d(\mathbf{k})|0\rangle, \tag{8.81}$$

and the higher particle states created by applying the creation operators and other quantities of interest can be derived in the standard manner.

As an example, let us determine the propagator for a right-handed fermion. We recall that the Feynman propagator is defined as the time ordered product (see, for example, (8.74))

$$\begin{aligned}iS_{F,R\alpha\beta}(x-y) &= \langle 0|T(\psi_{R\alpha}(x)\psi_{R\beta}^\dagger(y))|0\rangle \\ &= \theta(x^0-y^0)\langle 0|\psi_{R\alpha}(x)\psi_{R\beta}^\dagger(y)|0\rangle \\ &\quad -\theta(y^0-x^0)\langle 0|\psi_{R\beta}^\dagger(y)\psi_{R\alpha}(x)|0\rangle.\end{aligned} \quad (8.82)$$

Putting in the field decomposition in (8.77) and using the properties of the vacuum in (8.81) we obtain (we assume $k^0 = |\mathbf{k}|$ in the field decomposition)

$$\begin{aligned}iS_{F,R\alpha\beta}(x-y) &= \int \frac{d^3k\, d^3k'}{(2\pi)^3\sqrt{|\mathbf{k}||\mathbf{k}'|}} \\ &\quad \times \Big[\theta(x^0-y^0)e^{-ik\cdot x+ik'\cdot y}u_{R\alpha}(\mathbf{k})u_{R\beta}^\dagger(\mathbf{k}')\langle 0|c(\mathbf{k})c^\dagger(\mathbf{k}')|0\rangle \\ &\quad -\theta(y^0-x^0)e^{ik\cdot x-ik'\cdot y}v_{R\alpha}(\mathbf{k})v_{R\beta}^\dagger(\mathbf{k}')\langle 0|d(\mathbf{k}')d^\dagger(\mathbf{k})|0\rangle\Big] \\ &= \int \frac{d^3k}{(2\pi)^3|\mathbf{k}|}\Big[e^{-ik\cdot(x-y)}\theta(x^0-y^0)u_{R\alpha}(\mathbf{k})u_{R\beta}^\dagger(\mathbf{k}) \\ &\quad -e^{ik\cdot(x-y)}\theta(y^0-x^0)v_{R\alpha}(\mathbf{k})v_{R\beta}^\dagger(\mathbf{k})\Big]. \end{aligned} \quad (8.83)$$

Here we have used (8.80) in evaluating the vacuum expectation values. Dropping the spinor indices and using (3.172) this can also be written as

$$\begin{aligned}iS_{F,R}(x-y) &= \int \frac{d^3k}{(2\pi)^3}\frac{1}{4|\mathbf{k}|} \\ &\quad \times \Big[(|\mathbf{k}|\gamma^0-\boldsymbol{\gamma}\cdot\mathbf{k})\theta(x^0-y^0)e^{-i|\mathbf{k}|(x^0-y^0)+i\mathbf{k}\cdot(\mathbf{x}-\mathbf{y})} \\ &\quad -(|\mathbf{k}|\gamma^0-\boldsymbol{\gamma}\cdot\mathbf{k})\theta(y^0-x^0)e^{i|\mathbf{k}|(x^0-y^0)-i\mathbf{k}\cdot(\mathbf{x}-\mathbf{y})}\Big]\gamma^0(\mathbb{1}+\gamma_5) \\ &= \int \frac{d^3k}{(2\pi)^3}\frac{e^{i\mathbf{k}\cdot(\mathbf{x}-\mathbf{y})}}{4|\mathbf{k}|}\Big[(|\mathbf{k}|\gamma^0-\boldsymbol{\gamma}\cdot\mathbf{k})\theta(x^0-y^0)e^{-i|\mathbf{k}|(x^0-y^0)} \\ &\quad -(|\mathbf{k}|\gamma^0+\boldsymbol{\gamma}\cdot\mathbf{k})\theta(y^0-x^0)e^{i|\mathbf{k}|(x^0-y^0)}\Big]\gamma^0(\mathbb{1}+\gamma_5). \end{aligned} \quad (8.84)$$

8.8 MASSLESS DIRAC FIELDS

Let us evaluate the two terms in (8.84) separately. Using the integral representation for the theta function (see (6.100) and the limit $\epsilon \to 0$ is to be understood) we can write the first term in the integrand as

$$\frac{1}{4|\mathbf{k}|} \theta(x^0 - y^0)(|\mathbf{k}|\gamma^0 - \boldsymbol{\gamma} \cdot \mathbf{k})e^{-i|\mathbf{k}|(x^0-y^0)}$$

$$= -\int \frac{dk^0}{2\pi i} \frac{(|\mathbf{k}|\gamma^0 - \boldsymbol{\gamma} \cdot \mathbf{k})}{4|\mathbf{k}|} \frac{e^{-i(k^0+|\mathbf{k}|)(x^0-y^0)}}{k^0 + i\epsilon}$$

$$= -\int \frac{dk^0}{2\pi i} \frac{(|\mathbf{k}|\gamma^0 - \boldsymbol{\gamma} \cdot \mathbf{k})}{4|\mathbf{k}|} \frac{e^{-ik^0(x^0-y^0)}}{k^0 - |\mathbf{k}| + i\epsilon}$$

$$= -\int \frac{dk^0}{2\pi i} \frac{(k^0\gamma^0 - \boldsymbol{\gamma} \cdot \mathbf{k})}{4k^0} \frac{e^{-ik^0(x^0-y^0)}}{k^0 - |\mathbf{k}| + i\epsilon}$$

$$= -\int \frac{dk^0}{2\pi i} \frac{\not{k}}{4k^0} \frac{e^{-ik^0(x^0-y^0)}}{k^0 - |\mathbf{k}| + i\epsilon}. \tag{8.85}$$

Here we have used the fact that the integral has contributions only when $k^0 = |\mathbf{k}|$ and correspondingly have rewritten some of the terms. The second term in the integrand can also be simplified in a similar manner

$$-\frac{1}{4|\mathbf{k}|} \theta(y^0 - x^0)(|\mathbf{k}|\gamma^0 + \boldsymbol{\gamma} \cdot \mathbf{k})e^{i|\mathbf{k}|(x^0-y^0)}$$

$$= \int \frac{dk^0}{2\pi i} \frac{(|\mathbf{k}|\gamma^0 + \boldsymbol{\gamma} \cdot \mathbf{k})}{4|\mathbf{k}|} \frac{e^{i(k^0+|\mathbf{k}|)(x^0-y^0)}}{k^0 + i\epsilon}$$

$$= \int \frac{dk^0}{2\pi i} \frac{(|\mathbf{k}|\gamma^0 + \boldsymbol{\gamma} \cdot \mathbf{k})}{4|\mathbf{k}|} \frac{e^{ik^0(x^0-y^0)}}{k^0 - |\mathbf{k}| + i\epsilon}$$

$$= \int \frac{dk^0}{2\pi i} \frac{(k^0\gamma^0 + \boldsymbol{\gamma} \cdot \mathbf{k})}{4k^0} \frac{e^{ik^0(x^0-y^0)}}{k^0 - |\mathbf{k}| + i\epsilon}$$

$$= \int \frac{dk^0}{2\pi i} \frac{(-k^0\gamma^0 + \boldsymbol{\gamma} \cdot \mathbf{k})}{4k^0} \frac{e^{-ik^0(x^0-y^0)}}{k^0 + |\mathbf{k}| - i\epsilon}$$

$$= -\int \frac{dk^0}{2\pi i} \frac{\not{k}}{4k^0} \frac{e^{-ik^0(x^0-y^0)}}{k^0 + |\mathbf{k}| - i\epsilon}. \tag{8.86}$$

Therefore, adding the two results in (8.85) and (8.86) and substituting into (8.84) we obtain

$$iS_{F,R}(x-y) = \int \frac{d^4k}{(2\pi)^4} e^{-ik\cdot(x-y)} \frac{i\not{k}\gamma^0(\mathbb{1}+\gamma_5)}{4k^0}$$

$$\times \left(\frac{1}{k^0 - |\mathbf{k}| + i\epsilon} + \frac{1}{k^0 + |\mathbf{k}| - i\epsilon}\right)$$

$$= \int \frac{d^4k}{(2\pi)^4} e^{-ik\cdot(x-y)} \frac{i\not{k}\gamma^0(\mathbb{1}+\gamma_5)}{2(k^2 + i\epsilon)}. \tag{8.87}$$

It follows now that the fermion propagator in the momentum space has the form

$$iS_{F,R}(k) = \frac{i\not{k}\gamma^0(\mathbb{1}+\gamma_5)}{2(k^2+i\epsilon)} = \frac{i\not{k}\gamma^0 P_R}{k^2+i\epsilon}. \tag{8.88}$$

We note here that we have defined the propagator in (8.82) as the time ordered product of $\psi_R \psi_R^\dagger$ and not as the conventional time ordered product of $\psi_R \overline{\psi}_R$ (see, for example, (8.74)). The reason for this is that the adjoint field $\overline{\psi}_R$ has negative chirality. However, the conventional covariant propagator (defined as the time ordered product of $\psi_R \overline{\psi}_R$) can be obtained from (8.88) by simply multiplying a factor of γ^0 on the right which leads to

$$i\overline{S}_{F,R}(k) = iS_{F,R}(k)\gamma^0 = \frac{i\not{k}(\mathbb{1}-\gamma_5)}{2(k^2+i\epsilon)} = \frac{i\not{k} P_L}{k^2+i\epsilon}. \tag{8.89}$$

We note that although our discussion has been within the context of right-handed fermion fields, everything carries over in a straightforward manner to left-handed fermion fields.

8.9 Yukawa interaction

Let us next consider the theory which describes the interaction between a Dirac field and a charge neutral Klein-Gordon field. For

8.9 YUKAWA INTERACTION

example, such a theory can represent the interaction between protons (if assumed to be fundamental) and the π^0-meson as well as the self-interaction of the mesons. The Lorentz invariant Lagrangian density, in this case, has the form

$$\mathcal{L} = i\overline{\psi}\partial\!\!\!/\psi - m\overline{\psi}\psi + \frac{1}{2}\partial_\mu\phi\partial^\mu\phi - \frac{M^2}{2}\phi^2 - g\overline{\psi}\psi\phi - \frac{\lambda}{4!}\phi^4, \quad (8.90)$$

where we are assuming that the fermionic particle (say the proton) has mass m while the spin zero meson (e.g., the π^0 meson) has mass M. The trilinear coupling $g\overline{\psi}\psi\phi$ is manifestly Lorentz invariant (as we have seen in (3.86), $\overline{\psi}\psi$ is invariant under a Lorentz transformation and ϕ is a scalar field) and is known as the Yukawa interaction, since it leads to the Yukawa potential in the Born approximation and the coupling constant g represents the strength of the interaction. As in (6.36), we can separate the Lagrangian density (8.90) into a free part and an interaction part,

$$\mathcal{L} = \mathcal{L}_0 + \mathcal{L}_I, \quad (8.91)$$

where

$$\begin{aligned}
\mathcal{L}_0 &= i\overline{\psi}\partial\!\!\!/\psi - m\overline{\psi}\psi + \frac{1}{2}\partial_\mu\phi\partial^\mu\phi - \frac{M^2}{2}\phi^2, \\
\mathcal{L}_I &= -g\overline{\psi}\psi\phi - \frac{\lambda}{4!}\phi^4.
\end{aligned} \quad (8.92)$$

We note here that the π^0 meson is experimentally known to be a pseudoscalar meson (changes sign under a space reflection) which is not reflected in our trilinear interaction in (8.92). Namely, the correct parity invariant trilinear interaction for a pseudoscalar meson should be $(-g\overline{\psi}\gamma_5\psi\phi)$. However, we are going to ignore this aspect of the meson and also switch off the self-interactions of the meson field for simplicity, since here we are only interested in developing calculational methods in a quantum field theory. (The self-interaction can be switched off at the lowest order by setting $\lambda = 0$.) In this case, we can denote the interaction Lagrangian density as (remember everything is normal ordered)

$$\mathcal{L}_I = -g : \overline{\psi}\psi\phi := -\mathcal{H}_I. \tag{8.93}$$

As a result, in this theory the S-matrix will have the expansion (see (6.69), the adiabatic switch-off is assumed)

$$\begin{aligned}
S &= T\left(e^{-i\int_{-\infty}^{\infty} dt\, H_I}\right) = T\left(e^{-i\int d^4x\, \mathcal{H}_I}\right) \\
&= T\left(e^{-ig\int d^4x\, :\overline{\psi}\psi\phi:}\right) \\
&= \mathbb{1} - ig\int d^4x\, :\overline{\psi}(x)\psi(x)\phi(x): \\
&\quad -\frac{g^2}{2}\iint d^4x d^4x'\, T(:\overline{\psi}(x)\psi(x)\phi(x)::\overline{\psi}(x')\psi(x')\phi(x'):) \\
&\quad +\cdots.
\end{aligned} \tag{8.94}$$

Using Wick's theorem (see sections **6.5** and **6.6**), we can write (8.94) in terms of normal ordered products as

$$\begin{aligned}
S = {}& \mathbb{1} - ig\int d^4x\, :\overline{\psi}(x)\psi(x)\phi(x): \\
& -\frac{g^2}{2}\iint d^4x d^4x'\,\Big[\, :\overline{\psi}(x)\psi(x)\phi(x)\overline{\psi}(x')\psi(x')\phi(x'): \\
& + :\overline{\psi}(x)\psi(x)\overline{\psi}(x')\psi(x'): \contraction{}{\phi}{(x)}{\phi}\phi(x)\phi(x') \\
& + \contraction{}{\psi}{_\alpha(x)}{\overline{\psi}}\psi_\alpha(x)\overline{\psi}_\beta(x') :\overline{\psi}_\alpha(x)\psi_\beta(x')\phi(x)\phi(x'): \\
& - \contraction{}{\psi}{_\alpha(x')}{\overline{\psi}}\psi_\alpha(x')\overline{\psi}_\beta(x) :\psi_\beta(x)\overline{\psi}_\alpha(x')\phi(x)\phi(x'): \\
& + \contraction{}{\phi}{(x)}{\phi}\phi(x)\phi(x')\, \contraction{}{\psi}{_\alpha(x)}{\overline{\psi}}\psi_\alpha(x)\overline{\psi}_\beta(x') :\overline{\psi}_\alpha(x)\psi_\beta(x'): \\
& - \contraction{}{\phi}{(x)}{\phi}\phi(x)\phi(x')\, \contraction{}{\psi}{_\alpha(x')}{\overline{\psi}}\psi_\alpha(x')\overline{\psi}_\beta(x) :\psi_\beta(x)\overline{\psi}_\alpha(x'): \\
& - \contraction{}{\psi}{_\alpha(x)}{\overline{\psi}}\psi_\alpha(x)\overline{\psi}_\beta(x')\, \contraction{}{\psi}{_\beta(x')}{\overline{\psi}}\psi_\beta(x')\overline{\psi}_\alpha(x) :\phi(x)\phi(x'): \\
& - \contraction{}{\psi}{_\alpha(x)}{\overline{\psi}}\psi_\alpha(x)\overline{\psi}_\beta(x')\, \contraction{}{\psi}{_\beta(x')}{\overline{\psi}}\psi_\beta(x')\overline{\psi}_\alpha(x)\, \contraction{}{\phi}{(x)}{\phi}\phi(x)\phi(x')\Big] \\
& +\cdots\cdots.
\end{aligned} \tag{8.95}$$

8.9 YUKAWA INTERACTION

To see how perturbation theory works in the interacting quantum field theory, let us look at the lowest order (order g) term in the expansion (8.95). This can be written out in detail as

$$S^{(1)} = -ig \int d^4x \Big[\big(\overline{\psi}_\alpha^{(+)}(x)\psi_\alpha^{(+)}(x) - \psi_\alpha^{(-)}(x)\overline{\psi}_\alpha^{(+)}(x) $$
$$+ \overline{\psi}_\alpha^{(-)}(x)\psi_\alpha^{(+)}(x) + \overline{\psi}_\alpha^{(-)}(x)\psi_\alpha^{(-)}(x)\big)\phi^{(+)}(x)$$
$$+ \phi^{(-)}(x)\big(\overline{\psi}_\alpha^{(+)}(x)\psi_\alpha^{(+)}(x) - \psi_\alpha^{(-)}(x)\overline{\psi}_\alpha^{(+)}(x)$$
$$+ \overline{\psi}_\alpha^{(-)}(x)\psi_\alpha^{(+)}(x) + \overline{\psi}_\alpha^{(-)}(x)\psi_\alpha^{(-)}(x)\big)\Big]. \tag{8.96}$$

Each term in this expression leads to a distinct physical process. For example, just to get a feeling for what is involved, let us look at the first term in (8.96)

$$S_1^{(1)} = -ig \int d^4x\, \overline{\psi}_\alpha^{(+)}(x)\psi_\alpha^{(+)}(x)\phi^{(+)}(x). \tag{8.97}$$

Each positive energy operator in (8.97) contains an annihilation operator and, therefore, a nontrivial matrix element would result for the case

$$\langle 0|S_1^{(1)}|k,s;k',s';q\rangle, \tag{8.98}$$

where the initial state contains a proton with momentum k and spin s, an anti-proton with momentum k' and spin s' and a π^0 meson with momentum q and the final state contains no particles. (There can be other nontrivial matrix elements where the initial state has one more proton, anti-proton and π^0 meson than the final state.) To evaluate the matrix element (8.98) we substitute the field decomposition for each of the positive energy field operators and write

$$\langle 0|S_1^{(1)}|k,s;k',s';q\rangle$$

$$= -ig \sum_{\tilde{s},\tilde{s}'=\pm\frac{1}{2}} \iiiint d^4x \, \frac{d^3\tilde{p}\,d^3p\,d^3p'}{\sqrt{(2\pi)^3\,2\tilde{p}^0}} \sqrt{\frac{m}{(2\pi)^3 p^0}} \sqrt{\frac{m}{(2\pi)^3 p'^0}}$$

$$\times e^{-i(p+p'+\tilde{p})\cdot x} \, \bar{v}_\alpha(p',\tilde{s}')u_\alpha(p,\tilde{s})$$

$$\times \langle 0|d(\mathbf{p}',\tilde{s}')c(\mathbf{p},\tilde{s})a(\tilde{\mathbf{p}})c^\dagger(\mathbf{k},s)d^\dagger(\mathbf{k}',s')a^\dagger(\mathbf{q})|0\rangle$$

$$= -ig \sum_{\tilde{s},\tilde{s}'=\pm\frac{1}{2}} \iiint \frac{d^3\tilde{p}\,d^3p\,d^3p'}{\sqrt{(2\pi)^3\,2\tilde{p}^0}} \sqrt{\frac{m}{(2\pi)^3 p^0}} \sqrt{\frac{m}{(2\pi)^3 p'^0}}$$

$$\times (2\pi)^4 \delta^4(p+p'+\tilde{p}) \, \bar{v}_\alpha(p',\tilde{s}')u_\alpha(p,\tilde{s})$$

$$\times \delta_{\tilde{s}s}\,\delta^3(p-k)\,\delta_{\tilde{s}'s'}\,\delta^3(p'-k')\,\delta^3(\tilde{p}-q)$$

$$= -(2\pi)^4 ig\, \delta^4(k+k'+q)$$

$$\times \frac{1}{\sqrt{(2\pi)^3 2q^0}} \sqrt{\frac{m}{(2\pi)^3 k^0}} \sqrt{\frac{m}{(2\pi)^3 k'^0}}\, \bar{v}_\alpha(k',s')u_\alpha(k,s), \tag{8.99}$$

where we have used the appropriate (anti) commutation relations between the creation and the annihilation operators in evaluating the vacuum expectation value. The matrix element (8.99) represents the process where a proton, an anti-proton along with a π^0 meson are annihilated. The delta function merely ensures the overall conservation of energy and momentum in the process. We can similarly associate a physical process with every other term of $S^{(1)}$ in (8.96). (For completeness, we note here that this lowest order matrix element in (8.99) vanishes because energy-momentum conservation cannot be satisfied. In fact, note that since $k^0, k'^0, q^0 > 0$, energy conservation $\delta(k^0+k'^0+q^0)$ in (8.99) can be satisfied only if the energy of each of the particles vanishes which is not possible since the particles are massive. However, our goal in this section has been to describe the calculational methods in quantum field theory.)

Each term at order g^2 in the expansion in (8.95) also leads to a distinct physical process. For example, let us consider the second

8.9 YUKAWA INTERACTION

term in the expansion at order g^2 in (8.95)

$$S_2^{(2)} = -\frac{g^2}{2} \iint d^4x d^4x' \overline{\phi(x)\phi(x')} : \overline{\psi}(x)\psi(x)\overline{\psi}(x')\psi(x') :$$

$$= -\frac{g^2}{2} \iint d^4x d^4x' \, iG_F(x-x') : \overline{\psi}(x)\psi(x)\overline{\psi}(x')\psi(x') : . \quad (8.100)$$

A nontrivial matrix element of this operator would be obtained if the initial state contains a proton and an anti-proton of momenta k_1 and k_2 respectively and the final state also contains a proton and an anti-proton of momenta k_3 and k_4 respectively. In such a case, the matrix element will have the form

$$\langle k_4, s_4; k_3, s_3 | S_2^{(2)} | k_1, s_1; k_2, s_2 \rangle$$

$$= -\frac{ig^2}{2} \iint d^4x d^4x' \, G_F(x-x')$$

$$\times \langle k_4, s_4; k_3, s_3 | \overline{\psi}_\alpha^{(-)}(x)\psi_\alpha^{(-)}(x)\overline{\psi}_\beta^{(+)}(x')\psi_\beta^{(+)}(x')$$

$$+ \overline{\psi}_\alpha^{(-)}(x')\psi_\alpha^{(-)}(x')\overline{\psi}_\beta^{(+)}(x)\psi_\beta^{(+)}(x) | k_1, s_1; k_2, s_2 \rangle$$

$$= -ig^2 \iint d^4x d^4x' \, G_F(x-x')$$

$$\times \langle k_4, s_4; k_3, s_3 | \overline{\psi}_\alpha^{(-)}(x)\psi_\alpha^{(-)}(x)\overline{\psi}_\beta^{(+)}(x')\psi_\beta^{(+)}(x') | k_1, s_1; k_2, s_2 \rangle$$

$$= -ig^2 \iint d^4x d^4x' \, G_F(x-x') \, e^{i(k_3+k_4)\cdot x} \, e^{-i(k_1+k_2)\cdot x'}$$

$$\times \sqrt{\frac{m}{(2\pi)^3 k_1^0}} \sqrt{\frac{m}{(2\pi)^3 k_2^0}} \sqrt{\frac{m}{(2\pi)^3 k_3^0}} \sqrt{\frac{m}{(2\pi)^3 k_4^0}}$$

$$\times \overline{u}_\alpha(k_3, s_3) v_\alpha(k_4, s_4) \overline{v}_\beta(k_2, s_2) u_\beta(k_1, s_1). \quad (8.101)$$

Recalling that the Fourier transformation of the Green's function is given by (see (5.120))

$$G_F(x-x') = \int \frac{d^4q}{(2\pi)^4} \, e^{-iq\cdot(x-x')} G_F(q), \quad (8.102)$$

the matrix element (8.101) can be written as

$$\langle k_4, s_4; k_3, s_3 | S_2^{(2)} | k_1, s_1; k_2, s_2 \rangle$$

$$= -ig^2 \iiint \frac{d^4q}{(2\pi)^4} d^4x d^4x' \, G_F(q) e^{-i(q-k_3-k_4)\cdot x} e^{i(q-k_1-k_2)\cdot x'}$$

$$\times \sqrt{\frac{m}{(2\pi)^3 k_1^0}} \sqrt{\frac{m}{(2\pi)^3 k_2^0}} \sqrt{\frac{m}{(2\pi)^3 k_3^0}} \sqrt{\frac{m}{(2\pi)^3 k_4^0}}$$

$$\times \bar{u}_\alpha(k_3, s_3) v_\alpha(k_4, s_4) \bar{v}_\beta(k_2, s_2) u_\beta(k_1, s_1)$$

$$= -ig^2 \int d^4q \, (2\pi)^4 \delta^4(q - k_3 - k_4) \delta^4(q - k_1 - k_2) \, G_F(q)$$

$$\times \sqrt{\frac{m}{(2\pi)^3 k_1^0}} \sqrt{\frac{m}{(2\pi)^3 k_2^0}} \sqrt{\frac{m}{(2\pi)^3 k_3^0}} \sqrt{\frac{m}{(2\pi)^3 k_4^0}}$$

$$\times \bar{u}_\alpha(k_3, s_3) v_\alpha(k_4, s_4) \bar{v}_\beta(k_2, s_2) u_\beta(k_1, s_1)$$

$$= -(2\pi)^4 g^2 \, \delta^4(k_1 + k_2 - k_3 - k_4)$$

$$\times \sqrt{\frac{m}{(2\pi)^3 k_1^0}} \sqrt{\frac{m}{(2\pi)^3 k_2^0}} \sqrt{\frac{m}{(2\pi)^3 k_3^0}} \sqrt{\frac{m}{(2\pi)^3 k_4^0}}$$

$$\times \bar{u}_\alpha(k_3, s_3) v_\alpha(k_4, s_4) i G_F(k_1 + k_2) \bar{v}_\beta(k_2, s_2) u_\beta(k_1, s_1).$$

(8.103)

Physically this matrix element corresponds to a proton and an anti-proton annihilating to create a π^0 meson which subsequently pair produces a proton and an anti-proton. The delta function merely ensures the overall energy momentum conservation in the process.

8.10 Feynman diagrams

It is quite clear from the two simple examples that we have worked out in the last section that the evaluation of the S-matrix elements using the Wick expansion is quite tedious. Instead, Feynman developed a graphical method for evaluating these matrix elements which is in one to one correspondence with the Wick expansion. It is clear from looking at the expansion of the S-matrix that it contains terms

8.10 FEYNMAN DIAGRAMS

with a number of propagators each of which corresponds to the contraction of a pair of fields (Feynman Green's functions) and normal ordered products. The normal ordered products have to give nonvanishing matrix elements between the initial and the final state and, as we have seen, simply give rise to normalization factors as well as spinor functions for the particles in the initial and the final states. Thus, let us introduce a graphical representation for these two basic elements of Wick's expansion. Let us define the two kinds of Feynman propagators that are possible for our theory (the Yukawa theory), namely,

$$\text{-----}_{k}\text{-----} \;=\; iG_F(k) = \lim_{\epsilon \to 0} \frac{i}{k^2 - M^2 + i\epsilon},$$

$$\underset{\beta \quad k \quad \alpha}{\longrightarrow} \;=\; iS_{F,\alpha\beta}(k) = \lim_{\epsilon \to 0^+} \frac{i(\not{k} + m)_{\alpha\beta}}{k^2 - m^2 + i\epsilon}$$

$$= \lim_{\epsilon \to 0^+} \left(\frac{i}{\not{k} - m + i\epsilon}\right)_{\alpha\beta}. \qquad (8.104)$$

Several things are to be noted here. We note that the scalar (meson) Feynman propagator is an even function and corresponds to that of a charge neutral field. Therefore, the line representing this propagator has no direction associated with it. On the other hand, the Feynman propagator for the fermions is not an even function and corresponds to that of a (Dirac) field carrying charge. Hence, the line representing the fermion propagator has a direction associated with it, going from from $\overline{\psi}_\beta$ to ψ_α. (This is the same as saying that a fermionic particle moves in the direction shown or an anti-particle moves in the opposite direction.) It should also be remembered that the fermion propagator is a 4×4 matrix with α and β representing the (Dirac) matrix indices.

The propagators are completely independent of the initial and the final states and, therefore, occur as internal lines in a diagram. The initial and the final states, on the other hand, are represented by external lines. For example, an external boson line can represent either the annihilation or the creation of a particle at the interaction point (vertex), simply because we have a charge neutral scalar field and in such a case, the normalization factor for the external line (state) is denoted by ($k^0 = \sqrt{\mathbf{k}^2 + M^2}$)

$$\text{-----}\bullet_k = \frac{1}{\sqrt{(2\pi)^3 2k^0}}. \tag{8.105}$$

On the other hand, since the fermion field carries a charge (and, therefore, carries a direction for charge flow), a fermion external line with an arrow flowing into an interaction point (vertex) can represent either the annihilation of a particle or the creation of an anti-particle at that point and the external line factor, in such a case is given by (the simplest way to visualize this is to note that the end of a fermion line along the direction of the arrow corresponds to a ψ field which can either annihilate a particle or create an anti-particle)

$$\xrightarrow[k,s\ \alpha]{}\bullet \begin{cases} = \sqrt{\frac{m}{(2\pi)^3 k^0}}\ u_\alpha(k,s), & \text{(particle annihilation)}, \\ = \sqrt{\frac{m}{(2\pi)^3 k^0}}\ v_\alpha(k,s), & \text{(anti-particle creation)}. \end{cases} \tag{8.106}$$

However, a fermion external line with an arrow away from the interaction point (vertex) can represent the creation of a particle or the annihilation of an anti-particle (since it corresponds to the field $\overline{\psi}$) with an external line factor

$$\xleftarrow[k,s\ \alpha]{}\bullet \begin{cases} = \sqrt{\frac{m}{(2\pi)^3 k^0}}\ \overline{u}_\alpha(k,s), & \text{(particle creation)}, \\ = \sqrt{\frac{m}{(2\pi)^3 k^0}}\ \overline{v}_\alpha(k,s), & \text{(anti-particle annihilation)}. \end{cases} \tag{8.107}$$

Note that it is the interaction between the different fields which gives rise to nontrivial scattering. The interaction can be read out from the Lagrangian density $\mathcal{L}_I$ and, in a local quantum field theory, is represented as a vertex (point of interaction) with no meaning attached to the external lines. Thus, for the Yukawa interaction (8.93), we

8.10 FEYNMAN DIAGRAMS

have with the convention that all momenta are incoming at a vertex (basically, the interaction vertex is obtained from $(iS_I = i \int d^4x \, \mathcal{L}_I)$ by taking appropriate field derivatives and the factor of $(2\pi)^4$ arises from transforming to momentum space)

$$\text{(diagram)} = -(2\pi)^4 ig \, \delta^4(k_1 + k_2 + k_3) \, \delta_{\alpha\beta}. \tag{8.108}$$

These are the basic elements out of which Feynman diagrams are constructed by attaching internal and external lines to vertices. Furthermore, the rule for connecting vertices and lines is that the four momentum of every internal line (propagator) must be integrated over all possible values, with the normalization factor $\frac{1}{(2\pi)^4}$ for every momentum integration. With these rules, conventionally known as Feynman rules, we can show that every Feynman diagram corresponds to a unique physical process in one to one correspondence with Wick's expansion. (There is one other rule, namely, one has to have a factor of (-1) for every internal fermion loop. There are also some other subtleties like the symmetry factor associated with a diagram (graph), which we do not worry about at this point. We should also note for completeness that these are the Feynman rules in momentum space. We can also define analogous Feynman rules in coordinate space. However, calculations of S-matrix elements are simpler in momentum space.)

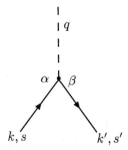

Figure 8.1: Lowest order Feynman diagram.

In this language, therefore, the physical process where a proton, an anti-proton and a π^0 meson are annihilated (which we considered in (8.99)) will correspond in the lowest order to evaluating the Feynman diagram in Fig. 8.1, where the external lines represent the respective physical particles that are annihilated. The value of this diagram can be obtained from the Feynman rules to be

$$\langle 0|S_1^{(1)}|k,s;k',s';q\rangle$$
$$= -(2\pi)^4 ig\delta^4(k+k'+q)\delta_{\alpha\beta}$$
$$\times \frac{1}{\sqrt{(2\pi)^3 2q^0}}\sqrt{\frac{m}{(2\pi)^3 k^0}}\, u_\alpha(k,s)\sqrt{\frac{m}{(2\pi)^3 k'^0}}\, \bar{v}_\beta(k',s')$$
$$= -(2\pi)^4 ig\, \delta^4(k+k'+q)$$
$$\times \frac{1}{\sqrt{(2\pi)^3 2q^0}}\sqrt{\frac{m}{(2\pi)^3 k^0}}\sqrt{\frac{m}{(2\pi)^3 k'^0}}\, \bar{v}_\alpha(k',s')u_\alpha(k,s). \tag{8.109}$$

This is exactly the value of the first order S-matrix element which we had evaluated in (8.99).

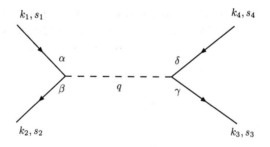

Figure 8.2: A second order Feynman diagram.

Let us next note that if we are interested in the second order process where a proton and an anti-proton annihilate and create a π^0 meson which subsequently pair creates a proton and an anti-proton, the corresponding lowest order Feynman diagram will be given by Fig. 8.2, with the external lines representing physical particles. The

8.10 FEYNMAN DIAGRAMS

value of the diagram can now be calculated using the Feynman rules to be

$$\langle k_4, s_4; k_3, s_3 | S_2^{(2)} | k_1, s_1; k_2, s_2 \rangle$$
$$= \int \frac{d^4q}{(2\pi)^4} (-(2\pi)^4 ig) \delta^4(k_1 + k_2 - q) \delta_{\alpha\beta} \sqrt{\frac{m}{(2\pi)^3 k_1^0}} \sqrt{\frac{m}{(2\pi)^3 k_2^0}}$$
$$\times u_\alpha(k_1, s_1) \bar{v}_\beta(k_2, s_2) i G_F(q) (-(2\pi)^4 ig) \delta^4(q - k_3 - k_4) \delta_{\gamma\delta}$$
$$\times \sqrt{\frac{m}{(2\pi)^3 k_3^0}} \sqrt{\frac{m}{(2\pi)^3 k_4^0}} \bar{u}_\gamma(k_3, s_3) v_\delta(k_4, s_4)$$
$$= -(2\pi)^4 g^2 \, \delta^4(k_1 + k_2 - k_3 - k_4)$$
$$\times \sqrt{\frac{m}{(2\pi)^3 k_1^0}} \sqrt{\frac{m}{(2\pi)^3 k_2^0}} \sqrt{\frac{m}{(2\pi)^3 k_3^0}} \sqrt{\frac{m}{(2\pi)^3 k_4^0}}$$
$$\times \bar{u}_\alpha(k_3, s_3) v_\alpha(k_4, s_4) i G_F(k_1 + k_2) \bar{v}_\beta(k_2, s_2) u_\beta(k_1, s_1).$$
(8.110)

Once again, this is exactly the second order S-matrix element which we had calculated from the Wick expansion in (8.103). This shows the power of the Feynman diagram method. However, it is worth emphasizing that whenever confusions are likely to arise in the Feynman method, it is always resolved by going back to Wick's expansion (of importance are concepts like the symmetry factor associated with a graph).

To conclude this section, let us observe that the same Feynman diagram can describe distinct physical processes depending on the external lines. For example, the Feynman diagram in Fig. 8.2 can describe the scattering of two protons through the exchange of a π^0 meson, namely, a proton with initial momentum k_1 scattering into a state with momentum k_2 by emitting a π^0 meson with momentum $q = k_1 - k_2$ which is captured by an initial proton of momentum k_4 scattering into a state with momentum k_3. If we are interested in only the basic diagram (and not the external line factors since the diagram can represent various distinct processes), from the Feynman rules we see that it will be given by (we are ignoring the overall energy-momentum conserving delta function $(2\pi)^4 \delta^4(k_1 + k_2 - k_3 - k_4)$)

$$-iV(q) = -ig^2 G_F(q)$$
$$= \lim_{\epsilon \to 0^+} -\frac{ig^2}{q^2 - M^2 + i\epsilon}, \quad q = k_1 - k_2, \tag{8.111}$$

where $V(q)$ represents the scattering amplitude in the Born approximation. Furthermore, if we are in the center of mass frame where

$$\mathbf{k}_1 = -\mathbf{k}_4, \quad \mathbf{k}_2 = -\mathbf{k}_3,$$
$$k_1^0 = k_4^0, \quad k_2^0 = k_3^0, \tag{8.112}$$

then, it follows from energy conservation that

$$k_1^0 = k_2^0, \quad q = (k_1 - k_2) = (0, \mathbf{k}_1 - \mathbf{k}_2). \tag{8.113}$$

Therefore, we see that $V(q) = V(\mathbf{q})$ and taking the Fourier transform of (8.113), we obtain

$$-iV(\mathbf{x}) = -ig^2 \int \frac{d^3q}{(2\pi)^3} e^{i\mathbf{q}\cdot\mathbf{x}} V(\mathbf{q})$$
$$= \lim_{\epsilon \to 0^+} -ig^2 \int \frac{d^3q}{(2\pi)^3} \frac{e^{i\mathbf{q}\cdot\mathbf{x}}}{-(\mathbf{q}^2 + M^2) + i\epsilon}$$
$$= \frac{ig^2}{(2\pi)^3} \int_0^\infty d|\mathbf{q}||\mathbf{q}|^2 \int_0^\pi d\theta \sin\theta \int_0^{2\pi} d\phi \frac{e^{i|\mathbf{q}||\mathbf{x}|\cos\theta}}{|\mathbf{q}|^2 + M^2}$$
$$= \frac{ig^2}{(2\pi)^2} \int_0^\infty d|\mathbf{q}| \frac{|\mathbf{q}|^2}{i|\mathbf{q}||\mathbf{x}|(|\mathbf{q}|^2 + M^2)} \left(e^{i|\mathbf{q}||\mathbf{x}|} - e^{-i|\mathbf{q}||\mathbf{x}|}\right)$$
$$= \frac{g^2}{(2\pi)^2|\mathbf{x}|} \int_{-\infty}^\infty dz \frac{ze^{iz|\mathbf{x}|}}{z^2 + M^2}$$
$$= \frac{g^2}{(2\pi)^2|\mathbf{x}|} (2\pi i) \left(\frac{iMe^{-M|\mathbf{x}|}}{2iM}\right)$$
$$= \frac{ig^2}{4\pi} \frac{e^{-M|\mathbf{x}|}}{|\mathbf{x}|}, \tag{8.114}$$

where we have used the method of residues to evaluate the integral (the poles of the integrand are along the imaginary axis and closing the contour in the upper half or in the lower half of the complex z-plane leads to the same result). We recall from studies in scattering theory in non-relativistic quantum mechanics that the potential in the coordinate space corresponds to the Fourier transform of the Born amplitude and, therefore, we obtain from (8.114) the potential between the fermions to be

$$V(\mathbf{x}) = -\frac{g^2}{4\pi} \frac{e^{-M|\mathbf{x}|}}{|\mathbf{x}|}. \tag{8.115}$$

We recognize this as the Yukawa potential (for the exchange of a particle with mass M) and this shows that the Yukawa interaction (8.93) leads to the Yukawa potential in the Born approximation.

8.11 References

1. S. Schweber, *Introduction to Relativistic Quantum Field Theory*, Row, Peterson, Evanston (1961).

2. J. D. Bjorken and S. Drell, *Relativistic Quantum Fields*, McGraw-Hill, New York, 1964.

3. P. Roman, *Introduction to Quantum Field Theory*, John Wliley, New York (1969).

4. C. Itzykson and J-B. Zuber, *Quantum Field Theory*, McGraw-Hill, New York, 1980.

5. N. N. Bogoliubov and D. V. Shirkov, *Introduction to the theory of Quantized Fields*, Nauka, Moscow (1984).

6. F. Gross, *Relativistic Quantum Mechanics and Field Theory*, John Wiley, New York (1993).

Chapter 9
Maxwell field theory

9.1 Maxwell's equations

The next field theory to consider logically is that of an Abelian gauge theory and the Maxwell field theory is a prototype of such a theory. As we know, this is described by the vector potential $A_\mu(x)$ which belongs to the $(\frac{1}{2},\frac{1}{2})$ representation of the Lorentz group (see section **4.2**). Thus, the dynamical field would naturally describe particles with spin 1. Such fields are commonly said to describe vector mesons (as opposed to (pseudo) scalar mesons described by the spin 0 Klein-Gordon fields) or gauge bosons if there is a gauge invariance associated with the corresponding theory.

Let us start with the classical Maxwell's equations in vacuum which are given by (remember $c=1$)

$$\boldsymbol{\nabla} \cdot \mathbf{E} = 0,$$
$$\boldsymbol{\nabla} \cdot \mathbf{B} = 0,$$
$$\boldsymbol{\nabla} \times \mathbf{E} = -\frac{\partial \mathbf{B}}{\partial t},$$
$$\boldsymbol{\nabla} \times \mathbf{B} = \frac{\partial \mathbf{E}}{\partial t}, \tag{9.1}$$

where $\mathbf{E}$ and $\mathbf{B}$ represent the electric and the magnetic fields respectively. It is more convenient from our point of view to rewrite Maxwell's equations (9.1) in a manifestly Lorentz covariant form. To that end, we note that we can solve the second and the third equations in (9.1) to write

$$\mathbf{B} = \nabla \times \mathbf{A},$$
$$\mathbf{E} = -\frac{\partial \mathbf{A}}{\partial t} - \nabla \Phi, \tag{9.2}$$

where $\mathbf{A}(x)$ and $\Phi(x)$ are known as the vector and the scalar potentials respectively. We can, of course, combine them into the four vector potential

$$A^\mu = (\Phi, \mathbf{A}), \qquad A_\mu = (\Phi, -\mathbf{A}). \tag{9.3}$$

If we now define the four dimensional curl of the four vector potential as

$$F_{\mu\nu} = \partial_\mu A_\nu - \partial_\nu A_\mu = -F_{\nu\mu}, \qquad \mu, \nu = 0, 1, 2, 3, \tag{9.4}$$

we note that

$$F_{0i} = \partial_0 A_i - \partial_i A_0 = E_i, \tag{9.5}$$

where

$$\mathbf{E} = (E_1, E_2, E_3), \tag{9.6}$$

denotes the electric field vector. Similarly,

$$F_{ij} = \partial_i A_j - \partial_j A_i = -\epsilon_{ijk} B_k, \tag{9.7}$$

where ϵ_{ijk} denotes the three dimensional Levi-Civita tensor and

$$\mathbf{B} = (B_1, B_2, B_3), \tag{9.8}$$

denotes the magnetic field vector. In other words, the six independent components of the tensor $F_{\mu\nu}$ are given by (three components each of) the electric and the magnetic fields. Therefore, the second rank anti-symmetric tensor $F_{\mu\nu}$ is also known as the field strength tensor.

It is now easy to check that the other two equations (the first and the fourth) in (9.1) are given by

$$\partial_\mu F^{\mu\nu} = 0. \tag{9.9}$$

For example, for $\nu = 0$, equation (9.9) takes the form ($F^{00} = 0$ by definition (9.4))

$$\partial_i F^{i0} = 0,$$
or, $\quad \nabla \cdot \mathbf{E} = 0,$ \hfill (9.10)

which coincides with the first of the equations in (9.1). Similarly, for $\nu = j$, equation (9.9) leads to

$$\partial_\mu F^{\mu j} = 0,$$
or, $\quad \partial_0 F^{0j} + \partial_i F^{ij} = 0,$

or, $\quad \partial^0 F_{0j} + \partial^i F_{ij} = 0,$

or, $\quad \dfrac{\partial E_j}{\partial t} + \partial^i \left(-\epsilon_{ijk} B_k\right) = 0,$

or, $\quad \dfrac{\partial E_j}{\partial t} = -\epsilon_{jik} \partial^i B_k,$

or, $\quad \dfrac{\partial \mathbf{E}}{\partial t} = \nabla \times \mathbf{B},$ \hfill (9.11)

which is the last equation in (9.1).

Thus, we see that the set of four Maxwell's equations (9.1) can be written in the manifestly covariant form as

$$\partial_\mu F^{\mu\nu} = \partial_\mu(\partial^\mu A^\nu - \partial^\nu A^\mu) = 0, \tag{9.12}$$

with the field strength tensor defined in (9.4) as

$$F_{\mu\nu} = \partial_\mu A_\nu - \partial_\nu A_\mu. \tag{9.13}$$

Furthermore, the gauge invariance of Maxwell's equations is obvious in this formulation. For example, we note that under a gauge transformation

$$A_\mu(x) \to A'_\mu(x) = A_\mu(x) + \partial_\mu \theta(x),$$
$$\text{or,} \quad \delta A_\mu(x) = A'_\mu(x) - A_\mu(x) = \partial_\mu \theta(x), \tag{9.14}$$

where $\alpha(x)$ is an arbitrary, real, local parameter of transformation,

$$\begin{aligned} F_{\mu\nu} \to F'_{\mu\nu} &= \partial_\mu A'_\nu - \partial_\nu A'_\mu \\ &= \partial_\mu (A_\nu + \partial_\nu \theta) - \partial_\nu (A_\mu + \partial_\mu \theta) \\ &= \partial_\mu A_\nu - \partial_\nu A_\mu = F_{\mu\nu}. \end{aligned} \tag{9.15}$$

In other words, the field strength tensor is unchanged under the gauge transformation (9.14) of the four vector potential $A_\mu(x)$. Consequently, Maxwell's equations (9.12) are also invariant under these transformations.

9.2 Canonical quantization

In trying to second quantize the Maxwell field theory, we have to treat the four vector potential $A_\mu(x)$, which denotes the dynamical variable of the theory, as an operator. But first, let us examine whether there exists a Lagrangian density which will lead to Maxwell's equations as its Euler-Lagrange equations. Let us consider the Lagrangian density

$$\begin{aligned} \mathcal{L} &= -\frac{1}{4} F_{\mu\nu} F^{\mu\nu} = -\frac{1}{4} (\partial_\mu A_\nu - \partial_\nu A_\mu)(\partial^\mu A^\nu - \partial^\nu A^\mu) \\ &= -\frac{1}{2} \partial_\mu A_\nu (\partial^\mu A^\nu - \partial^\nu A^\mu), \end{aligned} \tag{9.16}$$

which leads to

9.2 CANONICAL QUANTIZATION

$$\frac{\partial \mathcal{L}}{\partial A_\nu} = 0,$$

$$\begin{aligned}\frac{\partial \mathcal{L}}{\partial \partial_\mu A_\nu} &= -\frac{1}{2}\left(\partial^\mu A^\nu - \partial^\nu A^\mu\right) - \frac{1}{2}\partial^\mu A^\nu + \frac{1}{2}\partial^\nu A^\mu \\ &= -\left(\partial^\mu A^\nu - \partial^\nu A^\mu\right) = -F^{\mu\nu},\end{aligned} \quad (9.17)$$

so that the Euler-Lagrange equations, in this case, take the form

$$\frac{\partial \mathcal{L}}{\partial A_\nu} - \partial_\mu \frac{\partial \mathcal{L}}{\partial \partial_\mu A_\nu} = 0,$$

or, $\quad \partial_\mu \left(\partial^\mu A^\nu - \partial^\nu A^\mu\right) = \partial_\mu F^{\mu\nu} = 0, \quad (9.18)$

which coincides with the manifestly covariant Maxwell's equations (9.9). Note that a Lagrangian density

$$\mathcal{L} = \frac{1}{4} F_{\mu\nu} F^{\mu\nu} - \frac{1}{2} F_{\mu\nu} \left(\partial^\mu A^\nu - \partial^\nu A^\mu\right), \quad (9.19)$$

with both A_μ and $F_{\mu\nu}$ treated as independent dynamical variables would also lead to Maxwell's equations as its Euler-Lagrange equations, namely,

$$\frac{\partial \mathcal{L}}{\partial F_{\mu\nu}} = 0,$$

or, $\quad F_{\mu\nu} = \partial_\mu A_\nu - \partial_\nu A_\mu,$

$$\partial_\mu \frac{\partial \mathcal{L}}{\partial \partial_\mu A_\nu} = 0,$$

or, $\quad \partial_\mu F^{\mu\nu} = 0. \quad (9.20)$

However, we will work with the simpler Lagrangian density in (9.16) since the Lagrangian density (9.19) is equivalent to (9.16) when the field variable $F_{\mu\nu}$ is eliminated using the (first) auxiliary field equation in (9.20).

The Lagrangian density (9.16), indeed, gives Maxwell's equations as its Euler-Lagrange equations. It is manifestly Lorentz invariant.

Furthermore, since the field strength tensor $F_{\mu\nu}$ is gauge invariant (see (9.15)), this Lagrangian density is also invariant under the gauge transformation (9.14) of the vector potential. Such a theory which is invariant under a (local) gauge transformation is known as a gauge theory and $A_\mu(x)$ is correspondingly known as the gauge field. If we treat $A_\mu(x)$ as the independent field variable, we can define the associated conjugate momentum as (see (9.17))

$$\Pi^\mu(x) = \frac{\partial \mathcal{L}}{\partial \dot{A}_\mu(x)} = -F^{0\mu}, \tag{9.21}$$

and the first peculiarity of the Maxwell field theory is now obvious from (9.21), namely,

$$\Pi^0(x) = -F^{00} = 0. \tag{9.22}$$

In other words, the momentum conjugate to the field variable $A_0(x)$ vanishes. This is a consequence of the gauge invariance of the theory. As a result, we note that $A_0(x)$ which would have a nontrivial commutation relation only with $\Pi^0(x)$, now must commute with all the field variables in the theory. Therefore, we can think of $A_0(x)$ as a classical function (c-number function) – not as an operator – and without loss of generality we can set it equal to zero

$$A_0(x) = 0. \tag{9.23}$$

This is equivalent to saying that, for canonical quantization of such a system, we have to choose a gauge and this particular choice is known as the temporal (axial) gauge. Furthermore, in this gauge the nontrivial components of the conjugate momentum are given by

$$\Pi^i(x) = -F^{0i} = E_i = -(\partial^0 A^i - \partial^i A^0) = -\dot{A}^i(x). \tag{9.24}$$

The naive canonical quantization conditions, in this gauge, would then appear to be

$$\begin{aligned}
\left[A_i(x), A_j(y)\right]_{x^0=y^0} &= 0 = \left[\Pi^i(x), \Pi^j(y)\right]_{x^0=y^0}, \\
\left[A_i(x), \Pi^j(y)\right]_{x^0=y^0} &= i\delta_i^j \delta^3(x-y).
\end{aligned} \tag{9.25}$$

9.2 CANONICAL QUANTIZATION

However, this is not quite consistent which can be seen as follows. For example, in the gauge (9.23), the dynamical equations (see (9.18))

$$\partial_\mu(\partial^\mu A^\nu - \partial^\nu A^\mu) = 0, \tag{9.26}$$

lead to (for $\nu = 0$)

$$\partial_\mu(\partial^\mu A^0 - \partial^0 A^\mu) = 0,$$
or, $\quad \partial^0(\partial_\mu A^\mu) = 0,$
or, $\quad \partial^0(\nabla \cdot \mathbf{A}) = 0,$
or, $\quad \nabla \cdot \mathbf{A} = 0. \tag{9.27}$

Namely, in this gauge, the dynamical field variable $\mathbf{A}(x)$ must be transverse. (As a parenthetical remark, we note that even after imposing the gauge condition (9.23), the theory can be checked to have a residual gauge invariance under a time independent gauge transformation which preserves (9.23), namely,

$$\mathbf{A}(x) \to \mathbf{A}'(x) = \mathbf{A}(x) + \nabla\bar\theta(\mathbf{x}), \tag{9.28}$$

and the condition (9.27) can be thought of as the gauge choice for this residual invariance.) On the other hand, the canonical quantization condition (9.25) leads to

$$\left[A_i(x), \Pi^j(y)\right]_{x^0=y^0} = i\delta_i^j \delta^3(x-y),$$
or, $\quad \left[A_i(x), \dot{A}^j(y)\right]_{x^0=y^0} = -i\delta_i^j \delta^3(x-y), \tag{9.29}$

which does not satisfy the transversality condition (9.27). The proper quantization condition consistent with the transversality condition (9.27) is given by

$$\left[A_i(x), \dot{A}^j(y)\right]_{x^0=y^0} = -i\delta_{i\,\text{TR}}^j(x-y), \tag{9.30}$$

where the transverse delta function is formally defined as the nonlocal operator (whose coordinate representation is given by $\delta^j_{i\,\text{TR}}(x-y)$)

$$\delta^j_{i\,\text{TR}} = \delta^j_i + \partial_i\left(\frac{1}{\nabla^2}\right)\partial^j, \tag{9.31}$$

so that (remember that $\partial^i\partial_i = -\nabla^2$)

$$\begin{aligned}\partial^i\delta^j_{i\,\text{TR}} &= \partial^i\left(\delta^j_i + \partial_i\left(\frac{1}{\nabla^2}\right)\partial^j\right)\\ &= \partial^j + \partial^i\partial_i\left(\frac{1}{\nabla^2}\right)\partial^j = \partial^j - \partial^j = 0,\end{aligned} \tag{9.32}$$

and (9.30) is consistent with the transversality condition in (9.27). Here $\frac{1}{\nabla^2}$ stands for the inverse of the Laplacian operator and formally denotes the Green's function for ∇^2. We note here that relations such as (9.22), (9.23) and (9.27) constitute constraints on the dynamics of the theory and systems with constraints are known as constrained systems. As we have pointed out in connection with the discussions in the Dirac field theory, there exists a systematic procedure for deriving the correct quantization conditions for such systems, known as the Dirac method, which we will discuss in the next chapter. When quantized systematically through the Dirac method, the correct quantization condition (Dirac bracket) that results in the gauge (9.23) subject to (9.27) is given by (9.30). Let us note here that in momentum space, the transverse delta function in (9.30) has the explicit form

$$\begin{aligned}\delta^j_{i\,\text{TR}}(x-y) &= \left(\delta^j_i + \partial^x_i\partial^{jx}\left(\frac{1}{\nabla^2_x}\right)\right)\delta^3(x-y)\\ &= \int\frac{d^3k}{(2\pi)^3}\,e^{i\mathbf{k}\cdot(\mathbf{x}-\mathbf{y})}\left(\delta^j_i + \frac{k_ik^j}{\mathbf{k}^2}\right)\\ &= \int\frac{d^3k}{(2\pi)^3}\,e^{i\mathbf{k}\cdot(\mathbf{x}-\mathbf{y})}\,\delta^j_{i\,\text{TR}}(\mathbf{k}).\end{aligned} \tag{9.33}$$

In the gauge (9.23), the Hamiltonian density for the Maxwell field theory can be obtained to be

$$\begin{aligned}
\mathcal{H} &= \Pi^i \dot{A}_i - \mathcal{L} \\
&= -\Pi^i \Pi_i + \frac{1}{4} F_{\mu\nu} F^{\mu\nu} \\
&= -\Pi^i \Pi_i + \frac{1}{4} \left(2 F_{0i} F^{0i} + F_{ij} F^{ij} \right) \\
&= -\Pi^i \Pi_i + \frac{1}{2} \Pi_i \Pi^i + \frac{1}{4} (-\epsilon_{ijk} B_k)(-\epsilon_{ij\ell} B_\ell) \\
&= -\frac{1}{2} \Pi^i \Pi_i + \frac{1}{2} B_i B_i = \frac{1}{2} \left(\mathbf{\Pi}^2 + \mathbf{B}^2 \right) \\
&= \frac{1}{2} \left(\mathbf{E}^2 + \mathbf{B}^2 \right),
\end{aligned} \qquad (9.34)$$

and leads to the Hamiltonian

$$\begin{aligned}
H = \int \mathrm{d}^3 x\, \mathcal{H} &= \frac{1}{2} \int \mathrm{d}^3 x \left(\mathbf{\Pi}^2 + \mathbf{B}^2 \right) \\
&= \frac{1}{2} \int \mathrm{d}^3 x \left(\mathbf{E}^2 + \mathbf{B}^2 \right).
\end{aligned} \qquad (9.35)$$

Indeed this is what we know to be the Hamiltonian (energy) for the Maxwell theory even from studies in classical electrodynamics.

9.3 Field decomposition

In the temporal gauge (9.23) (see also (9.27))

$$A_0 = 0, \qquad \nabla \cdot \mathbf{A} = 0, \qquad (9.36)$$

which also lead to

$$\partial^\mu A_\mu = 0, \qquad (9.37)$$

Maxwell's equations (9.18) take the form

$$\partial_\mu F^{\mu\nu} = \partial_\mu (\partial^\mu A^\nu - \partial^\nu A^\mu) = \Box A^\nu = 0. \qquad (9.38)$$

As we have seen, for $\nu = 0$, the wave equation (9.38) vanishes identically (because of the choice of the gauge condition (9.23))

$$\Box A^0 \equiv 0. \tag{9.39}$$

On the other hand, for $\nu = i$, we have

$$\Box A^i = 0. \tag{9.40}$$

Thus, the three components of the vector potential, in this case, satisfy the Klein-Gordon equation for a massless particle (wave equation) and, therefore, will have plane wave solutions of the form

$$\mathbf{A}(x) \propto \boldsymbol{\epsilon}(\mathbf{k})\, e^{\mp ik\cdot x}, \tag{9.41}$$

with

$$k^0 = E_k = \sqrt{\mathbf{k}^2} = |\mathbf{k}|. \tag{9.42}$$

The vector $\boldsymbol{\epsilon}(\mathbf{k})$ represents the polarization (vector) of the plane wave solution travelling along $\hat{\mathbf{k}}$ and carries the vector nature of the vector potential. We note from (9.27) that, since the vector potential is transverse, we must have

$$\boldsymbol{\nabla} \cdot \mathbf{A}(x) = 0,$$
$$\text{or,} \quad \mathbf{k} \cdot \boldsymbol{\epsilon}(\mathbf{k}) = 0. \tag{9.43}$$

Namely, the polarization vector characterizing the vector potential must be transverse to the direction of propagation of the plane wave. There can be two such independent directions, for example, as shown in Fig. 9.1. We can choose the two independent polarization vectors to be orthonormal and, in general, complex satisfying

$$\mathbf{k} \cdot \boldsymbol{\epsilon}(\mathbf{k}, \lambda) = 0, \qquad \lambda = 1, 2,$$
$$\boldsymbol{\epsilon}^*(\mathbf{k}, \lambda) \cdot \boldsymbol{\epsilon}(\mathbf{k}, \lambda') = \delta_{\lambda\lambda'}. \tag{9.44}$$

9.3 FIELD DECOMPOSITION

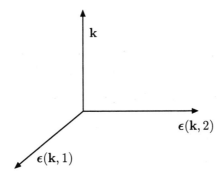

Figure 9.1: Two independent and orthogonal polarization vectors $\epsilon(\mathbf{k}, 1)$ and $\epsilon(\mathbf{k}, 2)$ transverse to the direction of propagation $\mathbf{k}$.

The vector nature of the field variable is now contained completely in the polarization vectors. It is also clear that any transverse vector (to the direction $\hat{\mathbf{k}}$) can now be expressed as a linear superposition of the two independent polarization vectors (9.44) (since they define a basis in the plane transverse to the direction of propagation).

Given this, we can decompose the vector field $\mathbf{A}(x)$ in the basis of the plane wave solutions as

$$\mathbf{A}(x) = \sum_{\lambda=1}^{2} \int \frac{\mathrm{d}^3 k}{\sqrt{(2\pi)^3 2k^0}} \big(\epsilon(\mathbf{k}, \lambda) \, e^{-ik\cdot x} \, a(\mathbf{k}, \lambda)$$
$$+ \epsilon^*(\mathbf{k}, \lambda) \, e^{ik\cdot x} \, a^\dagger(\mathbf{k}, \lambda) \big), \qquad (9.45)$$

where

$$k^0 = E_k = |\mathbf{k}|. \qquad (9.46)$$

Note that $a(\mathbf{k}, \lambda)$ and $a^\dagger(\mathbf{k}, \lambda)$ are operators and as we have defined, the field operator $\mathbf{A}(x)$ is Hermitian as it should be. This discussion is, therefore, quite analogous to that of the neutral Klein-Gordon field except for the presence of the polarization vectors.

Let us next note that

$$\int \frac{d^3x}{\sqrt{(2\pi)^3}} e^{ik\cdot x} \mathbf{A}(x) = \sum_{\lambda=1}^{2} \int\int \frac{d^3k'}{\sqrt{2k'^0}} \frac{d^3x}{(2\pi)^3}$$

$$\times \left(\boldsymbol{\epsilon}(\mathbf{k}',\lambda) e^{-i(k'-k)\cdot x} a(\mathbf{k}',\lambda) + \boldsymbol{\epsilon}^*(\mathbf{k}',\lambda) e^{i(k'+k)\cdot x} a^\dagger(\mathbf{k}',\lambda)\right)$$

$$= \sum_{\lambda=1}^{2} \int \frac{d^3k'}{\sqrt{2k'^0}} \left(\delta^3(\mathbf{k}-\mathbf{k}')\boldsymbol{\epsilon}(\mathbf{k}',\lambda) a(\mathbf{k}',\lambda)\right.$$

$$\left. + \delta^3(\mathbf{k}+\mathbf{k}')\boldsymbol{\epsilon}^*(\mathbf{k}',\lambda) e^{i(k^0+k'^0)t} a^\dagger(\mathbf{k}',\lambda)\right)$$

$$= \sum_{\lambda=1}^{2} \frac{1}{\sqrt{2k^0}} \left(\boldsymbol{\epsilon}(\mathbf{k},\lambda) a(\mathbf{k},\lambda) + \boldsymbol{\epsilon}^*(-\mathbf{k},\lambda) e^{2ik^0 t} a^\dagger(-\mathbf{k},\lambda)\right). \quad (9.47)$$

Similarly, we can show that

$$\int \frac{d^3x}{\sqrt{(2\pi)^3}} e^{ik\cdot x} \dot{\mathbf{A}}(x)$$

$$= -i \sum_{\lambda=1}^{2} \sqrt{\frac{k^0}{2}} \left(\boldsymbol{\epsilon}(\mathbf{k},\lambda) a(\mathbf{k},\lambda) - \boldsymbol{\epsilon}^*(-\mathbf{k},\lambda) e^{2ik^0 t} a^\dagger(-\mathbf{k},\lambda)\right). \quad (9.48)$$

From (9.47) and (9.48), we conclude that

$$i \int \frac{d^3x}{\sqrt{(2\pi)^3}} e^{ik\cdot x} \overleftrightarrow{\partial_0} \mathbf{A}(x) = \frac{2k^0}{\sqrt{2k^0}} \sum_{\lambda=1}^{2} \boldsymbol{\epsilon}(\mathbf{k},\lambda) a(\mathbf{k},\lambda),$$

$$\text{or,} \quad a(\mathbf{k},\lambda) = i \int \frac{d^3x}{\sqrt{(2\pi)^3 2k^0}} \left(e^{ik\cdot x} \overleftrightarrow{\partial_0} \mathbf{A}(x)\right) \cdot \boldsymbol{\epsilon}^*(\mathbf{k},\lambda), \quad (9.49)$$

where we have used the normalization for the polarization vectors in (9.44). Taking the Hermitian conjugate of (9.49) leads to

$$a^\dagger(\mathbf{k},\lambda) = -i \int \frac{d^3x}{\sqrt{(2\pi)^3 2k^0}} \left(e^{-ik\cdot x} \overleftrightarrow{\partial_0} \mathbf{A}(x)\right) \cdot \boldsymbol{\epsilon}(\mathbf{k},\lambda). \quad (9.50)$$

9.3 FIELD DECOMPOSITION

Equations (9.49) and (9.50) denote the two inversion formulae for the Maxwell field (analogous to (5.61) in the case of the Klein-Gordon field).

Given the quantization conditions (see (9.30)),

$$[A_i(x), A_j(y)]_{x^0=y^0} = 0 = [\dot{A}^i(x), \dot{A}^j(y)]_{x^0=y^0},$$
$$[A_i(x), \dot{A}^j(y)]_{x^0=y^0} = -i\delta^j_{i\,\mathrm{TR}}(x-y), \quad (9.51)$$

using (9.49) and (9.50), we can easily show that

$$[a(\mathbf{k}, \lambda), a(\mathbf{k}', \lambda')] = 0 = [a^\dagger(\mathbf{k}, \lambda), a^\dagger(\mathbf{k}', \lambda')],$$
$$[a(\mathbf{k}, \lambda), a^\dagger(\mathbf{k}', \lambda')] = \delta_{\lambda\lambda'}\delta^3(k-k'). \quad (9.52)$$

In other words, the operators $a(\mathbf{k}, \lambda)$ and $a^\dagger(\mathbf{k}, \lambda)$ behave exactly like the annihilation and the creation operators for a harmonic oscillator system.

The Hamiltonian and the momentum operators for the Maxwell theory can again be expressed in terms of the creation and the annihilation operators. The normal ordered forms for these operators can be shown to correspond to (see (5.77) and (5.97) and we do not show the normal ordering symbol explicitly)

$$H = \sum_{\lambda=1}^{2} \int d^3k \, E_k \, a^\dagger(\mathbf{k}, \lambda) a(\mathbf{k}, \lambda),$$
$$\mathbf{P} = \sum_{\lambda=1}^{2} \int d^3k \, \mathbf{k} \, a^\dagger(\mathbf{k}, \lambda) a(\mathbf{k}, \lambda), \quad (9.53)$$

where we recall from the Einstein relation (see (9.46)) that

$$E_k = |\mathbf{k}|. \quad (9.54)$$

The Hilbert space for this theory can be built up in the standard manner. If we denote the vacuum state of the theory by $|0\rangle$, then it satisfies

$$a(\mathbf{k},\lambda)|0\rangle = 0 = \langle 0|a^\dagger(\mathbf{k},\lambda),$$
$$H|0\rangle = 0 = \mathbf{P}|0\rangle. \tag{9.55}$$

The one particle state is then obtained to be ($\lambda = 1, 2$)

$$|k,\lambda\rangle = a^\dagger(\mathbf{k},\lambda)|0\rangle, \tag{9.56}$$

which satisfies

$$H|k,\lambda\rangle = E_k|k,\lambda\rangle,$$
$$\mathbf{P}|k,\lambda\rangle = \mathbf{k}|k,\lambda\rangle. \tag{9.57}$$

Therefore, the one particle state satisfies

$$(H^2 - \mathbf{P}^2)|k,\lambda\rangle = (E_k^2 - \mathbf{k}^2)|k,\lambda\rangle = 0. \tag{9.58}$$

This state, consequently, describes a one photon state and the index λ merely labels the polarization of the photon.

There are several things to note here. First of all, since the $\mathbf{A}(x)$ field is Hermitian like the charge neutral Klein-Gordon field, the particles of the theory, namely the photons, do not carry any electric charge. Second, the energy of the system is positive semi-definite and the Hilbert space is the standard harmonic oscillator space. Furthermore, there are two distinct one photon states corresponding to the two possible independent transverse polarizations a photon can have. We can, in fact, show explicitly that the polarization is directly related to the spin of photon. This, therefore, shows that even though the vector potential $A_\mu(x)$ has four independent field degrees of freedom, the true physical degrees of freedom are two in number. However, we achieved quantization, involving the true degrees of freedom, by eliminating the extra degrees of freedom through non-covariant gauge choices (see (9.23) and (9.27)). As a result, the canonically quantized description which involves only true degrees of freedom is not manifestly Lorentz covariant although the final result of any physical calculation will be. Later, we will also

9.3 FIELD DECOMPOSITION

quantize the Maxwell field theory in a manifestly covariant manner which will bring out some other interesting aspects of such a theory.

Let us note that if we choose the direction of propagation of the plane wave solution to be along the z-axis ($\mathbf{k} = k\hat{\mathbf{e}}_z$) and

$$\epsilon(\mathbf{k}, 1) = \hat{\mathbf{e}}_x, \qquad \epsilon(\mathbf{k}, 2) = \hat{\mathbf{e}}_y, \tag{9.59}$$

then, we can identify $a^\dagger(\mathbf{k}, 1)$ as creating a photon polarized along the x-axis and $a^\dagger(\mathbf{k}, 2)$ as creating a photon polarized along the y-axis. It is clear, in this language, that the operators

$$\begin{aligned} a_L^\dagger(\mathbf{k}) &= \frac{1}{\sqrt{2}}\left(a^\dagger(\mathbf{k}, 1) - ia^\dagger(\mathbf{k}, 2)\right), \\ a_R^\dagger(\mathbf{k}) &= \frac{1}{\sqrt{2}}\left(a^\dagger(\mathbf{k}, 1) + ia^\dagger(\mathbf{k}, 2)\right), \end{aligned} \tag{9.60}$$

would then create photons which are left and right circularly polarized respectively. (These one photon states are eigenstates of helicity with eigenvalue ± 1.) A simple way to see that these would represent the creation operators for left and right circularly polarized photons is to note that

$$\begin{aligned} \sum_{\lambda=1}^{2} \epsilon^*(\mathbf{k}, \lambda) a^\dagger(\mathbf{k}, \lambda) &= \hat{\mathbf{e}}_x a^\dagger(\mathbf{k}, 1) + \hat{\mathbf{e}}_y a^\dagger(\mathbf{k}, 2) \\ &= \frac{1}{2}(\hat{\mathbf{e}}_x + i\hat{\mathbf{e}}_y)(a^\dagger(\mathbf{k}, 1) - ia^\dagger(\mathbf{k}, 2)) \\ &\quad + \frac{1}{2}(\hat{\mathbf{e}}_x - i\hat{\mathbf{e}}_y)(a^\dagger(\mathbf{k}, 1) + ia^\dagger(\mathbf{k}, 2)) \\ &= \epsilon_L^*(\mathbf{k}) a_L^\dagger(\mathbf{k}) + \epsilon_R^*(\mathbf{k}) a_R^\dagger(\mathbf{k}), \end{aligned} \tag{9.61}$$

where we have used the familiar identification that

$$\epsilon_L(\mathbf{k}) = \frac{1}{\sqrt{2}}(\hat{\mathbf{e}}_x - i\hat{\mathbf{e}}_y), \qquad \epsilon_R(\mathbf{k}) = \frac{1}{\sqrt{2}}(\hat{\mathbf{e}}_x + i\hat{\mathbf{e}}_y), \tag{9.62}$$

describe respectively the left and the right circular polarization vectors.

9.4 Photon propagator

Let us note that in the temporal gauge (9.23) (see also (9.27)), if we define the two polarization vectors for $\lambda = 1, 2$ as four vectors

$$\epsilon^\mu(\mathbf{k}, \lambda) = (0, \boldsymbol{\epsilon}(\mathbf{k}, \lambda)), \tag{9.63}$$

and furthermore, for simplicity, if we choose the polarization vectors to be real as in (9.59) then we can write the field decomposition (9.45) as (recall (9.23))

$$A_\mu(x) = \sum_{\lambda=1}^{2} \int \frac{d^3k}{\sqrt{(2\pi)^3 2k^0}} \, \epsilon_\mu(\mathbf{k}, \lambda)(e^{-ik\cdot x} a(\mathbf{k}, \lambda) + e^{ik\cdot x} a^\dagger(\mathbf{k}, \lambda)), \tag{9.64}$$

with $k^0 = |\mathbf{k}|$. This allows us to decompose the field into its positive and negative energy parts as

$$A_\mu^{(+)}(x) = \sum_{\lambda=1}^{2} \int \frac{d^3k}{\sqrt{(2\pi)^3 2k^0}} \, \epsilon_\mu(\mathbf{k}, \lambda) e^{-ik\cdot x} a(\mathbf{k}, \lambda),$$

$$A_\mu^{(-)}(x) = \sum_{\lambda=1}^{2} \int \frac{d^3k}{\sqrt{(2\pi)^3 2k^0}} \, \epsilon_\mu(\mathbf{k}, \lambda) e^{ik\cdot x} a^\dagger(\mathbf{k}, \lambda). \tag{9.65}$$

From the commutation relations for $a(\mathbf{k}, \lambda)$ and $a^\dagger(\mathbf{k}, \lambda)$ in (9.52), it follows that the only nontrivial covariant commutation relations have the form (these are not at equal time)

$$[A_\mu^{(+)}(x), A_\nu^{(-)}(y)] = \sum_{\lambda,\lambda'=1}^{2} \frac{1}{(2\pi)^3} \int \frac{d^3k}{\sqrt{2k^0}} \frac{d^3k'}{\sqrt{2k'^0}}$$

$$\times \epsilon_\mu(\mathbf{k}, \lambda) \epsilon_\nu(\mathbf{k}', \lambda') e^{-ik\cdot x + ik'\cdot y} [a(\mathbf{k}, \lambda), a^\dagger(\mathbf{k}', \lambda')]$$

$$= \sum_{\lambda,\lambda'=1}^{2} \frac{1}{(2\pi)^3} \int \frac{d^3k}{\sqrt{2k^0}} \frac{d^3k'}{\sqrt{2k'^0}}$$

$$\times \epsilon_\mu(\mathbf{k}, \lambda) \epsilon_\nu(\mathbf{k}', \lambda') e^{-ik\cdot x + ik'\cdot y} \delta_{\lambda\lambda'} \delta^3(k - k')$$

$$\begin{aligned}
&= \sum_{\lambda=1}^{2} \frac{1}{(2\pi)^3} \int \frac{d^3k}{2k^0} \epsilon_\mu(\mathbf{k},\lambda)\epsilon_\nu(\mathbf{k},\lambda) e^{-ik\cdot(x-y)} \\
&= \frac{1}{2} \int \frac{d^3k}{(2\pi)^3} \frac{e^{-ik\cdot(x-y)}}{k^0} \left(\sum_{\lambda=1}^{2} \epsilon_\mu(\mathbf{k},\lambda)\epsilon_\nu(\mathbf{k},\lambda) \right) \\
&= -iG^{(+)}_{\mu\nu}(x-y),
\end{aligned}$$

$$[A^{(-)}_\mu(x), A^{(+)}_\nu(y)] = -[A^{(+)}_\nu(y), A^{(-)}_\mu(x)] = iG^{(+)}_{\nu\mu}(y-x)$$

$$\begin{aligned}
&= -\frac{1}{2} \int \frac{d^3k}{(2\pi)^3} \frac{e^{ik\cdot(x-y)}}{k^0} \left(\sum_{\lambda=1}^{2} \epsilon_\mu(\mathbf{k},\lambda)\epsilon_\nu(\mathbf{k},\lambda) \right) \\
&= -iG^{(-)}_{\mu\nu}(x-y), \quad\quad\quad (9.66)
\end{aligned}$$

which are the analogs of the positive and the negative energy Green's functions for the Maxwell field (see (5.143) and (5.145)).

It follows now that

$$\begin{aligned}
A_\mu(x)A_\nu(y) &= (A^{(+)}_\mu(x) + A^{(-)}_\mu(x))(A^{(+)}_\nu(y) + A^{(-)}_\nu(y)) \\
&= A^{(+)}_\mu(x)A^{(+)}_\nu(y) + A^{(+)}_\mu(x)A^{(-)}_\nu(y) \\
&\quad + A^{(-)}_\mu(x)A^{(+)}_\nu(y) + A^{(-)}_\mu(x)A^{(-)}_\nu(y) \\
&= :A_\mu(x)A_\nu(y): + [A^{(+)}_\mu(x), A^{(-)}_\nu(y)] \\
&= :A_\mu(x)A_\nu(y): -iG^{(+)}_{\mu\nu}(x-y) \\
&= :A_\mu(x)A_\nu(y): + \underbrace{A_\mu(x)A_\nu(y)}. \quad\quad (9.67)
\end{aligned}$$

Therefore, we obtain

$$\underbrace{A_\mu(x)A_\nu(y)} = \langle 0|A_\mu(x)A_\nu(y)|0\rangle = -iG^{(+)}_{\mu\nu}(x-y). \quad (9.68)$$

From this, we can define the Feynman propagator in the standard manner (see (6.107)), namely,

$$\langle 0|T(A_\mu(x)A_\nu(y))|0\rangle = iG_{F,\mu\nu}(x-y)$$
$$= \langle 0|\theta(x^0-y^0)A_\mu(x)A_\nu(y) + \theta(y^0-x^0)A_\nu(y)A_\mu(x)|0\rangle$$
$$= -i\theta(x^0-y^0)G^{(+)}_{\mu\nu}(x-y) - i\theta(y^0-x^0)G^{(+)}_{\nu\mu}(y-x)$$
$$= -i\theta(x^0-y^0)G^{(+)}_{\mu\nu}(x-y) + i\theta(y^0-x^0)G^{(-)}_{\mu\nu}(x-y)$$
$$= i(-\theta(x^0-y^0)G^{(+)}_{\mu\nu}(x-y) + \theta(y^0-x^0)G^{(-)}_{\mu\nu}(x-y)). \quad (9.69)$$

By definition, therefore, the Feynman Green's function has the form

$$G_{F,\mu\nu}(x-y)$$
$$= -\theta(x^0-y^0)G^{(+)}_{\mu\nu}(x-y) + \theta(y^0-x^0)G^{(-)}_{\mu\nu}(x-y)$$
$$= -i\int \frac{d^3k}{(2\pi)^3} \left(\sum_{\lambda=1}^{2} \epsilon_\mu(\mathbf{k},\lambda)\epsilon_\nu(\mathbf{k},\lambda)\right)$$
$$\times \frac{1}{2k^0}\left(\theta(x^0-y^0)e^{-ik\cdot(x-y)} + \theta(y^0-x^0)e^{ik\cdot(x-y)}\right)$$
$$= -i\int \frac{d^3k}{(2\pi)^3} e^{i\mathbf{k}\cdot(\mathbf{x}-\mathbf{y})} \left(\sum_{\lambda=1}^{2} \epsilon_\mu(\mathbf{k},\lambda)\epsilon_\nu(\mathbf{k},\lambda)\right)$$
$$\times \frac{1}{2k^0}\left(\theta(x^0-y^0)e^{-ik^0(x^0-y^0)} + \theta(y^0-x^0)e^{ik^0(x^0-y^0)}\right), \quad (9.70)$$

where we have changed $\mathbf{k} \to -\mathbf{k}$ in the second term and have used the fact that the polarization vectors are unaffected by this change of variables (see, for example, (9.59)). To simplify this expression, we use the integral representation for the step function given in (6.100) and write

$$\frac{1}{2k^0}\left(\theta(x^0-y^0)e^{-ik^0(x^0-y^0)} + \theta(y^0-x^0)e^{ik^0(x^0-y^0)}\right)$$
$$= \lim_{\epsilon\to 0^+} -\int \frac{dk'^0}{2\pi i} \frac{1}{2k^0}\left(\frac{e^{-i(k^0+k'^0)(x^0-y^0)}}{k'^0+i\epsilon} - \frac{e^{i(k^0-k'^0)(x^0-y^0)}}{k'^0-i\epsilon}\right)$$

9.4 Photon propagator

$$\begin{aligned}
&= \lim_{\epsilon \to 0^+} -\int \frac{\mathrm{d}k'^0}{2\pi i} \frac{e^{-ik'^0(x^0-y^0)}}{2k^0} \left(\frac{1}{k'^0 - k^0 + i\epsilon} - \frac{1}{k'^0 + k^0 - i\epsilon} \right) \\
&= \lim_{\epsilon \to 0^+} -\int \frac{\mathrm{d}k'^0}{2\pi i} \frac{e^{-ik'^0(x^0-y^0)}}{(k'^0)^2 - (|\mathbf{k}| - i\epsilon)^2} \\
&= \lim_{\epsilon \to 0^+} -\int \frac{\mathrm{d}k'^0}{2\pi i} \frac{e^{-ik'^0(x^0-y^0)}}{(k'^0)^2 - \mathbf{k}^2 + i\epsilon},
\end{aligned} \qquad (9.71)$$

where we have used the Einstein relation (9.46). Substituting this into (9.70) and denoting the variable of integration $k'^0 \to k^0$, we obtain

$$G_{F,\mu\nu}(x-y) = \lim_{\epsilon \to 0^+} \int \frac{\mathrm{d}^4 k}{(2\pi)^4} \frac{e^{-ik\cdot(x-y)}}{k^2 + i\epsilon} \left(\sum_{\lambda=1}^{2} \epsilon_\mu(\mathbf{k},\lambda)\epsilon_\nu(\mathbf{k},\lambda) \right). \qquad (9.72)$$

Therefore, we can identify the momentum space Feynman Green's function for the Maxwell (photon) field to be

$$\begin{aligned}
G_{F,\mu\nu}(k) &= \lim_{\epsilon \to 0^+} \frac{1}{k^2 + i\epsilon} \left(\sum_{\lambda=1}^{2} \epsilon_\mu(\mathbf{k},\lambda)\epsilon_\nu(\mathbf{k},\lambda) \right) \\
&= G_F(k) \left(\sum_{\lambda=1}^{2} \epsilon_\mu(\mathbf{k},\lambda)\epsilon_\nu(\mathbf{k},\lambda) \right),
\end{aligned} \qquad (9.73)$$

where $G_F(k)$ denotes the Feynman Green's function for a massless scalar field (see (5.140)).

The Feynman Green's function in (9.73) depends on the polarization sum and to evaluate the polarization sum, let us note that if we introduce two new four vectors

$$\begin{aligned}
\eta^\mu &= (1,0,0,0), \\
\hat{k}^\mu &= \frac{k^\mu - (k\cdot\eta)\eta^\mu}{\sqrt{(k\cdot\eta)^2 - k^2}} = \left(0, \frac{\mathbf{k}}{|\mathbf{k}|}\right),
\end{aligned} \qquad (9.74)$$

then, $\eta^\mu, \hat{k}^\mu$ and $\epsilon^\mu(\mathbf{k}, \lambda), \lambda = 1, 2$ define an orthonormal basis in four dimensions. Namely, they satisfy

$$\epsilon^\mu(\mathbf{k}, \lambda)\epsilon_\mu(\mathbf{k}, \lambda') = -\delta_{\lambda\lambda'},$$
$$\eta^\mu \eta_\mu = 1,$$
$$\hat{k}^\mu \hat{k}_\mu = -1,$$
$$\eta^\mu \epsilon_\mu(\mathbf{k}, \lambda) = 0 = \eta^\mu \hat{k}_\mu,$$
$$\hat{k}^\mu \epsilon_\mu(\mathbf{k}, \lambda) = 0. \tag{9.75}$$

Consequently, we can write the completeness relation for these basis vectors as

$$\sum_{\lambda=1}^{2} \frac{\epsilon_\mu(\mathbf{k}, \lambda)\epsilon_\nu(\mathbf{k}, \lambda)}{(\epsilon^\sigma(\mathbf{k}, \lambda)\epsilon_\sigma(\mathbf{k}, \lambda))} + \frac{\eta_\mu \eta_\nu}{(\eta^\sigma \eta_\sigma)} + \frac{\hat{k}_\mu \hat{k}_\nu}{(\hat{k}^\sigma \hat{k}_\sigma)} = \eta_{\mu\nu},$$

or, $\quad -\sum_{\lambda=1}^{2} \epsilon_\mu(\mathbf{k}, \lambda)\epsilon_\nu(\mathbf{k}, \lambda) + \eta_\mu \eta_\nu - \hat{k}_\mu \hat{k}_\nu = \eta_{\mu\nu},$

or, $\quad \sum_{\lambda=1}^{2} \epsilon_\mu(\mathbf{k}, \lambda)\epsilon_\nu(\mathbf{k}, \lambda) = -\eta_{\mu\nu} + \eta_\mu \eta_\nu - \hat{k}_\mu \hat{k}_\nu. \tag{9.76}$

Substituting this back into (9.73), we obtain the Feynman propagator for the photon field in the temporal gauge to be

$$iG_{F,\mu\nu}(k) = \lim_{\epsilon \to 0^+} -\frac{i}{k^2 + i\epsilon} (\eta_{\mu\nu} - \eta_\mu \eta_\nu + \hat{k}_\mu \hat{k}_\nu). \tag{9.77}$$

The Feynman propagator has a non-covariant look in this gauge. However, physical calculations carried out with this propagator do lead to Lorentz covariant results. The Feynman propagator, in this gauge, is manifestly transverse (see (9.75)), namely,

$$k^\mu G_{F,\mu\nu}(k) = 0, \tag{9.78}$$

and, as expected (since we are in the gauge $A_0 = 0$), also satisfies

$$G_{F,0\mu}(k) = 0, \tag{9.79}$$

which follows from (9.74).

9.5 Quantum electrodynamics

Quantum electrodynamics (QED) is the theory describing the interaction of electrons and positrons with the electromagnetic field. The Lagrangian density for this theory is given by

$$\begin{aligned}
\mathcal{L} &= -\frac{1}{4} F_{\mu\nu} F^{\mu\nu} + i\overline{\psi}\gamma^\mu D_\mu \psi - m\overline{\psi}\psi \\
&= -\frac{1}{4} F_{\mu\nu} F^{\mu\nu} + i\overline{\psi}\gamma^\mu (\partial_\mu + ieA_\mu) \psi - m\overline{\psi}\psi, \\
&= \mathcal{L}_0 + \mathcal{L}_I,
\end{aligned} \tag{9.80}$$

where

$$D_\mu \psi(x) = (\partial_\mu + ieA_\mu(x))\psi(x), \tag{9.81}$$

denotes the covariant derivative (see (7.83)) and we have separated the Lagrangian density in (9.80) into the free and the interaction parts as

$$\begin{aligned}
\mathcal{L}_0 &= -\frac{1}{4} F_{\mu\nu} F^{\mu\nu} + i\overline{\psi}\slashed{\partial}\psi - m\overline{\psi}\psi, \\
\mathcal{L}_I &= -e\overline{\psi}\gamma^\mu\psi A_\mu.
\end{aligned} \tag{9.82}$$

The total Lagrangian density (9.80) (or (9.82)) can be checked to be invariant under the finite local gauge transformations

$$\begin{aligned}
\psi(x) &\to \psi'(x) = e^{-i\theta(x)}\psi(x), \\
\overline{\psi}(x) &\to \overline{\psi}'(x) = \overline{\psi}(x)e^{i\theta(x)}, \\
A_\mu(x) &\to A'_\mu(x) = A_\mu(x) + \frac{1}{e}\partial_\mu \theta(x),
\end{aligned} \tag{9.83}$$

where $\theta(x)$ is the real and finite local parameter of transformation or under the infinitesimal form of the transformation (9.83) (with $\theta(x) = \epsilon(x)$ =infinitesimal and keeping terms to linear order in ϵ)

$$\begin{aligned}
\delta_\epsilon \psi(x) &= \psi'(x) - \psi(x) = -i\epsilon(x)\psi(x), \\
\delta_\epsilon \overline{\psi}(x) &= \overline{\psi}'(x) - \overline{\psi}(x) = i\overline{\psi}(x)\epsilon(x), \\
\delta_\epsilon A_\mu(x) &= A'_\mu(x) - A_\mu(x) = \frac{1}{e}\partial_\mu \epsilon(x),
\end{aligned} \quad (9.84)$$

where $\epsilon(x)$ denotes the infinitesimal local parameter of transformation. In fact, we have already seen that the Lagrangian density for the Maxwell theory is invariant under the gauge transformation in (9.83) (or (9.84)). Therefore, we need to examine the invariance of the minimally coupled Dirac Lagrangian density in (9.80). First, let us note that the covariant derivative in (9.81) transforms covariantly under a gauge transformation (see (7.85))

$$\begin{aligned}
D_\mu \psi(x) &\to D'_\mu \psi'(x) \\
&= \left(\partial_\mu + ieA'_\mu(x)\right)\psi'(x) \\
&= \left(\partial_\mu + ie\left(A_\mu(x) + \frac{1}{e}(\partial_\mu\theta(x))\right)\right)e^{-i\theta(x)}\psi(x) \\
&= e^{-i\theta(x)}\left(-i(\partial_\mu\theta(x)) + \partial_\mu + ieA_\mu(x) + i(\partial_\mu\theta(x))\right)\psi(x) \\
&= e^{-i\theta(x)}\left(\partial_\mu + ieA_\mu(x)\right)\psi(x) = e^{-i\theta(x)}D_\mu\psi(x). \quad (9.85)
\end{aligned}$$

It follows from this that under a (finite) gauge transformation, the minimally coupled Dirac Lagrangian density transforms as

$$\begin{aligned}
\mathcal{L}_{\text{Dirac}}(\psi, \overline{\psi}, A_\mu) &\to \mathcal{L}_{\text{Dirac}}(\psi', \overline{\psi}', A'_\mu) \\
&= i\overline{\psi}'\gamma^\mu D'_\mu \psi' - m\overline{\psi}'\psi' \\
&= i\overline{\psi}e^{i\theta(x)}\gamma^\mu e^{-i\theta(x)}D_\mu\psi - m\overline{\psi}e^{i\theta(x)}e^{-i\theta(x)}\psi \\
&= i\overline{\psi}\gamma^\mu D_\mu\psi - m\overline{\psi}\psi = \mathcal{L}_{\text{Dirac}}(\psi, \overline{\psi}, A_\mu). \quad (9.86)
\end{aligned}$$

9.5 QUANTUM ELECTRODYNAMICS

Namely, the minimally coupled Dirac Lagrangian density is also invariant under the gauge transformation (9.83) and, therefore, the complete QED Lagrangian density (9.80) is gauge invariant. In this theory, there is a conserved current,

$$J^\mu(x) = \overline{\psi}\gamma^\mu\psi, \quad \partial_\mu J^\mu(x) = 0. \tag{9.87}$$

In momentum space the current conservation takes the form

$$k_\mu J^\mu(k) = 0. \tag{9.88}$$

The Feynman rules for QED can be derived in the standard manner. We have already derived the fermion propagator as well as the fermion external line factors in (8.104) and in (8.106)-(8.107). In addition, in QED we have the photon propagator as well as the photon external line factor which take the forms,

$$\begin{aligned}
&= iG_{F,\mu\nu}(k) \\
&= \lim_{\epsilon \to 0^+} -\frac{i(\eta_{\mu\nu} - \eta_\mu\eta_\nu + \hat{k}_\mu\hat{k}_\nu)}{k^2 + i\epsilon},
\end{aligned} \tag{9.89}$$

$$= \frac{1}{\sqrt{(2\pi)^3 2k^0}}\, \epsilon_\mu(\mathbf{k},\lambda), \tag{9.90}$$

where we have chosen a real polarization vector as in (9.64) so that the external line in (9.90) can represent both the annihilation as well as the creation of a photon. We note here that the photon propagator is a gauge dependent quantity. The form of the propagator in (9.89) (or (9.77)) correponds to the temporal gauge (9.23) (see also (9.27)). The interaction vertex for QED can also be read out from the Lagrangian density (9.80)

$$= -(2\pi)^4 ie\, (\gamma^\mu)_{\beta\alpha}\, \delta^4(p+q+r). \tag{9.91}$$

9.6 Physical processes

With these rules, we can now calculate the S-matrix elements for various physical processes in QED and let us see how some of these calculations are carried out at low orders.

a) **Photon pair production:** For example, let us calculate the matrix element for an electron and a positron to annihilate and produce a pair of photons

$$e^-(k_1) + e^+(k_2) \rightarrow \gamma(k_3) + \gamma(k_4). \tag{9.92}$$

To lowest order there are two Feynman diagrams which will contribute to such a process and they are shown in Fig. 9.2. We note here that experimentally, we only measure two photons in the final state, but not the vertices (points) where each of the two photons was produced. Therefore, the amplitude for this process would simply be the sum of the amplitudes (quantum superposition principle) given by the two diagrams in Fig. 9.2. As a result, the S-matrix element for this process can be worked out to be (we are assuming that the final state photons are outgoing and have carried out the momentum integration for the internal line and are omitting the overall energy-momentum conserving delta function $\delta^4(k_1 + k_2 - k_3 - k_4)$)

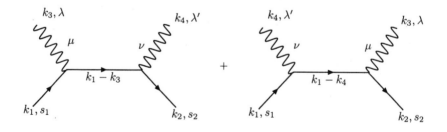

Figure 9.2: The two lowest order Feynman diagrams that contribute to the pair production of photons as well as to the Compton scattering.

9.6 PHYSICAL PROCESSES

$$\langle \gamma(k_3, \lambda), \gamma(k_4, \lambda') | S^{(2)} | e^-(k_1), e^+(k_2) \rangle$$
$$= \frac{1}{(2\pi)^4} \frac{1}{(2\pi)^6} \frac{1}{\sqrt{2k_3^0}} \frac{1}{\sqrt{2k_4^0}} \sqrt{\frac{m}{k_1^0}} \sqrt{\frac{m}{k_2^0}}$$
$$\times \epsilon_\mu(\mathbf{k}_3, \lambda) \epsilon_\nu(\mathbf{k}_4, \lambda') u_\alpha(k_1, s_1) \bar{v}_\beta(k_2, s_2)$$
$$\times \left[-(2\pi)^4 ie \, (\gamma^\nu)_{\beta\delta} \, iS_{F,\delta\gamma}(K)(-(2\pi)^4 ie) \, (\gamma^\mu)_{\gamma\alpha} \right.$$
$$\left. - (2\pi)^4 ie \, (\gamma^\mu)_{\beta\delta} \, iS_{F,\delta\gamma}(\tilde{K})(-(2\pi)^4 ie) \, (\gamma^\nu)_{\delta\alpha} \right]$$
$$= -\frac{ie^2}{(2\pi)^2} \frac{1}{\sqrt{2k_3^0}} \frac{1}{\sqrt{2k_4^0}} \sqrt{\frac{m}{k_1^0}} \sqrt{\frac{m}{k_2^0}}$$
$$\times \left[\bar{v}(k_2, s_2) \not{\epsilon}(\mathbf{k}_4, \lambda') \frac{1}{\not{K} - m + i\epsilon} \not{\epsilon}(\mathbf{k}_3, \lambda) u(k_1, s_1) \right.$$
$$\left. + \bar{v}(k_2, s_2) \not{\epsilon}(\mathbf{k}_3, \lambda) \frac{1}{\not{\tilde{K}} - m + i\epsilon} \not{\epsilon}(\mathbf{k}_4, \lambda') u(k_1, s_1) \right], \tag{9.93}$$

where, for simplicity, we have defined

$$K = k_1 - k_3, \qquad \tilde{K} = k_1 - k_4, \tag{9.94}$$

and the limit $\epsilon \to 0^+$ is understood.

The differential cross-section will be proportional to the absolute square of the matrix element. Furthermore, depending on whether we are interested in summing over all spin states of the fermions or not, the γ-algebra can be simplified and a simple formula for the differential cross-section can be obtained (as discussed in section **3.4**).

b) Compton scattering: The Feynman diagrams for the elastic scattering of an electron by a photon

$$e^-(k_1) + \gamma(k_2) \to e^-(k_3) + \gamma(k_4), \tag{9.95}$$

are given to the lowest order by the same diagrams as in Fig. 9.2 except for the labeling of the momenta. In this case, we will assume the final state electron and photon $(e^-(k_3), \gamma(k_4))$ to be outgoing. The amplitude differs from the case of the pair production of photons by only the external line factors, namely, instead of an electron and a positron being annihilated as was the case in (9.93), here we have the annihilation of an electron with momentum k_1 and the creation of an electron with momentum k_3 (the photon line factor in (9.90) is the same for the creation or the annihilation of a photon). Therefore, we can obtain the S-matrix element for Compton scattering directly from the results in (9.93) to be

$$\langle e^-(k_3), \gamma(k_4, \lambda') | S^{(2)} | e^-(k_1), \gamma(k_2, \lambda) \rangle$$

$$= -\frac{ie^2}{(2\pi)^2} \frac{1}{\sqrt{2k_2^0}} \frac{1}{\sqrt{2k_4^0}} \sqrt{\frac{m}{k_1^0}} \sqrt{\frac{m}{k_3^0}}$$

$$\times \left[\overline{u}(k_3, s_3) \not{\epsilon}(\mathbf{k}_4, \lambda') \frac{1}{\not{Q} - m + i\epsilon} \not{\epsilon}(\mathbf{k}_2, \lambda) u(k_1, s_1) \right.$$

$$\left. + \overline{u}(k_3, s_3) \not{\epsilon}(\mathbf{k}_2, \lambda) \frac{1}{\not{\tilde{K}} - m + i\epsilon} \not{\epsilon}(\mathbf{k}_4, \lambda') u(k_1, s_1) \right], \quad (9.96)$$

where as defined in (9.94) $\tilde{K} = k_1 - k_4$ and we have introduced

$$Q = k_1 + k_2. \tag{9.97}$$

c) **Möller scattering:** Möller scattering is the study of elastic scattering of two electrons

$$e^-(k_1) + e^-(k_2) \to e^-(k_3) + e^-(k_4). \tag{9.98}$$

To the lowest order, this process is described by the two Feynman diagrams shown in Fig. 9.3. Here, the internal line represents a photon propagator signifying that the electrons scatter by exchanging (emitting and absorbing) a photon. Thus, this amplitude will differ from (9.93) (or (9.96)) not only in the external line factors, but

9.6 PHYSICAL PROCESSES

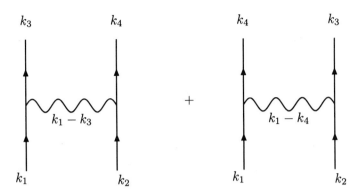

Figure 9.3: The two lowest order Feynman diagrams for Möller scattering.

also in the propagator of the diagram. This can be calculated in a straightforward manner as before. Assuming that the final state electrons are outgoing, carrying out the momentum integration for the internal line and omitting the overall energy-momentum conserving delta function $\delta^4(k_1 + k_2 - k_3 - k_4)$, the S-matrix element takes the form

$$\langle e^-(k_3), e^-(k_4)|S^{(2)}|e^-(k_1), e^-(k_2)\rangle$$
$$= \frac{ie^2}{(2\pi)^2} \sqrt{\frac{m}{k_1^0}} \sqrt{\frac{m}{k_2^0}} \sqrt{\frac{m}{k_3^0}} \sqrt{\frac{m}{k_4^0}}$$
$$\times \left[\overline{u}(k_4, s_4)\gamma^\mu u(k_2, s_2) \frac{(\eta_{\mu\nu} - \eta_\mu \eta_\nu + \hat{K}_\mu \hat{K}_\nu)}{K^2 + i\epsilon} \overline{u}(k_3, s_3)\gamma^\nu u(k_1, s_1) \right.$$
$$\left. + \overline{u}(k_3, s_3)\gamma^\mu u(k_2, s_2) \frac{(\eta_{\mu\nu} - \eta_\mu \eta_\nu + \tilde{\hat{K}}_\mu \tilde{\hat{K}}_\nu)}{\tilde{K}^2 + i\epsilon} \overline{u}(k_4, s_4)\gamma^\nu u(k_1, s_1) \right],$$
(9.99)

where K and $\tilde{K}$ are defined in (9.94).

d) **Bhabha scattering:** Finally, the Bhabha scattering can be thought of as the process

$$e^-(k_1) + e^+(k_2) \rightarrow e^-(k_3) + e^+(k_4). \tag{9.100}$$

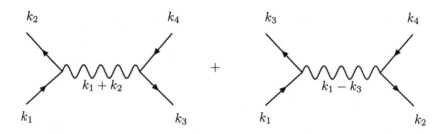

Figure 9.4: The two lowest order Feynman diagrams for Bhabha scattering.

The lowest order Feynman diagrams for this process are shown in Fig. 9.4. We see that topologically these diagrams are the same as in Fig. 9.3, but in the present case we will treat the final state electron and the positron as outgoing $(e^-(k_3), e^+(k_4))$. In this case, the first diagram in Fig. 9.4 represents the annihilation of an electron and a positron to create a photon which subsequently creates an electron and a positron. On the other hand, the second diagram simply describes the scattering of an electron and a positron by exchanging a photon. We can calculate this matrix element as before and omitting the overall delta function representing conservation of energy-momentum, we have

$$\langle e^-(k_3, s_3), e^+(k_4, s_4)|S^{(2)}|e^-(k_1, s_1), e^+(k_2, s_2)\rangle$$
$$= \frac{ie^2}{(2\pi)^2}\sqrt{\frac{m}{k_1^0}}\sqrt{\frac{m}{k_2^0}}\sqrt{\frac{m}{k_3^0}}\sqrt{\frac{m}{k_4^0}}$$
$$\times \left[\overline{u}(k_3,s_3)\gamma^\mu v(k_4,s_4)\frac{(\eta_{\mu\nu} - \eta_\mu\eta_\nu + \hat{Q}_\mu\hat{Q}_\nu)}{Q^2 + i\epsilon}\overline{v}(k_2,s_2)\gamma^\nu u(k_1,s_1)\right.$$
$$\left. + \overline{v}(k_2,s_2)\gamma^\mu v(k_4,s_4)\frac{(\eta_{\mu\nu} - \eta_\mu\eta_\nu + \hat{K}_\mu\hat{K}_\nu)}{K^2 + i\epsilon}\overline{u}(k_3,s_3)\gamma^\nu u(k_1,s_1)\right],$$
$$\tag{9.101}$$

9.7 Ward-Takahashi identity in QED

where as defined in (9.94) and (9.97) $K = k_1 - k_3$ and $Q = k_1 + k_2$. This gives a flavor of low order calculations in QED.

9.7 Ward-Takahashi identity in QED

Before we discuss more complicated symmetries, let us discuss very briefly the Ward-Takahashi identity in QED. In simple language this identity implies that there exists a relation between the fermion two point function (self-energy) and the vertex function in QED and the relation can be traced to the gauge invariance of the theory. Here we would only give a very simple derivation of this relation in QED. Such relations can be derived more systematically and more formally for more complicated theories like the non-Abelian gauge theories through the BRST (Becchi-Rouet-Stora-Tyutin) symmetry, which we will study later.

As we have seen, the Lagrangian density (9.80) for QED has the form

$$\mathcal{L} = -\frac{1}{4}F_{\mu\nu}F^{\mu\nu} + i\bar{\psi}\gamma^\mu \left(\partial_\mu + ieA_\mu\right)\psi - m\bar{\psi}\psi. \tag{9.102}$$

The structure of the photon propagator in the theory, of course, depends on the choice of gauge. However, independent of the choice of gauge, let us note that the fermion propagator in QED can be represented as

$$iS_F(p) = \frac{i}{\not{p} - m} = \underrightarrow{\quad p \quad}, \tag{9.103}$$

where the $i\epsilon$ factor in the denominator is understood. It follows from (9.103) that

$$\begin{aligned}
\frac{\partial}{\partial p^\mu} iS_F(p) &= \frac{\partial}{\partial p^\mu}\left(\frac{i}{\not{p} - m}\right) \\
&= \frac{i}{\not{p} - m}(-\gamma_\mu)\frac{1}{\not{p} - m} \\
&= \frac{i}{\not{p} - m}(i\gamma_\mu)\frac{i}{\not{p} - m} \\
&= -\frac{1}{e}\left(iS_F(p)\right)\Gamma_\mu\left(p, -p, 0\right)\left(iS_F(p)\right), \tag{9.104}
\end{aligned}$$

where $\Gamma_\mu(p,-p,0)$ denotes the three point vertex (9.91) in QED with a zero momentum photon (without the delta function and the factor of $(2\pi)^4$). As a result the right-hand side of (9.104) can be diagrammatically represented as

$$\frac{i}{\not{p}-m}i\gamma_\mu\frac{i}{\not{p}-m} = -\frac{1}{e}\quad\begin{array}{c}\vspace{-2pt}\\p\quad\raisebox{-2pt}{$\lessgtr$}\quad p\\k=0\end{array} \qquad (9.105)$$

$$= -\frac{1}{e}(iS_F(p))\Gamma_\mu(p,-p,k=0)(iS_F(p)).$$

Consequently, a diagrammatic representation for the basic identity (9.104) in QED resulting from the structure of the fermion propagator has the form

$$\frac{\partial}{\partial p^\mu}\xrightarrow{\quad p\quad} = -\frac{1}{e}\begin{array}{c}p\quad\raisebox{-2pt}{$\lessgtr$}\quad p\\k=0\end{array}. \qquad (9.106)$$

In this case, by inverting (9.104) we can obtain a relation between the fermion two point function and the three point vertex function as

$$\begin{aligned}\frac{1}{e}\Gamma_\mu(p,-p,k=0) &= -(iS_F(p))^{-1}\left(\frac{\partial}{\partial p^\mu}(iS_F(p))\right)(iS_F(p))^{-1}\\ &= \frac{\partial}{\partial p^\mu}(iS_F(p))^{-1}\\ &= -i\frac{\partial}{\partial p^\mu}S_F^{-1}(p),\end{aligned} \qquad (9.107)$$

which can also be written as

$$i\frac{\partial}{\partial p^\mu}S_F^{-1}(p) = -\frac{1}{e}\Gamma_\mu(p,-p,k=0). \qquad (9.108)$$

9.7 WARD-TAKAHASHI IDENTITY IN QED

This relates the three point vertex function to the fermion two point function (self-energy) in QED at the tree level and such relations are known as Ward-Takahashi identities (in non-Abelian gauge theories, such relations are called Slavnov-Taylor identities which we will discuss later). Such a relation can be seen to hold at any higher order in perturbation theory as well.

For example, let us look at the fermion self-energy at one loop. Using the graphical identity in (9.106), we can write

$$\frac{\partial}{\partial p^\mu} \quad \begin{array}{c}\text{(diagram)}\end{array} = -\frac{1}{e} \quad \begin{array}{c}\text{(diagram)}\end{array}_{q=0} \tag{9.109}$$

Here we are looking only at the proper vertex parts of the diagrams (namely, without the external lines). The diagram on the left-hand side in (9.109) represents the one loop correction to the fermion two point function or the self-energy (multiplied by a factor of i) whereas the right-hand side is the correction to the three point vertex function at one loop. Once again, we see that (9.108) holds for the one loop corrections to the amplitudes. (The trick in obtaining this relation in a simple manner lies in channeling the external momentum only through the fermion lines.)

At two loop order the fermion self energy diagrams have the forms shown in Fig. 9.5 (there are no external lines and we do not label the momenta for simplicity, but we understand that the external momentum is channelled only through the fermion lines)

Figure 9.5: Two loop corrections to the fermion self-energy in QED.

Using the basic graphical identity in (9.106) at this order we obtain

$$\frac{\partial}{\partial p^\mu}\left[\;\text{(diagram)}\;+\;\text{(diagram)}\;\right]$$

$$=-\frac{1}{e}\left[\;\text{(diagram)}\;+\;\text{(diagram)}\;+\;\text{(diagram)}\right.$$

$$\left.+\;\text{(diagram)}\;+\;\text{(diagram)}\;+\;\text{(diagram)}\;\right]$$

$$=-\frac{1}{e}\left[\;\text{(diagram)}\;+\;\text{(diagram)}\;+\;\text{(diagram)}\right.$$

$$\left.+\;\text{(diagram)}\;+\;\text{(diagram)}\;+\;\text{(diagram)}\;\right]. \qquad (9.110)$$

We recognize that the graphs in the bracket on the right-hand side of (9.110) correspond precisely to the two loop corrections to the vertex function. Thus, using (9.106) we can easily verify the validity of the identity (9.108) diagrammatically at every order in perturbation theory.

A simple and powerful consequence of the Ward-Takahashi identity can be derived as follows. We note that including quantum corrections (we will study renormalization in a later chapter), if the fermion self-energy and the vertex function change as

9.7 WARD-TAKAHASHI IDENTITY IN QED

$$S_F^{-1}(p, m=0) \longrightarrow Z_2 S_F^{-1}(p, m=0),$$
$$\Gamma_\mu \longrightarrow Z_1 \Gamma_\mu, \tag{9.111}$$

where Z_1, Z_2 are constants, then from (9.108) we conclude that these constants must be related as

$$Z_1 = Z_2. \tag{9.112}$$

This physically implies that if there are divergent parts in the fermion self-energy graphs at higher order, then they must equal those present in the vertex correction graphs at the same order (we will study these issues in more detail within the context of renormalization later).

That the Ward-Takahashi identity (9.108) is a consequence of gauge invariance can be seen heuristically in the following way. The Lagrangian density for QED including quantum corrections can be written as

$$\begin{aligned}\mathcal{L} &= -\frac{1}{4}F_{\mu\nu}F^{\mu\nu} + i\overline{\psi}\slashed{D}\psi - m\overline{\psi}\psi - \frac{A}{4}F_{\mu\nu}F^{\mu\nu} + iB\overline{\psi}\slashed{\partial}\psi \\ &\quad -eC\overline{\psi}\slashed{A}\psi - D\overline{\psi}\psi + \cdots \\ &= -\frac{1}{4}(1+A)F_{\mu\nu}F^{\mu\nu} + i(1+B)\overline{\psi}\slashed{\partial}\psi - e(1+C)\overline{\psi}\slashed{A}\psi \\ &\quad -(m+D)\overline{\psi}\psi + \cdots, \end{aligned} \tag{9.113}$$

where A, B, C, D are constants and the dots denote other structures that may be induced due to quantum corrections. Invariance under the local gauge transformations (9.83), on the other hand, requires that the kinetic energy part of the fermion and the photon interaction term must combine into the form of a covariant derivative (see (9.85))

$$ia\overline{\psi}\slashed{D}\psi = ia\overline{\psi}(\slashed{\partial} + ie\slashed{A})\psi. \tag{9.114}$$

Thus comparing with (9.113), we can identify

$$a = 1 + B = 1 + C, \tag{9.115}$$

which is another way of saying

$$Z_2 = 1 + B = 1 + C = Z_1. \tag{9.116}$$

We will discuss these topics in more detail when we study renormalization chapter **16**.

9.8 Covariant quantization of the Maxwell theory

As we have seen, canonical quantization of Maxwell's theory results in a lack of manifest Lorentz invariance. Let us recapitulate briefly how this arises. We recall that the Lagrangian density for Maxwell's theory is given by

$$\begin{aligned} \mathcal{L} &= -\frac{1}{4} F_{\mu\nu} F^{\mu\nu}, \\ F_{\mu\nu} &= \partial_\mu A_\nu - \partial_\nu A_\mu = -F_{\nu\mu}, \end{aligned} \tag{9.117}$$

where $F_{\mu\nu}$ denotes the field strength tensor. The canonical momenta conjugate to the field variables A_μ can be calculated from the Lagrangian density and take the forms (see (9.21) and (9.22))

$$\Pi^\mu = \frac{\partial \mathcal{L}}{\partial \dot{A}_\mu} = -F^{0\mu},$$

which implies the constraint

$$\Pi^0 = -F^{00} = 0. \tag{9.118}$$

As we have already seen in (9.15), the Lagrangian density (9.117) as well as the action for this theory are invariant under the gauge transformation $\delta A_\mu = \partial_\mu \alpha(x)$, since the field strength tensor $F_{\mu\nu}$ does not change under a shift of the vector potential by a gradient. As a result, we can choose a gauge condition and if we choose the gauge $\nabla \cdot \mathbf{A} = 0$ (in our earlier discussion we had chosen the temporal gauge $A_0 = 0$, the present choice of gauge is known as the Coulomb gauge), then the equations of motion (see (9.18))

9.8 Covariant quantization of the Maxwell theory

$$\partial_\mu F^{\mu\nu} = 0,$$

lead to (for $\nu = 0$)

$$\partial_\mu F^{\mu 0} = \partial_i F^{i0} = 0,$$

or, $\quad \partial_i \left(\partial^i A^0 - \partial^0 A^i\right) = 0,$

or, $\quad \nabla^2 A_0 = 0,$

or, $\quad A_0 = 0.$ (9.119)

On the other hand, if sources (charges and currents) are present, in the Coulomb gauge

$$\nabla \cdot \mathbf{A} = 0,$$

the dynamical equations

$$\partial_\mu F^{\mu\nu} = J^\nu,$$

would lead to

$$\partial_\mu F^{\mu 0} = J^0,$$

or, $\quad \nabla^2 A_0 = -J_0,$

or, $\quad A_0 = -\frac{1}{\nabla^2} J_0.$ (9.120)

In either case, the two transverse physical degrees of freedom describe the true dynamics of the theory while the time-like degree of freedom is merely related to the charge density. The canonical quantization can now be carried out for the two physical degrees of freedom, but we lose manifest Lorentz covariance in the process (because all components of A_μ are not being treated on a equal footing). Let us emphasize here that the final result for the calculation of any physical amplitude, however, remains manifestly Lorentz invariant in the canonical formalism. But, we lose manifest Lorentz covariance in the intermediate steps (which is highly desirable in any calculation). We recognize that this is a consequence of our choice of gauge.

The need for a choosing a gauge can be seen in an alternative manner as follows. (This is relevant in the discussion within the context of a path integral quantization of the system which we will discuss briefly in chapter **12**.) We note that we can write the Lagrangian density (9.117) for the Maxwell theory also as

$$\mathcal{L} = -\frac{1}{4} F_{\mu\nu} F^{\mu\nu} = \frac{1}{2} A_\mu P^{\mu\nu} A_\nu + \text{ total derivatives}, \qquad (9.121)$$

where

$$P^{\mu\nu} = \eta^{\mu\nu} \Box - \partial^\mu \partial^\nu. \qquad (9.122)$$

It follows from the explicit form of $P^{\mu\nu}$ that

$$\begin{aligned} P^{\mu\nu} P_\nu^{\ \lambda} &= (\eta^{\mu\nu} \Box - \partial^\mu \partial^\nu)\left(\delta_\nu^\lambda \Box - \partial_\nu \partial^\lambda\right) \\ &= \eta^{\mu\lambda} \Box^2 - \partial^\mu \partial^\lambda \Box - \partial^\mu \partial^\lambda \Box + \partial^\mu \partial^\lambda \Box \\ &= \Box \left(\eta^{\mu\lambda} \Box - \partial^\mu \partial^\lambda\right) \\ &= \Box P^{\mu\lambda}. \end{aligned} \qquad (9.123)$$

With a suitable normalization, we see that $P^{\mu\nu}$ can be thought of as a projection operator. (In fact, $\overline{P}^{\mu\nu} = \frac{1}{\Box} P^{\mu\nu}$ defines the normalized projection operator.) Furthermore, we note that

$$\begin{aligned} \partial_\mu P^{\mu\nu} &= \partial_\mu \left(\eta^{\mu\nu} \Box - \partial^\mu \partial^\nu\right) \\ &= (\partial^\nu \Box - \Box \partial^\nu) = 0 = \partial_\nu P^{\mu\nu}, \end{aligned} \qquad (9.124)$$

so that this is the transverse projection operator, namely, it projects on to the space of the components of any vector transverse (perpendicular) to the gradient operator ∂^μ. Consequently, the inverse of $P^{\mu\nu}$ does not exist (this also means that the determinant of $P^{\mu\nu}$ vanishes) and the Green's function and, therefore, the Feynman propagator for the theory cannot be defined. This implies that even if we can quantize the theory (say, in the naive path integral formalism), we cannot carry out calculations in perturbation theory.

We note here that whenever the determinant of the matrix of highest derivatives of the Lagrangian density vanishes, the system is singular and contains constraints among the field variables (as we will discuss in the next chapter). In such a case, without any further input, the Cauchy initial value problem cannot be uniquely solved, simply because the Green's function does not exist. Thus we see that the naive canonical quantization has unpleasant features in the case of Maxwell's theory since the fields are constrained and the momentum conjugate to A_0 vanishes. We can solve for the constraints and quantize only the true dynamical degrees of freedom. However, in this procedure we lose manifest Lorentz covariance since we single out the transverse degrees of freedom.

We can take an alternative approach. Namely, since we realize that the difficulties in quantization arise because of the singular nature of the Lagrangian density, we can try to modify the theory so as to make it nonsingular. Let us consider, for example, the Lagrangian density

$$\mathcal{L} = -\frac{1}{4} F_{\mu\nu} F^{\mu\nu} - \frac{1}{2} \left(\partial_\mu A^\mu \right)^2 + J^\mu A_\mu, \qquad (9.125)$$

where J^μ represents a conserved current

$$\partial_\mu J^\mu = 0. \qquad (9.126)$$

Here we have generalized Maxwell's theory to include a conserved current. But more than that we have also added a term $-\frac{1}{2} \left(\partial_\mu A^\mu \right)^2$ to Maxwell's Lagrangian density. (This formulation of the theory is due to Fermi and considered as a gauge choice, this gauge is known as the Feynman-Fermi gauge.) This additional term breaks gauge invariance and consequently leads to a nonsingular theory. But clearly the theory (9.125) would appear to be different from Maxwell's theory. Therefore, at this point there is no justification for adding this new term to the Lagrangian density. But to understand the issue better, let us look at the equations of motion following from the action in (9.125)

$$\partial_\mu \frac{\partial \mathcal{L}}{\partial \partial_\mu A_\nu} - \frac{\partial \mathcal{L}}{\partial A_\nu} = 0,$$

or, $\quad -\partial_\mu F^{\mu\nu} - \partial^\nu(\partial \cdot A) - J^\nu = 0,$

or, $\quad \partial_\mu F^{\mu\nu} + \partial^\nu(\partial \cdot A) = -J^\nu.$ \hfill (9.127)

Without the second term on the left-hand side in (9.127), this is just the Maxwell's equations in the presence of conserved sources. If we now write out the left-hand side explicitly, (9.127) takes the form

$$\partial_\mu \left(\partial^\mu A^\nu - \partial^\nu A^\mu \right) + \partial^\nu(\partial \cdot A) = -J^\nu,$$

or, $\quad \Box A^\nu = -J^\nu,$ \hfill (9.128)

or, $\quad \Box \partial \cdot A = -\partial \cdot J = 0.$ \hfill (9.129)

An alternate way to see this is to note that if we take the divergence of the equations of motion in (9.127), we have

$$\partial_\nu \partial_\mu F^{\mu\nu} + \Box(\partial \cdot A) = \partial_\nu J^\nu,$$

or, $\quad \Box(\partial \cdot A) = 0,$ \hfill (9.130)

where we have used the anti-symmetry of the field strength tensor as well as the conservation of J^μ.

Thus we see that although the presence of the term $-\frac{1}{2}\chi^2$ in the Lagrangian density (9.125) where

$$\chi = \partial \cdot A, \hfill (9.131)$$

seems to modify the theory, χ is a free field and, therefore, the presence of this additional term in the Lagrangian density would not change the physics of Maxwell's theory. Furthermore, we recognize that if we restrict classically to the initial value conditions

$$\chi = 0, \quad \frac{\partial \chi}{\partial t} = 0, \quad t = 0, \hfill (9.132)$$

9.8 COVARIANT QUANTIZATION OF THE MAXWELL THEORY

then $\chi = 0$ at all times and we recover the familiar Maxwell's theory. Thus, classically we can think of Maxwell's theory as described by the modified Lagrangian density (9.125) with the supplementary condition

$$\partial \cdot A = 0. \tag{9.133}$$

Let us now rewrite this modified Lagrangian density as

$$\begin{aligned} \mathcal{L} &= -\frac{1}{2}\partial_\mu A_\nu \left(\partial^\mu A^\nu - \partial^\nu A^\mu\right) - \frac{1}{2}\left(\partial_\mu A^\mu\right)^2 - J^\mu A_\mu \\ &= -\frac{1}{2}\partial_\mu A_\nu \partial^\mu A^\nu - J^\mu A_\mu + \text{total divergence.} \end{aligned} \tag{9.134}$$

It is now obvious that the coefficient matrix of highest derivatives is nonsingular in this case. We can define the canonical momenta as

$$\Pi^\mu = \frac{\partial \mathcal{L}}{\partial \dot{A}_\mu} = -\dot{A}^\mu. \tag{9.135}$$

Clearly now all components of the momenta are well defined without any constraint and hence we can quantize the theory as (Π^ν is conjugate to A_ν)

$$\begin{aligned} \left[A_\mu(x), \Pi^\nu(y)\right]_{x^0=y^0} &= i\delta_\mu^\nu \delta^3(x-y), \\ \text{or,} \quad \left[A_\mu(x), \dot{A}_\nu(y)\right]_{x^0=y^0} &= -i\eta_{\mu\nu}\delta^3(x-y), \end{aligned} \tag{9.136}$$

with all other commutators vanishing.

We now see that this quantization relation is exactly like the quantization condition for four distinct scalar fields (see (5.40) and note that we are being slightly sloppy here in the sense that the commutation relation should really involve field variables and their conjugate momenta)

$$\left[\phi(x), \dot{\phi}(y)\right]_{x^0=y^0} = i\delta^3(x-y), \tag{9.137}$$

except for one thing. Namely, the commutation relation between A_0 and $\dot{A}_0$ has a relative negative sign. This problem can, of course, be simply fixed by postulating that for the time component, the coordinate and the momenta exchange roles. However, that would be against the spirit of Lorentz covariance since we no longer treat all the components of A_μ on the same footing. But more serious than that is the fact that if we adopt the above convention, the Hamiltonian for Maxwell's theory would become unbounded from below.

Let us construct the Hamiltonian density for this theory and see whether the above quantization relations are consistent (assume, for simplicity, $J^\mu = 0$ and we will neglect the total divergence terms)

$$\begin{aligned}
\mathcal{H} &= \Pi^\mu \dot{A}_\mu - \mathcal{L} \\
&= -\Pi^\mu \Pi_\mu + \frac{1}{2}\partial_\mu A_\nu \partial^\mu A^\nu \\
&= -\Pi^\mu \Pi_\mu + \frac{1}{2}\Pi_\mu \Pi^\mu + \frac{1}{2}\partial_i A_\mu \partial^i A^\mu \\
&= -\frac{1}{2}\Pi_\mu \Pi^\mu + \frac{1}{2}\partial_i A_\mu \partial^i A^\mu.
\end{aligned} \tag{9.138}$$

It is worth pointing out at this point that the Hamiltonian density (and, therefore, the Hamiltonian) for this theory does not seem to be positive semi-definite since the time components ($\mu = 0$) contribute a negative amount. However, we note that if we calculate the commutator of the field variables with the Hamiltonian, we find

$$\begin{aligned}
[A_\mu(x), H] &= \left[A_\mu(x), \int d^3x' \left(-\frac{1}{2}\Pi_\nu(x')\Pi^\nu(x')\right.\right. \\
&\qquad \left.\left. + \frac{1}{2}\partial'_i A_\nu(x')\partial'^i A^\nu(x')\right)\right]_{x^0 = x'^0} \\
&= -\frac{1}{2}\int d^3x' \left[A_\mu(x), \Pi_\nu(x')\Pi^\nu(x')\right]_{x^0=x'^0} \\
&= -\int d^3x'\, i\eta_{\mu\nu}\delta^3(x-x')\Pi^\nu(x')\Big|_{x^0=x'^0} \\
&= -i\Pi_\mu(x) = i\dot{A}_\mu(x),
\end{aligned} \tag{9.139}$$

which is the correct Heisenberg equation of motion for the field variables A_μ and hence the quantization relations (9.136) are consistent.

We have, however, not made use of the supplementary condition (9.133) as yet. A little bit of thinking tells us that although in the classical theory we can impose the condition

$$\partial_\mu A^\mu = 0, \tag{9.140}$$

in the quantum theory, where A_μ's are operators, such a condition is hard to implement as an operator condition. It is easy to see that the commutation relation (9.136) would lead to

$$\left[A^\mu(x), \Pi^\nu(x')\right]_{x^0=x'^0} = i\eta^{\mu\nu}\delta^3(x-x'),$$
or, $\quad \partial_\mu \left[A^\mu(x), \Pi^\nu(x')\right]_{x^0=x'^0} = i\partial^\nu \delta^3(x-x'),$
or, $\quad \left[\partial_\mu A^\mu(x), \Pi^\nu(x')\right]_{x^0=x'^0} = i\partial^\nu \delta^3(x-x'). \tag{9.141}$

(This relation is essentially correct, but the derivation needs to be done carefully in a limiting manner $x^0 \to x'^0$.) We note that, whereas the left-hand side of the above expression would be zero if the supplementary condition (9.133) were to hold as an operator equation, the non vanishing right-hand side would lead to an inconsistency.

Thus, we can weaken the supplementary condition (9.133) and say that this condition is true only on the space of physical states of the theory, namely,

$$\partial_\mu A^\mu |\text{phys}\rangle = 0. \tag{9.142}$$

In other words, we can think of the supplementary condition as selecting out the subspace of the physical Hilbert space of the theory. (Note that since all field components are dynamical in this modified theory, the naive "Hilbert" space is much larger than the physical Hilbert space of the Maxwell theory consisting of only the transverse photon degrees of freedom. The supplementary condition (9.133) may be thought of as picking out the smaller physical subspace from the larger total vector space.) As we would see shortly, even (9.142) is a very stringent condition and we have to relax this further. We can see this already from the commutator relation (9.141). Basically, we note that if $|\psi\rangle$ represents a physical state, then

$$\langle\psi|\left[\partial \cdot A(x), \Pi^\nu(x')\right]|\psi\rangle\big|_{x^0=x'^0} = \langle\psi|i\partial^\nu\delta^3(x-x')|\psi\rangle,$$

or, $\quad 0 = i\partial^\nu\delta^3(x-x'),$ (9.143)

which is inconsistent.

To proceed further, let us make a plane wave expansion for the field variables

$$A_\mu(x) = \sum_\lambda \int d^3k \big[\epsilon_\mu(\mathbf{k},\lambda)f_k(x)a(\mathbf{k},\lambda)$$

$$+ \epsilon_\mu^*(\mathbf{k},\lambda)f_k^*(x)a^\dagger(\mathbf{k},\lambda)\big], \quad (9.144)$$

where

$$f_k(x) = \frac{1}{\sqrt{(2\pi)^3 2\omega}} e^{-ik\cdot x}, \qquad k_0 = \omega(\mathbf{k}) = |\mathbf{k}|. \quad (9.145)$$

Here $\epsilon_\mu(\mathbf{k},\lambda)$'s are the components of the polarization vector for a photon travelling along $\mathbf{k}$ and polarization index $\lambda = 0,1,2,3$. (Unlike the earlier discussion, here they are not required to be transverse.) They are normalized as

$$\epsilon_\mu(\mathbf{k},\lambda)\epsilon^{*\mu}(\mathbf{k},\lambda') = \eta^{\lambda\lambda'} = \begin{cases} 1, & \lambda = \lambda' = 0, \\ -1, & \lambda = \lambda' = i = 1,2,3. \end{cases}$$
(9.146)

As we have seen earlier (see (9.74)), we can always choose a basis such that the polarization vector for $\lambda = 0$ is time-like (recall η^μ), while that for $\lambda = 3$ is longitudinal (recall $\hat{k}^\mu$) and the other two polarization vectors are transverse to the momentum four vector.

At this point we can quantize the coefficients $a(\mathbf{k},\lambda)$ and $a^\dagger(\mathbf{k},\lambda)$ as annihilation and creation operators satisfying

$$\left[a(\mathbf{k},\lambda),a(\mathbf{k}',\lambda')\right] = 0 = \left[a^\dagger(\mathbf{k},\lambda),a^\dagger(\mathbf{k}',\lambda')\right],$$

$$\left[a(\mathbf{k},\lambda),a^\dagger(\mathbf{k}',\lambda')\right] = -\eta^{\lambda\lambda'}\delta^3(k-k'), \quad (9.147)$$

9.8 COVARIANT QUANTIZATION OF THE MAXWELL THEORY

where $\lambda, \lambda' = 0, 1, 2, 3$. Of course, we have to check that these conditions are consistent with the equal-time quantization conditions for the field variables (for simplicity, we do not write the equal-time arguments explicitly although they are assumed)

$$\left[A_\mu(x), \dot{A}_\nu(y)\right]$$

$$= \sum_{\lambda,\lambda'} \int d^3k d^3k' \left[\epsilon_\mu(\mathbf{k},\lambda) f_k(x) a(\mathbf{k},\lambda) + \epsilon_\mu^*(\mathbf{k},\lambda) f_k^*(x) a^\dagger(\mathbf{k},\lambda),\right.$$

$$\left. -i\omega' \left(\epsilon_\nu(\mathbf{k}',\lambda') f_{k'}(y) a(\mathbf{k}',\lambda') - \epsilon_\nu^*(\mathbf{k}',\lambda') f_{k'}^*(y) a^\dagger(\mathbf{k}',\lambda')\right)\right]$$

$$= \sum_{\lambda,\lambda'} \int d^3k d^3k' i\omega'$$

$$\times \left(\epsilon_\mu(\mathbf{k},\lambda) \epsilon_\nu^*(\mathbf{k}',\lambda') f_k(x) f_{k'}^*(y) \left[a(\mathbf{k},\lambda), a^\dagger(\mathbf{k}',\lambda')\right]\right.$$

$$\left. -\epsilon_\mu^*(\mathbf{k},\lambda) \epsilon_\nu(\mathbf{k}',\lambda') f_k^*(x) f_{k'}(y) \left[a^\dagger(\mathbf{k},\lambda), a(\mathbf{k}',\lambda')\right]\right)$$

$$= \sum_{\lambda,\lambda'} \int d^3k d^3k' (-i\omega' \eta^{\lambda\lambda'}) \delta^3(k-k')$$

$$\times \left(\epsilon_\mu(\mathbf{k},\lambda) \epsilon_\nu^*(\mathbf{k}',\lambda') f_k(x) f_{k'}^*(y) + \epsilon_\mu^*(\mathbf{k},\lambda) \epsilon_\nu(\mathbf{k}',\lambda') f_k^*(x) f_{k'}(y)\right)$$

$$= \sum_\lambda \int d^3k (-i\omega) \left(\frac{\epsilon_\mu(\mathbf{k},\lambda) \epsilon_\nu^*(\mathbf{k},\lambda)}{\epsilon_\sigma(\mathbf{k},\lambda) \epsilon^{*\sigma}(\mathbf{k},\lambda)} f_k(x) f_k^*(y)\right.$$

$$\left. +\frac{\epsilon_\mu^*(\mathbf{k},\lambda) \epsilon_\nu(\mathbf{k},\lambda)}{\epsilon_\sigma(\mathbf{k},\lambda) \epsilon^{*\sigma}(\mathbf{k},\lambda)} f_k^*(x) f_k(y)\right). \tag{9.148}$$

Using the completeness relation for the polarization vectors

$$\sum_\lambda \frac{\epsilon_\mu(\mathbf{k},\lambda) \epsilon_\nu^*(\mathbf{k},\lambda)}{\epsilon_\sigma(\mathbf{k},\lambda) \epsilon^{*\sigma}(\mathbf{k},\lambda)} = \eta_{\mu\nu}, \tag{9.149}$$

in (9.148), we obtain the equal-time commutator to be

$$\left[A_\mu(x), \dot{A}_\nu(y)\right] = \int d^3k (-i\omega) \eta_{\mu\nu} \left[f_k(x) f_k^*(y) + f_k^*(x) f_k(y)\right]$$

$$= -i\eta_{\mu\nu} \delta^3(x-y). \tag{9.150}$$

Thus the commutation relations for the creation and the annihilation operators are indeed consistent with the quantization condition for the field variables. However, it is strange to note that there are four independent photon degrees of freedom. We also note from (9.147) that although $a(\mathbf{k}, \lambda)$ and $a^\dagger(\mathbf{k}, \lambda)$ behave respectively like annihilation and creation operators for $\lambda = 1, 2, 3$, for the time-like component there is a relative negative sign in the commutation relation (9.147).

If we calculate the normal ordered Hamiltonian of the theory (recall $J^\mu = 0$), it has the form

$$:H: \;=\; \int d^3k \, \omega(\mathbf{k}) \left[-a^\dagger(\mathbf{k}, 0)a(\mathbf{k}, 0) + a^\dagger(\mathbf{k}, 1)a(\mathbf{k}, 1) \right.$$
$$\left. + a^\dagger(\mathbf{k}, 2)a(\mathbf{k}, 2) + a^\dagger(\mathbf{k}, 3)a(\mathbf{k}, 3) \right]. \qquad (9.151)$$

We can now develop the Hilbert space description for the photons in this theory which leads to some unusual features in the present case. For example, let us denote by $|0\rangle$ the vacuum state of the theory so that

$$\langle 0|0\rangle = 1, \quad a(\mathbf{k}, \lambda)|0\rangle = 0, \quad \lambda = 0, 1, 2, 3. \qquad (9.152)$$

Let us next consider the state with one time-like photon

$$|1, \lambda = 0\rangle = \int d^3k \, F(\mathbf{k}) a^\dagger(\mathbf{k}, \lambda = 0)|0\rangle, \qquad (9.153)$$

where $F(\mathbf{k})$ is a suitable smearing function such that

$$\int d^3k \, |F(\mathbf{k})|^2 < \infty. \qquad (9.154)$$

The norm of this one photon state can now be calculated

$$\langle 1, \lambda = 0 | 1, \lambda = 0 \rangle$$
$$= \int d^3k d^3k' F^*(\mathbf{k}) F(\mathbf{k}') \langle 0 | a(\mathbf{k}, 0) a^\dagger(\mathbf{k}', 0) | 0 \rangle$$
$$= \int d^3k d^3k' \ F^*(\mathbf{k}) F(\mathbf{k}') \langle 0 | a^\dagger(\mathbf{k}', 0) a(\mathbf{k}, 0) - \delta^3(k - k') | 0 \rangle$$
$$= -\int d^3k \ |F(\mathbf{k})|^2 < 0. \tag{9.155}$$

This shows that the vector space of the theory has an indefinite metric – the norm for the one time-like photon state is negative although states containing only space-like photons have positive norm. We also realize that this negative norm is a consequence of the negative sign in the commutator (9.147) for the annihilation and the creation operator for a time-like photon. Thus the natural modification that comes to mind is to interchange the roles of these operators for the time-like photon. However, this also leads to trouble as we see below.

Let us suppose that

$$a^\dagger(\mathbf{k}, 0) | 0 \rangle = 0, \tag{9.156}$$

so that the one time-like photon state is given by

$$|\mathbf{k}, 0\rangle = a(\mathbf{k}, 0) | 0 \rangle. \tag{9.157}$$

In this case, the normal ordered Hamiltonian will have the form (compare with (9.151))

$$: H : \ = \int d^3k \ \omega(\mathbf{k}) \Big[-a(\mathbf{k}, 0) a^\dagger(\mathbf{k}, 0) + a^\dagger(\mathbf{k}, 1) a(\mathbf{k}, 1)$$
$$+ a^\dagger(\mathbf{k}, 2) a(\mathbf{k}, 2) + a^\dagger(\mathbf{k}, 3) a(\mathbf{k}, 3) \Big], \tag{9.158}$$

and the energy of this one photon state is given by

$$\begin{aligned}
H|\mathbf{k},0\rangle &= Ha(\mathbf{k},0)|0\rangle = [H,a(\mathbf{k},0)]|0\rangle \\
&= \int d^3k'\, \omega(\mathbf{k}')[-a(\mathbf{k}',0)a^\dagger(\mathbf{k}',0), a(\mathbf{k},0)]|0\rangle \\
&= -\omega(\mathbf{k})a(\mathbf{k},0)|0\rangle = -\omega(\mathbf{k})|\mathbf{k},0\rangle, \quad\quad (9.159)
\end{aligned}$$

where we have used the fact that $H|0\rangle = 0$. Thus we see that if we exchange the roles of the creation and the annihilation operators for the time-like photon, the energy of the state with one time-like photon becomes negative. In fact it is obvious that with this definition, definite states of arbitrarily high negative energy values are possible and the Hamiltonian becomes unbounded from below.

So we are stuck with an indefinite metric space (with the standard interpretation for annihilation and creation operators for the time-like photon) and the normalization of states in this space is given by (we suppress the momentum label for simplicity)

$$\langle n_0, n_1, n_2, n_3 | m_0, m_1, m_2, m_3 \rangle$$
$$= (-1)^{n_0} \delta_{n_0 m_0} \delta_{n_1 m_1} \delta_{n_2 m_2} \delta_{n_3 m_3}, \quad\quad (9.160)$$

where n_λ, m_λ denote number of photons for each of the polarizations $\lambda = 0, 1, 2, 3$. Thus states containing an odd number of time-like photons have negative norm. The question that immediately comes to our mind is what happens to the probabilistic interpretation of the theory. In a free theory we may impose suitable conditions to prohibit any unwanted state. But in the presence of interactions such states may be excited. As we will see shortly in spite of the presence of negative norm states, physical results are well behaved.

This is seen by using the supplementary condition (subsidiary condition) which we still have not imposed. Namely, the physical states must satisfy

$$\partial_\mu A^\mu(x)|\text{phys}\rangle = 0. \quad\quad (9.161)$$

However, as discussed earlier this is too stringent a condition to allow any state of the radiation field. This not only demands that certain

9.8 COVARIANT QUANTIZATION OF THE MAXWELL THEORY

kinds of photons are not present in the physical state, but it also requires that those photons cannot be emitted. Gupta and Bleuler weakened the supplementary condition to have the form

$$\partial_\mu A^{\mu\,(+)}(x)|\text{phys}\rangle = 0, \tag{9.162}$$

where $\partial_\mu A^{\mu\,(+)}(x)$ is the positive frequency part of the divergence of the vector potential (Maxwell's field) and contains only the destruction operator or the annihilation operator. (This is commonly known as the Gupta-Bleuler quantization.) We remark here that since (see (9.129) or (9.130))

$$\Box \partial_\mu A^\mu(x) = 0, \tag{9.163}$$

$\partial_\mu A^\mu(x)$ is like a free scalar field. Therefore, it can be decomposed into positive and negative frequency parts uniquely in a relativistically invariant manner and this decomposition is preserved under time evolution. Recalling that

$$k_\mu \epsilon^\mu(\mathbf{k}, \lambda) = 0 \quad \text{for} \quad \lambda = 1, 2, \tag{9.164}$$

the Gupta-Bleuler supplementary condition leads to

$$\sum_\lambda k_\mu \epsilon^\mu(\mathbf{k}, \lambda) a(\mathbf{k}, \lambda)|\text{phys}\rangle = 0,$$

or, $\quad k_\mu \big(\epsilon^\mu(\mathbf{k}, 0) a(\mathbf{k}, 0) + \epsilon^\mu(\mathbf{k}, 3) a(\mathbf{k}, 3)\big)|\text{phys}\rangle = 0,$

or, $\quad \left(k_0 a(\mathbf{k}, \lambda = 0) - \dfrac{\mathbf{k}^2}{|\mathbf{k}|} a(\mathbf{k}, \lambda = 3)\right)|\text{phys}\rangle = 0,$

or, $\quad (a(\mathbf{k}, 0) - a(\mathbf{k}, 3))|\text{phys}\rangle = 0, \tag{9.165}$

where we have used the definitions and the properties of the polarization vectors in (9.74) as well as the fact that for positive energy photons $k_0 = \omega = |\mathbf{k}|$.

Let $V_\text{phys} = \{P\}$ denote the set of all states which satisfy the supplementary condition and let $|\psi\rangle$ represent such a state. Then, as we have seen, the supplementary condition (9.165) implies that

$$a(\mathbf{k},0)|\psi\rangle = a(\mathbf{k},3)|\psi\rangle,$$

or, $\quad \langle\psi|a^\dagger(\mathbf{k},0)a(\mathbf{k},0)|\psi\rangle = \langle\psi|a^\dagger(\mathbf{k},3)a(\mathbf{k},3)|\psi\rangle.$ \hfill (9.166)

This leads to the fact that the physical states must contain superpositions of states with an equal number of time-like and longitudinal photons. A general physical state is, of course, a superposition of different states of the form

$$|\psi\rangle = \sum_{n_0,n_1,n_2,n_3} C_{n_0,n_1,n_2,n_3}|n_0,n_1,n_2,n_3\rangle, \qquad (9.167)$$

where n_λ denotes the number of photons with polarization λ and the supplementary condition relates the number of time-like and longitudinal photon states as

$$\sqrt{n_0}\,C_{n_0,n_1,n_2,n_3-1} + \sqrt{n_3}\,C_{n_0-1,n_1,n_2,n_3} = 0. \qquad (9.168)$$

This can be seen as follows

$$\sum C_{n_0,n_1,n_2,n_3}\left(a(\mathbf{k},0) - a(\mathbf{k},3)\right)|n_0,n_1,n_2,n_3\rangle = 0,$$

or, $\quad \sum C_{n_0,n_1,n_2,n_3}\left(-\sqrt{n_0}\,|n_0-1,n_1,n_2,n_3\rangle \right.$

$\left. -\sqrt{n_3}\,|n_0,n_1,n_2,n_3-1\rangle\right) = 0,$

or, $\quad \sum \left(\sqrt{n_0}\,C_{n_0,n_1,n_2,n_3-1} + \sqrt{n_3}\,C_{n_0-1,n_1,n_2,n_3}\right)$

$\times |n_0-1,n_1,n_2,n_3-1\rangle = 0,$ \hfill (9.169)

which leads to the relation in (9.168). All the coefficients C_{n_0,n_1,n_2,n_3} in the expansion can be determined recursively from (9.168) with the condition

$$C_{0,n_1,n_2,0} = 1. \qquad (9.170)$$

9.8 COVARIANT QUANTIZATION OF THE MAXWELL THEORY

It follows now that for a fixed number n_1, n_2 of transverse photons, the physical states allowed by the supplementary condition have the forms

$$\begin{aligned}|\phi_0\rangle &= |0, n_1, n_2, 0\rangle, \\ |\phi_1\rangle &= |0, n_1, n_2, 1\rangle - |1, n_1, n_2, 0\rangle, \\ |\phi_3\rangle &= |0, n_1, n_2, 2\rangle - \sqrt{2}|1, n_1, n_2, 1\rangle + |2, n_1, n_2, 0\rangle, \end{aligned} \quad (9.171)$$

and so on where we have used the values of C_{n_0,n_1,n_2,n_3}'s determined recursively. Let us note that because of the negative norm of states containing an odd number of time-like photons, all such states except for $|\phi_0\rangle$ have zero norm. For example, we have

$$\begin{aligned} \langle\phi_1|\phi_1\rangle &= \langle 0, n_1, n_2, 1|0, n_1, n_2, 1\rangle + \langle 1, n_1, n_2, 0|1, n_1, n_2, 0\rangle \\ &= 1 - 1 = 0, \\ \langle\phi_2|\phi_2\rangle &= \langle 0, n_1, n_2, 2|0, n_1, n_2, 2\rangle + 2\langle 1, n_1, n_2, 1|1, n_1, n_2, 1\rangle \\ &\quad + \langle 2, n_1, n_2, 0|2, n_1, n_2, 0\rangle \\ &= 1 - 2 + 1 = 0, \end{aligned} \quad (9.172)$$

and so on. Thus although we can write a general physical state with a fixed number n_1, n_2 of transverse photons as a linear superposition of states of the form

$$|\phi\rangle = |\phi_0\rangle + b_1|\phi_1\rangle + b_2|\phi_2\rangle + \ldots, \quad (9.173)$$

because

$$\langle\phi|\phi\rangle = \langle\phi_0|\phi_0\rangle, \quad (9.174)$$

we can assume that in a truly physical state of the radiation field there are no time-like and longitudinal photons present. In other words,

$$n_0 = n_3 = 0, \quad (9.175)$$

in such states. If interactions are present time-like and longitudinal states may occur as intermediate states. However their effect cancels out in the final result. One way of heuristically saying this is that the probability for the emission of a longitudinal photon cancels out the negative probability for the emission of a time-like photon. A different way of saying this is to note that being orthogonal to every other state, such states decouple. We also point out here that the Gupta-Bleuler supplementary condition can be thought of as the quantum analog of the covariant Lorentz condition

$$\partial_\mu A^\mu(x) = 0, \tag{9.176}$$

on the space of physical states since

$$\langle \psi | \partial_\mu A^\mu | \psi \rangle = 0, \tag{9.177}$$

where $|\psi\rangle$ represents a physical state.

The physical subspace of the theory selected by the supplementary condition (9.165) contains states with positive semi-definite norm (negative norm states are eliminated by the supplementary condition or the physical state condition and, consequently, there is no problem with a probabilistic interpretation). Since the zero norm states are orthogonal to all the states including themselves, if we further mod out the states by the zero norm states, we have the true physical subspace of the theory where the norm of states is positive definite, namely,

$$\overline{V}_{\text{phys}} = \frac{V_{\text{phys}}}{V_0}, \tag{9.178}$$

where V_0 represents the set of states with zero norm. Without going into details we note here that the Feynman propagator for the photon in this theory has the simpler form

$$iG_{F,\mu\nu}(k) = \lim_{\epsilon \to 0^+} -\frac{i\eta_{\mu\nu}}{k^2 + i\epsilon}. \tag{9.179}$$

This is different from (9.77) or (9.89) simply because this can be thought of as quantizing the theory in a different gauge (Feynman-Fermi gauge). The physical S-matrix elements, however, are gauge independent and do not depend on the form of the photon propagator in a particular gauge. We will develop these ideas further when we study the covariant quantization of non-Abelian gauge theories in chapter **13**.

9.9 References

1. S. N. Gupta, Proceedings of the Physical Society (London) **A63**, 681 (1950).

2. K. Bleuler, Helvetica Physica Acta **23**, 567 (1950).

3. J. C. Ward, Physical Review **78**, 182 (1950).

4. Y. Takahashi, Nuovo Cimento **6**, 371 (1957).

5. S. Schweber, *Introduction to Relativistic Quantum Field Theory*, Row, Peterson, Evanston (1961).

6. J. D. Bjorken and S. Drell, *Relativistic Quantum Fields*, McGraw-Hill, New York, 1964.

7. P. Roman, *Introduction to Quantum Field Theory*, John Wliley, New York (1969).

8. C. Itzykson and J-B. Zuber, *Quantum Field Theory*, McGraw-Hill, New York, 1980.

9. N. N. Bogoliubov and D. V. Shirkov, *Introduction to the theory of Quantized Fields*, Nauka, Moscow (1984).

10. F. Gross, *Relativistic Quantum Mechanics and Field Theory*, John Wiley, New York (1993).

CHAPTER 10
Dirac method for constrained systems

10.1 Constrained systems

As we have seen in the study of the Dirac field theory in chapter **8** as well as the Maxwell field theory in chapter **9**, the dynamical phase space variables of these theories are not all independent, rather some of these variables have to satisfy constraints following from the structure of the theory. Such systems are known as constrained systems and the naive passage to the Hamiltonian description for such a system starting from the Lagrangian description fails. In this case, there is a systematic procedure due to Dirac which allows us to go from the Lagrangian description of a theory to the Hamiltonian description (and, thereby, carry out the quantization of a theory) and in this chapter we will discuss the Dirac method in some detail.

To appreciate the difficulties associated with constrained systems, let us consider a classical system of point particles described by the Lagrangian $L(q_i, \dot{q}_i), i = 1, 2, \cdots, N$. The Lagrangian is a function of N coordinates q_i as well as N velocities $\dot{q}_i$ which are assumed to be independent so that the configuration space of the theory is $2N$ dimensional. Given the Lagrangian of the theory, we define the momenta canonically conjugate to the coordinates as

$$p^i = \frac{\partial L}{\partial \dot{q}_i}. \tag{10.1}$$

Equation (10.1), in general, relates momenta to velocities (and coordinates) of the theory and if this relation is invertible, velocities can be expressed in terms of momenta (and coordinates). In this case, we can go from the configuration space of the system to the phase space

$$(q_i, \dot{q}_i) \to (q_i, p^i), \tag{10.2}$$

and uniquely define the Hamiltonian of the system through the Legendre transformation

$$H(q_i, p^i) = p^i \dot{q}_i - L(q_i, \dot{q}_i), \tag{10.3}$$

where summation over repeated indices is understood. The phase space spanned by the independent coordinates and momenta is also $2N$ dimensional with a (equal time) canonical Poisson bracket structure

$$\{q_i, q_j\} = 0 = \{p^i, p^j\},$$
$$\{q_i, p^j\} = \delta_i^j = -\{p^j, q_i\}, \tag{10.4}$$

which allows us to write the Poisson bracket between any two dynamical variables $A_1(q_i, p^i), A_2(q_i, p^i)$ as

$$\{A_1, A_2\} = \frac{\partial A_1}{\partial q_i} \frac{\partial A_2}{\partial p^i} - \frac{\partial A_1}{\partial p^i} \frac{\partial A_2}{\partial q_i}. \tag{10.5}$$

Using (10.4) or (10.5), it follows now that the $2N$ first order dynamical equations (evolution equations) for the system can be written in the Hamiltonian form

$$\begin{aligned}\dot{q}_i &= \{q_i, H\} = \frac{\partial H}{\partial p^i}, \\ \dot{p}^i &= \{p^i, H\} = -\frac{\partial H}{\partial q_i},\end{aligned} \tag{10.6}$$

which can be shown, using the definition (10.3), to be equivalent to the N second order Euler-Lagrange equations following from the Lagrangian description. In fact, the time evolution of any dynamical variable $A(q_i, p^i)$ in the phase space can be written in the Hamiltonian form

$$\dot{A}(q_i, p^i) = \{A(q_i, p^i), H\}. \tag{10.7}$$

This discussion can also be generalized to a classical system described by Grassmann (anti-commuting) variables. For example, if we have a classical system described by the Lagrangian $L(\theta_\alpha, \dot{\theta}_\alpha)$, $\alpha = 1, 2, \cdots M$ with

$$\theta_\alpha \theta_\beta = -\theta_\beta \theta_\alpha, \tag{10.8}$$

then, we can define the momenta conjugate to the coordinates as

$$\Pi^\alpha = \frac{\partial L}{\partial \dot{\theta}_\alpha}. \tag{10.9}$$

However, since the dynamical variables are now of Grassmann odd (fermionic) nature, we have to define the derivative in (10.9) and as discussed in chapter **8** (see (8.14)), we will choose the convention of taking derivatives from the left. If the relations (10.9) can be inverted to express velocities in terms of the momenta, we can go from the configuration space to the phase space

$$(\theta_\alpha, \dot{\theta}_\alpha) \to (\theta_\alpha, \Pi^\alpha), \tag{10.10}$$

through the Legendre transformation (this is consistent with the convention of left derivatives)

$$H(\theta_\alpha, \Pi^\alpha) = \dot{\theta}_\alpha \Pi^\alpha - L(\theta_\alpha, \dot{\theta}_\alpha) = -\Pi^\alpha \dot{\theta}_\alpha - L(\theta_\alpha, \dot{\theta}_\alpha). \tag{10.11}$$

The $2M$ dimensional phase space is endowed with a canonical Poisson bracket structure (with the convention of left derivatives)

$$\{\theta_\alpha, \theta_\beta\} = 0 = \{\Pi^\alpha, \Pi^\beta\},$$
$$\{\theta_\alpha, \Pi^\beta\} = -\delta_\alpha^\beta = \{\Pi^\beta, \theta_\alpha\}. \tag{10.12}$$

The Poisson bracket between any two dynamical variables can now be obtained from (10.12). However, unlike the bosonic case in (10.5), the structure of the Poisson bracket now depends on whether the dynamical variables are Grassmann even (bosonic) or Grassmann odd (fermionic). Thus, for example, with the convention of left derivatives, the Poisson bracket between two bosonic dynamical variables $B_1(\theta_\alpha, \Pi^\alpha), B_2(\theta_\alpha, \Pi^\alpha)$ takes the form

$$\begin{aligned}\{B_1, B_2\} &= -\frac{\partial B_1}{\partial \theta_\alpha}\{\theta_\alpha, \Pi^\beta\}\frac{\partial B_2}{\partial \Pi^\beta} - \frac{\partial B_1}{\partial \Pi^\alpha}\{\Pi^\alpha, \theta_\beta\}\frac{\partial B_2}{\partial \theta_\beta} \\ &= \left(\frac{\partial B_1}{\partial \theta_\alpha}\frac{\partial B_2}{\partial \Pi^\alpha} + \frac{\partial B_1}{\partial \Pi^\alpha}\frac{\partial B_2}{\partial \theta_\alpha}\right),\end{aligned} \qquad (10.13)$$

where we have used (10.12) as well as the fact that $\frac{\partial B}{\partial \theta_\alpha}$ and $\frac{\partial B}{\partial \Pi^\alpha}$ are Grassmann odd (fermionic) variables. In a similar manner, the Poisson brackets between a Grassmann even and a Grassmann odd variable as well as the bracket between two Grassmann odd variables can be obtained to correspond to

$$\begin{aligned}\{F, B\} &= -\left(\frac{\partial F}{\partial \theta_\alpha}\frac{\partial B}{\partial \Pi^\alpha} + \frac{\partial F}{\partial \Pi^\alpha}\frac{\partial B}{\partial \theta_\alpha}\right), \\ \{B, F\} &= \left(\frac{\partial B}{\partial \theta_\alpha}\frac{\partial F}{\partial \Pi^\alpha} + \frac{\partial B}{\partial \Pi^\alpha}\frac{\partial F}{\partial \theta_\alpha}\right), \\ \{F_1, F_2\} &= -\left(\frac{\partial F_1}{\partial \theta_\alpha}\frac{\partial F_2}{\partial \Pi^\alpha} + \frac{\partial F_1}{\partial \Pi^\alpha}\frac{\partial F_2}{\partial \theta_\alpha}\right).\end{aligned} \qquad (10.14)$$

Using (10.14) the $2M$ first order dynamical equations for the system can now be written in the Hamiltonian form as

$$\begin{aligned}\dot{\theta}_\alpha &= \{\theta_\alpha, H\} = -\frac{\partial H}{\partial \Pi^\alpha}, \\ \dot{\Pi}^\alpha &= \{\Pi^\alpha, H\} = -\frac{\partial H}{\partial \theta_\alpha},\end{aligned} \qquad (10.15)$$

which can be compared with (10.6).

The discussion so far assumes that the transformation to phase space in (10.2) or (10.10) is invertible so that all the velocities can be expressed uniquely in terms of independent momenta leading to the

10.1 Constrained systems

unique Hamiltonian of the system. However, the difficulty in passage to a Hamiltonian description arises when this transformation is non-invertible. To understand the difficulty, let us concentrate on the bosonic theory for simplicity. In this case, the transformation to phase space in (10.2) can be written in general in the matrix form as

$$\begin{pmatrix} q \\ p \end{pmatrix} = A \begin{pmatrix} q \\ \dot{q} \end{pmatrix} = \begin{pmatrix} \mathbb{1} & 0 \\ \tilde{b} & b \end{pmatrix} \begin{pmatrix} q \\ \dot{q} \end{pmatrix}, \tag{10.16}$$

where the elements $b, \tilde{b}$ denote $N \times N$ matrices. For conventional theories described by Lagrangians with quadratic velocity terms, it follows that

$$b = b(q), \qquad \tilde{b} = \tilde{b}(q). \tag{10.17}$$

The inverse of the matrix A in (10.16) can be easily seen to have the form

$$A^{-1} = \begin{pmatrix} \mathbb{1} & 0 \\ -b^{-1}\tilde{b} & b^{-1} \end{pmatrix}, \tag{10.18}$$

so that A^{-1} exists only if the matrix b is invertible. From (10.16) we note that

$$p^i = \tilde{b}^{ij}(q)q_j + b^{ij}(q)\dot{q}_j, \tag{10.19}$$

so that

$$b^{ij}(q) = \frac{\partial p^i}{\partial \dot{q}_j} = \frac{\partial^2 L}{\partial \dot{q}_j \partial \dot{q}_i}. \tag{10.20}$$

Therefore, it follows that the transformation is invertible (A^{-1} exists) if

$$\det \frac{\partial^2 L}{\partial \dot{q}_i \partial \dot{q}_j} \neq 0. \tag{10.21}$$

The naive passage to the Hamiltonian description can be carried out only in this case.

On the other hand, if the Lagrangian describing the theory satisfies

$$\det \frac{\partial^2 L}{\partial \dot{q}_i \partial \dot{q}_j} = 0, \tag{10.22}$$

then, the transformation (10.2) is not invertible. It follows from (10.20) that in this case not all the conjugate momenta can be thought of as independent variables leading to the fact that not all of N independent velocities can be expressed in terms of independent momenta. In other words, there exist relations or constraints between various dynamical variables and such systems are known as constrained systems. In this case, it is not clear *a priori* how to define the Hamiltonian uniquely. Furthermore, it is also clear that the naive canonical Poisson brackets in (10.4) need not satisfy the constraints of the theory and, therefore, may not represent the true brackets necessary for a Hamiltonian description of the system.

10.2 Dirac method and Dirac bracket

In a physical system with constraints, the passage to a Hamiltonian description is achieved systematically through the method of Dirac which we describe in this section. Let us assume that the $N \times N$ matrix

$$\frac{\partial^2 L}{\partial \dot{q}_i \partial \dot{q}_j}, \tag{10.23}$$

of the physical system is of rank $R < N$. In this case, therefore, we can determine only R of the N velocities in terms of coordinates and independent momenta as

$$\dot{q}_a = f_a(q_i, p^b), \qquad a, b = 1, 2, \cdots, R, \tag{10.24}$$

and the other $N - R$ velocities cannot be determined. As a result, we can write

10.2 Dirac method and Dirac bracket

$$p^a = g^a(q_i, \dot{q}_b), \quad a, b = 1, 2, \cdots, R,$$
$$p^\alpha = g^\alpha(q_i, p^a), \quad \alpha = R+1, R+2, \cdots, N. \quad (10.25)$$

It is clear that the second set of equations in (10.25) define $N - R$ constraints among dynamical variables which we denote by

$$\varphi^\alpha = p^\alpha - g^\alpha(q_i, p^a) = 0, \quad (10.26)$$

and we recognize that such constraints will reduce the dimensionality of the true phase space of the system. For example, we note that the constraints in (10.26) (following from the Lagrangian of the theory) are known as primary constraints of the theory and define a $2N-(N-R) = N+R$ dimensional hypersurface Γ_c in the naive $2N$ dimensional phase space Γ of the system. (If there are further constraints in the theory, the dimensionality of Γ_c would be further reduced.) Two dynamical variables F, G in Γ are said to be weakly equal, $F \approx G$, if they are equal on the constrained hypersurface Γ_c, namely, if

$$(F - G)\big|_{\Gamma_c} = 0. \quad (10.27)$$

As usual, we can define the canonical Hamiltonian of the theory as the Legendre transform

$$H_{\text{can}} = p^i \dot{q}_i - L(q_i, \dot{q}_i) = p^a \dot{q}_a + g^\alpha \dot{q}_\alpha - L(q_i, \dot{q}_i), \quad (10.28)$$

where we have used (10.25). We note from (10.28) that in spite of the fact that $\dot{q}_\alpha$ cannot be determined,

$$\frac{\partial H_{\text{can}}}{\partial \dot{q}_\alpha} = g^\alpha - \frac{\partial L}{\partial \dot{q}_\alpha} = g^\alpha - p^\alpha \approx 0, \quad (10.29)$$

so that on the constrained hypersurface Γ_c, the canonical Hamiltonian does not depend on velocities as we would expect, namely, $H_{\text{can}} = H_{\text{can}}(q_i, p^a)$. On the other hand, we also note that because of the non-invertibility of (10.2) (presence of constraints), the Hamiltonian of the theory is no longer unique. In fact, from the canonical

Hamiltonian of the theory, we can define the primary Hamiltonian associated with the system by incorporating the primary constraints of the theory (10.26) as (it is known as the primary Hamiltonian since it incorporates only the primary constraints in (10.26))

$$H_\text{p} = H_\text{can} + \lambda_\alpha \varphi^\alpha, \tag{10.30}$$

where λ_α denote undetermined Lagrange multipliers and it follows that

$$H_\text{p} \approx H_\text{can}. \tag{10.31}$$

Furthermore, with the canonical Poisson brackets in (10.4), we can write the Hamiltonian equations as

$$\begin{aligned} \dot{q}_i &\approx \{q_i, H_\text{p}\} = \frac{\partial H_\text{p}}{\partial p^i} = \frac{\partial (H_\text{can} + \lambda_\alpha \varphi^\alpha)}{\partial p^i}, \\ \dot{p}^i &\approx \{p^i, H_\text{p}\} = -\frac{\partial H_\text{p}}{\partial q_i} = -\frac{\partial (H_\text{can} + \lambda_\alpha \varphi^\alpha)}{\partial q_i}, \end{aligned} \tag{10.32}$$

and using (10.29), it follows that we can identify the Lagrange multipliers in (10.30) with velocities

$$\lambda_\alpha \approx \dot{q}_\alpha, \tag{10.33}$$

which remain undetermined. The time evolution of any dynamical variable F can now be written as

$$\dot{F} \approx \{F, H_\text{p}\} = \{F, H_\text{can} + \lambda_\alpha \varphi^\alpha\}. \tag{10.34}$$

The constraints of the theory should be invariant under time evolution and using (10.34) this leads to

$$\begin{aligned} \dot{\varphi}^\alpha &\approx \{\varphi^\alpha, H_\text{can} + \lambda_\beta \varphi^\beta\} \\ &= \{\varphi^\alpha, H_\text{can}\} + \lambda_\beta \{\varphi^\alpha, \varphi^\beta\} + \{\varphi^\alpha, \lambda_\beta\} \varphi^\beta \\ &\approx \{\varphi^\alpha, H_\text{can}\} + \lambda_\beta \{\varphi^\alpha, \varphi^\beta\} \approx 0, \end{aligned} \tag{10.35}$$

10.2 DIRAC METHOD AND DIRAC BRACKET

where we have used (10.26) in the last step. It is clear that (10.35) may determine some of the Lagrange multipliers or may lead to new constraints known as secondary constraints. Requiring the secondary constraints to be invariant under time evolution as in (10.35) (with H_p as the Hamiltonian) may determine some other Lagrange multipliers or generate newer constraints (tertiary etc). We continue with this process until all the constraints are determined to be evolution free. Let us denote all the constraints of the theory (primary, secondary, tertiary, ...) collectively as

$$\varphi^{\overline{\alpha}} \approx 0, \qquad \overline{\alpha} = 1, 2, \cdots, n < 2N, \tag{10.36}$$

which can be divided into two distinct classes.

1. The constraints which have weakly vanishing (see (10.27)) Poisson bracket with all the constraints (including themselves) are known as first class constraints and are denoted as

$$\psi^{\widetilde{\alpha}} \approx 0, \qquad \widetilde{\alpha} = 1, 2, \cdots, n_1. \tag{10.37}$$

2. On the other hand, the constraints which have at least one nonvanishing Poisson bracket with the constraints are known as second class constraints and are denoted as

$$\phi^{\widehat{\alpha}} \approx 0, \qquad \widehat{\alpha} = 1, 2, \cdots, 2n_2. \tag{10.38}$$

Here we have used the observation due to Dirac that there are an even number of second class constraints in a theory and we note from (10.36)-(10.38) that $n_1 + 2n_2 = n < 2N$.

It is also worth noting that in this process of determining all the constraints that are time independent (evolution free), all of the Lagrange multipliers may be completely determined or some of them may remain undetermined. In general, if there are primary constraints in the theory which are first class, then there remain undetermined Lagrange multipliers in the primary Hamiltonian after all the constraints have been determined. (The number of undetermined Lagrange multipliers equals the number of first class constraints among the primary constraints.)

The first class constraints have a special significance in that they are associated with gauge invariances (local invariances) in the theory. As we have seen in the case of Maxwell field theory, a consistent Hamiltonian description, in this case, requires gauge fixing conditions and we must choose as many gauge fixing conditions as there are first class contraints. The gauge fixing conditions are generally denoted as

$$\chi^{\widetilde{\alpha}} \approx 0, \qquad \widetilde{\alpha} = 1, 2, \cdots, n_1, \tag{10.39}$$

and are chosen such that they convert the first class constraints (10.37) into second class constraints. Thus, after gauge fixing we denote all the constraints of the theory (including the gauge fixing conditions), which are all second class, collectively as

$$\phi^A \approx 0, \qquad A = 1, 2, \cdots, 2(n_1 + n_2) < 2N, \tag{10.40}$$

so that the true phase space of the theory is $2(N - n_1 - n_2)$ dimensional.

Although we have identified all the constraints of the theory, we cannot yet impose them directly in the theory since the canonical Poisson brackets in (10.4) are not compatible with them. Namely, even though

$$\phi^A \approx 0, \tag{10.41}$$

the Poisson bracket of the constraints with any dynamical variable F does not, in general, vanish

$$\{F, \phi^A\} \not\approx 0. \tag{10.42}$$

This issue of incompatibility of the Poisson brackets can be addressed through the use of the Dirac brackets which are constructed as follows. First, let us note that we can construct the matrix of Poisson brackets of all the constraints in (10.40) as

10.2 Dirac method and Dirac bracket

$$\{\phi^A, \phi^B\} \approx C^{AB}. \tag{10.43}$$

This is an even dimensional anti-symmetric matrix and Dirac had shown that it is nonsingular so that its inverse C^{-1}_{AB} exists, namely,

$$C^{AD} C^{-1}_{DB} = \delta^A_B = C^{-1}_{BD} C^{DA}. \tag{10.44}$$

We note here parenthetically that in field theories (where the variables depend on space coordinates as well), the matrix of Poisson brackets in (10.43) will be coordinate dependent and the product in (10.44) will involve integration over the intermediate space coordinate. With this, we can now define the Dirac bracket between any two dynamical variables as

$$\{F, G\}_D = \{F, G\} - \{F, \phi^A\} C^{-1}_{AB} \{\phi^B, G\}, \tag{10.45}$$

which can be shown to have all the properties of a Poisson bracket (anti-symmetry, Jacobi identity) and in addition satisfies

$$\begin{aligned}
\{F, \phi^A\}_D &= \{F, \phi^A\} - \{F, \phi^B\} C^{-1}_{BD} \{\phi^D, \phi^A\} \\
&\approx \{F, \phi^A\} - \{F, \phi^B\} C^{-1}_{BD} C^{DA} \\
&= \{F, \phi^A\} - \{F, \phi^A\} = 0.
\end{aligned} \tag{10.46}$$

Namely, in contrast to the canonical Poisson brackets, the Dirac bracket is compatible with the constraints in the sense any dynamical variable of the theory has a vanishing Dirac bracket with a constraint. In fact, through the use of Lagrange brackets, we can show that the Dirac brackets are indeed the correct Poisson brackets of the theory subject to the constraints in (10.40). As a result, we can work with the Dirac brackets and set the constraints to zero in the theory (namely, in the definitions of the Hamiltonian, energy-momentum tensor and other observables of the theory since their Dirac bracket with any variable vanishes) and it is the Dirac brackets that can be quantized (promoted to commutators or anti-commutators) in a quantum theory. The Dirac bracket has the interesting property that it can be constructed in stages. Namely, when the number of constraints is large, rather than dealing with a large matrix of Poisson brackets in (10.43) and its inverse, we can equivalently choose a smaller set of even number of constraints and define an intermediate Dirac bracket and then construct the final Dirac bracket of the theory building on this structure. It is also worth emphasizing here that in dealing with field theories where the matrix C^{AB} is coordinate dependent, the inverse needs to be defined carefully with the appropriate boundary condition relevant for the problem.

10.3 Particle moving on a sphere

As a simple example of constrained systems, let us consider the motion of a point particle constrained to move on the surface of a n-dimensional sphere of radius unity. If we denote the coordinate of the particle as $q_i, i = 1, 2, \cdots, n$, then the coordinates are constrained to satisfy (the metric in this simple theory is Euclidean and, therefore, raising and lowering of indices can be carried out trivially)

$$q_i q_i = 1, \qquad (10.47)$$

where summation over repeated indices is understood. This simple system is an interesting example in the study of constrained quantization and is a prototype of the field theoretic model known as the nonlinear sigma model. We note that the dynamics of the system can be described by the Lagrangian

$$L = \frac{1}{2}\left(\dot{q}_i \dot{q}_i - F(q_i q_i - 1)\right), \qquad (10.48)$$

where F is a Lagrange multiplier field (an auxiliary field without independent dynamics) whose Euler-Lagrange equation

$$\frac{\partial L}{\partial F} = -\frac{1}{2}(q_i q_i - 1) = 0, \qquad (10.49)$$

gives the constraint (10.47) on the dynamics. We note that if we combine the dynamical variables q_i, F into $q_a = (q_i, F), a = 1, 2, \cdots, n + 1$, then the matrix of highest derivatives has the form

$$\frac{\partial^2 L}{\partial \dot{q}_a \partial \dot{q}_b} = \begin{pmatrix} \delta_{ij} & 0 \\ 0 & 0 \end{pmatrix}, \tag{10.50}$$

which is indeed singular (see (10.22)) reflecting the fact that there are constraints in the theory and the naive Hamiltonian description would not hold.

To understand the passage to the Hamiltonian description in this simple case, we note that the conjugate momenta of the theory are given by

$$p^i = \frac{\partial L}{\partial \dot{q}_i} = \dot{q}_i, \quad p_F = \frac{\partial L}{\partial \dot{F}} = 0. \tag{10.51}$$

Therefore, the theory has a primary constraint given by

$$\varphi^1 = p_F \approx 0. \tag{10.52}$$

The canonical Hamiltonian following from (10.48) and (10.51) has the form

$$H_{\text{can}} = p^i \dot{q}_i + p_F \dot{F} - L = \frac{1}{2} \left(p^i p^i + F(q_i q_i - 1) \right), \tag{10.53}$$

leading to the primary Hamiltonian (see (10.30))

$$H_{\text{p}} = H_{\text{can}} + \lambda_1 \varphi^1 = \frac{1}{2} \left(p^i p^i + F(q_i q_i - 1) \right) + \lambda_1 p_F. \tag{10.54}$$

The equal-time canonical Poisson brackets of the theory are given by

$$\{q_i, q_j\} = \{p^i, p^j\} = \{F, F\} = \{p_F, p_F\} = 0,$$
$$\{q_i, F\} = \{q_i, p_F\} = \{p^i, F\} = \{p^i, p_F\} = 0,$$
$$\{q_i, p^j\} = \delta_i^j, \quad \{F, p_F\} = 1. \tag{10.55}$$

With these we can now determine the time evolution of any dynamical variable. In particular, requiring the primary constraint to be independent of time leads to

$$\begin{aligned}
\dot{\varphi}^1 &\approx \{\varphi^1, H_\mathrm{p}\} \\
&= \{p_F, \frac{1}{2}(p^i p^i + F(q_i q_i - 1) + \lambda_1 p_F\} \\
&= -\frac{1}{2}(q_i q_i - 1) \approx 0.
\end{aligned} \qquad (10.56)$$

Let us denote this secondary constraint as

$$\varphi^2 = \frac{1}{2}(q_i q_i - 1) \approx 0. \qquad (10.57)$$

Requiring (10.57) to be time independent, we obtain

$$\begin{aligned}
\dot{\varphi}^2 &\approx \{\varphi^2, H_\mathrm{p}\} \\
&= \{\frac{1}{2}(q_i q_i - 1), \frac{1}{2}(p^j p^j + F(q_j q_j - 1) + \lambda_1 p_F)\} \\
&= q_i p^j \{q_i, p^j\} = q_i p^i \approx 0,
\end{aligned} \qquad (10.58)$$

which generates a new constraint

$$\varphi^3 = q_i p^i \approx 0. \qquad (10.59)$$

Requiring the new constraint (10.59) to be time independent, we obtain

$$\begin{aligned}
\dot{\varphi}^3 &\approx \{\varphi^3, H_\mathrm{p}\} \\
&= \{q_i p^i, \frac{1}{2}(p^j p^j + F(q_j q_j - 1) + \lambda_1 p_F)\} \\
&= p^i p^j \{q_i, p^j\} + q_i F q_j \{p^i, q_j\} = p^i p^i - F q_i q_i \\
&\approx p^i p^i - F \approx 0,
\end{aligned} \qquad (10.60)$$

10.3 PARTICLE MOVING ON A SPHERE

so that we can identify

$$\varphi^4 = p^i p^i - F \approx 0. \tag{10.61}$$

Requiring (10.61) to be time independent leads to

$$\begin{aligned}
\dot{\varphi}^4 &\approx \{\varphi^4, H_{\mathrm{p}}\} \\
&= \{p^i p^i - F, \frac{1}{2}(p^j p^j + F(q_j q_j - 1) + \lambda_1 p_F)\} \\
&= 2p^i F q_j \{p^i, q_j\} - \lambda_1 \{F, p_F\} = -2F q_i p^i - \lambda_1 \\
&\approx -\lambda_1 \approx 0,
\end{aligned} \tag{10.62}$$

which determines the Lagrange multiplier and the chain of constraints terminates.

It is easy to check that all the constraints of the theory (10.52), (10.57), (10.59) and (10.61) are second class. In fact, the nontrivial equal-time Poisson brackets between the constraints take the forms

$$\begin{aligned}
\{\varphi^1, \varphi^4\} &= \{p_F, p^i p^i - F\} = 1 \\
&= -\{\varphi^4, \varphi^1\}, \\
\{\varphi^2, \varphi^3\} &= \{\frac{1}{2}(q_i q_i - 1), q_j p^j\} = q_i q_j \{q_i, p^j\} = q_i q_i \approx 1 \\
&= -\{\varphi^3, \varphi^2\}, \\
\{\varphi^3, \varphi^4\} &= \{q_i p^i, p^j p^j - F\} = 2p^i p^j \{q_i, p^j\} = 2p^i p^i = 2\mathbf{p}^2 \\
&= -\{\varphi^4, \varphi^3\},
\end{aligned} \tag{10.63}$$

where we have identified $p^i p^i = \mathbf{p}^2$. Collecting all the constraints into $\phi^A = (\varphi^1, \varphi^2, \varphi^3, \varphi^4)$, $A = 1, 2, 3, 4$, we can calculate the matrix of the (equal-time) Poisson brackets of constraints

$$\{\phi^A, \phi^B\} = C^{AB}, \tag{10.64}$$

which takes the form

$$C^{AB} = \begin{pmatrix} 0 & 0 & 0 & 1 \\ 0 & 0 & 1 & 0 \\ 0 & -1 & 0 & 2\mathbf{p}^2 \\ -1 & 0 & -2\mathbf{p}^2 & 0 \end{pmatrix}. \tag{10.65}$$

The inverse of this matrix has the form

$$C^{-1}_{AB} = \begin{pmatrix} 0 & -2\mathbf{p}^2 & 0 & -1 \\ 2\mathbf{p}^2 & 0 & -1 & 0 \\ 0 & 1 & 0 & 0 \\ 1 & 0 & 0 & 0 \end{pmatrix}. \tag{10.66}$$

The Dirac brackets between any two dynamical variables can now be defined as (see (10.45))

$$\{F, G\}_D = \{F, G\} - \{F, \phi^A\} C^{-1}_{AB} \{\phi^B, G\}, \tag{10.67}$$

with C^{-1}_{AB} given in (10.66). In particular, the Dirac brackets between the fundamental dynamical variables take the forms

$$\{F, F\}_D = \{p_F, p_F\}_D = \{F, p_F\}_D = 0,$$

$$\{q_i, p_F\}_D = \{p^i, p_F\}_D = 0,$$

$$\{q_i, F\}_D = \{q_i, F\} - \{q_i, p^j p^j - F\} C^{-1}_{41} \{p_F, F\} = 2p^i,$$

$$\{p^i, F\}_D = \{p^i, F\} - \{p^i, \tfrac{1}{2}(q_j q_j - 1)\} C^{-1}_{21} \{p_F, F\} = -2q_i \mathbf{p}^2,$$

$$\{q_i, q_j\}_D = 0,$$

$$\{p^i, p^j\}_D = \{p^i, p^j\} - \{p^i, \tfrac{1}{2}(q_k q_k - 1)\} C^{-1}_{23} \{q_\ell p^\ell, p^j\}$$

$$\qquad - \{p^i, q_k p^k\} C^{-1}_{32} \{\tfrac{1}{2}(q_\ell q_\ell - 1), p^j\}$$

$$= -q_i p^j + p^i q_j,$$

$$\{q_i, p^j\}_D = \{q_i, p^j\} - \{q_i, q_k p^k\} C^{-1}_{32} \{\tfrac{1}{2}(q_\ell q_\ell - 1), p^j\}$$

$$= \delta^j_i - q_i q_j. \tag{10.68}$$

With the use of the Dirac brackets, we can set the constraints to zero so that the true independent dynamical variables are (q_i, p^i) (we note that $F = p^i p^i$ because of the constraint (10.61)) and the Hamiltonian for the theory is given by (see (10.54))

$$H_{\rm p} = \frac{1}{2} p^i p^i, \tag{10.69}$$

where we have used (10.62). This is the starting point for the quantization of this theory.

10.4 Relativistic particle

In the last section we studied a simple example of point particle dynamics which is constrained. The constraints of this theory were all second class. Let us next study the Hamiltonian description of a free relativistic massive particle which is described by the equation

$$m \frac{{\rm d}^2 x^\mu}{{\rm d}\tau^2} = 0, \tag{10.70}$$

where τ denoting the proper time is assumed to label the trajectory of the particle and m is the rest mass of the particle. We note that if we define the relativistic four velocity and momentum as

$$u^\mu = \frac{{\rm d} x^\mu}{{\rm d}\tau}, \qquad p^\mu = m u^\mu, \tag{10.71}$$

then the dynamical equation (10.70) can also be written as

$$\frac{{\rm d} p^\mu}{{\rm d}\tau} = 0. \tag{10.72}$$

The dynamical equation (10.70) can be obtained as the Euler-Lagrange equation following from the action (recall that $c = 1$)

$$S = m \int {\rm d}s = m \int {\rm d}\lambda \, (\dot{x}^\mu \dot{x}_\mu)^{\frac{1}{2}} = \int {\rm d}\lambda \, L, \tag{10.73}$$

where λ is a parameter labelling the trajectory and we have identified

$$\dot{x}^\mu = \frac{dx^\mu}{d\lambda}, \quad \dot{x}_\mu = \eta_{\mu\nu}\dot{x}^\nu. \tag{10.74}$$

We note that under a transformation

$$\lambda \to \xi = \xi(\lambda), \quad \dot{x}^\mu = \frac{dx^\mu}{d\lambda} = \frac{d\xi}{d\lambda}\frac{dx^\mu}{d\xi}, \tag{10.75}$$

so that

$$\begin{aligned} S &= m\int d\lambda \left(\frac{dx^\mu}{d\lambda}\frac{dx_\mu}{d\lambda}\right)^{\frac{1}{2}} \\ &\to m\int d\lambda \left(\frac{d\xi}{d\lambda}\frac{dx^\mu}{d\xi}\frac{d\xi}{d\lambda}\frac{dx_\mu}{d\xi}\right)^{\frac{1}{2}} \\ &= m\int d\lambda \frac{d\xi}{d\lambda}\left(\frac{dx^\mu}{d\xi}\frac{dx_\mu}{d\xi}\right)^{\frac{1}{2}} \\ &= m\int d\xi \left(\frac{dx^\mu}{d\xi}\frac{dx_\mu}{d\xi}\right)^{\frac{1}{2}} = S. \end{aligned} \tag{10.76}$$

Namely, the action (10.73) is invariant under the diffeomorphism (10.75) which is an example of a local gauge transformation much like in the Maxwell field theory. If we choose a gauge

$$\lambda = \tau, \tag{10.77}$$

where τ denotes the proper time associated with the particle, then (10.73) will lead to the dynamical equation (10.70) or (10.72) as the Euler-Lagrange equations. However, let us proceed with the Hamiltonian description without choosing a gauge at this point. We note from (10.73) that

$$P_{\mu\nu} = \frac{\partial^2 L}{\partial \dot{x}^\mu \dot{x}^\nu} = \frac{m}{\sqrt{\dot{x}^\rho \dot{x}_\rho}}\left(\eta_{\mu\nu} - \frac{\dot{x}_\mu \dot{x}_\nu}{\dot{x}^\sigma \dot{x}_\sigma}\right), \tag{10.78}$$

which leads to

10.4 RELATIVISTIC PARTICLE

$$\dot{x}^\mu P_{\mu\nu} = \frac{m}{\sqrt{\dot{x}^\rho \dot{x}_\rho}} \left(\dot{x}_\nu - \frac{\dot{x}^\mu \dot{x}_\mu \dot{x}_\nu}{\dot{x}^\sigma \dot{x}_\sigma} \right) = 0 = P_{\mu\nu} \dot{x}^\nu. \tag{10.79}$$

Therefore, $P_{\mu\nu}$ is a projection operator and

$$\det P_{\mu\nu} = \det \frac{\partial^2 L}{\partial \dot{x}^\mu \dot{x}^\nu} = 0, \tag{10.80}$$

as in the case of the Maxwell theory (see (9.124)) and the naive passage to the Hamiltonian description fails (see (10.22)).

Let us note that the action (10.73) can also be written in an alternate form that is more useful. For example, it is not clear how to take the $m = 0$ limit in (10.73) to describe the motion of a massless particle. We note that introducing an auxiliary variable g, we can rewrite the action (10.73) in the form

$$S = \int d\lambda\, L = \frac{1}{2} \int d\lambda\, \sqrt{g}\, \left(g^{-1} \dot{x}^\mu \dot{x}_\mu + m^2 \right), \tag{10.81}$$

where because of the diffeomorphism invariance of (10.73) we can think of g as the metric on the one dimensional manifold of the trajectory of the particle. Viewed in this manner, the action (10.81) is manifestly diffeomorphism invariant and the massless limit can now be taken in a straightforward manner. To see that the two actions (10.73) and (10.81) are equivalent, we note that the equation for the auxiliary field g takes the form

$$\frac{\partial L}{\partial g} = \frac{g^{-1/2}}{4} \left(-g^{-1} \dot{x}^\mu \dot{x}_\mu + m^2 \right) = 0,$$

or, $\quad g = \dfrac{\dot{x}^\mu \dot{x}_\mu}{m^2}.$ \hfill (10.82)

If we eliminate g in (10.81) using (10.82), then we obtain (10.73). However, it is more convenient for us to work with the action (10.81).

The two dynamical variables of the theory in (10.81) are x^μ and g. We can obtain the momenta conjugate to these variables to be

$$p_\mu = \frac{\partial L}{\partial \dot{x}^\mu} = g^{-1/2}\dot{x}_\mu, \quad p_g = \frac{\partial L}{\partial \dot{g}} = 0, \tag{10.83}$$

so that the primary constraint of the theory follows to be

$$\varphi^1 = p_g \approx 0. \tag{10.84}$$

The canonical Hamiltonian of the theory follows to be

$$H_{\text{can}} = p_\mu \dot{x}^\mu + p_g \dot{g} - L = \frac{\sqrt{g}}{2}\left(p^\mu p_\mu - m^2\right). \tag{10.85}$$

Adding the primary constraint (10.84), we obtain the primary Hamiltonian for the theory to correspond to

$$H_{\text{p}} = H_{\text{can}} + \lambda_1 \varphi^1 = \frac{\sqrt{g}}{2}\left(p^\mu p_\mu - m^2\right) + \lambda_1 p_g. \tag{10.86}$$

The equal time canonical Poisson brackets for the theory are given by

$$\{x^\mu, x^\nu\} = \{p_\mu, p_\nu\} = \{g, g\} = \{p_g, p_g\} = 0,$$
$$\{x^\mu, g\} = \{x^\mu, p_g\} = \{p_\mu, g\} = \{p_\mu, p_g\} = 0,$$
$$\{x^\mu, p_\nu\} = \delta^\mu_\nu, \quad \{g, p_g\} = 1. \tag{10.87}$$

With these we can now determine the evolution of the primary constraint (10.84) and requiring the constraint to be time independent, we obtain

$$\dot{\varphi}^1 \approx \{\varphi^1, H_{\text{p}}\} = \{p_g, \frac{\sqrt{g}}{2}\left(p^\mu p_\mu - m^2\right) + \lambda_1 p_g\}$$
$$= -\frac{1}{4\sqrt{g}}\left(p^\mu p_\mu - m^2\right) \approx 0, \tag{10.88}$$

which leads to the secondary constraint

10.4 RELATIVISTIC PARTICLE

$$\varphi^2 = \frac{1}{2}\left(p^\mu p_\mu - m^2\right) \approx 0. \tag{10.89}$$

We recognize that the secondary constraint (10.89) is the familiar Einstein relation for a massive relativistic particle (the multiplicative factor is for later calculational simplicity). Time evolution of the secondary constraint leads to

$$\dot{\varphi}^2 \approx \{\varphi^2, H_p\} = \{\frac{1}{2}\left(p^\mu p_\mu - m^2\right), \frac{\sqrt{g}}{2}\left(p^\nu p_\nu - m^2\right) + \lambda_1 p_g\}$$
$$= 0, \tag{10.90}$$

so that it is time independent and the chain of constraints terminates.

The two constraints of the theory can be easily checked to be first class which is associated with the fact that the theory has a local gauge invariance and is also reflected in the fact that the Lagrange multiplier λ_1 remains arbitrary (the primary constraint is first class). Let us choose the gauge fixing conditions

$$\chi^1 = g - \frac{1}{m^2} \approx 0,$$
$$\chi^2 = \lambda - x^0 \approx 0, \tag{10.91}$$

which render all the constraints to be second class. The non-vanishing Poisson brackets between the constraints are given by

$$\{\varphi^1, \chi^1\} = \{p_g, g - \frac{1}{m^2}\} = -1 = -\{\chi^1, \varphi^1\},$$
$$\{\varphi^2, \chi^2\} = \{\frac{1}{2}\left(p^\mu p_\mu - m^2\right), \lambda - x^0\} = p^0$$
$$= -\{\chi^2, \varphi^2\}. \tag{10.92}$$

As a result, if we combine all the constraints into $\phi^A = (\varphi^1, \varphi^2, \chi^1, \chi^2)$, then it follows from (10.92) that the matrix of Poisson brackets of constraints has the form

$$C^{AB} = \begin{pmatrix} 0 & 0 & -1 & 0 \\ 0 & 0 & 0 & p^0 \\ 1 & 0 & 0 & 0 \\ 0 & -p^0 & 0 & 0 \end{pmatrix}, \qquad (10.93)$$

whose inverse is given by

$$C^{-1}_{AB} = \begin{pmatrix} 0 & 0 & 1 & 0 \\ 0 & 0 & 0 & -\frac{1}{p^0} \\ -1 & 0 & 0 & 0 \\ 0 & \frac{1}{p^0} & 0 & 0 \end{pmatrix}. \qquad (10.94)$$

The equal time Dirac bracket between any two dynamical variables is given by (see (10.45))

$$\{F, G\}_D = \{F, G\} - \{F, \phi^A\} C^{-1}_{AB} \{\phi^B, G\}, \qquad (10.95)$$

and with (10.94) we can calculate the Dirac bracket between the fundamental variables to be

$$\{x^\mu, x^\nu\}_D = \{p_\mu, p_\nu\}_D = \{g, g\}_D = \{p_g, p_g\}_D = 0,$$
$$\{x^\mu, g\}_D = \{x^\mu, p_g\}_D = \{p_\mu, g\}_D = \{p_\mu, p_g\}_D = 0,$$
$$\{g, p_g\}_D = \{g, p_g\} - \{g, \phi^1\} C^{-1}_{13} \{\phi^3, p_g\}$$
$$= \{g, p_g\} - \{g, p_g\}\{g - \frac{1}{m^2}, p_g\} = 0,$$
$$\{x^\mu, p_\nu\}_D = \{x^\mu, p_\nu\} - \{x^\mu, \phi^2\} C^{-1}_{24} \{\phi^4, p_\nu\}$$
$$= \delta^\mu_\nu - \{x^\mu, \frac{1}{2}(p^\lambda p_\lambda - m^2)\}(-\frac{1}{p^0})\{\lambda - x^0, p_\nu\}$$
$$= \delta^\mu_\nu - \frac{p^\mu}{p^0} \delta^0_\nu. \qquad (10.96)$$

With the Dirac brackets we can set the constraints to zero in which case we see that the primary Hamiltonian (10.86) vanishes. (On the other hand, if we set the constraints strongly to zero, (10.89) in particular leads to $H = p^0 = \sqrt{\mathbf{p}^2 + m^2}$ which can be taken as the Hamiltonian.) Let us also note from (10.82) that the first of the gauge fixing conditions in (10.91) would imply

$$\dot{x}^\mu \dot{x}_\mu = \frac{\mathrm{d}x^\mu}{\mathrm{d}\lambda} \frac{\mathrm{d}x_\mu}{\mathrm{d}\lambda} = 1. \tag{10.97}$$

On the other hand, from the definition of the infinitesimal invariant length in the Minkowski space (see (1.20)), we have (remember that $c = 1$)

$$\mathrm{d}\tau^2 = \mathrm{d}x^\mu \mathrm{d}x_\mu,$$

$$\text{or,} \quad \frac{\mathrm{d}x^\mu}{\mathrm{d}\tau} \frac{\mathrm{d}x_\mu}{\mathrm{d}\tau} = 1. \tag{10.98}$$

It follows, therefore, that the first of the gauge fixing conditions corresponds to choosing $\lambda = \tau$.

10.5 Dirac field theory

As we have seen in (8.9), the free Dirac theory is described by the Lagrangian density

$$\mathcal{L} = i\overline{\psi}\slashed{\partial}\psi - m\overline{\psi}\psi, \tag{10.99}$$

where the adjoint spinor is given by

$$\overline{\psi} = \psi^\dagger \gamma^0. \tag{10.100}$$

The dynamical variables of the theory can be chosen to be $\psi_\alpha, \psi_\alpha^\dagger$, where $\alpha = 1, 2, 3, 4$ denote the Dirac indices and since the Lagrangian density is first order in derivatives, it is obvious that the matrix of second order dervatives vanishes and the system is singular (see (10.22)). This manifests in the definition of the conjugate momenta as (we have chosen the convention of left derivatives)

$$\Pi_\alpha^\dagger = \frac{\partial \mathcal{L}}{\partial \dot\psi_\alpha} = -i(\overline\psi \gamma^0)_\alpha = -i\psi_\alpha^\dagger,$$

$$\Pi_\alpha = \frac{\partial \mathcal{L}}{\partial \dot\psi_\alpha^\dagger} = 0. \tag{10.101}$$

(Note that $\Pi^\dagger$ is not the Hermitian conjugate of Π. Rather they correspond to the momenta conjugate to ψ and $\psi^\dagger$ respectively.) As a result, we obtain the two sets of primary constraints for the theory to be

$$\varphi_\alpha^\dagger = \Pi_\alpha^\dagger + i\psi_\alpha^\dagger \approx 0,$$

$$\rho_\alpha = \Pi_\alpha \approx 0, \tag{10.102}$$

which in this case correspond to fermionic constraints.

The canonical Hamiltonian density of the theory (consistent with the convention of left derivatives) is obtained to be (see (10.11))

$$\begin{aligned}\mathcal{H}_{\text{can}} &= -\Pi_\alpha^\dagger \dot\psi_\alpha + \dot\psi_\alpha^\dagger \Pi_\alpha - \mathcal{L} \\ &= i\psi_\alpha^\dagger \dot\psi_\alpha - i\psi_\alpha^\dagger \dot\psi_\alpha - i\overline\psi\gamma\cdot\nabla\psi + m\overline\psi\psi \\ &= -i\overline\psi\gamma\cdot\nabla\psi + m\overline\psi\psi, \end{aligned} \tag{10.103}$$

where we have used (10.101) in the intermediate steps. The canonical Hamiltonian is now obtained to be

$$H_{\text{can}} = \int d^3x\, \mathcal{H}_{\text{can}} = \int d^3x \left(-i\overline\psi\gamma\cdot\nabla\psi + m\overline\psi\psi\right). \tag{10.104}$$

Adding the primary constraints to the canonical Hamiltonian, we obtain the primary Hamiltonian for the Dirac field theory as (see (10.30))

$$H_p = H_{\text{can}} + \int d^3x \left(\phi_\alpha^\dagger \xi_\alpha + \lambda_\alpha^\dagger \rho_\alpha\right), \tag{10.105}$$

10.5 DIRAC FIELD THEORY

where the Lagrange multipliers $\xi_\alpha, \lambda_\alpha^\dagger$ in the present case denote fermionic variables.

The equal-time canonical Poisson brackets for the field variables take the forms (see (10.12))

$$\{\psi_\alpha(x), \Pi_\beta^\dagger(y)\} = -\delta_{\alpha\beta}\delta^3(x-y) = \{\psi_\alpha^\dagger(x), \Pi_\beta(y)\}, \quad (10.106)$$

with all other brackets vanishing. Using (10.106) (see also (10.13)-(10.14)), we can now calculate the time evolution of the primary constraints (10.102). For example,

$$\begin{aligned}\dot\phi_\alpha^\dagger(x) &\approx \{\phi_\alpha^\dagger(x), H_\mathrm{p}\} \\ &\approx \{\phi_\alpha^\dagger(x), H_\mathrm{can}\} + \int \mathrm{d}^3 y \big(\xi_\beta(y)\{\phi_\alpha^\dagger(x), \phi_\beta^\dagger(y)\} \\ &\quad -\lambda_\beta^\dagger(y)\{\phi_\alpha^\dagger(x), \rho_\beta(y)\}\big). \quad (10.107)\end{aligned}$$

Let us evaluate each of these terms separately. Using the fact that the primary Hamiltonian is independent of time, we can identify x^0 with the time variable of the field operators in H_p leading to (we do not show equal-time explicitly for simplicity)

$$\begin{aligned}&\{\phi_\alpha^\dagger(x), H_\mathrm{can}\} \\ &= \int \mathrm{d}^3 y\, \{\Pi_\alpha^\dagger(x) + i\psi_\alpha^\dagger(x), -i\overline{\psi}(y)\boldsymbol{\gamma}\cdot\boldsymbol{\nabla}_y\psi(y) + m\overline{\psi}(y)\psi(y)\} \\ &= \int \mathrm{d}^3 y\, (i\overline{\psi}(y)\boldsymbol{\gamma}\cdot\boldsymbol{\nabla}_y - m\overline{\psi}(y))_\beta\{\Pi_\alpha^\dagger(x), \psi_\beta(y)\} \\ &= \int \mathrm{d}^3 y\, (i\overline{\psi}(y)\boldsymbol{\gamma}\cdot\boldsymbol{\nabla}_y - m\overline{\psi}(y))_\beta\big(-\delta_{\alpha\beta}\delta^3(x-y)\big) \\ &= (i\boldsymbol{\nabla}\overline{\psi}(x)\cdot\boldsymbol{\gamma} + m\overline{\psi}(x))_\alpha,\end{aligned}$$

$$\begin{aligned}&\int \mathrm{d}^3 y\, \xi_\beta(y)\{\phi_\alpha^\dagger(x), \phi_\beta^\dagger(y)\} \\ &= \int \mathrm{d}^3 y\, \xi_\beta(y)\{\Pi_\alpha^\dagger(x) + i\psi_\alpha^\dagger(x), \Pi_\beta^\dagger(y) + i\psi_\beta^\dagger(y)\}\end{aligned}$$

$$= 0,$$

$$-\int d^3y\, \lambda_\beta^\dagger(y)\{\phi_\alpha^\dagger(x), \rho_\beta(y)\}$$

$$= -\int d^3y\, \lambda_\beta^\dagger(y)\{\Pi_\alpha^\dagger(x) + i\psi_\alpha^\dagger(x), \Pi_\beta(y)\}$$

$$= -\int d^3y\, \lambda_\beta^\dagger(y)\big(-i\delta_{\alpha\beta}\delta^3(x-y)\big)$$

$$= i\lambda_\alpha^\dagger(x). \tag{10.108}$$

Substituting these results into (10.107) and requiring the constraint to be independent of time leads to

$$\dot{\phi}_\alpha^\dagger(x) \approx \big(i\boldsymbol{\nabla}\overline{\psi}(x)\cdot\boldsymbol{\gamma} + m\overline{\psi}(x)\big)_\alpha + i\lambda_\alpha^\dagger(x) = 0, \tag{10.109}$$

which determines the Lagrange multiplier

$$\lambda_\alpha^\dagger = \big(-\boldsymbol{\nabla}\overline{\psi}(x)\cdot\boldsymbol{\gamma} + im\overline{\psi}(x)\big)_\alpha. \tag{10.110}$$

Similarly, requiring the second set of primary constraints to be time independent we obtain

$$\dot{\rho}_\alpha(x) \approx \{\rho_\alpha(x), H_\mathrm{p}\} = \{\Pi_\alpha(x), H_\mathrm{p}\}$$

$$= \int d^3y\, \{\Pi_\alpha(x), \psi_\beta^\dagger(y)\}\big(\gamma^0(-i\boldsymbol{\gamma}\cdot\boldsymbol{\nabla}_y + m)\psi(y) + i\xi(y)\big)_\beta$$

$$= \int d^3y\, \big(-\delta_{\alpha\beta}\delta^3(x-y)\big)\big(\gamma^0(-i\boldsymbol{\gamma}\cdot\boldsymbol{\nabla}_y + m)\psi(y) + i\xi(y)\big)_\beta$$

$$= -\big(\gamma^0(-i\boldsymbol{\gamma}\cdot\boldsymbol{\nabla} + m)\psi(x) + i\xi(x)\big)_\alpha$$

$$= 0, \tag{10.111}$$

which determines the other Lagrange multiplier

$$\xi_\alpha = \big(\gamma^0(\boldsymbol{\gamma}\cdot\boldsymbol{\nabla} + im)\psi(x)\big)_\alpha. \tag{10.112}$$

10.5 Dirac field theory

In this case, we see that requiring the two primary constraints to be time independent does not introduce any new constraint, rather it determines both the Lagrange multipliers in the primary Hamiltonian (10.105). Substituting (10.110) and (10.112) into the primary Hamiltonian, we obtain

$$\begin{aligned}
H_p &= \int d^3x \left(-i\overline{\psi}\gamma\cdot\nabla\psi + m\overline{\psi}\psi + \phi_\alpha^\dagger \xi_\alpha + \lambda_\alpha^\dagger \rho_\alpha\right) \\
&= \int d^3x \left(-i\overline{\psi}\gamma\cdot\nabla\psi + m\overline{\psi}\psi \right. \\
&\quad + (\Pi_\alpha^\dagger + i\psi_\alpha^\dagger)\left(\gamma^0(\gamma\cdot\nabla + im)\psi\right)_\alpha \\
&\quad \left. + \left(-\nabla\overline{\psi}\cdot\gamma + im\overline{\psi}\right)_\alpha \Pi_\alpha\right) \\
&= \int d^3x \left(\overline{\Pi}\gamma\cdot\nabla\psi + im\overline{\Pi}\psi + (-\nabla\overline{\psi}\cdot\gamma + im\overline{\psi})\Pi\right),
\end{aligned} \tag{10.113}$$

where, for simplicity of notation, we have defined

$$\overline{\Pi} = \Pi^\dagger \gamma^0. \tag{10.114}$$

Let us note next that the two sets of primary constraints (10.102), which constitute all the constraints of the theory, define second class constraints (this is consistent with the fact that there is no undetermined Lagrange multiplier in the theory). In fact, we have (at equal time)

$$\begin{aligned}
\{\phi_\alpha^\dagger(x), \phi_\beta^\dagger(y)\} &= \{\Pi_\alpha^\dagger(x) + i\psi_\alpha^\dagger(x), \Pi_\beta^\dagger(y) + i\psi_\beta^\dagger(y)\} = 0, \\
\{\phi_\alpha^\dagger(x), \rho_\beta(y)\} &= \{\Pi_\alpha^\dagger(x) + i\psi_\alpha^\dagger(x), \Pi_\beta(y)\} \\
&= -i\delta_{\alpha\beta}\delta^3(x-y) = \{\rho_\alpha(x), \phi_\beta^\dagger(y)\}, \\
\{\rho_\alpha(x), \rho_\beta(y)\} &= \{\Pi_\alpha(x), \Pi_\beta(y)\} = 0.
\end{aligned} \tag{10.115}$$

Thus, combining the constraints into $\Phi^A = (\phi_\alpha^\dagger, \rho_\alpha)$, we can write the matrix of Poisson brackets as

$$C(x,y) = \begin{pmatrix} \{\phi_\alpha^\dagger(x), \phi_\beta^\dagger(y)\} & \{\phi_\alpha^\dagger(x), \rho_\beta(y)\} \\ \{\rho_\alpha(x), \phi_\beta^\dagger(y)\} & \{\rho_\alpha(x), \rho_\beta(y)\} \end{pmatrix}$$

$$= -i \begin{pmatrix} 0 & \mathbb{1} \\ \mathbb{1} & 0 \end{pmatrix} \delta^3(x-y). \tag{10.116}$$

The inverse of the matrix is easily seen to be

$$C^{-1}(x,y) = i \begin{pmatrix} 0 & \mathbb{1} \\ \mathbb{1} & 0 \end{pmatrix} \delta^3(x-y). \tag{10.117}$$

The equal-time Dirac bracket of any two dynamical variables can now be easily defined (see (10.45))

$$\begin{aligned}
\{F(x), G(y)\}_D &= \{F(x), G(y)\} \\
&\quad - \iint d^3z\, d^3\bar{z}\, \{F(x), \Phi^A(z)\} C^{-1}_{AB}(z, \bar{z}) \{\Phi^B(\bar{z}), G(y)\} \\
&= \{F(x), G(y)\} \\
&\quad - \iint d^3z\, d^3\bar{z}\, \Big(\{F(x), \phi_\gamma^\dagger(z)\}(i\delta_{\gamma\delta}\delta^3(z-\bar{z}))\{\rho_\delta(\bar{z}), G(y)\} \\
&\qquad + \{F(x), \rho_\gamma(z)\}(i\delta_{\gamma\delta}\delta^3(z-\bar{z}))\{\phi_\delta^\dagger(\bar{z}), G(y)\} \Big) \\
&= \{F(x), G(y)\} \\
&\quad - i \int d^3z\, \Big(\{F(x), \Pi_\gamma^\dagger(z) + i\psi_\gamma^\dagger(z)\}\{\Pi_\gamma(z), G(y)\} \\
&\qquad + \{F(x), \Pi_\gamma(z)\}\{\Pi_\gamma^\dagger(z) + i\psi_\gamma^\dagger(z), G(y)\} \Big). \tag{10.118}
\end{aligned}$$

Using (10.118) we can calculate the Dirac bracket between the field variables. Recalling that constraints can be set to zero inside the Dirac bracket, we obtain

$$\{\psi_\alpha(x), \psi_\beta^\dagger(y)\}_\mathrm{D} = \{\psi_\alpha(x), \psi_\beta^\dagger(y)\}$$
$$-i\int \mathrm{d}^3z\{\psi_\alpha(x), \Pi_\gamma^\dagger(z) + i\psi_\gamma^\dagger(z)\}\{\Pi_\gamma(z), \psi_\beta^\dagger(y)\}$$
$$= -i\int \mathrm{d}^3z(-\delta_{\alpha\gamma}\delta^3(x-z))(-\delta_{\gamma\beta}\delta^3(z-y))$$
$$= -i\delta_{\alpha\beta}\delta^3(x-y), \tag{10.119}$$

where we have neglected the last term in (10.118) in evaluating this bracket because Π_γ has a vanishing Poisson bracket with ψ_α. Similarly, we can show that

$$\{\psi_\alpha(x), \psi_\beta(y)\}_\mathrm{D} = 0 = \{\psi_\alpha^\dagger(x), \psi_\beta^\dagger(y)\}_\mathrm{D}, \tag{10.120}$$

and these are the relations we have used in quantizing the Dirac theory (see (8.16)). Furthermore, we recall that when using the Dirac brackets, we can set the constraints to zero in the theory. Therefore, the Hamiltonian of the theory (10.113) becomes

$$H_\mathrm{p} = \int \mathrm{d}^3x \left(-i\overline{\psi}\boldsymbol{\gamma}\cdot\boldsymbol{\nabla}\psi + m\overline{\psi}\psi\right), \tag{10.121}$$

which is the Hamiltonian we have used in the quantization of the Dirac field theory (see (8.18)).

10.6 Maxwell field theory

Let us next consider the Maxwell field theory which as we know exhibits gauge invariance. As we have seen in (9.16), Maxwell's equations can be obtained as the Euler-Lagrange equations from the Lagrangian density

$$\mathcal{L} = -\frac{1}{4}F_{\mu\nu}F^{\mu\nu}, \quad \mu,\nu = 0,1,2,3, \tag{10.122}$$

where the field strength tensor (see (9.4))

$$F_{\mu\nu} = \partial_\mu A_\nu - \partial_\nu A_\mu = -F_{\nu\mu}, \qquad (10.123)$$

is the four dimensional curl of the four vector potential and contains the electric and the magnetic fields as its components (see (9.5) and (9.7))

$$F_{0i} = E_i, \quad F_{ij} = -\epsilon_{ijk} B_k, \quad i,j,k = 1,2,3. \qquad (10.124)$$

We also know that Maxwell's theory is invariant under gauge transformations

$$A_\mu(x) \to A'_\mu(x) = A_\mu(x) + \partial_\mu \alpha(x), \qquad (10.125)$$

where $\alpha(x)$ denotes the local parameter of gauge transformation.

As a result of the gauge invariance of the theory, the matrix of quadratic derivatives becomes singular (see (9.122)) and this manifests in the definition of the conjugate momenta as

$$\Pi^\mu = \frac{\partial \mathcal{L}}{\partial \dot{A}_\mu} = -F^{0\mu}, \qquad (10.126)$$

so that we have

$$\begin{aligned}\mathbf{\Pi} &= \mathbf{E} = -(\dot{\mathbf{A}} + \nabla A_0), \\ \Pi^0 &= 0.\end{aligned} \qquad (10.127)$$

This determines that the theory has a primary constraint given by

$$\varphi^1(x) = \Pi^0(x) \approx 0. \qquad (10.128)$$

We can now obtain the canonical Hamiltonian density of the theory to have the form

10.6 Maxwell field theory

$$\begin{aligned}\mathcal{H}_{\text{can}} &= \Pi^\mu \dot{A}_\mu - \mathcal{L} = -\mathbf{\Pi} \cdot \dot{\mathbf{A}} + \frac{1}{4} F_{\mu\nu} F^{\mu\nu} \\ &= -\mathbf{\Pi} \cdot (-\mathbf{\Pi} - \boldsymbol{\nabla} A_0) + \frac{1}{2}(-\mathbf{E}^2 + \mathbf{B}^2) \\ &= \frac{1}{2}(\mathbf{\Pi}^2 + \mathbf{B}^2) + \mathbf{\Pi} \cdot \boldsymbol{\nabla} A_0, \end{aligned} \quad (10.129)$$

where we have used (10.127). The canonical Hamiltonian, therefore, is given by

$$\begin{aligned} H_{\text{can}} &= \int \mathrm{d}^3 x\, \mathcal{H}_{\text{can}} \\ &= \int \mathrm{d}^3 x \left(\frac{1}{2}(\mathbf{\Pi}^2 + \mathbf{B}^2) - A_0 \boldsymbol{\nabla} \cdot \mathbf{\Pi} \right), \end{aligned} \quad (10.130)$$

where we have neglected a surface term (total divergence) in integrating the last term by parts. The primary Hamiltonian of the theory is now obtained by adding the primary constraint (10.128)

$$\begin{aligned} H_{\text{p}} &= H_{\text{can}} + \int \mathrm{d}^3 x\, \lambda_1 \varphi^1 \\ &= \int \mathrm{d}^3 x \left(\frac{1}{2}(\mathbf{\Pi}^2 + \mathbf{B}^2) - A_0 \boldsymbol{\nabla} \cdot \mathbf{\Pi} + \lambda_1 \Pi^0 \right), \end{aligned} \quad (10.131)$$

where λ_1 denotes the Lagrange multiplier of the theory.

The equal-time canonical Poisson brackets of the theory are given by

$$\{A_\mu(x), A_\nu(y)\} = 0 = \{\Pi^\mu(x), \Pi^\nu(y)\},$$
$$\{A_\mu(x), \Pi^\nu(y)\} = \delta_\mu^\nu \delta^3(x - y) = -\{\Pi^\nu(y), A_\mu(x)\}. \quad (10.132)$$

Using these we can calculate the time evolution of the primary constraint (10.128) and requiring the constraint to be time independent we obtain (the Poisson brackets are at equal-time which can be achieved using the time independence of H_{p})

$$\begin{aligned}
\dot\varphi^1(x) &\approx \{\varphi^1(x), H_{\mathrm p}\} \\
&= \int d^3y\, \{\Pi^0(x), \tfrac{1}{2}(\mathbf{\Pi}^2(y)+\mathbf{B}^2(y)) \\
&\quad - A_0(y)\mathbf{\nabla}\cdot\mathbf{\Pi}(y) + \lambda_1(y)\Pi^0(y)\} \\
&\approx -\int d^3y\, \{\Pi^0(x), A_0(y)\}\mathbf{\nabla}\cdot\mathbf{\Pi}(y) \\
&= \mathbf{\nabla}\cdot\mathbf{\Pi}(x) \approx 0.
\end{aligned} \qquad (10.133)$$

Therefore, we have a secondary constraint in the theory given by

$$\varphi^2(x) = \mathbf{\nabla}\cdot\mathbf{\Pi}(x) \approx 0. \qquad (10.134)$$

It can now be checked that the secondary constraint is time independent, namely,

$$\dot\varphi^2(x) \approx \{\varphi^2(x), H_{\mathrm p}\} = \{\mathbf{\nabla}\cdot\mathbf{\Pi}(x), H_{\mathrm p}\} \approx 0, \qquad (10.135)$$

so that the chain of constraints terminates.

We see that Maxwell's theory has two constraints

$$\varphi^1(x) = \Pi^0(x) \approx 0, \quad \varphi^2(x) = \mathbf{\nabla}\cdot\mathbf{\Pi}(x) \approx 0, \qquad (10.136)$$

and it is clear that both these constraints are first class constraints. This is consistent with our earlier observation that not all Lagrange multipliers in the primary Hamiltonian are determined when there are first class constraints present. In this case, as we have seen, the Lagrange multiplier λ_1 remains as yet undetermined (there is one primary constraint which is first class). Furthermore, first class constraints signal the presence of gauge invariances (local invariances) in the theory which we know very well in the case of Maxwell field theory. According to the Dirac procedure, in the presence of first class constraints we are supposed to add gauge fixing conditions to convert them into second class constraints. Let us choose

10.6 MAXWELL FIELD THEORY

$$\chi^1(x) = A_0(x) \approx 0, \quad \chi^2(x) = \boldsymbol{\nabla} \cdot \mathbf{A}(x) \approx 0, \tag{10.137}$$

as the gauge fixing conditions which clearly convert the constraints (10.136) into second class constraints. Therefore, relations (10.136) and (10.137) determine all the constraints of the theory.

Combining the first class constraints as well as the gauge fixing conditions into $\phi^A = (\varphi^\alpha, \chi^\alpha)$, $\alpha = 1, 2$, we can calculate the matrix of the second class constraints from the fact that the only nonvanishing equal-time Poisson brackets between the constraints are given by

$$\begin{aligned}
\{\varphi^1(x), \chi^1(y)\} &= \{\Pi^0(x), A_0(y)\} \\
&= -\delta^3(x-y) = -\{\chi^1(x), \varphi^1(y)\}, \\
\{\varphi^2(x), \chi^2(y)\} &= \{\boldsymbol{\nabla} \cdot \boldsymbol{\Pi}(x), \boldsymbol{\nabla} \cdot \mathbf{A}(y)\} \\
&= (\boldsymbol{\nabla}_x)_i (\boldsymbol{\nabla}_y)_j \{(\boldsymbol{\Pi})_i(x), (\boldsymbol{\Pi})_j(y)\} \\
&= (\boldsymbol{\nabla}_x)_i (\boldsymbol{\nabla}_y)_j (\delta_{ij} \delta^3(x-y)) \\
&= -\boldsymbol{\nabla}_x^2 \delta^3(x-y) = -\{\chi^2(x), \varphi^2(y)\}, \tag{10.138}
\end{aligned}$$

which leads to $(x^0 = y^0)$

$$C^{AB}(x,y) = \{\phi^A(x), \phi^B(y)\}$$

$$= \begin{pmatrix} 0 & 0 & -1 & 0 \\ 0 & 0 & 0 & -\boldsymbol{\nabla}_x^2 \\ 1 & 0 & 0 & 0 \\ 0 & \boldsymbol{\nabla}_x^2 & 0 & 0 \end{pmatrix} \delta^3(x-y). \tag{10.139}$$

The inverse of this matrix is easily calculated to be

$$C_{AB}^{-1}(x,y) = \begin{pmatrix} 0 & 0 & 1 & 0 \\ 0 & 0 & 0 & \boldsymbol{\nabla}_x^{-2} \\ -1 & 0 & 0 & 0 \\ 0 & -\boldsymbol{\nabla}_x^{-2} & 0 & 0 \end{pmatrix} \delta^3(x-y), \tag{10.140}$$

where, with the usual asymptotic condition that fields vanish at spatial infinity, the formal inverse of the Laplacian (Green's function) has the explicit form

$$\nabla_x^{-2} \delta^3(x-y) = \frac{1}{\nabla_x^2} \delta^3(x-y) = -\frac{1}{4\pi|\mathbf{x}-\mathbf{y}|}. \tag{10.141}$$

The equal-time Dirac bracket between any two dynamical variables can now be calculated from

$$\{F(x), G(y)\}_D = \{F(x), G(y)\}$$
$$- \iint d^3z\, d^3\bar{z}\, \{F(x), \phi^A(z)\} C_{AB}^{-1}(z,\bar{z}) \{\phi^B(\bar{z}), G(y)\}, \tag{10.142}$$

and, in particular, for the dynamical field variables, it leads to

$$\{A_\mu(x), A_\nu(y)\}_D = 0 = \{\Pi^\mu(x), \Pi^\nu(y)\}_D,$$
$$\{A_\mu(x), \Pi^\nu(y)\}_D = \{A_\mu(x), \Pi^\nu(y)\}$$
$$- \iint d^3z\, d^3\bar{z}\, \big(\{A_\mu(x), \Pi^0(z)\}(\delta^3(z-\bar{z}))\{A_0(\bar{z}), \Pi^\nu(y)\}$$
$$+ \{A_\mu(x), \nabla_z \cdot \mathbf{\Pi}(z)\}(\nabla_z^{-2}\delta^3(z-\bar{z}))\{\nabla_{\bar{z}} \cdot \mathbf{A}(\bar{z}), \Pi^\nu(y)\}\big)$$
$$= \delta_\mu^\nu \delta^3(x-y) - \delta_\mu^0 \delta_0^\nu \delta^3(x-y) - \iint d^3z\, d^3\bar{z}\, \delta_\mu^i \delta_j^\nu$$
$$\times ((\nabla_z)_i \delta^3(x-z))(\nabla_z^{-2}\delta^3(z-\bar{z}))((\nabla_{\bar{z}})^j \delta^3(\bar{z}-y))$$
$$= \left((\delta_\mu^\nu - \delta_\mu^0 \delta_0^\nu) + \delta_\mu^i \delta_j^\nu (\nabla_x)_i \frac{1}{\nabla_x^2}(\nabla_x)^j\right) \delta^3(x-y). \tag{10.143}$$

Since we can now set the constraints (10.136) and (10.137) to zero, the relevant Dirac brackets for the Maxwell field theory follow from (10.143) to be

$$\{A_i(x), A_j(y)\}_D = 0 = \{\Pi^i(x), \Pi^j(y)\}_D,$$
$$\{A_i(x), \Pi^j(y)\}_D = \left(\delta_i^j + (\nabla_x)_i \frac{1}{\nabla_x^2}(\nabla_x)^j\right)\delta^3(x-y)$$
$$= \delta_{i\,\text{TR}}^j(x-y), \tag{10.144}$$

which are the relations used in (9.30) (along with the first two of (9.25)) in quantizing Maxwell's theory. Furthermore, setting the constraints (10.136) to zero, we obtain the Hamiltonian for the theory (10.131) to be

$$H_\text{p} = \int \text{d}^3x \, \frac{1}{2} \left(\mathbf{\Pi}^2 + \mathbf{B}^2 \right), \tag{10.145}$$

which is the Hamiltonian we have used in the study of Maxwell's theory in chapter **9**.

10.7 References

1. P. A. M. Dirac, Canadian Journal of Mathematics **2**, 129 (1950); *ibid.* **3**, 1 (1951).

2. J. L. Anderson and P. G. Bergmann, Physical Review **83**, 1018 (1951).

3. P. A. M. Dirac, *Lectures on Quantum Mechanics*, Yeshiva University, New York (1964).

4. A. Hanson, T. Regge and C. Teitelboim, *Constrained Hamiltonian Systems*, Accademia Nazionale dei Lincei, Roma (1976).

5. K. Sundermeyer, *Constrained Dynamics*, Springer-Verlag, Berlin (1982).

6. J. Barcelos-Neto, A. Das and W. Scherer, Acta Physica Polonica **B18**, 269 (1987).

CHAPTER 11
Discrete symmetries

So far we have discussed mainly continuous symmetries of dynamical systems. In addition to continuous symmetries, however, there are a few discrete symmetries which play a fundamental role in physics. By definition, these are not continuous symmetries and, consequently, we cannot talk of an infinitesimal form for such transformations. In this chapter, we will describe three such important symmetries, namely, parity ($\mathcal{P}$), time reversal ($\mathcal{T}$) and charge conjugation ($\mathcal{C}$) where the first two correspond to space-time symmetry transformations while the last is an example of an internal symmetry transformation.

11.1 Parity

The simplest of the discrete transformations is known as parity or space inversion (also known as space reflection or mirror reflection) where one reflects the spatial coordinates through the origin

$$\mathbf{x} \xrightarrow{\mathcal{P}} -\mathbf{x}, \tag{11.1}$$

while t remains unchanged. Therefore, this represents a space-time transformation. Classically, the effect of a parity transformation can be thought of as choosing a left-handed coordinate system as opposed to the conventional right-handed one. Note that (in three space dimensions) space inversion cannot be obtained through any rotation and, therefore, it is not a continuous transformation.

We note that even classically, we can group objects into different categories depending on their behavior under parity or space inversion. It is clear that a vector would change sign in a left-handed coordinate system. Therefore,

$$\mathbf{A} \xrightarrow{P} -\mathbf{A},$$

$$\mathbf{A} \times \mathbf{B} \xrightarrow{P} (-\mathbf{A}) \times (-\mathbf{B}) = \mathbf{A} \times \mathbf{B},$$

$$\mathbf{A} \cdot \mathbf{B} \xrightarrow{P} (-\mathbf{A}) \cdot (-\mathbf{B}) = \mathbf{A} \cdot \mathbf{B},$$

$$\mathbf{A} \cdot (\mathbf{B} \times \mathbf{C}) \xrightarrow{P} (-\mathbf{A}) \cdot ((-\mathbf{B}) \times (-\mathbf{C})) = -\mathbf{A} \cdot (\mathbf{B} \times \mathbf{C}), \quad (11.2)$$

and so on. We see from (11.2) that even though a vector is expected to change sign under a parity transformation, the cross product of two vectors remains invariant and is correspondingly known as a pseudovector or an axial vector. Similarly, while a scalar (from the point of view of rotations, say, for example the length of a vector) is expected to remain unchanged under a parity transformation, we note from (11.2) that the volume (which behaves like a scalar under rotations) changes sign and is known as a pseudoscalar. This characterization carries over to the quantum domain as well. In particular, let us note the transformation properties of some of the well known classical variables under parity,

$$\mathbf{x} \xrightarrow{P} -\mathbf{x},$$

$$\mathbf{p} \xrightarrow{P} -\mathbf{p},$$

$$\mathbf{L} = \mathbf{x} \times \mathbf{p} \xrightarrow{P} (-\mathbf{x}) \times (-\mathbf{p}) = \mathbf{x} \times \mathbf{p} = \mathbf{L},$$

$$\frac{\mathbf{L} \cdot \mathbf{p}}{|\mathbf{p}|} \xrightarrow{P} \frac{\mathbf{L} \cdot (-\mathbf{p})}{|\mathbf{p}|} = -\frac{\mathbf{L} \cdot \mathbf{p}}{|\mathbf{p}|},$$

$$\mathbf{J}(\mathbf{x},t) \xrightarrow{P} -\mathbf{J}(-\mathbf{x},t),$$

$$J^0(\mathbf{x},t) = \rho(\mathbf{x},t) \xrightarrow{P} \rho(-\mathbf{x},t) = J^0(-\mathbf{x},t). \quad (11.3)$$

In quantum mechanics, these variables would, of course, be promoted to operators and they would satisfy the corresponding operator transformation properties.

We note from (11.1) that applying the parity transformation twice, returns the coordinates to their original value, namely,

$$\mathbf{x} \xrightarrow{P} -\mathbf{x} \xrightarrow{P} \mathbf{x}, \quad (11.4)$$

11.1 PARITY

so that that even classically parity operation defines a group with two elements (a group of order two), namely, $\mathbb{1}$ and $\mathcal{P}$ with

$$\mathcal{P}^2 = \mathbb{1}. \tag{11.5}$$

Defining a symmetry in a quantum system, of course, simply corresponds to asking whether this group can be carried over to quantum mechanics without any obstruction. The only difficulty we may apprehend is in the case of spinor states since they are double valued. But a systematic analysis, as we will see, shows that even that does not present any difficulty. Classically, we know that dynamical laws of physics (such as Newton's equation) do not depend on the handedness of the coordinate frame. In other words, the classical laws of physics are invariant under parity or space inversion. For a long time it was believed that the microscopic systems are also invariant under parity, but we know now that there exist processes in nature which do not respect parity.

11.1.1 Parity in quantum mechanics. For simplicity, let us consider a one dimensional quantum mechanical system. In this case, the parity transformation (11.1) would result in

$$\langle X \rangle \xrightarrow{\mathcal{P}} -\langle X \rangle,$$
$$\langle P \rangle \xrightarrow{\mathcal{P}} -\langle P \rangle, \tag{11.6}$$

where we have used the fact that classically vectors change sign under space reflection (see (11.2)) and that expectation values of quantum operators behave like classical objects (Ehrenfest theorem).

As is standard in quantum mechanics, we can analyze the parity transformation in (11.6) from two equivalent points of view. First, let us assume that under the parity transformation, quantum mechanical states change, but not the operators such that (11.6) holds. In this case, let us assume that $\mathcal{P}$ represents the operator which implements the action of parity on the quantum mechanical states. Namely, under space reflection an arbitrary quantum mechanical state transforms as

$$|\psi\rangle \to |\psi^{\mathcal{P}}\rangle = \mathcal{P}|\psi\rangle, \tag{11.7}$$

such that

$$\langle\psi|X|\psi\rangle \xrightarrow{\mathcal{P}} \langle\psi^{\mathcal{P}}|X|\psi^{\mathcal{P}}\rangle = \langle\psi|\mathcal{P}^\dagger X\mathcal{P}|\psi\rangle = -\langle\psi|X|\psi\rangle,$$

$$\langle\psi|P|\psi\rangle \xrightarrow{\mathcal{P}} \langle\psi^{\mathcal{P}}|P|\psi^{\mathcal{P}}\rangle = \langle\psi|\mathcal{P}^\dagger P\mathcal{P}|\psi\rangle = -\langle\psi|P|\psi\rangle. \tag{11.8}$$

Note that parity inverts space coordinates and, therefore, acting on the coordinate basis $\mathcal{P}$ must lead to

$$|x\rangle \xrightarrow{\mathcal{P}} |x^{\mathcal{P}}\rangle = \mathcal{P}|x\rangle = |-x\rangle. \tag{11.9}$$

It is now obvious from (11.9) that

$$\langle x^{\mathcal{P}}|y^{\mathcal{P}}\rangle = \langle x|\mathcal{P}^\dagger\mathcal{P}|y\rangle,$$

or, $\langle -x|-y\rangle = \langle x|\mathcal{P}^\dagger\mathcal{P}|y\rangle,$

or, $\delta(x-y) = \langle x|\mathcal{P}^\dagger\mathcal{P}|y\rangle, \tag{11.10}$

which leads to

$$\mathcal{P}^\dagger\mathcal{P} = \mathbb{1}. \tag{11.11}$$

In other words, the parity operator is unitary. Furthermore, since

$$\mathcal{P}|x\rangle = |-x\rangle,$$
$$\mathcal{P}^2|x\rangle = \mathcal{P}|-x\rangle = |x\rangle, \tag{11.12}$$

which shows that

$$\mathcal{P}^2 = \mathbb{1}. \tag{11.13}$$

Namely, $\mathcal{P}$ is an idempotent operator and the only eigenvalues of P are ± 1. Since $\mathcal{P}$ has real eigenvalues we conclude that the parity operator is Hermitian and thus we have

$$\mathcal{P}^\dagger = \mathcal{P} = \mathcal{P}^{-1},$$
$$\mathcal{P}^2 = \mathbb{1}. \tag{11.14}$$

To see how a scalar (not a multi-component) wave function transforms under parity, we note from (11.7) that

$$\begin{aligned} |\psi\rangle &\xrightarrow{\mathcal{P}} |\psi^\mathcal{P}\rangle = \mathcal{P}|\psi\rangle = \mathcal{P} \int dx\, |x\rangle\langle x|\psi\rangle \\ &= \mathcal{P} \int dx\, \psi(x)|x\rangle = \int dx\, \psi(x)\mathcal{P}|x\rangle \\ &= \int dx\, \psi(x)|-x\rangle, \end{aligned} \tag{11.15}$$

so that we obtain

$$\langle x|\psi^\mathcal{P}\rangle = \psi^\mathcal{P}(x) = \psi(-x). \tag{11.16}$$

Namely, under a parity transformation the wave function transforms as

$$\psi(x) \xrightarrow{\mathcal{P}} \psi^\mathcal{P}(x) = \psi(-x). \tag{11.17}$$

Let us note further that since the eigenvalues of $\mathcal{P}$ are ± 1, if $|\psi\rangle$ is an eigenstate of parity, then

$$\mathcal{P}|\psi\rangle = \pm|\psi\rangle. \tag{11.18}$$

In other words, for such a state

$$|\psi\rangle \xrightarrow{\mathcal{P}} |\psi^\mathcal{P}\rangle = \mathcal{P}|\psi\rangle = \pm|\psi\rangle,$$
$$\text{or,} \quad \psi(x) \xrightarrow{\mathcal{P}} \psi^\mathcal{P}(x) = \psi(-x) = \pm\psi(x). \tag{11.19}$$

Namely, an eigenstate of parity has an associated wave function which is either even or odd depending on the eigenvalue of the parity operator for that state. A general state of the system, however, does not have to be an eigenstate of the parity operator and, therefore, the corresponding wave function need not have any such symmetry behavior.

In the other point of view, we may assume that the states do not transform under the parity transformation, rather under a transformation only the operators change as (this point of view may be more relevant within the context of field theories)

$$\mathcal{O} \xrightarrow{\mathcal{P}} \mathcal{O}^\mathcal{P} = \mathcal{P}^\dagger \mathcal{O} \mathcal{P}, \tag{11.20}$$

such that (11.6) holds, namely,

$$\langle\psi|X|\psi\rangle \xrightarrow{\mathcal{P}} \langle\psi|X^\mathcal{P}|\psi\rangle = -\langle\psi|X|\psi\rangle,$$
$$\langle\psi|P|\psi\rangle \xrightarrow{\mathcal{P}} \langle\psi|P^\mathcal{P}|\psi\rangle = -\langle\psi|P|\psi\rangle. \tag{11.21}$$

It follows from this that

$$X \xrightarrow{\mathcal{P}} X^\mathcal{P} = \mathcal{P}^\dagger X \mathcal{P} = -X,$$
or, $\quad \mathcal{P} X + X \mathcal{P} = 0,$
or, $\quad [\mathcal{P}, X]_+ = 0, \tag{11.22}$

where we have used (11.14). Similarly, we can obtain that

$$[\mathcal{P}, P]_+ = 0. \tag{11.23}$$

In general, one can show that for any operator in this simple one dimensional theory,

$$\begin{aligned}
\mathcal{O}(X, P) \xrightarrow{\mathcal{P}} \mathcal{O}^\mathcal{P}(X, P) &= \mathcal{P}^\dagger \mathcal{O}(X, P) \mathcal{P} \\
&= \mathcal{O}(\mathcal{P}^\dagger X \mathcal{P}, \mathcal{P}^\dagger P \mathcal{P}) \\
&= \mathcal{O}(-X, -P). \tag{11.24}
\end{aligned}$$

11.1 PARITY

We conclude from our general discussion of symmetries (in earlier chapters) that a quantum mechanical theory would be parity invariant if the Hamiltonian of the theory remains invariant under the transformation, namely,

$$H(X,P) \xrightarrow{\mathcal{P}} \mathcal{P}^\dagger H(X,P)\mathcal{P} = H(-X,-P) = H(X,P),$$

or, $\quad [\mathcal{P}, H(X,P)] = 0.$ \hfill (11.25)

As a result, we see that if parity is a symmetry of a quantum mechanical system, $\mathcal{P}$ and H can be simultaneously diagonalized and the eigenstates of the Hamiltonian would carry specific parity quantum numbers. Therefore, in such a case, the eigenstates of the Hamiltonian would naturally decompose into even and odd states. It is obvious from simple systems like the one dimensional harmonic oscillator

$$H(X,P) = \frac{P^2}{2m} + \frac{1}{2}m\omega^2 X^2 = H(-X,-P), \quad (11.26)$$

that this is, indeed, what happens. (Recall that the Hermite polynomials, $H_n(x)$, which are eigenfunctions of the Hamiltonian are even or odd functions of x depending on the value of the index n.)

If the Hamiltonian H of a quantum mechanical system is time independent, the solution of the time dependent Schrödinger equation can be written as

$$|\psi(t)\rangle = e^{-iHt}|\psi(0)\rangle = U(t)|\psi(0)\rangle, \quad (11.27)$$

where $U(t)$ is know as the time evolution operator. If parity is a symmetry of the theory, then as we have seen

$$[\mathcal{P}, H] = 0, \quad (11.28)$$

and it follows from (11.27) that

$$[\mathcal{P}, U(t)] = 0. \quad (11.29)$$

As a result, we conclude that

$$\mathcal{P}|\psi(t)\rangle = \mathcal{P}U(t)|\psi(0)\rangle = U(t)\mathcal{P}|\psi(0)\rangle, \qquad (11.30)$$

so that an even parity state would continue to be an even state under time evolution just as an odd parity state would remain an odd state. This is another way to say that parity is conserved in such a theory.

Before we go on to the discussion of parity in quantum field theories, let us note that in this simple one dimensional quantum theory, we cannot construct any function of X and P which would represent the parity operator. This merely corresponds to the fact that classically there is no dynamical conserved quantity associated with this discrete symmetry. (Recall that Nöther's theorem holds for continuous symmetries.) However, a nontrivial representation of the parity operator is still possible and, in this one dimensional theory, takes the form

$$\mathcal{P} = \int dx \, |-x\rangle\langle x|. \qquad (11.31)$$

It is obvious from this definition that

$$\begin{aligned}
\mathcal{P}^\dagger &= \int dx \, |x\rangle\langle -x| = \int dx \, |-x\rangle\langle x| = \mathcal{P}, \\
\mathcal{P}^2 &= \int dx dy \, |-x\rangle\langle x|-y\rangle\langle y| \\
&= \int dx dy \, \delta(x+y)|-x\rangle\langle y| \\
&= \int dy \, |y\rangle\langle y| = \mathbb{1}, \\
|\psi^\mathcal{P}\rangle &= \mathcal{P}|\psi\rangle = \int dx \, |-x\rangle\langle x|\psi\rangle = \int dx \, \psi(x)|-x\rangle,
\end{aligned}$$

or, $\psi^\mathcal{P}(x) = \langle x|\psi^\mathcal{P}\rangle = \psi(-x),$

$$\mathcal{P}^\dagger X \mathcal{P} = \int dx dy \, |-y\rangle\langle y|X|-x\rangle\langle x|$$

11.1 PARITY

$$\begin{aligned}
&= \int \mathrm{d}x\mathrm{d}y\,(-x\delta(x+y))|-y\rangle\langle x| \\
&= -\int \mathrm{d}x\,x|x\rangle\langle x| = -\int \mathrm{d}x\,X|x\rangle\langle x| \\
&= -X\int \mathrm{d}x\,|x\rangle\langle x| = -X,
\end{aligned}$$

$$\begin{aligned}
\mathcal{P}^\dagger P \mathcal{P} &= \int \mathrm{d}x\mathrm{d}y\,|-y\rangle\langle y|P|-x\rangle\langle x| \\
&= \int \mathrm{d}x\mathrm{d}y\left(-i\frac{d}{dy}\delta(x+y)\right)|-y\rangle\langle x| \\
&= \int \mathrm{d}x\left(-i\frac{d}{\mathrm{d}x}|x\rangle\right)\langle x| = -P\int \mathrm{d}x\,|x\rangle\langle x| \\
&= -P, \quad (11.32)
\end{aligned}$$

so that it truly gives a representation of the parity operator. The important thing to note here is that such a representation is always nonlocal and not very useful from a practical point of view. On the other hand, the properties of the operator $\mathcal{P}$ such as in (11.14) are more important.

Let us now go over to relativistic quantum systems. As we have seen, such systems can be described consistently only in the language of quantized fields which are, in general, multi-component operators (recall the Dirac field or the Maxwell field). Therefore, we need to generalize our discussion of parity transformation to multi-component objects. We note that under a parity transformation a multi-component object would transform as

$$\psi_\alpha(\mathbf{x},t) \xrightarrow{\mathcal{P}} \psi_\alpha^\mathcal{P}(\mathbf{x},t) = \eta_\psi S_\alpha^\beta \psi_\beta(-\mathbf{x},t), \quad (11.33)$$

where the matrix S_α^β can, in principle, mix up the different components of the object. Namely, under a parity transformation, not only does $\mathbf{x} \to -\mathbf{x}$, but the components of the object may mix with one another. We have already seen this in the case of Lorentz transformations (see (3.35) for transformation of wavefunction). Furthermore, from the fact that two space inversions are equivalent to leaving the object unchanged, we are led to the requirement that

$$\eta_\psi^2 S^2 = \mathbb{1}. \tag{11.34}$$

In the language of quantized fields, the above relation represents an operator relation

$$\psi_\alpha(\mathbf{x}, t) \xrightarrow{\mathcal{P}} \psi_\alpha^{\mathcal{P}}(\mathbf{x}, t) = \mathcal{P}^\dagger \psi_\alpha(\mathbf{x}, t)\mathcal{P} = \eta_\psi S_\alpha^\beta \psi_\beta(-\mathbf{x}, t). \tag{11.35}$$

The parameter η_ψ is called the intrinsic parity of the field ψ_α and denotes the parity eigenvalue of the one particle state at rest. (Remember that parity and the momentum operators do not commute and, therefore, cannot have simultaneous eigenstates unless the eigenvalue of momentum vanishes.) Thus we see that η_ψ really measures the intrinsic behavior (and not the space part) of the wave function of a particle under space inversion.

11.1.2 Spin zero field. Let us begin with the spin zero field which is described by the Klein-Gordon equation

$$\Box \phi = \left(\frac{\partial^2}{\partial t^2} - \boldsymbol{\nabla}^2 \right) \phi(x) = 0, \quad \text{or,} \ (\Box + m^2)\phi(x) = 0, \tag{11.36}$$

depending on whether the field is massless or massive. In either of the cases, the equation is clearly invariant under space inversion. Consequently, if $\phi(\mathbf{x}, t)$ is a solution of the Klein-Gordon equation, so is $\phi(-\mathbf{x}, t)$. From our general discussion in (11.34) and (11.35), we conclude that in this case,

$$\phi(\mathbf{x}, t) \xrightarrow{\mathcal{P}} \phi^{\mathcal{P}}(\mathbf{x}, t) = \eta_\phi \phi(-\mathbf{x}, t), \tag{11.37}$$

with

$$\eta_\phi = \pm 1. \tag{11.38}$$

If $\eta_\phi = 1$ or the intrinsic parity of the spin zero field is positive, then

$$\phi^{\mathcal{P}}(\mathbf{x}, t) = \phi(-\mathbf{x}, t), \tag{11.39}$$

11.1 PARITY

and we say that such a field is a scalar field and that the associated particles are scalar particles. On the other hand, if the intrinsic parity of the spin zero field is negative or $\eta_\phi = -1$, then

$$\phi^P(\mathbf{x}, t) = -\phi(-\mathbf{x}, t), \tag{11.40}$$

and we say that such a field is a pseudoscalar field and that the associated particles are pseudoscalar particles (see, for example, (11.2)). In general, however, a field or a state may not have a well defined parity value (consider, for example, a one particle state not at rest). On the other hand, since parity commutes with angular momentum, we can expand such a field (or a state) in the basis of the angular momentum eigenstates, namely, the spherical harmonics as

$$\begin{aligned}\phi(\mathbf{x}, t) &= \sum_\ell \phi_\ell(r, \theta, \varphi, t) = \sum g_\ell(r, t) Y_{\ell,m}(\theta, \varphi), \\ \phi^P(\mathbf{x}, t) &= \sum_\ell \phi_\ell^P(r, \theta, \varphi, t).\end{aligned} \tag{11.41}$$

Furthermore, from the fact that

$$\mathbf{x} \xrightarrow{\mathcal{P}} -\mathbf{x}, \tag{11.42}$$

we have

$$r \xrightarrow{\mathcal{P}} r, \quad \theta \xrightarrow{\mathcal{P}} \pi - \theta, \quad \varphi \xrightarrow{\mathcal{P}} \pi + \varphi,$$
$$Y_{\ell,m}(\theta, \varphi) \xrightarrow{\mathcal{P}} Y_{\ell,m}(\pi - \theta, \pi + \varphi) = (-1)^\ell Y_{\ell,m}(\theta, \varphi), \tag{11.43}$$

so that

$$\phi_\ell^P(\mathbf{x}, t) = \eta_\phi \phi_\ell(-\mathbf{x}, t) = \eta_\phi (-1)^\ell \phi_\ell(\mathbf{x}, t). \tag{11.44}$$

In such a case, therefore, we can define the total parity for the state (with orbital angular momentum ℓ) as

$$\eta_{TOT} = \eta_\phi (-1)^\ell. \tag{11.45}$$

If parity is a symmetry of the theory, the total parity quantum number must be conserved in a physical process and this leads to the fact that for a decay (in the rest frame of A) of a spin zero particle into two spin zero particles

$$A \to B + C, \tag{11.46}$$

we must have

$$\eta_A = \eta_B \eta_C (-1)^\ell, \tag{11.47}$$

where η_A, η_B and η_C are the intrinsic parities of the particles A, B and C respectively and ℓ is the orbital angular momentum of the B-C system. It now follows that in a parity conserving theory, one spin zero particle can decay into two spin zero particles in an s-state ($\ell = 0$) only through the channels

$$\begin{aligned} S &\to S + S, \\ P &\to P + S, \\ S &\to P + P, \end{aligned} \tag{11.48}$$

whereas processes such as ($\ell = 0$)

$$\begin{aligned} S &\to P + S, \\ P &\to S + S, \\ P &\to P + P, \end{aligned} \tag{11.49}$$

will be forbidden.

Let us further note that for a complex spin zero field, if we have

11.1 PARITY

$$\phi^P(\mathbf{x},t) = \eta_\phi \phi(-\mathbf{x},t),$$
$$(\phi^P)^\dagger(\mathbf{x},t) = \eta_\phi^* \phi^\dagger(-\mathbf{x},t). \tag{11.50}$$

On the other hand, since η_ϕ is assumed to take only real values ± 1 (see (11.38)), the intrinsic parity of the field $\phi^\dagger$ is the same as that of the field ϕ. In quantum field theory ϕ and $\phi^\dagger$ are associated with particles and antiparticles. (ϕ destroys a particle or creates an antiparticle whereas $\phi^\dagger$ destroys an antiparticle or creates a particle, see discussion in sections **7.2** and **7.3**.) Thus we conclude that a charged spin zero particle and its antiparticle have the same intrinsic parity. This result is, in fact, quite general for the bosons and says that for bosons, particles and antiparticles would have the same intrinsic parity.

Let us next look at the electromagnetic current associated with a charged Klein-Gordon system. As we have seen in (7.29), the current density has the explicit form

$$J^\mu(\mathbf{x},t) = i\left(\phi^\dagger(\mathbf{x},t)\partial^\mu\phi(\mathbf{x},t) - \left(\partial^\mu\phi^\dagger(\mathbf{x},t)\right)\phi(\mathbf{x},t)\right). \tag{11.51}$$

Under a parity transformation

$$J^\mu(\mathbf{x},t) \xrightarrow{P} J^{P\mu}(\mathbf{x},t)$$
$$= i\left((\phi^P)^\dagger(\mathbf{x},t)\partial^\mu\phi^P(\mathbf{x},t) - (\partial^\mu(\phi^P)^\dagger(\mathbf{x},t))\phi^P(\mathbf{x},t)\right)$$
$$= i|\eta_\phi|^2\Big(\phi^\dagger(-\mathbf{x},t)\partial^\mu\phi(-\mathbf{x},t) - (\partial^\mu\phi^\dagger(-\mathbf{x},t))\phi(\mathbf{x},t)\Big), \tag{11.52}$$

which leads explicitly to (recall that $|\eta_\phi|^2 = 1$)

$$\mathbf{J}^P(\mathbf{x},t) = -\mathbf{J}(\mathbf{x},t), \quad J^{P0}(\mathbf{x},t) = J^0(-\mathbf{x},t), \tag{11.53}$$

which is the correct transformation for the current four vector as we have seen in (11.2).

11.1.3 Photon field.
From Maxwell's equations in the presence of sources (charges and currents)

$$\nabla \cdot \mathbf{E} = \rho = J^0,$$
$$\nabla \times \mathbf{B} = \frac{\partial \mathbf{E}}{\partial t} + \mathbf{J}, \tag{11.54}$$

we note that invariance of (11.54) under space reflection requires that under a parity transformation, we must have

$$\mathbf{E}(\mathbf{x}, t) \xrightarrow{\mathcal{P}} -\mathbf{E}(-\mathbf{x}, t),$$
$$\mathbf{B}(\mathbf{x}, t) \xrightarrow{\mathcal{P}} \mathbf{B}(-\mathbf{x}, t). \tag{11.55}$$

Namely, we see that whereas the electric field would transform like a vector, the magnetic field would behave like an axial vector if parity is a symmetry of the system. Furthermore, from the definitions of the electric and the magnetic fields in (9.2), namely,

$$\mathbf{E} = -\frac{\partial \mathbf{A}}{\partial t} - \nabla \phi = -\frac{\partial \mathbf{A}}{\partial t} - \nabla A^0,$$
$$\mathbf{B} = \nabla \times \mathbf{A}, \tag{11.56}$$

we conclude that under a parity transformation the components of the four vector potential would transform as

$$\mathbf{A}(\mathbf{x}, t) \xrightarrow{\mathcal{P}} -\mathbf{A}(-\mathbf{x}, t),$$
$$A^0(\mathbf{x}, t) \xrightarrow{\mathcal{P}} A^0(-\mathbf{x}, t). \tag{11.57}$$

This is very much like the transformations (11.53) of the components of the current four vector under a space reflection. In the quantum theory, we would then generalize these as operator transformations

$$\mathbf{A}(\mathbf{x}, t) \xrightarrow{\mathcal{P}} \mathbf{A}^{\mathcal{P}}(\mathbf{x}, t) = \eta_\mathbf{A} \mathbf{A}(-\mathbf{x}, t) = -\mathbf{A}(-\mathbf{x}, t),$$
$$A^0(\mathbf{x}, t) \xrightarrow{\mathcal{P}} A^{\mathcal{P} 0}(\mathbf{x}, t) = \eta_{A^0} A^0(-\mathbf{x}, t) = A^0(-\mathbf{x}, t). \tag{11.58}$$

11.1 PARITY

Note that a physical photon has only transverse degrees of freedom (namely, A^0 is nondynamical). From the parity transformation of the dynamical components, namely $\mathbf{A}$, we see that

$$\mathbf{A}^P(\mathbf{x}, t) = -\mathbf{A}(-\mathbf{x}, t), \tag{11.59}$$

from which we conclude that the intrinsic parity of the photon field (or the one photon state) is negative. In other words, the photon is a vector particle. From the form of the electromagnetic interaction Hamiltonian

$$H_{em} = \int d^3x \; J^\mu(\mathbf{x}, t) A_\mu(\mathbf{x}, t), \tag{11.60}$$

we note that under a parity transformation

$$\begin{aligned}
H_{em} &= \int d^3x \; J^\mu(\mathbf{x}, t) A_\mu(\mathbf{x}, t) \\
&\xrightarrow{\mathcal{P}} \int d^3x \; J^{P\mu}(\mathbf{x}, t) A_\mu^P(\mathbf{x}, t) \\
&= \int d^3x \left(J^{P0}(\mathbf{x}, t) A^{P0}(\mathbf{x}, t) - \mathbf{J}^P(\mathbf{x}, t) \cdot \mathbf{A}^P(\mathbf{x}, t) \right) \\
&= \int d^3x \left(J^0(-\mathbf{x}, t) A^0(-\mathbf{x}, t) - (-\mathbf{J}(\mathbf{x}, t)) \cdot (-\mathbf{A}(\mathbf{x}, t)) \right) \\
&= \int d^3x \; J^\mu(-\mathbf{x}, t) A_\mu(-\mathbf{x}, t) \\
&= \int d^3x \; J^\mu(\mathbf{x}, t) A_\mu(\mathbf{x}, t) = H_{em}, \tag{11.61}
\end{aligned}$$

where in the last step we have let $\mathbf{x} \to -\mathbf{x}$ under the integral. This shows that the electromagnetic interaction is invariant under parity and, therefore, parity (quantum number) must be conserved in electromagnetic processes.

11.1.4 Dirac field. Let us recall that the Dirac wave function or the Dirac field is described by a four component object. Thus according to our general discussion in (11.33) or (11.35), under a parity transformation

$$\psi_\alpha(\mathbf{x},t) \xrightarrow{\mathcal{P}} \psi_\alpha^\mathcal{P}(\mathbf{x},t) = \eta_\psi S_\alpha^\beta \psi_\beta(-\mathbf{x},t), \quad \alpha = 1,2,3,4. \tag{11.62}$$

In matrix notation, we can write the transformation as

$$\psi^\mathcal{P}(\mathbf{x},t) = \eta_\psi S \psi(-\mathbf{x},t), \tag{11.63}$$

with (see (11.34))

$$\eta_\psi^2 S^2 = \mathbb{1}. \tag{11.64}$$

Since we have assumed the parity eigenvalues to be ± 1, we can always choose

$$\eta_\psi = \pm 1, \quad S^2 = \mathbb{1}, \tag{11.65}$$

which will satisfy the above relation. With this choice (which also implies $\eta_\psi^* = \eta_\psi$) we note that

$$\begin{aligned}
\overline{\psi}^\mathcal{P}(\mathbf{x},t) &= (\psi^\mathcal{P})^\dagger(\mathbf{x},t)\gamma^0 = \eta_\psi^* \psi^\dagger(-\mathbf{x},t) S^\dagger \gamma^0 \\
&= \eta_\psi^* \psi^\dagger(-\mathbf{x},t)\gamma^0 \gamma^0 S^\dagger \gamma^0 \\
&= \eta_\psi \overline{\psi}(-\mathbf{x},t)\gamma^0 S^\dagger \gamma^0.
\end{aligned} \tag{11.66}$$

Let us recall that associated with the Dirac system is a conserved current (probability current or electromagnetic current, see (8.44))

$$J^\mu(\mathbf{x},t) = \overline{\psi}(\mathbf{x},t)\gamma^\mu \psi(\mathbf{x},t), \tag{11.67}$$

and under a parity transformation this would transform as

$$\begin{aligned}
J^\mu(\mathbf{x},t) \xrightarrow{\mathcal{P}} J^{\mathcal{P}\mu}(\mathbf{x},t) &= \overline{\psi}^\mathcal{P}(\mathbf{x},t)\gamma^\mu \psi^\mathcal{P}(\mathbf{x},t) \\
&= \eta_\psi^2 \overline{\psi}(-\mathbf{x},t)\gamma^0 S^\dagger \gamma^0 \gamma^\mu S \psi(-\mathbf{x},t) \\
&= \overline{\psi}(-\mathbf{x},t)\gamma^0 S^\dagger \gamma^0 \gamma^\mu S \psi(-\mathbf{x},t),
\end{aligned} \tag{11.68}$$

11.1 PARITY

where we have used the fact that $\eta_\psi^2 = 1$. On the other hand, we know the transformation properties of the current four vector under space inversion (see (11.53)), namely,

$$J^{P0}(\mathbf{x},t) = J^0(-\mathbf{x},t), \quad J^{Pi}(\mathbf{x},t) = -J^i(-\mathbf{x},t). \tag{11.69}$$

Comparing (11.68) and (11.69) for $\mu = 0$, we obtain

$$\overline{\psi}(-\mathbf{x},t)\gamma^0 S^\dagger S \psi(-\mathbf{x},t) = \overline{\psi}(-\mathbf{x},t)\gamma^0 \psi(-\mathbf{x},t),$$
$$\text{or,} \quad S^\dagger S = \mathbb{1}, \tag{11.70}$$

so that using (11.65) we have

$$S = S^\dagger = S^{-1}. \tag{11.71}$$

Similarly, comparing (11.68) and (11.69) for $\mu = i$, we obtain

$$\overline{\psi}(-\mathbf{x},t)\gamma^0 S^\dagger \gamma^0 \gamma^i S \psi(-\mathbf{x},t) = -\overline{\psi}(-\mathbf{x},t)\gamma^i \psi(-\mathbf{x},t),$$
$$\text{or,} \quad S = \gamma^0. \tag{11.72}$$

Thus we determine the transformation of the Dirac field (wave function) under a space reflection to be

$$\psi(\mathbf{x},t) \xrightarrow{\mathcal{P}} \psi^P(\mathbf{x},t) = \eta_\psi \gamma^0 \psi(-\mathbf{x},t),$$
$$\overline{\psi}(\mathbf{x},t) \xrightarrow{\mathcal{P}} \overline{\psi}^P(\mathbf{x},t) = \eta_\psi \overline{\psi}(-\mathbf{x},t)\gamma^0, \tag{11.73}$$

with

$$\eta_\psi = \pm 1. \tag{11.74}$$

It is now quite easy to see that if $\psi(\mathbf{x},t)$ is a solution of the Dirac equation, then the parity transformed function $\psi^P(\mathbf{x},t)$ also satisfies the Dirac equation in the new coordinates. Namely, if we define

$$y^\mu = (t, -\mathbf{x}),\tag{11.75}$$

then

$$\begin{aligned}\left(i\gamma^\mu \frac{\partial}{\partial y^\mu} - m\right)\psi^\mathcal{P}(y) &= \left(i\gamma^0 \frac{\partial}{\partial t} - i\gamma^i \frac{\partial}{\partial x^i} - m\right)\eta_\psi \gamma^0 \psi(x) \\ &= \eta_\psi \gamma^0 \left(i\gamma^0 \frac{\partial}{\partial t} + i\gamma^i \frac{\partial}{\partial x^i} - m\right)\psi(x) \\ &= \eta_\psi \gamma^0 \left(i\gamma^\mu \frac{\partial}{\partial x^\mu} - m\right)\psi(x) = 0.\end{aligned}\tag{11.76}$$

In other words, space inversion is a symmetry of the Dirac equation. This can also be seen by noting that parity is a symmetry of the free Dirac Lagrangian density (8.9). However, as we have discussed in section **3.6**, the two component Weyl fermions violate $\mathcal{P}$.

To determine the relative parity between Dirac particles and antiparticles, let us next analyze the decay of the positronium in the $^1S_0$ state where we use the standard spectroscopic notation $^{2S+1}L_J$ for the states (the positronium is a bound state of an electron and a positron much like the Hydrogen with a ground state energy equal to -6.8 eV)

$$e^- + e^+ \to \gamma + \gamma.\tag{11.77}$$

In the rest frame of the positronium, the two photons move with equal and opposite momentum. Let ϵ_1 and ϵ_2 denote the polarization vectors for the two photons while $\mathbf{k}$ represents their relative momentum. Since the electromagnetic interactions conserve parity, the scalar transition amplitude will have the general form

$$M_{i \to f} = (a\,\epsilon_1 \cdot \epsilon_2 + b\,\mathbf{k} \cdot (\epsilon_1 \times \epsilon_2))\,\psi_{\text{positronium}},\tag{11.78}$$

where 'a' and 'b' are scalar functions of momentum and $\psi_{\text{positronium}}$ represents the wave function for the positronium. The polarization planes of the photons in this decay can, in fact, be measured and it

11.1 PARITY

is experimentally observed that the two are perpendicular. Namely, for this decay we have

$$\epsilon_1 \cdot \epsilon_2 = 0. \tag{11.79}$$

Consequently, the transition amplitude for this decay takes the form

$$M_{i \to f} = (b\,\mathbf{k} \cdot (\epsilon_1 \times \epsilon_2))\,\psi_{\text{positronium}}. \tag{11.80}$$

Since the transition amplitude is a scalar under parity while

$$\mathbf{k} \cdot (\epsilon_1 \times \epsilon_2) \xrightarrow{\mathcal{P}} -\mathbf{k} \cdot (\epsilon_1 \times \epsilon_2), \tag{11.81}$$

we conclude that the positronium state must be a pseudoscalar under parity. On the other hand, in the rest frame of the positronium, the total parity of the state is given by

$$\eta_{\text{TOT}} = \eta_{e^-}\eta_{e^+}, \tag{11.82}$$

and for this state to represent a pseudoscalar state, we must have

$$\eta_{\text{TOT}} = \eta_{e^-}\eta_{e^+} = -1. \tag{11.83}$$

In other words, the electron and the positron must have relative negative intrinsic parity. This is, in fact, a very general result for fermions, namely, the relative intrinsic parity between Dirac particles and antiparticles is negative.

This result can be more directly seen when we study charge conjugation in the next section. However, it can also be seen in an intuitive manner from the structure of the spinor functions as follows. Let us recall from our earlier discussions that antiparticles are related to negative energy states. Let us choose two representative solutions (of positive and negative energy) of the Dirac equation in the rest frame (see (2.49))

$$\psi_{p^0>0} = \begin{pmatrix} 1 \\ 0 \\ 0 \\ 0 \end{pmatrix} \xrightarrow{P} \eta\gamma^0 \begin{pmatrix} 1 \\ 0 \\ 0 \\ 0 \end{pmatrix} = \eta \begin{pmatrix} 1 \\ 0 \\ 0 \\ 0 \end{pmatrix},$$

$$\psi_{p^0<0} = \begin{pmatrix} 0 \\ 0 \\ 1 \\ 0 \end{pmatrix} \xrightarrow{P} \eta\gamma^0 \begin{pmatrix} 0 \\ 0 \\ 1 \\ 0 \end{pmatrix} = -\eta \begin{pmatrix} 0 \\ 0 \\ 1 \\ 0 \end{pmatrix}. \qquad (11.84)$$

This analysis also brings out another feature involving the fermions, namely, since fermions always appear in pairs in any process (because of conservation of fermion number, angular momentum), the only meaningful quantity as far as the fermions are concerned is the relative parity of a pair of fermions.

Let us next calculate the transformation properties of the sixteen Dirac bilinears

$$\overline{\psi}\Gamma^{(\alpha)}\psi, \qquad (11.85)$$

introduced in (2.100) (or in section **3.3**) under parity. For example, we note that

$$\overline{\psi}(\mathbf{x},t)\psi(\mathbf{x},t) \xrightarrow{P} \overline{\psi}^P(\mathbf{x},t)\psi^P(\mathbf{x},t)$$
$$= |\eta|^2 \overline{\psi}(-\mathbf{x},t)\gamma^0\gamma^0\psi(-\mathbf{x},t)$$
$$= \overline{\psi}(-\mathbf{x},t)\psi(-\mathbf{x},t), \qquad (11.86)$$

which shows that this combination behaves like a scalar under parity. On the other hand,

$$\overline{\psi}(\mathbf{x},t)\gamma_5\psi(\mathbf{x},t) \xrightarrow{P} \overline{\psi}^P(\mathbf{x},t)\gamma_5\psi^P(\mathbf{x},t)$$
$$= |\eta|^2 \overline{\psi}(-\mathbf{x},t)\gamma^0\gamma_5\gamma^0\psi(-\mathbf{x},t)$$
$$= -\overline{\psi}(-\mathbf{x},t)\gamma_5\psi(-\mathbf{x},t), \qquad (11.87)$$

11.1 PARITY

so that this combination behaves like a pseudoscalar under parity. We have already seen that the combination $\overline{\psi}\gamma^\mu\psi$ behaves like a four vector, namely,

$$\overline{\psi}(\mathbf{x},t)\gamma^i\psi(\mathbf{x},t) \xrightarrow{P} -\overline{\psi}(-\mathbf{x},t)\gamma^i\psi(-\mathbf{x},t),$$

$$\overline{\psi}(\mathbf{x},t)\gamma^0\psi(\mathbf{x},t) \xrightarrow{P} \overline{\psi}(-\mathbf{x},t)\gamma^0\psi(-\mathbf{x},t). \tag{11.88}$$

It can similarly be checked that

$$\overline{\psi}(\mathbf{x},t)\gamma_5\gamma^i\psi(\mathbf{x},t) \xrightarrow{P} \overline{\psi}(-\mathbf{x},t)\gamma_5\gamma^i\psi(-\mathbf{x},t),$$

$$\overline{\psi}(\mathbf{x},t)\gamma_5\gamma^0\psi(\mathbf{x},t) \xrightarrow{P} -\overline{\psi}(-\mathbf{x},t)\gamma_5\gamma^0\psi(-\mathbf{x},t). \tag{11.89}$$

Namely, the combination $\overline{\psi}\gamma_5\gamma^\mu\psi$ behaves in an opposite way from the current four vector and consequently, it is known as an axial vector (or a pseudovector). Finally, we can also derive that

$$\overline{\psi}(\mathbf{x},t)\sigma^{0i}\psi(\mathbf{x},t) \xrightarrow{P} -\overline{\psi}(-\mathbf{x},t)\sigma^{0i}\psi(-\mathbf{x},t),$$

$$\overline{\psi}(\mathbf{x},t)\sigma^{ij}\psi(\mathbf{x},t) \xrightarrow{P} \overline{\psi}(-\mathbf{x},t)\sigma^{ij}\psi(-\mathbf{x},t), \tag{11.90}$$

so that the combination $\overline{\psi}\sigma^{\mu\nu}\psi$ behaves exactly like the field strength tensor $F^{\mu\nu}$ under space inversion and hence is known as a tensor.

The transformation properties of the bilinears are important in constructing relativistic theories. Thus, for example, it was thought sometime ago that the strong force between the nucleons was mediated through the Yukawa interaction of spin zero mesons. However, there are two possible Lorentz invariant interactions we can write down involving fermions and spin zero particles, namely

$$\overline{\psi}\psi\phi, \quad \text{or} \quad \overline{\psi}\gamma_5\psi\phi. \tag{11.91}$$

Since strong interactions conserve parity, only one of the forms of the interaction can be allowed depending on the intrinsic parity of the spin zero meson. As it turns out, the spin zero meson can be identified with the π-mesons which are known to be pseudoscalar mesons. Therefore, the only allowed interaction in this case has the form (recall also the discussion in section **8.8** after (8.92))

$$\bar{\psi}\gamma_5\psi\phi. \tag{11.92}$$

Conversely, if we know the interaction Lagrangian or Hamiltonian density for a given process, we can determine its behavior under a parity transformation which would then tell us whether parity will be conserved in processes mediated through such interactions.

11.2 Charge conjugation

To understand the discrete symmetry known as charge conjugation, let us go back to Maxwell's equations in the presence of charges and currents

$$\begin{aligned}
\nabla \cdot \mathbf{E} &= \rho, \\
\nabla \cdot \mathbf{B} &= 0, \\
\nabla \times \mathbf{E} &= -\frac{\partial \mathbf{B}}{\partial t}, \\
\nabla \times \mathbf{B} &= \frac{\partial \mathbf{E}}{\partial t} + \mathbf{J}.
\end{aligned} \tag{11.93}$$

We note that this set of equations is invariant under the discrete transformations

$$\rho \to -\rho, \quad \mathbf{J} \to -\mathbf{J}, \quad \mathbf{E} \to -\mathbf{E}, \quad \mathbf{B} \to -\mathbf{B}. \tag{11.94}$$

This discrete transformation can be thought of as charge reflection or charge conjugation (simply because reflecting the sign of the charge generates the transformations in (11.94)) and is denoted by $\mathcal{C}$. Furthermore, since

$$\begin{aligned}
\mathbf{E} &= -\frac{\partial \mathbf{A}}{\partial t} - \nabla A^0, \\
\mathbf{B} &= \nabla \times \mathbf{A},
\end{aligned} \tag{11.95}$$

classically we see that charge conjugation leads to

$$A_\mu \xrightarrow{\mathcal{C}} -A_\mu, \quad J_\mu \xrightarrow{\mathcal{C}} -J_\mu. \tag{11.96}$$

It is obvious that the electromagnetic interaction Hamiltonian

$$H_{em} = \int d^3x \ J^\mu A_\mu, \tag{11.97}$$

is invariant under this transformation. Let us also note here that like parity, two charge conjugation operations (reflecting the charge twice) would bring back any variable to itself.

In the quantum theory, we can generalize the classical result (11.96) by saying that

$$\begin{aligned} A_\mu(x) &\xrightarrow{\mathcal{C}} A_\mu^c(x) = \eta_A A_\mu(x) = -A_\mu(x), \\ J_\mu(x) &\xrightarrow{\mathcal{C}} J_\mu^c(x) = \eta_J J_\mu(x) = -J_\mu(x), \end{aligned} \tag{11.98}$$

and we say that the charge conjugation parity (or simply the charge parity) of the photon as well as the current is -1. Note that the transformation of charge conjugation does not change the space-time coordinates and, therefore, it is not a space-time transformation.

Let us also note that in a quantum theory, opposite charges are assigned to particles and antiparticles. Therefore, reflecting charges in a quantum theory is equivalent to saying that charge conjugation really interchanges particles and antiparticles. In other words, in a quantum theory, charge conjugation really interchanges every distinct quantum number between particles and antiparticles. For this reason, charge conjugation is sometimes also referred to as particle-antiparticle conjugation.

11.2.1 Spin zero field. Let us now analyze the complex Klein-Gordon field theory which describes charged spin zero particles. (As we have seen, a real Klein-Gordon theory describes charge neutral spin zero particles and, therefore, charge conjugation is trivial in this theory as we will discuss shortly.) We have seen that for such a theory we can define an electromagnetic current (7.29) as

$$J^\mu = i\left(\phi^\dagger \partial^\mu \phi - (\partial^\mu \phi^\dagger)\phi\right). \tag{11.99}$$

We note that if we treat $\phi(x)$ as a classical function (even though the functions are classical, we continue to use "†" instead of "∗" so that the passage to quantum theory will be straightforward), then

$$\phi(x) \xrightarrow{\mathcal{C}} \phi^c(x) = \eta_\phi \phi^\dagger(x),$$
$$\phi^\dagger(x) \xrightarrow{\mathcal{C}} (\phi^c)^\dagger(x) = \eta_\phi^* \phi(x), \tag{11.100}$$

with $|\eta_\phi|^2 = 1$, would define a transformation under which

$$\begin{aligned} J^\mu(x) &\xrightarrow{\mathcal{C}} i\left((\phi^c)^\dagger \partial^\mu \phi^c - (\partial^\mu(\phi^c)^\dagger)\phi^c\right) \\ &= i|\eta_\phi|^2\left(\phi \partial^\mu \phi^\dagger - (\partial^\mu \phi)\phi^\dagger\right) \\ &= -i\left(\phi^\dagger \partial^\mu \phi - (\partial^\mu \phi^\dagger)\phi\right) = -J^\mu(x). \end{aligned} \tag{11.101}$$

Thus the field transformations in (11.100) would define the appropriate charge conjugation transformation for the current. We have already discussed (see sections **7.2** and **7.3**) that $\phi(x)$ and $\phi^\dagger(x)$ can be thought of as being associated with particles and antiparticles (ϕ annihilates a particle/creates an antiparticle while $\phi^\dagger$ annihilates an antiparticle/creates a particle). Therefore, this further supports the fact that (11.100) is the appropriate charge conjugation transformation for complex scalar field.

As a result, under a charge conjugation, we have ($|\eta_\phi|^2 = 1$)

$$\begin{aligned} \phi(x) &\xrightarrow{\mathcal{C}} \eta_\phi \phi^\dagger(x), \\ \phi^\dagger(x) &\xrightarrow{\mathcal{C}} \eta_\phi^* \phi(x), \\ A_\mu(x) &\xrightarrow{\mathcal{C}} -A_\mu(x). \end{aligned} \tag{11.102}$$

It follows now that under such a transformation,

11.2 CHARGE CONJUGATION

$$F_{\mu\nu}(x) \xrightarrow{\mathcal{C}} -F_{\mu\nu}(x),$$

$$D_\mu \phi(x) = (\partial_\mu + ieA_\mu)\phi(x) \xrightarrow{\mathcal{C}} \eta_\phi (\partial_\mu - ieA_\mu)\phi^\dagger = \eta_\phi (D_\mu \phi)^\dagger,$$

$$(D_\mu \phi(x))^\dagger = (\partial_\mu - ieA_\mu)\phi^\dagger(x) \xrightarrow{\mathcal{C}} \eta_\phi^* (\partial_\mu + ieA_\mu)\phi = \eta_\phi^* D_\mu \phi.$$
(11.103)

Therefore, the Lagrangian density of charged, spin zero particles interacting with photons described by

$$\mathcal{L} = (D^\mu \phi)^\dagger (D_\mu \phi) - m^2 \phi^\dagger \phi - \frac{1}{4} F_{\mu\nu} F^{\mu\nu}, \tag{11.104}$$

is invariant under charge conjugation.

All of the above discussion carries over to the quantum theory if we are careful about the operator ordering of ϕ and $\phi^\dagger$. For example, we note that if we define the electromagnetic current in the quantum theory as in (11.99), then under the transformation (11.100) we would have

$$\begin{aligned} J^\mu(x) &= i\left(\phi^\dagger \partial^\mu \phi - (\partial^\mu \phi^\dagger)\phi\right) \xrightarrow{\mathcal{C}} J^{C\mu}(x) \\ &= i|\eta_\phi|^2 \left(\phi \partial^\mu \phi^\dagger - (\partial^\mu \phi)\phi^\dagger\right) \\ &= -i\left((\partial^\mu \phi)\phi^\dagger - \phi \partial^\mu \phi^\dagger\right) \\ &\neq -i\left(\phi^\dagger \partial^\mu \phi - (\partial^\mu \phi^\dagger)\phi\right), \end{aligned} \tag{11.105}$$

since the operators ϕ and $\phi^\dagger$ do not commute with $\partial_\mu \phi^\dagger$ and $\partial_\mu \phi$ respectively. As a result, it would appear as if the field transformations (11.100) do not lead to the correct transformation properties for the current operator. The problem, of course, can be traced to the fact that in going from a classical theory to the quantum theory, we have to choose an operator ordering which we have not taken care of in the above discussion. As is known, the operator ordering which works for the current involving a product of bosonic operators is to symmetrize the product (this can also be checked to be equivalent to normal ordering the product). If we symmetrize the products in the definition of the current, we obtain

$$J^\mu_{\text{sym}}(x) = \frac{i}{2} \left(\phi^\dagger \partial^\mu \phi + (\partial^\mu \phi)\phi^\dagger - (\partial^\mu \phi^\dagger)\phi - \phi \partial^\mu \phi^\dagger \right), \quad (11.106)$$

and it is easy to check that this current will transform properly (namely, will change sign) under charge conjugation. Let us note here that the symmetrized current differs from the original current merely by a constant which is infinite (this is the value of the commutator). But this has the interesting consequence that while (remember the Dirac vacuum picture)

$$\langle 0|J^\mu(x)|0\rangle \longrightarrow \infty, \quad (11.107)$$

the effect of the infinite constant in J^μ_{sym} in (11.106) leads to (this should be compared with the normal ordered charge defined in section **7.3**)

$$\langle 0|J^\mu_{\text{sym}}(x)|0\rangle = 0. \quad (11.108)$$

This is esthetically more pleasing and treats the vacuum in a more symmetrical way. Note that we can also write the symmetrized current as

$$J^\mu_{\text{sym}}(x) = \frac{1}{2} \left(J^\mu(x) - J^{c\mu}(x) \right), \quad (11.109)$$

which brings out the particle-antiparticle symmetry in a more direct way.

As we have discussed earlier, if a spin-zero field theory describes charge neutral particles which are their own antiparticles, it is described by a Hermitian field operator

$$\phi^\dagger(x) = \phi(x). \quad (11.110)$$

In such a case, under a charge conjugation

11.2 CHARGE CONJUGATION

$$\phi(x) \xrightarrow{\mathcal{C}} \phi^c(x) = \eta_\phi \phi^\dagger(x) = \eta_\phi \phi(x). \tag{11.111}$$

From the fact that two charge conjugation operations should return the field to itself, we expect

$$\eta_\phi^2 = 1,$$
$$\text{or,} \quad \eta_\phi = \pm 1. \tag{11.112}$$

In this case, the sign of the charge parity cannot be fixed from theory. But by analyzing various physical processes where charge conjugation is a symmetry or by analyzing the invariance of various interaction terms under charge conjugation, one can show that for all Hermitian spin zero fields that we know of (scalar or pseudoscalar)

$$\eta_\phi = 1, \tag{11.113}$$

so that in these cases

$$\phi(x) \xrightarrow{\mathcal{C}} \phi^c(x) = \phi(x). \tag{11.114}$$

Such fields or particles are known as self-charge-conjugate and simply describe particles that are their own antiparticles. As an example, let us note that in the decay

$$\pi^0 \to 2\gamma, \tag{11.115}$$

since the charge parity of the photon is (-1), the total charge parity of the final state is $(+1)$. If charge parity is to be conserved in this process (this corresponds to an electromagnetic process since photons are involved), we conclude that the π^0 meson must also have charge parity $(+1)$ and similar arguments hold for other mesons.

11.2.2 Dirac field.

Let us next analyze a system of Dirac fermions interacting with an electromagnetic field. If the fermions are minimally coupled to the photons, the equation of motion takes the form

$$(i\gamma^\mu (\partial_\mu + ieA_\mu) - m)\psi(x) = 0. \tag{11.116}$$

The adjoint equation is easily obtained to correspond to

$$\overline{\psi}(x)\left(i\gamma^\mu \left(\overleftarrow{\partial}_\mu - ieA_\mu\right) + m\right) = 0. \tag{11.117}$$

Taking the transpose of the adjoint equation (11.117), we obtain

$$\left(i(\gamma^\mu)^T (\partial_\mu - ieA_\mu) + m\right)\overline{\psi}^T = 0. \tag{11.118}$$

Let us recall that a positron which is the antiparticle of the electron is also a Dirac particle with the same mass but carries an opposite electric charge. Thus, interacting with an electromagnetic field, it would satisfy the equation

$$(i\gamma^\mu (\partial_\mu - ieA_\mu) - m)\psi^c(x) = 0, \tag{11.119}$$

where we have identified the charge conjugate function ψ^c as describing the positron. To see the relation between $\psi^c(x)$ and $\psi(x)$, let us note from (11.118) and (11.119) that the two equations are very similar in structure. In fact, we note that if we can find a matrix C satisfying

$$C(\gamma^\mu)^T C^{-1} = -\gamma^\mu, \tag{11.120}$$

from (11.118) we would have

$$C\left(i(\gamma^\mu)^T (\partial_\mu - ieA_\mu) + m\right)\overline{\psi}^T = 0,$$

or, $\quad \left(iC(\gamma^\mu)^T C^{-1} (\partial_\mu - ieA_\mu) + m\right) C\overline{\psi}^T = 0,$

or, $\quad (i\gamma^\mu (\partial_\mu - ieA_\mu) - m) C\overline{\psi}^T = 0, \tag{11.121}$

11.2 Charge Conjugation

which can be compared with (11.119). Therefore we conclude that if we can find a matrix C with the property

$$\begin{aligned} C(\gamma^\mu)^T C^{-1} &= -\gamma^\mu, \\ \text{or,} \quad C^{-1}\gamma^\mu C &= -(\gamma^\mu)^T, \end{aligned} \qquad (11.122)$$

we can identify (with $|\eta_\psi|^2 = 1$)

$$\psi^c(x) = \eta_\psi C \overline{\psi}^T. \qquad (11.123)$$

That such a matrix exists can be deduced from the fact that $-(\gamma^\mu)^T$ also satisfies the Clifford algebra, namely,

$$\left[-(\gamma^\mu)^T, -(\gamma^\mu)^T\right]_+ = 2\eta^{\mu\nu}\mathbb{1}, \qquad (11.124)$$

and hence from the generalized Pauli theorem (see (1.92)), it follows that there must exist a similarity transformation which relates $-(\gamma^\mu)^T$ and γ^μ. The important symmetry property that C must have is that it should be antisymmetric. (This property depends on the dimensionality of space-time. In 4-dimension $C^T = -C$.) To see this, let us assume that instead C is symmetric, namely,

$$C^T = C. \qquad (11.125)$$

In this case, from

$$C^{-1}\gamma^\mu C = -(\gamma^\mu)^T, \qquad (11.126)$$

we obtain

$$\gamma^\mu C = -C(\gamma^\mu)^T = -C^T(\gamma^\mu)^T = -(\gamma^\mu C)^T. \qquad (11.127)$$

Similarly, we can also show that

$$\sigma^{\mu\nu} C = -(\sigma^{\mu\nu} C)^T, \qquad (11.128)$$

so that if $C^T = C$, we can construct 10 linearly independent antisymmetric matrices. But these are 4×4 matrices and, therefore, there can be only six linearly independent antisymmetric matrices and C cannot be symmetric. Therefore, it follows that C must be antisymmetric,

$$C^T = -C. \tag{11.129}$$

Unlike the parity transformation, the structure of the matrix C depends on the representation of the Dirac matrices and, in the Pauli-Dirac representation that we have been using, it is easy to determine that

$$C = \gamma^0 \gamma^2, \tag{11.130}$$

so that it satisfies (it can be verified explicitly using the properties of the γ^μ matrices in our representation)

$$C = C^\dagger = C^{-1} = -C^T. \tag{11.131}$$

It now follows that

$$\psi^c = \eta_\psi C \overline{\psi}^T = \eta_\psi \gamma^0 \gamma^2 (\gamma^0)^T \psi^* = -\eta_\psi \gamma^2 \psi^*. \tag{11.132}$$

In general, however, if we do not restrict to any particular representation, we have (with $|\eta_\psi|^2 = 1, C^\dagger C = 1$)

$$\begin{aligned}\psi(x) &\xrightarrow{\mathcal{C}} \psi^c(x) = \eta_\psi C \overline{\psi}^T, \\ \overline{\psi}(x) &\longrightarrow \overline{\psi}^c(x) = -\eta_\psi^* \psi^T C^{-1}.\end{aligned} \tag{11.133}$$

The fermion fields, as quantum mechanical operators, satisfy anticommutation relations to reflect the Fermi-Dirac statistics that the underlying particles obey. In this case, the appropriate operator ordering for the current is antisymmetrization (this is equivalent to normal ordering in this case). Thus, given the classical current (see (8.44))

11.2 Charge Conjugation

$$J^\mu(x) = \overline{\psi}(x)\gamma^\mu\psi(x), \tag{11.134}$$

we can obtain the antisymmetrized current as

$$\begin{aligned}J^\mu_{\text{anti}}(x) &= \frac{1}{2}\left(\overline{\psi}_\alpha(x)\,(\gamma^\mu)_{\alpha\beta}\,\psi_\beta(x) - \psi_\beta(x)\,(\gamma^\mu)_{\alpha\beta}\,\overline{\psi}_\alpha(x)\right)\\ &= \frac{1}{2}\left(\overline{\psi}(x)\gamma^\mu\psi(x) - \psi^T(x)(\gamma^\mu)^T\overline{\psi}^T(x)\right).\end{aligned} \tag{11.135}$$

It can now be easily checked that under charge conjugation the antisymmetrized current transforms as

$$\begin{aligned}J^\mu_{\text{anti}}(x) &\xrightarrow{\mathcal{C}} \frac{1}{2}\left(\overline{\psi}^c(x)\gamma^\mu\psi^c(x) - (\psi^c)^T(x)(\gamma^\mu)^T(\overline{\psi}^c)^T(x)\right)\\ &= \frac{1}{2}|\eta_\psi|^2\left(-\psi^T(x)C^{-1}\gamma^\mu C\overline{\psi}^T(x) + \overline{\psi}\,C^T(\gamma^\mu)^T(C^{-1})^T\psi\right)\\ &= \frac{1}{2}\left(\psi^T(x)(\gamma^\mu)^T\overline{\psi}^T(x) + \overline{\psi}(x)\left(C^{-1}\gamma^\mu C\right)^T\psi(x)\right)\\ &= \frac{1}{2}\left(\psi^T(x)\gamma^{\mu T}\overline{\psi}^T(x) - \overline{\psi}(x)\gamma^\mu\psi(x)\right)\\ &= -J^\mu_{\text{anti}}(x), \end{aligned}\tag{11.136}$$

where we have used (11.122) in the intermediate steps. Therefore, this current transforms properly under charge conjugation. Let us note here again that the antisymmetrized current differs from the original current by an infinite constant whose effect is to make

$$\langle 0|J^\mu_{\text{anti}}(x)|0\rangle = 0, \tag{11.137}$$

whereas

$$\langle 0|J^\mu(x)|0\rangle \longrightarrow \infty. \tag{11.138}$$

Thus this current treats the vacuum more symmetrically which can also be seen from the fact that we can write the current in the form

$$J^\mu_{\text{anti}}(x) = \frac{1}{2}\left(J^\mu(x) - J^{c\mu}(x)\right). \tag{11.139}$$

Let us next calculate how the Dirac bilinears transform under charge conjugation. We have already seen in (11.122) that the charge conjugation matrix C satisfies

$$C^{-1}\gamma^\mu C = -(\gamma^\mu)^T. \tag{11.140}$$

It follows from this and the definition of γ_5 that

$$\begin{aligned}
C^{-1}\gamma_5 C &= C^{-1} i\gamma^0\gamma^1\gamma^2\gamma^3 C \\
&= iC^{-1}\gamma^0 C C^{-1}\gamma^1 C C^{-1}\gamma^2 C C^{-1}\gamma^3 C \\
&= i\left(-(\gamma^0)^T\right)\left(-(\gamma^1)^T\right)\left(-(\gamma^2)^T\right)\left(-(\gamma^3)^T\right) \\
&= i\left(\gamma^3\gamma^2\gamma^1\gamma^0\right)^T = i\left(\gamma^0\gamma^1\gamma^2\gamma^3\right)^T = \gamma_5^T.
\end{aligned} \tag{11.141}$$

Similarly, we obtain

$$\begin{aligned}
C^{-1}\gamma_5\gamma^\mu C &= C^{-1}\gamma_5 C C^{-1}\gamma^\mu C = \gamma_5^T\left(-(\gamma^\mu)^T\right) \\
&= -(\gamma^\mu\gamma_5)^T = (\gamma_5\gamma^\mu)^T, \\
C^{-1}\sigma^{\mu\nu} C &= C^{-1}\frac{i}{2}\left(\gamma^\mu\gamma^\nu - \gamma^\nu\gamma^\mu\right) C \\
&= \frac{i}{2}\left(C^{-1}\gamma^\mu C C^{-1}\gamma^\nu C - C^{-1}\gamma^\nu C C^{-1}\gamma^\mu C\right) \\
&= \frac{i}{2}\left((-(\gamma^\mu)^T)(-(\gamma^\nu)^T) - (-(\gamma^\nu)^T)(-(\gamma^\mu)^T)\right) \\
&= \frac{i}{2}\left((\gamma^\mu)^T(\gamma^\nu)^T - (\gamma^\nu)^T(\gamma^\mu)^T\right) \\
&= \frac{i}{2}\left(\gamma^\nu\gamma^\mu - \gamma^\mu\gamma^\nu\right)^T = -(\sigma^{\mu\nu})^T.
\end{aligned} \tag{11.142}$$

Therefore, if we define properly antisymmetrized Dirac bilinear combinations as

11.2 Charge Conjugation

$$\left(\overline{\psi}\Gamma^{(\alpha)}\psi\right)_{\text{anti}} \equiv \frac{1}{2}\left(\overline{\psi}\Gamma^{(\alpha)}\psi - \psi^T\left(\Gamma^{(\alpha)}\right)^T\overline{\psi}^T\right), \tag{11.143}$$

then it is clear that under charge conjugation the scalar combination would transform as

$$\begin{aligned}
\left(\overline{\psi}\psi\right)_{\text{anti}} &\xrightarrow{\mathcal{C}} \frac{1}{2}\left(\overline{\psi}^c\psi^c - (\psi^c)^T(\overline{\psi}^c)^T\right) \\
&= \frac{1}{2}|\eta_\psi|^2\left(\psi^T C^{-1}C\overline{\psi}^T - \overline{\psi}C^T\left(-(C^{-1})^T\psi\right)\right) \\
&= \frac{1}{2}\left(\overline{\psi}\psi - \psi^T\overline{\psi}^T\right) = \left(\overline{\psi}\psi\right)_{\text{anti}}. \tag{11.144}
\end{aligned}$$

(If we think for a moment, $(\overline{\psi}\psi)_{\text{anti}}$ really measures the sum of probability densities (number densities) for particles and antiparticles and that should not change under charge conjugation (particle $\leftrightarrow$ antiparticle).) We have already seen that

$$\left(\overline{\psi}\gamma^\mu\psi\right)_{\text{anti}} \xrightarrow{\mathcal{C}} -\left(\overline{\psi}\gamma^\mu\psi\right)_{\text{anti}}. \tag{11.145}$$

Similarly we can also show that

$$\begin{aligned}
\left(\overline{\psi}\gamma_5\psi\right)_{\text{anti}} &\xrightarrow{\mathcal{C}} \left(\overline{\psi}\gamma_5\psi\right)_{\text{anti}}, \\
\left(\overline{\psi}\gamma_5\gamma^\mu\psi\right)_{\text{anti}} &\xrightarrow{\mathcal{C}} \left(\overline{\psi}\gamma_5\gamma^\mu\psi\right)_{\text{anti}}, \\
\left(\overline{\psi}\sigma^{\mu\nu}\psi\right)_{\text{anti}} &\xrightarrow{\mathcal{C}} -\left(\overline{\psi}\sigma^{\mu\nu}\psi\right)_{\text{anti}}. \tag{11.146}
\end{aligned}$$

In general we can denote

$$\left(\overline{\psi}\Gamma^{(\alpha)}\psi\right)_{\text{anti}} \xrightarrow{\mathcal{C}} \epsilon_\alpha \left(\overline{\psi}\Gamma^{(\alpha)}\psi\right)_{\text{anti}}, \tag{11.147}$$

where the phase ϵ_α for the different classes has the form

$$\epsilon_\alpha = \begin{cases} 1, & \text{for } \alpha = S, P, A, \\ -1, & \text{for } \alpha = V, T. \end{cases} \tag{11.148}$$

Let us next recall from (11.73) that under parity a Dirac particle transforms as

$$\psi(\mathbf{x},t) \xrightarrow{\mathcal{P}} \psi^{\mathcal{P}}(\mathbf{x},t) = \eta^{\mathcal{P}}_{\psi}\gamma^0\psi(-\mathbf{x},t),$$
$$\overline{\psi}(\mathbf{x},t) \xrightarrow{\mathcal{P}} \overline{\psi}^{\mathcal{P}}(\mathbf{x},t) = \eta^{\mathcal{P}}_{\psi}\overline{\psi}(-\mathbf{x},t)\gamma^0. \qquad (11.149)$$

Since the antiparticle is described by

$$\psi^c(\mathbf{x},t) = \eta^{\mathcal{C}}_{\psi} C \overline{\psi}^T(\mathbf{x},t), \qquad (11.150)$$

under parity, the antiparticle will transform as

$$\begin{aligned}
\psi^c(\mathbf{x},t) \xrightarrow{\mathcal{P}} (\psi^c)^{\mathcal{P}}(\mathbf{x},t) &= \eta^{\mathcal{C}}_{\psi} C(\overline{\psi}^{\mathcal{P}})^T(\mathbf{x},t) \\
&= \eta^{\mathcal{C}}_{\psi}\eta^{\mathcal{P}}_{\psi} C(\overline{\psi}(-\mathbf{x},t)\gamma^0)^T \\
&= \eta^{\mathcal{C}}_{\psi}\eta^{\mathcal{P}}_{\psi} C\gamma^{0T} C^{-1} C(\gamma^0)^* \psi^*(-\mathbf{x},t) \\
&= -\eta^{\mathcal{P}}_{\psi}\eta^{\mathcal{C}}_{\psi}\gamma^0 C(\gamma^0)^T \psi^*(-\mathbf{x},t) \\
&= -\eta^{\mathcal{P}}_{\psi}\gamma^0 \left(\eta^{\mathcal{C}}_{\psi} C\overline{\psi}^T(-\mathbf{x},t)\right) \\
&= -\eta^{\mathcal{P}}_{\psi}\gamma^0 \psi^c(-\mathbf{x},t). \qquad (11.151)
\end{aligned}$$

Here we have used the fact that γ^0 is Hermitian in our metric. Thus if $\eta^{\mathcal{P}}_{\psi}$ represents the intrinsic parity of a Dirac particle, the antiparticle will have the intrinsic parity $-\eta^{\mathcal{P}}_{\psi}$ so that the relative particle-antiparticle parity will be given by

$$\eta^{\mathcal{P}}_{\psi}\left(-\eta^{\mathcal{P}}_{\psi}\right) = -(\eta^{\mathcal{P}}_{\psi})^2 = -1. \qquad (11.152)$$

Of course, we have already seen this from the analysis of the positronium decay in (11.83), but this derivation is simpler and more general.

Next, let us note that since the charge conjugate wave function is defined to be

$$\psi^c = \eta^{\mathcal{C}}_{\psi} C \overline{\psi}^T,$$

we obtain

11.2 CHARGE CONJUGATION

$$\begin{aligned}
\gamma_5 \psi^c &= \eta_\psi^c \gamma_5 C \gamma^{0T} \psi^* \\
&= \eta_\psi^c CC^{-1}\gamma_5 C(\gamma^0)^T \psi^* \\
&= \eta_\psi^c C\gamma_5^T(\gamma^0)^T \psi^* = -\eta_\psi^c C(\gamma^0)^T \gamma_5^T \psi^* \\
&= -\eta_\psi^c C(\gamma^0)^T \left(\gamma_5^\dagger \psi\right)^* = -\eta_\psi^c C(\gamma^0)^T (\gamma_5 \psi)^* \\
&= -\eta_\psi^c C(\gamma^0)^T ((\gamma_5 \psi)^\dagger)^T = -\eta_\psi^c C\overline{(\gamma_5 \psi)}^T.
\end{aligned} \quad (11.153)$$

As a result, we see that if

$$\gamma_5 \psi = \psi, \quad (11.154)$$

we have

$$\gamma_5 \psi^c = -\eta_\psi^c C\overline{\psi}^T = -\psi^c, \quad (11.155)$$

On the other hand, if

$$\gamma_5 \psi = -\psi, \quad (11.156)$$

then it follows that

$$\gamma_5 \psi^c = -\eta_\psi^c C\left(-\overline{\psi}^T\right) = \eta_\psi^c C\overline{\psi}^T = \psi^c. \quad (11.157)$$

In other words, the antiparticle of a particle with a given handedness has the opposite handedness. We have, of course, already seen this in the study of neutrinos in section **3.6** and have used this in the construction of the standard model in section **14.3**.

11.2.3 Majorana fermions. Let us recall that if a free Dirac particle satisfies the equation

$$(\not{p} - m)\psi = 0, \quad (11.158)$$

then from (11.119) we see that the antiparticle must also satisfy the same free equation, namely,

$$(\not{p} - m)\psi^c = 0, \quad (11.159)$$

where

$$\psi^c = \eta_\psi C \overline{\psi}^T. \tag{11.160}$$

We have also seen that there are bosons in nature which are their own antiparticles. These are charge neutral particles described by real quantum fields

$$\phi^\dagger = \phi. \tag{11.161}$$

We can now ask the corresponding question in the case of the Dirac particles, namely, whether there are also fermionic particles that are their own antiparticles and if so how does one describe them.

The answer to this question is, in fact, quite obvious. From the definition of charge conjugation in (11.123) we see that a Dirac particle which is its own antiparticle must satisfy

$$\psi = \psi^c = \eta_\psi C \overline{\psi}^T, \tag{11.162}$$

and would also satisfy the free Dirac equation

$$(\not{p} - m)\psi = (\not{p} - m)\psi^c = 0. \tag{11.163}$$

Such particles are known as self charge conjugate fermions or Majorana fermions. They can be massive and will have only two independent degrees of freedom (because of the constraint in (11.162)). Let us note from the definition of the antisymmetrized current in (11.135)

$$\begin{aligned} J^\mu_{\text{anti}}(x) &= \frac{1}{2}\left(\overline{\psi}(x)\gamma^\mu \psi(x) - \psi^T(x)\gamma^{\mu T}\overline{\psi}^T(x)\right) \\ &= \frac{1}{2}\left(\overline{\psi}(x)\gamma^\mu \psi(x) - \overline{\psi}^c(x)\gamma^\mu \psi^c(x)\right), \end{aligned} \tag{11.164}$$

that for Majorana fermions ($\psi = \psi^c$)

$$J^\mu_{\text{anti}}(x) = 0, \tag{11.165}$$

so that the Majorana fermions are charge neutral and hence cannot have any electromagnetic interaction.

11.2 CHARGE CONJUGATION

Secondly, as we have seen a particle with a definite handedness will lead to an antiparticle with the opposite handedness. Since a Majorana fermion is its own antiparticle, this implies that it cannot have a well defined handedness. Thus we cannot talk of a Majorana particle with a given helicity. (This is, however, not true if the space time dimensionality is not four. In particular for $d = 2$ mod 8, we can have Majorana-Weyl spinors.)

We have also seen that if the intrinsic parity of a Dirac particle is η_ψ^P, then the antiparticle will have the intrinsic parity $-(\eta_\psi^P)^*$ (although we had chosen the phase of the intrinsic parity for a Dirac fermion to be real, it can in principle be complex). For a Majorana fermion then, this would require

$$\eta_\psi^P = -(\eta_\psi^P)^*, \tag{11.166}$$

which can be satisfied only if the intrinsic parity of a Majorana fermion is imaginary.

Let us also note here that since both the Weyl and the Majorana fermions have only two degrees of freedom, we can find a simple relation between them. In fact, we note that if we define

$$\chi = \psi_L + (\psi_L)^c, \tag{11.167}$$

then

$$\chi^c = (\psi_L)^c + \psi_L = \chi, \tag{11.168}$$

so that we can express a Majorana fermion in terms of Weyl fermions. This then raises the question that since the two component Majorana fermion can be massive whether it is possible to have a mass for the two component Weyl fermion. To answer this, let us note that if we try to write a Lagrangian density for a massive Weyl fermion (say left-handed), the naive mass term would have the form

$$\mathcal{L} = i\overline{\psi}_L \slashed{\partial} \psi_L - m\overline{\psi}_L \psi_L. \tag{11.169}$$

However, the mass term would trivially vanish since

$$\overline{\psi}_L \psi_L = \overline{\psi}\frac{1}{2}(1+\gamma_5) \times \frac{1}{2}(1-\gamma_5)\psi = 0. \tag{11.170}$$

This is, of course, the essence of the statement that a Weyl fermion is massless. However, note that if we go beyond the conventional Dirac mass term in the Lagrangian density, we can write (we will choose $\eta_\psi^c = 1$)

$$\begin{aligned}\mathcal{L} &= i\overline{\psi}_L \slashed{\partial} \psi_L + m\overline{\psi^c_L}\psi_L \\ &= i\overline{\psi}_L \slashed{\partial}\psi_L - m\psi_L^T C^{-1}\psi_L.\end{aligned} \tag{11.171}$$

Clearly, in this case, the mass term is not zero and we can give a mass to the Weyl fermion. Such a mass therm is known as a Majorana mass term and in writing this term we have obviously given up something to which we will return in a moment.

Let us next ask whether there are any particles in nature which we can think of as Majorana particles. Obviously, the neutron and the neutrino are charge neutral and hence are prime candidates to be Majorana particles. But such an identification is not compatible with the experimental results, namely, if (here $\overline{n}$ represents the antineutron)

$$n = \overline{n}, \tag{11.172}$$

then this would imply violation of baryon number which is not seen at present. Similarly, if (once again $\overline{\nu}$ denotes the antineutrino)

$$\nu = \overline{\nu}, \tag{11.173}$$

this would imply violation of lepton number which is also not seen yet. In fact, note that any majorana particle would necessarily imply some sort of violation of a fermion number which is not seen in nature. It is now clear that the Majorana mass term would violate conservation of fermion number and this is what we give up in trying to write a mass term for the Weyl fermion. Obviously, if we require conservation of fermion number, a Weyl fermion cannot have mass.

11.2 CHARGE CONJUGATION

We can now ask why anyone would then like to study such particles or mass terms. The answer to this lies in the fact that while no one has really measured the mass of the neutrino, the direct experimental limits ($m_{\nu_e} < $ 2ev, $m_{\nu_\mu} < $ 190keV, $m_{\nu_\tau} < $ 18.2MeV) (the present trend is to quote the lowest mass bound along with the differences of mass squares for the neutrinos determined from the neutrino oscillation experiments) suggest that the neutrinos may indeed have small masses. The Majorana mass is one way to understand such small masses and can arise naturally in grand unified theories where both baryon and lepton numbers can be violated in small amounts. It is, of course, up to the experiments to decide whether this is really what happens. Another reason for the study of Majorana particles is that they are fundamental in the study of supersymmetry which we will not go into. Independent of whether supersymmetry is a symmetry of nature, its theoretical studies would require the study of the properties of such particles.

11.2.4 Eigenstates of charge conjugation. We have seen that the photon has odd charge conjugation parity. This, of course, means that a one photon state is an eigenstate of charge conjugation with eigenvalue -1. Similarly, we have also seen that a Hermitian spin zero field has charge parity $+1$. To understand the eigenstates of charge conjugation in general, let us note that if $\hat{Q}$ denotes the operator for charge (electric), then acting on an eigenstate it gives

$$\hat{Q}|Q,\ldots\rangle = Q|Q,\ldots\rangle, \tag{11.174}$$

where ... refer to other quantum numbers the state can depend on. It follows now that

$$\mathcal{C}\hat{Q}|Q,\ldots\rangle = Q\mathcal{C}|Q,\ldots\rangle = Q|-Q,\ldots\rangle. \tag{11.175}$$

On the other hand, we have

$$\hat{Q}\mathcal{C}|Q,\ldots\rangle = \hat{Q}|-Q,\ldots\rangle = -Q|-Q,\ldots\rangle, \tag{11.176}$$

so that comparing the two relations we obtain

$$\mathcal{C}\hat{Q} = -\hat{Q}\mathcal{C},$$
or, $\quad\left[\mathcal{C},\hat{Q}\right]_{+} = 0.$ \hfill (11.177)

In other words, the charge operator does not commute with the charge conjugation operator and, consequently, they cannot have simultaneous eigenstates unless the charge of the state vanishes. We have, of course, seen this already in the sense that the photon which does not carry charge has a well defined charge parity and so does a Hermitian spin zero field which describes charge neutral mesons. Let us note, however, that while all eigenstates of C must be charge neutral, not all charge neutral states would be eigenstates of charge conjugation operator. It is easy to see that

$$\begin{aligned}\mathcal{C}|n\rangle &= |\overline{n}\rangle,\\ \mathcal{C}|K^0\rangle &= |\overline{K}^0\rangle,\\ \mathcal{C}|\pi^- p\rangle &= |\pi^+ \overline{p}\rangle,\end{aligned} \qquad (11.178)$$

which clarifies the point. On the other hand, a state such as $|\pi^-\pi^+\rangle$ can be an eigenstate of $\mathcal{C}$. Note that

$$\mathcal{C}|\pi^-\pi^+\rangle = |\pi^+\pi^-\rangle. \qquad (11.179)$$

The final state simply corresponds to the initial state with the two particles interchanged. Thus if the state has well defined angular momentum quantum number, we would have

$$\mathcal{C}|\pi^-\pi^+,\ell\rangle = |\pi^+\pi^-,\ell\rangle = (-1)^\ell|\pi^-\pi^+,\ell\rangle, \qquad (11.180)$$

so that such a state would be an eigenstate of charge conjugation with eigenvalue $(-1)^\ell$. Similarly, for the positronium with orbital angular momentum and spin values ℓ, s respectively,

11.2 CHARGE CONJUGATION

$$\begin{aligned}
\mathcal{C}|e^-e^+;\ell,s\rangle &= |e^+e^-;\ell,s\rangle \\
&= -(-1)^\ell(-1)^{s+1}|e^-e^+;\ell,s\rangle \\
&= (-1)^{\ell+s}|e^-e^+;\ell,s\rangle.
\end{aligned} \qquad (11.181)$$

In other words, this is an eigenstate of charge conjugation with charge parity $(-1)^{\ell+s}$.

If charge conjugation is a symmetry of the theory, then the charge parity must be conserved. Thus if we look at the decay of the positronium to n photons, namely,

$$|e^-e^+;\ell,s\rangle \to n\gamma, \qquad (11.182)$$

then the conservation of charge parity would require (recall that the photon has charge parity -1)

$$(-1)^{\ell+s} = (-1)^n. \qquad (11.183)$$

This is, of course, consistent with our earlier discussion, namely, that the positronium in the $\ell = 0 = s$ state decays into two photons. Note, similarly, that upon requiring conservation of charge parity, the decay

$$\pi^0 \to n\gamma, \qquad (11.184)$$

would lead to the result that

$$1 = (-1)^n. \qquad (11.185)$$

It would, in particular, say that if charge conjugation is a symmetry, then

$$\pi^0 \not\to 3\gamma. \qquad (11.186)$$

Let us also note here that if we have an amplitude involving N photons shown in Fig. 11.1 which can, for example, describe the decay

$$n\gamma \to (N-n)\gamma, \tag{11.187}$$

then conservation of charge parity would require (electromagnetic interactions are invariant under charge conjugation)

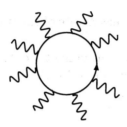

Figure 11.1: An amplitude with N photons.

$$\begin{aligned} (-1)^n &= (-1)^{N-n}, \\ \text{or,} \quad (-1)^N &= (-1)^{2n} = 1. \end{aligned} \tag{11.188}$$

In other words, any amplitude involving an odd number of photons must vanish if charge conjugation is a symmetry. This result goes under the name of Furry's theorem.

As we have seen, electromagnetic interactions are invariant under charge conjugation. Similarly, it is also known that charge conjugation symmetry holds true in strong interactions also. In fact, note that if a Hamiltonian is invariant under charge conjugation, then

$$\mathcal{C}^\dagger H \mathcal{C} = H, \tag{11.189}$$

and it follows from this that

$$\mathcal{C}^\dagger U \mathcal{C} = U, \tag{11.190}$$

where U denotes the time evolution operator. Therefore, if the strong interaction Hamiltonian respects charge conjugation symmetry, then we must have

11.2 Charge conjugation

$$\langle \pi^- p | U | \pi^- p \rangle = \langle \pi^- p | \mathcal{C}^\dagger U \mathcal{C} | \pi^- p \rangle = \langle \pi^+ \overline{p} | U | \pi^+ \overline{p} \rangle. \quad (11.191)$$

In other words, the scattering cross-section for $\pi^- p$ elastic scattering would be identical to the cross-section for $\pi^+ \overline{p}$ scattering which is experimentally observed.

Let us note here without going into details that while charge conjugation is a good symmetry for both strong and electromagnetic interactions, it is violated in weak processes just like parity. The simplest way to see this is to note that the weak current (see (14.100))

$$\begin{aligned}
J^+_{\mu WK} &= -\frac{g}{2\sqrt{2}} \overline{e}_{\rm L} \gamma_\mu (1-\gamma_5) \nu_{e\rm L} \xrightarrow{\mathcal{C}} -\frac{g}{2\sqrt{2}} \overline{e}^c_{\rm L} \gamma_\mu (1-\gamma_5) \nu^c_{e\rm L} \\
&= \frac{g}{2\sqrt{2}} (\eta^c_e)^* \eta^c_\nu e^T_{\rm L} C^{-1} \gamma_\mu (1-\gamma_5) C \overline{\nu}^T_{e\rm L} \\
&= -\frac{g}{2\sqrt{2}} (\eta^c_e)^* \eta^c_\nu e^T_{\rm L} \gamma^T_\mu (1-\gamma^T_5) \overline{\nu}^T_{e\rm L} \\
&= \frac{g}{2\sqrt{2}} (\eta^c_e)^* \eta^c_\nu \left(\overline{\nu}_{e\rm L} (1-\gamma_5) \gamma_\mu e_{\rm L} \right)^T \\
&= \frac{g}{2\sqrt{2}} (\eta^c_e)^* \eta^c_\nu \overline{\nu}_{e\rm L} \gamma_\mu (1+\gamma_5) e_{\rm L} \\
&\neq -\frac{g}{2\sqrt{2}} \eta_j \overline{\nu}_{e\rm L} \gamma_\mu (1-\gamma_5) e_{\rm L} = \eta_j J^-_{\mu WK}, \quad (11.192)
\end{aligned}$$

where η_J denotes a phase. Namely, the weak current does not have a well defined transformation property under charge conjugation and hence will violate $\mathcal{C}$-invariance.

Thus we see that $\mathcal{P}$ and $\mathcal{C}$ ($\mathcal{P}$ for parity) are conserved in strong and electromagnetic interactions but are violated in weak interactions. On the other hand the product operation $\mathcal{CP}$ appears to be a good symmetry in most weak processes. We can again check with the charged weak current that

$$\begin{aligned}
J^+_{\mu WK} &\xrightarrow{\mathcal{C}} \frac{g}{2\sqrt{2}} (\eta^c_e)^* \eta^c_\nu \overline{\nu}_{e\rm L} \gamma_\mu (1+\gamma_5) e_{\rm L} \\
&\xrightarrow{\mathcal{P}} \frac{g}{2\sqrt{2}} (\eta^c_e)^* \eta^c_\nu \overline{\nu}^\mathcal{P}_{e\rm L} \gamma_\mu (1+\gamma_5) e^\mathcal{P}_{\rm L}
\end{aligned}$$

$$\begin{aligned}
&= \frac{g}{2\sqrt{2}}(\eta_e^C)^* \eta_\nu^C (\eta_\nu^P)^* \eta_e^P \overline{\nu}_{eL} \gamma^0 \gamma_\mu (1-\gamma_5) \gamma^0 e_L \\
&= \frac{g}{2\sqrt{2}}(\eta_e^C)^* \eta_\nu^C (\eta_\nu^P)^* \eta_e^P \overline{\nu}_{eL} \gamma^0 \gamma_\mu \gamma^0 (1-\gamma_5) e_L \\
&= \eta_{\text{TOT}} \frac{g}{2\sqrt{2}} \overline{\nu}_{eL} \gamma^0 \gamma_\mu \gamma^0 (1-\gamma_5) e_L,
\end{aligned} \qquad (11.193)$$

where η_{TOT} denotes the total phase under the transformation. It follows from this that

$$\begin{aligned}
J^+_{0WK} &\xrightarrow{\mathcal{CP}} \eta_{\text{TOT}} \frac{g}{2\sqrt{2}} \overline{\nu}_{eL} \gamma_0 (1-\gamma_5) e_L = -\eta_{\text{TOT}} J^-_{0WK} \\
J^+_{iWK} &\xrightarrow{\mathcal{CP}} -\eta_{\text{TOT}} \frac{g}{2\sqrt{2}} \overline{\nu}_{eL} \gamma_i (1-\gamma_5) e_L = \eta_{\text{TOT}} J^-_{iWK}.
\end{aligned} \qquad (11.194)$$

This has well defined transformation properties under $\mathcal{CP}$ transformations and leads to $\mathcal{CP}$ invariance.

11.3 Time reversal

Classically, time reversal or time reflection corresponds to the space time transformation where we reverse only the sign of time, namely,

$$\mathbf{x} \xrightarrow{\mathcal{T}} \mathbf{x}, \quad t \xrightarrow{\mathcal{T}} -t. \qquad (11.195)$$

We note that the Newton's equation is invariant under time reversal since it is second order in the time derivative. Similarly, Maxwell's equations in the presence of sources

$$\begin{aligned}
\boldsymbol{\nabla} \cdot \mathbf{E}(x) &= \rho(x), \\
\boldsymbol{\nabla} \cdot \mathbf{B}(x) &= 0, \\
\boldsymbol{\nabla} \times \mathbf{E}(x) &= -\frac{\partial \mathbf{B}}{\partial t}, \\
\boldsymbol{\nabla} \times \mathbf{B}(x) &= \frac{\partial \mathbf{E}(x)}{\partial t} + \mathbf{J},
\end{aligned} \qquad (11.196)$$

are also form invariant if we define

11.3 TIME REVERSAL

$$\rho(\mathbf{x},t) \xrightarrow{\mathcal{T}} \rho^{\mathcal{T}}(\mathbf{x},t) = \rho(\mathbf{x},-t),$$
$$\mathbf{J}(\mathbf{x},t) \xrightarrow{\mathcal{T}} \mathbf{J}^{\mathcal{T}}(\mathbf{x},t) = -\mathbf{J}(\mathbf{x},-t),$$
$$\mathbf{E}(\mathbf{x},t) \xrightarrow{\mathcal{T}} \mathbf{E}^{\mathcal{T}}(\mathbf{x},t) = \mathbf{E}(\mathbf{x},-t),$$
$$\mathbf{B}(\mathbf{x},t) \xrightarrow{\mathcal{T}} \mathbf{B}^{\mathcal{T}}(\mathbf{x},t) = -\mathbf{B}(\mathbf{x},-t). \tag{11.197}$$

From the definition of the electric and the magnetic fields

$$\mathbf{E} = -\nabla A^0 - \frac{\partial \mathbf{A}}{\partial t},$$
$$\mathbf{B} = \nabla \times \mathbf{A}, \tag{11.198}$$

we determine that under time reversal

$$A^0(\mathbf{x},t) \xrightarrow{\mathcal{T}} A^{0\mathcal{T}}(\mathbf{x},t) = A^0(\mathbf{x},-t),$$
$$\mathbf{A}(\mathbf{x},t) \xrightarrow{\mathcal{T}} \mathbf{A}^{\mathcal{T}}(\mathbf{x},t) = -\mathbf{A}(\mathbf{x},-t). \tag{11.199}$$

The transformation of the $A^\mu(x)$ potentials under time reversal, therefore, is opposite from their behavior under parity in (11.57).

While some of the macroscopic equations are invariant under time reversal, we know that macroscopic physics is not. The diffusion equation which is first order in the time derivative is not symmetric under $t \to -t$. Macroscopically, we know that there is a sense of direction for time, namely, the entropy or disorder increases with time. However, in statistical mechanics we know that microscopic reversibility does hold. Thus we would like to investigate whether time reversal can be a symmetry of quantum mechanical systems. The first problem which we face, however, is that the time dependent Schrödinger equation is first order in the time derivative just like the diffusion equation and hence cannot be invariant under a naive extension of the time inversion operation to the quantum theory. Note that in quantum mechanics, symmetry transformations normally are implemented by linear operators which are also unitary. Thus assuming that time inversion is also implemented by such an operator $\mathcal{T}$, we will have

$$|\psi\rangle \xrightarrow{T} |\psi\rangle^T = T|\psi\rangle. \tag{11.200}$$

If we assume that the time independent Hamiltonian is also invariant under time inversion, namely,

$$[T, H] = 0, \tag{11.201}$$

then it follows that under a time inversion, the time dependent Schrödinger equation

$$i\frac{\partial |\psi\rangle}{\partial t} = H|\psi\rangle, \tag{11.202}$$

would lead to

$$\begin{aligned} iT\left(\frac{\partial |\psi\rangle}{\partial t}\right) &= TH|\psi\rangle, \\ \text{or,} \quad -i\frac{\partial}{\partial t}(T|\psi\rangle) &= H(T|\psi\rangle). \end{aligned} \tag{11.203}$$

Thus we see that in such a case $|\psi\rangle$ and $T|\psi\rangle$ satisfy different equations and hence this operation cannot define a symmetry of the time dependent Schrödinger equation.

However, the crucial difference between the diffusion equation and the Schrödiner equation is the fact that the time derivative in the quantum mechanical case comes multiplied with a factor of "i" which makes it possible to define a time inversion in the quantum theory which will be a symmetry of the system. Let us for the moment give up the assumption that the operator T is linear. In fact, let us assume that T is an antilinear (antiunitary) operator. By definition, an antilinear operator T satisfies

11.3 TIME REVERSAL

$$|\psi\rangle \xrightarrow{\mathcal{T}} |\psi\rangle^{\mathcal{T}} = \mathcal{T}|\psi\rangle,$$

$$\mathcal{T}(\alpha|\psi_1\rangle + \beta|\psi_2\rangle) = \alpha^*\mathcal{T}|\psi_1\rangle + \beta^*\mathcal{T}|\psi_2\rangle,$$

$$\langle\phi|\psi\rangle \xrightarrow{\mathcal{T}} {}^{\mathcal{T}}\langle\phi|\psi\rangle^{\mathcal{T}} = \langle\phi|\psi\rangle^* = \langle\psi|\phi\rangle,$$

$$\mathcal{T}i\mathcal{T}^\dagger = -i,$$

$$\mathcal{T}A\mathcal{T}^\dagger = A^{\mathcal{T}} = \eta_A A,$$

$$\mathcal{T}AB\mathcal{T}^\dagger = A^{\mathcal{T}}B^{\mathcal{T}} = \eta_A \eta_B AB, \tag{11.204}$$

where A and B are assumed to be Hermitian operators and η_A, η_B are phases. (We can also show that $\mathcal{T}^\dagger = \mathcal{T}^{-1}$, but we will not need this for our discussions.) Under such an antilinear (antiunitary) transformation, if the Hamiltonian is symmetric under time inversion, the dynamical equation would transform as

$$i\frac{\partial|\psi\rangle}{\partial t} = H|\psi\rangle,$$

would lead to

$$\mathcal{T}\left(i\frac{\partial|\psi\rangle}{\partial t}\right) = \mathcal{T}H|\psi\rangle,$$

$$\text{or,}\quad -i\frac{\partial\mathcal{T}|\psi\rangle}{\partial(-t)} = H\mathcal{T}|\psi\rangle,$$

$$\text{or,}\quad i\frac{\partial}{\partial t}(\mathcal{T}|\psi\rangle) = H(\mathcal{T}|\psi\rangle). \tag{11.205}$$

Therefore, with such a definition of $\mathcal{T}$, we see that $|\psi\rangle$ and $\mathcal{T}|\psi\rangle$ satisfy the same equation and hence such a transformation would define a symmetry of the time dependent Schrödinger equation.

Let us next assume that we are looking at a process which goes from the state $|\psi_i\rangle$ to the state $|\psi_f\rangle$. Thus the quantity we are interested in is the transition amplitude

$$\langle\psi_f|\psi_i\rangle. \tag{11.206}$$

In the time reversed process, on the other hand, the role of the initial and the final states are interchanged and the transition amplitude of interest would be

$$\langle \psi_i | \psi_f \rangle. \tag{11.207}$$

Thus, under time inversion, we expect

$$\langle \psi_f | \psi_i \rangle \xrightarrow{\mathcal{T}} \langle \psi_i | \psi_f \rangle, \tag{11.208}$$

which is indeed consistent with the definition of time inversion transformations given in (11.204) (see, in particular the third relation). In particular, let us note from (11.204) that

$$\psi(\mathbf{x}, t) = \langle \mathbf{x} | \psi(t) \rangle \xrightarrow{\mathcal{T}} \psi^{\mathcal{T}}(\mathbf{x}, t) = \langle \mathbf{x} | \psi(-t) \rangle^*$$
$$= \psi^*(\mathbf{x}, -t). \tag{11.209}$$

Let us derive the transformation properties of some of the operators under time inversion. Classically, we know that

$$\mathbf{x} \xrightarrow{\mathcal{T}} \mathbf{x}, \quad \mathbf{p} \xrightarrow{\mathcal{T}} -\mathbf{p}. \tag{11.210}$$

Thus from Ehrenfest's theorem we would expect the corresponding operators to transform as

$$\begin{aligned}
\mathcal{T} \mathbf{X} \mathcal{T}^\dagger &= \eta_\mathbf{X} \mathbf{X} = \mathbf{X}, \quad \eta_\mathbf{X} = 1, \\
\mathcal{T} \mathbf{P} \mathcal{T}^\dagger &= \eta_\mathbf{P} \mathbf{P} = -\mathbf{P}, \quad \eta_\mathbf{P} = -1.
\end{aligned} \tag{11.211}$$

It follows now that

$$\begin{aligned}
\mathcal{T} [X_i, P_j] \mathcal{T}^\dagger &= \mathcal{T} (X_i P_j - P_j X_i) \mathcal{T}^\dagger \\
&= X_i(-P_j) - (-P_j) X_i = -[X_i, P_j],
\end{aligned} \tag{11.212}$$

where we have used (11.204) as well as (11.211). It is reassuring to note that the canonical commutation relations are unchanged by the

11.3 TIME REVERSAL

time inversion transformation (note from (11.204) that $i \to -i$ under the anti-linear time reversal transformation). Let us note here that the naive extension of the time inversion transformation with a linear operator would have led to an inconsistency in the commutation relations.

Since the angular momentum operators can be written as

$$L_i = \epsilon_{ijk} X_j P_k,$$

we obtain

$$\mathcal{T} L_i \mathcal{T}^\dagger = \epsilon_{ijk} \mathcal{T} X_j P_k \mathcal{T}^\dagger = -\epsilon_{ijk} X_j P_k = -L_i. \tag{11.213}$$

Namely, under time inversion each component of the angular momentum operator would change sign which is consistent with the classical result. Therefore, it follows that

$$\mathcal{T} L^2 \mathcal{T}^\dagger = \mathcal{T} L_i L_i \mathcal{T}^\dagger = (-L_i)(-L_i) = L_i L_i = L^2, \tag{11.214}$$

so that the L^2 operator remains unchanged under time inversion. It follows now that for the angular momentum eigenstates satisfying

$$L^2 |\ell, m\rangle = \hbar^2 \ell(\ell+1) |\ell, m\rangle,$$
$$L_3 |\ell, m\rangle = \hbar m |\ell, m\rangle, \tag{11.215}$$

we have

$$L^2 \mathcal{T} |\ell, m\rangle = \mathcal{T} L^2 |\ell, m\rangle = \hbar^2 \ell(\ell+1)(\mathcal{T}|\ell, m\rangle),$$
$$L_3 \mathcal{T} |\ell, m\rangle = -\mathcal{T} L_3 |\ell, m\rangle = -\hbar m (\mathcal{T}|\ell, m\rangle). \tag{11.216}$$

We see that the state $\mathcal{T}|\ell, m\rangle$ is an eigenstate of L^2 and L_3 with eigenvalues $\hbar^2 \ell(\ell+1)$ and $-\hbar m$ respectively so that we can identify

$$\mathcal{T}|\ell, m\rangle = |\ell, m\rangle^\mathcal{T} = C_m |\ell, -m\rangle, \tag{11.217}$$

with

$$|C_m|^2 = 1. \tag{11.218}$$

By convention, the constant (phase) is chosen to be

$$C_m = (-1)^m = i^{2m}, \tag{11.219}$$

which follows because under time inversion (say for $m > 0$)

$$\begin{aligned}
Y_{\ell,m}(\theta, \phi) &= \langle \theta, \phi | \ell, m \rangle \\
&= (-1)^m \sqrt{\frac{2\ell+1}{4\pi} \frac{(\ell-|m|)!}{\ell+|m|)!}} P_{\ell,m}(\cos\theta) e^{im\phi} \\
&\xrightarrow{T} Y^*_{\ell,m}(\theta, \phi) \\
&= (-1)^m \sqrt{\frac{2\ell+1}{4\pi} \frac{(\ell-|m|)!}{(\ell+|m|)!}} P_{\ell,m}(\cos\theta) e^{-im\phi} \\
&= (-1)^m Y_{\ell,-m}(\theta, \phi), \tag{11.220}
\end{aligned}$$

where we have used the fact that the spherical harmonics for $m < 0$ are defined with a phase $+1$. Thus, we can indeed write

$$|\ell, m\rangle^T = \mathcal{T}|\ell, m\rangle = (i)^{2m}|\ell, -m\rangle. \tag{11.221}$$

It can be shown that the above relation holds true even for half integer angular momentum values, namely, in general, we can write

$$|j, m_j\rangle^T = \mathcal{T}|j, m_j\rangle = (i)^{2m_j}|j, -m_j\rangle. \tag{11.222}$$

11.3.1 Spin zero field and Maxwell's theory. We have already seen that under time reversal

$$A^\mu(\mathbf{x}, t) \xrightarrow{T} A^{\mu T}(\mathbf{x}, t) = \eta_{A^\mu} A^\mu(\mathbf{x}, -t), \tag{11.223}$$

where

$$\eta_{A^0} = 1, \qquad \eta_A = -1. \tag{11.224}$$

It follows, therefore, that

$$\begin{aligned}
F_{\mu\nu}(\mathbf{x},t) &= \partial_\mu A_\nu(\mathbf{x},t) - \partial_\nu A_\mu(\mathbf{x},t) \\
&\xrightarrow{T} \partial_\mu^T A_\nu^T(\mathbf{x},t) - \partial_\nu^T A_\mu^T(\mathbf{x},t),
\end{aligned} \tag{11.225}$$

so that

$$\begin{aligned}
F_{0i}(\mathbf{x},t) &\xrightarrow{T} F_{0i}^T(\mathbf{x}-t) = F_{0i}(\mathbf{x},-t), \\
F_{ij}(\mathbf{x},t) &\xrightarrow{T} F_{ij}^T(\mathbf{x},t) = -F_{ij}(\mathbf{x},-t).
\end{aligned} \tag{11.226}$$

Similarly, we have seen that under time reversal, the current four vector transforms as

$$J^\mu(\mathbf{x},t) \xrightarrow{T} J^{\mu T}(\mathbf{x},t) = \eta_{J^\mu} J^\mu(\mathbf{x},-t), \tag{11.227}$$

where

$$\eta_{J^0} = 1, \qquad \eta_J = -1. \tag{11.228}$$

The Maxwell theory is, of course, invariant under time reversal as we have already seen and as can be checked directly.

$$\begin{aligned}
S &= \int d^4 x \left(-\frac{1}{4} F_{\mu\nu} F^{\mu\nu} - J^\mu A_\mu \right) \\
&\xrightarrow{T} \int d^4 x \left(-\frac{1}{4} F_{\mu\nu}^T F^{\mu\nu T} - J^{\mu T} A_\mu^T \right) \\
&= \int d^4 x \left(-\frac{1}{4} F_{\mu\nu}(\mathbf{x},-t) F^{\mu\nu}(\mathbf{x},-t) - J^\mu(\mathbf{x},-t) A_\mu(\mathbf{x},-t) \right) \\
&= \int d^4 x \left(-\frac{1}{4} F_{\mu\nu}(x) F^{\mu\nu}(x) - J^\mu(x) A_\mu(x) \right) \\
&= S.
\end{aligned} \tag{11.229}$$

In general, for a multicomponent function or field we expect the behavior under time reversal to be

$$\psi_\alpha(x) \xrightarrow{T} \psi_\alpha^T(x) = \eta_\psi T_\alpha^\beta \psi_\beta^*(\mathbf{x}, -t), \tag{11.230}$$

where the matrix T mixes the components of the function and

$$|\eta_\psi|^2 = 1. \tag{11.231}$$

For example, for a spin zero field, we have

$$\phi(x) \xrightarrow{T} \phi^T(x) = \eta_\phi \phi^*(\mathbf{x}, -t). \tag{11.232}$$

Note that since the complex spin zero field satisfies

$$\begin{aligned}
(\Box + m^2)\phi(x) &= \left(\frac{\partial^2}{\partial t^2} - \nabla^2 + m^2\right)\phi(x) = 0, \\
(\Box + m^2)\phi^*(x) &= \left(\frac{\partial^2}{\partial t^2} - \nabla^2 + m^2\right)\phi^*(x) = 0,
\end{aligned} \tag{11.233}$$

it is clear that in this case

$$\begin{aligned}
(\Box + m^2)\phi^T(x) &= \left(\frac{\partial^2}{\partial t^2} - \nabla^2 + m^2\right)\phi^T(x) \\
&= \left(\frac{\partial^2}{\partial t^2} - \nabla^2 + m^2\right)\eta_\phi \phi^*(\mathbf{x}, -t) \\
&= 0.
\end{aligned} \tag{11.234}$$

That is, the Klein-Gordon equation is invariant under time inversion. Furthermore, note that under time inversion (assume classical behavior)

$$\begin{aligned}
J^\mu(x) &= i\left(\phi^*(x)\partial^\mu \phi(x) - (\partial^\mu \phi^*(x))\phi(x)\right) \\
&\xrightarrow{T} i\left(\phi^{T*}(x)\partial^\mu \phi^T(x) - (\partial^\mu \phi^{T*}(x))\phi^T(x)\right) \\
&= i\left(\phi(\mathbf{x}, -t)\partial^\mu \phi^*(\mathbf{x}, -t) - (\partial^\mu \phi(\mathbf{x}, -t))\phi^*(\mathbf{x}, -t)\right) \\
&= -i\left(\phi^*(\mathbf{x}, -t)\partial^\mu \phi(\mathbf{x}, -t) - (\partial^\mu \phi^*(\mathbf{x}, -t))\phi(\mathbf{x}, -t)\right).
\end{aligned} \tag{11.235}$$

11.3 TIME REVERSAL

Therefore, we obtain

$$J^0(\mathbf{x},t) \xrightarrow{T} -i\left(\phi^*(\mathbf{x},-t)\frac{\partial}{\partial t}\phi(\mathbf{x},-t) - (\frac{\partial}{\partial t}\phi^*(\mathbf{x},-t))\phi(\mathbf{x},-t)\right)$$
$$= J^0(\mathbf{x},-t),$$
$$J^i(\mathbf{x},t) \xrightarrow{T} -i\left(\phi^*(\mathbf{x},-t)\partial^i\phi(\mathbf{x},-t) - (\partial^i\phi^*(\mathbf{x},-t))\phi(\mathbf{x},-t)\right)$$
$$= -J^i(\mathbf{x},-t), \tag{11.236}$$

which is, of course, the correct behavior of the current four-vector under time inversion as we have already seen in (11.197). (Here we have used the fact that $J^0(\mathbf{x},-t)$ is defined with $-\frac{\partial}{\partial t}$.) Thus we conclude that the spin zero charged particles interacting with photons can be described by a theory which will be time reversal invariant.

11.3.2 Dirac fields. From our general discussion (see (11.230)), we expect that under time reversal

$$\psi(\mathbf{x},t) \xrightarrow{T} \psi^T(\mathbf{x},t) = \eta_\psi T\psi^*(\mathbf{x},-t). \tag{11.237}$$

To determine when this function will be a solution of the Dirac equation (so that time reversal can be a symmetry of the system) we note that the Dirac equation

$$(i\gamma^\mu \partial_\mu - m)\psi(x) = 0, \tag{11.238}$$

leads to (upon complex conjugation)

$$(-i\gamma^{\mu*}\partial_\mu - m)\psi^*(x) = 0,$$
$$\text{or,} \quad \left(-i\gamma^{0*}\frac{\partial}{\partial t} - i\gamma^{i*}\frac{\partial}{\partial x^i} - m\right)\psi^*(\mathbf{x},t) = 0,$$
$$\text{or,} \quad \left(-i\gamma^{0*}\frac{\partial}{\partial(-t)} - i\gamma^{i*}\frac{\partial}{\partial x^i} - m\right)\psi^*(\mathbf{x},-t) = 0,$$
$$\text{or,} \quad \left(i\gamma^{0*}\frac{\partial}{\partial t} - i\gamma^{i*}\frac{\partial}{\partial x^i} - m\right)\psi^*(\mathbf{x},-t) = 0. \tag{11.239}$$

On the other hand, if $\psi^T(\mathbf{x},t)$ were to be a solution of the Dirac equation, we should have

$$(i\gamma^\mu \partial_\mu - m)\psi^T(x) = 0,$$

or, $\quad \left(i\gamma^0 \dfrac{\partial}{\partial t} + i\gamma^i \dfrac{\partial}{\partial x^i} - m\right)\psi^T(x) = 0,$

or, $\quad \left(i\gamma^0 \dfrac{\partial}{\partial t} + i\gamma^i \dfrac{\partial}{\partial x^i} - m\right)T\psi^*(\mathbf{x},-t) = 0. \quad (11.240)$

Therefore, comparing the two equations we see that if we can find a matrix T such that

$$\begin{aligned} T\gamma^{0*}T^{-1} &= \gamma^0, \\ T\gamma^{i*}T^{-1} &= -\gamma^i, \end{aligned} \quad (11.241)$$

then the time reversed wave function

$$\psi^T(x) = \eta_\psi T\psi*(\mathbf{x},-t), \quad (11.242)$$

would satisfy the same Dirac equation as $\psi(\mathbf{x},t)$.

We note that the matrices $\gamma^{\mu*}$ satisfy the Clifford algebra

$$\{\gamma^{\mu*},\gamma^{\nu*}\} = 2\eta^{\mu\nu}\mathbb{1}, \quad (11.243)$$

and the matrices $\tilde{\gamma}^\mu$ defined as $(\tilde{\gamma}^\mu = \gamma^{\mu\dagger})$

$$\begin{aligned} \tilde{\gamma}^0 &= \gamma^0, \\ \tilde{\gamma}^i &= -\gamma^i, \end{aligned} \quad (11.244)$$

also satisfy the Clifford algebra

$$\{\tilde{\gamma}^\mu, \tilde{\gamma}^\nu\} = 2\eta^{\mu\nu}\mathbb{1}. \quad (11.245)$$

11.3 TIME REVERSAL

Thus both $\gamma^{\mu*}$ and $\tilde{\gamma}^\mu$ must be related to γ^μ through similarity transformations (by Pauli's fundamental theorem). They must also be related to each other through a similarity transformation. Thus we can identify

$$T\gamma^{\mu*}T^{-1} = \tilde{\gamma}^\mu, \tag{11.246}$$

leading to

$$\begin{aligned} T\gamma^{0*}T^{-1} &= \gamma^0, \\ T\gamma^{i*}T^{-1} &= -\gamma^i. \end{aligned} \tag{11.247}$$

To determine the explicit form of T, let us invert these relations and write

$$\begin{aligned} T^{-1}\gamma^0 T &= \gamma^{0*} = \left(\gamma^{0\dagger}\right)^T = \gamma^{0T}, \\ T^{-1}\gamma^i T &= -\gamma^{i*} = -\left(\gamma^{i\dagger}\right)^T = \gamma^{iT}. \end{aligned} \tag{11.248}$$

Thus we see that the inverse relation can be written as

$$T^{-1}\gamma^\mu T = \gamma^{\mu T}. \tag{11.249}$$

Let us also recall from (11.122) that

$$C\gamma^{\mu T}C^{-1} = -\gamma^\mu. \tag{11.250}$$

Combining the two results we obtain

$$\begin{aligned} CT^{-1}\gamma^\mu TC^{-1} &= C\gamma^{\mu T}C^{-1} = -\gamma^\mu, \\ \text{or,} \quad \left(CT^{-1}\right)\gamma^\mu\left(CT^{-1}\right)^{-1} &= -\gamma^\mu. \end{aligned} \tag{11.251}$$

In other words, the matrix CT^{-1} anticommutes with all the γ^μ's and hence must be a multiple of γ_5. Choosing

$$CT^{-1} = -\gamma_5 = -i\gamma^0\gamma^1\gamma^2\gamma^3, \tag{11.252}$$

and using the definition (11.130)

$$C = \gamma^0\gamma^2, \tag{11.253}$$

we determine (in our representation)

$$T = i\gamma^1\gamma^3. \tag{11.254}$$

Note that with this choice, the T matrix satisfies

$$T = T^\dagger = T^{-1}. \tag{11.255}$$

Furthermore, as in the case of charge conjugation, we can show that in four space-time dimensions, independent of any representation (the superscript denotes transposition)

$$(T)^T = -T, \tag{11.256}$$

so that the time inversion matrix in the spinor space is antisymmetric.

Therefore we see that the time inversion transformations

$$\begin{aligned}
\psi(\mathbf{x},t) &\xrightarrow{\mathcal{T}} \psi^{\mathcal{T}}(\mathbf{x},t) = \eta_\psi T \psi*(\mathbf{x},-t) \\
&= \eta_\psi T \gamma^{0T}\overline{\psi}^T(\mathbf{x},-t) = \eta_\psi \gamma^0 T \overline{\psi}^T(\mathbf{x},-t), \\
\overline{\psi}(\mathbf{x},t) &\xrightarrow{\mathcal{T}} \overline{\psi}^{\mathcal{T}}(\mathbf{x},t) = \eta_\psi^* \psi^T(\mathbf{x},-t) T^{-1}\gamma^0,
\end{aligned} \tag{11.257}$$

with $(|\eta_\psi|^2 = 1)$ defines a symmetry of the Dirac equation. Let us next analyze the behavior of the current four-vector under time inversion

11.3 TIME REVERSAL

$$\begin{aligned}
J^\mu(x) &= \overline{\psi}(x)\gamma^\mu\psi(x) \\
&\xrightarrow{\mathcal{T}} \overline{\psi}^T(x)\gamma^\mu\psi^T(x) \\
&= |\eta_\psi|^2 \psi^T(\mathbf{x},-t) T^{-1}\gamma^0\gamma^\mu\gamma^0 T \overline{\psi}^T(\mathbf{x},-t) \\
&= \psi^T(\mathbf{x},-t) T^{-1}\gamma^0\gamma^\mu\gamma^0 T \overline{\psi}^T(\mathbf{x},-t), \quad (11.258)
\end{aligned}$$

so that we have

$$\begin{aligned}
J^0(x) &\xrightarrow{\mathcal{T}} \psi^T(\mathbf{x},-t) T^{-1}\gamma^0 T \overline{\psi}^T(\mathbf{x},-t) \\
&= \psi^T(\mathbf{x},-t) \gamma^{0T} \overline{\psi}^T(\mathbf{x},-t) \\
&= \left(\overline{\psi}(\mathbf{x},-t)\gamma^0\psi(\mathbf{x},-t)\right)^T = J^0(\mathbf{x},-t), \\
J^i(\mathbf{x},t) &\xrightarrow{\mathcal{T}} \psi^T(\mathbf{x},-t) T^{-1}\gamma^0\gamma^i\gamma^0 T \overline{\psi}^T(\mathbf{x},-t) \\
&= -\psi^T(\mathbf{x},-t) \gamma^{iT} \overline{\psi}^T(\mathbf{x},-t) \\
&= -\left(\overline{\psi}(\mathbf{x},-t)\gamma^i\psi(\mathbf{x},-t)\right)^T = -J^i(\mathbf{x},-t), \quad (11.259)
\end{aligned}$$

which is the behavior of the current four vector under a time inversion and we can identify the transformations in (11.257) to correspond to time inversion for a Dirac system. Furthermore, we also see that a Dirac system interacting with photons would be invariant under such a transformation.

Before we examine the transformation properties of Dirac bilinears under time inversion, let us note that since

$$T^{-1}\gamma^\mu T = \gamma^{\mu T}, \quad (11.260)$$

(this is very much like the charge conjugation in (11.122) except for the -1 sign), we have

$$\begin{aligned}
T^{-1}\gamma_5 T &= \gamma_5^T, \\
T^{-1}\gamma_5 \gamma^\mu T &= \gamma_5^T \gamma^{\mu T} = (\gamma^\mu \gamma_5)^T = -(\gamma_5 \gamma^\mu)^T, \\
T^{-1}\sigma^{\mu\nu} T &= \frac{i}{2} T^{-1} \left(\gamma^\mu \gamma^\nu - \gamma^\nu \gamma^\mu \right) T \\
&= \frac{i}{2} \left(\gamma^{\mu T} \gamma^{\nu T} - \gamma^{\nu T} \gamma^{\mu T} \right) \\
&= \frac{i}{2} \left(\gamma^\nu \gamma^\mu - \gamma^\mu \gamma^\nu \right)^T = -\sigma^{\mu\nu T}. \quad (11.261)
\end{aligned}$$

The behavior of the Dirac bilinears under time reversal can now be obtained in a simple manner,

$$\begin{aligned}
\overline{\psi}(\mathbf{x},t)\psi(\mathbf{x},t) &\xrightarrow{\mathcal{T}} \overline{\psi}^T(\mathbf{x},t)\psi^T(\mathbf{x},t) \\
&= \psi^T(\mathbf{x},-t) T^{-1}\gamma^0 \gamma^0 T \overline{\psi}^T(\mathbf{x},-t) \\
&= \left(\overline{\psi}(\mathbf{x},-t)\psi(\mathbf{x},-t) \right)^T \\
&= \overline{\psi}(\mathbf{x},-t)\psi(\mathbf{x},-t), \\
\overline{\psi}(\mathbf{x},t)\gamma_5 \psi(\mathbf{x},t) &\xrightarrow{\mathcal{T}} \overline{\psi}^T(\mathbf{x},t)\gamma_5 \psi^T(\mathbf{x},t) \\
&= \psi^T(\mathbf{x},-t) T^{-1}\gamma^0 \gamma_5 \gamma^0 T \overline{\psi}^T(\mathbf{x},-t) \\
&= -\left(\overline{\psi}(\mathbf{x},-t)\gamma_5 \psi(\mathbf{x},-t) \right)^T \\
&= -\overline{\psi}(\mathbf{x},-t)\gamma_5 \psi(\mathbf{x},-t). \quad (11.262)
\end{aligned}$$

Similarly, we can show that

$$\begin{aligned}
\overline{\psi}(\mathbf{x},t)\gamma_5 \psi^0 \psi(\mathbf{x},t) &\xrightarrow{\mathcal{T}} \overline{\psi}(\mathbf{x},-t)\gamma_5 \gamma^0 \psi(\mathbf{x},-t), \\
\overline{\psi}(\mathbf{x},t)\gamma_5 \psi^i \psi(\mathbf{x},t) &\longrightarrow -\overline{\psi}(\mathbf{x},-t)\gamma_5 \gamma^i \psi(\mathbf{x},-t). \quad (11.263)
\end{aligned}$$

Namely, the axial vector (current) behaves exactly the same way as the vector (current) under a time inversion. In particular, the transformation of $J_5^0 = -\psi^\dagger \gamma_5 \psi$ shows that under time inversion the helicity does not change. We could have expected this from the definition (see (3.121))

$$h = \frac{\mathbf{s}\cdot\mathbf{p}}{|\mathbf{p}|} \xrightarrow{\mathcal{T}} \frac{(-\mathbf{s})\cdot(-\mathbf{p})}{|\mathbf{p}|} = \frac{\mathbf{s}\cdot\mathbf{p}}{|\mathbf{p}|} = h. \tag{11.264}$$

We can also show that

$$\overline{\psi}(\mathbf{x},t)\sigma^{0i}\psi(\mathbf{x},t) \xrightarrow{\mathcal{T}} \overline{\psi}(\mathbf{x},-t)\sigma^{0i}\psi(\mathbf{x},-t),$$
$$\overline{\psi}(\mathbf{x},t)\sigma^{ij}\psi(\mathbf{x},t) \xrightarrow{\mathcal{T}} -\overline{\psi}(\mathbf{x},-t)\sigma^{ij}\psi(\mathbf{x},-t). \tag{11.265}$$

In other words, the tensor bilinears have precisely the same behavior under time reversal as the electromagnetic field strength tensor $F^{\mu\nu}$. It is also easy to see now that under time reversal, the chirality of a spinor does not change, namely,

$$\begin{aligned}\gamma_5\psi^{\mathcal{T}} &= \eta_\psi\gamma_5 T\psi^* = \eta_\psi T\gamma_5^T\psi^* \\ &= \eta_\psi T\left(\gamma_5^\dagger\psi\right)^* = \eta_\psi T\left(\gamma_5\psi\right)^*,\end{aligned} \tag{11.266}$$

so that a left-handed spinor remains left-handed under time reversal and a right-handed spinor remains right-handed.

11.3.3 Consequences of $\mathcal{T}$ invariance. Let us recall that under time reversal, a state with given angular momentum transforms as (see (11.222))

$$|\alpha,\mathbf{p},j,m_j\rangle \xrightarrow{\mathcal{T}} (i)^{2m_j}|\alpha,-\mathbf{p},j,-m_j\rangle, \tag{11.267}$$

where α stands for all the other quantum numbers of the state. Let us also note that if we are looking at a reaction of the form

$$a+b \longrightarrow c+d, \tag{11.268}$$

then the transition amplitude can be represented as

$$\langle f|S|i\rangle, \tag{11.269}$$

where $|i\rangle$ stands for the initial state, $|f\rangle$ for the final state and S denotes the S-matrix or the time evolution matrix. On the other hand, if we are looking at the reciprocal reaction

$$c+d \longrightarrow a+b, \tag{11.270}$$

then the transition amplitude can be represented as

$$\langle i|S|f\rangle, \tag{11.271}$$

where we are still using the same notation, namely,

$$|i\rangle = |a,b\rangle, \qquad |f\rangle = |c,d\rangle. \tag{11.272}$$

Since the time evolution operator is not Hermitian, namely,

$$S^\dagger \neq S, \tag{11.273}$$

it does not follow *a priori* that

$$|\langle i|S|f\rangle|^2 = |\langle f|S|i\rangle|^2. \tag{11.274}$$

On the other hand, experimentally this is seen to hold in many processes. This is known as the principle of detailed balance and has been used, for example, to obtain the spin of π-meson etc. It is worth emphasizing here that the detailed balance does not say that the rates for the two reactions

$$a+b \longrightarrow c+d,$$

and

$$c+d \longrightarrow a+b, \tag{11.275}$$

11.3 TIME REVERSAL

have to be the same. In fact, from Fermi's Golden rule we know that the rate for a process $|i\rangle \to |f\rangle$ is given by

$$W_{i \to f} \propto \frac{1}{\hbar} |\langle f|S|i\rangle|^2 \rho_f, \qquad (11.276)$$

where ρ_f denotes the density of states for the final state which may not be the same for the two processes.

Let us next try to understand, theoretically, when the principle of detailed balance may hold. Let us assume that we are dealing with a system where the interaction Hamiltonian can be treated as a perturbation. In this case, we can write (for distinct $|i\rangle$ and $|f\rangle$ states)

$$\langle f|S|i\rangle = -2\pi i \langle f|H_{\text{int}}|i\rangle \delta(E_i - E_f), \qquad (11.277)$$

and furthermore, if the interaction Hamiltonian is Hermitian, then we have

$$\begin{aligned}
|\langle f|S|i\rangle| &= 2\pi |\langle f|H_{\text{int}}|i\rangle| \delta(E_i - E_f) \\
&= 2\pi |\langle i|H_{\text{int}}|f\rangle| \delta(E_i - E_f) \\
&= |\langle i|S|f\rangle|.
\end{aligned} \qquad (11.278)$$

In this case, therefore, the principle of detailed balance will hold.

Let us next assume that we are dealing with a system whose Hamiltonian is invariant under time reversal. It follows then (in the operator formulation) that the evolution operator or the S-matrix must also be invariant under time inversion. That is,

$$\mathcal{T} S \mathcal{T}^\dagger = S. \qquad (11.279)$$

Let us choose the initial and the final states to correspond to

$$\begin{aligned}
|i\rangle &= |\alpha, \mathbf{p}_i, m_i\rangle, \\
|f\rangle &= |\beta, \mathbf{p}_f, m_f\rangle,
\end{aligned} \qquad (11.280)$$

where m_i and m_f denote the spin projections for the initial and the final state respectively (corresponding to the same angular momentum). Under time reversal, as we have seen in (11.222)

$$\begin{aligned}
\mathcal{T}|i\rangle &= (i)^{2m_i}|\alpha, -\mathbf{p}_i, -m_i\rangle, \\
\mathcal{T}|f\rangle &= (i)^{2m_f}|\beta, -\mathbf{p}_f, -m_f\rangle,
\end{aligned} \qquad (11.281)$$

so that in this case, time reversal invariance would imply that

$$\begin{aligned}
\langle f|S|i\rangle &= \langle \beta, \mathbf{p}_f, m_f|S|\alpha, \mathbf{p}_i, m_i\rangle \\
&= \langle \beta, \mathbf{p}_f, m_f|\mathcal{T}^{-1}\mathcal{T}S\mathcal{T}^{-1}\mathcal{T}|\alpha, \mathbf{p}_i, m_i\rangle \\
&= (i)^{2m_i-2m_f}\langle \alpha, -\mathbf{p}_i, -m_i|S|\beta, -\mathbf{p}_f, -m_f\rangle, \quad (11.282)
\end{aligned}$$

where we have used (the third) relation in (11.204). Since the transitions can only be from integer spin states to integer spin states or half integer spin states to half integer spin states, the phase factor can only take values ± 1 and we have

$$\begin{aligned}
\langle \beta, \mathbf{p}_f, m_f|S|\alpha, \mathbf{p}_i, m_i\rangle &= \pm\langle \alpha, -\mathbf{p}_i, -m_i|S|\beta, -\mathbf{p}_f, -m_f\rangle, \\
\text{or,} \quad |\langle \beta, \mathbf{p}_f, m_f|S|\alpha, \mathbf{p}_i, m_i\rangle| &= |\langle \alpha, -\mathbf{p}_i, -m_i|S|\beta, -\mathbf{p}_f, -m_f\rangle|.
\end{aligned}$$
$$(11.283)$$

This is known as the reciprocity relation and holds for any system where time reversal invariance is a symmetry. But this is not the same as the detailed balance. Let us note that if, in addition, the system is invariant under parity, we have

$$\begin{aligned}
&|\langle \beta, \mathbf{p}_f, m_f|S|\alpha, \mathbf{p}_i, m_i\rangle| \\
&= |\langle \alpha, -\mathbf{p}_i, m_i|\mathcal{P}^{-1}\mathcal{P}S\mathcal{P}^{-1}\mathcal{P}|\beta, -\mathbf{p}_f, m_f\rangle| \\
&= |\langle \alpha, \mathbf{p}_i, -m_i|S|\beta, \mathbf{p}_f, -m_f\rangle|.
\end{aligned} \qquad (11.284)$$

Therefore, if we square this relation and sum over all the spin projections m_i, m_f, we obtain

11.3 TIME REVERSAL

$$\sum_{m_i,m_f} |\langle \beta, \mathbf{p}_f, m_f|S|\alpha, \mathbf{p}_i, m_i\rangle|^2$$
$$= \sum_{m_i,m_f} |\langle \alpha, \mathbf{p}_i, m_i|S|\beta, \mathbf{p}_f, m_f\rangle|^2. \tag{11.285}$$

This is very similar to the principle of detailed balance but not quite. This relation is known as the principle of semi-detailed balance and holds true in every system for which $\mathcal{P}$ and $\mathcal{T}$ are symmetries.

Note that if we are looking for a subclass of processes where

$$m_i = m_f = m, \tag{11.286}$$

then from $\mathcal{P}$ and $\mathcal{T}$ invariance we have

$$\langle \beta, \mathbf{p}_f, m|S|\alpha, \mathbf{p}_i, m\rangle = \langle \alpha, \mathbf{p}_i, -m|S|\beta, \mathbf{p}_f, -m\rangle. \tag{11.287}$$

Furthermore, since rotations are assumed to be a symmetry of physical systems, we can make a rotation on the right-hand side to bring the state $|-m\rangle \to |m\rangle$. Thus in this case we will have

$$\langle \beta, \mathbf{p}_f, m|S|\alpha, \mathbf{p}_i, m\rangle = \langle \alpha, \mathbf{p}_i, m|S|\beta, \mathbf{p}_f, m\rangle \tag{11.288}$$

In this case, therefore, the principle of detailed balance will hold.

Note that in weak interactions, $\mathcal{P}$ is violated – but the strength of the interaction Hamiltonian is assumed to be weak. On the other hand, strong and electromagnetic interactions are assumed to be $\mathcal{P}$ and $\mathcal{T}$ symmetric. Therefore, it is obvious that the principle of detailed balance or the semi-detailed balance must hold in practically all processes.

11.3.4 Electric dipole moment of neutron. Time reversal invariance holds extremely well in strong and electromagnetic interactions. The stringent limits on $\mathcal{T}$-invariance in electromagnetic interactions comes from the measurement of the electric dipole moment of the neutron. Let us recall that although the neutron is electrically neutral, it has a magnetic dipole moment. Let us assume that it also has an electric dipole moment. Since there is no natural axis for the neutron other than the spin axis, the electric dipole moment would be parallel to its spin. Thus the neutron would interact with an electromagnetic field through the interaction Hamiltonian (there is no minimal coupling, this is the Pauli coupling)

$$H_{\text{int}} = \mu_e \, \boldsymbol{\sigma} \cdot \mathbf{E} + \mu_m \, \boldsymbol{\sigma} \cdot \mathbf{B}. \tag{11.289}$$

Let us also note here that the only natural dimension associated with the neutron is its size which is of the order of $1F = 10^{-13}$ cm. Thus the magnitude of any dipole moment (electric) will be of the order of

$$\mu_e \simeq ed \simeq e\,10^{-13} \text{ cm} = 10^{-13} \text{ e-cm}. \tag{11.290}$$

We have already seen that under parity and time reversal

$$\mathbf{E} \xrightarrow{\mathcal{P}} -\mathbf{E}, \qquad \mathbf{E} \xrightarrow{\mathcal{T}} \mathbf{E},$$
$$\mathbf{B} \xrightarrow{\mathcal{P}} \mathbf{B}, \qquad \mathbf{B} \xrightarrow{\mathcal{T}} -\mathbf{B},$$
$$\boldsymbol{\sigma} \xrightarrow{\mathcal{P}} \boldsymbol{\sigma}, \qquad \boldsymbol{\sigma} \xrightarrow{\mathcal{T}} -\boldsymbol{\sigma}, \tag{11.291}$$

so that we have

$$\boldsymbol{\sigma} \cdot \mathbf{E} \xrightarrow{\mathcal{P}} -\boldsymbol{\sigma} \cdot \mathbf{E}, \qquad \boldsymbol{\sigma} \cdot \mathbf{E} \xrightarrow{\mathcal{T}} -\boldsymbol{\sigma} \cdot \mathbf{E},$$
$$\boldsymbol{\sigma} \cdot \mathbf{B} \xrightarrow{\mathcal{P}} \boldsymbol{\sigma} \cdot \mathbf{B}, \qquad \boldsymbol{\sigma} \cdot \mathbf{B} \xrightarrow{\mathcal{T}} \boldsymbol{\sigma} \cdot \mathbf{B}. \tag{11.292}$$

In other words, whereas the interaction involving the magnetic dipole moment is invariant under $\mathcal{P}$ and $\mathcal{T}$, the electric dipole interaction changes sign under each one of the transformations. Consequently, if electromagnetic interactions are invariant under $\mathcal{P}$ and $\mathcal{T}$, then the electric dipole moment of the neutron must vanish. Experiments have obtained extremely small limits on μ_e,

$$(\mu_e)_{\text{exp}} < 0.63 \times 10^{-25} \text{ e-cm}. \tag{11.293}$$

This basically measures the extent to which $\mathcal{P}$ and $\mathcal{T}$ invariances hold in electromagnetic interactions.

11.4 CPT theorem

We have seen that some of the discrete symmetries appear to be violated in weak interactions. We can naturally ask whether there exists any combination of these symmetries which will be a symmetry of the weak Hamiltonian. The CPT theorem is basically an answer to this question and in essence says that any physical Hamiltonian would be invariant under the combined operation of P, C and T in any order even though the individual transformations may not be a symmetry of the system. Rather than going into the technical proof of this theorem, let us discuss the obvious consequences of this theorem.

11.4.1 Equality of mass for particles and antiparticles. Let us consider the state of a particle at rest which satisfies

$$H|\alpha, m, \mathbf{s}\rangle = m|\alpha, m, \mathbf{s}\rangle, \tag{11.294}$$

where m denotes the mass of the particle, $\mathbf{s}$ its spin and α denotes the other quantum numbers. The operation of CP, of course, takes a particle to the antiparticle state. Thus under the CPT transformation,

$$|\alpha, m, \mathbf{s}\rangle \xrightarrow{CPT} \eta\langle\overline{\alpha}, \overline{m}, -\mathbf{s}|, \tag{11.295}$$

where $\overline{m}, \overline{\alpha}$ denote the parameters associated with the antiparticle (in this section, quantities with bars would refer to antiparticle parameters). Thus we see that

$$\begin{aligned} m &= \langle\alpha, m, \mathbf{s}|H|\alpha, m, \mathbf{s}\rangle \\ &= \langle\alpha, m, \mathbf{s}|(CPT)^{-1}(CPT)H(CPT)^{-1}(CPT)|\alpha, m, \mathbf{s}\rangle \\ &= \langle\overline{\alpha}, \overline{m}, -\mathbf{s}|(CPT)H(CPT)^{-1}|\overline{\alpha}, \overline{m}, -\mathbf{s}\rangle. \end{aligned} \tag{11.296}$$

If the Hamiltonian is CPT invariant, then it follows that

$$m = \langle\overline{\alpha}, \overline{m}, -\mathbf{s}|H|\overline{\alpha}, \overline{m}, -\mathbf{s}\rangle = \overline{m}. \tag{11.297}$$

In other words, the equality of particle and antiparticle masses is a consequence of $\mathcal{CPT}$ invariance. Even though C-invariance is violated in weak interactions, it is CPT invariance which guarantees that, for example, K^0 and $\overline{K}^0$ have the same mass. In fact, the experimental limit on the mass difference between the $K^0-\overline{K}^0$ provides the best limit on $\mathcal{CPT}$ invariance.

11.4.2 Electric charge for particles and antiparticles. Let $\hat{Q}$ denote the operator which measures the charge of a state. As we have seen

$$\hat{Q} \xrightarrow{\mathcal{P}} \hat{Q}, \qquad \hat{Q} \xrightarrow{\mathcal{C}} -\hat{Q}, \qquad \hat{Q} \xrightarrow{\mathcal{T}} \hat{Q}, \qquad (11.298)$$

so that

$$\hat{Q} \xrightarrow{\mathcal{CPT}} -\hat{Q}. \qquad (11.299)$$

(Recall $\hat{Q} = \int d^3x\, \overline{\psi}\gamma^0\psi$ for a Dirac particle.)

Thus for a one particle state with charge q at rest, we have

$$\begin{aligned} q &= \langle \alpha, q, \mathbf{s} | \hat{Q} | \alpha, q, \mathbf{s} \rangle \\ &= \langle \alpha, q, \mathbf{s} | (\mathcal{CPT})^{-1}(\mathcal{CPT})\hat{Q}(\mathcal{CPT})^{-1}(\mathcal{CPT}) | \alpha, q, \mathbf{s} \rangle \\ &= \langle \overline{\alpha}, \overline{q}, -\mathbf{s} | (-\hat{Q}) | \overline{\alpha}, \overline{q}, -\mathbf{s} \rangle \\ &= -\overline{q}. \end{aligned} \qquad (11.300)$$

Namely, the electric charge for particles is equal in magnitude but opposite in sign from that of the antiparticles. In this way, one can show that all the quantum numbers of antiparticles are equal in magnitude but opposite in sign from those of the particles.

11.4.3 Equality of lifetimes for particles and antiparticles. Since the weak interactions violate parity, let us decompose the weak Hamiltonian into a sum of Hermitian terms as (basically into a parity even and a parity odd part)

$$H_{WK} = H_+ + H_-, \qquad (11.301)$$

11.4 $\mathcal{CPT}$ THEOREM

where

$$\mathcal{P}H_{\pm}\mathcal{P}^{-1} = \pm H_{\pm}. \tag{11.302}$$

In such a case, the transition probability from an initial state $|i\rangle$ to a final state $|f\rangle$ would be given by

$$\begin{aligned}
|\langle f|S|i\rangle|^2 &\propto |\langle f|H_{WK}|i\rangle|^2 \\
&= |\langle f|H_+ + H_-|i\rangle|^2 \\
&= |\langle f|H_+|i\rangle|^2 + |\langle f|H_-|i\rangle|^2. \tag{11.303}
\end{aligned}$$

The cross terms would have a pseudoscalar character and would vanish since the reaction rate is a scalar quantity.

Let us consider the decay channel of a particle

$$|\alpha, j, m_j\rangle \longrightarrow |\beta, j, m_j\rangle, \tag{11.304}$$

and the corresponding antiparticle decay channel

$$|\overline{\alpha}, j, m_j\rangle \longrightarrow |\overline{\beta}, j, m_j\rangle. \tag{11.305}$$

We know that

$$\begin{aligned}
&\langle \beta, j, m_j|H_{\pm}|\alpha, j, m_j\rangle \\
&= \langle \beta, j, m_j|(\mathcal{CPT})^{-1}(\mathcal{CPT})H_{\pm}(\mathcal{CPT})^{-1}(\mathcal{CPT})|\alpha, j, m_j\rangle \\
&= (i)^{2m_j - 2m_j}\langle \overline{\alpha}, j, -m_j|H_{\pm}|\overline{\beta}, j, -m_j\rangle \\
&= \langle \overline{\alpha}, j, -m_j|H_{\pm}|\overline{\beta}, j, -m_j\rangle, \tag{11.306}
\end{aligned}$$

where we have used the fact that $H_{\pm}$ would be individually invariant under $\mathcal{CPT}$ according to the $\mathcal{CPT}$ theorem. Furthermore, using the rotational invariance of physical Hamiltonians, we can rewrite this relation as

$$\langle \beta, j, m_j|H_{\pm}|\alpha, j, m_j\rangle = \langle \overline{\alpha}, j, m_j|H_{\pm}|\overline{\beta}, j, m_j\rangle. \tag{11.307}$$

Since the Hamiltonians $H_\pm$ are Hermitian we also have

$$\langle \beta, j, m_j | H_\pm | \alpha, j, m_j \rangle = \langle \overline{\beta}, j, m_j | H_\pm | \overline{\alpha}, j, m_j \rangle^*. \tag{11.308}$$

Here we have neglected an overall phase factor (depending on α, β) which is not relevant for the subsequent discussion.

From $\mathcal{CPT}$ invariance, therefore, we have obtained

$$|\langle \beta, j, m_j | H_\pm | \alpha, j, m_j \rangle|^2 = |\langle \overline{\beta}, j, m_j | H_\pm | \overline{\alpha}, j, m_j \rangle|^2, \tag{11.309}$$

which leads to a relation between the total life times (τ, and $\overline{\tau}$) of particles and antiparticles, namely

$$\frac{1}{\tau} = \frac{2\pi}{\hbar} \sum_\beta \left[|\langle \beta, j, m_j | H_+ | \alpha, j, m_j \rangle|^2 + |\langle \beta, j, m_j | H_- | \alpha, j, m_j \rangle|^2 \right] \rho_\beta$$

$$= \frac{2\pi}{\hbar} \sum_{\overline{\beta}} \left[|\langle \overline{\beta}, j, m_j | H_+ | \overline{\alpha}, j, m_j \rangle|^2 + |\langle \overline{\beta}, j, m_j | H_- | \overline{\alpha}, j, m_j \rangle|^2 \right] \rho_{\overline{\beta}}$$

$$= \frac{1}{\overline{\tau}}. \tag{11.310}$$

Here we have assumed that the phase space for the decay products of the antiparticle is the same as that for the decay products of the particle. This follows trivially from the equality of particle and antiparticle masses. However, we note here that the individual decay channels may have different life times for particle and anti-particle.

There are many other interesting consequences of the $\mathcal{CPT}$ theorem including the connection between spin and statistics which we cannot go into. Let us simply note here that since $\mathcal{CPT}$ must be conserved in any physical theory and we know that $\mathcal{CP}$ is violated in weak interactions by a small amount, it follows that the weak interactions must violate $\mathcal{T}$-invariance by a small amount as well so that $\mathcal{CPT}$ will hold.

11.5 References

1. S. Okubo, Physical Review **109**, 984 (1958).

2. S. Schweber, *Introduction to Relativistic Quantum Field Theory*, Row, Peterson, Evanston (1961).

3. J. D. Bjorken and S. Drell, *Relativistic Quantum Fields*, McGraw-Hill, New York, 1964.

4. J. J. Sakurai, *Invariance Principles and Elementary Particles*, Princeton University Press, Princeton (1964).

5. R. F. Streater and A. S. Wightman, *PCT, Spin and Statistics and All That*, Benjamin, New York (1964).

6. S. Gasiorowicz, *Elementary Particle Physics*, John Wiley, New York (1966).

7. P. Roman, *Introduction to Quantum Field Theory*, John Wliley, New York (1969).

8. C. Itzykson and J-B. Zuber, *Quantum Field Theory*, McGraw-Hill, New York, 1980.

9. F. Gross, *Relativistic Quantum Mechanics and Field Theory*, John Wiley, New York (1993).

CHAPTER 12

Yang-Mills theory

12.1 Non-Abelian gauge theories

So far, we have talked about the simplest of gauge theories, namely, the Maxwell theory which is based on the Abelian gauge group $U(1)$. We will now study gauge theories based on more complicated symmetries. Such theories belonging to non-Abelian (non-commutative) gauge groups are commonly known as Yang-Mills theories and are fundamental building blocks in the construction of physical theories. Let us recall that gauge field theories necessarily arise when we try to promote a global symmetry to a local symmetry. For example, let us quickly review how the gauge fields come into the theory in the case of quantum electrodynamics (QED).

Let us start with the Dirac Lagrangian density for a free fermion (8.9)

$$\mathcal{L} = i\overline{\psi}\partial\!\!\!/\psi - m\overline{\psi}\psi, \tag{12.1}$$

where ψ provides a representation of $U(1)$ (namely, it is a complex field). As we have seen in (8.37) and (8.38), this Lagrangian density has the global symmetry

$$\begin{aligned}\psi &\to e^{-i\theta}\psi, \\ \overline{\psi} &\to \overline{\psi}e^{i\theta},\end{aligned} \tag{12.2}$$

where θ is a real constant parameter and infinitesimally, we have

$$\begin{aligned}\delta\psi &= -i\epsilon\psi, \\ \delta\overline{\psi} &= i\epsilon\overline{\psi},\end{aligned} \tag{12.3}$$

with $\theta = \epsilon =$ infinitesimal. The internal symmetry in this case arises from the fact that ψ is a complex field variable while the Lagrangian density which is a functional of ψ is Hermitian. Consequently, the phase of ψ remains arbitrary.

If we now want to promote the symmetry to be local, namely $\epsilon = \epsilon(x)$ (or $\theta = \theta(x)$), we note that

$$\begin{aligned}
\delta\psi &= -i\epsilon(x)\psi, \\
\delta\overline{\psi} &= i\epsilon(x)\overline{\psi}, \\
\delta\left(\partial_\mu\psi\right) &= \partial_\mu(\delta\psi) = \partial_\mu(-i\epsilon(x)\psi) \\
&= -i\left((\partial_\mu\epsilon(x)) + \epsilon(x)\partial_\mu\right)\psi(x).
\end{aligned} \qquad (12.4)$$

Consequently, we have

$$\begin{aligned}
\delta\mathcal{L} &= i\delta\overline{\psi}\partial\!\!\!/\psi + i\overline{\psi}\gamma^\mu\partial_\mu\delta\psi - m\delta\overline{\psi}\psi - m\overline{\psi}\delta\psi \\
&= -\epsilon(x)\overline{\psi}\partial\!\!\!/\psi + \overline{\psi}\gamma^\mu\left((\partial_\mu\epsilon(x)) + \epsilon(x)\partial_\mu\right)\psi \\
&\quad - im\epsilon(x)\overline{\psi}\psi + im\epsilon(x)\overline{\psi}\psi, \\
&= (\partial_\mu\epsilon(x))\overline{\psi}\gamma^\mu\psi.
\end{aligned} \qquad (12.5)$$

Neither the Lagrangian density (12.1) nor the corresponding action is invariant under the local phase transformation in (12.4) since $\partial_\mu\psi$ does not transform covariantly and there is no term in the Lagrangian density (12.1) whose variation would cancel the $\partial_\mu\epsilon(x)$ term.

As we have seen earlier (see (9.81)), the way out of this difficulty is to define a covariant derivative $D_\mu\psi$ such that it transforms covariantly under the local transformation, namely,

$$\delta\left(D_\mu\psi\right) = -i\epsilon(x)D_\mu\psi(x). \qquad (12.6)$$

Clearly this can be achieved if we introduce a new field variable $A_\mu(x)$ (gauge field) and define the covariant derivative as

$$D_\mu\psi(x) = \left(\partial_\mu + ieA_\mu(x)\right)\psi(x), \qquad (12.7)$$

with the transformation property for the additional field

$$\delta A_\mu(x) = \frac{1}{e}\partial_\mu\epsilon(x). \tag{12.8}$$

We recognize (12.7) as describing the minimal coupling of the charged fermions to the electromagnetic field and as we have seen in (9.86), the Lagrangian density

$$\mathcal{L} = i\overline{\psi}\gamma^\mu D_\mu\psi(x) - m\overline{\psi}\psi, \tag{12.9}$$

is invariant under the infinitesimal local gauge transformations

$$\begin{aligned}\delta\psi &= -i\epsilon(x)\psi(x),\\ \delta\overline{\psi} &= i\epsilon(x)\overline{\psi}(x),\\ \delta A_\mu &= \frac{1}{e}\partial_\mu\epsilon(x).\end{aligned} \tag{12.10}$$

However, the field A_μ known as the gauge field, is nondynamical in this theory. To introduce the kinetic energy part for the gauge field (in order to give it dynamics) in a gauge invariant manner, we note that since the gauge field transforms longitudinally (by a gradient), the curl of the field would be unchanged by such a transformation. That is, if we define a field strength tensor

$$F_{\mu\nu} = \partial_\mu A_\nu - \partial_\nu A_\mu = -F_{\nu\mu}, \tag{12.11}$$

under the gauge transformation (12.10), this would transform as

$$\delta F_{\mu\nu} = \partial_\mu \delta A_\nu - \partial_\nu \delta A_\mu = \frac{1}{e}(\partial_\mu\partial_\nu\epsilon - \partial_\nu\partial_\mu\epsilon) = 0. \tag{12.12}$$

Hence the field strength tensor is gauge invariant and the gauge invariant Lagrangian density for the dynamical part of the gauge field can be written as (quadratic in derivatives)

$$\mathcal{L}_{\text{gauge}} = -\frac{1}{4}F_{\mu\nu}F^{\mu\nu}. \tag{12.13}$$

The total Lagrangian density for QED is then given by (see also (9.80))

$$\mathcal{L}_{\text{QED}} = -\frac{1}{4}F_{\mu\nu}F^{\mu\nu} + i\overline{\psi}\gamma^{\mu}D_{\mu}\psi - m\overline{\psi}\psi, \tag{12.14}$$

and has the local gauge invariance described in (12.10).

Let us now generalize these ideas to theories with more complicated symmetries. Let us consider a free Dirac theory described by the Lagrangian density (containing several complex fermion fields)

$$\mathcal{L} = i\overline{\psi}_k \partial\!\!\!/\psi_k - m\overline{\psi}_k\psi_k, \qquad k = 1, 2, \ldots, \dim R, \tag{12.15}$$

where k is an internal symmetry index. Namely, we assume that ψ_k belongs to a nontrivial representation R of some internal symmetry group G and $\dim R$ denotes the dimensionality of the representation. This Lagrangian density is invariant under the global phase transformations

$$\begin{aligned}\psi_k &\rightarrow (U\psi)_k = \left(e^{-i\theta^a T^a}\psi\right)_k, \qquad a = 1, 2, \cdots, \dim G,\\ \overline{\psi}_k &\rightarrow \left(\overline{\psi}U^{\dagger}\right)_k = \left(\overline{\psi}U^{-1}\right)_k = \left(\overline{\psi}e^{i\theta^a T^a}\right)_k,\end{aligned} \tag{12.16}$$

where θ^a denotes the real global parameter of transformation and infinitesimally we have

$$\begin{aligned}\delta\psi_k &= -i\epsilon^a T^a_{k\ell}\psi_\ell, & a = 1, 2, \ldots, \dim G,\\ \delta\overline{\psi}_k &= i\epsilon^a \overline{\psi}_\ell (T^a)_{\ell k},\end{aligned} \tag{12.17}$$

with $\theta^a = \epsilon^a = $ infinitesimal. Here the T^a's represent the generators of the internal symmetry group G and are assumed to be Hermitian. Furthermore, $\dim G$ represents the dimensionality of the symmetry group. (We note here parenthetically that $\dim G = n^2 - 1$ for the group $SU(n)$.) The generators (matrices) T^a satisfy the Lie algebra of the Lie group G (this is necessary for the transformations to form a group)

12.1 NON-ABELIAN GAUGE THEORIES

$$[T^a, T^b] = if^{abc}T^c, \qquad a,b,c = 1,2,\ldots,\dim G, \qquad (12.18)$$

where the real constants f^{abc} which are completely antisymmetric (antisymmetric under the exchange of any pair of indices) are known as the structure constants of the symmetry group. (For semi-simple groups the structure constants can always be chosen to be completely antisymmetric. For example, we already know from the study of angular momenta, the structure constants of the group $SU(2)$ can be identified with $f^{abc} = \epsilon^{abc}, a, b, c = 1, 2, 3$ while for the group $SU(3)$ the structure constants are more complicated and the nontrivial ones are given by $f^{123} = 1, f^{147} = f^{246} = f^{257} = f^{345} = \frac{1}{2}, f^{156} = f^{367} = -\frac{1}{2}, f^{458} = f^{678} = \frac{\sqrt{3}}{2}$.) When the generators of the symmetry group do not commute (namely, the structure constants are nontrivial), the symmetry group is known as a non-Abelian group. It is easy to see that the Lagrangian density is invariant under the transformations (12.17) (or (12.16)), namely,

$$\begin{aligned}
\delta \mathcal{L} &= \delta\overline{\psi}_k(i\slashed{\partial} - m)\psi_k + \overline{\psi}_k(i\slashed{\partial} - m)\delta\psi_k \\
&= i\epsilon^a \overline{\psi}_\ell (T^a)_{\ell k}(i\slashed{\partial} - m)\psi_k - i\epsilon^a \overline{\psi}_k(i\slashed{\partial} - m)T^a_{k\ell}\psi_\ell \\
&= i\epsilon^a \overline{\psi}_k (T^a)_{k\ell}(i\slashed{\partial} - m)\psi_\ell - i\epsilon^a \overline{\psi}_k(i\slashed{\partial} - m)T^a_{k\ell}\psi_\ell \\
&= 0. \qquad (12.19)
\end{aligned}$$

Here we have used the fact that the symmetry generators are Hermitian, $(T^a)^\dagger = T^a$ as well as the fact that the internal symmetry generators commute with the Dirac gamma matrices since the two act on different spaces.

Let us now try to make this symmetry a local symmetry of our theory. That is, let us consider the infinitesimal local phase transformations

$$\begin{aligned}
\delta\psi_k &= -i\epsilon^a(x)T^a_{k\ell}\psi_\ell, \\
\delta\overline{\psi}_k &= i\epsilon^a(x)\overline{\psi}_\ell T^a_{\ell k}. \qquad (12.20)
\end{aligned}$$

Once again, we note that the ordinary derivative acting on the fermion field does not transform covariantly under this transformation, namely,

$$\delta\left(\partial_{\mu}\psi_{k}\right) = \partial_{\mu}\left(-i\epsilon^{a}(x)T^{a}_{k\ell}\psi_{\ell}\right)$$
$$= -i\left(\partial_{\mu}\epsilon^{a}(x)\right)T^{a}_{k\ell}\psi_{\ell} - i\epsilon^{a}(x)T^{a}_{k\ell}\partial_{\mu}\psi_{\ell}, \quad (12.21)$$

so that under the local transformation (12.20),

$$\delta\mathcal{L} = \delta\overline{\psi}_{k}(i\slashed{\partial} - m)\psi_{k} + \overline{\psi}_{k}(i\slashed{\partial} - m)\delta\psi_{k}$$
$$= (\partial_{\mu}\epsilon^{a}(x))\overline{\psi}_{k}\gamma^{\mu}(T^{a})_{k\ell}\psi_{\ell}. \quad (12.22)$$

As in the case of the $U(1)$ theory, there is no other term in the Lagrangian density whose variation would cancel the $\partial_{\mu}\epsilon^{a}(x)$ term in (12.22). Hence we define a covariant derivative $(D_{\mu}\psi)_{k}$ such that under an infinitesimal local transformation

$$\delta\left(D_{\mu}\psi\right)_{k} = -i\epsilon^{a}(x)T^{a}_{k\ell}\left(D_{\mu}\psi\right)_{\ell}. \quad (12.23)$$

Introducing a new field (gauge field), we write the covariant derivative (in the matrix notation) as

$$D_{\mu}\psi = (\partial_{\mu} + igA_{\mu})\psi, \quad (12.24)$$

where g denotes the appropriate coupling constant and

$$A_{\mu} = T^{a}A^{a}_{\mu}. \quad (12.25)$$

This is the generalization of the minimal coupling (in QED) to the present case and the T^{a}'s are the generators of the symmetry group belonging to the same representation as the fermions. We can compare this with QED where the generator is the identity matrix $\mathbb{1}$ which commutes with every operator. We note that the number of gauge fields (that are needed to define a covariant derivative) is the same as $\dim G$, the dimension of the Lie group. We can also write the covariant derivative in (12.24) explicitly as

$$D_{\mu}\psi_{k} = \partial_{\mu}\psi_{k} + igT^{a}_{k\ell}A^{a}_{\mu}\psi_{\ell} = \left(\partial_{\mu}\delta_{k\ell} + igT^{a}_{k\ell}A^{a}_{\mu}\right)\psi_{\ell}. \quad (12.26)$$

12.1 NON-ABELIAN GAUGE THEORIES

Noting that we would like the covariant derivative to transform covariantly under an infinitesimal local transformation,

$$\delta \left(D_\mu \psi\right)_k = -i\epsilon^a T^a_{k\ell} \left(D_\mu \psi\right)_\ell, \tag{12.27}$$

we conclude from (12.26) that we should have

$$\begin{aligned}
igT^a_{k\ell}\delta\left(A^a_\mu \psi_\ell\right) &= \delta(D_\mu \psi_k) - \partial_\mu \delta \psi_k \\
&= -i\epsilon^a T^a_{k\ell}(D_\mu \psi)_\ell - \partial_\mu \delta \psi_k \\
&= -i\epsilon^a T^a_{k\ell}(\partial_\mu \psi_\ell + igT^b_{\ell m}A^b_\mu \psi_m) - \partial_\mu\left(-i\epsilon^a T^a_{k\ell}\psi_\ell\right) \\
&= -i\epsilon^a T^a_{k\ell}(\partial_\mu \psi_\ell + igT^b_{\ell m}A^b_\mu \psi_m) \\
&\quad + i\left(\partial_\mu \epsilon^a\right) T^a_{k\ell}\psi_\ell + i\epsilon^a T^a_{k\ell}\partial_\mu \psi_\ell \\
&= i\left(\partial_\mu \epsilon^a\right)T^a_{k\ell}\psi_\ell + g\epsilon^a T^a_{k\ell}T^b_{\ell m}A^b_\mu \psi_m.
\end{aligned} \tag{12.28}$$

We note that the left-hand side in (12.28) can be written in the form

$$\begin{aligned}
igT^a_{k\ell}\delta\left(A^a_\mu\psi_\ell\right) &= igT^a_{k\ell}\delta A^a_\mu \psi_\ell + igT^a_{k\ell}A^a_\mu \delta \psi_\ell \\
&= igT^a_{k\ell}\delta A^a_\mu \psi_\ell + igT^a_{k\ell}A^a_\mu\left(-i\epsilon^b\right)T^b_{\ell m}\psi_m \\
&= igT^a_{k\ell}\delta A^a_\mu \psi_\ell + g\epsilon^a T^b_{k\ell}A^b_\mu T^a_{\ell m}\psi_m.
\end{aligned} \tag{12.29}$$

Substituting this into (12.28), we obtain

$$\begin{aligned}
igT^a_{k\ell}&\delta A^a_\mu \psi_\ell \\
&= i\left(\partial_\mu \epsilon^a\right)T^a_{k\ell}\psi_\ell + g\epsilon^a \left(T^a T^b\right)_{km}A^b_\mu \psi_m \\
&\quad - g\epsilon^a \left(T^b T^a\right)_{km}A^b_\mu \psi_m \\
&= i\left(\partial_\mu \epsilon^a\right)T^a_{k\ell}\psi_\ell + g\epsilon^a \left[T^a, T^b\right]_{km}A^b_\mu \psi_m \\
&= i\left(\partial_\mu \epsilon^a\right)T^a_{k\ell}\psi_\ell + ig\epsilon^a f^{abc}\left(T^c\right)_{km}A^b_\mu \psi_m.
\end{aligned} \tag{12.30}$$

This can also be written as

$$ig T^a_{k\ell}\psi_\ell \left(\delta A^a_\mu - \frac{1}{g}\partial_\mu \epsilon^a + f^{abc} A^b_\mu \epsilon^c\right) = 0, \tag{12.31}$$

which determines the transformation for the gauge field to be

$$\delta A^a_\mu = \frac{1}{g}\left(\partial_\mu \epsilon^a - g f^{abc} A^b_\mu \epsilon^c\right) = \frac{1}{g}\partial_\mu \epsilon^a - f^{abc} A^b_\mu \epsilon^c. \tag{12.32}$$

Thus the Lagrangian density

$$\mathcal{L} = i\overline{\psi}_k \gamma^\mu (D_\mu \psi)_k - m\overline{\psi}_k \psi_k, \tag{12.33}$$

with the covariant derivative defined in (12.26) is invariant under the infinitesimal local transformations

$$\begin{aligned}
\delta \psi_k &= -i\epsilon^a(x) T^a_{k\ell} \psi_\ell(x), \\
\delta \overline{\psi}_k &= i\epsilon^a(x) \overline{\psi}_\ell T^a_{\ell k}, \\
\delta A^a_\mu &= \frac{1}{g}\partial_\mu \epsilon^a - f^{abc} A^b_\mu \epsilon^c.
\end{aligned} \tag{12.34}$$

If we use the definition (see (12.25))

$$A_\mu = T^a A^a_\mu, \qquad a = 1, 2, \ldots, \dim G, \tag{12.35}$$

where the gauge fields A_μ are matrices belonging to the same representation of the group as the fermions (since the generators T^a belong to this representation), we can write the Lagrangian density in (12.33) in terms of matrices as (of course, the Lagrangian density is a scalar)

$$\mathcal{L} = i\overline{\psi}\gamma^\mu D_\mu \psi - m\overline{\psi}\psi, \tag{12.36}$$

with normal product rules for matrices. We can now define a unitary representation of the group as describing the finite symmetry transformation matrix (see (12.16))

12.1 Non-Abelian gauge theories

$$U = e^{-i\theta^a(x)T^a}, \qquad U^\dagger = U^{-1}; \tag{12.37}$$

so that we can write the finite transformations (corresponding to (12.34)) for the field variables in the matrix form as

$$\begin{aligned}
\psi &\to U\psi, \\
\overline{\psi} &\to \overline{\psi} U^{-1}, \\
A_\mu &\to U A_\mu U^{-1} - \frac{1}{ig} (\partial_\mu U) U^{-1}.
\end{aligned} \tag{12.38}$$

To construct the Lagrangian density for the dynamical part of the gauge field in a gauge invariant manner, we note that under the gauge transformation (12.38), the tensor representing the Abelian field strength (see (12.11)) would transform as

$$\begin{aligned}
f_{\mu\nu} &= \partial_\mu A_\nu - \partial_\nu A_\mu \\
&\to \partial_\mu \left[U A_\nu U^{-1} - \frac{1}{ig} (\partial_\nu U) U^{-1} \right] \\
&\quad - \partial_\nu \left[U A_\mu U^{-1} - \frac{1}{ig} (\partial_\mu U) U^{-1} \right] \\
&= \partial_\mu \left(U A_\nu U^{-1} \right) - \frac{1}{ig} (\partial_\mu \partial_\nu U) U^{-1} - \frac{1}{ig} (\partial_\nu U) (\partial_\mu U^{-1}) \\
&\quad - \partial_\nu \left(U A_\mu U^{-1} \right) + \frac{1}{ig} (\partial_\nu \partial_\mu U) U^{-1} + \frac{1}{ig} (\partial_\mu U) (\partial_\nu U^{-1}) \\
&= \frac{1}{ig} (\partial_\mu U) (\partial_\nu U^{-1}) - \frac{1}{ig} (\partial_\nu U) (\partial_\mu U^{-1}) + \partial_\mu U A_\nu U^{-1} \\
&\quad + U \partial_\mu A_\nu U^{-1} + U A_\nu \partial_\mu U^{-1} - \partial_\nu U A_\mu U^{-1} \\
&\quad - U \partial_\nu A_\mu U^{-1} - U A_\mu \partial_\nu U^{-1} \\
&= \frac{1}{ig} (\partial_\mu U \partial_\nu U^{-1} - \partial_\nu U \partial_\mu U^{-1}) + \partial_\mu U A_\nu U^{-1} \\
&\quad + U A_\nu \partial_\mu U^{-1} - \partial_\nu U A_\mu U^{-1} - U A_\mu \partial_\nu U^{-1} \\
&\quad + U (\partial_\mu A_\nu - \partial_\nu A_\mu) U^{-1}. \tag{12.39}
\end{aligned}$$

Namely, we see that unlike in the case of QED, here $f_{\mu\nu} = \partial_\mu A_\nu - \partial_\nu A_\mu$ is neither invariant nor does it have a simple transformation under the gauge transformation (12.38). Let us also note that under the gauge transformation

$$\begin{aligned}
ig\,[A_\mu, A_\nu] &= ig\,(A_\mu A_\nu - A_\nu A_\mu) \\
&\rightarrow ig\bigg[\left(UA_\mu U^{-1} - \frac{1}{ig}(\partial_\mu U)U^{-1}\right)\left(UA_\nu U^{-1} - \frac{1}{ig}(\partial_\nu U)U^{-1}\right) \\
&\quad - \left(UA_\nu U^{-1} - \frac{1}{ig}(\partial_\nu U)U^{-1}\right)\left(UA_\mu U^{-1} - \frac{1}{ig}(\partial_\mu U)U^{-1}\right)\bigg] \\
&= ig\bigg[UA_\mu A_\nu U^{-1} - \frac{1}{ig}\left(\partial_\mu U A_\nu U^{-1} - UA_\mu \partial_\nu U^{-1}\right) \\
&\quad + \frac{1}{g^2}\partial_\mu U \partial_\nu U^{-1} - UA_\nu A_\mu U^{-1} \\
&\quad + \frac{1}{ig}\left(\partial_\nu U A_\mu U^{-1} - UA_\nu \partial_\mu U^{-1}\right) - \frac{1}{g^2}\partial_\nu U \partial_\mu U^{-1}\bigg] \\
&= igU\,[A_\mu, A_\nu]\,U^{-1} - \frac{1}{ig}\left(\partial_\mu U \partial_\nu U^{-1} - \partial_\nu U \partial_\mu U^{-1}\right) \\
&\quad - [\partial_\mu U A_\nu U^{-1} + UA_\nu \partial_\mu U^{-1} - \partial_\nu U A_\mu U^{-1} - UA_\mu \partial_\nu U^{-1}].
\end{aligned}$$
(12.40)

Thus comparing (12.39) and (12.40), it is clear that, in the present case, if we define the field strength tensor as

$$F_{\mu\nu} = \partial_\mu A_\nu - \partial_\nu A_\mu + ig\,[A_\mu, A_\nu],\tag{12.41}$$

then under the gauge transformation (12.38),

$$F_{\mu\nu} \rightarrow UF_{\mu\nu}U^{-1},\tag{12.42}$$

so that $F_{\mu\nu}$ transforms covariantly. It is now easy to construct the gauge invariant Lagrangian density for the dynamical part of the gauge field as (quadratic in derivatives)

12.1 NON-ABELIAN GAUGE THEORIES

$$\mathcal{L}_{\text{gauge}} = -\frac{1}{2}\text{Tr } F_{\mu\nu}F^{\mu\nu} = -\frac{1}{4}F^a_{\mu\nu}F^{\mu\nu\,a}, \tag{12.43}$$

where a particular normalization for the trace of the generators is assumed (namely, $T^aT^b = \frac{1}{2}\delta^{ab}$, see (12.59) and the comments following that equation). From the cyclicity of trace this Lagrangian density is easily seen to be invariant under the gauge transformations (12.38). In components, the field strength tensor (12.41) is easily seen to take the form,

$$F_{\mu\nu} = \partial_\mu A_\nu - \partial_\nu A_\mu + ig[A_\mu, A_\nu],$$

or, $\quad F^a_{\mu\nu}T^a = (\partial_\mu A^a_\nu - \partial_\nu A^a_\mu)T^a + igA^b_\mu A^c_\nu [T^b, T^c]$

$$= (\partial_\mu A^a_\nu - \partial_\nu A^a_\mu)T^a - gf^{abc}A^b_\mu A^c_\nu T^a,$$

or, $\quad F^a_{\mu\nu} = \partial_\mu A^a_\nu - \partial_\nu A^a_\mu - gf^{abc}A^b_\mu A^c_\nu = -F^a_{\nu\mu}, \tag{12.44}$

so that the gauge invariant Lagrangian density for the gauge field can also be written in terms of component fields as

$$\begin{aligned}\mathcal{L}_{\text{gauge}} &= -\frac{1}{4}F^a_{\mu\nu}F^{\mu\nu\,a} \\ &= -\frac{1}{4}(\partial_\mu A^a_\nu - \partial_\nu A^a_\mu - gf^{abc}A^b_\mu A^c_\nu) \\ &\quad \times (\partial^\mu A^{\nu a} - \partial^\nu A^{\mu a} - gf^{apq}A^{\mu p}A^{\nu q}).\end{aligned} \tag{12.45}$$

The complete Lagrangian density for QCD (quantum chromodynamics, more generally for fermions interacting with a non-Abelian gauge field, QCD corresponds to the particular case when $G = SU(3)$) is, therefore, given by

$$\mathcal{L}_{\text{QCD}} = -\frac{1}{4}F^a_{\mu\nu}F^{\mu\nu\,a} + i\overline{\psi}_k\gamma^\mu(D_\mu\psi)_k - m\overline{\psi}_k\psi_k, \tag{12.46}$$

where we can set $m = 0$ if the fermions (quarks) are massless. This Lagrangian density is invariant under the non-Abelian gauge symmetry transformations (12.38) which have the infinitesimal form (see (12.34))

$$\delta\psi_k = -i\epsilon^a(x)(T^a)_{k\ell}\psi_\ell,$$
$$\delta\overline{\psi}_k = i\epsilon^a(x)\overline{\psi}_\ell(T^a)_{\ell k},$$
$$\delta A^a_\mu = \frac{1}{g}\partial_\mu \epsilon^a - f^{abc}A^b_\mu \epsilon^c. \tag{12.47}$$

Let us discuss briefly some of the properties of the Lie algebra of the symmetry group G. The algebra of the Hermitian generators, as we have noted in (12.18), has the form

$$[T^a, T^b] = if^{abc}T^c, \tag{12.48}$$

and the Jacobi identity associated with this algebra is given by

$$[[T^a, T^b], T^c] + [[T^c, T^a], T^b] + [[T^b, T^c], T^a] = 0. \tag{12.49}$$

Using (12.48) it can be seen that the Jacobi identity (12.49) imposes a restriction on the structure constants of the group of the form

$$if^{abp}[T^p, T^c] + if^{cap}[T^p, T^b] + if^{bcp}[T^p, T^a] = 0,$$
$$\text{or,} \quad f^{abp}f^{pcq}T^q + f^{cap}f^{pbq}T^q + f^{bcp}f^{paq}T^q = 0,$$
$$\text{or,} \quad f^{abp}f^{pcq} + f^{cap}f^{pbq} + f^{bcp}f^{paq} = 0. \tag{12.50}$$

From the structure of the Lie algebra we know that we can write the finite group elements in a unitary representation as (recall finite rotations and the angular momentum operators and see (12.37) as well)

$$U(x) = e^{i\theta^a(x)T^a}. \tag{12.51}$$

If we can divide the generators of the non-Abelian Lie algebra into two non-Abelian subsets such that $f^{abc} = 0$ when one index is in one set and another index is in the second set, then the Lie algebra breaks up into two commuting non-Abelian subalgebras. In this case

12.1 NON-ABELIAN GAUGE THEORIES

the group G is a direct product of two independent non-Abelian Lie groups. A non-Abelian Lie group that cannot be so factorized is called a simple Lie group. A direct product of simple Lie groups is called semi-simple. In all our discussions, we will assume that the symmetry group G is simple.

For any representation of a simple Lie group we can write

$$\text{Tr } T^a T^b = C_2 \delta^{ab}. \tag{12.52}$$

We note that C_2 is a normalization constant which determines the values of the structure constants. It depends on the representation but not on the indices a and b. To prove this let us note that we can always diagonalize the tensor $\text{Tr}(T^a T^b)$ such that (this is a symmetric real matrix in the "ab" space)

$$\text{Tr}\left(T^a T^b\right) = \begin{cases} 0 & \text{if} \quad a \neq b, \\ K_a & \text{if} \quad a = b. \end{cases} \tag{12.53}$$

Using the cyclicity property of the trace, let us next note that the quantity

$$\begin{aligned} h^{abc} &= \text{Tr}\left([T^a, T^b] T^c\right) \\ &= \text{Tr}\left(T^a T^b T^c\right) - \text{Tr}\left(T^b T^a T^c\right), \end{aligned} \tag{12.54}$$

is completely antisymmetric in all its indices. Furthermore, using the commutation relation (12.18) we obtain

$$\begin{aligned} h^{abc} &= \text{Tr}\left(i f^{abp} T^p T^c\right) \\ &= i f^{abp} \text{Tr}\left(T^p T^c\right) \\ &= i f^{abp} K_p \delta^{pc} \\ &= i K_c f^{abc}, \end{aligned} \tag{12.55}$$

where the index c is fixed (and not summed). On the other hand, we note that

$$\begin{aligned}
h^{acb} &= \operatorname{Tr}\left([T^a, T^c]T^b\right) \\
&= if^{acp} \operatorname{Tr}\left(T^p T^b\right) \\
&= if^{acp} K_p \delta^{pb} \\
&= iK_b f^{acb} = -iK_b f^{abc},
\end{aligned} \qquad (12.56)$$

with the index b fixed. However, since h^{abc} is completely antisymmetric, we have

$$h^{acb} = -h^{abc}. \qquad (12.57)$$

Using this and comparing (12.55) and (12.56) we conclude that

$$K_b = K_c = K. \qquad (12.58)$$

This shows that we can write

$$\operatorname{Tr}\left(T^a T^b\right) = C_2 \delta^{ab}, \qquad (12.59)$$

where the constant C_2 depends only on the representation. (It is chosen to be $\frac{1}{2}$ for the fundamental representation to which the fermions belong in $SU(n)$ and this is the normalization used in (12.43). Furthermore, we note here that it is (12.59) which is used to show that the structure constants for a semi-simple group are completely antisymmetric in the indices.) We note that if we write

$$\operatorname{Tr} T^a T^b = T(R)\, \delta^{ab}, \qquad a,b = 1, 2, \cdots, \dim G, \qquad (12.60)$$

then $T(R)$ is known as the index of the representation R. Similarly, we have

$$(T^a T^a)_{mn} = C(R)\, \delta_{mn}, \qquad m, n = 1, 2, \cdots, \dim R, \qquad (12.61)$$

12.1 NON-ABELIAN GAUGE THEORIES

where $C(R)$ is known as the Casimir of the representation R. The two are clearly related as (this is easily seen by taking trace in the respective spaces)

$$T(R) \dim G = C(R) \dim R. \tag{12.62}$$

We can determine the generators of the group in various representations much like in the case of angular momentum. However, a particular representation that is very important as well as useful is given by

$$\left(T^a\right)_{bc} = -if^{abc}. \tag{12.63}$$

This is consistent with the hermiticity requirement for the generators, namely,

$$\begin{aligned}\left(T^{a\dagger}\right)_{bc} &= \left(\left(T^a\right)_{cb}\right)^* \\ &= \left(-if^{acb}\right)^* = if^{acb} = -if^{abc} = \left(T^a\right)_{bc}.\end{aligned} \tag{12.64}$$

Furthermore, we can easily check that this representation satisfies the Lie algebra,

$$\begin{aligned}\left[T^a, T^b\right]_{cq} &= \left(T^a T^b - T^b T^a\right)_{cq} \\ &= \left(T^a\right)_{cp}\left(T^b\right)_{pq} - \left(T^b\right)_{cp}\left(T^a\right)_{pq} \\ &= \left(-if^{acp}\right)\left(-if^{bpq}\right) - \left(-if^{bcp}\right)\left(-if^{apq}\right) \\ &= -f^{acp}f^{bpq} + f^{bcp}f^{apq} \\ &= -f^{cap}f^{pbq} - f^{bcp}f^{paq} \\ &= f^{abp}f^{pcq} \\ &= if^{abp}\left(-if^{pcq}\right) \\ &= if^{abp}\left(T^p\right)_{cq}, \end{aligned} \tag{12.65}$$

where we have used the anti-symmetry of the structure constants as well as (12.50). As a result, we conclude that the identification

$$\left(T^a_{(\text{adj})}\right)_{bc} = -if^{abc}, \tag{12.66}$$

indeed defines a representation of the Lie algebra known as the adjoint representation.

Let us next look at the transformation for the gauge fields in (12.34)

$$\begin{aligned}\delta A^a_\mu &= \frac{1}{g}\left(\partial_\mu \epsilon^a - gf^{abc}A^b_\mu \epsilon^c\right) \\ &= \frac{1}{g}\left(\partial_\mu \epsilon^a + gf^{bac}A^b_\mu \epsilon^c\right) \\ &= \frac{1}{g}\left(\partial_\mu \epsilon^a + ig\left(-if^{bac}\right)A^b_\mu \epsilon^c\right) \\ &= \frac{1}{g}\left(\partial_\mu \epsilon^a + ig\left(T^b_{(\text{adj})}\right)_{ac}A^b_\mu \epsilon^c\right),\end{aligned} \tag{12.67}$$

so that we can equivalently write (see, for example, (12.26))

$$\delta A^a_\mu = \frac{1}{g}\left(D^{(\text{adj})}_\mu \theta\right)^a, \tag{12.68}$$

which shows that the gauge field, A_μ, transforms according to the adjoint representation of the group. This can also be seen from the transformation of the field strength tensor in (12.42), which infinitesimally has the form

$$\begin{aligned}F_{\mu\nu} &\to UF_{\mu\nu}U^{-1} \\ &= F_{\mu\nu} + i\epsilon^b\left[F_{\mu\nu}, T^b\right],\end{aligned} \tag{12.69}$$

so that

12.1 NON-ABELIAN GAUGE THEORIES

$$\delta F_{\mu\nu} = i\epsilon^b [F_{\mu\nu}, T^b]$$
$$= i\epsilon^b F^a_{\mu\nu}[T^a, T^b],$$
or, $\quad \delta F^a_{\mu\nu} T^a = i\epsilon^b F^a_{\mu\nu}(if^{abc}T^c),$
or, $\quad \delta F^a_{\mu\nu} = -f^{abc} F^b_{\mu\nu} \epsilon^c$
$$= -i(-if^{cab})\epsilon^c F^b_{\mu\nu}$$
$$= -i(T^c_{(\mathrm{adj})})_{ab} \epsilon^c F^b_{\mu\nu}$$
$$= -i\epsilon^c (T^c_{(\mathrm{adj})})_{ab} F^b_{\mu\nu}. \tag{12.70}$$

Comparing this with the transformation of the fermions in (12.34) we conclude that the field strength $F_{\mu\nu}$ as well as the gauge field A_μ transform according to the adjoint representation of the group. (It does not matter what representation the matter fields belong to, the gauge field must transform in the adjoint representation.) For completeness, let us note here that the covariant derivative in the adjoint representation (12.67)

$$\left(D_\mu^{(\mathrm{adj})}\epsilon\right)^a = \partial_\mu \epsilon^a - gf^{abc} A^b_\mu \epsilon^c, \tag{12.71}$$

can also be written in the matrix form as

$$D_\mu^{(\mathrm{adj})} \epsilon = \partial_\mu \epsilon + ig[A_\mu, \epsilon], \tag{12.72}$$

where $A_\mu = A^a_\mu T^a$ and $\epsilon = \epsilon^a T^a$ (with T^a in any representation).

Let us now concentrate only on the gauge field (Yang-Mills) part of the Lagrangian density

$$\begin{aligned}
\mathcal{L}_{\mathrm{YM}} &= -\frac{1}{4} F^a_{\mu\nu} F^{\mu\nu\, a} \\
&= -\frac{1}{4}\left(\partial_\mu A^a_\nu - \partial_\nu A^a_\mu - gf^{abc} A^b_\mu A^c_\nu\right) \\
&\quad \times \left(\partial^\mu A^{\nu a} - \partial^\nu A^{\mu a} - gf^{apq} A^{\mu p} A^{\nu q}\right).
\end{aligned} \tag{12.73}$$

Let us note the following features of this Lagrangian density. For example, let us scale the field variables as

$$A_\mu^a \;\to\; \frac{1}{g} A_\mu^a.$$

It follows then that
$$F_{\mu\nu}^a \;\to\; \frac{1}{g}\left(\partial_\mu A_\nu^a - \partial_\nu A_\mu^a - f^{abc} A_\mu^b A_\nu^c\right),$$
which leads to
$$\begin{aligned}\mathcal{L}_{\text{YM}} \;\to\; &-\frac{1}{4g^2}\left(\partial_\mu A_\nu^a - \partial_\nu A_\mu^a - f^{abc} A_\mu^b A_\nu^c\right) \\ &\times \left(\partial^\mu A^{\nu a} - \partial^\nu A^{\mu a} - f^{apq} A^{\mu p} A^{\nu q}\right).\end{aligned} \qquad (12.74)$$

Thus we note that in this theory, the coupling constant can be scaled out and written as an overall multiplicative factor in the Lagrangian density. The other feature to note is that unlike the photon field, here the gauge fields have self interaction (they interact with themselves) and, therefore, the pure Yang-Mills theory is an interacting theory unlike the Maxwell theory. Physically we understand this in the following way. In the present case, the gauge fields carry the charge of the non-Abelian symmetry group (they have a nontrivial symmetry index) in contrast to the photon field which is chargeless. Since the gauge fields couple to any field (particle) carrying charge of the symmetry group, in the case of non-Abelian symmetry they must couple to themselves and possess self interactions.

12.2 Canonical quantization of Yang-Mills theory

Let us discuss the canonical quantization of Yang-Mills theory in the same spirit as the discussion of Maxwell's theory in chapter **9**. We note that the Lagrangian density for the Yang-Mills theory is given by (12.43) (or (12.45))

$$\mathcal{L}_{\text{YM}} = -\frac{1}{2}\text{Tr}\, F_{\mu\nu} F^{\mu\nu} = -\frac{1}{4} F_{\mu\nu}^a F^{\mu\nu\, a}, \qquad (12.75)$$

where the field strength tensor is given by (see (12.41) and (12.44))

$$\begin{aligned} F_{\mu\nu} &= \partial_\mu A_\nu - \partial_\nu A_\mu + ig[A_\mu, A_\nu], \\ F_{\mu\nu}^a &= \partial_\mu A_\nu^a - \partial_\nu A_\mu^a - g f^{abc} A_\mu^b A_\nu^c. \end{aligned} \qquad (12.76)$$

12.2 CANONICAL QUANTIZATION OF YANG-MILLS THEORY

As in the case of the Abelian theory (Maxwell theory), we can identify the non-Abelian electric and magnetic fields from (12.76) as

$$F^a_{0i} = \partial_0 A^a_i - \partial_i A^a_0 - gf^{abc} A^b_0 A^c_i = E^a_i,$$
$$F^a_{ij} = \partial_i A^a_j - \partial_j A^a_i - gf^{abc} A^b_i A^c_j = -\epsilon_{ijk} B^a_k. \tag{12.77}$$

The Lagrangian density (12.75) can also be written explicitly in terms of the non-Abelian electric and magnetic field strength tensors (12.77) as

$$\mathcal{L}_{\text{YM}} = \text{Tr}\left(E_i E_i - \frac{1}{2} F_{ij} F^{ij}\right) = \frac{1}{2} E^a_i E^a_i - \frac{1}{4} F^a_{ij} F^{ij\,a}. \tag{12.78}$$

The Euler-Lagrange equation for the theory are obtained from (12.75) and have the form

$$\partial_\mu \frac{\partial \mathcal{L}}{\partial \partial_\mu A^a_\nu} - \frac{\partial \mathcal{L}}{\partial A^a_\nu} = 0,$$

or, $\quad -\partial_\mu F^{\mu\nu\,a} - \left(-gf^{abc} A^b_\mu F^{\mu\nu\,c}\right) = 0,$

or, $\quad (D_\mu F^{\mu\nu})^a = \partial_\mu F^{\mu\nu\,a} - gf^{abc} A^b_\mu F^{\mu\nu\,c} = 0, \tag{12.79}$

where the covariant derivative is defined to be in the adjoint representation of the group (see (12.67) and (12.68)). In the matrix notation (see (12.72)), we can write the Euler-Lagrange equations as

$$D_\mu F^{\mu\nu} = \partial_\mu F^{\mu\nu} + ig[A_\mu, F^{\mu\nu}] = 0. \tag{12.80}$$

Let us note that the coefficient matrix of highest derivatives in (12.75) is the transverse projection operator just like in the case of Maxwell's theory. Therefore, the theory is singular implying that there are constraints which is evident from the fact that the Yang-Mills Lagrangian density is invariant under the infinitesimal gauge transformation (see (12.68) and we are suppressing the symbol "(adj)" to denote the covariant derivative in the adjoint representation for simplicity)

$$\delta A_\mu^a = \frac{1}{g} \left(D_\mu \epsilon(x)\right)^a. \tag{12.81}$$

In fact, the constrained structure of the theory is already obvious in the Euler-Lagrange equation of motion (12.79) for $\nu = 0$, namely,

$$D_i F^{i0\,a} = 0, \tag{12.82}$$

which is a constraint relation.

Let us obtain the momenta canonically conjugate to the field variables A_μ^a from (12.75) and this leads to

$$\Pi^{\mu\,a}(x) = \frac{\partial \mathcal{L}}{\partial \dot{A}_\mu^a(x)} = -F^{0\mu\,a}(x). \tag{12.83}$$

Noting that the field strength $F_{\mu\nu}^a$ is antisymmetric in the indices μ, ν, we have

$$\Pi^{0\,a}(x) = -F^{00\,a}(x) = 0, \tag{12.84}$$

as in the case of Maxwell's theory and

$$\Pi^{i\,a}(x) = -F^{0i\,a} = E_i^a(x), \tag{12.85}$$

where the non-Abelian electric field $E_i^a(x)$ is defined in (12.77). Thus, we see that we have only $3N$ canonical momenta where $N = \dim G$. The momentum conjugate to A_0^a does not exist. This implies that A_0^a is like a c-number quantity which commutes with every other operator in the theory. Thus we can choose a gauge condition and set it equal to zero. Namely, we choose

$$A_0^a(x) = 0, \quad \text{or,} \quad A_0(x) = 0. \tag{12.86}$$

If a gauge field configuration does not satisfy this condition we can always make a suitable gauge transformation so that the transformed field would satisfy the condition. For example, requiring that

$$A_0' = UA_0U^{-1} - \frac{1}{ig}(\partial_0 U)U^{-1} = 0,$$

or, $\quad U^{-1}\partial_0 U = igA_0,$ \hfill (12.87)

we can determine the parameter of transformation and we see that a gauge field can always be transformed to satisfy (12.86).

Let us note that with the gauge condition (12.86) the Lagrangian density (12.75) takes the form

$$\begin{aligned}\mathcal{L}_{\text{YM}} &= \text{Tr}\left(E_i E_i - \frac{1}{2}F_{ij}F^{ij}\right) = \frac{1}{2}E_i^a E_i^a - \frac{1}{4}F_{ij}^a F^{ij\,a} \\ &= \text{Tr}\left(\dot{A}_i \dot{A}_i - \frac{1}{2}F_{ij}F^{ij}\right) = \frac{1}{2}\dot{A}_i^a \dot{A}_i^a - \frac{1}{4}F_{ij}^a F^{ij\,a},\end{aligned}$$ (12.88)

where we have used the fact that in the gauge (12.86) the non-Abelian electric field in (12.77) takes the form

$$E_i^a = \dot{A}_i^a, \quad \text{or,} \quad E_i = \dot{A}_i. \tag{12.89}$$

From our discussion in the case of the Maxwell theory, we expect the physical dynamical degrees of freedom of a massless gauge field to be transverse. However, in the present case, we note from (12.82) that

$$D_i E_i = \partial_i E_i + ig[A_i, E_i] = 0, \tag{12.90}$$

so that neither the gauge field A_i nor the electric field E_i is transverse (recall that $E_i = \dot{A}_i$). On the other hand, let us also note that the potential term in (12.88) (namely, the term quadratic in the non-Abelian magnetic field) is invariant under the field redefinition (gauge transformation)

$$A_i = S\bar{A}_i S^{-1} - \frac{1}{ig}(\partial_i S)S^{-1} = S\left(\bar{A}_i - \frac{1}{ig}S^{-1}(\partial_i S)\right)S^{-1}, \tag{12.91}$$

so that there is a cyclic variable in the theory that needs to be separated. We note that the relation (12.91) can be inverted to give

$$\bar{A}_i = S^{-1}A_iS + \frac{1}{ig}S^{-1}\partial_iS = S^{-1}\Big(A_i + \frac{1}{ig}(\partial_iS)S^{-1}\Big)S. \quad (12.92)$$

Keeping in mind that we would like the dynamical variables to be transverse, let us require the new field variable $\bar{A}_i$ to be transverse, namely,

$$\partial_i\bar{A}_i = 0. \quad (12.93)$$

Imposing this condition in (12.91), we obtain

$$\begin{aligned}
\partial_iA_i &= \partial_i\big(S\bar{A}_iS^{-1}\big) - \frac{1}{ig}\partial_i\big((\partial_iS)S^{-1}\big) \\
&= (\partial_iS)\bar{A}_iS^{-1} + S\bar{A}_i(\partial_iS^{-1}) - \frac{1}{ig}\partial_i\big((\partial_iS)S^{-1}\big) \\
&= (\partial_iS)\big(S^{-1}A_i - \frac{1}{ig}(\partial_iS^{-1})\big) \\
&\quad + \big(A_iS + \frac{1}{ig}(\partial_iS)\big)(\partial_iS^{-1}) - \frac{1}{ig}\partial_i\big((\partial_iS)S^{-1}\big) \\
&= -\frac{1}{ig}\partial_i\big((\partial_iS)S^{-1}\big) - [A_i,(\partial_iS)S^{-1}] \\
&= -\frac{1}{ig}D_i(A)\big((\partial_iS)S^{-1}\big), \quad (12.94)
\end{aligned}$$

where we have used (12.92) in the intermediate step and the covariant derivative in the adjoint representation is defined for a matrix function G as (see (12.72))

$$D_i(A)G = \partial_iG + ig[A_i,G]. \quad (12.95)$$

This shows that if we can find a gauge transformation which satisfies (12.94), then we can define a new field variable $\bar{A}_i$ satisfying the transversality condition (12.93).

We note next from the field redefinition in (12.91) that

12.2 CANONICAL QUANTIZATION OF YANG-MILLS THEORY

$$\begin{aligned}
\dot{A}_i &= \partial_0\left(S\bar{A}_i S^{-1} - \frac{1}{ig}(\partial_i S)S^{-1}\right) \\
&= \dot{S}\bar{A}_i S^{-1} + S\dot{\bar{A}}_i S^{-1} - S\bar{A}_i S^{-1}\dot{S}S^{-1} \\
&\quad - \frac{1}{ig}\left((\partial_i \dot{S})S^{-1} - (\partial_i S)S^{-1}\dot{S}S^{-1}\right) \\
&= S\left(\dot{\bar{A}}_i - [\bar{A}_i, S^{-1}\dot{S}] - \frac{1}{ig}(\partial_i(S^{-1}\dot{S}))\right)S^{-1} \\
&= S\left(\dot{\bar{A}}_i - \frac{1}{ig}D_i(\bar{A})(S^{-1}\dot{S})\right)S^{-1} \\
&= S\left(\dot{\bar{A}}_i - \frac{1}{ig}D_i(\bar{A})f\right)S^{-1}, \quad (12.96)
\end{aligned}$$

where we have used $\partial_0 S^{-1} = -S^{-1}\dot{S}S^{-1}$. Here $D_i(\bar{A})$ is the covariant derivative in the adjoint representation (see (12.95)) with respect to the gauge field $\bar{A}_i$ and we have identified

$$f = S^{-1}\dot{S}. \quad (12.97)$$

From (12.96) we note that in the gauge (12.86) we can write

$$E_i = \dot{A}_i = S\left(\dot{\bar{A}}_i - \frac{1}{ig}D_i(\bar{A})f\right)S^{-1} = S\bar{E}_i S^{-1}, \quad (12.98)$$

where we can identify

$$\bar{E}_i = \dot{\bar{A}}_i - \frac{1}{ig}D_i(\bar{A})f. \quad (12.99)$$

Furthermore, decomposing $\bar{E}_i$ into its transverse and longitudinal components as

$$\bar{E}_i = \bar{E}_i^{\mathrm{T}} + \partial_i \phi, \quad (12.100)$$

where

$$\partial_i \overline{E}_i^{\mathrm{T}} = 0, \qquad \partial_i \overline{E}_i = \partial_i \partial_i \phi = \boldsymbol{\nabla}^2 \phi, \tag{12.101}$$

we conclude from the definition in (12.99) that

$$\overline{E}_i^{\mathrm{T}} = \dot{\overline{A}}_i, \qquad \phi = -\frac{1}{ig}\frac{1}{\boldsymbol{\nabla}^2}\partial_i D_i(\overline{A})f. \tag{12.102}$$

Let us next note that under the field redefinition (12.91),

$$\begin{aligned}
D_i(A)E_i &= \partial_i E_i + ig[A_i, E_i] \\
&= \partial_i\bigl(S\overline{E}_i S^{-1}\bigr) + ig[S(\overline{A}_i - \tfrac{1}{ig}S^{-1}(\partial_i S))S^{-1}, S\overline{E}_i S^{-1}] \\
&= (\partial_i S)\overline{E}_i S^{-1} + S(\partial_i \overline{E}_i)S^{-1} - S\overline{E}_i S^{-1}(\partial_i S)S^{-1} \\
&\quad + ig S[\overline{A}_i - \tfrac{1}{ig}S^{-1}(\partial_i S), \overline{E}_i]S^{-1} \\
&= S(\partial_i \overline{E}_i + ig[\overline{A}_i, \overline{E}_i])S^{-1} \\
&= S(D_i(\overline{A})\overline{E}_i)S^{-1},
\end{aligned} \tag{12.103}$$

where we have used (12.91) and (12.98). As a result of this relation, the constraint equation (12.90) takes the form

$$\begin{aligned}
&D_i(\overline{A})\overline{E}_i = \partial_i \overline{E}_i + ig[\overline{A}_i, \overline{E}_i] = 0, \\
\text{or,} \quad & \partial_i\partial_i\phi + ig[\overline{A}_i, (\overline{E}_i^{\mathrm{T}} + \partial_i\phi)] = 0, \\
\text{or,} \quad & \partial_i\partial_i\phi + ig\partial_i[\overline{A}_i, \phi] + ig[\overline{A}_i, \overline{E}_i^{\mathrm{T}}] = 0, \\
\text{or,} \quad & \partial_i\bigl(\partial_i\phi + ig[\overline{A}_i, \phi]\bigr) = -ig[\overline{A}_i, \overline{E}_i^{\mathrm{T}}] = \rho, \\
\text{or,} \quad & \partial_i D_i(\overline{A})\phi = \rho,
\end{aligned} \tag{12.104}$$

where we have used $\partial_i \overline{A}_i = 0$ in the intermediate step. We can invert this relation to determine

12.2 Canonical quantization of Yang-Mills theory

$$\phi = \frac{1}{\partial_i D_i(\bar{A})} \rho, \tag{12.105}$$

so that from (12.100) we obtain

$$\overline{E}_i = \overline{E}_i^T + \partial_i \phi = \dot{\bar{A}}_i + \partial_i \frac{1}{\partial_j D_j(\bar{A})} \rho. \tag{12.106}$$

Using the definition in (12.91), (12.98) and the decomposition in (12.106) as well as the cyclicity of trace, the Lagrangian density (12.88) takes the form

$$\begin{aligned}
\mathcal{L}_{\text{YM}} &= \operatorname{Tr}\left(E_i E_i - \frac{1}{2} F_{ij}(A) F^{ij}(A)\right) \\
&= \operatorname{Tr}\left(\overline{E}_i \overline{E}_i - \frac{1}{2} F_{ij}(\bar{A}) F^{ij}(\bar{A})\right) \\
&= \operatorname{Tr}\left((\dot{\bar{A}}_i + \partial_i \frac{1}{\partial_j D_j(\bar{A})} \rho)(\dot{\bar{A}}_i + \partial_i \frac{1}{\partial_k D_k(\bar{A})} \rho) \right. \\
&\quad \left. - \frac{1}{2} F_{ij}(\bar{A}) F^{ij}(\bar{A})\right) \\
&= \operatorname{Tr}\left(\dot{\bar{A}}_i \dot{\bar{A}}_i - \frac{1}{2} F_{ij}(\bar{A}) F^{ij}(\bar{A}) \right. \\
&\quad \left. - \rho \frac{1}{\partial_j D_j(\bar{A})} \nabla^2 \frac{1}{\partial_k D_k(\bar{A})} \rho\right) \\
&= \frac{1}{2} \dot{\bar{A}}_i^a \dot{\bar{A}}_i^a - \frac{1}{4} F_{ij}(\bar{A}) F^{ij}(\bar{A}) \\
&\quad - \frac{1}{2} \rho^a \left(\frac{1}{\partial_j D_j(\bar{A})} \nabla^2 \frac{1}{\partial_k D_k(\bar{A})} \rho\right)^a, \tag{12.107}
\end{aligned}$$

where we have neglected total divergence terms in the intermediate steps (note also that because of (12.93), $\partial_i D_i(\bar{A}) = D_i(\bar{A}) \partial_i$ which has been used together with integration by parts).

We have isolated the dynamical variables of the theory to be $\bar{A}_i^a$ which are transverse and the conjugate momenta can be determined from (12.107) to be

$$\overline{\Pi}^{ia} = \frac{\partial \mathcal{L}_{\text{YM}}}{\partial \dot{\bar{A}}_i^a} = \dot{\bar{A}}_i^a. \tag{12.108}$$

The equal-time canonical Poisson brackets for the theory can now be obtained as in the case of the Maxwell theory

$$\{\bar{A}_i^a(x), \overline{\Pi}^{jb}(y)\} = \delta^{ab} \delta_{i\,\text{TR}}^{\;j}(x-y), \tag{12.109}$$

with all other brackets vanishing. Here the transverse delta function corresponds to the one already defined in the case of Maxwell's theory in (9.31). We note that the Lagrangian density in (12.107) has exactly the same form as in the Abelian theory except for the last term. The last term is like the long range Coulomb interaction term in the case of QED (in an Abelian theory where the structure constants vanish and the adjoint covariant derivative reduces to an ordinary derivative, this last term has the form $\rho \frac{1}{\nabla^2} \rho$ which corresponds to the long range Coulomb interaction). Here we see that since the gauge fields are self interacting, even in the absence of other matter fields there is a long range interaction. Let us note here that the long range behavior of the interaction (and, therefore, of the theory) is, therefore, controlled by the eigenvalues of the operator $\partial_i D_i(\bar{A})$. Gribov has shown that the operator $\partial_i D_i(\bar{A})$ does possess zero modes or eigenvectors with zero eigenvalue. Thus the long range behavior of this theory is still not well defined in the Coulomb gauge. This is related to the question of gauge fixing, namely the Coulomb gauge does not uniquely define the gauge fields. This can be seen from the fact that if we have a field configuration $\bar{A}_i^a$ which is transverse, i.e.

$$\partial_i \bar{A}_i^a = 0, \tag{12.110}$$

then we can make a time independent infinitesimal gauge transformation (under which the theory is invariant)

$$\bar{A}_i'^a = \bar{A}_i^a + \frac{1}{g} \left(D_i(\bar{A}) \bar{\epsilon} \right)^a, \tag{12.111}$$

such that

$$\partial_i \bar{A}_i'^a = 0, \tag{12.112}$$

provided $\bar{\epsilon}^a$ corresponds to a zero mode of $\partial_i D_i(\bar{A})$. This shows that the Coulomb gauge does not really specify the gauge field configuration uniquely. In other words, the hypersurface defining the gauge choice intersects the gauge orbits more than once. We can show that this nonuniqueness, known as the Gribov ambiguity, occurs in all gauges other than the axial gauge. There now exist prescriptions to take care of this problem. But let us note from the form of $\bar{A}_i'^{a}$ in (12.111) that this has a singularity at $g = 0$. (It is of the form $O(\frac{1}{g})$.) Hence such gauge field configurations can only be reached nonperturbatively. As long as we are within perturbation theory we can neglect such configurations. The other way of saying this is that within the framework of perturbation theory we are interested in infinitesimal gauge transformations. On the other hand, Gribov ambiguity is a phenomenon associated with large gauge transformations and is an important issue since it leads to nonperturbative effects. However, we would not worry about it within the context of perturbation theory.

The canonical quantization of Yang-Mills theories does not have manifest covariance very much like the Maxwell theory. Therefore, we can also try to quantize the non-Abelian gauge theory covariantly, very much along the lines of the Abelian theory that we have discussed in chapter **9**, namely, by modifying the theory. Thus, for example, let us modify the theory (12.75) as (see (9.125) with $J^\mu = 0$)

$$\mathcal{L} = -\frac{1}{4} F^a_{\mu\nu} F^{\mu\nu\, a} - \frac{1}{2} (\partial_\mu A^{\mu a})^2 . \tag{12.113}$$

The additional term clearly breaks gauge invariance and, consequently, makes the theory nonsingular, much like in the Maxwell theory. However, in the present case, there are serious differences from Maxwell's theory. For example, we recall that in QED (see (9.129))

$$\Box (\partial \cdot A) = 0, \tag{12.114}$$

namely, $\partial \cdot A$ behaves like a free field and, therefore, classically we can impose the condition

$$\partial \cdot A = 0, \tag{12.115}$$

which then translates to the Gupta-Bleuler condition on the physical states

$$\partial \cdot A^{(+)}|\text{phys}\rangle = 0. \tag{12.116}$$

In the case of the Yang-Mills theory, however, even classically the equations of motion are given by

$$\partial_\mu F^{\mu\nu\,a} - gf^{abc} A^b_\mu F^{\mu\nu\,c} + \partial^\nu \left(\partial \cdot A^a\right) = 0. \tag{12.117}$$

Contracting with ∂_ν, we obtain

$$\Box \left(\partial \cdot A^a\right) = gf^{abc}\partial_\nu \left(A^b_\mu F^{\mu\nu\,c}\right) \neq 0. \tag{12.118}$$

Thus, in contrast to the Abelian theory, we note that $(\partial \cdot A^a)$ does not behave like a free field and, consequently, the additional term in (12.113) has truly modified the theory. Furthermore, since $(\partial \cdot A^a)$ is not a free field, it cannot be uniquely decomposed into a positive and a negative frequency part (in a time invariant manner), nor can we think of the supplementary condition

$$\partial \cdot A^{a(+)}|\text{phys}\rangle = 0, \tag{12.119}$$

in a physically meaningful manner, since it is not invariant under time evolution. (Namely, the physical subspace would keep changing with time which is not desirable.) Correspondingly the naive analog of the Gupta-Bleuler condition for non-Abelian gauge theories does not seem to exist. In other words, just modifying the Lagrangian density as in (12.113) does not seem to be sufficient in contrast to the Abelian gauge theory. Therefore, we need to analyze the question of modifying the theory in a more systematic and detailed manner. We would see next how we can derive intuition on this important question from the path integral quantization of the theory.

12.3 Path integral quantization of gauge theories

Path integrals provide an alternative to the canonical quantization of field theories and are particularly useful in studying complicated

12.3 PATH INTEGRAL QUANTIZATION OF GAUGE THEORIES

gauge theories. Since this is a complete topic in itself, we will not go into the systematic details of this description, rather, we will only go over some of the essential concepts involved in such a description.

Let us recall that the primary goal in the study of relativistic field theories is to calculate scattering matrix elements which can be derived from the vacuum to vacuum transition amplitude (namely, from the "in" vacuum to the "out" vacuum, see section **6.4** for the definitions of "in" and "out" states) in the presence of interactions. In scalar and fermion field theories where the relation between the Hamiltonian and the Lagrangian is conventional the vacuum to vacuum transition amplitude can be written as a path integral. For example, for a real scalar field interacting only with an external source, the path integral description is given by (recall $\hbar = 1$)

$$\langle 0|0\rangle^J = Z[J] = e^{iW[J]} = N\int \mathcal{D}\phi\, e^{i(S[\phi]+\int d^4x\, J(x)\phi(x))}$$
$$= N\int \mathcal{D}\phi\, e^{iS^{(J)}[\phi]}, \quad (12.120)$$

where N is a normalization constant, $J(x)$ is the external source to which the scalar field is coupled and

$$S[\phi] = \int d^4x\, \frac{1}{2}(\partial_\mu\phi\partial^\mu\phi - m^2\phi^2). \quad (12.121)$$

Furthermore, $Z[J]$ and $W[J]$ are known as generating functionals for the Green's functions of the theory. A functional is roughly defined as a function of a function and it is clear that the action of a field theory is naturally a functional. The functional dependence of a quantity on a variable is generally denoted by a square bracket. Like the derivative of a function, we can also define functional derivatives in a simple manner through the relation (in four dimensions)

$$\frac{\delta F[\phi(x)]}{\delta \phi(y)} = \lim_{\epsilon \to 0^+} \frac{F[\phi(x) + \epsilon\delta^4(x-y)] - F[\phi(x)]}{\epsilon}, \quad (12.122)$$

and various Green's functions (n-point functions) are obtained from the generating functionals $Z[J]$ and $W[J]$ through functional differentiation.

The integration in (12.120) is known as a functional integration and is defined as follows. Let us divide the entire space-time into infinitesimal cells labelled by "i" of volume δV_i and define

$$\phi_i = \frac{1}{\delta V_i} \int_{\delta V_i} \mathrm{d}^4 x \, \phi(x). \tag{12.123}$$

Clearly in the limit of vanishing volume we recover

$$\lim_{\delta V_i \to 0} \phi_i = \phi(x), \tag{12.124}$$

and in terms of these discretized field variables, the functional integration is defined (up to a normalization constant) as

$$\int \mathcal{D}\phi = \prod_i \mathrm{d}\phi_i. \tag{12.125}$$

With these basics a simple functional integral for a quadratic action such as in (12.120) can be evaluated in a straightforward manner. Basically, this is a generalization of the Gaussian integral to the functional space and leads to

$$Z[J] = e^{iW[J]}$$
$$= N \left(\det \left(\partial_\mu \partial^\mu + m^2 \right)^{-\frac{1}{2}} \right) e^{-\frac{i}{2} \int \mathrm{d}^4 x \mathrm{d}^4 y \, J(x) G(x-y) J(y)}, \tag{12.126}$$

where the Green's function $G(x - y)$ is formally the inverse of the operator in the quadratic part of the action, namely,

$$G = (-\partial_\mu \partial^\mu - m^2)^{-1}. \tag{12.127}$$

(We note here that a Gaussian integral for fermions leads to positive powers of determinants in contrast to the negative powers for bosons which is related to the fact that fermion loops have a negative sign associated with them.) The determinant in (12.126) is independent of field variables and is a constant (possibly divergent) and can be incorporated into the normalization constant N. When we have an interacting theory (not quadratic in the field variables) the functional integral, in general, cannot be obtained in a closed form. In this case, we evaluate the functional integral perturbatively (expanding the interaction term in the exponent so that the functional integral becomes a series of integrals corresponding to different moments of a Gaussian integration) and the perturbative expansion coincides with the perturbative expansion of amplitudes in the conventional canonical quantized field theory. The advantage of using the path integral description lies in the fact that there are no operators in (12.120), everything is a classical function.

With this brief introduction to path integral description, let us go back to the Maxwell theory

$$\mathcal{L}^{(J)} = -\frac{1}{4}F_{\mu\nu}F^{\mu\nu} + J^\mu A_\mu, \tag{12.128}$$

where J^μ represents a conserved current (source). In this case, the generating functional in the path integral formalism is given by

$$Z[J_\mu] = e^{iW[J_\mu]} = N \int \mathcal{D}A_\mu e^{iS^{(J)}[A_\mu]} \tag{12.129}$$

$$= N \int \mathcal{D}A_\mu \, e^{i[\frac{1}{2}(A_\mu, P^{\mu\nu}A_\nu)+(J^\mu, A_\mu)]}, \tag{12.130}$$

where N is a normalization constant and (see (9.122)) we have identified

$$P^{\mu\nu}(x-y) = (\eta^{\mu\nu}\Box - \partial^\mu\partial^\nu)\delta^4(x-y),$$

$$(J^\mu, A_\mu) = \int \mathrm{d}^4x \, J^\mu(x)A_\mu(x). \tag{12.131}$$

This is a Gaussian functional integral and we can evaluate this using our earlier result in (12.126)

$$Z[J_\mu] = e^{iW[J_\mu]} = N\left(\det(-P^{\mu\nu})\right)^{-\frac{1}{2}} e^{-\frac{i}{2}\left(J^\mu, P_{\mu\nu}^{-1} J^\nu\right)}, \tag{12.132}$$

where we have used the notation in (12.131). However, as we have seen before (see (9.123) and (9.124)), the operator $P^{\mu\nu}$ is a projection operator for transverse photons. The longitudinal vectors k_μ (or ∂_μ) are its eigenvectors with zero eigenvalue. Clearly therefore, the determinant of $P^{\mu\nu}$ vanishes. This implies that the generating functional does not exist in this case. (The operator possesses zero modes and, consequently, the inverse of the matrix cannot be defined either.)

The source of the difficulty is not hard to see. The Lagrangian density for Maxwell's theory is invariant under the gauge transformation

$$A_\mu \to A_\mu^{(\theta)} = U A_\mu U^{-1} - \frac{1}{ie}(\partial_\mu U) U^{-1}, \tag{12.133}$$

where

$$U(\theta) = e^{-i\theta(x)}. \tag{12.134}$$

We can immediately see that for a $U(1)$ gauge group, we can write the transformation as

$$A_\mu \to A_\mu^{(\theta)} = A_\mu + \frac{1}{e}\partial_\mu \theta(x), \tag{12.135}$$

which is the familiar gauge transformation for Maxwell's theory. This is a $U(1)$ symmetry and the symmetry is noncompact since the parameter of transformation $\theta(x)$ can take any real value. (As we have seen the form of transformation for the gauge fields (12.133) in terms of $U(\theta)$ is quite general and holds for the non-Abelian theories as well.)

For a fixed A_μ, all the $A_\mu^{(\theta)}$'s that are obtained by making a gauge transformations with all possible $\theta(x)$ are said to lie on an "orbit" in the group space. The action S, on the other hand, is constant on such orbits. Therefore, the generating functional, even in the absence of any sources, is proportional to the "volume" of the orbits denoted by

$$\int \prod_x d\theta(x). \tag{12.136}$$

(In the non-Abelian case, this should be replaced by the group invariant Haar measure $\prod_x dU(x)$.) This is an infinite factor (which is one of the reasons for the divergence in the naive evaluation of the functional integral) and must be extracted out before doing any calculations. The method for extracting this factor out of the path integral is due to Faddeev and Popov and relies on the method of gauge fixing. We recognize that we should not integrate over all gauge field configurations because they are not really distinct. Rather we should integrate over each orbit only once.

The way this is carried out is by choosing a hypersurface which intersects each orbit only once, namely, if

$$F[A_\mu(x)] = 0, \tag{12.137}$$

defines the hypersurface which intersects the gauge orbits once, then even if A_μ does not satisfy the condition, we can find a gauge transformed $A_\mu^{(\theta)}$ which does and

$$F[A_\mu^{(\theta)}(x)] = 0, \tag{12.138}$$

has a unique solution for some $\theta(x)$. In this way, we pick up a representative gauge field from each gauge orbit. This procedure is known as gauge fixing and the condition

$$F[A_\mu(x)] = 0, \tag{12.139}$$

is known as the gauge condition (or gauge fixing condition). Thus, for example, here are a few of the familiar gauge fixing conditions

$$\begin{aligned} F[A_\mu(x)] = \partial_\mu A^\mu(x) &= 0, \quad \text{the Lorentz/Landau gauge,} \\ \nabla \cdot \mathbf{A}(x) &= 0, \quad \text{the Coulomb gauge,} \\ A_0(x) &= 0, \quad \text{the temporal gauge,} \\ A_3(x) &= 0, \quad \text{the axial gauge,} \end{aligned} \tag{12.140}$$

and so on. Physical quantities are, of course, gauge invariant and do not depend on the choice of the hypersurface (gauge). We can already see the need for gauge fixing from the fact that because the action is gauge invariant so is the generating functional (if sources are transformed appropriately in the non-Abelian case). Therefore, it would lead only to gauge invariant Green's functions. On the other hand, we know from ordinary perturbation theory that the Green's functions are, in general, gauge dependent (recall, for example, the photon propagator) although the physical S-matrix (the scattering matrix) elements are gauge independent. Thus we have to fix a gauge without which even the Cauchy initial value problem cannot be solved. (Only physical quantities need to be gauge independent.)

To extract out the infinite gauge volume factor, let us do the following trick due to Faddeev and Popov. Let us define

$$\Delta_{\text{FP}}^{-1}[A_\mu] = \int \prod_x d\theta(x) \, \delta\!\left(F[A_\mu^{(\theta)}(x)]\right). \tag{12.141}$$

This can also be written as

$$\Delta_{\text{FP}}[A_\mu] \int \prod_x d\theta(x) \, \delta\!\left(F[A_\mu^{(\theta)}(x)]\right) = 1, \tag{12.142}$$

which can, therefore, be thought of as a completeness relation. (The integration measure should be $dU(x)$ which is essential in the case of non-Abelian theories.) Note that the quantity $\Delta_{\text{FP}}[A_\mu]$, known as the Faddeev-Popov determinant, is gauge invariant which can be seen as follows. Let us make a gauge transformation $A_\mu \to A_\mu^{(\theta')}$ in (12.141). Then

$$\begin{aligned}
\Delta_{\text{FP}}^{-1}[A_\mu^{(\theta')}] &= \int \prod_x d\theta(x) \, \delta\!\left(F[A_\mu^{(\theta+\theta')}(x)]\right) \\
&= \int \prod_x d\theta(x) \, \delta\!\left(F[A_\mu^{(\theta)}(x)]\right) \\
&= \Delta_{\text{FP}}^{-1}[A_\mu].
\end{aligned} \tag{12.143}$$

This follows from the fact that the measure in the group space is invariant under a gauge transformation. That is (for the Abelian group a translation of the transformation parameter corresponds to a gauge transformation)

$$\int \mathrm{d}\theta(x) = \int \mathrm{d}\left(\theta(x) + \theta'(x)\right). \tag{12.144}$$

In the non-Abelian case, we should have the Haar measure which is gauge invariant, namely,

$$\int \mathrm{d}(UU') = \int \mathrm{d}U. \tag{12.145}$$

Remembering that $\Delta_{\rm FP}[A_\mu]$ is gauge invariant we can now insert this identity factor into the generating functional to write

$$Z[J_\mu] = N \int \mathcal{D}A_\mu \Delta_{\rm FP}[A_\mu] \int \prod_x \mathrm{d}\theta(x)\, \delta\!\left(F[A_\mu^{(\theta)}(x)]\right) e^{iS^{(J)}[A_\mu]}. \tag{12.146}$$

Furthermore, let us make an inverse gauge transformation

$$A_\mu \to A_\mu^{(-\theta)}, \tag{12.147}$$

under which the generating functional takes the form

$$\begin{aligned}
Z[J_\mu] &= N \int \mathcal{D}A_\mu \Delta_{\rm FP}[A_\mu] \int \prod_x \mathrm{d}\theta(x)\, \delta\!\left(F[A_\mu(x)]\right) e^{iS^{(J)}[A_\mu]} \\
&= N\!\left(\int \prod_x \mathrm{d}\theta(x)\right)\!\int \mathcal{D}A_\mu \Delta_{\rm FP}[A_\mu]\, \delta\!\left(F[A_\mu(x)]\right) e^{iS^{(J)}[A_\mu]} \\
&= N \int \mathcal{D}A_\mu \Delta_{\rm FP}[A_\mu]\, \delta\!\left(F[A_\mu(x)]\right) e^{iS^{(J)}[A_\mu]}, \tag{12.148}
\end{aligned}$$

where the (infinite) gauge volume has been factored out and absorbed into the normalization constant N of the path integral.

Therefore, this gives the correct functional form for the generating functional. However, we still have to determine what $\Delta_{\rm FP}[A_\mu]$ is. To do this let us note from (12.141) that

$$\Delta_{\text{FP}}^{-1}[A_\mu] = \int \prod_x d\theta(x)\, \delta\!\left(F[A_\mu^{(\theta)}(x)]\right)$$

$$= \int \prod_x dF\, \delta\!\left(F[A_\mu^{(\theta)}(x)]\right)\left(\det \frac{\delta\theta}{\delta F}\right)$$

$$= \left.\det \frac{\delta\theta}{\delta F}\right|_{F[A_\mu^{(\theta)}(x)]=0}. \tag{12.149}$$

We note that since $\Delta_{\text{FP}}^{-1}[A_\mu]$ is gauge invariant we can make an inverse gauge transformation to make $F[A_\mu(x)] = 0$ in the above derivation. On the other hand, for gauge fields which satisfy the condition

$$F[A_\mu(x)] = 0, \tag{12.150}$$

we have

$$\theta(x) = 0. \tag{12.151}$$

Thus, using (12.149)-(12.151) we determine

$$\Delta_{\text{FP}}[A_\mu] = \det \left(\frac{\delta F[A_\mu^{(\theta)}(x)]}{\delta \theta(y)}\right)_{\theta=0}. \tag{12.152}$$

The Faddeev-Popov determinant can, therefore, be thought of as the Jacobian that goes with a given gauge choice. We see that the Faddeev-Popov determinant can be calculated simply by restricting to infinitesimal gauge transformations (since we take $\theta = 0$ at the end). (Here we can completely ignore the problem of Gribov ambiguity associated with large gauge transformations.)

We can further generalize our derivation by noting that a general equation of the hypersurface has the form (physical results are not sensitive to the choice of the hypersurface)

$$F[A_\mu(x)] = f(x), \tag{12.153}$$

12.3 Path integral quantization of gauge theories

where $f(x)$ is independent of A_μ. Then we can insert the identity

$$\Delta_{\text{FP}}[A_\mu] \int \prod_x d\theta(x)\ \delta\Big(F[A_\mu^{(\theta)}(x)] - f(x)\Big) = 1, \qquad (12.154)$$

into the functional integral. The Faddeev-Popov determinant is unchanged by this modification because $f(x)$ does not depend on $A_\mu(x)$. Thus the generating functional in this case is given by

$$Z[J_\mu] = N \int \mathcal{D}A_\mu \Delta_{\text{FP}}[A_\mu]\delta\Big(F[A_\mu(x)] - f(x)\Big) e^{iS^{(J)}[A_\mu]}.$$
$$(12.155)$$

Following 't Hooft, we can now do the following (also known as the 't Hooft trick). We note that physical quantities are independent of $f(x)$. Hence we can multiply the generating functional by a weight factor and integrate over all $f(x)$. Thus, the generating functional becomes

$$\begin{aligned}
Z[J_\mu] &= N \int \mathcal{D}A_\mu \Delta_{\text{FP}}[A_\mu] \\
&\quad \times \int \mathcal{D}f \delta\Big(F[A_\mu(x)] - f(x)\Big) e^{-\frac{i}{2\xi}\int d^4x (f(x))^2} e^{iS^{(J)}[A_\mu]} \\
&= N \int \mathcal{D}A_\mu \Delta_{\text{FP}}[A_\mu] e^{i\left[S^{(J)}[A_\mu] - \frac{1}{2\xi}\int d^4x (F[A_\mu(x)])^2\right]} \\
&= N \int \mathcal{D}A_\mu \Delta_{\text{FP}}[A_\mu] e^{i(S^{(J)} + S_{\text{GF}})}, \qquad (12.156)
\end{aligned}$$

where we have defined a gauge fixing action as

$$S_{\text{GF}} = \int d^4x\ \mathcal{L}_{\text{GF}} = -\frac{1}{2\xi} \int d^4x\ (F[A_\mu(x)])^2, \qquad (12.157)$$

and ξ is known as the gauge fixing parameter.

We note further that since

$$\Delta_{\text{FP}}[A_\mu] = \det\left(\frac{\delta F[A_\mu^{(\theta)}(x)]}{\delta \theta(y)}\right)_{\theta=0}, \qquad (12.158)$$

we can write this as

$$\begin{aligned}
\Delta_{\text{FP}}[A_\mu] &= \det\left(\frac{\delta F[A_\mu^\theta(x)]}{\delta\theta(y)}\right)_{\theta=0} \\
&= \int \mathcal{D}\bar{c}\,\mathcal{D}c\; e^{-i\left(\bar{c},\left(\frac{\delta F}{\delta\theta}\right)_{\theta=0}c\right)} \\
&= \int \mathcal{D}\bar{c}\,\mathcal{D}c\; e^{-i\int d^4x\,d^4y\,\bar{c}(x)\left(\frac{\delta F[A_\mu^\theta(x)]}{\delta\theta(y)}\right)_{\theta=0}c(y)} \\
&= \int \mathcal{D}\bar{c}\,\mathcal{D}c\; e^{iS_{\text{ghost}}}, \quad\quad (12.159)
\end{aligned}$$

where

$$S_{\text{ghost}} = -\int d^4x\,d^4y\; \bar{c}(x)\left(\frac{\delta F[A_\mu^\theta(x)]}{\delta\theta(y)}\right)_{\theta=0} c(y). \quad (12.160)$$

Here we have introduced two independent fictitious fields $c(x)$ and $\bar{c}(x)$, known as ghost fields, to write the Faddeev-Popov determinant in the form of a ghost action (action involving ghost fields). We note here that this is possible (since we have a determinant with a non-negative power) only if the ghost fields $c(x)$ and $\bar{c}(x)$ anticommute (ghost fields have the same Lorentz structure as the parameters of transformation, but opposite statistics.), i.e.,

$$\begin{aligned}
[c(x), c(y)]_+ &= 0, \\
[\bar{c}(x), \bar{c}(y)]_+ &= 0, \\
[c(x), \bar{c}(y)]_+ &= 0. \quad\quad (12.161)
\end{aligned}$$

Thus although these fields behave as scalar fields under Lorentz transformations, they obey anticommutation rules. These fields are known as Faddeev-Popov ghosts and as is obvious from their anticommutation relations, graphs involving these fictitious particles in closed loops must have an additional (-1) factor just like the fermions. Thus the generating functional now takes the form

$$Z[J_\mu] = e^{iW[J_\mu]} = N\int \mathcal{D}A_\mu \mathcal{D}\bar{c}\,\mathcal{D}c\; e^{iS_{\text{eff}}^{(J)}[A_\mu,c,\bar{c}]}, \quad (12.162)$$

where

$$S^{(J)}_{\text{eff}}[A_\mu, c, \bar{c}] = S^{(J)}[A_\mu] + S_{\text{GF}} + S_{\text{ghost}}$$
$$= \int d^4x \, \mathcal{L}^{(J)}_{\text{eff}}[A_\mu, c, \bar{c}]. \tag{12.163}$$

Thus we can summarize what we have done so far. To do covariant perturbation theory in the path integral formalism, we start with a gauge invariant Lagrangian density and add to it a gauge fixing Lagrangian density determined by the gauge fixing condition that we want to work with. Of course, this modifies the starting theory. We then add a ghost Lagrangian density which is determined by an infinitesimal gauge variation of the gauge fixing condition and this is expected to compensate for the change in the theory due to the gauge fixing term (recall from (12.142) that both these actions arose from inserting a factor of unity into the functional integral). It is worth noting at this point that we have modified our starting Lagrangian density by a series of formal manipulations. It is, of course, our responsibility to show that the physical interpretation of the theory has not changed, that the S-matrix we obtain is independent of the gauge choice and is unitary and that this formulation leads naturally to the Gupta-Bleuler states as physical states. We would show this in the next chapter. However, for the moment let us emphasize that the purpose of a gauge fixing Lagrangian density is to break gauge invariance, namely, the gauge fixing Lagrangian density should make the quadratic part of the effective Lagrangian density nonsingular. Otherwise it fails its purpose and will not correspond to an acceptable gauge fixing condition.

Let us now look at a simple gauge fixing condition in Maxwell's theory, for example, the covariant condition (see (12.153))

$$F[A_\mu(x)] = \partial_\mu A^\mu(x) = f(x). \tag{12.164}$$

In this case, the gauge fixing Lagrangian density (see (12.157)) has the form

$$\mathcal{L}_{\text{GF}} = -\frac{1}{2\xi}(F[A_\mu(x)])^2 = -\frac{1}{2\xi}(\partial_\mu A^\mu(x))^2. \tag{12.165}$$

It is clear that this provides longitudinal components to the quadratic terms in fields and hence breaks gauge invariance. To obtain the corresponding ghost Lagrangian density for this gauge choice, we note that

$$F[A_\mu^{(\theta)}(x)] = \partial_\mu A^{(\theta)\mu}(x) = \partial_\mu \left(A^\mu(x) + \frac{1}{e} \partial^\mu \theta(x) \right), \qquad (12.166)$$

so that (see (12.152))

$$\left. \frac{\delta F[A_\mu^{(\theta)}(x)]}{\delta \theta(y)} \right|_{\theta=0} = \frac{1}{e} \Box_x \delta^4(x-y). \qquad (12.167)$$

Absorbing the factor $\frac{1}{e}$ into the normalization factor (alternatively scaling the ghost fields), the ghost action (12.160) for this gauge choice is obtained to be

$$\begin{aligned}
S_{\text{ghost}} &= -\int d^4x\, d^4y\, \bar{c}(x) \left. \frac{\delta F[A_\mu^\theta(x)]}{\delta \theta(y)} \right|_{\theta=0} c(y) \\
&= -\int d^4x\, d^4y\, \bar{c}(x) \left(\Box_x \delta^4(x-y) \right) c(y) \\
&= \int d^4x\, \partial_\mu \bar{c}(x) \partial^\mu c(x) = \int d^4x\, \mathcal{L}_{\text{ghost}}, \qquad (12.168)
\end{aligned}$$

where we have neglected total divergence terms.

Thus our effective Lagrangian density for the Maxwell theory with the choice of a covariant gauge fixing condition becomes

$$\mathcal{L}_{\text{eff}}^{(J)} = -\frac{1}{4} F_{\mu\nu} F^{\mu\nu} - \frac{1}{2\xi} (\partial_\mu A^\mu)^2 + \partial_\mu \bar{c} \partial^\mu c + J^\mu A_\mu. \qquad (12.169)$$

We note here that the ghost fields are noninteracting in the case of the Maxwell theory in flat space-time and, therefore, we may neglect them and then for $\xi = 1$ we recognize that our effective Lagrangian density (12.169) is nothing other than Maxwell's theory in the Feynman-Fermi gauge (9.125). In non-Abelian gauge theories, however, the ghost fields are interacting and have to be present. This explains why the naive modification in (12.113) of the non-Abelian gauge theory failed to be sufficient unlike in the Maxwell theory (namely, the ghost Lagrangian density was missing). Furthermore, since the ghost fields and the ghost action are really necessary for the unitarity of the S-matrix, we cannot neglect them even if they are noninteracting particularly when we are doing calculations at finite temperature. It is also true that when Maxwell's theory is coupled to a gravitational field, the ghost fields automatically couple to the geometry also. Hence omitting the ghost Lagrangian density in such a case would lead to incorrect results.

One way of looking at the ghost fields is as if they are there to subtract out the unphysical field degrees of freedom. For example, the A_μ field has four field degrees of freedom. On the other hand each of the ghost fields, being a scalar, has only one field degree of freedom (we will discuss the question of hermiticity of the ghost fields in the next chapter). Hence we can think of the effective Lagrangian density as having two $(4-2\times 1 = 2)$ effective field degrees of freedom which is the correct number of physical dynamical components as we have already seen within the context of canonical quantization of Maxwell's theory. This naive counting works pretty well as we will see later. (The ghost degrees of freedom subtract because they have the unphysical statistics, namely, they anticommute even though they are scalar fields.)

Let us now go back to the effective Lagrangian density (12.169) and ask how we can recover the Lorentz gauge from the Lorentz like covariant gauge condition in (12.164). To do this let us rewrite the effective Lagrangian density (12.169) as

$$\mathcal{L}_{\text{eff}}^{(J)} = -\frac{1}{4}F_{\mu\nu}F^{\mu\nu} + \frac{\xi}{2}F^2 - F(\partial_\mu A^\mu) + \partial_\mu \bar{c}\partial^\mu c + J^\mu A_\mu. \quad (12.170)$$

Here we have introduced an auxiliary field F which does not have any dynamics. The equation of motion for this field leads to the gauge fixing condition and elimination of this field through its equations of motion leads to the familiar gauge fixing Lagrangian density in (12.169). On the other hand, we see that the equation of motion for F is given by

$$\xi F(x) = \partial_\mu A^\mu(x), \quad (12.171)$$

and, therefore, in the limit $\xi \to 0$, this equation leads to the Lorentz condition (or the Landau gauge)

$$\partial_\mu A^\mu(x) = 0. \tag{12.172}$$

Therefore, the effective Lagrangian density in the Lorentz gauge has the form

$$\mathcal{L}_{\text{eff}}^{(J)} = \lim_{\xi \to 0} -\frac{1}{4}F_{\mu\nu}F^{\mu\nu} - \frac{1}{2\xi}(\partial_\mu A^\mu)^2 + \partial_\mu \bar{c}\partial^\mu c + J^\mu A_\mu, \tag{12.173}$$

which can also be written equivalently as (see (12.170) with $\xi = 0$)

$$\mathcal{L}_{\text{eff}}^{(J)} = -\frac{1}{4}F_{\mu\nu}F^{\mu\nu} - F(\partial_\mu A^\mu) + \partial_\mu \bar{c}\partial^\mu c + J^\mu A_\mu, \tag{12.174}$$

and describes Maxwell's theory in the Lorentz gauge.

From our discussions of the Abelian gauge theories thus far, we see that there are two distinct quantization procedures. First, we have the operator quantization and again there appear to be two distinct possibilities in this case. Namely, we can quantize the Abelian gauge theory canonically. Here we explicitly eliminate the unphysical (dependent) field degrees of freedom and then quantize the physical (independent) field degrees of freedom. The Hilbert space contains only photon states of physical polarization. However, in the process of eliminating the dependent field degrees of freedom we lose manifest Lorentz covariance. The second possibility is to use the Gupta-Bleuler quantization method to quantize the theory in a manifestly covariant manner. Here we modify the theory so that all the field degrees of freedom are independent. We maintain manifest Lorentz covariance and quantize all components of the field as independent variables. Thus the vector space that we work with in this case is much larger than the physical Hilbert space. We select out the physical Hilbert space by imposing supplementary conditions on the state vectors in a Lorentz covariant way. In this case, the larger vector space of the theory contains states of indefinite norm and when we do perturbation theory in this formalism, in the intermediate states we find time-like photon states which contribute negatively whereas the longitudinal states contribute an equal positive amount. As a

12.3 PATH INTEGRAL QUANTIZATION OF GAUGE THEORIES

result, their contributions cancel out and effectively we are left with only two physical transverse degrees of freedom.

The second method of quantizing the Abelian gauge theory is through the method of path integrals. In this formalism the field variables are treated as classical variables. The generating functional for physical Green's functions is given by

$$\begin{aligned} Z[J_\mu] &= N \int \mathcal{D}A_\mu^{\mathrm{T}} e^{iS^{(J)}[A_\mu^{\mathrm{T}}]} \\ &\neq N \int \mathcal{D}A_\mu \delta(\partial \cdot A(x)) e^{iS^{(J)}[A_\mu]}, \end{aligned} \quad (12.175)$$

where A_μ^{T} denotes the transverse physical degrees of freedom. The correct description for the generating functional, according to Faddeev and Popov, is given by

$$\begin{aligned} Z[J_\mu] &= N \int \mathcal{D}A_\mu^{\mathrm{T}} e^{iS^{(J)}[A_\mu^{\mathrm{T}}]} \\ &= N \int \mathcal{D}A_\mu \Delta_{\mathrm{FP}}[A_\mu] \delta(\partial \cdot A(x))\, e^{iS^{(J)}[A_\mu]}. \end{aligned} \quad (12.176)$$

In this way, the Faddeev-Popov determinant or the ghost action can be thought of as the Jacobian in transforming from the physical dynamical field variables to all components of the field variables through the constraint relation. In the path integral formalism, we also modify the theory (as in the Gupta-Bleuler method) so that there are no dependent variables and all components of the field contribute to any Green's function of the theory. However, in this formalism there is no reference to the Hilbert space of the theory. Rather the extra contributions (from the unphysical degrees of freedom) are cancelled by the Faddeev-Popov determinant which we can write as a ghost action (namely, the ghost contributions cancel those from the unphysical gauge field degrees of freedom).

In summary, we note that gauge invariance puts a very strong constraint on the structure of the Lagrangian density for the gauge field. In particular, the coefficient matrix of the quadratic terms in the Lagrangian density is singular and, therefore, non-invertible. As a result, if we take $\mathcal{L}_{\mathrm{inv}}$, as the Lagrangian density describing the dynamics of the gauge field theory, then we cannot define propagators and the entire philosophy of doing perturbative calculations with Feynman diagrams breaks down. In order to circumvent this difficulty, we add to the gauge invariant Lagrangian density a term which breaks gauge invariance and thereby allows us to define the propagator for the gauge field. Such a term is called a gauge fixing term and any term which makes the coefficient matrix of the quadratic terms (in the action) nonsingular and maintains various global symmetries of the theory is allowed for this purpose. On the other hand, adding a gauge fixing Lagrangian density changes the theory, in general, and to compensate for that we have to add a corresponding Lagrangian density for the ghost fields following the prescription of Faddeev and Popov.

Let us now apply these ideas to the study of the non-Abelian gauge theory. As we have seen, the Lagrangian density for the gauge fields is given by (see (12.75))

$$\mathcal{L}_{\text{inv}} = -\frac{1}{4} F_{\mu\nu}^a F^{\mu\nu a}, \tag{12.177}$$

which is invariant under the infinitesimal gauge transformation

$$A_\mu^a(x) \to A_\mu^{(\epsilon)a}(x) = A_\mu^a(x) + \frac{1}{g} D_\mu \epsilon^a(x), \tag{12.178}$$

where $\epsilon^a(x)$ is the infinitesimal parameter of transformation. Here the covariant derivative in the adjoint representation as well as the non-Abelian field strength tensors are defined as (see (12.67) and (12.76))

$$\begin{aligned} D_\mu \epsilon^a(x) &= \partial_\mu \epsilon^a(x) - g f^{abc} A_\mu^b(x) \epsilon^c(x), \\ F_{\mu\nu}^a(x) &= \partial_\mu A_\nu^a(x) - \partial_\nu A_\mu^a(x) - g f^{abc} A_\mu^b(x) A_\nu^c(x), \end{aligned} \tag{12.179}$$

with g denoting the coupling constant of the theory.

The standard covariant gauge fixing (12.164), in the non-Abelian gauge theory, consists of adding to the invariant Lagrangian density a gauge fixing Lagrangian density of the form

12.3 PATH INTEGRAL QUANTIZATION OF GAUGE THEORIES

$$\mathcal{L}_{\text{GF}} = -\frac{1}{2\xi}\left(\partial_\mu A^{\mu a}(x)\right)^2, \tag{12.180}$$

which corresponds to a gauge fixing condition of the form

$$F^a[A_\mu(x)] = \partial_\mu A^{\mu a}(x) = f^a(x). \tag{12.181}$$

Here ξ represents a real arbitrary constant parameter known as the gauge fixing parameter. Following the prescription of Faddeev and Popov, we can write the ghost action corresponding to this gauge choice as (see (12.160) and note that since ϵ^a is an infinitesimal parameter, it is redundant to set it equal to zero at the end since $F^a[A_\mu^{(\epsilon)}(x)]$ is linear in ϵ^a)

$$\begin{aligned}S_{\text{ghost}} &= \int \mathrm{d}^4 x\, \mathcal{L}_{\text{ghost}} \\ &= -\int \mathrm{d}^4 x \mathrm{d}^4 y\, \overline{c}^a(x) \frac{\delta F^a[A_\mu^{(\epsilon)}(x)]}{\delta \epsilon^b(y)} c^b(y). \end{aligned} \tag{12.182}$$

We note that for the covariant gauge choice (12.181) that we are using, we can write

$$F^a[A_\mu^{(\epsilon)}(x)] = \partial^\mu A_\mu^{(\epsilon)a}(x) = \partial^\mu \left(A_\mu^a(x) + \frac{1}{g} D_\mu \epsilon^a(x)\right), \tag{12.183}$$

so that we have (see (12.122))

$$\begin{aligned}\frac{\delta F^a[A_\mu^{(\epsilon)}(x)]}{\delta \epsilon^b(y)} &= \frac{1}{g} \partial_x^\mu D_{x\mu}^{ab} \delta^4(x-y) \\ &= \partial_x^\mu \left(\partial_{x\mu} \delta^{ab} - g f^{acb} A_\mu^c(x)\right) \delta^4(x-y). \end{aligned} \tag{12.184}$$

Consequently, rescaling the ghost fields (to absorb the factor of $\frac{1}{g}$) we can write the ghost Lagrangian density for this choice of gauge fixing to be

$$\begin{aligned}
S_{\text{ghost}} &= \int d^4x\, \mathcal{L}_{\text{ghost}} \\
&= -\int d^4x\, d^4y\, \bar{c}^a(x) \left(\partial_x^\mu D^{ab}_{x\mu} \delta^4(x-y)\right) c^b(y) \\
&= \int d^4x\, \partial^\mu \bar{c}^a(x)\, (D_\mu c(x))^a\,,
\end{aligned} \qquad (12.185)$$

where we have dropped total derivative terms (surface terms).

With all these modifications, the total Lagrangian density for the non-Abelian gauge theory can be written in this covariant gauge as

$$\begin{aligned}
\mathcal{L}_{\text{TOT}} &= \mathcal{L}_{\text{inv}} + \mathcal{L}_{\text{GF}} + \mathcal{L}_{\text{ghost}} \\
&= -\frac{1}{4} F^a_{\mu\nu} F^{\mu\nu a} - \frac{1}{2\xi}(\partial_\mu A^{\mu a})^2 + \partial^\mu \bar{c}^a\, (D_\mu c)^a\,.
\end{aligned} \qquad (12.186)$$

We note that the ghost fields in the present case are interacting unlike in Maxwell's theory and, therefore, cannot be neglected even in flat space-time. As we have mentioned earlier, the gauge fixing and the ghost Lagrangian densities modify the original theory in a compensating manner which allows us to define the Feynman rules of the theory and carry out perturbative calculations. In a deeper sense, the gauge fixing and the ghost Lagrangian densities, in the path integral formulation, merely correspond to a multiplicative factor of unity (see (12.142)) which does not change the physical content of the theory.

12.4 Path integral quantization of tensor fields

The method due to Faddeev and Popov gives a simple recipe for quantizing gauge theories in the path integral formalism. However, in some cases we have to work through the details of this analysis rather carefully in order to obtain the correct result. As an example of how we should be careful in carrying out the Faddeev-Popov analysis in complicated gauge theories, let us consider the gauge theory of the antisymmetric tensor field $A_{\mu\nu}(x)$

12.4 Path integral quantization of tensor fields

$$A_{\mu\nu}(x) = -A_{\nu\mu}(x), \qquad \mu, \nu = 0, 1, 2, 3. \tag{12.187}$$

Such tensor fields have been studied in connection with the question of confinement (of quarks in QCD) and are known as Kalb-Ramond fields. Let us consider only the free Lagrangian density for this field defined by

$$\mathcal{L} = -\frac{1}{6} F_{\mu\nu\lambda} F^{\mu\nu\lambda}, \tag{12.188}$$

where the field strength tensor corresponds to the totally antisymmetric tensor

$$F_{\mu\nu\lambda} = \partial_{[\mu} A_{\nu\lambda]} = \partial_\mu A_{\nu\lambda} + \partial_\nu A_{\lambda\mu} + \partial_\lambda A_{\mu\nu}. \tag{12.189}$$

Naive counting shows that $A_{\mu\nu}$ has six field degrees of freedom. However, we also note that this Lagrangian density is invariant under the gauge transformation

$$\delta A_{\mu\nu}(x) = \partial_\mu \theta_\nu(x) - \partial_\nu \theta_\mu(x), \tag{12.190}$$

so that not all field variables are independent. This can be seen from the definition of the conjugate momenta

$$\Pi^{\mu\nu}(x) = \frac{\partial \mathcal{L}}{\partial \dot{A}_{\mu\nu}(x)} = -F^{0\mu\nu}(x). \tag{12.191}$$

The independent field variables of the theory are A_{0i}, A_{ij} and we note that

$$\Pi^{\mu\nu} = -\Pi^{\nu\mu} = -F^{0\mu\nu}, \tag{12.192}$$

namely, the canonical momenta are antisymmetric (like the field variables). Furthermore, since $F_{\mu\nu\lambda}$ is completely antisymmetric in all the indices, we conclude that

$$\Pi^{0\mu} = -F^{00\mu} = 0,$$
$$\text{or,} \quad \Pi^{0i} = 0, \quad i = 1,2,3. \tag{12.193}$$

Since these momenta identically vanish, the corresponding field variables are like c-number quantities and we can choose the gauge condition

$$A_{0i}(x) = 0. \tag{12.194}$$

With these conditions we have

$$\Pi_{ij} = -F_{0ij} = -\dot{A}_{ij}, \tag{12.195}$$

and the Lagrangian density (12.188) takes the form

$$\begin{aligned}\mathcal{L} &= -\frac{1}{6}F_{\mu\nu\lambda}F^{\mu\nu\lambda} = -\frac{1}{6}\left[3F_{0ij}F^{0ij} + F_{ijk}F^{ijk}\right] \\ &= -\frac{1}{6}\left[3\Pi_{ij}\Pi^{ij} + F_{ijk}F^{ijk}\right], \quad i,j,k = 1,2,3. \end{aligned} \tag{12.196}$$

Thus it would seem that the theory has truly three degrees of freedom.

However, we note that the theory still possesses a residual (static) gauge invariance under the transformation

$$\delta A_{ij} = \partial_i \theta_j(\mathbf{x}) - \partial_j \theta_i(\mathbf{x}), \tag{12.197}$$

so that we can impose a Coulomb gauge condition of the form

$$\partial_i A_{ij}(x) = 0. \tag{12.198}$$

This would seem like three constraints and hence we would naively conclude that this theory has no dynamical degrees of freedom. However, on closer inspection we notice that the Coulomb gauge condition (12.198) actually represents only two independent conditions. We can see this by writing out the gauge condition explicitly

12.4 Path integral quantization of tensor fields

$$\partial_1 A_{1j} + \partial_2 A_{2j} + \partial_3 A_{3j} = 0. \tag{12.199}$$

Thus, the three conditions corresponding to $j = 1, 2, 3$ are

$$\begin{aligned}
\partial_2 A_{21} + \partial_3 A_{31} &= 0, \\
\partial_1 A_{12} + \partial_3 A_{32} &= 0, \\
\partial_1 A_{13} + \partial_2 A_{23} &= 0.
\end{aligned} \tag{12.200}$$

It is now clear that any two of the three conditions lead to the third condition so that there are only two independent conditions. Hence the theory has truly one physical degree of freedom and the antisymmetric tensor field describes a scalar field (it is a gauge theory describing a scalar degree of freedom). The fact that the theory has only one degree of freedom can also be seen in the following manner. Since $F_{\mu\nu\lambda}$ represent a totally antisymmetric third rank tensor in four space-time dimensions satisfying the Bianchi identity (see (12.189)),

$$\partial_\rho F_{\mu\nu\lambda} - \partial_\lambda F_{\rho\mu\nu} + \partial_\nu F_{\lambda\rho\mu} - \partial_\mu F_{\nu\lambda\rho} = 0, \tag{12.201}$$

we can represent them also as (this is the dual of the field strength tensor in four dimensions)

$$F_{\mu\nu\lambda}(x) = \frac{1}{\sqrt{2}} \epsilon_{\mu\nu\lambda}{}^\rho \partial_\rho \phi(x), \tag{12.202}$$

where $\phi(x)$ represents a real scalar field. In this case, the Lagrangian-density (12.188) takes the form

$$\begin{aligned}
\mathcal{L} &= -\frac{1}{6} F_{\mu\nu\lambda} F^{\mu\nu\lambda} = -\frac{1}{6} \times \frac{1}{2} \epsilon_{\mu\nu\lambda}{}^\rho \epsilon^{\mu\nu\lambda}{}_\sigma \partial_\rho \phi \partial^\sigma \phi \\
&= -\frac{1}{12} \times (-6\delta^\rho_\sigma) \partial_\rho \phi \partial^\sigma \phi = \frac{1}{2} \partial_\rho \phi \partial^\rho \phi.
\end{aligned} \tag{12.203}$$

We recognize this as the Lagrangian density for a free, massless real scalar field describing a single degree of freedom.

Let us now look at the path integral quantization of this theory. According to our earlier discussions, the starting gauge invariant Lagrangian density has the form

$$\mathcal{L} = -\frac{1}{6} F_{\mu\nu\lambda} F^{\mu\nu\lambda}, \tag{12.204}$$

and we add to it the covariant gauge fixing Lagrangian density (ξ is the gauge fixing parameter)

$$\mathcal{L}_{\text{GF}} = -\frac{1}{2\xi} (\partial_\mu A^{\mu\nu})^2, \tag{12.205}$$

corresponding to the choice of gauge condition

$$F^\nu[A_{\mu\lambda}(x)] = \partial_\mu A^{\mu\nu}(x) = f^\nu(x). \tag{12.206}$$

To determine the Lagrangian density for the ghosts we note that

$$\begin{aligned}
F^\nu[A^{(\theta)}_{\mu\lambda}(x)] &= \partial_\mu A^{(\theta)\mu\nu}(x) \\
&= \partial_\mu(A^{\mu\nu}(x) + \partial^\mu \theta^\nu(x) - \partial^\nu \theta^\mu(x)),
\end{aligned} \tag{12.207}$$

so that we have

$$\begin{aligned}
\frac{\delta F^\nu[A^{(\theta)}_{\mu\sigma}(x)]}{\delta \theta^\lambda(y)}\bigg|_{\theta=0} &= \partial_{x\mu} \left(\delta^\nu_\lambda \partial^\mu_x - \delta^\mu_\lambda \partial^\nu_x\right) \delta^4(x-y) \\
&= \left(\delta^\nu_\lambda \Box_x - \partial_{x\lambda} \partial^\nu_x\right) \delta^4(x-y).
\end{aligned} \tag{12.208}$$

This is a matrix with two vector indices. Hence to write the determinant as an action we need ghost fields which carry a vector index and we have

12.4 Path integral quantization of tensor fields

$$\begin{aligned}
S_{\text{ghost}} &= -\int d^4x d^4y\, \bar{c}_\nu(x) \frac{\delta F^\nu[A^{(\theta)}_{\mu\sigma}(x)]}{\delta \theta^\lambda(y)} c^\lambda(y) \\
&= -\int d^4x d^4y\, \bar{c}_\nu(x) \left((\delta^\nu_\lambda \Box_x - \partial_{x\lambda}\partial^\nu_x) \delta^4(x-y)\right) c^\lambda(y) \\
&= \frac{1}{2}\int d^4x\, (\partial_\mu \bar{c}_\nu(x) - \partial_\nu \bar{c}_\mu(x))(\partial^\mu c^\nu(x) - \partial^\nu c^\mu(x)) \\
&= \int d^4x\, \mathcal{L}_{\text{ghost}}.
\end{aligned} \quad (12.209)$$

Therefore, the effective Lagrangian density for the theory appears to be given by

$$\begin{aligned}
\mathcal{L}_{\text{eff}} &= \mathcal{L} + \mathcal{L}_{\text{GF}} + \mathcal{L}_{\text{ghost}} \\
&= -\frac{1}{6} F_{\mu\nu\lambda} F^{\mu\nu\lambda} - \frac{1}{2\xi}(\partial_\mu A^{\mu\nu})^2 \\
&\quad + \frac{1}{2}(\partial_\mu \bar{c}_\nu - \partial_\nu \bar{c}_\mu)(\partial^\mu c^\nu - \partial^\nu c^\mu).
\end{aligned} \quad (12.210)$$

Let us now try our naive counting of degrees of freedom of the theory. The field variable $A_{\mu\nu}$ has six degrees of freedom. The ghost fields being vectors have four degrees of freedom each and since they satisfy anticommutation relations, they subtract out field degrees of freedom. Thus the effective Lagrangian density appears to have

$$6 - 2 \times 4 = 6 - 8 = -2, \quad (12.211)$$

effective field degrees of freedom. This does not seem right (does not agree with the analysis from canonical quantization) and we will show now that this is a consequence of our careless application of the Faddeev-Popov procedure.

Let us start with the generating functional

$$Z[J_{\mu\nu}] = N \int \mathcal{D}A_{\mu\nu}\, e^{iS^{(J)}[A_{\mu\nu}]}. \quad (12.212)$$

The gauge symmetry allows us to choose a gauge condition and we have chosen a Lorentz like gauge condition.

$$F^\nu[A_{\mu\lambda}(x)] = \partial_\mu A^{\mu\nu}(x) = f^\nu(x). \tag{12.213}$$

Therefore, we introduce the identity element as (see (12.154))

$$\Delta_{\text{FP}}[A_{\mu\nu}] \int \prod_x d\theta^\sigma(x) \delta\big(F^\nu[A_{\mu\lambda}^{(\theta)}(x)] - f^\nu(x)\big) = 1, \tag{12.214}$$

where, as we have seen (see (12.152)),

$$\begin{aligned}\Delta_{\text{FP}}[A_{\mu\nu}] &= \det \frac{\delta F^\nu[A_{\mu\lambda}^{(\theta)}(x)]}{\delta \theta^\sigma(y)}\bigg|_{\theta=0} \\ &= \det\big((\Box_x \delta_\sigma^\nu - \partial_x^\nu \partial_{x\sigma})\delta^4(x-y)\big).\end{aligned} \tag{12.215}$$

Thus inserting this identity element into the functional integral, the generating functional can be written as

$$\begin{aligned}Z[J_{\mu\nu}] = N \int \mathcal{D}A_{\mu\nu} \mathcal{D}\bar{c}^\mu \mathcal{D}c^\nu \\ \times \delta\left(F^\nu[A_{\mu\lambda}(x)] - f^\nu(x)\right) e^{i\left(S^{(J)} + S_{\text{ghost}}\right)},\end{aligned} \tag{12.216}$$

where as we have seen before in (12.209)

$$S_{\text{ghost}} = \frac{1}{2} \int d^4x \, (\partial_\mu \bar{c}_\nu - \partial_\nu \bar{c}_\mu)(\partial^\mu c^\nu - \partial^\nu c^\mu). \tag{12.217}$$

In the naive application of the quantization procedure, we would use the 't Hooft trick to write

$$\begin{aligned}\int \mathcal{D}f^\mu \, \delta\left(F^\nu[A_{\mu\lambda}(x)] - f^\nu(x)\right) e^{-\frac{i}{2\xi} \int d^4x \, f_\nu(x) f^\nu(x)} \\ = e^{-\frac{i}{2\xi} \int d^4x \, F^\nu[A_{\mu\lambda}(x)] F_\nu[A_{\mu\lambda}(x)]}.\end{aligned} \tag{12.218}$$

12.4 PATH INTEGRAL QUANTIZATION OF TENSOR FIELDS

However, we note that the gauge condition

$$\partial_\mu A^{\mu\nu}(x) = f^\nu(x), \tag{12.219}$$

implies that

$$\partial_\nu \partial_\mu A^{\mu\nu} = \partial_\nu f^\nu(x) = 0. \tag{12.220}$$

That is, the function $f^\nu(x)$ has to be transverse and if we neglect to take this fact into account we may get an incorrect result. We take into account the transverse nature of $f^\mu(x)$ exactly like the Maxwell field and we apply the 't Hooft trick by integrating over a transverse weight factor

$$\int \mathcal{D}f^\mu \delta\left(F^\nu[A_{\mu\lambda}(x)] - f^\nu(x)\right) e^{-\frac{i}{2\xi} \int d^4x d^4y\, f_\mu(x)\overline{P}^{\mu\nu}(x-y)f_\nu(y)}, \tag{12.221}$$

where $\overline{P}^{\mu\nu}(x-y)$ is the normalized transverse projection operator defined earlier (see, for example, (9.123)) as

$$\overline{P}^{\mu\nu}(x-y) = \left(\eta^{\mu\nu} - \frac{\partial_x^\mu \partial_x^\nu}{\Box_x}\right) \delta^4(x-y). \tag{12.222}$$

We see that because of the transverse projection operator, the action for f_μ in (12.221) has a gauge invariance (just like Maxwell's theory) of the form

$$\delta f_\mu(x) = \partial_\mu \theta(x). \tag{12.223}$$

Thus one has to use a gauge fixing condition and we choose again a covariant gauge condition

$$\overline{F}[f_\mu(x)] = \frac{1}{\Box^{\frac{1}{2}}} \partial_\mu f^\mu(x) = f(x), \tag{12.224}$$

so that we can write

$$\overline{\Delta}_{\rm FP}\left[f_\mu\right] \int \prod_x d\theta(x)\delta\left(\overline{F}[f_\mu^{(\theta)}(x)] - f(x)\right) = 1, \qquad (12.225)$$

where

$$\overline{\Delta}_{\rm FP}\left[f_\mu\right] = \det\Big(\frac{\delta \overline{F}[f_\mu^{(\theta)}(x)]}{\delta\theta(y)}\Big)_{\theta=0} = \left(\det \Box_x \delta^4(x-y)\right)^{\frac{1}{2}}. \qquad (12.226)$$

Thus using the 't Hooft trick we can write

$$\int \mathcal{D}f^\mu \delta\left(F^\nu[A_{\mu\lambda}(x)] - f^\nu(x)\right) e^{-\frac{i}{2\xi}(f_\mu, \overline{P}^{\mu\nu} f_\nu)}$$

$$\times \int \mathcal{D}f \overline{\Delta}_{\rm FP}\left[f_\mu\right] \delta\left(\overline{F}[f_\mu(x)] - f(x)\right) e^{-\frac{i}{2\xi}(f,f)}$$

$$= \int \mathcal{D}f^\mu \,\delta\left(F^\nu[A_{\mu\lambda}(x)] - f^\nu(x)\right) e^{-\frac{i}{2\xi}(f_\mu, \overline{P}^{\mu\nu} f_\nu)}$$

$$\times \overline{\Delta}_{\rm FP}\left[f_\mu\right] e^{-\frac{i}{2\xi}(\overline{F}(f_\mu), \overline{F}(f_\mu))}$$

$$= \int \mathcal{D}f^\mu \delta\left(F^\nu[A_{\mu\lambda}(x)] - f^\nu(x)\right) e^{-\frac{i}{2\xi}(f_\mu, f^\mu)} \overline{\Delta}_{\rm FP}[f_\mu]. \qquad (12.227)$$

Furthermore, remembering that (see (12.226))

$$\overline{\Delta}_{\rm FP}\left[f_\mu\right] = \left(\det \Box_x \delta^4(x-y)\right)^{\frac{1}{2}}$$

$$= \int \mathcal{D}c(x)\, e^{i\overline{S}_{\rm ghost}},$$

we determine

$$\overline{S}_{\rm ghost} = -\frac{1}{2}(c, \Box c) = -\frac{1}{2}\int d^4x\, c(x)\Box c(x). \qquad (12.228)$$

Substituting this into (12.227) the functional integral becomes

$$\int \mathcal{D}f^\mu \mathcal{D}c\, \delta\left(F^\nu[A_{\mu\lambda}(x)] - f^\nu(x)\right) e^{-\frac{i}{2\xi}(f_\mu, f^\mu) - \frac{i}{2}(c, \Box c)}$$

$$= \int \mathcal{D}c\, e^{-\frac{i}{2\xi}(F^\nu[A_{\mu\lambda}], F_\nu[A_{\mu\lambda}]) - \frac{i}{2}(c, \Box c)}. \qquad (12.229)$$

12.4 Path integral quantization of tensor fields

Note here that the field $c(x)$ is a real anticommuting scalar field. This is different from the usual Faddeev-Popov ghosts in the sense that the Faddeev-Popov ghosts seem to come in pairs. The ghosts of the present form are known as Nielsen ghosts. In flat space-time this ghost Lagrangian density can be seen to give a total divergence and, therefore, may be neglected, but in the presence of gravitation it cannot be written as a total divergence and is quite relevant.

Thus using this modified 't Hooft weighting factor, the generating functional takes the form

$$Z[J_{\mu\nu}]$$
$$= N \int \mathcal{D}A_{\mu\nu} \mathcal{D}\bar{c}_\mu \mathcal{D}c_\nu \mathcal{D}c \; e^{i\left(S^J + S_{\text{ghost}} + \bar{S}_{\text{ghost}}\right)} e^{-\frac{i}{2\xi}(F_\mu[A_{\lambda\sigma}], F^\mu[A_{\lambda\sigma}])}$$
$$= N \int \mathcal{D}A_{\mu\nu} \mathcal{D}\bar{c}_\mu \mathcal{D}c_\nu \mathcal{D}c \; e^{i\left(S^J + S_{\text{GF}} + S_{\text{ghost}} + \bar{S}_{\text{ghost}}\right)}, \qquad (12.230)$$

where

$$S_{\text{GF}} = -\frac{1}{2\xi} \int d^4x \; F_\mu[A_{\lambda\sigma}(x)] F^\mu[A_{\lambda\sigma}(x)]$$
$$= -\frac{1}{2\xi} \int d^4x \, (\partial_\mu A^{\mu\nu}(x)) \left(\partial^\lambda A_\lambda{}^\nu\right). \qquad (12.231)$$

Therefore, we can write the effective Lagrangian density as

$$\mathcal{L}_{\text{eff}}^{(J)} = -\frac{1}{6} F_{\mu\nu\lambda} F^{\mu\nu\lambda} + J^{\mu\nu} A_{\mu\nu} - \frac{1}{2\xi} (\partial_\mu A^{\mu\nu}) \left(\partial^\lambda A_{\lambda\nu}\right)$$
$$+ \frac{1}{2} (\partial_\mu \bar{c}_\nu - \partial_\nu \bar{c}_\mu)(\partial^\mu c^\nu - \partial^\nu c^\mu) - \frac{1}{2} c \Box c. \qquad (12.232)$$

Counting the field degrees of freedom we see that the effective number of degrees of freedom seems to be

$$6 - 2 \times 4 - 1 = -3. \qquad (12.233)$$

This is again not right and the agreement with canonical quantization seems to be worse than before.

We notice at this point that although we have fixed up the gauge invariance of the gauge fixing Lagrangian density, the ghost Lagrangian density also possesses a gauge invariance, namely, under

$$\delta c_\mu(x) = \partial_\mu \lambda(x),$$

and

$$\delta \bar{c}_\mu(x) = \partial_\mu \bar{\lambda}(x), \tag{12.234}$$

where $\lambda(x)$ and $\bar{\lambda}(x)$ are anticommuting scalar parameters, the ghost Lagrangian density $\mathcal{L}_{\text{ghost}}$ is invariant. We can again use the Faddeev-Popov trick and write

$$\widetilde{\Delta}_{\text{FP}}[c_\mu] \int \prod_x \mathrm{d}\lambda(x)\, \delta\left(\widetilde{F}[c_\mu^{(\lambda)}(x)] - \widetilde{f}(x)\right) = 1, \tag{12.235}$$

and

$$\widetilde{\bar{\Delta}}_{\text{FP}}[\bar{c}_\mu] \int \prod_x \mathrm{d}\bar{\lambda}(x)\, \delta\left(\widetilde{\bar{F}}[\bar{c}_\mu^{(\bar{\lambda})}(x)] - \widetilde{\bar{f}}(x)\right) = 1. \tag{12.236}$$

Let us choose for simplicity Lorentz like gauge conditions

$$\widetilde{F} = \partial_\mu c^\mu(x) = \widetilde{f}(x), \tag{12.237}$$

and

$$\widetilde{\bar{F}} = \partial_\mu \bar{c}^\mu(x) = \widetilde{\bar{f}}(x). \tag{12.238}$$

We are now ready to calculate the Faddeev-Popov determinants associated with these gauge fixing conditions,

$$\begin{aligned}
\widetilde{\Delta}_{\text{FP}}^{-1}[c_\mu] &= \int \prod_x \mathrm{d}\lambda(x) \delta(\widetilde{F}[c_\mu^{(\lambda)}(x)] - \widetilde{f}(x)) \\
&= \int \prod_x \mathrm{d}\widetilde{F}\, \delta(\widetilde{F}[c_\mu^{(\lambda)}(x)] - \widetilde{f}(x)) \det \frac{\delta \widetilde{F}[c_\mu^{(\lambda)}(x)]}{\delta \lambda(y)} \\
&= \det \left. \frac{\delta \widetilde{F}[c_\mu^{(\lambda)}(x)]}{\delta \lambda(y)} \right|_{\lambda=0}, \tag{12.239}
\end{aligned}$$

12.4 Path integral quantization of tensor fields

so that

$$\widetilde{\Delta}_{\text{FP}}[c_\mu] = \left[\det \left.\frac{\delta \widetilde{F}[c_\mu^{(\lambda)}(x)]}{\delta \lambda(y)}\right|_{\lambda=0}\right]^{-1}$$
$$= \left(\det \Box_x \delta^4(x-y)\right)^{-1}, \qquad (12.240)$$

where we have used the fact that for anticommuting variables, the Jacobian for a change of variables is the inverse of the determinant. We note that contrary to the usual case, the Faddeev-Popov term is an inverse determinant and, consequently, in the present case we can write

$$\widetilde{\Delta}_{\text{FP}}[c_\mu] = \left(\det \Box_x \delta^4(x-y)\right)^{-1}$$
$$= \int \mathcal{D}\widetilde{\bar{c}}\,\mathcal{D}\widetilde{c}\, e^{-i\left(\widetilde{\bar{c}},\Box \widetilde{c}\right)}. \qquad (12.241)$$

We note here that since we are writing an inverse determinant as an action, the ghost fields $\widetilde{c}$ and $\widetilde{\bar{c}}$ behave like commuting scalars. Similarly we can show that

$$\widetilde{\Delta}_{\text{FP}}[\bar{c}_\mu] = \left(\det \Box_x \delta^4(x-y)\right)^{-1}$$
$$= \int \mathcal{D}\widetilde{\bar{\bar{c}}}\,\mathcal{D}\widetilde{\bar{c}}\, e^{-i\left(\widetilde{\bar{\bar{c}}},\Box \widetilde{\bar{c}}\right)}. \qquad (12.242)$$

The gauge conditions (12.237) and (12.238) correspond to fermionic conditions, namely, $\widetilde{f}$ and $\widetilde{\bar{f}}$ are Grassmann (fermionic) functions. Therefore, the 't Hooft trick needs to be carried out rather carefully. Namely, we cannot use weight factors of the forms

$$e^{-\frac{i}{2\chi}\int d^4x\, \widetilde{f}(x)\widetilde{f}(x)}, \qquad e^{-\frac{i}{2\chi}\int d^4x\, \widetilde{\bar{f}}(x)\widetilde{\bar{f}}(x)}, \qquad (12.243)$$

since the exponents vanish (because of the fermionic nature of the variables). Rather, the appropriate weight factor in this case would correspond to

$$e^{-\frac{i}{2\chi} \int d^4x \, \tilde{\bar{f}}(x) \, \tilde{f}(x)}. \tag{12.244}$$

Thus putting in these identity elements into the functional integral for the generating functional and using the 't Hooft trick with the weight factor (12.244) the generating functional takes the form

$$Z\left[J_{\mu\nu}\right] = N \int \mathcal{D}A_{\mu\nu} \mathcal{D}\bar{c}_\mu \mathcal{D}c_\nu \mathcal{D}c \mathcal{D}\tilde{\bar{c}} \mathcal{D}\tilde{c} \mathcal{D}\tilde{\bar{\tilde{c}}} \mathcal{D}\tilde{\tilde{c}} \, e^{iS_{\text{eff}}}, \tag{12.245}$$

where

$$S_{\text{eff}} = \int d^4x \, \mathcal{L}_{\text{eff}}, \tag{12.246}$$

and

$$\begin{aligned}\mathcal{L}_{\text{eff}} &= -\frac{1}{6} F_{\mu\nu\lambda} F^{\mu\nu\lambda} - \frac{1}{2\xi} (\partial_\mu A^{\mu\nu})^2 + J^{\mu\nu} A_{\mu\nu} \\ &+ \frac{1}{2} (\partial_\mu \bar{c}_\nu - \partial_\nu \bar{c}_\mu)(\partial^\mu c^\nu - \partial^\nu c^\mu) - \frac{1}{2} c \Box c \\ &- \frac{1}{2\chi} (\partial_\nu \bar{c}^\nu)(\partial_\mu c^\mu) - \tilde{\bar{c}} \Box \tilde{c} - \tilde{\bar{\tilde{c}}} \Box \tilde{\tilde{c}}. \end{aligned} \tag{12.247}$$

We are now ready to count the number of effective field degrees of freedom in the theory,

$$A_{\mu\nu} : 6 \text{ degrees of freedom}, \tag{12.248}$$

$\left.\begin{array}{l}c_\mu : -4 \text{ degrees of freedom,} \\ \bar{c}_\mu : -4 \text{ degrees of freedom,} \\ c : -1 \text{ degrees of freedom,}\end{array}\right\}$ anticommuting,

$\left.\begin{array}{l}\tilde{c} : 1 \text{ degrees of freedom,} \\ \tilde{\bar{c}} : 1 \text{ degrees of freedom,} \\ \tilde{\tilde{c}} : 1 \text{ degrees of freedom,} \\ \tilde{\bar{\tilde{c}}} : 1 \text{ degrees of freedom,}\end{array}\right\}$ commuting.

The ghosts $\tilde{c}$, $\tilde{\bar{c}}$, $\tilde{\tilde{c}}$ and $\tilde{\bar{\tilde{c}}}$ being commuting scalars contribute positively to the counting of the number of degrees of freedom. The other way of saying this is that these are ghosts of the ghosts c_μ and $\bar{c}_\mu$ respectively and hence they contribute just the opposite way from c_μ and $\bar{c}_\mu$. Thus counting the degrees of freedom, we see that effectively the theory has

$$6 - 2 \times 4 - 1 + 4 \times 1 = 6 - 8 - 1 + 4 = 1, \qquad (12.249)$$

degree of freedom. This matches exactly with the counting of the degrees of freedom from the canonical quantization.

12.5 References

1. C. N. Yang and R. L. Mills, Physical Review **96**, 191 (1954).

2. R. Utiyama, Physical Review **101**, 1597 (1956).

3. A. R. Hibbs and R. P. Feynman. *Quantum Mechanics and Path Integrals*, McGraw-Hill, New York (1965).

4. L. D. Faddeev and V. N. Popov, Physics Letters **25B**, 29 (1967).

5. L. D. Faddeev, Theoretical and Mathematical Physics **1**, 1 (1970).

6. R. N. Mohapatra, Physical Review **D4**, 378 (1971).

7. G. 't Hooft and M. Veltman, Nuclear Physics **B44**, 189 (1972).

8. G. 't Hooft and M. Veltman, *Diagrammar*, CERN preprint (1973).

9. V. N. Gribov, Nuclear Physics **B139**, 1 (1978).

10. W. Siegel, Physics Letters **93B**, 170 (1980).

11. K. Huang, *Qualks, Leptons and Gauge Fields*, World Scientific, Singapore (1982).

12. C. Itzykson and J-B. Zuber, *Quantum Field Theory*, McGraw-Hill, New York, 1980.

13. F. Gross, *Relativistic Quantum Mechanics and Field Theory*, John Wiley, New York (1993).

14. A. Das, *Finite Temperature Field Theory*, World Scientific, Singapore (1997).

15. A. Das, *Field Theory:A Path Integral Approach*, World Scientific, Singapore (2006).

CHAPTER 13
BRST invariance and its consequences

13.1 BRST symmetry

As we have seen in (12.186), the total Lagrangian density for the non-Abelian gauge theory (Yang-Mills theory) has the form

$$\mathcal{L}_{\text{TOT}} = \mathcal{L}_{\text{inv}} + \mathcal{L}_{\text{GF}} + \mathcal{L}_{\text{ghost}}, \tag{13.1}$$

which for the covariant gauge in (12.181) is given by

$$\mathcal{L}_{\text{TOT}} = -\frac{1}{4} F^a_{\mu\nu} F^{\mu\nu a} - \frac{1}{2\xi} \left(\partial^\mu A^a_\mu \right)^2 + \partial^\mu \bar{c}^a \left(D_\mu c \right)^a. \tag{13.2}$$

Here ξ represents the arbitrary gauge fixing parameter and the covariant derivative in the adjoint representation is defined as (see (12.67))

$$(D_\mu c)^a = \partial_\mu c^a - g f^{abc} A^b_\mu c^c, \tag{13.3}$$

where g denotes the coupling constant of the theory. The total Lagrangian density has been gauge fixed and, therefore, does not have the gauge invariance (12.34) of the original theory. However, the total Lagrangian density, with gauge fixing and ghost terms, develops a global fermionic symmetry which, in some sense, remembers the gauge invariance of the original theory. It is easy to check that the total Lagrangian density (13.2) is invariant under the global transformations

$$\delta A_\mu^a = \frac{\omega}{g}(D_\mu c)^a,$$

$$\delta c^a = \frac{\omega}{2} f^{abc} c^b c^c,$$

$$\delta \bar{c}^a = -\frac{\omega}{g\xi}\left(\partial^\mu A_\mu^a\right), \tag{13.4}$$

where ω is an arbitrary anti-commuting constant parameter of the global transformations. The invariance of the Lagrangian density (13.2) can be seen by first noting that (recall that the ghost fields are Grassmann variables)

$$\begin{aligned}
\delta\left(D_\mu c^a\right) &= D_\mu \delta c^a - g f^{abc} \delta A_\mu^b c^c \\
&= \frac{\omega}{2} D_\mu\left(f^{abc} c^b c^c\right) - \omega f^{abc}\left(D_\mu c^b\right) c^c = 0, \\
\delta\left(\frac{1}{2} f^{abc} c^b c^c\right) &= f^{abc} \delta c^b c^c = \frac{\omega}{2} f^{abc} f^{bpq} c^p c^q c^c \\
&= \frac{\omega}{6}\left(f^{abc} f^{bpq} + f^{abp} f^{bqc} + f^{abq} f^{bcp}\right) c^p c^q c^c = 0. \tag{13.5}
\end{aligned}$$

Here we have used the Jacobi identity for the symmetry algebra (see (12.50)). Similarly, we obtain

$$\delta\left(\partial^\mu A_\mu^a\right) = \frac{\omega}{g} \partial^\mu D_\mu c^a = 0, \tag{13.6}$$

when the ghost equation of motion is used. This shows that under two successive transformations of the kind (13.4) we have

$$\delta_2 \delta_1 \phi^a = 0, \tag{13.7}$$

for the fields $\phi^a = A_\mu^a, c^a, \bar{c}^a$ independent of the parameters of transformations where $\delta_{1,2}$ correspond to transformations with the parameters $\omega_{1,2}$ respectively. We note that the nilpotency of the transformations holds off-shell only for the fields A_μ^a, c^a, while for $\bar{c}^a$ it is true only on-shell (namely, only when the ghost equation of motion is used).

13.1 BRST SYMMETRY

The invariance of the Lagrangian density (13.2) under the transformations (13.4) can now be easily checked. First, we note that the transformation for A_μ^a can really be thought of as an infinitesimal gauge transformation (see (12.34)) with the parameter $\epsilon^a(x) = \omega c^a(x)$ and, therefore, the invariant Lagrangian density is trivially invariant under these transformations, namely,

$$\delta \mathcal{L}_{\text{inv}} = 0. \tag{13.8}$$

Consequently, we need to worry only about the changes in the gauge fixing and the ghost Lagrangian densities which lead to

$$\begin{aligned}
\delta \left(\mathcal{L}_{\text{GF}} + \mathcal{L}_{\text{ghost}} \right) &= -\frac{1}{\xi} \left(\partial^\nu A_\nu^a \right) \left(\partial^\mu \delta A_\mu^a \right) + \left(\partial^\mu \delta \overline{c}^a \right) D_\mu c^a \\
&= -\frac{\omega}{g\xi} \left(\partial^\nu A_\nu^a \right) \partial^\mu D_\mu c^a - \frac{\omega}{g\xi} \partial^\mu \left(\partial^\nu A_\nu^a \right) D_\mu c^a \\
&= -\partial^\mu \left(\frac{\omega}{g\xi} \left(\partial^\nu A_\nu^a \right) D_\mu c^a \right),
\end{aligned} \tag{13.9}$$

so that the action is invariant. In this derivation, we have used the fact that $\delta \left(D_\mu c \right)^a = 0$ which we have seen in (13.5). This shows that the action for the total Lagrangian density (13.2) is invariant under the global transformations (13.4) with an anti-commuting constant parameter. This is known as the BRST (Becchi-Rouet-Stora-Tyutin) transformation for a gauge theory and arises when the gauge fixing and the ghost Lagrangian densities have been added to the original gauge invariant Lagrangian density. The present formulation of the BRST symmetry, however, is slightly unpleasant in the sense that the nilpotency of the anti-ghost field transformation holds only on-shell. Generally, this is a reflection of the fact that the theory is lacking in some auxiliary field variables and once the correct auxiliary fields are incorporated the symmetry algebra will close off-shell (without the use of equations of motion).

As we have seen earlier within the context of the Abelian gauge theory in (12.170) we can write the gauge fixing Lagrangian density by introducing an auxiliary field. Since this will also be quite useful for our latter discussions, let us recall that the gauge fixing Lagrangian density in (13.2) can also be written equivalently as

$$\mathcal{L}_{\text{GF}} = \frac{\xi}{2} F^a F^a + (\partial^\mu F^a) A_\mu^a, \tag{13.10}$$

where F^a is an auxiliary field. (The form of the gauge fixing Lagrangian density in (13.10) differs from that in (12.170) by a total divergence, but it is this form that is very useful as we will see.) It is clear from the form of $\mathcal{L}_{\text{GF}}$ in (13.10) that the equation of motion for the auxiliary field takes the form

$$\xi F^a = \partial^\mu A_\mu^a, \tag{13.11}$$

and when we eliminate F^a from the Lagrangian density using this equation, we recover the original gauge fixing Lagrangian density (up to a total divergence term). Among other things $\mathcal{L}_{\text{GF}}$ as written above allows us to take such gauge choices as the Landau gauge which corresponds to simply taking the limit $\xi = 0$. The total Lagrangian density can now be written as

$$\begin{aligned}\mathcal{L}_{\text{TOT}} &= \mathcal{L}_{\text{inv}} + \mathcal{L}_{\text{GF}} + \mathcal{L}_{\text{ghost}} \\ &= -\frac{1}{4} F_{\mu\nu}^a F^{\mu\nu a} + \frac{\xi}{2} F^a F^a + \partial^\mu F^a A_\mu^a + \partial^\mu \bar{c}^a (D_\mu c)^a.\end{aligned} \tag{13.12}$$

In this case, the BRST transformations in (13.4) take the form

$$\begin{aligned}\delta A_\mu^a &= \frac{\omega}{g} (D_\mu c)^a, \\ \delta c^a &= \frac{\omega}{2} f^{abc} c^b c^c, \\ \delta \bar{c}^a &= -\frac{\omega}{g} F^a, \\ \delta F^a &= 0,\end{aligned} \tag{13.13}$$

and it is straightforward to check that these transformations are nilpotent off-shell, namely,

$$\delta_2 \delta_1 \phi^a = 0, \tag{13.14}$$

13.1 BRST SYMMETRY

for all the field variables $\phi^a = A_\mu^a, F^a, c^a, \bar{c}^a$. Therefore, F^a represents the missing auxiliary field that we had alluded to earlier.

We note that $\mathcal{L}_{\text{inv}}$ is invariant under the BRST transformation as we had argued earlier in (13.8) and the auxiliary field F^a does not transform at all which leads to

$$\begin{aligned}
\delta\mathcal{L}_{\text{TOT}} &= \delta\left(\mathcal{L}_{\text{GF}} + \mathcal{L}_{\text{ghost}}\right) \\
&= \partial^\mu F^a \delta A_\mu^a + \partial^\mu \delta\bar{c}^a \left(D_\mu c\right)^a \\
&= \frac{\omega}{g} \partial^\mu F^a \left(D_\mu c\right)^a - \frac{\omega}{g} \partial^\mu F^a \left(D_\mu c\right)^a = 0.
\end{aligned} \qquad (13.15)$$

Unlike in the formulation of BRST variations without the auxiliary field in (13.4), here we see that the total Lagrangian density is invariant under the BRST transformations (as opposed to the Lagrangian density changing by a total divergence in (13.9)).

In some sense the BRST transformations, which define a residual global symmetry of the full theory, replace the original gauge invariance of the theory and play a fundamental role in the study of non-Abelian gauge theories. There is also a second set of fermionic transformations involving the anti-ghost fields of the form

$$\begin{aligned}
\bar{\delta} A_\mu^a &= \frac{\bar{\omega}}{g} \left(D_\mu \bar{c}\right)^a, \\
\bar{\delta} c^a &= \frac{\bar{\omega}}{g} \left(F^a + g f^{abc} c^b \bar{c}^c\right), \\
\bar{\delta} \bar{c}^a &= \frac{\bar{\omega}}{2} f^{abc} \bar{c}^b \bar{c}^c, \\
\bar{\delta} F^a &= -\bar{\omega} f^{abc} F^b \bar{c}^c,
\end{aligned} \qquad (13.16)$$

which can also be easily checked to leave the total Lagrangian density (13.12) invariant. These are known as the anti-BRST transformations. However, since these do not lead to any new constraint on the structure of the theory beyond what the BRST invariance provides, we will not pursue this symmetry further in our discussions. We note here that the BRST and the anti-BRST transformations are not quite symmetric in the ghost and the anti-ghost fields which is a reflection of the asymmetric manner in which these fields occur in the ghost Lagrangian density in (13.2) or (13.12). Without going into details, we note here that these fermionic symmetries arise naturally in a superspace formulation of gauge theories.

In addition to these two anti-commuting symmetries, the total Lagrangian density (13.12) is also invariant under the infinitesimal bosonic global symmetry transformations

$$\delta c^a = \epsilon c^a,$$
$$\delta \bar{c}^a = -\epsilon \bar{c}^a, \qquad (13.17)$$

with all other fields remaining inert. Here ϵ represents a constant, commuting infinitesimal parameter and the generator of this symmetry transformation merely corresponds to the ghost number operator, namely, it corresponds to the operator that counts the ghost number of the fields. This is known as the ghost scaling symmetry of the theory. (The fact that these transformations are like scale transformations and not like phase transformations, which is normally associated with the number operator, has to do with the particular hermiticity properties that the ghost and the anti-ghost fields satisfy for a consistent covariant quantization of the theory which we will discuss in the next section.)

13.2 Covariant quantization of Yang-Mills theory

The presence of the BRST symmetry and the ghost scaling symmetry in the gauge fixed Yang-Mills theory leads to many interesting consequences. For example, it allows us to carry out covariant quantization of the non-Abelian gauge theory. Let us recall that the vector space of the full theory in the covariant gauge (12.181), as we have emphasized several times by now, contains many more states than the physical states alone. Therefore, the physical Hilbert space needs to be properly selected for a discussion of physical questions associated with the non-Abelian gauge theory. We have already seen that the naive Gupta-Bleuler quantization does not work in the non-Abelian case. We recall that the physical space must be selected in such a way that it remains invariant under the time evolution of the system. In the covariant gauge in Maxwell's theory, for example, we have seen that the states in the physical space are selected as the ones satisfying the Gupta-Bleuler condition (see (9.162))

$$\partial^\mu A_\mu^{(+)}(x)|\text{phys}\rangle = 0, \tag{13.18}$$

where the superscript, "$_{(+)}$", stands for the positive frequency part of the field. We recognize that even though this looks like one condition, in reality it is an infinite number of conditions since it has to hold for every value of the coordinates. In the Abelian theory, the Gupta-Bleuler condition works because $\partial^\mu A_\mu$ satisfies the free Klein-Gordon equation (9.130) in the covariant gauge (12.164) and hence the physical space so selected remains invariant under time evolution. The corresponding operator in a non-Abelian theory, as we have seen in (12.118), does not satisfy a free equation and hence it is not a suitable operator for identifying the physical subspace in a time invariant manner. On the other hand, the generators of the BRST symmetry, Q_{BRST}, and the ghost scaling symmetry, Q_c, are conserved and hence can be used to define a physical Hilbert space which would remain invariant under the time evolution of the system. (Q_{BRST} and Q_c are the charges constructed from the Nöther current for the respective transformations whose explicit forms can be obtained from the Nöther procedure and will be derived below.) Thus, we can identify the physical space of states of the gauge theory as satisfying (we note that at this point this only defines a subspace of the total vector space and we still have to show that this subspace indeed coincides with the physical Hilbert space)

$$\begin{aligned} Q_{\text{BRST}}|\text{phys}\rangle &= 0, \\ Q_c|\text{phys}\rangle &= 0. \end{aligned} \tag{13.19}$$

Note that even in the case of Maxwell's theory, the conditions in (13.19) would appear to correspond to only two conditions and not an infinite number of conditions as we have seen is the case with the Gupta-Bleuler condition in (13.18). It is, therefore, not clear *a priori* if the conditions (13.19) are sufficient to reproduce even the Gupta-Bleuler condition in the case of the Abelian theory (namely, whether they can reduce the vector space sufficiently enough to coincide with the physical space).

To see that these conditions indeed lead to the Gupta-Bleuler condition in Maxwell's theory, let us note that the Nöther current densities associated with the BRST transformation as well as the ghost scaling transformation have the forms (recall that we use left derivatives for Grassmann variables and that ω is an anticommuting parameter)

$$J_{\text{BRST}}^{(\omega)\mu}(x) = \frac{\partial \mathcal{L}_{\text{TOT}}}{\partial \partial_\mu A_\nu^a}\delta A_\nu^a + \frac{\partial \mathcal{L}_{\text{TOT}}}{\partial \partial_\mu F^a}\delta F^a + \delta\bar{c}^a \frac{\partial \mathcal{L}_{\text{TOT}}}{\partial \partial_\mu \bar{c}^a} + \delta c^a \frac{\partial \mathcal{L}_{\text{TOT}}}{\partial \partial_\mu c^a}$$

$$= -F^{\mu\nu a}\delta A_\nu^a + \delta\bar{c}^a (D^\mu c)^a - \delta c^a (\partial^\mu \bar{c}^a)$$

$$= -\frac{\omega}{g}\left(F^{\mu\nu a}(D_\nu c)^a + F^a (D^\mu c)^a + \frac{g}{2} f^{abc} (\partial^\mu \bar{c}^a) c^b c^c\right),$$

$$J_c^{\mu(\epsilon)} = \delta\bar{c}^a \frac{\partial \mathcal{L}_{\text{TOT}}}{\partial \partial_\mu \bar{c}^a} + \delta c^a \frac{\partial \mathcal{L}_{\text{TOT}}}{\partial \partial_\mu c^a}$$

$$= -\epsilon\bar{c}^a (D^\mu c)^a - \epsilon c^a \partial^\mu \bar{c}^a = \epsilon\left((\partial^\mu \bar{c}^a) c^a - \bar{c}^a (D^\mu c)^a\right), \quad (13.20)$$

where we have used the fact that the auxiliary field does not transform under the BRST transformations and the fields A_μ^a, F^a are inert under the scaling of ghost fields. From this, we can obtain the BRST as well as the ghost scaling current densities without the parameters of transformation to correspond to

$$J_{\text{BRST}}^\mu = F^{\mu\nu a}(D_\nu c)^a + F^a (D^\mu c)^a + \frac{g}{2} f^{abc} (\partial^\mu \bar{c}^a) c^b c^c,$$

$$J_c^\mu = (\partial^\mu \bar{c}^a) c^a - \bar{c}^a (D^\mu c)^a. \quad (13.21)$$

The corresponding conserved charges can also be obtained from these current densities and take the forms

$$Q_{\text{BRST}} = \int d^3x\, J_{\text{BRST}}^0$$

$$= \int d^3x \left(F^{0i\,a}(D_i c)^a + F^a (D^0 c)^a + \frac{g}{2} f^{abc} \dot{\bar{c}}^a c^b c^c\right)$$

$$= \int d^3x \left(-(\partial^0 F^a) c^a + F^a (D^0 c)^a + \frac{g}{2} f^{abc} \dot{\bar{c}}^a c^b c^c\right),$$

$$Q_c = \int d^3x\, J_c^0 = \int d^3x \left(\dot{\bar{c}}^a c^a - \bar{c}^a (D^0 c)^a\right), \quad (13.22)$$

13.2 COVARIANT QUANTIZATION OF YANG-MILLS THEORY

where we have integrated by parts the first term (in Q_{BRST}) and have used the equation of motion for the gauge field

$$D_\mu F^{\mu\nu a} = -\partial^\nu F^a, \tag{13.23}$$

following from (13.12). In particular, we note that for the Abelian theory where $f^{abc} = 0$ and there is no internal index "a", the BRST and the ghost scaling charge operators can be obtained from (13.22) to correspond to

$$\begin{aligned} Q_{\text{BRST}} &= \int \mathrm{d}^3 x \left(-\dot{F} c + F \dot{c} \right) = \int \mathrm{d}^3 x \, F \overleftrightarrow{\partial^0} c, \\ Q_c &= \int \mathrm{d}^3 x \left(\dot{\overline{c}} c - \overline{c} \dot{c} \right) = -\int \mathrm{d}^3 x \, \overline{c} \overleftrightarrow{\partial^0} c. \end{aligned} \tag{13.24}$$

If we use the field decomposition for the fields and normal order the charges (so that the annihilation/positive frequency operators are to the right of the creation/negative frequency operators), the BRST charge has the explicit form (we do not show the normal ordering explicitly)

$$Q_{\text{BRST}} = i \int \mathrm{d}^3 k \left(c^{(-)}(-\mathbf{k}) F^{(+)}(\mathbf{k}) - F^{(-)}(-\mathbf{k}) c^{(+)}(\mathbf{k}) \right). \tag{13.25}$$

Let us note that the condition (13.19)

$$Q_c |\text{phys}\rangle = 0, \tag{13.26}$$

implies that the physical states must have (net) zero ghost number. In principle, this allows for states containing an equal number of ghost and anti-ghost particles. Thus, denoting the physical states of the theory as

$$|\text{phys}\rangle = |A_\mu\rangle \otimes |n, \overline{n}\rangle, \tag{13.27}$$

where $n, \bar{n}$ denote the (equal) number of ghost and anti-ghost particles, we note that if the physical states have to further satisfy the condition

$$\begin{aligned}
&Q_{\text{BRST}}|\text{phys}\rangle \\
&= i\int d^3k \left(c^{(-)}(-\mathbf{k})F^{(+)}(\mathbf{k}) - F^{(-)}(-\mathbf{k})c^{(+)}(\mathbf{k})\right)|A_\mu\rangle \otimes |n, \bar{n}\rangle \\
&= 0,
\end{aligned} \tag{13.28}$$

then this implies that

$$c^{(+)}(\mathbf{k})|n, \bar{n}\rangle = 0, \quad \text{and} \quad F^{(+)}(\mathbf{k})|A_\mu\rangle = 0. \tag{13.29}$$

Namely, the physical states should have no ghost particles (the number of ghost and anti-ghost particles have to be the same by the other physical condition),

$$|\text{phys}\rangle = |A_\mu\rangle \otimes |0, \bar{0}\rangle = |A_\mu\rangle, \tag{13.30}$$

and must further satisfy (with $k^0 = |\mathbf{k}|$)

$$F^{(+)}(\mathbf{k})|\text{phys}\rangle = 0 = \left(k_\mu A^{\mu(+)}(\mathbf{k})\right)|\text{phys}\rangle, \tag{13.31}$$

where we have used the equation of motion for the auxiliary field (see (13.11)). This is precisely the Gupta-Bleuler condition in momentum space and this derivation shows how a single condition can give rise to an infinite number of conditions (in this case, for every momentum mode $\mathbf{k}$). Thus, we feel confident that the physical state conditions in (13.19) are the right ones even for the non-Abelian theory.

To investigate systematically whether the physical state conditions in (13.19) really select out the subspace of physical states, we note from the form of the Lagrangian density (13.12) that we can obtain the canonical momenta conjugate to various field variables of the theory to correspond to

13.2 COVARIANT QUANTIZATION OF YANG-MILLS THEORY

$$\Pi^{ia} = \frac{\partial \mathcal{L}_{\text{TOT}}}{\partial \dot{A}_i^a} = -F^{0ia},$$

$$\Pi^a = \frac{\partial \mathcal{L}_{\text{TOT}}}{\partial \dot{F}^a} = A_0^a,$$

$$\Pi_c^a = \frac{\partial \mathcal{L}_{\text{TOT}}}{\partial \dot{c}^a} = -\dot{\bar{c}}^a,$$

$$\Pi_{\bar{c}}^a = \frac{\partial \mathcal{L}_{\text{TOT}}}{\partial \dot{\bar{c}}^a} = (D_0 c)^a. \tag{13.32}$$

Here we have used left derivatives for the anti-commuting ghost fields. In particular, we see that in this formulation with the auxiliary field, A_0^a plays the role of the momentum conjugate to F^a. The equal-time canonical (anti) commutation relations for the theory can now be written as ($\hbar = 1$)

$$\left[A_i^a(\mathbf{x}, t), \Pi^{jb}(\mathbf{y}, t) \right] = i\delta^{ab} \delta_i^j \delta^3(x-y),$$

$$\left[F^a(\mathbf{x}, t), \Pi^b(\mathbf{y}, t) \right] = i\delta^{ab} \delta^3(x-y),$$

$$\left[c^a(\mathbf{x}, t), \Pi_c^b(\mathbf{y}, t) \right]_+ = i\delta^{ab} \delta^3(x-y),$$

$$\left[\bar{c}^a(\mathbf{x}, t), \Pi_{\bar{c}}^b(\mathbf{y}, t) \right]_+ = i\delta^{ab} \delta^3(x-y). \tag{13.33}$$

The hermiticity conditions for the ghost fields which arise out of various consistency conditions (for example, the Lagrangian density and the conserved charges have to be Hermitian, the (anti) commutation relations (13.33) have to satisfy the proper hermiticity properties etc.) are given by

$$(c^a)^\dagger = c^a,$$

$$(\bar{c}^a)^\dagger = -\bar{c}^a. \tag{13.34}$$

With this choice, the BRST parameter ω is seen to be anti-Hermitian

$$\omega^\dagger = -\omega, \tag{13.35}$$

and we note that the currents (13.21) and, therefore, the conserved charges (13.22) are Hermitian with the assigned hermiticity conditions (13.34) for the ghost fields. The conserved charges (13.22) can now be expressed in terms of the fields and the conjugate momenta as (before integrating the first term by parts in Q_{BRST}, see (13.22))

$$Q_{\text{BRST}} = -\int d^3x \left(\Pi^{ia}(D_i c)^a - F^a \Pi_{\bar{c}}^a + \frac{g}{2} f^{abc} \Pi_c^a c^b c^c \right),$$

$$Q_c = -\int d^3x \left(\Pi_c^a c^a + \bar{c}^a \Pi_{\bar{c}}^a \right), \quad (13.36)$$

and we can calculate the algebra of charges as well as various other relations of interest using the (anti) commutation relations (13.33) for the field variables. We note, for example, that (this basically describes the behavior of the ghost fields under a scaling (13.17))

$$[c^a(x), Q_c] = -ic^a(x),$$
$$[\bar{c}^a(x), Q_c] = i\bar{c}^a(x). \quad (13.37)$$

Therefore, we see that we can think of iQ_c as the ghost number operator (with the ghost number for $\bar{c}^a$ being negative).

The algebra of the conserved charges also follows in a straightforward manner from (13.36) and (13.33)

$$[Q_{\text{BRST}}, Q_{\text{BRST}}]_+ = 2Q_{\text{BRST}}^2 = 0,$$
$$[Q_c, Q_c] = 0,$$
$$[Q_{\text{BRST}}, Q_c] = -iQ_{\text{BRST}}. \quad (13.38)$$

This algebra can also be derived directly from the transformation laws for the field variables in (13.13) and (13.17). The first equation in (13.38) simply reiterates in an operator language the fact that the BRST transformations are nilpotent (see (13.14)). The second equation implies that the ghost scaling symmetry is Abelian. The third is a statement of the fact that the BRST charge carries a ghost number of unity. An immediate consequence of the algebra in (13.38) is that

13.2 COVARIANT QUANTIZATION OF YANG-MILLS THEORY

$$Q_{\text{BRST}}\, e^{\pi Q_c} = e^{\pi(Q_c - i)} Q_{\text{BRST}} = -e^{\pi Q_c} Q_{\text{BRST}},$$

or, $\quad \left[Q_{\text{BRST}}, e^{\pi Q_c}\right]_+ = 0,$ \hfill (13.39)

which is quite useful (particularly) in dealing with gauge theories at finite temperature.

As is clear from our earlier discussions, the total vector space V of the complete gauge theory (13.12) contains various unphysical states as well as states with negative norm in addition to the physical states. Consequently, the metric of this space and the inner product become indefinite and a probabilistic description of the quantum theory is lost unless we can restrict to a suitable subspace with a positive definite inner product. As we have discussed earlier, we can select out a subspace V_{phys} by requiring that states in this space are annihilated by Q_{BRST}. Namely, for every $|\Psi\rangle \in V_{\text{phys}}$, we have

$$Q_{\text{BRST}}|\Psi\rangle = 0. \tag{13.40}$$

We emphasize again that such an identification of the physical subspace is invariant under the time evolution of the system since Q_{BRST} commutes with the Hamiltonian since the BRST transformations define a symmetry of the full theory,

$$[Q_{\text{BRST}}, H] = 0. \tag{13.41}$$

We note that there are two possible kinds of states which will satisfy the physical state condition (13.40). First, if a state $|\Psi\rangle$ cannot be written as

$$|\Psi\rangle \neq Q_{\text{BRST}}|\tilde{\Psi}\rangle, \tag{13.42}$$

for some $|\tilde{\Psi}\rangle$ and still satisfies the physical state condition, then, it is truly a BRST singlet (invariant) state. The field operator which creates such a state must necessarily commute with Q_{BRST} and, therefore, must represent a truly gauge invariant field variable, namely, (we assume that the vacuum state is a BRST singlet (invariant) state)

$$Q_{\text{BRST}}|\Psi\rangle = Q_{\text{BRST}}\Psi|0\rangle = [Q_{\text{BRST}}, \Psi]|0\rangle = 0, \tag{13.43}$$

where we have assumed that the operator Ψ creates the state $|\Psi\rangle$ from vacuum, namely,

$$\Psi|0\rangle = |\Psi\rangle. \tag{13.44}$$

This implies that

$$[Q_{\text{BRST}}, \Psi] = 0, \tag{13.45}$$

and transverse fields would represent such operators which are gauge invariant asymptotically. Such states would, therefore, correspond to truly physical states of the theory and since gauge invariant degrees of freedom have physical (non-negative) commutation relations, such states would have positive norm. (Note that the auxiliary field does not transform under a BRST transformation and, therefore, also satisfies the above relation. However, as we have seen in (13.13), it can be written as the BRST variation of the anti-ghost field and, therefore, has a different character.)

The second class of states which would satisfy the physical state condition (13.40) can be written in the form

$$|\Psi\rangle = Q_{\text{BRST}}|\tilde{\Psi}\rangle. \tag{13.46}$$

We see that the physical state condition is trivially satisfied in this case because of the nilpotency of Q_{BRST} (see (13.38)). The nilpotency of Q_{BRST} also implies that all such states would have zero norm because (Q_{BRST} is Hermitian with our choice of hermiticity conditions (13.34) for the ghost fields.)

$$\langle\Psi|\Psi\rangle = \langle\tilde{\Psi}|Q_{\text{BRST}}Q_{\text{BRST}}|\tilde{\Psi}\rangle = 0. \tag{13.47}$$

Such a state would also be orthogonal to any physical state (either of the first or the second kind) satisfying the physical state condition

13.2 COVARIANT QUANTIZATION OF YANG-MILLS THEORY

(13.40) because if $|\Psi\rangle$ represents a physical state (of either kind) while $|\Psi'\rangle$ is a state of the second kind, then

$$\langle \Psi'|\Psi\rangle = \langle \tilde{\Psi}'|Q_{\text{BRST}}|\Psi\rangle = 0, \tag{13.48}$$

which follows from the physical state condition. If we denote all the states of V_{phys} of the second kind by V_0, then, this would contain all the zero norm states which would be orthogonal to V_{phys} itself. The true physical states, which satisfy the physical state condition and are of the first kind can, therefore, be identified as belonging to the quotient space $\overline{V}_{\text{phys}} = \frac{V_{\text{phys}}}{V_0}$ and will have positive definite norm.

Every state in V_{phys} can, of course, be decomposed into states containing fixed numbers of unphysical particles. Thus, defining $P^{(n)}$ as the projection operator onto the n-unphysical particle sector, we have

$$\sum_{n=0}^{\infty} P^{(n)} = \mathbb{1},$$

$$P^{(n)} P^{(m)} = \delta_{nm} P^{(n)}. \tag{13.49}$$

Here, by unphysical, we mean states containing quanta of the ghost fields, the longitudinal components of A_μ^a and the auxiliary fields. We recall from (13.13) that the auxiliary field can be written as a BRST variation of the anti-ghost field as (note that Q_{BRST} is the generator of the BRST transformations and generates the transformations in (13.13) through (anti) commutation with the appropriate field variables)

$$F^a \sim [Q_{\text{BRST}}, \overline{c}^a]_+. \tag{13.50}$$

It is clear from the definition in (13.49) that $P^{(0)}$ projects onto the truly physical states of the theory (namely, with no unphysical particles) and from the completeness of the projection operators in (13.49), we note that we can write

$$P^{(0)} = \mathbb{1} - P', \tag{13.51}$$

where

$$P' = \sum_{n=1}^{\infty} P^{(n)}. \tag{13.52}$$

We can, of course, construct the actual forms of all the projection operators explicitly, but this is not very illuminating. What is interesting, however, is the fact that $P^{(0)}$ projects onto the true physical states and, therefore, must be gauge invariant and will commute with Q_{BRST}. It follows, then, that P' must also commute with Q_{BRST}. Namely, we have

$$\begin{aligned} \left[Q_{\text{BRST}}, P^{(0)}\right] &= 0, \\ \left[Q_{\text{BRST}}, P'\right] &= 0. \end{aligned} \tag{13.53}$$

However, the difference between the two lies in the fact that P' must necessarily involve unphysical fields since it projects onto unphysical states and, therefore, cannot be truly gauge invariant. Consequently, it must have the form (for some fermionic operator R, whose explicit form is not important for our discussions)

$$P' = [Q_{\text{BRST}}, R]_+, \tag{13.54}$$

so that

$$\left[Q_{\text{BRST}}, P'\right] = \left[Q_{\text{BRST}}, [Q_{\text{BRST}}, R]_+\right] = 0, \tag{13.55}$$

which follows from the nilpotency of Q_{BRST}. In fact, we can even show that for every $P^{(n)}$, $n \geq 1$, we can write

$$P^{(n)} = \left[Q_{\text{BRST}}, R^{(n)}\right]_+. \tag{13.56}$$

It is now easy to show that $P^{(n)}$, for $n \geq 1$, and, therefore, P' project onto the zero norm space V_0 when acting on states $|\Psi\rangle$

which satisfy the physical state condition (13.40). For example, let $|\Psi\rangle, |\Psi'\rangle \in V_{\text{phys}}$, then,

$$\begin{aligned} \langle \Psi' | P^{(n)} | \Psi \rangle &= \langle \Psi' | \left[Q_{\text{BRST}}, R^{(n)} \right]_+ | \Psi \rangle = 0, \\ \langle \Psi' | P' | \Psi \rangle &= \langle \Psi' | [Q_{\text{BRST}}, R]_+ | \Psi \rangle = 0, \end{aligned} \qquad (13.57)$$

which follows from the physical state condition (13.40) (as well as the hermiticity of Q_{BRST}). Any vector $|\Psi\rangle \in V_{\text{phys}}$ can now be written as

$$|\Psi\rangle = P^{(0)}|\Psi\rangle + P'|\Psi\rangle, \qquad (13.58)$$

and the norm of any such state is, then, obtained to be

$$\langle \Psi | \Psi \rangle = \langle \Psi | P^{(0)} | \Psi \rangle \geq 0. \qquad (13.59)$$

This, therefore, shows that V_{phys}, defined by the physical state condition, has a positive semi-definite norm as we should have for a physical vector space and the value of the norm depends on the truly physical component of the state. This also makes it clear that the norm of a state $|\overline{\Psi}\rangle \in \overline{V}_{\text{phys}} = \frac{V_{\text{phys}}}{V_0}$ would be positive definite. This would correspond to the true physical subspace of the total vector space. This completes the covariant quantization of the non-Abelian theory and shows that the generalization of the supplementary condition (Gupta-Bleuler condition) to the non-Abelian theory can be achieved as a consequence of the BRST symmetry of the theory and is given by (13.19).

13.3 Unitarity

The BRST invariance of a gauge theory is quite important from yet another consideration. It leads to a formal proof of unitarity of the theory when restricted to the subspace of the physical Hilbert space. We will only outline the proof of unitarity in this section. The question of the unitarity of the S-matrix in a gauge theory can be formulated in the following manner. We note that, with the hermiticity assignments for the ghost fields in (13.34), the total Lagrangian density (13.12) is Hermitian so that the S-matrix of the theory is formally unitary, namely,

$$S^\dagger S = SS^\dagger = \mathbb{1}. \tag{13.60}$$

Furthermore, the S-matrix is BRST invariant since the full theory is, so that

$$[Q_{\text{BRST}}, S] = 0. \tag{13.61}$$

Given these the question that we would like to understand is whether one can define an operator S_{phys} which would act and correspond to the S-matrix in the physical subspace of states of the theory and which will also be unitary. This question can be systematically analyzed as follows.

Let us note here from the discussions of the last section that for any $|\Psi\rangle, |\Phi\rangle \in V_{\text{phys}}$ and any two operators A and B acting on V_{phys} (i.e., any two operators which do not take us out of V_{phys})

$$\begin{aligned}\langle\Psi|P^{(0)}|\Phi\rangle &= \langle\Psi|(\mathbb{1}-P')|\Phi\rangle = \langle\Psi|\Phi\rangle, \\ \langle\Psi|AP^{(0)}B|\Phi\rangle &= \langle\Psi|A(\mathbb{1}-P')B|\Phi\rangle = \langle\Psi|AB|\Phi\rangle,\end{aligned} \tag{13.62}$$

which follow from (13.57). Furthermore, we note that given any state $|\Psi\rangle \in V_{\text{phys}}$, we can define a unique state $|\overline{\Psi}\rangle \in \overline{V}_{\text{phys}}$ as

$$|\overline{\Psi}\rangle = P^{(0)}|\Psi\rangle, \tag{13.63}$$

where we note that in the space $\overline{V}_{\text{phys}}$, $P^{(0)}$ acts as the identity operator, namely,

$$P^{(0)}|\overline{\Psi}\rangle = \left(P^{(0)}\right)^2|\Psi\rangle = P^{(0)}|\Psi\rangle = |\overline{\Psi}\rangle. \tag{13.64}$$

The two states, $|\Psi\rangle$ and $|\overline{\Psi}\rangle$, differ only by a zero norm state which is orthogonal to every state in V_{phys} so that the inner product of any two such states is the same

13.3 UNITARITY

$$\langle\overline{\Psi}|\overline{\Phi}\rangle = \langle\Psi|(P^{(0)})^2|\Phi\rangle = \langle\Psi|P^{(0)}|\Phi\rangle = \langle\Psi|\Phi\rangle, \tag{13.65}$$

where the last line follows from the first of the conditions in (13.62). Given this, it is clear that we can define the S-matrix which acts on the physical space $\overline{V}_{\text{phys}}$ as satisfying

$$P^{(0)}S = S_{\text{phys}}P^{(0)}, \tag{13.66}$$

such that for $|\Psi\rangle \in V_{\text{phys}}$, $|\overline{\Psi}\rangle = P^{(0)}|\Psi\rangle \in \overline{V}_{\text{phys}}$ as defined in (13.63),

$$S_{\text{phys}}|\overline{\Psi}\rangle = S_{\text{phys}}P^{(0)}|\Psi\rangle = P^{(0)}S|\Psi\rangle = \overline{S|\Psi\rangle}. \tag{13.67}$$

It is clear that since S is BRST invariant, it will leave the space V_{phys} invariant and hence S_{phys} will not take a state out of $\overline{V}_{\text{phys}}$. Furthermore, we now have

$$\begin{aligned}\langle\overline{\Psi}|S^\dagger_{\text{phys}}S_{\text{phys}}|\overline{\Phi}\rangle &= \langle\Psi|P^{(0)}S^\dagger_{\text{phys}}S_{\text{phys}}P^{(0)}|\Phi\rangle \\ &= \langle\Psi|S^\dagger P^{(0)}S|\Phi\rangle \\ &= \langle\Psi|S^\dagger S|\Phi\rangle = \langle\Psi|\Phi\rangle = \langle\overline{\Psi}|\overline{\Phi}\rangle,\end{aligned} \tag{13.68}$$

where we have used (13.66) and (13.62) as well as the formal unitarity of S (13.60). It follows from (13.68) that

$$S^\dagger_{\text{phys}}S_{\text{phys}} = \mathbb{1}. \tag{13.69}$$

In other words, the physical state condition (13.40) which naturally follows from the BRST invariance of the theory, also automatically leads to a formal proof of unitarity of the S-matrix in the subspace of the truly physical states of the theory. In a similar manner, we can also show that

$$S_{\text{phys}}S^\dagger_{\text{phys}} = \mathbb{1}. \tag{13.70}$$

As another consequence of the BRST invariance of the theory, we can show that the gauge fixing and the ghost Lagrangian densities lead to no physical consequences. This is particularly important since (as we have emphasized before) we have modified the starting theory through a series of formal manipulations and we should show that this has not changed the physical results of the theory. To show this, we note that these extra terms in the Lagrangian density (namely, the gauge fixing and the ghost Lagrangian densities) can, in fact, be written as a BRST variation (with the parameter of transformation taken out), namely,

$$\begin{aligned}\mathcal{L}_{\text{GF}} + \mathcal{L}_{\text{ghost}} &= \frac{\xi}{2} F^a F^a + (\partial^\mu F^a) A_\mu^a + \partial^\mu \bar{c}^a (D_\mu c)^a \\ &= g\delta\left(-\frac{\xi}{2} \bar{c}^a F^a - \partial^\mu \bar{c}^a A_\mu^a\right) \\ &= g\left[Q_{\text{BRST}}, \left(-\frac{\xi}{2}\bar{c}^a F^a - \partial^\mu \bar{c}^a A_\mu^a\right)\right]_+. \end{aligned} \quad (13.71)$$

Here we have used the fact that the BRST charge is the generator of the BRST transformations so that the transformations for any fermionic operator is generated through the anti-commutator of the operator with the generator. It now follows from the physical state condition (13.40) that

$$\begin{aligned}&\langle \text{phys}|\left(\mathcal{L}_{\text{GF}} + \mathcal{L}_{\text{ghost}}\right)|\text{phys}'\rangle \\ &= -g\langle\text{phys}|\left[Q_{\text{BRST}}, \left(\frac{\xi}{2}\bar{c}^a F^a + \partial^\mu \bar{c}^a A_\mu^a\right)\right]_+ |\text{phys}'\rangle \\ &= 0. \end{aligned} \quad (13.72)$$

This shows that the terms added to modify the original Lagrangian density have no contribution to the physical matrix elements of the theory. We can also show that all the physical matrix elements of the theory are independent of the choice of the gauge fixing parameter ξ in the following manner (the BRST variation denoted is with the parameter of transformation taken out)

$$\frac{\partial}{\partial \xi}\langle 0|0\rangle^J = \frac{\partial Z[J]}{\partial \xi} = \frac{i}{2}\langle 0|\int d^4x\, F^a F^a |0\rangle^J$$
$$= -\frac{ig}{2}\int d^4x\, \langle 0|\delta\left(\bar{c}^a F^a\right)|0\rangle^J$$
$$= -\frac{ig}{2}\int d^4x\, \langle 0|[Q_{\text{BRST}}, \bar{c}^a F^a]_+|0\rangle^J = 0, \quad (13.73)$$

where we have used the fact that the vacuum belongs to the physical Hilbert space of the theory and as such is annihilated by the BRST charge.

13.4 Slavnov-Taylor identity

The BRST invariance of the full theory leads to many relations between various scattering amplitudes of the theory. These are known as the Ward-Takahashi identities (in the Abelian case) or the Slavnov-Taylor identities (in the non-Abelian case) of the theory and are quite essential in establishing the renormalizability of gauge theories. We have already given a simple diagrammatic derivation of such identities in the case of QED in section **9.7**. However, such a simple diagrammatic derivation does not carry over to non-Abelian gauge theories and they are best described systematically within the context of path integrals which we will do next.

However, before we go into the actual derivation, let us recapitulate briefly some of the essential concepts from path integrals (a detailed discussion is beyond the scope of these lectures). For simplicity, let us consider the self-interacting field theory of a real scalar field coupled to an external source described by the action

$$S^J[\phi] = S[\phi] + \int d^4x\, J(x)\phi(x), \quad (13.74)$$

where $S[\phi]$ denotes the dynamical action of the self-interacting scalar field. In this case, as we have seen in (12.120), the vacuum functional is given by

$$\langle 0|0\rangle^J = Z[J] = e^{iW[J]} = N\int \mathcal{D}\phi\, e^{iS^J[\phi]}. \quad (13.75)$$

As we have already mentioned in the last chapter, $Z[J]$ and $W[J]$ are known as the generating functionals for the Green's functions of the theory. For example,

$$\langle 0|T\big(\phi(x_1)\cdots\phi(x_n)\big)|0\rangle = \frac{(-i)^n}{Z[J]}\frac{\delta^n Z[J]}{\delta J(x_1)\cdots\delta J(x_n)}\bigg|_{J=0}, \qquad (13.76)$$

defines the n-point time ordered Green's function of the theory. On the other hand,

$$\langle 0|T\big(\phi(x_1)\cdots\phi(x_n)\big)|0\rangle^c$$
$$= (-i)^{n-1}\frac{\delta^n W[J]}{\delta J(x_1)\cdots\delta J(x_n)}\bigg|_{J=0}, \qquad (13.77)$$

leads to the connected n-point time ordered Green's functions of the theory. To clarify this distinction further, let us note here that each term in the calculation of the Green's function in (13.76) and (13.77) in the path integral formalism corresponds to a unique Feynman diagram in perturbation theory. In general, the n-point Green's functions would involve Feynman diagrams which consist of parts that are not connected and $Z[J]$ (through (13.76)) includes contributions of all diagrams (including the ones that are not connected) to a Green's function. On the other hand, the Green's functions calculated from $W[J]$ (the logarithm of $Z[J]$) through (13.77) involve only Feynman diagrams where all the parts of the diagram are connected. This is more fundamental since the general Green's functions can be constructed in terms of these. In particular we note that

$$\langle 0|\phi(x)|0\rangle^J = \phi_c(x) = \frac{\delta W[J]}{\delta J(x)} = \frac{(-i)}{Z[J]}\frac{\delta Z[J]}{\delta J(x)}, \qquad (13.78)$$

is known as the classical field (see (7.49)) and is a functional of J. The connected two point time ordered Green's function which also corresponds to the Feynman propagator of the theory is similarly obtained from (13.77)

$$iG_{\mathrm{F}}(x-y) = \langle 0|T\big(\phi(x)\phi(y)\big)|0\rangle^c$$
$$= (-i)\frac{\delta^2 W[J]}{\delta J(x)\delta J(y)}\bigg|_{J=0} = \frac{(-i)^2}{Z[J]}\frac{\delta^2 Z[J]}{\delta J(x)\delta J(y)}\bigg|_{J=0}. \qquad (13.79)$$

13.4 SLAVNOV-TAYLOR IDENTITY

The connected n-point time ordered Green's functions consist of connected Feynman diagrams with external lines (propagators). However, the more fundamental concept in perturbation theory is the diagram without the external propagators known as the vertex function. The diagrams that contribute to the n-point vertex function and which cannot be separated into two disconnected diagrams by cutting a single internal line (propagator) of the diagram lead to what are known as the 1PI (one particle irreducible) vertex functions or the proper vertex functions of the theory. These are quite fundamental in the study of quantum field theory and the generating functional which generates such vertex functions is constructed as follows. Let us Legendre transform $W[J]$ as (this is like going from the Lagrangian to the Hamiltonian)

$$\Gamma[\phi_c] = W[J] - \int d^4x\, J(x)\phi_c(x). \tag{13.80}$$

It can now be checked using (13.78) that

$$\frac{\delta \Gamma[\phi_c]}{\delta \phi_c(x)}$$
$$= \int d^4y \left[\frac{\delta W[J]}{\delta J(y)} \frac{\delta J(y)}{\delta \phi_c(x)} - \frac{\delta J(y)}{\delta \phi_c(x)} \phi_c(y) - \delta^4(x-y) J(y) \right]$$
$$= -J(y). \tag{13.81}$$

The generating functional $\Gamma[\phi_c]$ is known as the effective action of the theory (including all quantum corrections) and generates the proper (1PI) vertex functions of the theory.

With these basic ideas from path integrals, let us consider the effective Lagrangian density (at the lowest order or tree level) which consists of $\mathcal{L}_{\text{TOT}}$ for the Yang-Mills theory (13.12) as well as source terms as follows

$$\begin{aligned}\mathcal{L}_{\text{eff}} &= \mathcal{L}_{\text{TOT}} + J^{\mu a} A^a_\mu + J^a F^a + i\left(\overline{\eta}^a c^a - \overline{c}^a \eta^a\right) \\ &\quad + K^{\mu a}\left(D_\mu c\right)^a + K^a \left(\frac{g}{2} f^{abc} c^b c^c\right).\end{aligned} \tag{13.82}$$

Here, we have not only introduced sources for all the field variables in the theory, but we have also added sources $(K^{\mu a}, K^a)$ for the composite variations under the BRST transformation (namely, for the variations in (13.13) of the fields A_μ^a, c^a which are nonlinear in the field variables). It is worth noting here that the source $K^{\mu a}$ is of fermionic nature (in addition to the fermionic sources $\eta^a, \bar\eta^a$). The usefulness of this will become clear shortly. Denoting all the fields and the sources generically by A and J respectively, we can write the generating functional for the theory as

$$\langle 0|0\rangle^J = Z[J] = e^{iW[J]} = N \int \mathcal{D}A \, e^{i\int d^4x \, \mathcal{L}_{\text{eff}}}. \tag{13.83}$$

The vacuum expectation values of operators, in the presence of sources, can now be written as

$$\langle 0|A_\mu^a|0\rangle^J = \langle A_\mu^a\rangle^J = A_\mu^{(c)\,a} = \frac{\delta W[J]}{\delta J^{\mu a}},$$

$$\langle 0|F^a|0\rangle^J = \langle F^a\rangle^J = F^{(c)a} = \frac{\delta W[J]}{\delta J^a},$$

$$\langle 0|c^a|0\rangle^J = \langle c^a\rangle^J = c^{(c)a} = (-i)\frac{\delta W[J]}{\delta \bar\eta^a},$$

$$\langle 0|\bar c^a|0\rangle^J = \langle \bar c^a\rangle^J = \bar c^{(c)a} = (-i)\frac{\delta W[J]}{\delta \eta^a},$$

$$\langle 0|\left(D_\mu c\right)^a|0\rangle^J = \langle (D_\mu c)^a\rangle^J = \frac{\delta W[J]}{\delta K^{\mu a}},$$

$$\langle 0|\left(\frac{g}{2}f^{abc}c^b c^c\right)|0\rangle^J = \langle \left(\frac{g}{2}f^{abc}c^b c^c\right)\rangle^J = \frac{\delta W[J]}{\delta K^a}. \tag{13.84}$$

Here, we have assumed the convention of left derivatives for the anti-commuting fields. The fields $A^{(c)}$ are known as the classical fields (see, for example, (7.49)) and in what follows we will ignore the superscript (c) for notational simplicity.

The effective Lagrangian density is no longer invariant under the BRST transformations (13.13) when the external sources are held fixed. In fact, recalling that $\mathcal{L}_{\text{TOT}}$ in (13.82) is BRST invariant and that the BRST transformations are nilpotent, we obtain the change

13.4 SLAVNOV-TAYLOR IDENTITY

in $\mathcal{L}_{\text{eff}}$ to be (remember that the parameter of the BRST transformation is anti-commuting)

$$\begin{aligned}
\delta \mathcal{L}_{\text{eff}} &= J^{\mu a}\delta A^a_\mu + J^a \delta F^a + i(\overline{\eta}^a \delta c^a - \delta \overline{c}^a \eta^a) \\
&= \frac{\omega}{g}\left[J^{\mu a}(D_\mu c)^a + i\left(-\frac{g}{2}f^{abc}\overline{\eta}^a c^b c^c + F^a \eta^a\right)\right].
\end{aligned} \quad (13.85)$$

We note that the generating functional is defined by integrating over all possible field configurations. Therefore, if we redefine the fields under the path integral as,

$$A \to A + \delta_{\text{BRST}} A, \quad (13.86)$$

the generating functional should be invariant (namely, since it does not depend on the field variables, it should not change under any field redefinition). This immediately leads from (13.85) to (see also (13.78))

$$\begin{aligned}
\delta Z[J] = 0 &= N \int \mathcal{D}A \left(i \int d^4x\, \delta \mathcal{L}_{\text{eff}}\right) e^{i \int d^4x\, \mathcal{L}_{\text{eff}}} \\
&= \frac{\omega}{g} \int d^4x \left(J^{\mu a}(x)\frac{\delta Z}{\delta K^{\mu a}(x)} - i\overline{\eta}^a(x)\frac{\delta Z}{\delta K^a(x)}\right. \\
&\qquad\qquad \left. + i\eta^a(x)\frac{\delta Z}{\delta J^a(x)}\right).
\end{aligned} \quad (13.87)$$

The functional integration measure can be easily checked to be invariant under such a fermionic transformation and using (13.75) we note that (13.87) can also be written as

$$\int d^4x \left(J^{\mu a}(x)\frac{\delta W}{\delta K^{\mu a}(x)} - i\overline{\eta}^a(x)\frac{\delta W}{\delta K^a(x)} + i\eta^a(x)\frac{\delta W}{\delta J^a(x)}\right) = 0. \quad (13.88)$$

This is the master equation from which we can derive all the identities relating the connected Green's functions of the theory by taking functional derivatives with respect to sources. It is here that the usefulness of the sources for the composite BRST variations becomes evident.

Most often, however, we are interested in the identities satisfied by the proper (1PI) vertices of the theory. These can be obtained by passing from the generating functional for the connected Green's functions $W[J]$ to the generating functional for the proper vertices $\Gamma[A]$ through the Legendre transformation involving the field variables of the theory (see (13.80) and the field variables are really the classical fields and we are dropping the superscript (c) for simplicity), we have

$$\Gamma[A,K] = W[J,K] - \int \mathrm{d}^4x \left(J^{\mu a} A_\mu^a + J^a F^a + i\left(\overline{\eta}^a c^a - \overline{c}^a \eta^a\right)\right), \tag{13.89}$$

where K stands generically for the sources for the composite BRST variations. From the definition of the generating functional for the proper vertices (see (13.81)), it is clear that

$$\begin{aligned}
\frac{\delta \Gamma}{\delta A_\mu^a} &= -J^{\mu a}, \\
\frac{\delta \Gamma}{\delta F^a} &= -J^a, \\
\frac{\delta \Gamma}{\delta c^a} &= i\overline{\eta}^a, \\
\frac{\delta \Gamma}{\delta \overline{c}^a} &= i\eta^a, \\
\frac{\delta \Gamma}{\delta K_\mu^a} &= \frac{\delta W}{\delta K_\mu^a}, \\
\frac{\delta \Gamma}{\delta K^a} &= \frac{\delta W}{\delta K^a}.
\end{aligned} \tag{13.90}$$

Using these definitions, we see that we can rewrite the master equation (13.88) in terms of the generating functional for the proper vertices as

13.4 SLAVNOV-TAYLOR IDENTITY

$$\int d^4x \left(\frac{\delta \Gamma}{\delta A^a_\mu(x)} \frac{\delta \Gamma}{\delta K^{\mu a}(x)} + \frac{\delta \Gamma}{\delta c^a(x)} \frac{\delta \Gamma}{\delta K^a(x)} - F^a(x) \frac{\delta \Gamma}{\delta \bar{c}^a(x)} \right) = 0.$$
(13.91)

This is the master equation from which we can derive all the relations between various (1PI) proper vertices resulting from the BRST invariance of the theory by taking functional derivatives with respect to (classical) fields. This is essential in proving the renormalizability of gauge theories. Thus, for example, let us note that we can write the master identity (13.91) in the momentum space as

$$\int d^4k \left(\frac{\delta \Gamma}{\delta A^a_\mu(-k)} \frac{\delta \Gamma}{\delta K^{\mu a}(k)} + \frac{\delta \Gamma}{\delta c^a(-k)} \frac{\delta \Gamma}{\delta K^a(k)} \right.$$
$$\left. - F^a(k) \frac{\delta \Gamma}{\delta \bar{c}^a(k)} \right) = 0.$$
(13.92)

Taking derivative of this with respect to $\frac{\delta^2}{\delta F^b(p) \delta c^c(-p)}$ and setting all field variables to zero gives

$$\frac{\delta^2 \Gamma}{\delta F^b(p) \delta A^a_\mu(-p)} \frac{\delta^2 \Gamma}{\delta c^c(-p) \delta K^{\mu a}(p)} - \frac{\delta^2 \Gamma}{\delta c^c(-p) \delta \bar{c}^b(p)} = 0. \quad (13.93)$$

A simple analysis of this relation shows that the mixed two point vertex function involving the fields F-A_μ is related to the two point function for the ghost fields and, consequently, the counter terms (quantum corrections) should satisfy such a relation. The BRST invariance, in this way, is very fundamental in the study of gauge theories as far as renormalizability and gauge independence of physical observables are concerned. For example, the gauge dependence of the effective potential in a gauge theory (with scalar fields) as well as the gauge independence of the physical poles of the propagator can be obtained systematically from the Nielsen identities which, like the Slavnov-Taylor identities, follow from the BRST invariance of the theory. (We would discuss renormalization of field theories in a later chapter.)

13.5 Feynman rules

Understanding gauge fixing for gauge theories in the path integral formalism is quite essential in developing the perturbation theory for gauge theories. We note that since gauge fixing breaks gauge invariance, it renders the theory nonsingular making it possible to define the propagator of the theory and derive the Feynman rules for the non-Abelian gauge theory. These are the essential elements in carrying out any perturbative calculation in a quantum field theory. Let us recall that the total Lagrangian density after gauge fixing (in the covariant gauge) has the form (13.2)

$$\mathcal{L}_{\text{TOT}} = -\frac{1}{4}F^a_{\mu\nu}F^{\mu\nu a} - \frac{1}{2\xi}\left(\partial^\mu A^a_\mu\right)^2 + \partial^\mu \bar{c}^a D_\mu c^a. \tag{13.94}$$

The propagators of the theory can be derived from the free part of the Lagrangian density which is quadratic in the field variables

$$\begin{aligned}\mathcal{L}_Q &= -\frac{1}{4}\left(\partial_\mu A^a_\nu - \partial_\nu A^a_\mu\right)\left(\partial^\mu A^{\nu a} - \partial^\nu A^{\mu a}\right) \\ &\quad - \frac{1}{2\xi}\left(\partial^\mu A^a_\mu\right)^2 + \partial^\mu \bar{c}^a \partial_\mu c^a \\ &= \frac{1}{2}A^a_\mu\left(\eta^{\mu\nu}\Box - \left(1 - \frac{1}{\xi}\right)\partial^\mu\partial^\nu\right)A^a_\nu - \bar{c}^a\Box c^a \\ &= \frac{1}{2}A^a_\mu O^{\mu\nu\,ab} A^b_\nu + \bar{c}^a M^{ab} c^b, \end{aligned} \tag{13.95}$$

where we have neglected total divergence terms and have identified

$$O^{\mu\nu\,ab} = \delta^{ab}\left(\eta^{\mu\nu}\Box - \left(1 - \frac{1}{\xi}\right)\partial^\mu\partial^\nu\right), \quad M^{ab} = -\delta^{ab}\Box. \tag{13.96}$$

The propagators for the gauge and the ghost fields are the inverses $O^{-1\,ab}_{\mu\nu}$ and $M^{-1\,ab}$ respectively of the two point functions (up to multiplicative factors). (They are the Green's functions of the free theory.) The inverses of the two point functions are easily calculated in the momentum space where the quadratic operators in (13.96) take the forms

13.5 FEYNMAN RULES

$$M^{ab}(p) = \delta^{ab}p^2,$$
$$O^{\mu\nu\,ab}(p) = -\delta^{ab}\left(\eta^{\mu\nu}p^2 - \left(1 - \frac{1}{\xi}\right)p^\mu p^\nu\right). \tag{13.97}$$

The inverse of $M^{ab}(p)$ in (13.97) is quite simple (it is like the propagator of a massless scalar field)

$$M^{-1\,ab}(p) = \frac{\delta^{ab}}{p^2}, \tag{13.98}$$

while the derivation of the inverse of $O^{\mu\nu\,ab}(p)$ is a bit more involved. Therefore, let us derive this inverse systematically. Let us note that the Green's function $O^{-1\,ab}_{\mu\nu}(p)$ is a symmetric second rank tensor (of rank 2) and with all the Lorentz structures available $(p_\mu, \eta_{\mu\nu})$, we can parameterize the most general form of the inverse as

$$O^{-1\,ab}_{\mu\nu}(p) = \delta^{ab}\left(\alpha\,\eta_{\mu\nu} + \beta\,\frac{p_\mu p_\nu}{p^2}\right), \tag{13.99}$$

where α, β are arbitrary parameters to be determined (they are not necessarily constants and can be Lorentz invariant functions of the momentum). With this parameterization, we note that the inverse is defined to satisfy (repeated indices are summed)

$$O^{\mu\nu\,ab}(p)O^{-1\,bc}_{\nu\lambda}(p) = \delta^{ac}\delta^\mu_\lambda,$$
$$\text{or,} \quad \left(\eta^{\mu\nu}p^2 - \left(1 - \frac{1}{\xi}\right)p^\mu p^\nu\right)O^{-1\,ac}_{\nu\lambda}(p) = -\delta^{ac}\delta^\mu_\lambda. \tag{13.100}$$

Substituting the parameterization (13.99) for the inverse and carrying out the multiplication explicitly we have

$$\left(\eta^{\mu\nu}p^2 - \left(1 - \frac{1}{\xi}\right)p^\mu p^\nu\right)\left(\alpha\eta_{\nu\lambda} + \beta\,\frac{p_\nu p_\lambda}{p^2}\right) = -\delta^\mu_\lambda,$$
$$\text{or,} \quad \alpha p^2 \delta^\mu_\lambda - \alpha\left(1 - \frac{1}{\xi}\right)p^\mu p_\lambda + \beta\,p^\mu p_\lambda - \beta\left(1 - \frac{1}{\xi}\right)p^\mu p_\lambda = -\delta^\mu_\lambda,$$
$$\text{or,} \quad (\alpha p^2 + 1)\delta^\mu_\lambda - \left(\alpha\left(1 - \frac{1}{\xi}\right) - \frac{\beta}{\xi}\right)p^\mu p_\lambda = 0. \tag{13.101}$$

Setting the coefficients of each distinct Lorentz structure to zero, it follows that

$$\alpha = -\frac{1}{p^2},$$

$$\alpha\left(1 - \frac{1}{\xi}\right) - \frac{\beta}{\xi} = 0,$$

or, $\quad \beta = \alpha(\xi - 1) = -\frac{1}{p^2}(\xi - 1),$ \hfill (13.102)

so that we can write the inverse in (13.99) as

$$O^{-1\,ab}_{\mu\nu}(p) = -\frac{\delta^{ab}}{p^2}\left(\eta_{\mu\nu} + (\xi - 1)\frac{p_\mu p_\nu}{p^2}\right). \tag{13.103}$$

With these, we can write the generating functional for the free theory as (involving only the quadratic terms of the action as well as sources for the fields, see (12.126))

$$Z_0 = N e^{\left[-\frac{i}{2}\left(J^{\mu a}(-p), O^{-1\,ab}_{\mu\nu}(p) J^{\nu b}(p)\right) + i\left(\overline{\eta}^a(-p), M^{-1\,ab}(p)\eta^b(p)\right)\right]}, \tag{13.104}$$

where $(\,,\,)$ represents the integral over p and we deduce from this that (see (13.79), sometimes the gauge propagator is also denoted by $D^{ab}_{\mu\nu}(p)$)

$$\begin{aligned}
iG^{ab}_{\mu\nu}(p) &= \frac{(-i)^2}{Z_0}\frac{\delta^2 Z_0}{\delta J^{\mu a}(-p)\delta J^{\nu b}(p)}\bigg|_{J^{\mu a}=\eta^a=\overline{\eta}^a=0} \\
&= (-1)\left(-iO^{-1\,ab}_{\mu\nu}(p)\right) = iO^{-1\,ab}_{\mu\nu}(p) \\
&= -\frac{i\delta^{ab}}{p^2}\left(\eta_{\mu\nu} + (\xi - 1)\frac{p_\mu p_\nu}{p^2}\right).
\end{aligned} \tag{13.105}$$

This defines the Feynman propagator (we do not explicitly write the subscript "F" and the $i\epsilon$ in the denominator for simplicity) for the gauge field which clearly depends on the gauge fixing parameter ξ.

13.5 FEYNMAN RULES

When $\xi = 0$, i.e., when we are in the Landau gauge this propagator is transverse, while for $\xi = 1$, the gauge is known as the Feynman gauge where the propagator has a much simpler form which is more suitable for perturbative calculations. Similarly we note that (once again we suppress the subscript "F" and the $i\epsilon$ prescription which is assumed)

$$\begin{aligned} iG^{ab}(p) &= \frac{(-i)^2}{Z_0}\frac{\delta^2 Z_0}{\delta\overline{\eta}^a(-p)\delta\eta^b(p)}\bigg|_{J^{\mu a}=\eta^a=\overline{\eta}^a=0} \\ &= (-1)\left(-iM^{-1ab}(p)\right) = iM^{-1ab}(p) \\ &= \frac{i\delta^{ab}}{p^2}, \end{aligned} \qquad (13.106)$$

which defines the ghost propagator (which is sometimes also denoted by $D^{ab}(p)$). The propagators can be diagrammatically represented as

$$\begin{aligned} &= iG^{ab}_{\mu\nu}(p) &&(13.107) \\ &= -\frac{i\delta^{ab}}{p^2}\left(\eta_{\mu\nu} + (\xi-1)\frac{p_\mu p_\nu}{p^2}\right), \\ &= iG^{ab}(p) = \frac{i\delta^{ab}}{p^2}. &&(13.108) \end{aligned}$$

We can now write the complete generating functional as

$$Z[J] = \exp\left[i\int d^4x\, \mathcal{L}_{\text{int}}\left(\frac{1}{i}\frac{\delta}{\delta J}\right)\right]Z_0[J], \qquad (13.109)$$

where we have used J generically for all the sources and the interaction Lagrangian density can now be identified with

$$\begin{aligned} \mathcal{L}_{\text{int}} &= \mathcal{L}_{\text{TOT}} - \mathcal{L}_{\text{Q}} \\ &= gf^{abc}\partial_\mu A^a_\nu A^{\mu b}A^{\nu c} - \frac{g^2}{4}f^{abp}f^{cdp}A^a_\mu A^b_\nu A^{\mu c}A^{\nu d} \\ &\quad - gf^{abc}A^a_\mu \partial^\mu \overline{c}^c c^b. \end{aligned} \qquad (13.110)$$

Let us now derive the interaction vertices for the theory which are more useful in the momentum space (where Feynman diagram calculations are primarily carried out). We note that we can write the interaction action in the momentum space as

$$\begin{aligned}
S_{\text{int}} &= \int d^4x\, \mathcal{L}_{\text{int}} \\
&= (2\pi)^4 g f^{abc} \int d^4p_1 d^4p_2 d^4p_3\, \delta^4(p_1+p_2+p_3) \\
&\quad \times (-ip_{1\mu}) A^a_\nu(p_1) A^{\mu b}(p_2) A^{\nu c}(p_3) \\
&\quad -\frac{(2\pi)^4 g^2}{4} f^{abp} f^{cdp} \int d^4p_1 d^4p_2 d^4p_3 d^4p_4 \\
&\quad \times \delta^4(p_1+p_2+p_3+p_4) A^a_\mu(p_1) A^b_\nu(p_2) A^{\mu c}(p_3) A^{\nu d}(p_4) \\
&\quad -(2\pi)^4 g f^{abc} \int d^4p_1 d^4p_2 d^4p_3\, \delta^4(p_1+p_2+p_3) \\
&\quad \times (-ip_3^\mu) A^a_\mu(p_1) \bar{c}^c(p_3) c^b(p_2).
\end{aligned} \qquad (13.111)$$

The tree level 3-point and 4-point interaction vertices of the theory follow from this to correspond to (the restriction | denotes setting all field variables to zero)

$$\begin{aligned}
V^{\mu\nu\lambda\,abc}(p_1,p_2,p_3) &= i \frac{\delta^3 S}{\delta A^a_\mu(p_1)\delta A^b_\nu(p_2)\delta A^c_\lambda(p_3)}\bigg| \\
&= i \frac{\delta^3 S_{\text{int}}}{\delta A^a_\mu(p_1)\delta A^b_\nu(p_2)\delta A^c_\lambda(p_3)}\bigg| \\
&= (2\pi)^4 g f^{abc} \delta^4(p_1+p_2+p_3) \\
&\quad \times \left[\eta^{\mu\nu}(p_1-p_2)^\lambda + \eta^{\nu\lambda}(p_2-p_3)^\mu + \eta^{\lambda\mu}(p_3-p_1)^\nu\right],
\end{aligned} \qquad (13.112)$$

$$\begin{aligned}
V^{\mu\nu\lambda\rho\,abcd}(p_1,p_2,p_3,p_4) &= i \frac{\delta^4 S}{\delta A^a_\mu(p_1)\delta A^b_\nu(p_2)\delta A^c_\lambda(p_3)\delta A^d_\rho(p_4)}\bigg| \\
&= i \frac{\delta^4 S_{\text{int}}}{\delta A^a_\mu(p_1)\delta A^b_\nu(p_2)\delta A^c_\lambda(p_3)\delta A^d_\rho(p_4)}\bigg|
\end{aligned}$$

$$\begin{aligned}
= & -(2\pi)^4 ig^2 \delta^4(p_1+p_2+p_3+p_4) \Big[f^{abp}f^{cdp}\big(\eta^{\mu\lambda}\eta^{\nu\rho}-\eta^{\mu\rho}\eta^{\nu\lambda}\big) \\
& + f^{acp}f^{dbp}\big(\eta^{\mu\rho}\eta^{\nu\lambda}-\eta^{\mu\nu}\eta^{\lambda\rho}\big) \\
& + f^{adp}f^{bcp}\big(\eta^{\mu\nu}\eta^{\lambda\rho}-\eta^{\mu\lambda}\eta^{\nu\rho}\big) \Big],
\end{aligned}$$

(13.113)

$$\begin{aligned}
V^{\mu\,abc}(p_1,p_2,p_3) &= i\frac{\delta^3 S}{\delta A^a_\mu(p_1)\delta c^b(p_2)\delta \bar{c}^c(p_3)}\bigg| \\
&= i\frac{\delta^3 S_{\text{int}}}{\delta A^a_\mu(p_1)\delta c^b(p_2)\delta \bar{c}^c(p_3)}\bigg| \\
&= -(2\pi)^4 g f^{abc} p_3^\mu \, \delta^4(p_1+p_2+p_3).
\end{aligned}$$

(13.114)

Therefore, with (13.112)-(13.114) we can represent all the interaction vertices of the theory as

$$= V^{\mu\nu\lambda\,abc}(p_1,p_2,p_3),$$

$$= V^{\mu\nu\lambda\rho\,abcd}(p_1,p_2,p_3,p_4),$$

$$= V^{\mu\,abc}(p_1,p_2,p_3). \quad (13.115)$$

All the momenta in the above diagrams are assumed to be incoming and the arrow in the ghost diagram shows the direction of flow of the ghost number. The dot over the ghost line denotes where the derivative acts in the interaction term in the Lagrangian density (namely, the $\bar{c}^c$ field, see for example (13.114)). These represent all the Feynman rules for the Yang-Mills theory and perturbative calculations can now be carried out using these Feynman rules.

13.6 Ghost free gauges

In quantizing a non-Abelian gauge theory, we find that a gauge fixing term necessarily modifies the theory and thereby requires the addition of a ghost Lagrangian density which as we have argued is necessary to balance the modification induced by the gauge fixing term. It is interesting to ask if there exist gauge conditions in a non-Abelian theory where the ghost degrees of freedom may not be important much like in the Abelian theory. If possible, this would, of course, simplify the perturbative calculations enormously primarily because the number of diagrams to evaluate will be much smaller (namely, the diagrams with ghosts will be absent). A class of gauge conditions of the form

$$n \cdot A^a(x) = f^a(x), \qquad n^2 \neq 0 \quad \text{or} \quad n^2 = 0, \tag{13.116}$$

where n^μ is an arbitrary vector does indeed achieve this. Such a class of gauge choices is conventionally known as ghost free gauges. When $n^2 = 0$, such a gauge is known as the light-cone gauge, for $n^2 = 1$ (normalized) it is called the temporal gauge while for $n^2 = -1$ it is known as the axial gauge. We will now discuss, from two distinct points of view, in what sense these gauges become ghost free in a non-Abelian gauge theory.

Let us first discuss this from the diagrammatic point of view. In a general axial-like gauge (13.116), we recall (see (12.153) and (12.160)) that the total Lagrangian density for a non-Abelian gauge theory takes the form

$$\mathcal{L}_{\text{TOT}} = -\frac{1}{4} F^a_{\mu\nu} F^{\mu\nu a} - \frac{1}{2\xi} (n \cdot A^a)^2 - \bar{c}^a n \cdot D c^a. \tag{13.117}$$

13.6 Ghost free gauges

With such a choice of gauge the gauge propagator can be calculated to have the form

$$iG^{ab}_{\mu\nu} = i\delta^{ab}\left[\frac{1}{p^2}\left(-\eta_{\mu\nu} + \frac{n_\mu p_\nu + n_\nu p_\mu}{(n\cdot p)} - \frac{n^2 p_\mu p_\nu}{(n\cdot p)^2}\right)\right.$$
$$\left. -\xi \frac{p_\mu p_\nu}{(n\cdot p)^2}\right], \qquad (13.118)$$

so that we have (independent of whether $n^2 = 0$ or $n^2 \neq 0$)

$$n^\mu G^{ab}_{\mu\nu} = -\xi\delta^{ab}\frac{p_\nu}{(n\cdot p)}, \qquad n^\mu n^\nu G^{ab}_{\mu\nu} = -\xi\delta^{ab}. \qquad (13.119)$$

Similarly, the ghost propagator in the axial-like gauge (13.116) in this theory has the form

$$iG^{ab} = \frac{\delta^{ab}}{n\cdot p}, \qquad (13.120)$$

and the antighost-ghost-gauge vertex involves a multiplicative factor of n^μ. Therefore, the integrand of any diagram involving ghosts, either open lines or closed loops, will have the structure (we are not writing the possible multiplicative Lorentz factors involving n^μ which do not effect the integral and we are evaluating the integral in D dimensions)

Figure 13.1: Examples of diagrams with open and closed ghost lines.

$$\int \mathrm{d}^D k \, \frac{1}{(n \cdot k)(n \cdot (k+p_1)) \ldots (n \cdot (k+p_1+\cdots+p_m))}$$

$$= \int_0^1 \prod_{i=0}^m \mathrm{d}x_i \int \mathrm{d}^D k \, \frac{\delta(1-x_0-x_1-\cdots-x_m)}{\left(n \cdot (k+x_1 p_1 + \cdots + x_m p_m)\right)^{m+1}}$$

$$= \int_0^1 \prod_{i=0}^m \mathrm{d}x_i \int \mathrm{d}^D k \, \frac{\delta(1-x_0-x_1-\cdots-x_m)}{(n \cdot k)^{m+1}}. \quad (13.121)$$

Here we have used the Feynman parameterization to combine the denominators (to be discussed in the next chapter) and have shifted the variable of integration in the last step to bring it to the simpler form. Using the fact that the integral is Lorentz invariant, we can rewrite this also as

$$\int_0^1 \prod_{i=0}^m \mathrm{d}x_i \, \frac{\delta(1-x_0-x_1-\cdots-x_m)}{(n^2)^{(m+1)/2}} \int \frac{\mathrm{d}^D k}{(k^2)^{(m+1)/2}} = 0, \quad (13.122)$$

and the vanishing of this integral (in the last step) is easily seen using dimensional regularization (that we will discuss in chapter **15**, namely, in (15.10)). This shows that all the diagrams involving ghost interactions can be regularized to zero so that the ghost degrees of freedom are irrelevant with such a choice of gauge.

An alternative way to see this is to note that the generating functional in the axial-like gauge (13.116) has the form

$$Z = N \int \mathcal{D}A_\mu^a \mathcal{D}f^a \, \det(n \cdot D^{ab}) \delta(n \cdot A^a - f^a) G[f^a] e^{iS}$$

$$= N \int \mathcal{D}A_\mu^a \mathcal{D}f^a \, \det(n \cdot \partial \delta^{ab} + g f^{abc} n \cdot A^c)$$

$$\times \delta(n \cdot A^a - f^a) G[f^a] e^{iS}. \quad (13.123)$$

Here $G[f^a]$ is an arbitrary functional of f^a which we normally choose to correspond to the 't Hooft weight factor

$$G[f^a] = e^{-\frac{i}{2\xi}\int d^4x\, f^a f^a}. \tag{13.124}$$

However, since the generating functional does not depend on the functional form of $G[f^a]$ (up to irrelevant multiplicative constants), let us choose

$$G[f^a] = \left[\det\left(n\cdot\partial\delta^{ab} + gf^{abc}f^c\right)\right]^{-1} e^{-\frac{i}{2\xi}\int d^4x\, f^a f^a}. \tag{13.125}$$

In this case, we can write the generating functional as

$$\begin{aligned}
Z &= N\int \mathcal{D}A^a_\mu \mathcal{D}f^a \det\left(n\cdot\partial\delta^{ab} + gf^{abc}n\cdot A^c\right)\delta(n\cdot A^a - f^a) \\
&\quad \times \left[\det\left(n\cdot\partial\delta^{ab} + gf^{abc}f^c\right)\right]^{-1} e^{-\frac{i}{2\xi}\int d^4x\, f^a f^a} e^{iS} \\
&= N\int \mathcal{D}A^a_\mu \mathcal{D}f^a\, \delta(n\cdot A^a - f^a)\, e^{-\frac{i}{2\xi}\int d^4x\, f^a f^a} e^{iS} \\
&= N\int \mathcal{D}A^a_\mu\, e^{iS - \frac{i}{2\xi}\int d^4x\, (n\cdot A^a)^2}. \tag{13.126}
\end{aligned}$$

This shows that in this gauge the ghosts decouple completely and can be absorbed into the normalization factor. Note that such a decoupling works only because n^μ (in $n\cdot A^a$ inside the determinant) is a multiplicative operator. It would not work in say, the covariant gauge. The axial-like gauge choices (13.116) in a non-Abelian gauge theory, therefore, correspond to ghost free gauges very much like in the Abelian gauge theories.

13.7 References

1. J. C. Taylor, Nuclear Physics **B33**, 436 (1971).

2. A. A. Slavnov, Theoretical and Mathematical Physics **10**, 99 (1972).

3. C. Becchi, A. Rouet and R. Stora, Physics Letters **52B**, 344 (1974).

4. C. Becchi, A. Rouet and R. Stora, Communications in Mathematical Physics **42**, 127 (1975).

5. T. Kugo and I. Ojima, Progress of Theoretical Physics **60**, 1869 (1978).

6. T. Kugo and I. Ojima, Progress of Theoretical Physics Supplement No. **66** (1979).

7. J. Frenkel, Physical Review **D13**, 2325 (1979).

8. G. Leibbrandt, Reviews of Modern Physics **59**, 1067 (1987).

9. A. Das, *Finite Temperature Field Theory*, World Scientific, Singapore (1997).

CHAPTER 14
Higgs phenomenon and the standard model

In chapters **9**, **12** and **13** we have studied gauge theories (both Abelian and non-Abelian) in great detail. One of the striking features of these theories is that the gauge fields are massless simply because the action for a mass term for the gauge field in the Lagrangian density

$$\frac{M^2}{2} A_\mu A^\mu, \quad \text{or} \quad \frac{M^2}{2} A^a_\mu A^{\mu a}, \tag{14.1}$$

is not invariant under the gauge transformation (9.14) or (12.34) respectively. As we have mentioned earlier gauge fields can be thought of as the carriers of physical forces. Therefore, a massless gauge field is completely consistent with the observed fact that electromagnetic forces are long ranged. However, we also know of physical forces in nature (such as the weak force) that are short ranged and this would seem to suggest (intuitively) that the gauge fields associated with such forces may be massive. Therefore, in this chapter we would discuss this important question of massive gauge fields concluding with the standard model of the electroweak interactions.

14.1 Stückelberg formalism

Let us consider the Lagrangian density for a charge neutral spin 1 field A_μ given by

$$\mathcal{L} = -\frac{1}{4} F_{\mu\nu} F^{\mu\nu} + \frac{M^2}{2} A_\mu A^\mu, \tag{14.2}$$

where the field strength tensor is defined as in the Maxwell theory

$$F_{\mu\nu} = \partial_\mu A_\nu - \partial_\nu A_\mu. \tag{14.3}$$

The Lagrangian density (14.2) describes a massive photon which can be seen as follows. The Euler-Lagrange equation following from (14.2) has the form

$$\partial_\mu \frac{\partial \mathcal{L}}{\partial \partial_\mu A_\nu} - \frac{\partial \mathcal{L}}{\partial A_\nu} = 0,$$

or, $\quad -\partial_\mu F^{\mu\nu} - M^2 A^\nu = 0,$

or, $\quad \partial_\mu F^{\mu\nu} + M^2 A^\nu = 0. \tag{14.4}$

Contracting (14.4) with ∂_ν and using the anti-symmetry of the field strength tensor, we obtain

$$\partial \cdot A = 0, \tag{14.5}$$

and substituting this back into the Euler-Lagrange equation in (14.4) yields

$$\partial_\mu (\partial^\mu A^\nu - \partial^\nu A^\mu) + M^2 A^\nu = 0,$$

or, $\quad \left(\Box + M^2\right) A^\nu = 0, \tag{14.6}$

Equation (14.6) together with the constraint (14.5) defines the Proca equation (and (14.2) denotes the Proca Lagrangian density) and it is clear that this system of equations defines a massive spin 1 field theory (a massive photon). Note that the field variable A_μ has four field degrees of freedom while the constraint (14.5) eliminates one degree of freedom leaving us with three field degrees of freedom which is the correct number of dynamical field degrees of freedom for a massive spin 1 field. As has been mentioned earlier, we also note here that the theory (14.2) does not have a gauge invariance (because of the mass term) unlike the Maxwell theory. The propagator for this theory can be worked out (see, for example, section **13.5**) in a simple manner and in momentum space has the form

14.1 Stückelberg formalism

$$iG_{\mu\nu}(p) = -\frac{i}{p^2 - M^2}\left(\eta_{\mu\nu} - \frac{p_\mu p_\nu}{M^2}\right). \tag{14.7}$$

Thus, we see that the propagator in this theory does not fall off fast enough for large values of the momenta unlike in all other theories that we have studied so far. This leads to difficulties in establishing renormalizability of the (interacting) Proca theory for massive photons. The limit $M \to 0$ of the Proca theory is clearly quite subtle.

As a result of these difficulties, Stückelberg considered the following generalized Lagrangian density

$$\begin{aligned}\mathcal{L} &= -\frac{1}{4}F_{\mu\nu}F^{\mu\nu} + \frac{M^2}{2}\left(A_\mu + \frac{1}{M}\partial_\mu\chi\right)\left(A^\mu + \frac{1}{M}\partial^\mu\chi\right) \\ &= -\frac{1}{4}F_{\mu\nu}F^{\mu\nu} + \frac{M^2}{2}A_\mu A^\mu + \frac{1}{2}\partial_\mu\chi\partial^\mu\chi + MA^\mu\partial_\mu\chi,\end{aligned} \tag{14.8}$$

which, in addition to the spin 1 field, also contains a dynamical charge neutral spin zero field (real scalar field) which mixes with A_μ. In contrast to the Proca theory (14.2), the Stückelberg theory (14.8) was constructed to be invariant under a Maxwell-like gauge transformation

$$\begin{aligned}A_\mu(x) &\to A_\mu(x) + \frac{1}{e}\partial_\mu\theta(x), \\ \chi(x) &\to \chi(x) - \frac{M}{e}\theta(x),\end{aligned} \tag{14.9}$$

where e denotes the coupling constant of QED (this can be set to unity, but we have kept it for consistency with earlier discussion, say in (9.83) as well as for discussions in connection with the Higgs phenomenon in the next section). The field strength tensor is, of course, invariant under the gauge transformation (14.9). To see the invariance of the full theory, it is sufficient to note that under the transformation (14.9)

$$\begin{aligned}A_\mu + \frac{1}{M}\partial_\mu\chi &\to A_\mu + \frac{1}{e}\partial_\mu\theta + \frac{1}{M}\left(\partial_\mu\chi - \frac{M}{e}\partial_\mu\theta\right) \\ &= A_\mu + \frac{1}{M}\partial_\mu\chi.\end{aligned} \tag{14.10}$$

Because of the gauge invariance of the theory, we can choose a gauge and if we choose the unitary gauge

$$\chi(x) = 0, \tag{14.11}$$

which can be achieved by the choice of the gauge transformation parameter

$$\theta(x) = \frac{e}{M}\chi(x), \tag{14.12}$$

then the Lagrangian density (14.8) takes the form

$$\mathcal{L} = -\frac{1}{4}F_{\mu\nu}F^{\mu\nu} + \frac{M^2}{2}A_\mu A^\mu. \tag{14.13}$$

We recognize this to be the Proca theory (14.2) and this makes it clear that the Proca theory can be thought of as the gauge fixed Stückelberg theory. However, in this unitary gauge, the gauge propagator (see also (14.7))

$$iG_{\mu\nu}(p) = -\frac{i}{p^2 - M^2}\left(\eta_{\mu\nu} - \frac{p_\mu p_\nu}{M^2}\right), \tag{14.14}$$

has the unpleasant features described earlier. This is typical of the behavior of theories in a unitary gauge. On the other hand, if we choose a covariant gauge condition which leads to a gauge fixing Lagrangian density of the form (see, for example, section **12.3** and we note that this is known as the 't Hooft gauge which we will discuss in the next section)

$$-\frac{1}{2\xi}\left(\partial_\mu A^\mu - \xi M\chi\right)^2, \tag{14.15}$$

this would induce a ghost action, but would lead to propagators that are well behaved at high momentum. As a result, the there is no difficulty in establishing renormalizability in such a gauge. Furthermore, the massless limit $M \to 0$ is now straightforward.

14.1 STÜCKELBERG FORMALISM

All of this analysis can be carried over to non-Abelian gauge theories as well as to the Einstein theory of gravitation. We will now briefly describe the Stückelberg formalism for non-Abelian gauge theories. Let us consider the Lagrangian density

$$\begin{aligned}\mathcal{L} = \mathrm{Tr}\Big(&-\frac{1}{2}F_{\mu\nu}F^{\mu\nu} \\ &+ M^2\big(A_\mu + \frac{1}{ig}(\partial_\mu V)V^{-1}\big)\big(A^\mu + \frac{1}{ig}(\partial^\mu V)V^{-1}\big)\Big),\end{aligned} \quad (14.16)$$

where the trace is over the fundamental representation of the group, say, $SU(n)$ (as we have discussed earlier in section **12.1**). Here the field strength tensor and the operator V are defined as

$$\begin{aligned} F_{\mu\nu} &= \partial_\mu A_\nu - \partial_\nu A_\mu + ig\,[A_\mu, A_\nu], \\ V &= e^{\frac{ig}{M}\chi}, \end{aligned} \quad (14.17)$$

where χ denotes a matrix valued scalar field. This Lagrangian density can be easily checked to be invariant under the gauge transformations (see, for example, (12.38) and (12.41))

$$\begin{aligned} A_\mu &\to UA_\mu U^{-1} - \frac{1}{ig}(\partial_\mu U)U^{-1}, \\ V &\to UV. \end{aligned} \quad (14.18)$$

As we have already seen in chapter **12**, the Lagrangian density for the gauge field is invariant under the transformation (14.18). To see that the complete Lagrangian density is also invariant, it is sufficient to note that

$$A_\mu + \frac{1}{ig}(\partial_\mu V)V^{-1}$$

$$\rightarrow UA_\mu U^{-1} - \frac{1}{ig}(\partial_\mu U)U^{-1} + \frac{1}{ig}(\partial_\mu UV)(UV)^{-1}$$

$$= UA_\mu U^{-1} - \frac{1}{ig}(\partial_\mu U)U^{-1}$$

$$+ \frac{1}{ig}\left((\partial_\mu U)U^{-1} + U(\partial_\mu V)V^{-1}U^{-1}\right)$$

$$= U\left(A_\mu + \frac{1}{ig}(\partial_\mu V)V^{-1}\right)U^{-1}. \tag{14.19}$$

Using the cyclicity under trace it is now straightforward to see that the full theory is invariant under the gauge transformation in (14.18). If we choose the unitary gauge condition

$$V = \mathbb{1}, \tag{14.20}$$

which can be achieved by the choice of the gauge transformation matrix

$$U(x) = V^{-1}(x), \tag{14.21}$$

then the Lagrangian density (14.16) becomes

$$\mathcal{L} = -\frac{1}{2}\operatorname{Tr} F_{\mu\nu}F^{\mu\nu} + M^2 \operatorname{Tr} A_\mu A^\mu. \tag{14.22}$$

This can be thought of as a generalization of the Proca theory (14.2) to the non-Abelian case and describes a massive spin 1 field belonging to $SU(n)$. In this gauge, as before, the propagator has the form (recall that the quadratic part of the Lagrangian density has the similar form both in the Abelian as well as in the non-Abelian theories)

$$iG^{ab}_{\mu\nu}(p) = -\frac{i\delta^{ab}}{p^2 - M^2}\left(\eta_{\mu\nu} - \frac{p_\mu p_\nu}{M^2}\right), \tag{14.23}$$

with the unpleasant features discussed earlier that are characteristics of the unitary gauge choice. On the other hand, if we choose a covariant gauge fixing condition (a generalization of (14.15)), the propagator will have the correct asymptotic behavior for large values of the momenta.

14.2 Higgs phenomenon

In section **7.5** we studied the phenomena of spontaneous break down of a global symmetry in the self-interacting complex scalar field theory where the mass term for the scalar field had the "wrong" sign. In that case, we saw that one of the real components of the scalar field becomes massless while the other picks up a mass (with the right sign). This is known as the Nambu-Goldstone phenomena and the massless scalar field is known as the Nambu-Goldstone boson. Let us next ask what happens if there is a spontaneous breakdown of a local symmetry. For simplicity we consider again the self-interacting theory of a complex scalar field (as in section **7.5**), but now interacting with an Abelian gauge field as well (scalar QED). The Lagrangian density for such a theory is given by (with $\lambda > 0$, see also section **7.6**)

$$\mathcal{L} = -\frac{1}{4}F_{\mu\nu}F^{\mu\nu} + (D_\mu \phi)^\dagger (D^\mu \phi) + m^2 (\phi^\dagger \phi) - \frac{\lambda}{4}(\phi^\dagger \phi)^2, \quad (14.24)$$

where the theory has a "wrong" sign for the mass term and the covariant derivative is defined as in (7.83)

$$D_\mu \phi = \partial_\mu \phi + ieA_\mu \phi. \quad (14.25)$$

The Lagrangian density (14.24) can be checked easily (see (7.84) and (7.85)) to be invariant under the infinitesimal local transformations

$$\begin{aligned} \delta\phi &= -i\epsilon(x)\phi(x), \\ \delta\phi^\dagger &= i\epsilon(x)\phi^\dagger(x), \\ \delta A_\mu &= \frac{1}{e}\partial_\mu \epsilon(x). \end{aligned} \quad (14.26)$$

The gauge boson in this theory is massless because the gauge symmetry (14.26) does not allow a mass term for the gauge field. On the other hand, since the mass term for the scalar field has the "wrong" sign, as we have discussed in section **7.5**, the normal vacuum for which

$$\langle\phi\rangle = \langle\phi^\dagger\rangle = 0, \tag{14.27}$$

does not correspond to the true vacuum. (As we have seen in section **7.5**, this actually corresponds to the local maximum of the potential.) Rather the true vacuum of the theory satisfies (see (7.64), there is actually an infinity of minima of the potential, this is just one of them)

$$\sigma_c = \langle\sigma\rangle = \frac{2m}{\sqrt{\lambda}}, \qquad \chi_c = \langle\chi\rangle = 0, \tag{14.28}$$

where (see also (7.42))

$$\phi = \frac{1}{\sqrt{2}}(\sigma + i\chi). \tag{14.29}$$

For further analysis, let us rewrite the Lagrangian density (14.24) in terms of the σ, χ fields

$$\begin{aligned}\mathcal{L} &= -\frac{1}{4}F_{\mu\nu}F^{\mu\nu} + (D_\mu\phi)^\dagger(D^\mu\phi) + m^2\phi^\dagger\phi - \frac{\lambda}{4}\left(\phi^\dagger\phi\right)^2 \\ &= -\frac{1}{4}F_{\mu\nu}F^{\mu\nu} + \partial_\mu\phi^\dagger\partial^\mu\phi - ieA^\mu\phi^\dagger\overleftrightarrow{\partial_\mu}\phi \\ &\quad + e^2A_\mu A^\mu\phi^\dagger\phi + m^2\phi^\dagger\phi - \frac{\lambda}{4}\left(\phi^\dagger\phi\right)^2 \\ &= -\frac{1}{4}F_{\mu\nu}F^{\mu\nu} + \frac{1}{2}\partial_\mu\sigma\partial^\mu\sigma + \frac{1}{2}\partial_\mu\chi\partial^\mu\chi + \frac{m^2}{2}\left(\sigma^2+\chi^2\right) \\ &\quad - eA^\mu\chi\overleftrightarrow{\partial_\mu}\sigma + \frac{e^2}{2}A_\mu A^\mu\left(\sigma^2+\chi^2\right) - \frac{\lambda}{16}\left(\sigma^2+\chi^2\right)^2. \end{aligned} \tag{14.30}$$

Let us now rewrite the theory around the true vacuum by shifting the field variables as (see (7.77))

14.2 HIGGS PHENOMENON

$$\sigma \to \sigma + \langle \sigma \rangle = \sigma + \frac{2m}{\sqrt{\lambda}} = \sigma + v, \qquad \chi \to \chi + \langle \chi \rangle = \chi, \quad (14.31)$$

where we have identified the vacuum expectation value (vev) of the field with

$$\langle \sigma \rangle = v = \frac{2m}{\sqrt{\lambda}}. \quad (14.32)$$

In this case, the Lagrangian density (14.30) becomes

$$\begin{aligned}
\mathcal{L} &= -\frac{1}{4} \left(\partial_\mu A_\nu - \partial_\nu A_\mu \right) \left(\partial^\mu A^\nu - \partial^\nu A^\mu \right) - e A^\mu \chi \overleftrightarrow{\partial_\mu} (\sigma + v) \\
&\quad + \frac{e^2}{2} A_\mu A^\mu ((\sigma+v)^2 + \chi^2) + \frac{1}{2} \partial_\mu \sigma \partial^\mu \sigma + \frac{1}{2} \partial_\mu \chi \partial^\mu \chi \\
&\quad + \frac{m^2}{2} ((\sigma+v)^2 + \chi^2) - \frac{\lambda}{16} ((\sigma+v)^2 + \chi^2)^2 \\
&= -\frac{1}{4} \left(\partial_\mu A_\nu - \partial_\nu A_\mu \right) \left(\partial^\mu A^\nu - \partial^\nu A^\mu \right) + \frac{1}{2} \partial_\mu \sigma \partial^\mu \sigma \\
&\quad + \frac{1}{2} \partial_\mu \chi \partial^\mu \chi - m^2 \sigma^2 + \frac{m^2 v^2}{4} + \frac{e^2 v^2}{2} A_\mu A^\mu + e v A^\mu \partial_\mu \chi \\
&\quad - e A^\mu \chi \overleftrightarrow{\partial_\mu} \sigma + \frac{e^2}{2} A_\mu A^\mu \left(\sigma^2 + \chi^2 + 2v\sigma \right) \\
&\quad - \frac{\lambda}{16} (\sigma^4 + 2\sigma^2 \chi^2 + \chi^4 + 4v\sigma^3 + 4v\sigma \chi^2), \quad (14.33)
\end{aligned}$$

where we have used the form of v in (14.32) to simplify some of the terms. Let us look at only the terms in (14.33) which are quadratic in the field variables

$$\begin{aligned}
\mathcal{L}_Q &= -\frac{1}{4} \left(\partial_\mu A_\nu - \partial_\nu A_\mu \right) \left(\partial^\mu A^\nu - \partial^\nu A^\mu \right) + \frac{e^2 v^2}{2} A_\mu A^\mu \\
&\quad + \frac{1}{2} \partial_\mu \chi \partial^\mu \chi + e v A^\mu \partial_\mu \chi + \frac{1}{2} \partial_\mu \sigma \partial^\mu \sigma - m^2 \sigma^2. \quad (14.34)
\end{aligned}$$

We note here that the χ field appears to be massless in this theory as in (7.79), but it now mixes with the gauge field A_μ so that we have to diagonalize the theory in order to determine the true spectrum of the theory. (The Lagrangian density in (14.34), ignoring the σ field should be compared with the Stückelberg Lagrangian density (14.8).) To achieve this, let us define

$$B_\mu = A_\mu + \frac{1}{ev}\partial_\mu \chi, \tag{14.35}$$

then we have

$$\begin{aligned}
\partial_\mu B_\nu - \partial_\nu B_\mu &= \partial_\mu\left(A_\nu + \frac{1}{ev}\partial_\nu \chi\right) - \partial_\nu\left(A_\mu + \frac{1}{ev}\partial_\mu \chi\right) \\
&= \partial_\mu A_\nu - \partial_\nu A_\mu, \\
B_\mu B^\mu &= \left(A_\mu + \frac{1}{ev}\partial_\mu \chi\right)\left(A^\mu + \frac{1}{ev}\partial^\mu \chi\right) \\
&= A_\mu A^\mu + \frac{2}{ev}A^\mu \partial_\mu \chi + \frac{1}{e^2 v^2}\partial_\mu \chi \partial^\mu \chi. \quad (14.36)
\end{aligned}$$

Thus in terms of this new field variable (14.35) the quadratic Lagrangian density (14.34) takes the form

$$\begin{aligned}
\mathcal{L}_Q &= -\frac{1}{4}(\partial_\mu B_\nu - \partial_\nu B_\mu)(\partial^\mu B^\nu - \partial^\nu B^\mu) + \frac{e^2 v^2}{2}B_\mu B^\mu \\
&\quad + \frac{1}{2}\partial_\mu \sigma \partial^\mu \sigma - m^2 \sigma^2. \tag{14.37}
\end{aligned}$$

We see from (14.37) that the quadratic part of the shifted Lagrangian density is completely diagonalized. As in (7.79), it describes a massive scalar field σ with mass

$$m_\sigma^2 = 2m^2, \tag{14.38}$$

and a massive gauge field B_μ with mass (this is the analog of the unitary gauge in (14.11))

$$m_B^2 = M^2 = e^2 v^2 = \frac{4e^2 m^2}{\lambda}. \tag{14.39}$$

14.2 HIGGS PHENOMENON 593

The other real component χ of the spin 0 boson - the Nambu-Goldstone boson - has completely disappeared from the spectrum of the theory. This is consistent with the unitarity of the theory (from the point of view of field degrees of freedom). In fact, what has happened is that the original gauge field which was massless had only two physical degrees of freedom just like the photon. (We can think of the physical degrees of freedom as the two transverse components.) However, the Goldstone mode of the complex scalar field, namely, the massless mode has combined with the massless gauge field to give it an additional degree of freedom. Hence the gauge field has become massive with three degrees of freedom without violating unitarity. This phenomenon of the gauge field acquiring a mass by absorbing a Goldstone particle is known as the Higgs phenomenon and was investigated independently by Higgs; Brout and Englert; Guralnik, Hagen and Kibble. The scalar field responsible for this phenomenon is conventionally known as the Higgs field.

At this point we may ask why the Goldstone theorem fails in this case, namely, why is there no massless particle even though we have a spontaneous breakdown of symmetry. The answer to this question lies in the fact that if a gauge theory is written in a manifestly Lorentz invariant way, as we have seen (recall Gupta-Bleuler quantization in **9.8**) the metric of the Hilbert space becomes indefinite. That is there can arise states in the Hilbert space with negative norm. Intuitively, we can see this happening in the following way. The field A_μ is a four vector and hence must have four independent polarization states. The physical states, however, consist only of transverse polarization. Thus the other two states must have cancelling effect with each other as we have seen earlier. Thus some states must have negative norm in such a description.

A crucial assumption in the Goldstone theorem is that the theory should be manifestly Lorentz invariant as well as the metric of the Hilbert space should be positive semi-definite. When we are dealing with a gauge theory, it is the incompatibility of these two conditions that avoids the Goldstone theorem. This is in a way a double benefit in the sense that we do not have to worry about the absence of massless bosons in nature and furthermore, we have a mechanism for giving masses to gauge bosons without violating renormalizability and unitarity of the theory. It is, of course, this second aspect that is important in developing a gauge theory of the weak interactions.

Although the unitary gauge is useful for the counting of true dynamical degrees of freedom in the theory, as pointed out in the last section, it is not the right gauge to study renormalizability of the theory. To introduce the gauge choice, commonly known as the 't Hooft gauge or the R_ξ gauge, that is more useful for this purpose let us proceed as follows. First we note that the unshifted theory (14.30) is invariant under the infinitesimal gauge transformations (see (7.84))

$$\begin{aligned}
\delta\sigma(x) &= \epsilon(x)\chi(x), \\
\delta\chi(x) &= -\epsilon(x)\sigma(x), \\
\delta A_\mu(x) &= \frac{1}{e}\partial_\mu\epsilon(x).
\end{aligned} \quad (14.40)$$

The shifted theory (14.33) which reflects a spontaneous breakdown of the local symmetry (14.40) nonetheless is invariant the shifted local transformations (recall that it is the vacuum of the theory that breaks the symmetry)

$$\begin{aligned}
\delta\sigma(x) &= \epsilon(x)\chi(x), \\
\delta\chi(x) &= -\epsilon(x)(v+\sigma(x)) = -\epsilon(x)\left(\frac{M}{e}+\sigma(x)\right), \\
\delta A_\mu(x) &= \frac{1}{e}\partial_\mu\epsilon(x),
\end{aligned} \quad (14.41)$$

where M is defined in (14.39).

The presence of the local symmetry (14.41) in the theory allows us to choose a gauge fixing Lagrangian density which we would prefer to be manifestly Lorentz invariant and to simultaneously diagonalize the quadratic Lagrangian density (14.34) (namely, eliminate the mixing term between the A_μ and the χ fields). This is achieved by the 't Hooft gauge fixing which is described by the gauge fixing Lagrangian density

$$\begin{aligned}
\mathcal{L}_{\text{GF}} &= -\frac{1}{2\xi}(\partial^\mu A_\mu - \xi ev\chi)^2 = -\frac{1}{2\xi}(\partial^\mu A_\mu - \xi M\chi)^2, \\
&= -\frac{1}{2\xi}(\partial^\mu A_\mu)^2 + M(\partial^\mu A_\mu)\chi - \frac{\xi M^2}{2}\chi^2,
\end{aligned} \quad (14.42)$$

14.2 Higgs phenomenon

which can be compared with (14.15). It is clear that the mixing terms in (14.34) together with the cross terms in (14.42) combine into a total divergence and, thereby, disappear from the action. As a result, the propagators for the A_μ and the χ fields in this gauge are given respectively by

$$iG_{\mu\nu}(p) = -\frac{i}{p^2 - M^2}\left(\eta_{\mu\nu} - (1 - \xi)\frac{p_\mu p_\nu}{p^2 - \xi M^2}\right),$$

$$iG_\chi(p) = \frac{i}{p^2 - \xi M^2}, \qquad (14.43)$$

and we note that in this gauge the propagators fall off for large values of the momenta unlike in the unitary gauge.

The gauge fixing Lagrangian density (14.42) can also be written with an auxiliary field as (see (12.170))

$$\mathcal{L}_{\text{GF}} = \frac{\xi}{2} F^2 - F(\partial \cdot A - \xi M \chi), \qquad (14.44)$$

and the choice of this gauge fixing leads to a ghost Lagrangian density of the form (scaling out a factor of $\frac{1}{e}$)

$$\mathcal{L}_{\text{ghost}} = \partial^\mu \bar{c} \partial_\mu c - \xi M^2 \bar{c} c - \xi M e \sigma \bar{c} c. \qquad (14.45)$$

We see that unlike in the conventional covariant gauge fixing (12.164), the 't Hooft gauge choice in (14.42) (or (14.44)) leads to an interacting ghost Lagrangian density.

The complete Lagrangian density in this gauge has the form

$$\mathcal{L}_{\text{TOT}} = \mathcal{L} + \mathcal{L}_{\text{GF}} + \mathcal{L}_{\text{ghost}}, \qquad (14.46)$$

and can be easily checked to be invariant under the BRST transformations

$$\begin{aligned}
\delta A_\mu &= \frac{1}{e}\omega\partial_\mu c, \\
\delta\sigma &= \omega\chi c, \\
\delta\chi &= -\omega\left(\frac{M}{e}+\sigma\right)c, \\
\delta c &= 0, \\
\delta\bar{c} &= -\omega F, \\
\delta F &= 0.
\end{aligned} \qquad (14.47)$$

Following the discussions in section **13.4** (see, for example, (13.91)) we can now derive the master equation for the 1PI generating functional following from this invariance which takes the form

$$\frac{\delta\Gamma}{\delta A_\mu(x)}\partial_\mu c(x) + \frac{\delta\Gamma}{\delta\sigma(x)}\frac{\delta\Gamma}{\delta P_\sigma(x)} + \frac{\delta\Gamma}{\delta\chi(x)}\frac{\delta\Gamma}{\delta P_\chi(x)} - F(x)\frac{\delta\Gamma}{\delta\bar{c}(x)} = 0, \qquad (14.48)$$

where P_σ and P_χ denote respectively the sources for the composite BRST variations of the fields σ and χ in (14.47). This is the starting point for establishing the renormalizability of the gauge theory with a spontaneously broken symmetry (Higgs theory).

14.3 The standard model

The standard model or the gauge theory of electroweak interactions was built on the results of years of earlier work on weak interactions. Therefore, let us review some of these essential earlier results before constructing the standard model (known as the Weinberg-Salam-Glashow theory).

At low energies, weak interactions (for example, the β decay of the neutron etc.) are described quite well by a specific form of the Fermi theory which involves four fermion interactions in the form of a current-current interaction Hamiltonian density

$$\mathcal{H}_I = \frac{G_F}{\sqrt{2}} J_\mu^\dagger J^\mu, \qquad (14.49)$$

where the current is a sum of the hadronic and leptonic currents of the form

$$\begin{aligned} J_\mu &= J_\mu^{(\text{had})} + J_\mu^{(\text{lept})} \\ &= \overline{\psi}_p \gamma_\mu (1-\gamma_5)\psi_n + \overline{\psi}_\nu \gamma_\mu (1-\gamma_5)\psi_e + \cdots, \end{aligned} \quad (14.50)$$

with $\psi_n, \psi_p, \psi_e, \psi_\nu$ denoting the fermion fields associated with neutron, proton, electron and neutrino respectively. The coupling constant (interaction strength) G_F is known as the Fermi coupling and is determined from low energy experiments to have the value

$$G_\text{F} = 1.17 \times 10^{-5} (\text{GeV})^{-2}. \quad (14.51)$$

The weak interactions are, therefore, quite weak and were also known to be extremely short ranged.

It was known that all the weakly interacting particles (leptons and hadrons) can be described by multiplets belonging to the weak isospin group of $SU(2)$. As a result, the currents in (14.50) have a weak isospin structure and are also known as the weak isospin currents. Furthermore, it is clear from the $V - A$ (vector minus axial vector) structure of the currents in (14.50) that only the left-handed components of the fermion fields participate in the weak interactions because of which the weak isospin group can be identified with the group $SU_\text{L}(2)$ (meaning that the symmetry transformation changes only the left-handed components of the fields (see (3.139) and (3.140) for a definition of right and left-handed particles/fields). (The $V - A$ structure is connected with the fact that parity is violated maximally in weak interactions.) In addition to the weak isospin quantum number, it was also known that we can assign an additive ($U(1)$) quantum number known as the weak hypercharge to all the weakly interacting particles such that the charge of any particle can be written as

$$Q = I_{3\text{L}} + \frac{Y}{2}, \quad (14.52)$$

where Q denotes the electric charge of the particle, $I_{3\text{L}}$ the quantum number associated with the 3-component of the weak isospin generator and Y its weak hypercharge quantum number. Experimentally it was known that weak interactions violate both weak isospin and weak hypercharge quantum numbers, but in such a way that the electric charge is conserved in any weak process.

Although the $V - A$ theory with four fermion interactions works quite well at low energy, it is not renormalizable. (The nonrenormalizable character is already manifest in the fact that the coupling constant of the theory (14.51) is dimensional with inverse dimensions of mass (energy).) Therefore, the $V - A$ theory cannot be the fundamental theory underlying weak interactions (forces). On the other hand, we have already seen that gauge theories can describe physical forces and do not have any problem of renormalizability (we will discuss renormalizability and renormalizability of gauge theories in a later chapter). As a result we may try to formulate the weak interactions as a gauge theory such that it reduces to the $V - A$ theory in the low energy limit. Furthermore, since weak interactions are short ranged, a gauge theoretic description would naturally involve massive gauge bosons. We have seen in the last section that gauge bosons can become massive through the Higgs mechanism which would involve the spontaneous breakdown of some local symmetry. As we have discussed weak interactions violate both the weak isospin and the weak hypercharge symmetries and, therefore, it is logical to associate the Higgs mechanism to the spontaneous breakdown of these symmetries. As a result, it is clear that we should look for a gauge theory of weak interactions based on the local symmetry group

$$SU_\mathrm{L}(2) \times U_\mathrm{Y}(1), \tag{14.53}$$

where $SU_\mathrm{L}(2)$ and $U_\mathrm{Y}(1)$ denote respectively the weak isospin and the weak hypercharge symmetry groups and we would like this symmetry group to spontaneously breakdown (at the weak interaction scale) in a way such that only the electromagnetic symmetry survives as the true symmetry of the low energy theory. Namely, we would like the gauge theory to describe the symmetry behavior ("SSB" denotes spontaneous symmetry breakdown)

$$SU_\mathrm{L}(2) \times U_\mathrm{Y}(1) \xrightarrow{\text{SSB}} U_\mathrm{EM}(1). \tag{14.54}$$

With these basics, let us construct the standard model in several steps.

14.3 THE STANDARD MODEL

14.3.1 Field content. Let us first discuss the field content of the standard model. The standard model is expected to describe the weak and electromagnetic interactions of all fundamental particles that participate in the weak processes. These include both hadrons and leptons as we have already mentioned. However, although the leptons are believed to be fundamental particles, we now know that the constituents of hadronic matter are the quarks. Thus, unlike the $V - A$ theory (see (14.50) where the hadronic current was given in terms of neutron and proton fields) the matter fields in the standard model consist of quarks and leptons and they come in three families. However, for simplicity we will consider the matter to consist of only one family of leptons, namely, the electron and its neutrino. The quarks as well as the other families can be introduced in a straightforward manner. We note that the electron is massive and, therefore, its field can be decomposed into a left-handed part and a right-handed part. In contrast, the neutrino is experimentally known to be essentially left-handed (see section **3.6**). The matter fields associated with the electron family can be grouped as multiplets of $SU_{\rm L}(2)$ as

$$L = \begin{pmatrix} \nu_e \\ e \end{pmatrix}_{\rm L}, \qquad e_{\rm R} = R, \tag{14.55}$$

where the left- and the right-handed components of a spinor field are defined by (see also section **3.7**)

$$\psi = \psi_{\rm L} + \psi_{\rm R}, \quad \psi_{\rm L} = \frac{1}{2}(\mathbb{1} - \gamma_5)\psi, \quad \psi_{\rm R} = \frac{1}{2}(\mathbb{1} + \gamma_5)\psi. \tag{14.56}$$

The left-handed multiplets of the standard model are doublets of $SU_{\rm L}(2)$ while the right-handed particles correspond to singlets. Each multiplet of $SU_{\rm L}(2)$ has a unique value of the weak hypercharge quantum number (corresponding to $U_{\rm Y}(1)$) and for the leptonic matter fields, the left-handed doublets have $Y = -1$ while for the right-handed singlets carry a hypercharge of $Y = -2$. The other lepton families (μ and the τ families) can be introduced through an obvious generalization. The quark families can also be described by such structures. The only difference is that since all the quark fields are massive, there will be right-handed singlet components for every quark field (as opposed to the absence of $\nu_{\rm R}$ in the leptonic sector).

(The hypercharge quantum numbers for the quarks are also different from those for leptons in order to be compatible with (14.52).)

The gauge fields which are expected to correspond to the carriers of forces are determined from the local symmetry structure of the theory. If we assume (14.53) to describe the local (gauge) symmetry of the theory, then the corresponding gauge fields are determined to be

$$SU_{\rm L}(2) \; : \; 3 \text{ gauge fields,} \quad W_\mu^a, \quad a = 1, 2, 3,$$
$$U_{\rm Y}(1) \; : \; 1 \text{ gauge field,} \quad Y_\mu, \tag{14.57}$$

so that the theory should contain four gauge fields in total.

Furthermore, if we would like the gauge bosons to be massive, the local symmetry of the electroweak group must be spontaneously broken through the Higgs mechanism and this requires that the theory should also contain scalar Higgs fields which would be described by multiplets of $SU_{\rm L}(2)$. As we will see shortly, the minimal multiplet of Higgs fields that can achieve the spontaneous breakdown of the symmetry (as in (14.54)) is given by a doublet of charged fields,

$$\text{minimal Higgs field} : \phi = \begin{pmatrix} \phi^+ \\ \phi^0 \end{pmatrix}, \quad (\phi^+)^\dagger = \phi^-, \quad (\phi^0)^\dagger = \overline{\phi}^0. \tag{14.58}$$

Since the Higgs multiplet is charged, it carries a nontrivial weak hypercharge quantum number given by $Y = 1$ which is, in fact, the opposite of that for the left-handed lepton doublet.

Let us now write down the quantum numbers for the matter and the Higgs field and verify explicitly that (14.52) holds,

14.3 THE STANDARD MODEL

Fields	I_{3L}	$\frac{Y}{2}$	$I_{3L} + \frac{Y}{2}$	Q
$\begin{pmatrix} \nu_e \\ e \end{pmatrix}_L$	$\frac{1}{2}$ $-\frac{1}{2}$	$-\frac{1}{2}$ $-\frac{1}{2}$	$\frac{1}{2} - \frac{1}{2} = 0$ $-\frac{1}{2} - \frac{1}{2} = -1$	0 -1
e_R	0	-1	$0 - 1 = -1$	-1
$\begin{pmatrix} \phi^+ \\ \phi^0 \end{pmatrix}$	$\frac{1}{2}$ $-\frac{1}{2}$	$\frac{1}{2}$ $\frac{1}{2}$	$\frac{1}{2} + \frac{1}{2} = 1$ $-\frac{1}{2} + \frac{1}{2} = 0$	1 0

(14.59)

This shows that all the quantum numbers assigned to various fields are consistent with the relation (14.52).

14.3.2 Lagrangian density. It is now a straightforward matter to write down the complete Lagrangian density for the theory which has the local gauge invariance $SU_L(2) \times U_Y(1)$ and we will do this in several steps. First, let us note that the gauge invariant Lagrangian density for the gauge fields follows from our earlier discussions (see (9.16) and (12.45)) to correspond to

$$\mathcal{L}_G = -\frac{1}{4} F^a_{\mu\nu} F^{\mu\nu\, a} - \frac{1}{4} F_{\mu\nu} F^{\mu\nu}, \qquad a = 1, 2, 3, \tag{14.60}$$

where the field strength tensors for the Abelian as well as the non-Abelian gauge fields are defined as (see (9.4) and (12.44))

$$\begin{aligned} F_{\mu\nu} &= \partial_\mu Y_\nu - \partial_\nu Y_\mu, \\ F^a_{\mu\nu} &= \partial_\mu W^a_\nu - \partial_\nu W^a_\mu - g\epsilon^{abc} W^b_\mu W^c_\nu. \end{aligned} \tag{14.61}$$

We are assuming that g and g' are the coupling constants for the gauge interactions associated with the groups $SU_L(2)$ and $U_Y(1)$ respectively. (The Abelian gauge fields are not self-interacting which is why g' does not appear in the Lagrangian density for the gauge fields, but it will be present in the matter Lagrangian density. Note that ϵ^{abc} denotes the structure constant of $SU_L(2)$ as we know from the study of angular momentum.) The Lagrangian density (14.60) is clearly invariant under the infinitesimal gauge transformations of $SU_L(2)$ and $U_Y(1)$ defined by

$$\delta W_\mu^a(x) = \frac{1}{g}(D_\mu \epsilon(x))^a = \frac{1}{g}\bigl(\partial_\mu \epsilon^a(x) - g\epsilon^{abc}W_\mu^b(x)\epsilon^c(x)\bigr),$$

$$\delta Y_\mu(x) = \frac{1}{g'}\partial_\mu \epsilon(x), \qquad (14.62)$$

where $\epsilon^a(x), a = 1,2,3$ and $\epsilon(x)$ denote the local infinitesimal parameters of symmetry transformations associated with the groups $SU_L(2)$ and $U_Y(1)$ respectively. Since the two symmetry groups are commuting (direct products), the gauge field W_μ^a for the group $SU_L(2)$ does not transform under the group $U_Y(1)$ and *vice versa*. An alternative way to see this is to note that the fields W_μ^a do not carry weak hypercharge quantum number just as Y_μ does not carry weak isospin quantum number and, therefore, they are inert under the corresponding transformations.

The gauge invariant Lagrangian density for the matter fields (leptons) can be obtained through minimal coupling, we simply have to remember that there are two local symmetries present in the theory. Since the left-handed fermions belong to an isospin doublet while the right-handed fermions are isospin singlets, the Lagrangian density for the leptons takes the form

$$\mathcal{L}_f = i\overline{L}\gamma^\mu\left(\partial_\mu + ig\frac{\sigma}{2}\cdot \mathbf{W}_\mu + \frac{ig'}{2}Y_\mu\right)L + i\overline{R}\gamma^\mu\left(\partial_\mu + ig'Y_\mu\right)R.$$

$$(14.63)$$

Here we have used the fact that the generators of angular momentum $(SU(2))$ are related to the Pauli matrices as $\frac{1}{2}\sigma$ in the fundamental representation to which the left-handed doublets belong. (We note here that within the context of isospin, sometimes it is conventional to denote the generators as $\frac{\tau}{2}$ by making the identification $\sigma = \tau$. We have also used a vector notation here to represent the three components of the isospin vectors which is often used interchangeably.) The other thing to note from the structure of (14.63) is that the right-handed fermions couple twice as strongly to the Y_μ

field as the left-handed ones simply reflecting the Y quantum numbers for the two fields. Note also that the $SU_\mathrm{L}(2)$ invariance forbids a mass term for the lepton fields (the electron and the neutrino are, therefore, massless in this theory to begin with). The minimally coupled fermion Lagrangian density (14.63) can be easily checked to be invariant under the infinitesimal local transformations

$$\begin{aligned}\delta L(x) &= -i\frac{\sigma}{2}\cdot\epsilon(x)L(x) - \frac{i}{2}\epsilon(x)L(x),\\ \delta R(x) &= -i\epsilon(x)R(x),\end{aligned} \qquad (14.64)$$

together with the transformations in (14.62). We note that we have combined the three components of $\epsilon^a(x), a = 1, 2, 3$ into a vector $\epsilon(x)$ in (14.64). The transformations (14.64) show explicitly that the right-handed fermions do not change under the $SU_\mathrm{L}(2)$ transformations.

Besides the gauge fields and the leptons, we also need a Lagrangian density for the scalar Higgs fields which are expected to lead to a spontaneous breakdown of the local symmetries. The minimally coupled Lagrangian density consistent with the assignment of the $SU_\mathrm{L}(2)$ and $U_\mathrm{Y}(1)$ quantum numbers takes the form

$$\mathcal{L}_\mathrm{H} = \left(\left(\partial_\mu + ig\frac{\sigma}{2}\cdot\mathbf{W}_\mu - \frac{ig'}{2}Y_\mu\right)\phi\right)^\dagger\left(\partial^\mu + ig\frac{\sigma}{2}\cdot\mathbf{W}^\mu - \frac{ig'}{2}Y^\mu\right)\phi. \qquad (14.65)$$

We note that since the hypercharge of the Higgs doublet is just the opposite of that of the left-handed lepton doublet, the coupling to the $U_\mathrm{Y}(1)$ gauge field is correspondingly with the opposite sign. The Lagrangian density (14.65) can be checked to be invariant under the infinitesimal local transformations

$$\delta\phi(x) = -i\frac{\sigma}{2}\cdot\epsilon(x)\phi(x) + \frac{i}{2}\epsilon(x)\phi(x), \qquad (14.66)$$

together with the gauge field transformations in (14.62).

A mass term as well as a potential for the Higgs field can be introduced consistent with the $SU_\mathrm{L}(2)$ symmetry. In fact, since we would like the potential for the Higgs field to lead to the spontaneous breakdown of the local symmetries we choose it to have the form (see (14.24), it is worth noting that in spite of the similarity in their forms for the potential, here the Higgs field is represented by an isospin doublet)

$$V(\phi, \phi^\dagger) = -m^2 \phi^\dagger \phi + \frac{\lambda}{4}(\phi^\dagger \phi)^2, \quad \lambda > 0, \tag{14.67}$$

which is invariant under the local transformations in (14.66). In addition, of course, the fermions can have Yukawa interactions with the scalar fields which can be described by the Lagrangian density

$$\mathcal{L}_Y = -h(\overline{R}\phi^\dagger L + \overline{L}\phi R), \tag{14.68}$$

where h denotes the Yukawa coupling and the Lagrangian density (14.68) can be checked to be invariant under the transformations (14.64) and (14.66). As we will see shortly, the Yukawa interaction leads to masses for fermions after the spontaneous breakdown of the symmetry.

Therefore, collecting all the terms in (14.60), (14.63), (14.65), (14.67) and (14.68), the complete weak interaction Lagrangian density describing just one family of leptons (electron family) can be written as

$$\mathcal{L} = \mathcal{L}_G + \mathcal{L}_f + \mathcal{L}_H + \mathcal{L}_Y - V(\phi, \phi^\dagger), \tag{14.69}$$

which is invariant under the infinitesimal local gauge transformations

$$\begin{aligned} \delta L(x) &= -i\frac{\sigma}{2} \cdot \epsilon(x) L(x) - \frac{i}{2} \epsilon(x) L(x), \\ \delta R(x) &= -i\epsilon(x) R(x), \\ \delta \phi(x) &= -i\frac{\sigma}{2} \cdot \epsilon(x) \phi(x) + \frac{i}{2} \epsilon(x) \phi(x), \\ \delta W_\mu^a(x) &= \frac{1}{g}(D_\mu \epsilon(x))^a = \frac{1}{g}\left(\partial_\mu \epsilon^a(x) - g\epsilon^{abc} W_\mu^b(x) \epsilon^c(x)\right), \\ \delta Y_\mu(x) &= \frac{1}{g'} \partial_\mu \epsilon(x). \end{aligned} \tag{14.70}$$

14.3 THE STANDARD MODEL

14.3.3 Spontaneous symmetry breaking. We know that both $SU_L(2)$ and $U_Y(1)$ symmetries are violated in weak interactions (the associated quantum numbers are not conserved in weak processes). Furthermore, the weak interactions are short ranged so that we would like the gauge bosons to be massive. Both these issues can be addressed simultaneously if the local symmetries are spontaneously broken by the Higgs phenomenon and this is what we discuss next. We have chosen the Higgs potential (14.67) in the form

$$V(\phi, \phi^\dagger) = -m^2 \phi^\dagger \phi + \frac{\lambda}{4}(\phi^\dagger \phi)^2, \qquad \lambda > 0, \tag{14.71}$$

so that the minimum of the potential occurs at

$$\begin{aligned}\frac{\partial V}{\partial \phi} &= \phi^\dagger \left(-m^2 + \frac{\lambda}{2}(\phi^\dagger \phi) \right) = 0, \\ \frac{\partial V}{\partial \phi^\dagger} &= \left(-m^2 + \frac{\lambda}{2}(\phi^\dagger \phi) \right) \phi = 0.\end{aligned} \tag{14.72}$$

As we have seen in the last section, there are two solutions to (14.72), namely,

$$\begin{aligned}\langle \phi \rangle &= \langle \phi^\dagger \rangle = 0, \\ \langle \phi^\dagger \phi \rangle &= \frac{2m^2}{\lambda}.\end{aligned} \tag{14.73}$$

Furthermore, we have noted earlier in section **7.5** (and also in section **14.2**) that the first solution in (14.73) represents a local maximum while the second solution corresponds to the true minimum of the potential. Let us choose the solution for the minimum to be of the form (recall that there is an infinity of possible minima lying on a circle, furthermore, the factor of $\frac{1}{\sqrt{2}}$ below is a consequence of the definition in (14.29))

$$\langle \phi \rangle = \begin{pmatrix} 0 \\ \frac{v}{\sqrt{2}} \end{pmatrix}, \qquad v = \frac{2m}{\sqrt{\lambda}}. \tag{14.74}$$

Therefore, as we have discussed earlier, to determine the spectrum of particles we need to expand the theory around the true ground state (vacuum) by shifting the scalar field as

$$\phi \to \phi + \langle \phi \rangle = \phi + \begin{pmatrix} 0 \\ \frac{v}{\sqrt{2}} \end{pmatrix}. \tag{14.75}$$

However, before shifting the fields let us rewrite the Lagrangian densities for the Higgs and the Yukawa sector of the theory for simplicity (see (14.65), (14.67) and (14.68)). (Basically, this is the sector of the theory that involves the scalar field and will be affected by the shift, the other parts will not change under the shift.) We note that we can write explicitly in the matrix form

$$\begin{aligned}\frac{\sigma}{2} \cdot \mathbf{W}_\mu \phi &= \frac{1}{2} \begin{pmatrix} W_\mu^3 & W_\mu^1 - i W_\mu^2 \\ W_\mu^1 + i W_\mu^2 & -W_\mu^3 \end{pmatrix} \\ &= \frac{1}{2} \begin{pmatrix} W_\mu^3 & \sqrt{2} W_\mu^+ \\ \sqrt{2} W_\mu^- & -W_\mu^3 \end{pmatrix}, \end{aligned} \tag{14.76}$$

where we have defined

$$W_\mu^\pm = \frac{1}{\sqrt{2}} (W_\mu^1 \mp i W_\mu^2). \tag{14.77}$$

As a result, it follows that

$$\begin{aligned}\frac{\sigma}{2} \cdot \mathbf{W}_\mu &= \frac{1}{2} \begin{pmatrix} W_\mu^3 & \sqrt{2} W_\mu^+ \\ \sqrt{2} W_\mu^- & -W_\mu^3 \end{pmatrix} \begin{pmatrix} \phi^+ \\ \phi^0 \end{pmatrix} \\ &= \begin{pmatrix} \frac{1}{2} W_\mu^3 \phi^+ + \frac{1}{\sqrt{2}} W_\mu^+ \phi^0 \\ \frac{1}{\sqrt{2}} W_\mu^- \phi^+ - \frac{1}{2} W_\mu^3 \phi^0 \end{pmatrix}. \end{aligned} \tag{14.78}$$

Using these as well as the explicit doublet representation of the Higgs field in (14.58) we obtain

14.3 THE STANDARD MODEL

$$\mathcal{L}_H + \mathcal{L}_Y - V(\phi, \phi^\dagger)$$

$$= \left(\left(\partial_\mu + ig\frac{\sigma}{2}\cdot\mathbf{W}_\mu - \frac{ig'}{2}Y_\mu\right)\phi\right)^\dagger\left(\partial^\mu + ig\frac{\sigma}{2}\cdot\mathbf{W}^\mu - \frac{ig'}{2}Y^\mu\right)\phi$$

$$-h\left(\overline{R}\phi^\dagger L + \overline{L}\phi R\right) + m^2\phi^\dagger\phi - \frac{\lambda}{4}(\phi^\dagger\phi)^2$$

$$= \left(\left(\partial_\mu + \frac{ig'}{2}Y_\mu - \frac{ig}{2}W_\mu^3\right)\phi^- - \frac{ig}{\sqrt{2}}W_\mu^-\overline{\phi}^0\right)$$

$$\times\left(\left(\partial^\mu - \frac{ig'}{2}Y^\mu + \frac{ig}{2}W^{\mu 3}\right)\phi^+ + \frac{ig}{\sqrt{2}}W^{\mu+}\phi^0\right)$$

$$+\left(\left(\partial_\mu + \frac{ig'}{2}Y_\mu + \frac{ig}{2}W_\mu^3\right)\overline{\phi}^0 - \frac{ig}{\sqrt{2}}W_\mu^+\phi^-\right)$$

$$\times\left(\left(\partial^\mu - \frac{ig'}{2}Y^\mu - \frac{ig}{2}W^{\mu 3}\right)\phi^0 + \frac{ig}{\sqrt{2}}W^{\mu-}\phi^+\right)$$

$$-h\left(\overline{e}_R(\phi^-\nu_{eL} + \overline{\phi}^0 e_L) + (\overline{\nu}_{eL}\phi^+ + \overline{e}_L\phi^0)e_R\right)$$

$$+m^2(\phi^-\phi^+ + \overline{\phi}^0\phi^0) - \frac{\lambda}{4}(\phi^-\phi^+ + \overline{\phi}^0\phi^0)^2. \qquad (14.79)$$

The expansion around (14.74) can be done equivalently by letting

$$\begin{aligned}\phi^0 &\to \phi^0 + \frac{v}{\sqrt{2}}, \\ \overline{\phi}^0 &\to \overline{\phi}^0 + \frac{v}{\sqrt{2}},\end{aligned} \qquad (14.80)$$

under which the Lagrangian density (14.79) in the Higgs and the Yukawa sector becomes

$$\mathcal{L}_H + \mathcal{L}_Y - V(\phi, \phi^\dagger)$$

$$= \left(\left(\partial_\mu + \frac{ig'}{2}Y_\mu - \frac{ig}{2}W_\mu^3\right)\phi^- - \frac{ig}{\sqrt{2}}W_\mu^-\left(\overline{\phi}^0 + \frac{v}{\sqrt{2}}\right)\right)$$

$$\times \left(\left(\partial^\mu - \frac{ig'}{2} Y^\mu + \frac{ig}{2} W^{\mu 3} \right) \phi^+ + \frac{ig}{\sqrt{2}} W^{\mu +} \left(\phi^0 + \frac{v}{\sqrt{2}} \right) \right)$$

$$+ \left(\left(\partial_\mu + \frac{ig'}{2} Y_\mu + \frac{ig}{2} W_\mu^3 \right) \left(\overline{\phi}^0 + \frac{v}{\sqrt{2}} \right) - \frac{ig}{\sqrt{2}} W_\mu^+ \phi^- \right)$$

$$\times \left(\left(\partial^\mu - \frac{ig'}{2} Y^\mu - \frac{ig}{2} W^{\mu 3} \right) \left(\phi^0 + \frac{v}{\sqrt{2}} \right) + \frac{ig}{\sqrt{2}} W^{\mu -} \phi^+ \right)$$

$$- h \bar{e}_R \nu_{eL} \phi^- - h \bar{e}_R e_L \left(\overline{\phi}^0 + \frac{v}{\sqrt{2}} \right) - h \bar{\nu}_{eL} e_R \phi^+$$

$$- h \bar{e}_L e_R \left(\phi^0 + \frac{v}{\sqrt{2}} \right) + m^2 \left(\phi^- \phi^+ + \left(\overline{\phi}^0 + \frac{v}{\sqrt{2}} \right) \left(\phi^0 + \frac{v}{\sqrt{2}} \right) \right)$$

$$- \frac{\lambda}{4} \left(\phi^- \phi^+ + \left(\overline{\phi}^0 + \frac{v}{\sqrt{2}} \right) \left(\phi^0 + \frac{v}{\sqrt{2}} \right) \right)^2. \tag{14.81}$$

The spectrum of the theory can be obtained from the quadratic part of the Lagrangian density and the quadratic part of (14.81) gives

$$\partial_\mu \phi^- \partial^\mu \phi^+ + \partial_\mu \overline{\phi}^0 \partial^\mu \phi^0 + \frac{igv}{2} W_\mu^+ \partial^\mu \phi^- - \frac{igv}{2} W_\mu^- \partial^\mu \phi^+$$

$$+ \frac{g^2 v^2}{4} W_\mu^+ W^{\mu -} - \frac{iv}{2\sqrt{2}} (g' Y_\mu + g W_\mu^3) \partial^\mu (\overline{\phi}^0 - \phi^0)$$

$$+ \frac{v^2}{8} \left(g' Y_\mu + g W_\mu^3 \right) \left(g' Y^\mu + g W^{\mu 3} \right)$$

$$- \frac{hv}{\sqrt{2}} (\bar{e}_R e_L + \bar{e}_L e_R) - \frac{\lambda v^2}{8} (\phi^0 + \overline{\phi}^0)^2. \tag{14.82}$$

We note that the field combinations $(\phi^0 + \overline{\phi}^0)$, $(g' Y_\mu + g W_\mu^3)$ as well as the fields $W_\mu^\pm$ appear to have become massive. However, since there is mixing between various fields, we cannot truly determine the spectrum of the theory until we diagonalize the quadratic part of the Lagrangian density. Adding the quadratic terms from the gauge and the fermion Lagrangian densities in (14.60) and (14.63) respectively, the complete quadratic Lagrangian density is obtained to be

14.3 THE STANDARD MODEL

$$\begin{aligned}
\mathcal{L}_Q =\ & -\frac{1}{2}\left(\partial_\mu W_\nu^+ - \partial_\nu W_\mu^+\right)\left(\partial^\mu W^{\nu-} - \partial^\nu W^{\mu-}\right) \\
& -\frac{1}{4}\left(\partial_\mu W_\nu^3 - \partial_\nu W_\mu^3\right)\left(\partial^\mu W^{\nu 3} - \partial^\nu W^{\mu 3}\right) \\
& -\frac{1}{4}\left(\partial_\mu Y_\nu - \partial_\nu Y_\mu\right)\left(\partial^\mu Y^\nu - \partial^\nu Y^\mu\right) + i\bar{e}_\mathrm{L}\slashed{\partial} e_\mathrm{L} + i\bar{\nu}_{e\mathrm{L}}\slashed{\partial}\nu_{e\mathrm{L}} \\
& + i\bar{e}_\mathrm{R}\slashed{\partial} e_\mathrm{R} + \partial_\mu\phi^-\partial^\mu\phi^+ + \partial_\mu\bar{\phi}^0\partial^\mu\phi^0 + \frac{g^2 v^2}{4}W_\mu^+ W^{\mu-} \\
& + \frac{igv}{2}\left(W_\mu^+ \partial^\mu\phi^- - W_\mu^- \partial^\mu\phi^+\right) \\
& - \frac{iv}{2\sqrt{2}}\left(g'Y_\mu + gW_\mu^3\right)\partial^\mu\left(\bar{\phi}^0 - \phi^0\right) \\
& + \frac{v^2}{8}\left(g'Y_\mu + gW_\mu^3\right)\left(g'Y^\mu + gW^{\mu 3}\right) \\
& - \frac{hv}{\sqrt{2}}(\bar{e}_\mathrm{R} e_\mathrm{L} + \bar{e}_\mathrm{L} e_\mathrm{R}) - \frac{\lambda v^2}{8}(\phi^0 + \bar{\phi}^0)^2. \quad (14.83)
\end{aligned}$$

It is now a straightforward matter to diagonalize the Lagrangian density (14.83). Let us define

$$\begin{aligned}
Z_\mu &= \frac{g'}{\sqrt{g^2+g'^2}}Y_\mu + \frac{g}{\sqrt{g^2+g'^2}}W_\mu^3 \\
&= \sin\theta_\mathrm{W} Y_\mu + \cos\theta_\mathrm{W} W_\mu^3, \\
A_\mu &= \frac{g}{\sqrt{g^2+g'^2}}Y_\mu - \frac{g'}{\sqrt{g^2+g'^2}}W_\mu^3 \\
&= \cos\theta_\mathrm{W} Y_\mu - \sin\theta_\mathrm{W} W_\mu^3, \quad (14.84)
\end{aligned}$$

where θ_W is known as the Weinberg angle (also called the weak mixing angle). We can invert the relations in (14.84) to write

$$\begin{aligned}
W_\mu^3 &= Z_\mu \cos\theta_\mathrm{W} - A_\mu \sin\theta_\mathrm{W}, \\
Y_\mu &= Z_\mu \sin\theta_\mathrm{W} + A_\mu \cos\theta_\mathrm{W}. \quad (14.85)
\end{aligned}$$

With this, it is easy to check that

$$\begin{aligned}(\partial_\mu W_\nu^3 - \partial_\nu W_\mu^3)&(\partial^\mu W^{\nu 3} - \partial^\nu W^{\mu 3}) \\ + (\partial_\mu Y_\nu - \partial_\nu Y_\nu)&(\partial^\mu Y^\nu - \partial^\nu Y^\mu) \\ = (\partial_\mu Z_\nu - \partial_\nu Z_\mu)&(\partial^\mu Z^\nu - \partial^\nu Z^\mu) \\ + (\partial_\mu A_\nu - \partial_\nu A_\mu)&(\partial^\mu A^\nu - \partial^\nu A^\mu).\end{aligned} \tag{14.86}$$

Similarly, we can define (this is equivalent to defining $\phi^0 = \frac{1}{\sqrt{2}}(\sigma + i\chi)$, $\overline{\phi}^0 = \frac{1}{\sqrt{2}}(\sigma - i\chi)$, see, for example, (14.29))

$$\begin{aligned}\sigma &= \frac{1}{\sqrt{2}}(\overline{\phi}^0 + \phi^0), \\ \chi &= \frac{i}{\sqrt{2}}(\overline{\phi}^0 - \phi^0),\end{aligned} \tag{14.87}$$

so that we can write

$$\partial_\mu \overline{\phi}^0 \partial^\mu \phi^0 = \frac{1}{2}\partial_\mu\sigma\partial^\mu\sigma + \frac{1}{2}\partial_\mu\chi\partial^\mu\chi. \tag{14.88}$$

Furthermore, if we define the parameters

$$\begin{aligned}M_W &= \frac{gv}{2}, \\ M_Z &= \frac{\sqrt{g^2 + g'^2}\, v}{2} = \frac{M_W}{\cos\theta_W}, \\ m_e &= \frac{hv}{\sqrt{2}}, \\ M_H^2 &= \frac{\lambda v^2}{2} = \frac{\lambda}{2}\left(\frac{4m^2}{\lambda}\right) = 2m^2,\end{aligned} \tag{14.89}$$

then together with (14.84)-(14.89), we can write the quadratic part of the Lagrangian density (14.83) as

14.3 THE STANDARD MODEL

$$\begin{aligned}\mathcal{L}_Q &= -\frac{1}{2}\left(\partial_\mu W_\nu^+ - \partial_\nu W_\mu^+\right)\left(\partial^\mu W^{\nu-} - \partial^\nu W^{\mu-}\right)\\&\quad -\frac{1}{4}\left(\partial_\mu Z_\nu - \partial_\nu Z_\mu\right)\left(\partial^\mu Z^\nu - \partial^\nu Z^\mu\right)\\&\quad -\frac{1}{4}\left(\partial_\mu A_\nu - \partial_\nu A_\mu\right)\left(\partial^\mu A^\nu - \partial^\nu A^\mu\right)\\&\quad + i\bar{e}_L \slashed{\partial} e_L + i\bar{e}_R \slashed{\partial} e_R + i\bar{\nu}_{eL}\slashed{\partial}\nu_{eL}\\&\quad + \partial_\mu \phi^- \partial^\mu \phi^+ + \frac{1}{2}\partial_\mu \sigma \partial^\mu \sigma + \frac{1}{2}\partial_\mu \chi \partial^\mu \chi - \frac{M_H^2}{2}\sigma^2\\&\quad + iM_W(W_\mu^+ \partial^\mu \phi^- - W_\mu^- \partial^\mu \phi^+) + M_W^2 W_\mu^+ W^{\mu-}\\&\quad - M_Z Z_\mu \partial^\mu \chi + \frac{M_Z^2}{2} Z_\mu Z^\mu - m_e(\bar{e}_R e_L + \bar{e}_L e_R)\\&= -\frac{1}{2}\left(\partial_\mu W_\nu^+ - \partial_\nu W_\mu^+\right)\left(\partial^\mu W^{\nu-} - \partial^\nu W^{\mu-}\right)\\&\quad + M_W^2\left(W_\mu^+ - \frac{i}{M_W}\partial_\mu \phi^+\right)\left(W^{\mu-} + \frac{i}{M_W}\partial^\mu \phi^-\right)\\&\quad - \frac{1}{4}\left(\partial_\mu Z_\nu - \partial_\nu Z_\mu\right)\left(\partial^\mu Z^\nu - \partial^\nu Z^\mu\right)\\&\quad + \frac{M_Z^2}{2}\left(Z_\mu - \frac{1}{M_Z}\partial_\mu \chi\right)\left(Z^\mu - \frac{1}{M_Z}\partial^\mu \chi\right)\\&\quad - \frac{1}{4}\left(\partial_\mu A_\nu - \partial_\nu A_\mu\right)\left(\partial^\mu A^\nu - \partial^\nu A^\mu\right)\\&\quad + \frac{1}{2}\partial_\mu \sigma \partial^\mu \sigma - \frac{M_H^2}{2}\sigma^2 + \bar{e}(i\slashed{\partial} - m_e)e + i\bar{\nu}_{eL}\slashed{\partial}\nu_{eL}, \quad (14.90)\end{aligned}$$

where we have combined the two chirality states e_L, e_R to define the massive electron field as (see (14.56))

$$e = e_L + e_R. \qquad (14.91)$$

If we now redefine the (massive) gauge fields as

$$W^{\pm}_{\mu} \mp \frac{i}{M_W} \partial_\mu \phi^{\pm} \;\to\; W^{\pm}_{\mu},$$

$$Z_\mu - \frac{1}{M_Z} \partial_\mu \chi \;\to\; Z_\mu, \qquad (14.92)$$

then the quadratic Lagrangian density (14.90) becomes completely diagonal with the form

$$\begin{aligned}\mathcal{L}_Q &= -\frac{1}{2}\left(\partial_\mu W^+_\nu - \partial_\nu W^+_\mu\right)\left(\partial^\mu W^{\nu-} - \partial^\nu W^{\mu-}\right) + M_W^2 W^+_\mu W^{\mu-} \\ &\quad - \frac{1}{4}\left(\partial_\mu Z_\nu - \partial_\nu Z_\mu\right)\left(\partial^\mu Z^\nu - \partial^\nu Z^\mu\right) + \frac{M_Z^2}{2} Z_\mu Z^\mu \\ &\quad - \frac{1}{4}\left(\partial_\mu A_\nu - \partial_\nu A_\mu\right)\left(\partial^\mu A^\nu - \partial^\nu A^\mu\right) + \frac{1}{2}\partial_\mu \sigma \partial^\mu \sigma - \frac{1}{2} M_H^2 \sigma^2 \\ &\quad + \bar{e}(i\slashed{\partial} - m_e)e + i\bar{\nu}_{eL}\slashed{\partial}\nu_{eL}. \end{aligned} \qquad (14.93)$$

The quadratic Lagrangian density (14.93) is now completely diagonalized and shows some very desirable features. Although the starting theory had a massless electron, we see that after spontaneous breakdown of symmetry the electron has become massive with mass

$$m_e = \frac{hv}{\sqrt{2}}, \qquad (14.94)$$

while the neutrino remains massless as we would like. The Higgs particle has the usual mass $M_H = \sqrt{2}m$ (see (14.38)). The charged vector (spin 1) bosons $W^{\pm}_\mu$ are massive with mass

$$M_W = \frac{gv}{2}. \qquad (14.95)$$

One of the two neutral vector bosons, namely, Z_μ is also massive with mass

14.3 THE STANDARD MODEL

$$M_Z = \frac{\sqrt{g^2 + g'^2}\, v}{2} = \frac{M_W}{\cos\theta_W}, \qquad (14.96)$$

where the Weinberg angle θ_W is defined by (see (14.84))

$$\begin{aligned}\sin\theta_W &= \frac{g'}{\sqrt{g^2 + g'^2}}, \\ \cos\theta_W &= \frac{g}{\sqrt{g^2 + g'^2}}.\end{aligned} \qquad (14.97)$$

There is another neutral vector boson in the theory (14.93), namely, A_μ which is massless. This suggests that there is still a residual unbroken gauge symmetry in the theory which can possibly be identified with the low energy electromagnetic gauge symmetry $U_{\mathrm{EM}}(1)$. That this is indeed true can be seen as follows.

Let us look at the minimally coupled fermion Lagrangian density (14.63)

$$\begin{aligned}\mathcal{L}_{\mathrm{f}} &= i\overline{L}\gamma^\mu\left(\partial_\mu + ig\frac{\sigma}{2}\cdot\mathbf{W}_\mu + \frac{ig'}{2}Y_\mu\right)L + i\overline{R}\gamma^\mu\left(\partial_\mu + ig'Y_\mu\right)R \\ &= i\begin{pmatrix}\overline{\nu}_{e\mathrm{L}} & \overline{e}_{\mathrm{L}}\end{pmatrix}\gamma^\mu\begin{pmatrix}(\partial_\mu + \frac{ig'}{2}Y_\mu + \frac{ig}{2}W^3_\mu)\nu_{e\mathrm{L}} + \frac{ig}{\sqrt{2}}W^+_\mu e_{\mathrm{L}} \\ (\partial_\mu + \frac{ig'}{2}Y_\mu - \frac{ig}{2}W^3_\mu)e_{\mathrm{L}} + \frac{ig}{\sqrt{2}}W^-_\mu \nu_{e\mathrm{L}}\end{pmatrix} \\ &\quad + i\overline{e}_{\mathrm{R}}\gamma^\mu\left(\partial_\mu + ig'Y_\mu\right)e_{\mathrm{R}} \\ &= i\overline{\nu}_{e\mathrm{L}}\gamma^\mu\left(\left(\partial_\mu + \frac{ig'}{2}Y_\mu + \frac{ig}{2}W^3_\mu\right)\nu_{e\mathrm{L}} + \frac{ig}{\sqrt{2}}W^+_\mu e_{\mathrm{L}}\right) \\ &\quad + i\overline{e}_{\mathrm{L}}\gamma^\mu\left(\left(\partial_\mu + \frac{ig'}{2}Y_\mu - \frac{ig}{2}W^3_\mu\right)e_{\mathrm{L}} + \frac{ig}{\sqrt{2}}W^-_\mu \nu_{e\mathrm{L}}\right) \\ &\quad + i\overline{e}_{\mathrm{R}}\gamma^\mu\left(\partial_\mu + ig'Y_\mu\right)e_{\mathrm{R}} \\ &= i\overline{\nu}_{e\mathrm{L}}\slashed{\partial}\nu_{e\mathrm{L}} + i\overline{e}_{\mathrm{L}}\slashed{\partial}e_{\mathrm{L}} + i\overline{e}_{\mathrm{R}}\slashed{\partial}e_{\mathrm{R}} \\ &\quad - \frac{g}{\sqrt{2}}W^+_\mu \overline{\nu}_{e\mathrm{L}}\gamma^\mu e_{\mathrm{L}} - \frac{g}{\sqrt{2}}W^-_\mu \overline{e}_{\mathrm{L}}\gamma^\mu \nu_{e\mathrm{L}} \\ &\quad - \frac{g}{2}W^3_\mu\left(\overline{\nu}_{e\mathrm{L}}\gamma^\mu \nu_{e\mathrm{L}} - \overline{e}_{\mathrm{L}}\gamma^\mu e_{\mathrm{L}}\right)\end{aligned}$$

$$-\frac{g'}{2}Y_\mu\left(\bar{\nu}_{eL}\gamma^\mu\nu_{eL}+\bar{e}_L\gamma^\mu e_L+2\bar{e}_R\gamma^\mu e_R\right)$$

$$=i\bar{\nu}_{eL}\slashed{\partial}\nu_{eL}+i\bar{e}_L\slashed{\partial}e_L+i\bar{e}_R\slashed{\partial}e_R$$

$$-\frac{g}{\sqrt{2}}\left(W_\mu^+\bar{\nu}_{eL}\gamma^\mu e_L+W_\mu^-\bar{e}_L\gamma^\mu\nu_{eL}\right)$$

$$-\frac{g}{2}\left(Z_\mu\cos\theta_W-A_\mu\sin\theta_W\right)\left(\bar{\nu}_{eL}\gamma^\mu\nu_{eL}-\bar{e}_L\gamma^\mu e_L\right)$$

$$-\frac{g'}{2}(Z_\mu\sin\theta_W+A_\mu\cos\theta_W)(\bar{\nu}_{eL}\gamma^\mu\nu_{eL}+\bar{e}_L\gamma^\mu e_L+2\bar{e}_R\gamma^\mu e_R)$$

$$=i\bar{\nu}_{eL}\slashed{\partial}\nu_{eL}+i\bar{e}\slashed{\partial}e$$

$$-\frac{g}{\sqrt{2}}\left(W_\mu^+\bar{\nu}_{eL}\gamma^\mu e_L+W_\mu^-\bar{e}_L\gamma^\mu\nu_{eL}\right)$$

$$-\frac{Z_\mu}{2\sqrt{g^2+g'^2}}\left[g^2\left(\bar{\nu}_{eL}\gamma^\mu\nu_{eL}-\bar{e}_L\gamma^\mu e_L\right)\right.$$
$$\left.+g'^2\left(\bar{\nu}_{eL}\gamma^\mu\nu_{eL}+\bar{e}_L\gamma^\mu e_L+2\bar{e}_R\gamma^\mu e_R\right)\right]$$

$$-\frac{gg'}{\sqrt{g^2+g'^2}}A_\mu\bar{e}\gamma^\mu e. \qquad(14.98)$$

We note from (14.98) that the massless neutral vector boson A_μ couples only to the charged fermions as the photon should and we can identify the electric charge as

$$e=\frac{gg'}{\sqrt{g^2+g'^2}}=g\sin\theta_W=g'\cos\theta_W. \qquad(14.99)$$

Thus, indeed we can think of the spontaneous symmetry breaking as leaving a residual $U_{EM}(1)$ gauge invariance as the low energy symmetry of the electroweak theory. We also note from (14.98) that the standard model has, in addition to the correct electromagnetic current, both charged as well as neutral weak currents given by

$$J_\mu^- = -\frac{g}{\sqrt{2}}\bar{\nu}_{eL}\gamma_\mu e_L = -\frac{g}{2\sqrt{2}}\bar{\nu}_{eL}\gamma_\mu\left(1-\gamma_5\right)e_L,$$

$$J_\mu^+ = -\frac{g}{\sqrt{2}}\bar{e}_L\gamma_\mu\nu_{eL} = -\frac{g}{2\sqrt{2}}\bar{e}_L\gamma_\mu\left(1-\gamma_5\right)\nu_{eL},$$

$$J_\mu^0 = -\frac{1}{2\sqrt{g^2+g'^2}}\bigl[g^2\left(\bar\nu_{eL}\gamma^\mu\nu_{eL} - \bar e_L\gamma^\mu e_L\right)$$
$$+g'^2\left(\bar\nu_{eL}\gamma^\mu\nu_{eL} + \bar e_L\gamma^\mu e_L + 2\bar e_R\gamma^\mu e_R\right)\bigr]. \quad (14.100)$$

We recall that the $V-A$ theory describes only charged current interactions. If we look at the charged current interaction terms in (14.98),

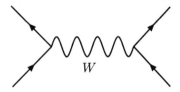

Figure 14.1: Charged current-current interaction in the lowest order.

they would lead to a low energy effective current-current interation Hamiltonian density in the lowest order of the form (basically, two charged currents interact through the charged W gauge boson propagator (exchange) as shown in Fig. 14.1 and the propagator reduces to a multiplicative factor of $\frac{1}{M_W^2}$ in the low energy limit $\left(p^2 \ll M_W^2\right)$)

$$\mathcal{H}_I = \frac{g^2}{8M_W^2}\bar\nu_{eL}\gamma_\mu\left(1-\gamma_5\right)e_L\bar e_L\gamma^\mu\left(1-\gamma_5\right)\nu_{eL}. \quad (14.101)$$

This has exactly the form of the $V-A$ four-fermion interaction in (14.49) (for just one family of leptons) and comparing the two we determine

$$\frac{G_F}{\sqrt{2}} = \frac{g^2}{8M_W^2}, \quad (14.102)$$

which leads to

$$M_W^2 = \frac{g^2}{4\sqrt{2}G_F} = \frac{e^2}{4\sqrt{2}G_F\sin^2\theta_W}. \quad (14.103)$$

Experimental measurements lead to a value of the Weinberg angle as

$$\sin^2\theta_W \simeq 0.23, \tag{14.104}$$

and using the values of other constants (see, for example, (14.51) and recall that we have set $\hbar = c = 1$)

$$G_F \simeq 10^{-5} \, (\text{GeV})^{-2},$$

$$\frac{e^2}{4\pi} \simeq \frac{1}{137},$$

we can now determine

$$M_W = \left(\frac{e^2}{4\sqrt{2} G_F \sin^2\theta_W}\right)^{\frac{1}{2}} \simeq 80 \text{ GeV},$$

$$M_Z = \frac{M_W}{\cos\theta_W} \simeq 90 \text{ GeV}. \tag{14.105}$$

These gauge bosons have been discovered and their masses are determined very close to the theoretically predicted values. Furthermore, the standard model differs from the conventional Fermi theory in (14.49) in that it predicts weak neutral currents (in addition to the conventional charged weak currents) and hence processes where a neutral heavy vector boson (Z_μ) is exchanged. Such processes have also been experimentally observed. The standard model compares very well with the experimental results. However, the Weinberg-Salam-Glashow theory does not predict a unique mass for the Higgs particle since its mass depends on the quartic coupling λ, although bounds on the value of its mass can be put from various other arguments. Search for Higgs particles is a top priority in many experiments at the upcoming LHC. The standard model also suffers from the fact that the neutrino is massless in this theory while we now believe that the neutrino may have a small mass (from neutrino oscillation experiments). Once again these can be accomodated into extensions of the standard model that we will not go into here.

14.4 References

1. A. Proca, J. de Phys. et le Radium **7**, 347 (1936).

14.4 References

2. E. C. G. Stückelberg, Helvetica Physica Acta **30**, 209 (1957).

3. E. C. G. Sudarshan and R. E. Marshak, *Proceedings of Padua-Venice conference on mesons and newly discovered particles*, (1957); Physical Review **109**, 1860 (1958).

4. R. P. Feynman and M. Gell-Mann, Physical Review **109**, 193 (1958).

5. F. Englert and R. Brout, Physical Review Letters **13**, 321 (1964).

6. G. S. Guralnik, C. R. Hagen and T. W. B. Kibble, Physical Review Letters **13**, 585 (1964).

7. P. W. Higgs, Physical Review **145**, 1156 (1966).

8. S. Weinberg, Physical Review Letters **19**, 1264 (1967).

9. G. 't Hooft, Nuclear Physics **B35**, 167 (1971).

10. J. Bernstein, Reviews of Modern Physics **46**, 7 (1974).

11. H. Ruegg and M. Ruiz-Altaba, International Journal of Modern Physics **A19**, 3265 (2004).

CHAPTER 15
Regularization of Feynman diagrams

15.1 Introduction

So far we have discussed the basic structures of various quantum field theories. We have also carried out a few calculations of simple Feynman diagrams describing physical processes in section **9.6**. The simple diagrams in **9.6** are known as tree diagrams where the momenta of all the internal propagators are uniquely determined (through energy-momentum conserving delta functions) in terms of the external momenta (in the diagrams in **9.6** there was only one internal propagator). However, as we go to higher orders in perturbation, the Feynman diagrams become topologically more complicated and may contain internal propagators whose momenta are not uniquely determined in terms of the external momenta. Instead some of the internal propagators may involve additional momenta (loop momenta) that are integrated over all possible values. In such a case, the evaluation of Feynman diagrams may lead to divergences depending on the structure of the integral that is being evaluated. As a result, we have to find a way of extracting meaningful results from such a quantum field theory. This procedure is known, in general, as renormalization of a quantum field theory and involves several steps. We will study this question in a systematic manner developing the necessary ideas in this chapter as well as in the next.

As we have discussed earlier (see section **13.4**), any scattering matrix element (S matrix element) in a given quantum field theory can be built out of 1PI (one particle irreducible) graphs. We recall that an 1PI graph is defined to be a (vertex) graph which cannot be separated into two disconnected graphs by cutting a single internal line. For example, let us look at one of the simplest interacting field theories, namely, the scalar ϕ^4 theory described by the Lagrangian density (see (6.36))

$$\mathcal{L} = \frac{1}{2}\partial_\mu\phi\partial^\mu\phi - \frac{M^2}{2}\phi^2 - \frac{\lambda}{4!}\phi^4, \tag{15.1}$$

where the coupling constant λ is assumed to be positive. The Feynman rules for this theory are given by

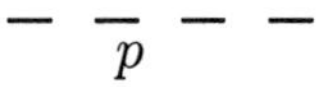

$$= iG_F(p) = \frac{i}{p^2 - M^2},$$

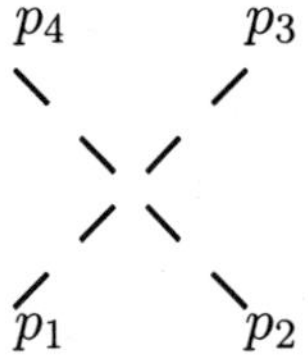

$$= -(2\pi)^4 i\lambda \delta^4(p_1 + p_2 + p_3 + p_4), \tag{15.2}$$

where all the momenta in the vertex are assumed to be incoming and we have not written the "$i\epsilon$" term in the propagator explicitly although it should be understood. With these Feynman rules, we can calculate any perturbative amplitude in the ϕ^4 theory. In fact, given these Feynman rules, we can construct the graphs in Fig. 15.1 (which topologically involve loops and, therefore, are called loop diagrams)

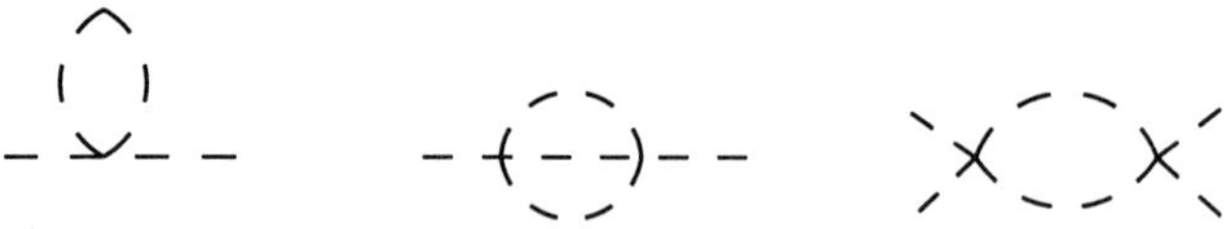

Figure 15.1: A few examples of 1PI diagrams in the ϕ^4 theory.

which are 1PI graphs whereas the graph in Fig. 15.2, for example, is not. Clearly the 1PI diagrams are fundamental since any other diagram can be built using them as basic elements. Therefore, in studying quantum field theories at higher orders, it is sufficient to concentrate on the 1PI graphs.

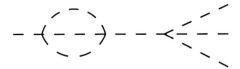

Figure 15.2: An example of a diagram in the ϕ^4 theory which is not 1PI.

As we have mentioned, in evaluating scattering amplitudes particularly at the loop level (where the topology of the diagram involves loops as in Fig. 15.1 and Fig. 15.2), we invariably run into divergences. Namely, the integrals over the arbitrary loop momenta may not, in general, be convergent. In such a case, we are forced to give a meaning to the Feynman amplitudes in some manner and this process is known as regularization of the diagrams. Since the 1PI graphs are fundamental, to study the divergence structure of any amplitude (graph), it is sufficient to study the divergence behavior of only the 1PI graphs. In the discussions below we will introduce a number of regularization schemes that are commonly used in studying divergent Feynman amplitudes.

15.2 Loop expansion

As we have mentioned repeatedly, problems with divergences arise in a quantum field theory at the loop level. Therefore, let us discuss the notion of loop expansion in a quantum field theory in this section. Let us note that if our theory is described by a Lagrangian density such that (although we are assuming the basic field variable to be a scalar field, this discussion applies to any quantum field theory)

$$\mathcal{L}(\phi, \partial_\mu \phi, a) = a^{-1} \mathcal{L}(\phi, \partial_\mu \phi), \tag{15.3}$$

where a is a constant parameter, then it is clear that each interaction vertex in the theory will have a factor a^{-1} multiplied to it, while the propagator of the theory which is the inverse of the tree level two point function will have a factor "a" associated with it. Thus if we are considering a proper vertex graph (1PI graph), and P is the power of "a" associated with this graph, then it follows easily that

$$P = I - V, \qquad (15.4)$$

where I denotes the number of internal lines and V the number of interaction vertices in the graph.

On the other hand the number of loops in a graph is defined as the number of independent momenta which we have to integrate over (after all the energy-momentum conserving delta functions have been taken care of). Realizing that with every internal line there is a momentum which is integrated and that each vertex has a delta function enforcing the conservation of energy and momentum at that vertex, the number of loops (independent momenta) in such a graph, therefore, is given by

$$L = I - V + 1, \qquad (15.5)$$

where the term (+1) reflects the fact that every 1PI vertex function has an overall energy-momentum conserving delta function associated with it. From the two relations in (15.4) and (15.5), we can eliminate $I - V$ to obtain

$$L = P + 1. \qquad (15.6)$$

In other words, the power of "a" in a diagram in such a theory also determines the number of loops in the diagram and can, therefore, be thought of as the parameter of expansion in the number of loops.

Let us recall that in the path integral description of a quantum field theory, the generating functional is defined as (see, for example, (12.120), we have set $\hbar = 1$ throughout our discussion, but for a clear understanding of the loop expansion we are going to restore $\hbar$ in this section only)

$$Z[J] = N \int \mathcal{D}\phi \, e^{\frac{i}{\hbar} S^J}, \qquad (15.7)$$

where $\hbar$ represents the Planck's constant and we see that it can be thought of as an overall multiplicative parameter like "a" in (15.3).

As a result, expanding in powers of $\hbar$ in a quantum field theory also corresponds to expanding in the number of loops as described in (15.6).

We are usually accustomed to the idea of perturbation in powers of the coupling constant in the theory. Loop expansion is a different perturbative expansion. The reason why such an expansion is advantageous can be seen from the fact that since this parameter multiplies the whole action, the expansion is unaffected by how one divides the total action into a free part and an interaction part. This is quite important, for example, when we study theories where there is spontaneous breakdown of a symmetry. In such a case, as we have seen (see, for example, sections **7.5** and **14.2**), the scalar field picks up a vacuum expectation value which depends on the coupling constant of the theory. Therefore, when we perturb around the true vacuum (by shifting the field variable by its vacuum expectation value), the entire perturbation expansion has to be rearranged which is nontrivial. On the other hand, the loop expansion is unaffected by shifting of fields since the parameter of expansion is a multiplicative factor in front of the total action and as a result the loop expansion is quite useful. In addition the small value of $\hbar$ makes it a legitimate expansion (perturbation) parameter.

15.3 Cut-off regularization

To get an idea of how divergences arise in a quantum field theory at the loop level, let us calculate the simplest nontrivial graph in the ϕ^4 theory (15.1) which gives the one loop correction to the two point function. Factoring out an overall momentum conserving delta function $(2\pi)^4 \delta^4(p_1 - p_2)$ and identifying $p_1 = p_2 = p$ we have (we will always factor out the overall momentum conserving delta function along with the factor of $(2\pi)^4$)

$$= I = -\frac{i\lambda}{2} \int \frac{\mathrm{d}^4 k}{(2\pi)^4} \frac{i}{k^2 - M^2}$$

$$= -\frac{i\lambda}{2}\int \frac{id^4k_E}{(2\pi)^4}\frac{i}{-k_E^2-M^2}$$

$$= -\frac{i\lambda}{2(2\pi)^4}\int d^4k_E \frac{1}{k_E^2+M^2}$$

$$= -\frac{i\lambda}{2(2\pi)^4}\int d\Omega \int_0^\infty k_E^3 dk_E \frac{1}{k_E^2+M^2}$$

$$= -\frac{i\lambda}{2(2\pi)^4}(2\pi^2)\int_0^\infty \frac{1}{2}dk_E^2 \frac{k_E^2}{k_E^2+M^2}$$

$$= -\frac{i\lambda}{32\pi^2}\int_0^\infty dy \frac{y+M^2-M^2}{y+M^2}$$

$$= -\frac{i\lambda}{32\pi^2}\int_0^\infty dy\left(1-\frac{M^2}{y+M^2}\right). \tag{15.8}$$

Here we have used the fact that the Feynman diagram in (15.8) has an associated symmetry factor of $\frac{1}{2}$ and have rotated the momentum to Euclidean space by letting $k_0 \to ik_{0E}$ in the intermediate step to facilitate the evaluation of the integral. Furthermore, we have used the value of the angular integral in four dimensions which can be easily determined as

$$\begin{aligned}\int d\Omega &= \int_0^\pi d\theta_1 \sin^2\theta_1 \int_0^\pi d\theta_2 \sin\theta_2 \int_0^{2\pi} d\theta_3 \\ &= \int_0^\pi d\theta_1 \frac{1}{2}(1-\cos 2\theta_1)\int_{-1}^1 d\cos\theta_2 \times (2\pi) \\ &= 2\pi^2.\end{aligned} \tag{15.9}$$

We see from (15.8) that this integral is divergent and, therefore, we define (regularize) this integral by cutting off the momentum integration at some higher value Λ and then taking the limit $\Lambda \to \infty$ at the end of the calculation. (This is the reason for the name cut-off regularization.) Since large values of momentum correspond to small values of coordinate separation, such divergences are also known as ultraviolet divergences. With this regularization, the two point function in (15.8) takes the form

15.3 CUT-OFF REGULARIZATION

$$\begin{aligned}
I &= \lim_{\Lambda\to\infty} -\frac{i\lambda}{32\pi^2}\left[\int_0^{\Lambda^2}dy - M^2\int_0^{\Lambda^2}\frac{dy}{y+M^2}\right]\\
&= \lim_{\Lambda\to\infty} -\frac{i\lambda}{32\pi^2}\left[\Lambda^2 - M^2\ln(y+M^2)\Big|_0^{\Lambda^2}\right]\\
&= \lim_{\Lambda\to\infty} -\frac{i\lambda}{32\pi^2}\left[\Lambda^2 - M^2\ln\left(\frac{\Lambda^2+M^2}{M^2}\right)\right]\\
&= \lim_{\Lambda\to\infty} -\frac{i\lambda}{32\pi^2}\left[\Lambda^2 - M^2\ln\frac{\Lambda^2}{M^2} + O\left(\frac{1}{\Lambda^2}\right)\right]. \quad (15.10)
\end{aligned}$$

The result at the end should be rotated back to the Minkowski space. However, since there is no momentum vector in the final result in (15.10) (and m, Λ are scalar parameters), in this case this is also the result in the Minkowski space.

We add here for completeness that the cut-off can also be implemented through a weight factor in the following manner. We note that we can write (15.8) in a regularized manner as

$$\begin{aligned}
I &= -\frac{i\lambda}{32\pi^2}\int_0^\infty dy\left(1-\frac{M^2}{y+M^2}\right)\\
&= \lim_{\tilde\Lambda\to\infty} -\frac{i\lambda}{32\pi^2}\int_0^\infty dy\left(1-\frac{M^2}{y+M^2}\right)e^{-\frac{y}{\tilde\Lambda^2}}\\
&= \lim_{\tilde\Lambda\to\infty} -\frac{i\lambda}{32\pi^2}\left[\tilde\Lambda^2 + M^2\left(\gamma+\ln\frac{M^2}{\tilde\Lambda^2}+O(\frac{1}{\tilde\Lambda^2})\right)\right]\\
&= \lim_{\tilde\Lambda\to\infty} -\frac{i\lambda}{32\pi^2}\left[\tilde\Lambda^2 + M^2\left(\gamma-\ln\frac{\tilde\Lambda^2}{M^2}+O(\frac{1}{\tilde\Lambda^2})\right)\right], \quad (15.11)
\end{aligned}$$

where

$$\gamma = \lim_{n\to\infty}\left(\sum_{k=1}^n\frac{1}{k}-\ln n\right) = 0.577215665, \quad (15.12)$$

denotes the Euler's constant (also known as the Euler-Mascheroni constant and sometimes denoted by C) and we have used the standard tables of integrals (see, for example, Gradshteyn and Ryzhik 3.352.4 and 8.214.1). This result coincides with (15.10) with the identification

$$\tilde{\Lambda}^2 = \Lambda^2 - M^2 \gamma. \tag{15.13}$$

Furthermore, we also note here that the diagram in (15.8) can be evaluated without rotating the momentum into Euclidean space as follows

$$\begin{aligned} I &= -\frac{i\lambda}{2} \int \frac{d^4k}{(2\pi)^4} \frac{i}{k^2 - M^2 + i\epsilon} \\ &= \frac{\lambda}{2(2\pi)^4} \int d^3k dk_0 \frac{1}{(k_0 - (E_k - i\epsilon))(k_0 + E_k - i\epsilon)} \\ &= \frac{\lambda}{2(2\pi)^4} (-2\pi i) \int d^3k \frac{1}{2E_k} = -\frac{i\lambda}{32\pi^3} \int d^3k \frac{1}{E_k}, \end{aligned} \tag{15.14}$$

where we have identified (see, for example, (5.47)) $E_k = \sqrt{|\mathbf{k}|^2 + M^2}$. Since the integrand in (15.14) does not depend on the angular variables, the integration over the angles can be carried out, which gives (4π) in three dimensions. Furthermore, defining $y = |\mathbf{k}|$, the integral takes the form

$$\begin{aligned} I &= -\frac{i\lambda}{32\pi^3} (4\pi) \int_0^\infty dy \frac{y^2}{\sqrt{y^2 + M^2}} \\ &= \lim_{\bar{\Lambda} \to \infty} -\frac{i\lambda}{8\pi^2} \int_0^{\bar{\Lambda}} dy \left(\sqrt{y^2 + M^2} - \frac{M^2}{\sqrt{y^2 + M^2}} \right) \\ &= \lim_{\bar{\Lambda} \to \infty} -\frac{i\lambda}{8\pi^2} \frac{1}{2} \left[y\sqrt{y^2 + M^2} - M^2 \ln\left(y + \sqrt{y^2 + M^2}\right) \right]_0^{\bar{\Lambda}} \\ &= \lim_{\bar{\Lambda} \to \infty} -\frac{i\lambda}{16\pi^2} \left[\bar{\Lambda}\sqrt{\bar{\Lambda}^2 + M^2} - M^2 \ln \frac{\bar{\Lambda} + \sqrt{\bar{\Lambda}^2 + M^2}}{M} \right] \\ &= \lim_{\bar{\Lambda} \to \infty} -\frac{i\lambda}{32\pi^2} \left[2\bar{\Lambda}^2 + M^2 - M^2 \ln\left(\frac{4\bar{\Lambda}^2}{M^2}\right) + O(\frac{1}{\bar{\Lambda}^2}) \right]. \end{aligned} \tag{15.15}$$

15.3 CUT-OFF REGULARIZATION

Here we have used the standard table of integrals in the intermediate step (see, for example, Gradshteyn and Ryzhik, 2.271.2 and 2.271.4). This result coincides with (15.10) if we identify

$$\bar{\Lambda}^2 = \frac{1}{2}(\Lambda^2 - M^2(1 - \ln 2)). \tag{15.16}$$

This derivation makes it clear that the evaluation of the amplitude is much easier in the Euclidean space.

The integral in (15.8) can also be evaluated in an alternative manner which is sometimes useful and so let us discuss the alternate method as well. We note that the integral in Euclidean space has the form

$$\begin{aligned}
I &= -\frac{i\lambda}{2(2\pi)^4} \times 2\pi^2 \int_0^\infty \frac{1}{2} dk_E^2 \, \frac{k_E^2}{k_E^2 + M^2} \\
&= -\frac{i\lambda}{32\pi^2} \int_0^\infty dy \, \frac{y}{y + M^2} \\
&= -\frac{i\lambda}{32\pi^2} \iint_0^\infty dy d\tau \, y \, e^{-\tau(y+M^2)} \\
&= -\frac{i\lambda}{32\pi^2} \int_0^\infty d\tau \, \frac{1}{\tau^2} \, \Gamma(2) \, e^{-\tau M^2} \\
&= -\frac{i\lambda}{32\pi^2} \int_0^\infty \frac{d\tau}{\tau^2} e^{-\tau M^2}. \tag{15.17}
\end{aligned}$$

Here $\Gamma(2)$ denotes the gamma function and the parameter τ is known as the Schwinger parameter (or the proper time parameter). We note that the integrand in (15.17) is well behaved for large values of τ, but is divergent at the lower limit $\tau = 0$. Thus, the ultraviolet divergence in (15.10) has been transformed into an infrared divergence in the Schwinger parameter τ. This integral can be regularized by cutting off the integral at some lower limit $\frac{1}{\Lambda^2}$ as

$$\begin{aligned}
I &= \lim_{\Lambda \to \infty} -\frac{i\lambda}{32\pi^2} \int_0^\infty \frac{d\tau}{\tau^2} e^{-\tau M^2 - \frac{1}{\Lambda^2 \tau}} \\
&= \lim_{\Lambda \to \infty} -\frac{i\lambda}{32\pi^2} \times 2 \left(\frac{1}{\Lambda^2 M^2}\right)^{-\frac{1}{2}} K_{-1}\left(\frac{2M}{\Lambda}\right)
\end{aligned}$$

$$\begin{aligned}
&= \lim_{\Lambda\to\infty} -\frac{i\lambda}{32\pi^2} \times (2\Lambda M) K_1\left(\frac{2M}{\Lambda}\right) \\
&= \lim_{\Lambda\to\infty} -\frac{i\lambda}{32\pi^2} \times (2\Lambda M)\left[\left(\frac{2M}{\Lambda}\right)^{-1} + \frac{M}{\Lambda}\ln\frac{M}{\Lambda} + \cdots\right] \\
&= \lim_{\Lambda\to\infty} -\frac{i\lambda}{32\pi^2}\left[\Lambda^2 + M^2\ln\frac{M^2}{\Lambda^2} + O\left(\frac{1}{\Lambda^2}\right)\right] \\
&= \lim_{\Lambda\to\infty} -\frac{i\lambda}{32\pi^2}\left[\Lambda^2 - M^2\ln\frac{\Lambda^2}{M^2} + O\left(\frac{1}{\Lambda^2}\right)\right], \quad (15.18)
\end{aligned}$$

where $K_m(z)$ denotes the modified Bessel function of the second kind of order m and we have used the standard tables of integral (see, for example, Gradshteyn and Ryzhik 3.471.9) to evaluate the integral over τ. Equation (15.18) can be compared with (15.10) and we see that both ways of evaluating the integral lead to the same result. We note that since the cutoff Λ has to be actually taken to infinity in the final result, this graph is divergent. However, regularizing the integral brings out the divergence structure of the Feynman graph. For example, we note from (15.10) (or (15.18)) that the self-energy diagram at one loop contains a term that diverges quadratically as well as a term which is logarithmically divergent. It is also clear from this simple calculation that the nature of the divergence depends on the dimensionality of space-time (we are restricting ourselves to four space-time dimensions, but quantum field theories can be defined in any number of space-time dimensions). Without going into details, we note that the result in (15.18) can also be obtained by simply cutting off the τ integral at the lower limit $\frac{1}{\Lambda^2}$ (without the regularizing exponential factor).

Let us next look at another one loop graph, namely, the one loop diagram for the vertex correction shown in Fig. 15.3. For simplicity we would consider all the incoming momenta to vanish. This diagram is also divergent (in four dimensions) and with the cut-off regularization leads to (we assume $p_1 = p_2 = p_3 = p_4 = 0$ for simplicity so that each graph contributes the same amount and also because we are interested in looking at only the divergence structure of the graph)

15.3 Cut-off regularization

Figure 15.3: One loop correction to the four point vertex function.

$$
\begin{aligned}
I &= \frac{3(-i\lambda)^2}{2} \int \frac{d^4k}{(2\pi)^4} \frac{i}{k^2-M^2} \frac{i}{k^2-M^2} \\
&= \frac{3\lambda^2}{2(2\pi)^4} \int i d^4 k_E \left(\frac{1}{-k_E^2 - M^2}\right)^2 \\
&= \frac{3i\lambda^2}{2(2\pi)^4} \int d^4 k_E \frac{1}{(k_E^2 + M^2)^2} \\
&= \frac{3i\lambda^2}{2(2\pi)^4} (2\pi^2) \int_0^\infty k_E^3 \, dk_E \frac{1}{(k_E^2+M^2)^2} \\
&= \frac{3i\lambda^2}{32\pi^2} \int_0^\infty dy \frac{y}{(y+M^2)^2} \\
&= \frac{3i\lambda^2}{32\pi^2} \int_0^\infty dy \frac{y + M^2 - M^2}{(y+M^2)^2} \\
&= \frac{3i\lambda^2}{32\pi^2} \left[\lim_{\Lambda \to \infty} \int_0^{\Lambda^2} \frac{dy}{y+M^2} - M^2 \int_0^\infty \frac{dy}{(y+M^2)^2} \right] \\
&= \frac{3i\lambda^2}{32\pi^2} \left[\lim_{\Lambda \to \infty} \ln(y+M^2) \Big|_0^{\Lambda^2} + \frac{M^2}{y+M^2} \Big|_0^\infty \right] \\
&= \lim_{\Lambda \to \infty} \frac{3i\lambda^2}{32\pi^2} \left[\ln \frac{\Lambda^2 + M^2}{M^2} - 1 \right] \\
&= \lim_{\Lambda \to \infty} \frac{3i\lambda^2}{32\pi^2} \left[\ln \frac{\Lambda^2}{M^2} - 1 + O\left(\frac{1}{\Lambda^2}\right) \right]. \quad (15.19)
\end{aligned}
$$

There are several things to note from this derivation. First we have factored out an overall momentum conserving delta function along with the factor of $(2\pi)^4$. The second factor of $(2\pi)^4$ that comes from the second vertex has cancelled with the factor of $\frac{1}{(2\pi)^4}$ in the integration over the momentum of one of the internal propagators. Finally, each of these diagrams has an associated symmetry factor of $\frac{1}{2}$ and we have used the value of the angular integral from (15.9) in the intermediate step. Thus we see that this graph is also divergent (in four dimensions), but the divergence in this case is logarithmic. (This can also be seen from the fact that the integral in (15.19) is given by $-\frac{\partial}{\partial M^2}$ of that in (15.8) up to a multiplicative factor.)

Let us note here that these divergences are there in the quantum field theory besides the zero point energy or the ground state energy divergence which, as we have seen earlier, can be removed by normal ordering the Hamiltonian of the theory. This is, of course, equivalent to saying that since we only measure differences in the energy levels, we are free to redefine the value of the ground state energy to zero. However, the present divergences are nontrivial and are present even after normal ordering of the theory. In fact, any given quantum field theory, in general, is plagued with divergences at every (loop) order in perturbation theory (unless we are in lower dimensions). The divergence structure of any diagram can be determined simply by counting the powers of momentum in the numerator and in the denominator. Thus, for example, for the first graph in (15.8) we have four powers of momentum in the numerator coming from the momentum integration while there are two powers of momentum in the propagator. Consequently, we see that there is a net $4 - 2 = 2$ powers of momentum in the numerator reflected in the fact that the highest degree of divergence of the diagram is quadratic. Similarly, for the diagrams in Fig. 15.3, we have four powers of momentum in the numerator coming from the momentum integration while there are four powers of momentum in the denominator coming from two propagators. As a result, there is a net $4 - 4 = 0$ power of momentum in the numerator reflected in the logarithmic divergence of the diagram.

Renormalization is the process of redefining these infinities that we encounter in perturbation theory. Very broadly it consists of two essential parts. First we introduce a regularization procedure (scheme) which gives a meaning to the divergent Feynman integrals by isolating the divergent parts of the diagram. In the examples we have studied in (15.10) and (15.19), the cutoff regularizes and defines

15.3 CUT-OFF REGULARIZATION

the integrals. But in this process the amplitude (and the theory) becomes cutoff dependent. The second part of renormalization consists of removing the cutoff dependence (or regularization dependence) of the theory which we will take up in the next chapter.

15.3.1 Calculation in the Yukawa theory. Let us note that any Feynman diagram (amplitude), in general, depends analytically on the external momenta. In the previous examples, particularly in the second example (15.19), the lack of dependence on the external momenta is simply due to our special choice of the external momenta to be all vanishing. But let us examine the dependence of diagrams on external momenta (and this will be important in studying renormalization) in another theory, namely, the theory with a Yukawa coupling described by the Lagrangian density

$$\mathcal{L} = i\overline{\psi}\not{\partial}\psi - m\overline{\psi}\psi + \frac{1}{2}\partial_\mu\phi\partial^\mu\phi - \frac{M^2}{2}\phi^2 - g\overline{\psi}\psi\phi - \frac{\lambda}{4!}\phi^4. \quad (15.20)$$

This theory describes the interaction between a fermion and a scalar field and the Feynman rules for this theory are given by

$$-\!\!-\!\!-\!\!-\!\!-\!\!-\!\!-\!\!- \quad = \quad iG_F(p) = \frac{i}{p^2 - M^2},$$
$$p$$

$$\xrightarrow{} \quad = \quad iS_F(p) = \frac{i}{\not{p} - m} = \frac{i(\not{p} + m)}{p^2 - m^2},$$
$$p$$

$$\quad = \quad -(2\pi)^4 i g \delta^4(p_1 + p_2 + p_3),$$

$$\quad = \quad -(2\pi)^4 i \lambda \delta^4(p_1 + p_2 + p_3 + p_4), \quad (15.21)$$

where the arrows in the vertices denote the fact that momenta are all incoming. The lowest order graph involving the fermions in this theory gives rise to the Yukawa potential (as we have seen in section **8.9**) which is why the cubic coupling involving fermions is called the Yukawa coupling.

Before we proceed to calculate higher order diagrams in this theory, let us note that at higher orders in perturbation theory, diagrams necessarily involve more internal propagators. Therefore, it would be useful to find a way to combine denominators and this is achieved by Feynman's formula as follows. Suppose we would like to write $\frac{1}{AB}$ as a single factor, then we note that

$$\begin{aligned}
\int_0^1 \frac{\mathrm{d}x}{(xA + (1-x)B)^2} &= \int_0^1 \frac{\mathrm{d}x}{(x(A-B) + B)^2} \\
&= -\frac{1}{A-B} \frac{1}{x(A-B)+B} \Big|_0^1 \\
&= -\frac{1}{A-B} \left[\frac{1}{A} - \frac{1}{B} \right] \\
&= \frac{1}{AB}.
\end{aligned} \qquad (15.22)$$

Therefore, we see that we can combine two denominators in a simple manner as

$$\begin{aligned}
\frac{1}{AB} &= \int_0^1 \frac{\mathrm{d}x}{(xA + (1-x)B)^2} \\
&= \iint_0^1 \mathrm{d}x_1 \mathrm{d}x_2 \frac{\delta(1 - x_1 - x_2)}{(x_1 A + x_2 B)^2}.
\end{aligned} \qquad (15.23)$$

It is worth emphasizing here that this simple combination formula holds if both the factors $\frac{1}{A}$ and $\frac{1}{B}$ have the same analyticity property (namely, the $i\epsilon$ term with the same sign). On the other hand when the two factors have opposite analytic behavior ($i\epsilon$ terms with opposite sign), then this formula needs to be generalized (which is important in studying finite temperature field theories). In quantum field theory at zero temperature, however, (15.23) holds and is quite

15.3 CUT-OFF REGULARIZATION

useful. Generalization of (15.23) to include more denominators is straightforward and has the form

$$\prod_{i=1}^{n} \frac{1}{A_i} = \int_0^1 \prod_{i=1}^{n} \mathrm{d}x_i \, \frac{\delta(1 - x_1 - \cdots - x_n)}{(\sum_{i=1}^{n} x_i A_i)^n}, \tag{15.24}$$

which we have already used in (13.121).

Equipped with the Feynman combination formula, let us next calculate the two simplest nontrivial graphs involving loops in this theory. The fermion self-energy at one loop has the form

$$\begin{aligned}
&= (-ig)^2 \int \frac{\mathrm{d}^4 k}{(2\pi)^4} \frac{i}{k^2 - M^2} \frac{i}{(\not{k} + \not{p}) - m} \\
&= g^2 \int \frac{\mathrm{d}^4 k}{(2\pi)^4} \frac{1}{k^2 - M^2} \frac{(\not{k} + \not{p}) + m}{(k + p)^2 - m^2} \\
&= g^2 \int \frac{i\mathrm{d}^4 k_E}{(2\pi)^4} \frac{1}{-k_E^2 - M^2} \frac{-(\not{k}_E + \not{p}_E) + m}{-(k_E + p_E)^2 - m^2} \\
&= -\frac{ig^2}{(2\pi)^4} \int \mathrm{d}^4 k_E \, \frac{(\not{k}_E + \not{p}_E) - m}{(k_E^2 + M^2)\left((k_E + p_E)^2 + m^2\right)}. \tag{15.25}
\end{aligned}$$

Here we have rotated the momenta as well as the gamma matrices to Euclidean space as $\gamma^0 \to i\gamma_E^0$ so that

$$\not{k} = \gamma^0 k^0 - \boldsymbol{\gamma} \cdot \mathbf{k} = -\gamma_E^0 k_E^0 - \boldsymbol{\gamma} \cdot \mathbf{k} = -\not{k}_E. \tag{15.26}$$

Using the Feynman combination formula (15.23), the integral in (15.25) can be written as (for simplicity we factor out the multiplicative factor $-\frac{ig^2}{(2\pi)^4}$ which we will put back at the end of the calculation)

$$= \int d^4 k_E \int_0^1 dx \frac{(\slashed{k}_E + \slashed{p}_E) - m}{\left[x((k_E + p_E)^2 + m^2) + (1-x)(k_E^2 + M^2)\right]^2}$$

$$= \int d^4 k_E \int_0^1 dx \frac{(\slashed{k}_E + \slashed{p}_E) - m}{(k_E^2 + 2x k_E \cdot p_E + x p_E^2 + x m^2 + (1-x) M^2)^2}$$

$$= \int_0^1 dx \int d^4 k_E \frac{(\slashed{k}_E + \slashed{p}_E) - m}{((k_E + x p_E)^2 + x(1-x) p_E^2 + x m^2 + (1-x) M^2)^2}$$

$$= \int_0^1 dx \int d^4 k_E \frac{\slashed{k}_E + (1-x)\slashed{p}_E - m}{(k_E^2 + x(1-x) p_E^2 + x m^2 + (1-x) M^2)^2}$$

$$= \int_0^1 dx ((1-x)\slashed{p}_E - m) \int_0^\infty (2\pi^2) k_E^3 \, dk_E \frac{1}{(k_E^2 + Q^2)^2}, \quad (15.27)$$

where we have defined

$$Q^2 = x(1-x) p_E^2 + x m^2 + (1-x) M^2. \quad (15.28)$$

In (15.27) we have shifted the variable of integration $k_E \to k_E - x p_E$ in the intermediate step after which the term $\slashed{k}_E$ in the numerator vanishes because of anti-symmetry.

Furthermore, the k_E integral can be carried out to give (we now put back the overall multiplicative factor $-\frac{ig^2}{(2\pi)^4}$)

$$= -\frac{ig^2}{16\pi^2} \int_0^1 dx \, ((1-x)\slashed{p}_E - m) \int_0^\infty dy \frac{y}{[y + Q^2]^2}$$

$$= -\frac{ig^2}{16\pi^2} \int_0^1 dx \, ((1-x)\slashed{p}_E - m) \left[\ln\left(\frac{\Lambda^2 + Q^2}{Q^2}\right) - 1\right]$$

$$= -\frac{ig^2}{16\pi^2} \int_0^1 dx \, ((1-x)\slashed{p}_E - m)$$
$$\times \left[\ln \frac{\Lambda^2 + x(1-x) p_E^2 + x m^2 + (1-x) M^2}{x(1-x) p_E^2 + x m^2 + (1-x) M^2} - 1\right]$$

$$= \frac{ig^2}{16\pi^2} \int_0^1 dx \, ((1-x)\slashed{p} + m)$$

15.3 CUT-OFF REGULARIZATION

$$\times \left[\ln \frac{\Lambda^2 - x(1-x)p^2 + xm^2 + (1-x)M^2}{-x(1-x)p^2 + xm^2 + (1-x)M^2} - 1 \right]$$

$$= f(p, m, M, \Lambda, g), \tag{15.29}$$

where it is understood that the limit $\Lambda \to \infty$ is to be taken. We also note here that we have used the integral from (15.19) in the intermediate step and have rotated back to Minkowski space in the final step.

Similarly the scalar self-energy graph is given by

$$= -\mathrm{Tr} \left[(-ig)^2 \int \frac{\mathrm{d}^4 k}{(2\pi)^4} \frac{i}{\slashed{k} - m} \frac{i}{(\slashed{k} + \slashed{p}) - m} \right]$$

$$= -g^2 \mathrm{Tr} \int \frac{\mathrm{d}^4 k}{(2\pi)^4} \frac{(\slashed{k} + m)((\slashed{k} + \slashed{p}) + m)}{(k^2 - m^2)((k+p)^2 - m^2)}. \tag{15.30}$$

Here the overall negative sign reflects the fact that we are evaluating a graph with a fermion loop. Furthermore, the trace is a consequence of the fact that the Dirac matrix indices coming from the two fermion propagators are being summed over (there are no free Dirac indices in the diagram). Using the trace identities

$$\begin{aligned} \mathrm{Tr}\, \mathbb{1} &= 4, \\ \mathrm{Tr}\, \slashed{A} &= 0, \\ \mathrm{Tr}\, \slashed{A}\slashed{B} &= 4A \cdot B, \end{aligned} \tag{15.31}$$

the self energy graph in (15.30) takes the form

$$
\begin{aligned}
&= -\frac{g^2}{(2\pi)^4} \int d^4k \frac{4(k \cdot (k+p)) + m^2)}{(k^2 - m^2)((k+p)^2 - m^2)} \\
&= -\frac{4g^2}{(2\pi)^4} \int i d^4k_E \frac{(-k_E \cdot (k_E + p_E)) + m^2}{(-k_E^2 - m^2)(-(k_E + p_E)^2 - m^2)} \\
&= \frac{4ig^2}{(2\pi)^4} \int d^4k_E \frac{(k_E \cdot (k_E + p_E)) - m^2}{(k_E^2 + m^2)((k_E + p_E)^2 + m^2)} \\
&= \frac{4ig^2}{(2\pi)^4} \int d^4k_E \\
&\quad \times \int_0^1 dx \frac{(k_E \cdot (k_E + p_E)) - m^2}{(x((k_E + p_E)^2 + m^2) + (1-x)(k_E^2 + m^2))^2} \\
&= \frac{4ig^2}{(2\pi)^4} \int_0^1 dx \int d^4k_E \frac{(k_E \cdot (k_E + p_E)) - m^2}{((k_E + xp_E)^2 + x(1-x)p_E^2 + m^2)^2} \\
&= \frac{4ig^2}{(2\pi)^4} \int_0^1 dx \int d^4k_E \frac{(k_E^2 - x(1-x)p_E^2 - m^2)}{(k_E^2 + x(1-x)p_E^2 + m^2)^2} \\
&= \frac{4ig^2}{(2\pi)^4} \int_0^1 dx\, 2\pi^2 \int_0^\infty k_E^3\, dk_E \frac{(k_E^2 - Q^2)}{(k_E^2 + Q^2)^2} \\
&= \frac{4ig^2 \pi^2}{(2\pi)^4} \int_0^1 dx \int_0^\infty dy \frac{y(y - Q^2)}{(y + Q^2)^2} \\
&= \frac{ig^2}{4\pi^2} \int_0^1 dx \int_0^\infty dy\, y \left[\frac{1}{y + Q^2} - \frac{2Q^2}{(y + Q^2)^2} \right] \\
&= \frac{ig^2}{4\pi^2} \int_0^1 dx \int_0^\infty dy \left[1 - \frac{Q^2}{y + Q^2} - \frac{2Q^2}{y + Q^2} + \frac{2Q^4}{(y + Q^2)^2} \right] \\
&= \lim_{\Lambda \to \infty} \frac{ig^2}{4\pi^2} \int_0^1 dx \left[\int_0^{\Lambda^2} dy - 3Q^2 \int_0^{\Lambda^2} \frac{dy}{y + Q^2} \right. \\
&\qquad \left. + 2Q^4 \int_0^\infty \frac{dy}{(y + Q^2)^2} \right] \\
&= \lim_{\Lambda \to \infty} \frac{ig^2}{4\pi^2} \int_0^1 dx \left[\Lambda^2 - 3Q^2 \ln \frac{\Lambda^2 + Q^2}{Q^2} + 2Q^2 \right]
\end{aligned}
$$

15.3 CUT-OFF REGULARIZATION

$$= \lim_{\Lambda \to \infty} \frac{ig^2}{4\pi^2} \int_0^1 dx \left[\Lambda^2 - 3\overline{Q}^2 \ln \frac{\Lambda^2 + \overline{Q}^2}{\overline{Q}^2} + 2\overline{Q}^2 \right]$$

$$= f(p, m, g, \Lambda), \tag{15.32}$$

where we have identified $Q^2 = x(1-x)p_E^2 + m^2$ while $\overline{Q}^2 = -x(1-x)p^2 + m^2$ represents the function rotated to Minkowski space.

These two simple calculations show that Feynman amplitudes are functions of external momenta. In fact, a Feynman amplitude with n external lines is an analytic function of $(n-1)$ external momentum variables. This is because the overall momentum conservation eliminates one of the momentum variables. In the examples worked out above it is clear that one can Taylor expand the amplitudes around zero momentum of the external lines. Thus, for example, for the fermion self-energy in (15.29), we can write

$$= \frac{ig^2}{16\pi^2} \int_0^1 dx \, ((1-x)\not{p} + m)$$

$$\times \left[\ln \frac{\Lambda^2 - x(1-x)p^2 + xm^2 + (1-x)M^2}{-x(1-x)p^2 + xm^2 + (1-x)M^2} - 1 \right]$$

$$= \frac{ig^2}{16\pi^2} \int_0^1 dx \left[m \left(\ln \frac{\Lambda^2 + xm^2 + (1-x)M^2}{xm^2 + (1-x)M^2} - 1 \right) \right.$$

$$+ (1-x)\not{p} \left(\ln \frac{\Lambda^2 + xm^2 + (1-x)M^2}{xm^2 + (1-x)M^2} - 1 \right)$$

$$\left. + \text{ finite terms as } \Lambda \to \infty \right]$$

$$= \frac{ig^2}{16\pi^2} \int_0^1 dx \left[((1-x)\not{p} + m) \ln \frac{\Lambda^2}{xm^2 + (1-x)M^2} \right.$$

$$\left. + \text{finite terms as } \Lambda \to \infty \right]. \tag{15.33}$$

Similarly, for the scalar self-energy in (15.32) we can write

$$= \lim_{\Lambda \to \infty} \frac{ig^2}{4\pi^2} \int_0^1 dx \left[\Lambda^2 - 3\overline{Q}^2 \ln \frac{\Lambda^2 + \overline{Q}^2}{\overline{Q}^2} + 2\overline{Q}^2 \right]$$

$$= \lim_{\Lambda \to \infty} \frac{ig^2}{4\pi^2} \int_0^1 dx \left[\Lambda^2 - 3m^2 \ln \frac{\Lambda^2 + m^2}{m^2} + 2m^2 \right.$$

$$\left. + 3x(1-x)p^2 \ln \frac{\Lambda^2 + m^2}{m^2} \right.$$

$$\left. + \text{finite terms as } \Lambda \to \infty \right]$$

$$= \frac{ig^2}{4\pi^2} \int_0^1 dx \left[\Lambda^2 - 3m^2 \ln \frac{\Lambda^2}{m^2} + 3x(1-x)p^2 \ln \frac{\Lambda^2}{m^2} \right.$$

$$\left. + \text{finite terms as } \Lambda \to \infty \right]. \tag{15.34}$$

Thus we see explicitly that any Feynman amplitude can be Taylor expanded so that the divergent parts can be separated out as local functions (independent of momenta). Let us note here that if the theory is massless, then expanding around zero external momenta would be disastrous. This is because of the infrared divergences of the theory which can be seen in the above examples by setting $m = 0$. To avoid this problem, in massless theories we Taylor expand the amplitudes around a nonzero but finite value of the external momentum.

15.4 Pauli-Villars regularization

Although regularizing a Feynman diagram by cutting off contributions from large values of momentum seems natural, it is clear that such a regularization violates manifest Poincaré invariance of the quantum theory. Furthermore, a cut-off can lead to violation of gauge invariance in gauge theories by giving a mass to the gauge boson. Similarly, a lattice regularization which regularizes a theory by defining it on a discrete space-time lattice (we will not go into a detailed discussion of lattice regularization), while quite useful, leads to a lack of manifest rotational invariance (although in the continuum limit this symmetry is recovered). Therefore, it is useful to find

15.4 PAULI-VILLARS REGULARIZATION

a covariant regularization scheme which also respects gauge invariance. The Pauli-Villars regularization provides such a regularization scheme.

To discuss the Pauli-Villars regularization scheme, let us consider QED (quantum electrodynamics) in the Feynman gauge (see (9.80) and (9.125))

$$\mathcal{L} = -\frac{1}{4}F_{\mu\nu}F^{\mu\nu} + i\overline{\psi}\slashed{D}\psi - m\overline{\psi}\psi - \frac{1}{2}(\partial_\mu A^\mu)^2, \qquad (15.35)$$

which leads to the Feynman rules

$$\begin{aligned}
& = iG_{F,\mu\nu}(p) = -\frac{i\eta_{\mu\nu}}{p^2},\\
& = iS_F(p) = \frac{i(\slashed{p}+m)}{p^2 - m^2},\\
& = -(2\pi)^4 ie\gamma^\mu \, \delta^4(p+q+r). \quad (15.36)
\end{aligned}$$

With these Feynman rules, we can now calculate various 1PI amplitudes in QED. For example, in this theory the fermion self-energy at one loop takes the form (as usual we factor out the overall energy-momentum conserving delta function along with a factor of $(2\pi)^4$)

$$= (-ie)^2 \int \frac{d^4k}{(2\pi)^4} \gamma^\mu \frac{i(\slashed{k}+\slashed{p}+m)}{(k+p)^2 - m^2} \gamma^\nu \left(-i\frac{\eta_{\mu\nu}}{k^2 - \mu^2}\right)$$

$$\begin{aligned}
&= -\frac{e^2}{(2\pi)^4} \int d^4k \, \frac{-2(\slashed{k}+\slashed{p})+4m}{((k+p)^2-m^2)(k^2-\mu^2)} \\
&= -\frac{e^2}{(2\pi)^4} \int id^4k_E \, \frac{2(\slashed{k}_E+\slashed{p}_E)+4m}{((k_E+p_E)^2+m^2)(k_E^2+\mu^2)} \\
&= -\frac{ie^2}{(2\pi)^4} \int dx \int d^4k_E \\
&\quad \times \frac{2(\slashed{k}_E+\slashed{p}_E)+4m}{\left((k_E+xp_E)^2+x(1-x)p_E^2+xm^2+(1-x)\mu^2\right)^2} \\
&= -\frac{ie^2}{(2\pi)^4} \int dx\, d^4k_E \, \frac{2\slashed{k}_E+2(1-x)\slashed{p}_E+4m}{(k_E^2+Q^2(m,\mu))^2} \\
&= -\frac{ie^2}{16\pi^2} \int dx\, dk_E^2 \, \frac{k_E^2\left(2(1-x)\slashed{p}_E+4m\right)}{(k_E^2+Q^2(m,\mu))^2} \\
&= -\frac{ie^2}{8\pi^2} \int dx\, ((1-x)\slashed{p}_E+2m) \\
&\quad \times \int_0^\infty dk_E^2 \left(\frac{1}{k_E^2+Q^2(m,\mu)} - \frac{Q^2(m,\mu)}{(k_E^2+Q^2(m,\mu))^2}\right) \\
&= -\frac{ie^2}{8\pi^2} \int dx((1-x)\slashed{p}_E+2m)\left(\int_0^\infty \frac{dk_E^2}{k_E^2+Q^2(m,\mu)} - 1\right),
\end{aligned}$$
(15.37)

where we have defined

$$Q^2(m,\mu) = x(1-x)p_E^2 + xm^2 + (1-x)\mu^2, \tag{15.38}$$

and have introduced a mass μ for the photon (to regulate infrared divergences) which can be taken to zero at the end of the calculations. We have also used gamma matrix identities in (2.111) and (2.112) as well as (15.9) for the angular integral. (The Dirac gamma matrices have also been rotated to Euclidean space as discussed in (15.26).)

The momentum integral is, of course, divergent (logarithmically by power counting) as is the case in the earlier calculation of the fermion self-energy with the Yukawa coupling (see (15.27)). In the

15.4 Pauli-Villars regularization

Pauli-Villars regularization, we define the theory with a minimal coupling of the photon to another fermion which we assume to be heavy as well as introduce a second massive photon which couples to the normal as well as the heavy fermions. Namely, let us add to the Lagrangian density of QED (15.35) another gauge invariant term of the form

$$\mathcal{L}' = \frac{1}{4}F'_{\mu\nu}F'^{\mu\nu} - \frac{\Lambda^2}{2}A'_\mu A'^\mu - i\overline{\psi}'\slashed{D}(A+A')\psi'$$
$$+\Lambda\overline{\psi}'\psi' + e\overline{\psi}\slashed{A}'\psi, \qquad (15.39)$$

where we assume that $\Lambda \gg m, \mu$ and ψ', A'_μ are fictitious heavy fields introduced to regularize diagrams. (The covariant derivative in (15.39) is defined with the gauge field combination $A_\mu + A'_\mu$.) We note that the sign of the Lagrangian density for these fictitious fields is opposite to that of the standard fields and hence they act as ghost fields and subtract out contributions. The idea is to take $\Lambda \to \infty$ at the end and in that limit the heavy fields do not propagate and, therefore, decouple. However, these additional interacting fields give an additional contribution to the fermion self-energy (15.37) coming from the diagram where a heavy photon is being exchanged. The contribution from the heavy photon to the self-energy would be exactly the same as what we have already calculated in (15.37) except for a sign and $\mu \to \Lambda$. Therefore, without doing any further calculation, we can write the additional contribution to the fermion self-energy as

$$I' = \frac{ie^2}{8\pi^2}\int dx\,((1-x)\slashed{p}_E + 2m)\left(\int_0^\infty \frac{dk_E^2}{k_E^2 + Q^2(m,\Lambda)} - 1\right), \qquad (15.40)$$

so that the effective regularized fermion self-energy can be written as

$$I^{(\text{reg})} = I + I' = -\frac{ie^2}{8\pi^2}\int dx\,((1-x)\slashed{p}_E + 2m)\ln\frac{Q^2(m,\Lambda)}{Q^2(m,\mu)}, \qquad (15.41)$$

which is finite for any given value of Λ. The final result can now be rotated back to Minkowski space and has the form

$$I^{(\text{reg})} = \frac{ie^2}{8\pi^2} \int dx \; ((1-x)\slashed{p} - 2m) \ln \frac{\overline{Q}^2(m,\Lambda)}{\overline{Q}^2(m,\mu)}, \qquad (15.42)$$

where $\overline{Q}^2(m,\mu) = -x(1-x)p^2 + xm^2 + (1-x)\mu^2$.

We can also calculate the one loop photon self-energy in QED in a similar manner. The Feynman rules (15.36) lead to

$$= i\Pi_{\mu\nu}(p)$$

$$= -(-ie)^2 \operatorname{Tr} \int \frac{d^4k}{(2\pi)^4} \gamma_\mu \frac{i(\slashed{k}+m)}{k^2 - m^2} \gamma_\nu \frac{i(\slashed{k}+\slashed{p}+m)}{(k+p)^2 - m^2}$$

$$= -\frac{4e^2}{(2\pi)^4} \int d^4k \frac{k_\mu(k+p)_\nu + k_\nu(k+p)_\mu - \eta_{\mu\nu} k \cdot (k+p) + m^2 \eta_{\mu\nu}}{(k^2 - m^2)((k+p)^2 - m^2)}$$

$$= -\frac{4e^2}{(2\pi)^4} \int d^4k \int dx$$
$$\times \left[\frac{k_\mu(k+p)_\nu + k_\nu(k+p)_\mu - \eta_{\mu\nu}(k \cdot (k+p) - m^2)}{((k+xp)^2 + x(1-x)p^2 - m^2)^2} \right]$$

$$= -\frac{4e^2}{(2\pi)^4} \int d^4k \int dx$$
$$\times \left[\frac{2k_\mu k_\nu - 2x(1-x)p_\mu p_\nu - \eta_{\mu\nu}(k^2 - (x(1-x)p^2 + m^2))}{(k^2 + \overline{Q}^2(m))^2} \right]$$

$$= -\frac{4e^2}{(2\pi)^4} \int dx \int d^4k$$
$$\times \left[\frac{2x(1-x)(\eta_{\mu\nu}p^2 - p_\mu p_\nu) + 2k_\mu k_\nu - \eta_{\mu\nu}(k^2 + \overline{Q}^2(m))}{(k^2 + \overline{Q}^2(m))^2} \right], \qquad (15.43)$$

15.4 Pauli-Villars regularization

where we have used the gamma matrix identities in (2.115) and (2.116) and have defined

$$\overline{Q}^2(m) = x(1-x)p^2 - m^2. \tag{15.44}$$

The overall negative sign in (15.43) reflects the fact that we are evaluating a fermion loop.

To evaluate the momentum integral, we can rotate (15.43) to Euclidean space $((k_0, p_0) \to i(k_{0E}, p_{0E}), (k^2, p^2) \to -(k_E^2, p_E^2), \eta_{00} \to \delta_{00}, \Pi_{00} \to (i)^2 \Pi_{00}^E, \Pi_{0i} \to (i)\Pi_{0i}^E)$ and we have

$$i\Pi_{\mu\nu}^E = -\frac{4ie^2}{(2\pi)^4} \int dx \int d^4 k_E$$

$$\times \left[\frac{2x(1-x)(\delta_{\mu\nu} p_E^2 - p_{\mu E} p_{\nu E}) + 2k_{\mu E} k_{\nu E} - \delta_{\mu\nu}(k_E^2 + Q^2(m))}{(k_E^2 + Q^2(m))^2} \right], \tag{15.45}$$

where $Q^2(m) = x(1-x)p_E^2 + m^2$. We can now integrate each of the terms in (15.45) individually. For terms involving $k_{\mu E} k_{\nu E}$ we can use symmetric integration $(k_{\mu E} k_{\nu E} \to \frac{1}{4}\delta_{\mu\nu} k_E^2)$ and in this way we have

$$\int d^4 k_E \frac{2k_{\mu E} k_{\nu E}}{(k_E^2 + Q^2(m))^2} = \int d^4 k_E \frac{\frac{1}{2}\delta_{\mu\nu} k_E^2}{(k_E^2 + Q^2(m))^2}$$

$$= \pi^2 \int dy \frac{y^2}{(y + Q^2(m))^2} \frac{1}{2}\delta_{\mu\nu}$$

$$= \frac{\pi^2}{2}\delta_{\mu\nu} \int dy \frac{(y+Q^2(m))^2 - 2(y+Q^2(m))Q^2(m) + (Q^2(m))^2}{(y+Q^2(m))^2}$$

$$= \frac{\pi^2}{2}\delta_{\mu\nu} \int dy \left[1 - \frac{2Q^2(m)}{y+Q^2(m)} + \frac{(Q^2(m))^2}{(y+Q^2(m))^2} \right],$$

$$\int d^4 k_E \frac{1}{k_E^2 + Q^2(m)} (-\delta_{\mu\nu})$$

$$= -\pi^2 \delta_{\mu\nu} \int dy \frac{y}{y + Q^2(m)}$$

$$= -\pi^2 \delta_{\mu\nu} \int dy \left[1 - \frac{Q^2(m)}{y + Q^2(m)}\right],$$

so that we have

$$\int d^4 k_E \left(\frac{2k_{\mu E} k_{\nu E}}{(k_E^2 + Q^2(m))^2} - \frac{\delta_{\mu\nu}}{k_E^2 + Q^2(m)}\right)$$

$$= -\frac{\pi^2}{2} \delta_{\mu\nu} \int dy \left[1 - \frac{(Q^2(m))^2}{(y + Q^2(m))^2}\right]. \tag{15.46}$$

Here we have used (15.9) in the intermediate steps. Putting this back into (15.45), we obtain for the self-energy of the photon (in Euclidean space),

$$i\Pi_{\mu\nu}^E = -\frac{ie^2}{4\pi^2} \int dx \int_0^\infty dy \bigg[\frac{2x(1-x)\left(\delta_{\mu\nu} p_E^2 - p_{\mu E} p_{\nu E}\right) y}{(y + Q^2(m))^2}$$

$$+ \delta_{\mu\nu} \left(-\frac{1}{2} + \frac{(Q^2(m))^2}{2(y + Q^2(m))^2}\right)\bigg]$$

$$= -\frac{ie^2}{4\pi^2} \int dx \int_0^\infty dy \bigg[2x(1-x)\left(\delta_{\mu\nu} p_E^2 - p_{\mu E} p_{\nu E}\right)$$

$$\times \left(\frac{1}{y + Q^2(m)} - \frac{Q^2(m)}{(y + Q^2(m))^2}\right)$$

$$+ \delta_{\mu\nu} \left(-\frac{1}{2} + \frac{(Q^2(m))^2}{2(y + Q^2(m))^2}\right)\bigg]$$

$$= -\frac{ie^2}{4\pi^2} \int dx \bigg[2x(1-x)(\delta_{\mu\nu} p_E^2 - p_{\mu E} p_{\nu E})\left(\int_0^\infty \frac{dy}{y + Q^2(m)} - 1\right)$$

$$+ \frac{1}{2} \delta_{\mu\nu} \left(-\int_0^\infty dy + Q^2(m)\right)\bigg]. \tag{15.47}$$

This integral, as we see, is quadratically divergent. However, recalling that the photon has coupling to fictitious heavy fermions,

15.4 PAULI-VILLARS REGULARIZATION

we will have another contribution coming from the heavy fermion loop with exactly the same contribution except for a sign and $m \to \Lambda$. As a result, the regularized photon self-energy will have the form

$$i\Pi^E_{\mu\nu}(p_E, m) - i\Pi^E_{\mu\nu}(p_E, \Lambda)$$
$$= -\frac{ie^2}{4\pi^2} \int dx \left[2x(1-x) \left(\delta_{\mu\nu} p_E^2 - p_{\mu E} p_{\nu E}\right) \ln \frac{Q^2(\Lambda)}{Q^2(m)} \right.$$
$$\left. -\frac{1}{2}\delta_{\mu\nu} \left(Q^2(\Lambda) - Q^2(m)\right) \right]. \qquad (15.48)$$

This shows that the diagram is finite for any given value of Λ. However, this has the unpleasant feature that the photon self-energy has developed a mass term (term proportional to $\delta_{\mu\nu}$, see, for example, section **14.1**) which would violate gauge invariance. Such a term can be cancelled by adding another set of heavy fermion fields. Let us assume that we have two sets of heavy ghost fermions with masses Λ_1, Λ_2 and charges $\sqrt{c_1}e, \sqrt{c_2}e$ respectively. Then the regularized photon amplitude with these two sets of heavy fermions will have the form

$$i\Pi^{E(\text{reg})}_{\mu\nu}(p_E) = i\Pi^E_{\mu\nu}(p_E, m) - ic_1 \Pi^E_{\mu\nu}(p_E, \Lambda_1) - ic_2 \Pi^E_{\mu\nu}(p_E, \Lambda_2)$$
$$= -\frac{ie^2}{4\pi^2} \int dx \left[2x(1-x) \left(\delta_{\mu\nu} p_E^2 - p_{\mu E} p_{\nu E}\right) \right.$$
$$\times \left(\int_0^\infty dy \left(\frac{1}{y + Q^2(m)} - \frac{c_1}{y + Q^2(\Lambda_1)} - \frac{c_2}{y + Q^2(\Lambda_2)} \right) \right.$$
$$\left. -(1 - c_1 - c_2) \right)$$
$$-\frac{1}{2}\delta_{\mu\nu} \left((1 - c_1 - c_2) \left(\int_0^\infty dy \right) \right.$$
$$\left.\left. -\left(Q^2(m) - c_1 Q^2(\Lambda_1) - c_2 Q^2(\Lambda_2)\right) \right) \right]$$
$$= -\frac{ie^2}{4\pi^2} \int dx \left[2x(1-x) \left(\delta_{\mu\nu} p_E^2 - p_{\mu E} p_{\nu E}\right) \right.$$

$$\times \left(\int_0^\infty dy \left(\frac{1}{y+Q^2(m)} - \frac{c_1}{y+Q^2(\Lambda_1)} - \frac{c_2}{y+Q^2(\Lambda_2)} \right) \right.$$

$$\left. -(1-c_1-c_2) \right)$$

$$-\frac{1}{2}\delta_{\mu\nu}\left((1-c_1-c_2)\left(\int_0^\infty dy\right)\right.$$

$$\left.-(1-c_1-c_2)x(1-x)p_E^2 - (m^2 - c_1\Lambda_1^2 - c_2\Lambda_2^2)\right)\bigg]. \quad (15.49)$$

It is clear, therefore, that if the parameters satisfy

$$c_1 + c_2 = 1, \quad c_1\Lambda_1^2 + c_2\Lambda_2^2 = m^2, \tag{15.50}$$

then, the quadratic divergences as well as the logarithmic divergences in (15.49) would cancel leading to

$$i\Pi_{\mu\nu}^{E(\text{reg})} = -\frac{ie^2}{2\pi^2}\left(\delta_{\mu\nu}p_E^2 - p_{\mu E}p_{\nu E}\right)\int dx\, x(1-x)$$

$$\times \left[c_1 \ln Q^2(\Lambda_1) + c_2 \ln Q^2(\Lambda_2) - \ln Q^2(m)\right], (15.51)$$

which represents a gauge invariant photon self-energy (it is manifestly transverse). Rotating this back to Minkowski space we obtain

$$i\Pi_{\mu\nu}^{(\text{reg})} = -\frac{ie^2}{2\pi^2}\left(\eta_{\mu\nu}p^2 - p_\mu p_\nu\right)\int dx\, x(1-x)$$

$$\times \left[c_1 \ln \overline{Q}^2(\Lambda_1) + c_2 \ln \overline{Q}^2(\Lambda_2) - \ln \overline{Q}^2(m)\right]. (15.52)$$

We note here that even though we are working in the Feynman gauge, since the diagram for the photon self-energy in (15.43) does not involve the photon propagator, the result (15.52) holds for any covariant gauge with an arbitrary value of the gauge fixing parameter ξ.

In general, Pauli-Villars regularization involves introducing, in a gauge invariant manner, a set of heavy fields with masses Λ_i and charges ($\sqrt{c_i}e$) with $i = 1, 2, \cdots, n$ depending on the nature of the divergence in the diagram such that the regularized amplitude

$$I^{(\text{reg})} = I(p, m) - \sum_{i=1}^{n} c_i I_i(p, \Lambda_i), \tag{15.53}$$

with the conditions on the parameters given by

$$\sum_i c_i = 1, \quad \sum_i c_i \Lambda_i^2 = m^2, \quad \cdots, \tag{15.54}$$

will be finite and gauge invariant.

15.5 Dimensional regularization

From the analysis of the divergence structure of amplitudes in the last two sections it is clear that the divergences are functions of the dimension of space time. For example, the integral

$$\int \frac{\mathrm{d}^4 k}{(k^2 - m^2)\left((k+p)^2 - M^2\right)}, \tag{15.55}$$

is logarithmically divergent in four dimensions. However, in less than four dimensions it is finite. Thus we see that a method of regularizing Feynman integrals can very well be to analytically continue the integral into n dimensions where it is well defined (well behaved). This procedure is known as dimensional regularization. Furthermore, since gauge invariance is independent of the number of space time dimensions, it is by construction a gauge invariant regularization scheme. In fact, it is quite useful in studying non-Abelian gauge theories for which the Pauli-Villars regularization does not work. It is worth pointing out here that the Pauli-Villars regularization is sufficient to regularize all the Feynman amplitudes in an Abelian gauge theory in a gauge invariant way. In non-Abelian theories, however, there are graphs of the form shown in Fig. 15.4 which cannot be regularized by the Pauli-Villars method. (Very roughly speaking the Pauli-Villars regularization is a gauge invariant regularization but not a gauge covariant regularization which is why it fails in non-Abelian theories.) We should mention here that dimensional regularization has its own problems which we will discuss at the end of this section, but it is worth pointing out here that manipulations with dimensional regularization are extremely simple which is why it is quite useful.

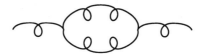

Figure 15.4: One of the one loop graphs contributing to the gluon self-energy.

In dimensional regularization we analytically continue the theory to n dimensions. For example, if we are looking at the ϕ^4 theory described by the Lagrangian density

$$\mathcal{L} = \frac{1}{2}\partial_\mu \phi \partial^\mu \phi - \frac{M^2}{2}\phi^2 - \frac{\lambda}{4!}\phi^4, \tag{15.56}$$

we can carry out the (canonical) dimensional analysis in n dimensions to determine

$$\begin{aligned}
[\phi] &= \frac{n-2}{2}, \\
[M] &= 1, \\
[\lambda] &= 4 - n.
\end{aligned} \tag{15.57}$$

Therefore, the coupling constant for the quartic interaction in (15.56) carries a dimension and introducing an arbitrary mass scale μ we can then write the Lagrangian density (15.56) for the theory in n dimensions as

$$\mathcal{L} = \frac{1}{2}\partial_\mu \phi \partial^\mu \phi - \frac{M^2}{2}\phi^2 - \frac{\lambda \mu^{4-n}}{4!}\phi^4, \tag{15.58}$$

so that the coupling constant λ is rendered dimensionless. The Feynman rules for the theory (15.58) are given by

$$\text{---} \!\!\! \underset{p}{\text{---}} \!\!\! \text{---} \quad = \frac{i}{p^2 - M^2},$$

$$\underset{\substack{p_1 \quad p_2}}{\overset{p_4 \quad p_3}{\times}} \quad = -(2\pi)^4 i\lambda \mu^{4-n} \delta^n(p_1 + p_2 + p_3 + p_4).$$

(15.59)

We need to be careful about manipulating tensors in n dimensions. For example, since each vector index takes n values, it follows that in n dimensions we have

$$\eta^\mu_\mu = n. \tag{15.60}$$

Furthermore, to calculate amplitudes in n dimensions, we need to evaluate the basic integral

$$\begin{aligned} I &= \int \frac{d^n k}{(2\pi)^n} \frac{1}{(k^2 + 2k \cdot p - M^2)^\alpha} \\ &= \frac{i}{(2\pi)^n} \int d^n k_E \frac{(-1)^\alpha}{(k_E^2 + 2k_E \cdot p_E + M^2)^\alpha} \\ &= \frac{i(-1)^\alpha}{(2\pi)^n} \int d^n k_E \frac{1}{((k_E + p_E)^2 - p_E^2 + M^2)^\alpha} \\ &= \frac{i(-1)^\alpha}{(2\pi)^n} \int d^n k_E \frac{1}{(k_E^2 + Q^2)^\alpha}, \end{aligned} \tag{15.61}$$

where we have rotated to Euclidean space, shifted the variable of integration and have defined $Q^2 = M^2 - p_E^2$.

To evaluate this integral, let us note that the integrand is spherically symmetric (independent of angular coordinates) and, therefore, we can separate out the angular part of the integral

$$\mathrm{d}^n k_E = k_E^{n-1} \mathrm{d}k_E \mathrm{d}\Omega,$$

where, in n dimensions, we can write

$$\mathrm{d}\Omega = \mathrm{d}\theta_1 \sin^{n-2}\theta_1 \mathrm{d}\theta_2 \sin^{n-3}\theta_2 \ldots \mathrm{d}\theta_{n-1}, \quad n \geq 2. \quad (15.62)$$

The integral over the angles $\int \mathrm{d}\Omega$ can be evaluated from the simple Gaussian integral as follows. Let us consider the basic n dimensional Gaussian integral whose value is given by

$$\int \mathrm{d}^n k_E \, e^{-\frac{k_E^2}{2}} = (2\pi)^{\frac{n}{2}}. \quad (15.63)$$

On the other hand, we can evaluate the integral in (15.63) in spherical coordinates as

$$\begin{aligned}
(2\pi)^{\frac{n}{2}} &= \int \mathrm{d}^n k_E \, e^{-\frac{k_E^2}{2}} \\
&= \int \mathrm{d}\Omega \int_0^\infty \mathrm{d}k_E \, k_E^{n-1} \, e^{-\frac{k_E^2}{2}} \\
&= \int \mathrm{d}\Omega \int_0^\infty \frac{1}{2}\mathrm{d}k_E^2 \, (k_E^2)^{\frac{n}{2}-1} \, e^{-\frac{k_E^2}{2}} \\
&= \int \mathrm{d}\Omega \, 2^{\frac{n}{2}-1} \int_0^\infty \mathrm{d}y \, y^{\frac{n}{2}-1} \, e^{-y} \\
&= 2^{\frac{n}{2}-1} \Gamma\left(\frac{n}{2}\right) \int \mathrm{d}\Omega,
\end{aligned} \quad (15.64)$$

where we have defined $y = \frac{k_E^2}{2}$ in the intermediate step and have used the definition of the gamma function. This determines the value of the angular integral to be

$$\int \mathrm{d}\Omega = \frac{(2\pi)^{\frac{n}{2}}}{2^{\frac{n}{2}-1}\Gamma\left(\frac{n}{2}\right)} = \frac{2\pi^{\frac{n}{2}}}{\Gamma\left(\frac{n}{2}\right)}. \quad (15.65)$$

Let us compare this with known results in lower dimensions that we are familiar with. We see that

15.5 Dimensional regularization

$$n = 2: \quad \int d\Omega = \int d\theta = 2\pi,$$

$$n = 3: \quad \int d\Omega = \int d\theta_1 \sin\theta_1 d\theta_2 = 4\pi,$$

$$n = 4: \quad \int d\Omega = \int d\theta_1 \sin^2\theta_1 d\theta_2 \sin\theta_2 d\theta_3 = 2\pi^2, \quad (15.66)$$

which agree with what we already know from explicitly doing these integrals.

Thus, for the basic integral (15.61) we have

$$\begin{aligned}
I &= \int \frac{d^n k}{(2\pi)^n} \frac{1}{(k^2 + 2k \cdot p - M^2)^\alpha} \\
&= \frac{i(-1)^\alpha}{(2\pi)^n} \int d\Omega \int_0^\infty dk_E \, k_E^{n-1} \frac{1}{(k_E^2 + Q^2)^\alpha} \\
&= \frac{i(-1)^\alpha}{(2\pi)^n} \frac{2\pi^{\frac{n}{2}}}{\Gamma\left(\frac{n}{2}\right)} \int_0^\infty dk_E \, k_E^{n-1} \frac{1}{(k_E^2 + Q^2)^\alpha} \\
&= \frac{i(-1)^\alpha}{(2\pi)^n} \frac{2\pi^{\frac{n}{2}}}{\Gamma\left(\frac{n}{2}\right)} \int_0^\infty d\left(\frac{k_E}{Q}\right) \left(\frac{k_E}{Q}\right)^{n-1} \frac{Q^n}{Q^{2\alpha}} \frac{1}{\left(1 + \frac{k_E^2}{Q^2}\right)^\alpha} \\
&= \frac{i(-1)^\alpha}{(2\pi)^n} \frac{\pi^{\frac{n}{2}}}{\Gamma\left(\frac{n}{2}\right)} \frac{1}{(Q^2)^{\alpha - \frac{n}{2}}} \times 2 \int_0^\infty dt \, t^{n-1} (1+t^2)^{-\alpha}. \quad (15.67)
\end{aligned}$$

Let us recall that the beta function is defined as

$$B(p, q) = \frac{\Gamma(p)\Gamma(q)}{\Gamma(p+q)} = 2 \int_0^\infty dt \, t^{2p-1} \left(1+t^2\right)^{-p-q}. \quad (15.68)$$

Therefore, we see that with the identification

$$p = \frac{n}{2}, \quad q = \alpha - p = \alpha - \frac{n}{2}, \quad (15.69)$$

the integral (15.67) can be written as

$$
\begin{aligned}
I &= \frac{i(-1)^\alpha}{(2\pi)^n} \frac{\pi^{\frac{n}{2}}}{\Gamma\left(\frac{n}{2}\right)} \frac{1}{(Q^2)^{\alpha-\frac{n}{2}}} \frac{\Gamma\left(\frac{n}{2}\right)\Gamma\left(\alpha-\frac{n}{2}\right)}{\Gamma(\alpha)} \\
&= \frac{i\pi^{\frac{n}{2}}}{(2\pi)^n} \frac{(-1)^\alpha}{\Gamma(\alpha)} \frac{\Gamma\left(\alpha-\frac{n}{2}\right)}{(Q^2)^{\alpha-\frac{n}{2}}}.
\end{aligned}
\tag{15.70}
$$

Rotating back to Minkowski space, we see that this gives us our basic integral as

$$
\begin{aligned}
I &= \int \frac{\mathrm{d}^n k}{(2\pi)^n} \frac{1}{(k^2 + 2k\cdot p - M^2)^\alpha} \\
&= \frac{i\pi^{\frac{n}{2}}}{(2\pi)^n} \frac{(-1)^\alpha}{\Gamma(\alpha)} \frac{\Gamma\left(\alpha-\frac{n}{2}\right)}{(p^2 + M^2)^{\alpha-\frac{n}{2}}}.
\end{aligned}
\tag{15.71}
$$

This basic integral generates all other integrals that we need for evaluating amplitudes in n dimensions and, in fact, any other formula can be obtained from (15.71) by differentiation. For example, using (15.71) we note that we can write (this result can also be obtained by shifting the variable of integration)

$$
\begin{aligned}
I_\mu &= \int \frac{\mathrm{d}^n k}{(2\pi)^n} \frac{k_\mu}{(k^2 + 2k\cdot p - M^2)^\alpha} \\
&= -\frac{1}{2(\alpha-1)} \frac{\partial}{\partial p^\mu} \int \frac{\mathrm{d}^n k}{(2\pi)^n} \frac{1}{(k^2 + 2k\cdot p - M^2)^{\alpha-1}} \\
&= -\frac{1}{2(\alpha-1)} \frac{\partial}{\partial p^\mu} \left[\frac{i\pi^{\frac{n}{2}}}{(2\pi)^n} \frac{(-1)^{\alpha-1}}{\Gamma(\alpha-1)} \frac{\Gamma\left(\alpha-1-\frac{n}{2}\right)}{(p^2 + M^2)^{\alpha-1-\frac{n}{2}}} \right] \\
&= -\frac{i\pi^{\frac{n}{2}}}{(2\pi)^n} \frac{(-1)^{\alpha-1}}{2(\alpha-1)\Gamma(\alpha-1)} \frac{\left(-2\left(\alpha-1-\frac{n}{2}\right)p_\mu\right)}{(p^2 + M^2)^{\alpha-\frac{n}{2}}} \\
&\quad \times \Gamma\left(\alpha-1-\frac{n}{2}\right) \\
&= \frac{i\pi^{\frac{n}{2}}}{(2\pi)^n} \frac{(-1)^{\alpha-1}}{\Gamma(\alpha)} \frac{p_\mu}{(p^2 + M^2)^{\alpha-\frac{n}{2}}} \Gamma\left(\alpha-\frac{n}{2}\right).
\end{aligned}
\tag{15.72}
$$

Similarly we can obtain

15.5 Dimensional regularization

$$I_{\mu\nu} = \int \frac{d^n k}{(2\pi)^n} \frac{k_\mu k_\nu}{(k^2 + 2k \cdot p - M^2)^\alpha}$$

$$= \frac{i\pi^{\frac{n}{2}}}{(2\pi)^n} \frac{(-1)^\alpha}{\Gamma(\alpha)} \frac{1}{(p^2 + M^2)^{\alpha - \frac{n}{2}}}$$

$$\times \left[p_\mu p_\nu \Gamma\left(\alpha - \frac{n}{2}\right) - \frac{1}{2}\eta_{\mu\nu} (p^2 + M^2) \Gamma\left(\alpha - 1 - \frac{n}{2}\right) \right], \quad (15.73)$$

and so on.

With these basic integration formulae, let us now calculate various amplitudes in the ϕ^4 theory (15.58). First of all, the one loop scalar self-energy takes the form (see also (15.8))

$$= -\frac{i\lambda\mu^{4-n}}{2} \int \frac{d^n k}{(2\pi)^n} \frac{i}{k^2 - M^2}$$

$$= \frac{\lambda\mu^{4-n}}{2} \frac{i\pi^{\frac{n}{2}}}{(2\pi)^n} \frac{(-1)}{\Gamma(1)} \frac{\Gamma\left(1 - \frac{n}{2}\right)}{(M^2)^{1-\frac{n}{2}}}$$

$$= -\frac{i\lambda\mu^{4-n}}{2} \frac{\pi^{\frac{n}{2}}}{(2\pi)^n} \frac{\Gamma\left(1 - \frac{n}{2}\right)}{(M^2)^{1-\frac{n}{2}}}. \quad (15.74)$$

Let us next set $n = 4 - \epsilon$, i.e., we are analytically continuing away from four dimensions (to a lower dimension where the integral is well defined). At the end of our calculations we should, of course, take the limit $n \to 4$ which translates to $\epsilon \to 0$. With this, the amplitude (15.74) becomes

$$= -\frac{i\lambda\mu^\epsilon}{2} \frac{\pi^{\frac{n}{2}}}{(2\pi)^n} \frac{\Gamma\left(1 - 2 + \frac{\epsilon}{2}\right)}{(M^2)^{1-2+\frac{\epsilon}{2}}}$$

$$= -\frac{i\lambda\mu^\epsilon}{2} \frac{\pi^{\frac{n}{2}}}{(2\pi)^n} \frac{\Gamma\left(-1 + \frac{\epsilon}{2}\right)}{(M^2)^{-1+\frac{\epsilon}{2}}}. \tag{15.75}$$

Let us next work out some of the identities involving the gamma functions that will be useful to us. In the limit $\epsilon \to 0$, we have

$$\begin{aligned}
\Gamma\left(3 - \frac{n}{2}\right) &= \Gamma\left(1 + \frac{\epsilon}{2}\right) \simeq 1 - \frac{\epsilon}{2}\gamma, \\
\Gamma\left(2 - \frac{n}{2}\right) &= \Gamma\left(\frac{\epsilon}{2}\right) = \frac{2}{\epsilon}\Gamma\left(1 + \frac{\epsilon}{2}\right) \\
&\simeq \frac{2}{\epsilon}\left(1 - \frac{\epsilon}{2}\gamma\right) = \frac{2}{\epsilon} - \gamma, \\
\Gamma\left(1 - \frac{n}{2}\right) &= \Gamma\left(-1 + \frac{\epsilon}{2}\right) = \frac{\Gamma\left(\frac{\epsilon}{2}\right)}{-1 + \frac{\epsilon}{2}} \\
&\simeq -\left(1 + \frac{\epsilon}{2}\right)\left(\frac{2}{\epsilon} - \gamma\right) \simeq -\frac{2}{\epsilon} + (\gamma - 1), \tag{15.76}
\end{aligned}$$

where γ denotes the Euler's constant defined in (15.12). Similarly, we have

$$\begin{aligned}
\frac{\pi^{\frac{n}{2}}}{(2\pi)^n} &= \frac{1}{(4\pi)^{\frac{n}{2}}} = \frac{1}{(4\pi)^{2-\frac{\epsilon}{2}}} \\
&\simeq \frac{1}{16\pi^2}\left(1 + \frac{\epsilon}{2}\ln 4\pi\right). \tag{15.77}
\end{aligned}$$

Using (15.76) as well as (15.77), the scalar self-energy (15.75) takes the form

15.5 Dimensional regularization

$$\begin{aligned}
&= -\frac{i\lambda\mu^\epsilon}{2} \frac{\pi^{\frac{n}{2}}}{(2\pi)^n} M^2 M^{-\epsilon} \Gamma\left(-1+\frac{\epsilon}{2}\right) \\
&= -\frac{i\lambda}{2} \frac{\pi^{\frac{n}{2}}}{(2\pi)^n} M^2 \Gamma\left(-1+\frac{\epsilon}{2}\right) \left(\frac{M^2}{\mu^2}\right)^{-\frac{\epsilon}{2}} \\
&\simeq -\frac{i\lambda}{2} \frac{M^2}{16\pi^2} \left(1+\frac{\epsilon}{2}\ln 4\pi\right) \left(-\frac{2}{\epsilon}+(\gamma-1)\right) \left(1-\frac{\epsilon}{2}\ln\frac{M^2}{\mu^2}\right) \\
&= -\frac{i\lambda}{2} \frac{M^2}{16\pi^2} \left(-\frac{2}{\epsilon}+(\gamma-1)-\ln 4\pi\right) \left(1-\frac{\epsilon}{2}\ln\frac{M^2}{\mu^2}\right) \\
&\simeq -\frac{i\lambda M^2}{32\pi^2} \left(-\frac{2}{\epsilon}+\ln\frac{M^2}{\mu^2}+(\gamma-1)-\ln 4\pi\right) \\
&= -\frac{i\lambda M^2}{32\pi^2} \left(-\frac{2}{\epsilon}+\ln\frac{M^2}{4\pi\mu^2}+(\gamma-1)\right). \quad (15.78)
\end{aligned}$$

We note here that for $M^2 = 0$, this graph would be regularized to zero in dimensional regularization which is the type of argument used in (13.122).

Similarly, we can calculate the one loop correction to the vertex function shown in Fig. 15.3 and, for simplicity, we will set all the external momenta to vanish as we had done earlier in (15.19). In this case, the amplitude in n dimensions takes the form

$$\begin{aligned}
&= \frac{3}{2}(-i\lambda\mu^\epsilon)^2 \int \frac{d^n k}{(2\pi)^n} \left(\frac{i}{k^2-M^2}\right)^2 \\
&= \frac{3\lambda^2\mu^{2\epsilon}}{2} \int \frac{d^n k}{(2\pi)^n} \frac{1}{(k^2-M^2)^2}
\end{aligned}$$

$$= \frac{3\lambda^2 \mu^{2\epsilon}}{2} \frac{i}{(4\pi)^{\frac{n}{2}}} \frac{(-1)^2}{\Gamma(2)} \frac{\Gamma\left(2 - \frac{n}{2}\right)}{(M^2)^{2-\frac{n}{2}}}$$

$$\simeq \frac{3i\lambda^2 \mu^\epsilon}{2} \frac{1}{16\pi^2} \left(1 + \frac{\epsilon}{2} \ln 4\pi\right) \left(\frac{2}{\epsilon} - \gamma\right) \left(\frac{M^2}{\mu^2}\right)^{-\frac{\epsilon}{2}}$$

$$\simeq \frac{3i\lambda^2 \mu^\epsilon}{32\pi^2} \left(\frac{2}{\epsilon} - \gamma + \ln 4\pi\right) \left(1 - \frac{\epsilon}{2} \ln \frac{M^2}{\mu^2}\right)$$

$$\simeq \frac{3i\lambda^2 \mu^\epsilon}{32\pi^2} \left(\frac{2}{\epsilon} - \ln \frac{M^2}{\mu^2} - \gamma + \ln 4\pi\right)$$

$$= \frac{3i\lambda^2 \mu^\epsilon}{32\pi^2} \left(\frac{2}{\epsilon} - \ln \frac{M^2}{4\pi\mu^2} - \gamma\right), \tag{15.79}$$

where we have used (15.76) and (15.77). We note that these amplitudes are well behaved (regularized) for any finite value of ϵ. However, as we take the limit $\epsilon \to 0$ (to go to four dimensions), the amplitudes diverge.

15.5.1 Calculations in QED. Let us next calculate one loop amplitudes in QED described by the Lagrangian density in the Feynman gauge (see (15.35))

$$\mathcal{L} = -\frac{1}{4} F_{\mu\nu} F^{\mu\nu} + i\overline{\psi} \slashed{D} \psi - m\overline{\psi}\psi - \frac{1}{2} (\partial_\mu A^\mu)^2. \tag{15.80}$$

In n dimensions, the canonical dimensions of various fields can be easily determined to be (we are assuming $n = 4 - \epsilon$)

$$[A_\mu] = \frac{n-2}{2} = 1 - \frac{\epsilon}{2},$$
$$[\psi] = [\overline{\psi}] = \frac{n-1}{2} = \frac{3}{2} - \frac{\epsilon}{2},$$
$$[e] = \frac{\epsilon}{2}. \tag{15.81}$$

As a result, to make the coupling constant dimensionless, we let

$$e \to e \mu^{\frac{\epsilon}{2}}, \tag{15.82}$$

15.5 Dimensional regularization

where μ is an arbitrary mass scale so that the covariant derivative in the theory is understood to have the form

$$D_\mu \psi = \left(\partial_\mu + ie\mu^{\frac{\epsilon}{2}} A_\mu\right)\psi. \tag{15.83}$$

As a result, the Feynman rules of the theory in $n = 4 - \epsilon$ dimensions are given by

$$\begin{aligned}
&= iG_{F,\mu\nu}(p) = -\frac{i\eta_{\mu\nu}}{p^2}, \\
&= iS_F(p) = \frac{i(\not{p}+m)}{p^2 - m^2}, \\
&= -(2\pi)^4 ie\mu^{\frac{\epsilon}{2}} \gamma^\mu \, \delta^4(p+q+r).
\end{aligned} \tag{15.84}$$

With these Feynman rules, we can calculate the fermion self-energy at one loop which gives (see also (15.37) and the discussion following that equation)

$$\begin{aligned}
&= \left(-ie\mu^{\frac{\epsilon}{2}}\right)^2 \int \frac{d^n k}{(2\pi)^n} \gamma^\mu \frac{i\left((\not{k}+\not{p})+m\right)}{(k+p)^2 - m^2} \gamma^\nu \left(-\frac{i\eta_{\mu\nu}}{k^2}\right) \\
&= -e^2 \mu^\epsilon \int \frac{d^n k}{(2\pi)^n} \frac{\gamma^\mu \left((\not{k}+\not{p})+m\right)\gamma_\mu}{k^2 \left((k+p)^2 - m^2\right)}.
\end{aligned} \tag{15.85}$$

We note here that if we use the algebra of the gamma matrices (1.79) in n dimensions as well as (15.60) we obtain

$$\gamma_\mu \gamma^\mu = n,$$
$$\gamma_\mu \not{k} \gamma^\mu = (2-n)\not{k}. \tag{15.86}$$

Using this, the one loop fermion self energy in (15.85) takes the form

$$= -e^2 \mu^\epsilon \int \frac{\mathrm{d}^n k}{(2\pi)^n} \frac{(2-n)(\not{k}+\not{p}) + nm}{k^2 ((k+p)^2 - m^2)} \tag{15.87}$$

$$= -e^2 \mu^\epsilon \int \mathrm{d}x \frac{\mathrm{d}^n k}{(2\pi)^n} \frac{(2-n)(\not{k}+\not{p}) + nm}{((k+xp)^2 + x(1-x)p^2 - xm^2)^2}$$

$$= -e^2 \mu^\epsilon \int \mathrm{d}x \frac{\mathrm{d}^n k}{(2\pi)^n} \frac{(2-n)(\not{k}+(1-x)\not{p}) + nm}{(k^2 - Q^2)^2}$$

$$= -e^2 \mu^\epsilon \int \mathrm{d}x \frac{\mathrm{d}^n k}{(2\pi)^n} \frac{(2-n)(1-x)\not{p} + nm}{(k^2 - Q^2)^2} \tag{15.88}$$

$$= -e^2 \mu^\epsilon \int \mathrm{d}x \, ((2-n)(1-x)\not{p} + nm) \frac{i}{(4\pi)^{\frac{n}{2}}} \frac{(-1)^2}{\Gamma(2)} \frac{\Gamma\left(2 - \frac{n}{2}\right)}{(Q^2)^{2-\frac{n}{2}}}$$

$$\simeq \frac{ie^2}{8\pi^2} \int \mathrm{d}x \, ((1-x)\not{p} - 2m) \left(1 - \frac{\epsilon}{2}\right) \left(1 + \frac{\epsilon}{2}\ln 4\pi\right)$$
$$\times \left(\frac{2}{\epsilon} - \gamma\right) \left(1 - \frac{\epsilon}{2}\ln \frac{Q^2}{\mu^2}\right)$$

$$\simeq \frac{ie^2}{8\pi^2} \int \mathrm{d}x \, ((1-x)\not{p} - 2m) \left(1 - \frac{\epsilon}{2}\right) \left(\frac{2}{\epsilon} - \gamma + \ln 4\pi\right)$$
$$\times \left(1 - \frac{\epsilon}{2}\ln \frac{Q^2}{\mu^2}\right)$$

$$\simeq \frac{ie^2}{8\pi^2} \int \mathrm{d}x \, ((1-x)\not{p} - 2m) \left(\frac{2}{\epsilon} - \ln \frac{Q^2}{\mu^2} - \gamma - 1 + \ln 4\pi\right)$$

$$= \frac{ie^2}{8\pi^2} \int \mathrm{d}x \, ((1-x)\not{p} - 2m)$$
$$\times \left(\frac{2}{\epsilon} - \ln \frac{xm^2 - x(1-x)p^2}{4\pi\mu^2} - \gamma - 1\right), \tag{15.89}$$

15.5 DIMENSIONAL REGULARIZATION

where we have defined $Q^2 = -x(1-x)p^2 + xm^2$ and have used (15.76) as well as (15.77). This can be compared with (15.42).

Similarly the photon self energy at one loop can also be calculated to have the form (the overall negative sign is because of the fermion loop)

$$\begin{aligned}
& \quad\quad\text{(diagram)} \quad = i\Pi_{\mu\nu}(p) \\
&= -\left(-ie\mu^{\frac{\epsilon}{2}}\right)^2 \int \frac{d^n k}{(2\pi)^n} \, \text{Tr} \, \gamma_\mu \frac{i(\slashed{k}+m)}{k^2-m^2} \gamma_\nu \frac{i(\slashed{k}+\slashed{p}+m)}{(k+p)^2-m^2} \\
&= -e^2 \mu^\epsilon \int \frac{d^n k}{(2\pi)^n} \, \frac{\text{Tr} \, \gamma_\mu(\slashed{k}+m)\gamma_\nu(\slashed{k}+\slashed{p}+m)}{(k^2-m^2)\left((k+p)^2-m^2\right)}. \quad (15.90)
\end{aligned}$$

Using the n-dimensional identities (see (2.115) and (2.116)),

$$\begin{aligned}
&\text{Tr} \, \mathbb{1} = n, \\
&\text{Tr} \, \slashed{A} = 0, \\
&\text{Tr} \, \slashed{A}\slashed{B} = nA \cdot B, \\
&\text{Tr} \, \slashed{A}\slashed{B}\slashed{C} = 0, \\
&\text{Tr} \, \gamma_\mu\gamma_\nu\gamma_\lambda\gamma_\rho = n\left(\eta_{\mu\nu}\eta_{\lambda\rho} - \eta_{\mu\lambda}\eta_{\nu\rho} + \eta_{\mu\rho}\eta_{\nu\lambda}\right), \quad (15.91)
\end{aligned}$$

the photon self-energy (15.90) becomes (we factor out $(-e^2\mu^\epsilon)$ for simplicity and will restore this later)

$$\begin{aligned}
&= \int \frac{d^n k}{(2\pi)^n} \frac{n(k_\mu(k+p)_\nu - \eta_{\mu\nu} k \cdot (k+p) + k_\nu(k+p)_\mu + m^2\eta_{\mu\nu})}{(k^2-m^2)((k+p)^2-m^2)} \\
&= n \int dx \frac{d^n k}{(2\pi)^n} \frac{k_\mu(k+p)_\nu + k_\nu(k+p)_\mu - \eta_{\mu\nu}(k \cdot (k+p) - m^2)}{((k+xp)^2 + x(1-x)p^2 - m^2)^2} \\
&= n \int dx \frac{d^n k}{(2\pi)^n}
\end{aligned}$$

$$\times \frac{2k_\mu k_\nu - 2x(1-x)p_\mu p_\nu - \eta_{\mu\nu}(k^2 - x(1-x)p^2 - m^2)}{(k^2 - Q^2)^2}$$

$$= n \int dx \frac{d^n k}{(2\pi)^n}$$

$$\times \frac{\eta_{\mu\nu}(x(1-x)p^2 + m^2) - 2x(1-x)p_\mu p_\nu + 2k_\mu k_\nu - \eta_{\mu\nu}k^2}{(k^2 - Q^2)^2}, \quad (15.92)$$

where we have defined $Q^2 = -x(1-x)p^2 + m^2$.

The integral in (15.92) contains three different kinds of terms which can be integrated using the standard formulae in (15.71)-(15.73). Let us look at the terms without any k in the numerator and this takes the form (we now put back the factor $(-e^2\mu^\epsilon)$)

$$= -ne^2\mu^\epsilon \int dx \frac{d^n k}{(2\pi)^n} \frac{\eta_{\mu\nu}(x(1-x)p^2 + m^2) - 2x(1-x)p_\mu p_\nu}{(k^2 - Q^2)^2}$$

$$= -ne^2\mu^\epsilon \int dx \left(\eta_{\mu\nu}\left(x(1-x)p^2 + m^2\right) - 2x(1-x)p_\mu p_\nu\right)$$

$$\times \frac{i}{(4\pi)^{\frac{n}{2}}} \frac{(-1)^2}{\Gamma(2)} \frac{\Gamma\left(2 - \frac{n}{2}\right)}{(Q^2)^{2-\frac{n}{2}}}$$

$$\simeq -\frac{ie^2}{4\pi^2} \int dx \left(\eta_{\mu\nu}(x(1-x)p^2 + m^2) - 2x(1-x)p_\mu p_\nu\right)$$

$$\times \left(1 - \frac{\epsilon}{4}\right)\left(1 + \frac{\epsilon}{2}\ln 4\pi\right)\left(\frac{2}{\epsilon} - \gamma\right)\left(1 - \frac{\epsilon}{2}\ln\frac{Q^2}{\mu^2}\right)$$

$$\simeq -\frac{ie^2}{4\pi^2} \int dx \left(\eta_{\mu\nu}\left(x(1-x)p^2 + m^2\right) - 2x(1-x)p_\mu p_\nu\right)$$

$$\times \left(\frac{2}{\epsilon} - \gamma - \frac{1}{2} + \ln 4\pi\right)\left(1 - \frac{\epsilon}{2}\ln\frac{Q^2}{\mu^2}\right)$$

$$\simeq -\frac{ie^2}{4\pi^2} \int dx \left[\left(\frac{2}{\epsilon} - \ln\frac{Q^2}{4\pi\mu^2} - \gamma - \frac{1}{2}\right)\right.$$

$$\left. \times \left(\eta_{\mu\nu}\left(x(1-x)p^2 + m^2\right) - 2x(1-x)p_\mu p_\nu\right)\right], \quad (15.93)$$

where we have used (15.76) as well as (15.77). Let us next look at the terms with $k_\mu k_\nu$ in the numerator separately which leads to

15.5 Dimensional regularization

$$= -ne^2\mu^\epsilon \int dx \frac{d^n k}{(2\pi)^n} \frac{2k_\mu k_\nu}{(k^2 - Q^2)^2}$$

$$= -2ne^2\mu^\epsilon \int dx \frac{1}{2}\eta_{\mu\nu} \frac{i}{(4\pi)^{\frac{n}{2}}} \frac{(-1)^{2-1}\,\Gamma\left(1-\frac{n}{2}\right)}{\Gamma(2)} \frac{1}{(Q^2)^{1-\frac{n}{2}}}$$

$$\simeq \frac{ie^2 \eta_{\mu\nu}}{4\pi^2} \int dx\, Q^2 \left(1 - \frac{\epsilon}{4}\right)\left(1 + \frac{\epsilon}{2}\ln 4\pi\right)$$

$$\times \left(-\frac{2}{\epsilon} + (\gamma - 1)\right)\left(1 - \frac{\epsilon}{2}\ln\frac{Q^2}{\mu^2}\right)$$

$$\simeq \frac{ie^2}{4\pi^2}\eta_{\mu\nu} \int dx\, Q^2 \left(-\frac{2}{\epsilon} + \gamma - \frac{1}{2} - \ln 4\pi\right)\left(1 - \frac{\epsilon}{2}\ln\frac{Q^2}{\mu^2}\right)$$

$$\simeq \frac{ie^2}{4\pi^2}\eta_{\mu\nu} \int dx\, Q^2 \left(-\frac{2}{\epsilon} + \ln\frac{Q^2}{\mu^2} + \gamma - \frac{1}{2} - \ln 4\pi\right)$$

$$= \frac{ie^2}{4\pi^2}\eta_{\mu\nu} \int dx\, (-x(1-x)p^2 + m^2)$$

$$\times \left(-\frac{2}{\epsilon} + \ln\frac{Q^2}{4\pi\mu^2} + \gamma - \frac{1}{2}\right). \tag{15.94}$$

Similarly, the term with k^2 in the numerator leads to

$$= -ne^2\mu^\epsilon \int dx \frac{d^n k}{(2\pi)^n} \frac{-\eta_{\mu\nu} k^2}{(k^2 - Q^2)^2}$$

$$= ne^2\mu^\epsilon \eta_{\mu\nu} \int dx \frac{n}{2} \frac{i}{(4\pi)^{\frac{n}{2}}} \frac{(-1)^{2-1}\,\Gamma\left(1-\frac{n}{2}\right)}{\Gamma(2)} \frac{1}{(Q^2)^{1-\frac{n}{2}}}$$

$$\simeq -\frac{ie^2}{4\pi^2}\eta_{\mu\nu} \int dx\, Q^2 \left(1 - \frac{\epsilon}{4}\right)\left(2 - \frac{\epsilon}{2}\right)\left(1 + \frac{\epsilon}{2}\ln 4\pi\right)$$

$$\times \left(-\frac{2}{\epsilon} + \gamma - 1\right)\left(1 - \frac{\epsilon}{2}\ln\frac{Q^2}{\mu^2}\right)$$

$$\simeq -\frac{ie^2}{4\pi^2}\eta_{\mu\nu} \int dx\, Q^2\, (2 + \epsilon(-1 + \ln 4\pi))$$

$$\times \left(-\frac{2}{\epsilon} + \ln\frac{Q^2}{\mu^2} + \gamma - 1\right)$$

$$\simeq -\frac{ie^2}{4\pi^2}\eta_{\mu\nu}\int \mathrm{d}x\, Q^2 \left(-\frac{4}{\epsilon}+2-2\ln 4\pi+2\ln\frac{Q^2}{\mu^2}+2\gamma-2\right)$$

$$= -\frac{ie^2}{4\pi^2}\eta_{\mu\nu}\int \mathrm{d}x\, \left(-x(1-x)p^2+m^2\right)$$

$$\times \left(-\frac{4}{\epsilon}+2\ln\frac{Q^2}{4\pi\mu^2}+2\gamma\right). \tag{15.95}$$

Adding the contributions from (15.93)-(15.95), the photon self energy in (15.92) takes the final form

$$= -\frac{ie^2}{2\pi^2}(\eta_{\mu\nu}p^2-p_\mu p_\nu)\int \mathrm{d}x\, x(1-x)\left(\frac{2}{\epsilon}-\ln\frac{Q^2}{4\pi\mu^2}-\gamma-\frac{1}{2}\right). \tag{15.96}$$

This shows that the photon self energy graph is completely transverse which is consistent with gauge invariance (see also (15.52)). The photon does not acquire a mass term and we note here that even though we have chosen to work with the Feynman gauge, since the photon self-energy does not involve the photon propagator, this result holds in any covariant gauge with the gauge fixing term ξ arbitrary. Since there is no longitudinal term in the photon self-energy, the gauge fixing Lagrangian density does not receive any quantum correction which can also be seen from the BRST identities for QED.

Let us next calculate the one loop amplitude corresponding to vertex correction (charge renormalization). For simplicity we will put the momentum of the external photon equal to zero.

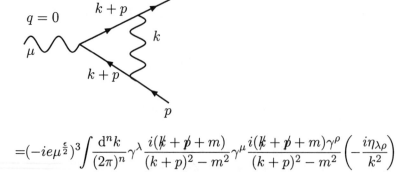

$$= -e^3 \mu^{\frac{3\epsilon}{2}} \int \frac{d^n k}{(2\pi)^n} \frac{\gamma^\lambda(\slashed{k}+\slashed{p}+m)\gamma^\mu(\slashed{k}+\slashed{p}+m)\gamma_\lambda}{k^2\left((k+p)^2-m^2\right)^2}. \tag{15.97}$$

Using the n-dimensional identities (see also (2.113))

$$\gamma^\lambda \slashed{A} \gamma^\mu \slashed{A} \gamma_\lambda = (2-n)(2A^\mu \slashed{A} - \gamma^\mu A^2), \tag{15.98}$$

as well as (15.86) we have

$$\begin{aligned}
&\gamma^\lambda(\slashed{k}+\slashed{p}+m)\gamma^\mu(\slashed{k}+\slashed{p}+m)\gamma_\lambda \\
&= \gamma^\lambda(\slashed{k}+\slashed{p})\gamma^\mu(\slashed{k}+\slashed{p})\gamma_\lambda + m\gamma^\lambda(\gamma^\mu(\slashed{k}+\slashed{p})+(\slashed{k}+\slashed{p})\gamma^\mu)\gamma_\lambda \\
&\quad + m^2 \gamma^\lambda \gamma^\mu \gamma_\lambda \\
&= (2-n)(2(\slashed{k}+\slashed{p})(k+p)^\mu - (k+p)^2 \gamma^\mu) + 2nm(k+p)^\mu \\
&\quad + (2-n)m^2 \gamma^\mu \\
&= -(2-n)\gamma^\mu((k+p)^2 - m^2) \\
&\quad + 2((2-n)(\slashed{k}+\slashed{p}) + nm)(k+p)^\mu. \tag{15.99}
\end{aligned}$$

Therefore, we can write the amplitude (15.97) as

$$= -e^3 \mu^{\frac{3\epsilon}{2}} \int \frac{d^n k}{(2\pi)^n}$$

$$\times \frac{-(2-n)\gamma^\mu((k+p)^2 - m^2) + 2((2-n)(\slashed{k}+\slashed{p}) + nm)(k+p)^\mu}{k^2\left((k+p)^2-m^2\right)^2}$$

$$= -e^3 \mu^{\frac{3\epsilon}{2}} \int \frac{d^n k}{(2\pi)^n} \left[\frac{-(2-n)\gamma^\mu}{k^2((k+p)^2 - m^2)} \right.$$

$$\left. + \frac{2\left((2-n)(\slashed{k}+\slashed{p}) + nm\right)(k+p)^\mu}{k^2((k+p)^2 - m^2)^2} \right]. \tag{15.100}$$

Let us next use (15.23) as well as the identity

$$\frac{1}{AB^2} = -\frac{\partial}{\partial B}\frac{1}{AB} = -\frac{\partial}{\partial B}\int dx \frac{1}{((1-x)A+xB)^2}$$
$$= \int dx \frac{2x}{((1-x)A+xB)^3}, \qquad (15.101)$$

to combine the denominators in (15.100) which leads to

$$\frac{1}{k^2((k+p)^2-m^2)} = \int dx \frac{1}{((k+xp)^2-Q^2)^2},$$
$$\frac{1}{k^2((k+p)^2-m^2)^2} = \int dx \frac{2x}{((k+xp)^2-Q^2)^3}, \qquad (15.102)$$

where we have defined $Q^2 = -x(1-x)p^2 + xm^2$. Using this in (15.100) and shifting the variable of integration, we obtain for the vertex correction

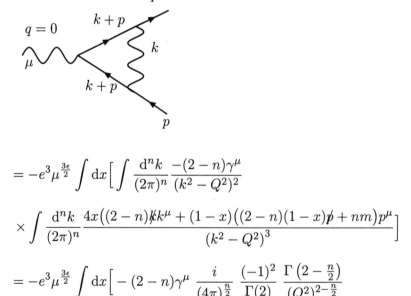

$$= -e^3\mu^{\frac{3\epsilon}{2}} \int dx \Bigg[\int \frac{d^n k}{(2\pi)^n} \frac{-(2-n)\gamma^\mu}{(k^2-Q^2)^2}$$
$$\times \int \frac{d^n k}{(2\pi)^n} \frac{4x\big((2-n)\slashed{k}k^\mu + (1-x)((2-n)(1-x)\slashed{p}+nm)p^\mu\big)}{(k^2-Q^2)^3}\Bigg]$$
$$= -e^3\mu^{\frac{3\epsilon}{2}} \int dx \Bigg[-(2-n)\gamma^\mu \frac{i}{(4\pi)^{\frac{n}{2}}} \frac{(-1)^2}{\Gamma(2)} \frac{\Gamma\left(2-\frac{n}{2}\right)}{(Q^2)^{2-\frac{n}{2}}}$$

15.5 Dimensional regularization

$$+4x(2-n)\gamma^\mu \frac{i}{(4\pi)^{\frac{n}{2}}} \frac{(-1)^3}{\Gamma(3)} \left(-\frac{1}{2}\right) \frac{\Gamma\left(2-\frac{n}{2}\right)}{(Q^2)^{2-\frac{n}{2}}}$$

$$+4x(1-x)\left((2-n)(1-x)\slashed{p}+nm\right)p^\mu \frac{i}{(4\pi)^{\frac{n}{2}}} \frac{(-1)^3}{\Gamma(3)} \frac{\Gamma\left(3-\frac{n}{2}\right)}{(Q^2)^{3-\frac{n}{2}}}\Bigg]$$

$$= -ie^3\mu^{\frac{3\epsilon}{2}} \int dx \frac{\Gamma\left(2-\frac{n}{2}\right)}{(4\pi)^{\frac{n}{2}}} \left[-\frac{(2-n)(1-x)\gamma^\mu}{(Q^2)^{2-\frac{n}{2}}}\right.$$

$$\left. -\frac{(4-n)x(1-x)\left((2-n)(1-x)\slashed{p}+nm\right)p^\mu}{(Q^2)^{3-\frac{n}{2}}}\right] \quad (15.103)$$

$$= \frac{ie^3\mu^{\frac{\epsilon}{2}}}{16\pi^2} \int dx \left[(1-x)\gamma^\mu \left(1+\frac{\epsilon}{2}\ln 4\pi\right)\left(\frac{2}{\epsilon}-\gamma\right)(-2+\epsilon)\right.$$

$$\times \left(1-\frac{\epsilon}{2}\ln\frac{Q^2}{\mu^2}\right)$$

$$\left. +\frac{x(1-x)(-2(1-x)\slashed{p}+4m)p^\mu}{-x(1-x)p^2+xm^2} \epsilon\left(\frac{2}{\epsilon}-\gamma\right)\right]$$

$$= -\frac{ie^3\mu^{\frac{\epsilon}{2}}}{8\pi^2} \int dx(1-x)\left[\gamma^\mu\left(\frac{2}{\epsilon}-\ln\frac{Q^2}{4\pi\mu^2}-\gamma-1\right)\right.$$

$$\left. +\frac{2((1-x)\slashed{p}-2m)p^\mu}{-(1-x)p^2+m^2}\right]. \quad (15.104)$$

It is worth noting from this derivation that at every step if we compare (15.87), (15.88) and (15.89) with (15.100), (15.103) and (15.104) respectively, we see that the fermion self-energy and the vertex function are related as (see also (9.109) and recall that the coupling constant in the present case is $e\mu^{\frac{\epsilon}{2}}$)

$$\frac{\partial}{\partial p^\mu} \underset{p \quad k+p \quad p}{\overset{k}{\frown}} = -\frac{1}{e\mu^{\frac{\epsilon}{2}}} \underset{p \quad k+p \underset{q=0}{\lessgtr} k+p \quad p}{\overset{k}{\frown}},$$

$$(15.105)$$

which we recognize as the Ward-Takahashi identity. (Note that the divergent parts as well as the finite parts satisfy the identity which demonstrates that dimensional regularization preserves gauge invariance.) The power of dimensional regularization is quite obvious from these calculations. It is gauge invariant and extremely simple to manipulate with. However, it has its own drawbacks. It is not naively applicable if the calculation involves quantities that typically exist only in four dimensions. For example, we know that

$$\gamma_5 = i\gamma^0\gamma^1\gamma^2\gamma^3 = -\frac{i}{4!}\epsilon_{\mu\nu\lambda\rho}\gamma^\mu\gamma^\nu\gamma^\lambda\gamma^\rho, \tag{15.106}$$

is defined only in four dimensions (both γ_5 and the Levi-Civita tensor $\epsilon_{\mu\nu\lambda\rho}$ are manifestly four dimensional quantities). Analytic continuation of these to other dimensions is nontrivial. Hence if the theory or the physical quantity of interest involves such objects, naive dimensional regularization leads to incorrect answers. Such quantities often occur in physics like in the chiral anomaly which is related to the life time of the π meson decaying through

$$\pi^0 \to 2\gamma. \tag{15.107}$$

There are several ways to address this issue with the γ_5 matrix. We will discuss this further in section **16.6** where we discuss chiral anomaly.

In addition to the regularization methods that we have discussed in this chapter, there exist several other important regularization schemes, for example, the higher derivative method, the point splitting method, the ζ-function regularization etc. The lattice regularization is also quite useful. However, we will not go into the details of these other methods here. We simply note here that we select a particular regularization scheme depending on the theory (or the problem) that we are analyzing.

15.6 References

1. J. Schwinger, Physical Review **74**, 1439 (1948).

15.6 References

2. R. Feynman, Physical Review **76**, 769 (1949).

3. W. Pauli and F. Villars, Reviews of Modern Physics **21**, 434 (1949).

4. J. Schwinger, Physical Review **82**, 664 (1951).

5. C. G. Bollini and J. J. Giambiaggi, Nuovo Cimento **12B**, 20 (1972).

6. G. 't Hooft and M. Veltman, Nuclear Physics **B44**, 189 (1972).

7. G. 't Hooft and M. Veltman, *Diagrammar*, CERN preprint (1973).

8. N. N. Bogoliubov and D. V. Shirkov, *Introduction to the theory of Quantized Fields*, Nauka, Moscow (1984).

CHAPTER 16
Renormalization theory

16.1 Superficial degree of divergence

In the last chapter, we have seen that loop diagrams in various theories become divergent and need to be regularized. Furthermore, we have also seen that since Feynman amplitudes are analytic functions of external momenta, we can expand them around some reference momentum value (in massive theories conventionally chosen to be zero) so that the regularized divergent parts can be isolated as local terms. The next step in understanding the process of renormalization is to devise a method for determining which Feynman diagrams in a theory will be divergent as well as the nature of the divergence (without actually evaluating the integrals) and this is achieved through the notion of the superficial degree of divergence of a graph. To define this let us start with the Lagrangian density for a given theory

$$\mathcal{L} = \mathcal{L}_0 + \sum_i \mathcal{L}_i, \tag{16.1}$$

where $\mathcal{L}_0$ denotes the sum of the free Lagrangian densities for all the field variables in the theory and each $\mathcal{L}_i$ representing some interaction is a monomial in the basic field variables and derivatives. In units of $\hbar = c = 1$ which we have been using, the action for the theory (16.1) is dimensionless. As a result, the Lagrangian density must have canonical dimension 4 in these units (since we are in four space time dimensions, in n dimensions it should be n). In our units, it is easy to check that

$$[M] = [L]^{-1} = 1. \tag{16.2}$$

Let us note from the form of the free Lagrangian density for a real scalar field

$$\mathcal{L}_0^{(\phi)} = \frac{1}{2}\partial_\mu\phi\partial^\mu\phi - \frac{M^2}{2}\phi^2, \tag{16.3}$$

that since

$$[\partial_\mu] = [L]^{-1} = 1, \tag{16.4}$$

and

$$\left[\mathcal{L}_0^{(\phi)}\right] = 4, \tag{16.5}$$

we have

$$[\phi] = 1. \tag{16.6}$$

Namely, the canonical dimension of a scalar field (in our units and in four dimensions) is 1. (The mass term also has the correct dimension with this assignment.) Similarly the free fermion Lagrangian density

$$\mathcal{L}_0^{(\psi)} = i\overline{\psi}\slashed{\partial}\psi - m\overline{\psi}\psi, \tag{16.7}$$

leads to

$$[\psi] = [\overline{\psi}] = \frac{3}{2}. \tag{16.8}$$

In fact, we can show that boson fields in general have canonical dimension 1 as long as we ignore gravitation (even gravitation can be included in this category depending on what we consider as the basic field variable) while that of the fermion fields is $\frac{3}{2}$ in four space time dimensions.

Let us further introduce the notations f_i, b_i and d_i to denote respectively the number of fermions, bosons and derivatives at an interaction vertex following from the interaction Lagrangian density $\mathcal{L}_i$. Thus, for example, for the Yukawa interaction

$$\mathcal{L}_Y = g\overline{\psi}\psi\phi, \tag{16.9}$$

16.1 SUPERFICIAL DEGREE OF DIVERGENCE

we have

$$f = 2, \ b = 1, \ d = 0. \tag{16.10}$$

On the other hand, the ϕ^4 interaction

$$\mathcal{L}_I = \frac{\lambda}{4!} \phi^4, \tag{16.11}$$

leads to

$$f = 0, \ b = 4, \ d = 0. \tag{16.12}$$

Similarly, for an interaction of the form

$$\mathcal{L}_I = h\overline{\psi}\gamma_\mu\psi\partial^\mu\phi, \quad \text{or,} \quad \kappa\overline{\psi}\sigma^{\mu\nu}\psi F_{\mu\nu}, \tag{16.13}$$

we have

$$f = 2, \ b = 1, \ d = 1, \tag{16.14}$$

and so on.

With these notations we are now ready to introduce the notion of the superficial degree of divergence of a Feynman graph. This is defined to be the difference between the number of momenta in the numerator arising from the loop integrations as well as derivative couplings and the number of momenta in the denominator arising from the propagators. For example, we can determine easily the superficial degree of divergence of the following Feynman graphs as

$$D = 4 - 2 = 2,$$

$$D = 4 - 2 \times 2 = 0,$$

$$D = 4 - 2 - 1 = 1,$$

$$D = 4 - 2 \times 1 = 2,$$

$$D = 8 - 3 \times 2 = 2,$$

$$D = 4 - 2 - 2 \times 1 = 0,$$

$$D = 8 - 5 \times 2 = -2. \qquad (16.15)$$

If $D = 0$ for a diagram, we say that the diagram is superficially logarithmically divergent. Similarly, diagrams with $D = 1$ or 2 are known

16.1 SUPERFICIAL DEGREE OF DIVERGENCE

respectively to have superficial linear or quadratic divergences. If $D < 0$ for a graph, then the Feynman diagram is said to be superficially convergent. The meaning of the adjective "superficial" becomes clear once we look at the last graph in (16.15). Although this graph is superficially convergent, it is actually divergent since it contains a subgraph which is divergent. However, as we will see it is the notion of superficial degree of divergence which is useful in the study of renormalization theory.

Rather than calculating the superficial degree of divergence for each graph, let us develop a general formula for the superficial degree of divergence of any connected Feynman graph. Let us consider a Feynman diagram with

$$
\begin{aligned}
B &= \text{number of external bosons lines,} \\
I_B &= \text{number of internal boson lines,} \\
F &= \text{number of external fermion lines,} \\
I_F &= \text{number of internal fermion lines,} \\
n_i &= \text{number of vertices of the ith type from $\mathcal{L}_i$.}
\end{aligned}
\tag{16.16}
$$

There exist topological relations between these numbers. Since a vertex of $\mathcal{L}_i$ has b_i boson lines attached to it and since each of these lines can become either an external boson line or an internal boson line we must have

$$B + 2I_B = \sum_i n_i b_i. \tag{16.17}$$

The factor $2I_B$ reflects the fact that it takes two boson lines at two different vertices to form a propagator (internal boson line). Similarly, we have

$$F + 2I_F = \sum_i n_i f_i. \tag{16.18}$$

The superficial degree of divergence is easily seen to be given by

$$D = \sum_i n_i d_i + 2I_B + 3I_F - 4\sum_i n_i + 4. \tag{16.19}$$

The first term in (16.19) simply says that each derivative at a vertex gives rise to a momentum in the numerator and if there are n_i vertices of the ith type in a graph, this would lead to a power of momentum in the numerator $n_i d_i$ (which must be summed over all possible types of vertices). With each internal boson line is associated a momentum integration and a propagator. Thus each internal boson line effectively leads to two powers of momentum in the numerator. Similarly each internal fermion line adds three powers of the momentum to the numerator. At each vertex, however, energy-momentum has to be conserved and since a four dimensional delta function (expressing conservation of energy-momentum) has dimension -4, each vertex takes away four powers of momentum except for an overall delta function which is necessary for the overall energy-momentum conservation. The last two terms reflect this.

Using (16.17) and (16.18) and eliminating I_B and I_F from (16.19) we obtain

$$\begin{aligned} D &= \sum_i n_i d_i + \sum_i n_i b_i - B + \frac{3}{2}\sum_i n_i f_i - \frac{3}{2}F - 4\sum_i n_i + 4 \\ &= 4 - B - \frac{3}{2}F + \sum_i n_i \left(d_i + b_i + \frac{3}{2}f_i - 4\right) \\ &= 4 - B - \frac{3}{2}F + \sum_i n_i \delta_i, \end{aligned} \tag{16.20}$$

where we have defined

$$\delta_i = d_i + b_i + \frac{3}{2}f_i - 4. \tag{16.21}$$

This is known as the index of divergence of the interaction Lagrangian density $\mathcal{L}_i$ and can also be expressed as

$$\delta_i = \dim \mathcal{L}_i - 4, \tag{16.22}$$

where dim $\mathcal{L}_i$ is calculated only from the dimensions of the field variables and derivatives and not from any dimensionful parameters. In all theories we consider, the interaction Lagrangian density has dimension 4 and hence

$$\delta_i = 0. \tag{16.23}$$

In such cases (16.20) reduces to

$$D = 4 - B - \frac{3}{2}F. \tag{16.24}$$

Namely, in such cases the superficial degree of divergence is completely determined by the number of external lines in the graph. Let us note here that the superficial degree of divergence is a function of the number of space-time dimensions (we are working in four dimensions).

We can check the relation (16.24) against our explicit power counting calculations in (16.15),

$$D = 4 - 2 - 0 = 2,$$

$$D = 4 - 4 - 0 = 0,$$

$$D = 4 - 0 - \frac{3}{2} \times 2 = 1,$$

$$D = 4 - 2 - 0 = 2,$$

$$D = 4 - 2 - 0 = 2,$$

$$D = 4 - 1 - \frac{3}{2} \times 2 = 0,$$

$$D = 4 - 6 - 0 = -2, \qquad (16.25)$$

and they agree completely.

From the form of the formula for the superficial degree of divergence, it is clear that only a few graphs in a theory would have non-negative superficial degree of divergence. Thus for example, if we are considering the ϕ^4 theory, then only the following 1PI functions would be superficially divergent.

B	$D = 4 - B$
0	4
1	3
2	2
3	1
4	0

The zero point function contributes only to the zero point energy which can be eliminated by normal ordering the theory. The one point function is in principle divergent. However, such graphs do not exist in the ϕ^4 theory since the theory has the discrete symmetry

$$\phi \to -\phi, \qquad (16.26)$$

and the one point function violates this symmetry. Similarly, although the three point function can be superficially divergent it does not exist because of this discrete symmetry. Thus, only two graphs are superficially divergent, namely the 2-point and the 4-point functions. In the last chapter we have explicitly calculated the two point as well as the four point functions at one loop in the ϕ^4 theory where we have seen that these graphs are indeed divergent.

We have to develop a systematic way of making these graphs finite. Any higher point 1PI graph can, of course, contain these graphs as subgraphs and even though these graphs are superficially convergent, they will in fact be divergent. However, corresponding to each such graph we can define a skeleton graph. Thus for example, in the ϕ^4 theory, the graph for the six point function in Fig. 16.1

Figure 16.1: A superficially convergent graph with a divergent subgraph.

has the skeleton graph shown in Fig. 16.2.

Figure 16.2: The skeleton graph associated with Fig. 16.1.

Namely, the skeleton graph of a superficially convergent graph is a graph where no divergent subgraph can be found (divergent subgraphs shrunk to a point). The skeleton graph of a superficially convergent graph is, therefore, by definition convergent. The full graph can be obtained from the skeleton graph by making insertion of the two point and the four point functions at appropriate places. However, if we have a method of making these graphs, namely, $\Gamma^{(2)}$, and $\Gamma^{(4)}$, finite and regularization independent through some renormalization procedure, then the same procedure would make all the n point functions finite and regularization independent.

Let us next look at QED described by the Lagrangian density

$$\mathcal{L} = -\frac{1}{4}F_{\mu\nu}F^{\mu\nu} + i\overline{\psi}\slashed{D}\psi - m\overline{\psi}\psi - \frac{1}{2\xi}(\partial_\mu A^\mu)^2,$$

$$D_\mu \psi = (\partial_\mu + ieA_\mu)\psi. \tag{16.27}$$

The Feynman rules for the theory (in the Feynman gauge with $\xi = 1$) are

$$= iG_{F,\mu\nu}(p) = -\frac{i\eta_{\mu\nu}}{p^2},$$

$$= iS_F(p) = \frac{i(\slashed{p}+m)}{p^2 - m^2},$$

$$= -(2\pi)^4 ie\gamma^\mu \, \delta^4(p+q+r). \tag{16.28}$$

Let us now analyze the divergence structure of graphs in this theory.

B	F	$D = 4 - B - \frac{3}{2}F$
0	0	4
1	0	3
2	0	2
3	0	1
4	0	0
0	1	$\frac{5}{2}$
0	2	1
1	2	0

The zero point graphs are neglected because they can be taken care of by normal ordering. The 1 point boson graphs do not exist because they violate Lorentz invariance as well as gauge invariance. Similarly the 3 point boson graphs also do not exist. (Vanishing of photon graphs with an odd number of photons is a consequence of the symmetry of charge conjugation $\mathcal{C}$, also known as the Furry's theorem (see (11.188)).) The four photon graph, in this analysis, would appear to be superficially divergent, but it is actually finite because of gauge invariance. In fact, in gauge theories the actual degree of divergence may be softer than the naive power counting because of constraints coming from gauge invariance. Fermion graphs with only an even number of fermion lines can be nonzero because of Lorentz invariance (as well as fermion number conservation). Therefore, there are only three possible superficially divergent graphs, namely, the fermion and the photon self-energy as well as the fermion interaction vertex with the photon, and if we can somehow make these finite in a regularization independent manner, the theory will be well defined.

16.2 A brief history of renormalization

Dyson laid the foundations for the systematic study of renormalization in two papers in 1949 where he studied the renormalization of QED. He basically used the Schwinger-Dyson equations (to be discussed in the next section) and showed that most of the divergences can be absorbed into a redefinition of parameters of the theory. The class of graphs which he could not incorporate into his study are known as overlapping divergent graphs of the type shown in Fig. 16.3. Salam showed how the overlapping divergences can be handled in any theory.

Figure 16.3: The overlapping divergent graph in the fermion self-energy at two loops.

In QED things were a bit simpler because of the Ward-Takahashi identities (which relate various amplitudes, see for example, section **9.7**) and renormalization of QED was thought to be straightforward. However, Yang and Mills noticed that at the 14th order (in the coupling), the photon self-energy graph shown in Fig. 16.4 does lead to overlapping divergence and needed further prescription to handle this graph as Ward-Takahashi identities do not restrict the photon self-energy. (It was believed earlier that the photon self energy cannot have overlapping divergences.) Yang and Mills solved the problem of the photon self-energy and together with Salam's work as well as the subsequent work of Weinberg, the question of overlapping divergence was considered to be solved. (This is within the framework of integral equations for Green's functions and skeleton expansions.)

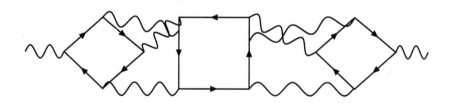

Figure 16.4: An overlapping divergent graph in the photon self-energy at the 14th order.

There are mainly two equivalent renormalization methods known as the multiplicative renormalization and the BPHZ renormalization. In multiplicative renormalization, we calculate 1PI Feynman amplitudes in perturbation theory (as we have done in the last chapter) until we encounter a divergent graph. The amplitude is then regularized with a momentum cut off (or dimensional regularization or any other regularization). The regularized amplitude is then Taylor expanded about zero external momenta (for massive theories) so as to separate out the local divergent parts. We then add counter terms to the Lagrangian density to exactly cancel these divergences. We continue calculating with this modified Lagrangian density and add more counter terms whenever faced with new divergences. This procedure renders the theory finite.

Let us consider a specific theory to see how this procedure works. We have seen that in the ϕ^4 theory only the two point and the four point functions are divergent. Thus we add counter terms to cancel these divergences. For example, at one loop we obtain the counter terms from calculating $\Gamma^{(2)}$ and $\Gamma^{(4)}$ as shown in Fig. 16.5 so that the divergent contributions from the one loop graph and the counter terms cancel and render the amplitude finite as shown in Fig. 16.6. The counter terms are clearly of one loop order (this is counted by the power of $\hbar$ in the coefficient which we have set to unity, but should be understood).

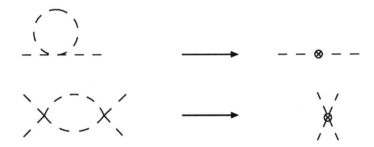

Figure 16.5: One loop counter terms for the two point and the four point functions in the ϕ^4 theory.

Including these one loop counter terms, at two loops the self energy graphs would, therefore, have the form shown in Fig. 16.7. These would be divergent and hence we have to add two loop counter terms for the two point function to cancel the new divergences at this order. Similarly the four point function would also need counter terms at the two loop level. However, no other 1PI graph would be divergent at this order. As an example, let us look at the 6 point function at two loops in Fig. 16.1 whose superficial degree of divergence is -2 but which is divergent because of a divergent subgraph. However, with the modified Lagrangian density (i.e., with one loop counter

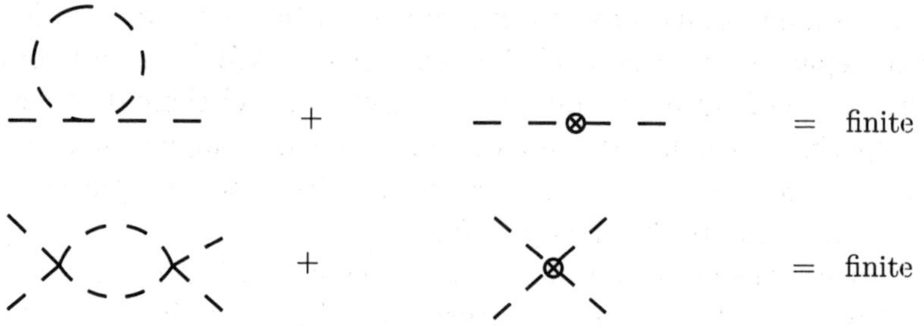

16 RENORMALIZATION THEORY

$$\text{(loop diagram)} + \text{(counter term)} = \text{finite}$$

$$\text{(loop diagram)} + \text{(counter term)} = \text{finite}$$

Figure 16.6: The sum of the graphs and the counter terms make the amplitudes finite.

$$\text{(diagrams)}$$

Figure 16.7: Two loop self-energy graphs including the one loop counter terms in the ϕ^4 theory.

terms) there exist another two loop graph in this theory shown in Fig. 16.8 (this is of two loop order because the counter term is of one loop order).

Figure 16.8: Two loop 6-point function graph with the one loop vertex counter term in the ϕ^4 theory.

It is clear that since the counter term cancels the divergence of the one loop four point function, the sum of the two graphs in Fig. 16.1 and Fig. 16.8 is convergent. The point which this analysis brings out is that the only counter terms we really need to make a theory finite are for the graphs whose superficial degree of divergence is non-negative (this is why the concept of superficial divergence is important within the context of renormalization). The renormaliza-

16.2 A BRIEF HISTORY OF RENORMALIZATION

tion procedure works because the structure of the counter terms is the same as the terms in the original Lagrangian density and hence the divergence analysis remains the same even after adding counter terms. Furthermore, since the counter terms have the same form as the terms in the original Lagrangian density, they can be simply absorbed into a redefinition of the original parameters in the theory.

For example, in the ϕ^4 theory we have (C.T. stands for counter terms)

$$\begin{aligned}\mathcal{L} &= \frac{1}{2}\partial_\mu\phi\partial^\mu\phi - \frac{M^2}{2}\phi^2 - \frac{\lambda}{4!}\phi^4 + \text{C.T.} \\ &= \frac{1}{2}\partial_\mu\phi\partial^\mu\phi - \frac{M^2}{2}\phi^2 - \frac{\lambda}{4!}\phi^4 \\ &\quad + \frac{A}{2}\partial_\mu\phi\partial^\mu\phi - \frac{B}{2}\phi^2 - \frac{C}{4!}\phi^4.\end{aligned} \quad (16.29)$$

The constants A, B and C receive contributions from various loops and are regularization dependent. We can now combine the counter terms with the original terms in the Lagrangian density as

$$\mathcal{L} = \frac{1}{2}(1+A)\partial_\mu\phi\partial^\mu\phi - \frac{1}{2}\left(M^2+B\right)\phi^2 - \frac{1}{4!}(\lambda+C)\phi^4, \quad (16.30)$$

and define

$$\begin{aligned}\phi_0 &= (1+A)^{\frac{1}{2}}\phi = Z^{\frac{1}{2}}\phi, \\ M_0^2 &= \left(M^2+B\right)(1+A)^{-1} \\ &= \left(M^2+B\right)Z^{-1} = M^2 Z_M Z^{-1}, \\ \lambda_0 &= (\lambda+C)(1+A)^{-2} = (\lambda+C)Z^{-2} = \lambda Z_1 Z^{-2}. \end{aligned} \quad (16.31)$$

(The mass renormalization takes this form only if the regularization procedure does not introduce any mass parameter.) With this redefinition, therefore, the Lagrangian density becomes

$$\mathcal{L} = \frac{1}{2}(1+A)\partial_\mu\phi\partial^\mu\phi - \frac{1}{2}\left(M^2+B\right)\phi^2 - \frac{1}{4!}(\lambda+C)\phi^4$$
$$= \frac{1}{2}\partial_\mu\phi_0\partial^\mu\phi_0 - \frac{M_0^2}{2}\phi_0^2 - \frac{\lambda_0}{4!}\phi_0^4. \tag{16.32}$$

We see that this Lagrangian density in (16.32) has the same form as the one we started out with. However, our new field variable as well as the parameters m_0, λ_0 have become regularization dependent. These are known as the bare field and the bare parameters of the theory. (Sometimes they are also referred to as the unrenormalized field and the unrenormalized parameters of the theory and are denoted respectively by ϕ_u, M_u, λ_u. We will use these notations interchangeably.) However, the renormalized fields and parameters are finite. Furthermore, if we calculate with the bare Lagrangian density and bare parameters, then the amplitudes will be finite in terms of the renormalized variables. Let us also note here that the renormalized parameters we are talking about are not the conventional renormalized parameters since our graphs are expanded around zero external momenta. However, these renormalized parameters are related to the usual renormalized parameters through renormalization group equations which we will study later.

Since we have already calculated the one loop amplitudes in the scalar theory, let us indicate the one loop renormalization of the ϕ^4 theory. The only divergences in one loop come from $\Gamma^{(2)}, \Gamma^{(4)}$ which in the cut-off regularization have the forms (see (15.10) and (15.19))

$$-\!\!\!\bigvee\!\!\!- \;=\; -\frac{i\lambda}{32\pi^2}\left[\Lambda^2 - M^2\ln\frac{\Lambda^2+M^2}{M^2}\right],$$

$$\times\!-\!\times \;=\; \frac{3i\lambda^2}{32\pi^2}\left[\ln\frac{\Lambda^2+M^2}{M^2}-1\right]. \tag{16.33}$$

Thus, we see that to one loop the parameters of the counter terms in (16.29) are

16.2 A BRIEF HISTORY OF RENORMALIZATION

$$A = 0,$$
$$B = -\frac{\lambda}{32\pi^2}\left[\Lambda^2 - M^2 \ln\frac{\Lambda^2 + M^2}{M^2}\right],$$
$$C = \frac{3\lambda^2}{32\pi^2}\left[\ln\frac{\Lambda^2 + M^2}{M^2}\right], \qquad (16.34)$$

which lead to the one loop definition of the bare fields and parameters as

$$\phi_0 = (1+A)^{\frac{1}{2}}\phi = Z^{\frac{1}{2}}\phi = \phi,$$
$$M_0^2 = (M^2 + B)(1+A)^{-1}$$
$$= \left[M^2\left(1 + \frac{\lambda}{32\pi^2}\ln\frac{\Lambda^2 + M^2}{M^2}\right) - \frac{\lambda\Lambda^2}{32\pi^2}\right],$$
$$\lambda_0 = (\lambda + C)(1+A)^{-2} = \lambda Z_1$$
$$= \lambda\left[1 + \frac{3\lambda}{32\pi^2}\ln\frac{\Lambda^2 + M^2}{M^2}\right]. \qquad (16.35)$$

On the other hand, we have also seen that in the dimensional regularization, we can write

$$\mathcal{L} = \frac{1}{2}\partial_\mu\phi\partial^\mu\phi - \frac{M^2}{2}\phi^2 - \mu^\epsilon\frac{\lambda}{4!}\phi^4 + \frac{A}{2}\partial_\mu\phi\partial^\mu\phi - \frac{B}{2}\phi^2 - \mu^\epsilon\frac{C}{4!}\phi^4$$
$$= \frac{1}{2}(1+A)\partial_\mu\phi\partial^\mu\phi - \frac{1}{2}(M^2 + B)\phi^2 - \mu^\epsilon\frac{(\lambda+C)}{4!}\phi^4$$
$$= \frac{1}{2}\partial_\mu\phi_0\partial^\mu\phi_0 - \frac{M_0^2}{2}\phi_0^2 - \frac{\lambda_0}{4!}\phi_0^4, \qquad (16.36)$$

with (see (15.78) and (15.79))

$$A = 0, \quad B = \frac{\lambda M^2}{32\pi^2}\frac{2}{\epsilon}, \quad C = \frac{3\lambda^2}{32\pi^2}\frac{2}{\epsilon}, \qquad (16.37)$$

so that we have

$$Z = 1, \quad Z_M = \left(1 + \frac{\lambda}{32\pi^2}\frac{2}{\epsilon}\right), \quad Z_1 = \left(1 + \frac{3\lambda}{32\pi^2}\frac{2}{\epsilon}\right), \quad (16.38)$$

and in this case, we have defined

$$\lambda_0 = \mu^\epsilon \lambda Z_1 Z^{-2}. \tag{16.39}$$

Our calculation so far has been at the one loop. However, at n-loops, the divergence structure and, therefore, the counter terms in the dimensional regularization, in general, have the form

$$\sum_{m=-\infty}^{m=n} \frac{a_m}{\epsilon^m}, \tag{16.40}$$

where a_m represents constants. At higher loops, we calculate amplitudes using the counter terms already present at lower orders (namely, we also include diagrams coming from counter terms at lower order).

Similarly, in QED in the covariant gauge in the dimensional regularization we can write the Lagrangian density of QED in the covariant gauge with counter terms as

$$\begin{aligned}
\mathcal{L} &= -\frac{1}{4}F_{\mu\nu}F^{\mu\nu} + i\overline{\psi}\slashed{\partial}\psi - m\overline{\psi}\psi - e\mu^{\frac{\epsilon}{2}}\overline{\psi}\slashed{A}\psi - \frac{1}{2\xi}(\partial_\mu A^\mu)^2 \\
&\quad - \frac{A}{4}F_{\mu\nu}F^{\mu\nu} + iB\overline{\psi}\slashed{\partial}\psi - C\overline{\psi}\psi - \mu^{\frac{\epsilon}{2}}D\overline{\psi}\slashed{A}\psi \\
&= -\frac{1}{4}(1+A)F_{\mu\nu}F^{\mu\nu} + i(1+B)\overline{\psi}\slashed{\partial}\psi - (m+C)\overline{\psi}\psi \\
&\quad - (e+D)\mu^{\frac{\epsilon}{2}}\overline{\psi}\slashed{A}\psi - \frac{1}{2\xi}(\partial_\mu A^\mu)^2 \\
&= -\frac{1}{4}F^{(0)}_{\mu\nu}F^{\mu\nu(0)} + i\overline{\psi}^{(0)}\slashed{\partial}\psi^{(0)} - m_0\overline{\psi}^{(0)}\psi^{(0)} \\
&\quad - e_0\overline{\psi}^{(0)}\slashed{A}^{(0)}\psi^{(0)} - \frac{1}{2\xi_0}\left(\partial_\mu A^{(0)\mu}\right)^2,
\end{aligned}$$

where the bare fields and the bare parameters are defined as

16.2 A BRIEF HISTORY OF RENORMALIZATION

$$A_\mu^{(0)} = (1+A)^{\frac{1}{2}} A_\mu = Z_3^{\frac{1}{2}} A_\mu,$$

$$\psi^{(0)} = (1+B)^{\frac{1}{2}} \psi = Z_2^{\frac{1}{2}} \psi,$$

$$m_0 = (m+C)(1+B)^{-1} = (m+C) Z_2^{-1} = m Z_m Z_2^{-1},$$

$$e_0 = \mu^{\frac{\epsilon}{2}} (e+D)(1+B)^{-1}(1+A)^{-\frac{1}{2}}$$

$$= \mu^{\frac{\epsilon}{2}} (e+D) Z_2^{-1} Z_3^{-\frac{1}{2}} = \mu^{\frac{\epsilon}{2}} e Z_1 Z_2^{-1} Z_3^{-\frac{1}{2}},$$

$$\xi_0 = \xi(1+A) = \xi Z_3. \tag{16.41}$$

We have explicitly determined at one loop that (see (15.89), (15.96) and (15.104))

$$A = -\frac{e^2}{12\pi^2} \frac{2}{\epsilon}, \quad Z_3 = \left(1 - \frac{e^2}{12\pi^2} \frac{2}{\epsilon}\right),$$

$$B = -\frac{e^2}{16\pi^2} \frac{2}{\epsilon}, \quad Z_2 = \left(1 - \frac{e^2}{16\pi^2} \frac{2}{\epsilon}\right),$$

$$C = -\frac{e^2 m}{4\pi^2} \frac{2}{\epsilon}, \quad Z_m = \left(1 - \frac{e^2}{4\pi^2} \frac{2}{\epsilon}\right),$$

$$D = -\frac{e^3}{16\pi^2} \frac{2}{\epsilon}, \quad Z_1 = \left(1 - \frac{e^2}{16\pi^2} \frac{2}{\epsilon}\right). \tag{16.42}$$

The Ward identities of the theory imply that $Z_1 = Z_2$ which is explicitly seen at one loop (see section **9.7** as well as the discussion after (15.104)). This, in turn, implies that $e_0 = e\mu^{\frac{\epsilon}{2}} Z_3^{-\frac{1}{2}}$. This is interesting because it says that the renormalization of charge depends only on the photon wave function renormalization. As a result, the charge (coupling) of any fermion is renormalized exactly in the same manner. This is commonly known as the universality of charge which is a consequence of the Ward-Takahashi identity of the theory.

The theories which have the property that all the divergences can be absorbed into a redefinition of the parameters of the theory are known as renormalizable theories. It is clear that only theories whose index of divergence $\delta_i \leq 0$ would be renormalizable. This is because since

$$D = 4 - B - \frac{3}{2}F + \sum_i n_i \delta_i, \tag{16.43}$$

if $\delta_i > 0$ then when we go to higher and higher orders of interaction, we generate divergent graphs with more and more external lines. That would correspond to adding to the Lagrangian density counter terms which cannot be absorbed into a redefinition of the existing parameters of the theory.

Motivated by the notion of counter terms, Bogoliubov and Parasiuk developed a recursive subtraction scheme. However, one of their intermediate theorems was not true which was corrected by Hepp. Hence the method is known as the BPH method. Subsequently, Zimmermann provided a solution to their equation thereby extending it to what is known as the BPHZ method which represents the second renormalization method. The BPHZ method generalizes quite nicely to non-Abelian gauge theories and, therefore, we will discuss this in detail in section **16.4**. Here we simply summarize the method briefly.

The BPHZ method corresponds to defining forests associated with each graph in the following manner. Corresponding to each Feynman graph, if there are renormalization parts in the graph (subgraphs with superficial degree of divergence non-negative), then we draw boxes around the renormalization parts in various ways such that no boxes ever overlap. A particular laying down of boxes is called a forest F. The elements or boxes in a forest F are denoted by γ. Associated with each graph there is a set of forests corresponding to all possible ways of laying down boxes around renormalization parts. A forest may be empty and it is all right to draw a box around the entire graph provided the graph is a renormalization part. As an example let us look at the complete set of forests shown in Fig. 16.9 for the two loop graph of $\Gamma^{(4)}$ in the ϕ^4 theory. It should be emphasized that the boxes contain only the renormalization parts and not any propagator external to the renormalization part. Furthermore, the graphs are considered functions of the internal loop momenta as well as the external momenta. However, the loop momenta are not integrated yet (namely, we are working with the integrand of the amplitude).

We now define a Taylor operator t^γ which acts on a boxed subgraph γ and replaces it by its Taylor expansion in the external mo-

16.2 A BRIEF HISTORY OF RENORMALIZATION 689

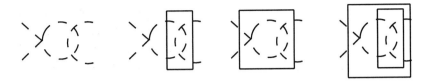

Figure 16.9: Forest diagrams associated with the simple two loop vertex correction diagram in the ϕ^4 theory.

mentum variables about zero four-momentum out to order $D(\gamma)$ which denotes the superficial degree of divergence of γ. For example, if γ is a $\Gamma^{(4)}$, then $D(\gamma) = 0$ and t^γ simply evaluates the graph γ at zero external momentum. If γ denotes the two point function $\Gamma^{(2)}$, then $(D(\gamma) = 2$ and k denotes the generic internal momentum)

$$t^\gamma \Gamma^{(2)}(k,p) = \Gamma^{(2)}(k,0) + \frac{1}{2!} p^\mu p^\nu \frac{\partial^2 \Gamma^{(2)}(k,p)}{\partial p^\mu \partial p^\nu} \bigg|_{p=0}. \qquad (16.44)$$

Here we have neglected the linear term in the Taylor expansion since it would vanish upon integration because of antisymmetry. For a convergent graph H, we have $t^\gamma H = 0$. With the notion of the Taylor operator, we can obtain a renormalized Feynman integrand $R(G)$ for the graph G given by the expression

$$R(G) = (1 - t^G) \sum_{F \in \Phi} \prod_{\gamma \in F} (-t^\gamma) I(G), \qquad (16.45)$$

where $I(G)$ represents the integrand of the graph G and Φ represents the complete set of disjoint and nested forests associated with the graph (these concepts will be discussed in section **16.4**). We should only remember that when nested boxes exist then the t^γ operation should be carried from inside out. Then the assertion is that the renormalized Feynman graph is convergent. Note that the BPHZ method subtracts out the divergent parts in the integrand itself so that the renormalized integral is finite. We would see in detail how this works in section **16.4**.

Weinberg's theorem. The integral of a Feynman graph G is absolutely convergent if the superficial degree of divergence D_H is negative for every subgraph H of G including the case when $H = G$. This theorem is extremely important in the study of renormalization.

16.3 Schwinger-Dyson equation

Let us consider the ϕ^3 theory described by the Lagrangian density

$$\mathcal{L} = \frac{1}{2}\partial_\mu\phi\partial^\mu\phi - \frac{m^2}{2}\phi^2 - \frac{g}{3!}\phi^3. \tag{16.46}$$

The Euler-Lagrange equation for the theory is given by

$$-\frac{\delta S}{\delta \phi} = F[\phi] = \left(\partial_\mu\partial^\mu + m^2\right)\phi + \frac{g}{2}\phi^2 = 0. \tag{16.47}$$

This describes the dynamics of the system at the tree level. However, when we include quantum corrections, this equation modifies and the modified equation can be obtained from the generating functional for the theory as follows. We recall that the generating functional for the theory in the presence of an external source is defined as (see (12.120))

$$Z[J] = e^{iW[J]} = N \int \mathcal{D}\phi \, e^{i(S[\phi] + \int J\phi)}. \tag{16.48}$$

We note that since the generating functional does not depend on the field variable (we are integrating over all field configurations), under an arbitrary field redefinition $\phi \to \phi + \delta\phi$ inside the path integral, the generating functional would be stationary leading to

$$\delta Z = 0 = N \int \mathcal{D}\phi \left(F[\phi] - J\right) e^{i(S[\phi] + \int J\phi)}, \tag{16.49}$$

where the Euler-Lagrange operator F is defined in (16.47). Working out in detail, (16.49) leads to

16.3 Schwinger-Dyson equation

$$e^{iW[J]}\left(F\left[-i\frac{\delta}{\delta J}\right] - J\right)e^{iW[J]} = 0,$$

or, $\quad F\left[\frac{\delta W}{\delta J} - i\frac{\delta}{\delta J}\right] - J = 0,$

or, $\quad F\left[\phi_c - i\frac{\delta}{\delta J}\right] - J = 0.$ (16.50)

Here ϕ_c denotes the classical field which is defined in (13.78).

For the ϕ^3 theory described by (16.46), this can be written explicitly as

$$(\partial_\mu \partial^\mu + m^2)\phi_c + \frac{g}{2}\left(\phi_c - i\frac{\delta}{\delta J}\right)\left(\phi_c - i\frac{\delta}{\delta J}\right) - J = 0,$$

or, $\quad (\partial_\mu \partial^\mu + m^2)\phi_c + \frac{g}{2}\phi_c^2 - i\frac{g}{2}\frac{\delta \phi_c}{\delta J} - J = 0,$

or, $\quad (\partial_\mu \partial^\mu + m^2)\phi_c + \frac{g}{2}\phi_c^2 + \frac{ig}{2}\left(\frac{\delta^2 \Gamma}{\delta \phi_c^2}\right)^{-1} + \frac{\delta \Gamma}{\delta \phi_c} = 0,$ (16.51)

where $\Gamma[\phi_c]$ corresponds to the effective action defined in (13.80). If we now take the functional derivative of (16.51) with respect to ϕ_c and set $\phi_c = 0$, we obtain

$$(\partial_\mu \partial^\mu + m^2)\mathbb{1} + \left.\frac{\delta^2 \Gamma}{\delta \phi_c^2}\right| = -\frac{ig}{2}\frac{\delta}{\delta \phi_c}\left.\left(\frac{\delta^2 \Gamma}{\delta \phi_c^2}\right)^{-1}\right|$$

or, $\quad \Sigma = \frac{ig}{2}\left(\frac{\delta^2 \Gamma}{\delta \phi_c^2}\right)^{-1}\left(\frac{\delta^3 \Gamma}{\delta \phi_c^3}\right)\left.\left(\frac{\delta^2 \Gamma}{\delta \phi_c^2}\right)^{-1}\right|,$ (16.52)

where the restriction "|" stands for setting $\phi_c = 0$ and Σ denotes the complete self-energy the ϕ field (the self-energy is defined to be the complete two point function minus the tree level contribution). Equation (16.52) can be represented graphically as in Fig. 16.10 where the line with the blob on the left-hand side represents the complete self-energy while the lines with blobs as well as the vertex with a blob on the right-hand side denote the full propagators and the full vertex of the theory including quantum corrections to all

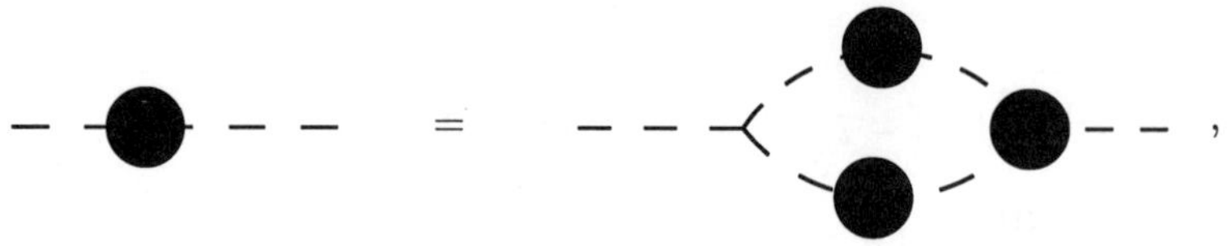

Figure 16.10: Schwinger-Dyson equation for the two point function.

orders. (We are using a very compact notation where integrations over intermediate coordinates/momenta are suppressed.)

Similarly, by taking the second functional derivative of (16.51) with respect to ϕ_c and setting $\phi_c = 0$, we obtain

$$\frac{\delta^3 \Gamma}{\delta \phi_c^3}\bigg| = -g - ig \left(\frac{\delta^2 \Gamma}{\delta \phi_c^2}\right)^{-1} \left(\frac{\delta^3 \Gamma}{\delta \phi_c^3}\right) \left(\frac{\delta^2 \Gamma}{\delta \phi_c^2}\right)^{-1} \left(\frac{\delta^3 \Gamma}{\delta \phi_c^3}\right) \left(\frac{\delta^2 \Gamma}{\delta \phi_c^2}\right)^{-1}\bigg|$$
$$+ \frac{ig}{2} \left(\frac{\delta^2 \Gamma}{\delta \phi_c^2}\right)^{-1} \left(\frac{\delta^4 \Gamma}{\delta \phi_c^4}\right) \left(\frac{\delta^2 \Gamma}{\delta \phi_c^2}\right)^{-1}\bigg|, \qquad (16.53)$$

which can be represented graphically as in Fig. 16.11.

Figure 16.11: Schwinger-Dyson equation for the three point function.

The equations in (16.52) and (16.53) are integral equations (they are written in a compact notation here) for the two point and the three point functions. Such equations (and for the higher point functions) are known as the Schwinger-Dyson equations.

16.4 BPHZ renormalization

We have seen in detail how multiplicative renormalization works. To see in some detail how the BPHZ method works, let us consider the ϕ^3 theory in six dimensions described by the Lagrangian density

16.4 BPHZ RENORMALIZATION

$$\mathcal{L} = \frac{1}{2}\partial_\mu\phi\partial^\mu\phi - \frac{M^2}{2}\phi^2 - \frac{g}{3!}\phi^3. \tag{16.54}$$

This is a simple quantum field theory that is quite helpful in understanding various features of renormalization. Furthermore, since we are interested in the ultraviolet behavior of the theory, let us set $M = 0$ for simplicity.

In six dimensions, the counting of the canonical dimension gives

$$[\phi] = 2. \tag{16.55}$$

Thus, for this theory, the superficial degree of divergence for any graph is given by

B	$D = 6 - 2B$
0	6
1	4
2	2
3	0

and we see that the one point, the 2-point and the 3-point amplitudes are superficially divergent. (Note that unlike the ϕ^4 theory, here the Lagrangian density is not invariant under $\phi \to -\phi$.) Although the zero point amplitude is divergent, we can remove this divergence by normal ordering the theory. Furthermore, in dimensional regularization which we will be using, the one point amplitude vanishes (in the massless theory) as can be seen easily from

$$\sim \int \frac{d^n k}{(2\pi)^n} \frac{1}{k^2} = \lim_{M \to 0} \int \frac{d^n k}{(2\pi)^n} \frac{1}{k^2 - M^2}$$

$$\sim \lim_{M \to 0} (M^2)^{\frac{n}{2}-1} \to 0, \tag{16.56}$$

when $n \to 6$. Therefore, only the 2-point and the 3-point amplitudes are divergent which can be made finite with conventional counter terms. Thus, this theory should be renormalizable.

In order to see how renormalization works in some detail and, in particular, how the overlapping divergences are handled, let us calculate some amplitudes beyond the simple one loop. Using dimensional regularization we analytically continue to n dimensions and define $\epsilon = 6 - n$. In this case, the naive dimensional analysis gives

$$[\phi] = \frac{(n-2)}{2}, \qquad [g] = n - \frac{3(n-2)}{2} = \frac{(6-n)}{2} = \frac{\epsilon}{2}, \qquad (16.57)$$

so that we can introduce, as before, an arbitrary mass scale to make the coupling constant dimensionless, namely,

$$g \to g\mu^{\frac{\epsilon}{2}}. \qquad (16.58)$$

In this case, the one loop self-energy can be evaluated as

$$= \frac{1}{2}(-ig\mu^{\frac{\epsilon}{2}})^2 \int \frac{d^n k}{(2\pi)^n} \frac{(i)^2}{k^2(k+p)^2}$$

$$= \frac{g^2 \mu^\epsilon}{2} \int dx \int \frac{d^n k}{(2\pi)^n} \frac{1}{((k+xp)^2 + x(1-x)p^2)^2}$$

$$= \frac{g^2 \mu^\epsilon}{2} \int dx \frac{(-1)^2}{\Gamma(2)} \frac{i\pi^{\frac{n}{2}}}{(2\pi)^n} \frac{\Gamma\left(2 - \frac{n}{2}\right)}{(-x(1-x)p^2)^{2-\frac{n}{2}}}$$

$$= \frac{ig^2 \mu^\epsilon}{2} \frac{1}{(4\pi)^{\frac{n}{2}}} \int dx \frac{\Gamma\left(-1 + \frac{\epsilon}{2}\right)}{(-x(1-x)p^2)^{2-\frac{n}{2}}}$$

$$= \frac{ig^2 \mu^\epsilon}{2} \frac{1}{(4\pi)^{\frac{n}{2}}} \frac{\Gamma\left(-1 + \frac{\epsilon}{2}\right)}{(-p^2)^{2-\frac{n}{2}}} \int dx \frac{1}{(x(1-x))^{2-\frac{n}{2}}}. \qquad (16.59)$$

16.4 BPHZ RENORMALIZATION

This form is quite suitable to carry out two loop calculations which we will do shortly. However, to understand the structure of the self energy at one loop, let us simplify this.

As we have discussed earlier (see (15.76)),

$$2 - \frac{n}{2} = 2 - \frac{6-\epsilon}{2} = 2 - 3 + \frac{\epsilon}{2} = -1 + \frac{\epsilon}{2},$$

$$\Gamma\left(-1 + \frac{\epsilon}{2}\right) \simeq \left(-\frac{2}{\epsilon} + (\gamma - 1)\right),$$

$$\frac{1}{(4\pi)^{\frac{n}{2}}} \simeq \frac{1}{(4\pi)^3}\left(1 + \frac{\epsilon}{2} \ln 4\pi\right),$$

$$\frac{1}{(4\pi)^{\frac{n}{2}}} \Gamma\left(-1 + \frac{\epsilon}{2}\right) \simeq \frac{1}{(4\pi)^3}\left(1 + \frac{\epsilon}{2} \ln 4\pi\right)\left(-\frac{2}{\epsilon} + (\gamma - 1)\right)$$

$$\simeq \frac{1}{(4\pi)^3}\left(-\frac{2}{\epsilon} + (\gamma - 1) - \ln 4\pi\right),$$

$$\int_0^1 dx \, \frac{1}{(x(1-x))^{2-\frac{n}{2}}} = \int_0^1 dx \, (x(1-x))^{1-\frac{\epsilon}{2}}$$

$$\simeq \int_0^1 dx \, x(1-x) \left(1 - \frac{\epsilon}{2} \ln x(1-x)\right)$$

$$= \int_0^1 dx \, x(1-x) \left(1 - \epsilon \ln x\right)$$

$$= \frac{1}{6}\left(1 + \frac{5\epsilon}{6}\right). \tag{16.60}$$

Using these, we obtain the one loop self energy of the theory in (16.59) to be

$$\simeq -\frac{ig^2 \mu^\epsilon}{2(4\pi)^3} \frac{p^2}{6}\left(1 + \frac{5\epsilon}{6}\right)\left(-\frac{2}{\epsilon} + (\gamma - 1) - \ln 4\pi\right)\left(1 - \frac{\epsilon}{2} \ln(-p^2)\right)$$

$$\simeq \frac{ig^2 p^2}{12(4\pi)^3} \left(\frac{2}{\epsilon} - \ln \frac{(-p^2)}{\mu^2} - \left(\gamma - 1 - \frac{5}{3}\right) + \ln 4\pi \right)$$

$$\simeq \frac{ig^2 p^2}{12(4\pi)^3} \left(\frac{2}{\epsilon} - \ln \frac{(-p^2)}{4\pi\mu^2} - \left(\gamma - \frac{8}{3}\right) \right). \tag{16.61}$$

This leads to the one loop counter term

$$- - -\otimes - - - = -\frac{ig^2 p^2}{12(4\pi)^3} \frac{2}{\epsilon}. \tag{16.62}$$

Similarly, we can calculate the three point amplitude at one loop as (there is a second graph with $p_1 \leftrightarrow p_2$ which contributes to the amplitude, but since the BPHZ method subtracts out divergences in every graph, we will concentrate on graphs)

$$= \left(-ig\mu^{\frac{\epsilon}{2}}\right)^3 \int \frac{\mathrm{d}^n k}{(2\pi)^n} \frac{(i)^3}{k^2(k+p)^2(k+p+p_1)^2}$$

$$= g^3 \mu^{\frac{3\epsilon}{2}} \, 2! \int \frac{\mathrm{d}^n k}{(2\pi)^n} \mathrm{d}x_1 \mathrm{d}x_2 \mathrm{d}x_3$$

$$\times \frac{\delta(1 - x_1 - x_2 - x_3)}{\left(x_1 k^2 + x_2(k+p+p_1)^2 + x_3(k+p)^2\right)^3}. \tag{16.63}$$

Here we have used the generalized Feynman combination formula

$$\prod_{i=1}^{p} \frac{1}{(A_i)^{n_i}} = \frac{\Gamma(\sum_i n_i)}{\prod_i \Gamma(n_i)} \int \mathrm{d}x_1 \cdots \mathrm{d}x_p \, \delta(1 - x_1 - \cdots - x_p)$$

$$\times \frac{\left(\prod_{i=1}^{p} x_i^{n_i - 1}\right)}{\left(\sum_i x_i A_i\right)^{n_1 + \cdots + n_p}}. \tag{16.64}$$

16.4 BPHZ RENORMALIZATION

Let us next simplify the denominator of the integrand as

$$x_1 k^2 + x_2 (k+p+p_1)^2 + x_3 (k+p)^2$$
$$= (x_1 + x_2 + x_3) k^2 + 2k \cdot (x_3 p + x_2 (p+p_1))$$
$$\quad + x_3 p^2 + x_2 (p+p_1)^2$$
$$= k^2 + 2k \cdot ((1-x_1)p + x_2 p_1) + (1-x_1)p^2$$
$$\quad + x_2 p_1^2 + 2x_2 p \cdot p_1$$
$$= (k + (1-x_1)p + x_2 p_1)^2 - (1-x_1)^2 p^2 - x_2^2 p_1^2$$
$$\quad - 2x_2(1-x_1) p \cdot p_1 + (1-x_1)p^2 + x_2 p_1^2 + 2x_2 p \cdot p_1$$
$$= (k + (1-x_1)p + x_2 p_1)^2 + x_1(1-x_1)p^2 + x_2(1-x_2)p_1^2$$
$$\quad + 2x_1 x_2 p \cdot p_1$$
$$= (k + (1-x_1)p + x_2 p_1)^2 + x_1(1-x_1)Q^2, \qquad (16.65)$$

where we have identified

$$\begin{aligned} x_1(1-x_1)Q^2 &= x_1(1-x_1)p^2 + x_2(1-x_2)p_1^2 \\ &\quad + 2x_1 x_2 p \cdot p_1, \end{aligned} \qquad (16.66)$$

where we have used the constraint from the delta function in the intermediate steps. Therefore, by shifting the variable of integration we obtain the value of the integral in (16.63) to be

$$= 2g^3 \mu^{\frac{3\epsilon}{2}} \int_0^1 dx_1 \int_0^{1-x_1} dx_2 \, \frac{d^n k}{(2\pi)^n} \, \frac{1}{(k^2 + x_1(1-x_1)Q^2)^3}$$

$$
\begin{aligned}
&= 2g^3\mu^{\frac{3\epsilon}{2}} \int_0^1 dx_1 \int_0^{1-x_1} dx_2 \frac{i(-1)^3}{\Gamma(3)(4\pi)^{\frac{n}{2}}} \frac{\Gamma\left(3-\frac{n}{2}\right)}{(-x_1(1-x_1)Q^2)^{3-\frac{n}{2}}} \\
&= -\frac{ig^3\mu^{\frac{3\epsilon}{2}}}{(4\pi)^{\frac{n}{2}}} \int_0^1 dx_1 \int_0^{1-x_1} dx_2 \frac{\Gamma\left(3-\frac{n}{2}\right)}{(-x_1(1-x_1)Q^2)^{3-\frac{n}{2}}} \quad (16.67)\\
&\simeq -\frac{ig^3\mu^{\frac{\epsilon}{2}}}{(4\pi)^3} \int_0^1 dx_1 \int_0^{1-x_1} dx_2 \left(\frac{2}{\epsilon} - \gamma\right) \\
&\qquad \times \left(1 - \frac{\epsilon}{2} \ln \frac{(-x_1(1-x_1)Q^2)}{4\pi\mu^2}\right) \\
&\simeq -\frac{ig^3\mu^{\frac{\epsilon}{2}}}{(4\pi)^3} \int_0^1 dx_1 \int_0^{1-x_1} dx_2 \left(\frac{2}{\epsilon} - \ln \frac{(-x_1(1-x_1)Q^2)}{4\pi\mu^2} - \gamma\right).
\end{aligned}
$$
(16.68)

In this derivation we have used the fact that because of the δ function involving the Feynman parameters, we have $x_3 = 1 - x_1 - x_2$ and

$$
\begin{aligned}
x_3 &= 1 \quad \Rightarrow x_1 = x_2 = 0, \\
x_3 &= 0 \quad \Rightarrow x_2 = 1 - x_1,
\end{aligned}
\quad (16.69)
$$

so that

$$
\int dx_1 dx_2 dx_3 \, \delta(1 - x_1 - x_2 - x_3) = \int_0^1 dx_1 \int_0^{1-x_1} dx_2. \quad (16.70)
$$

Furthermore, defining a new variable (this would be quite useful for the calculation of the two loop self-energy which we will do shortly)

$$
x_2 = (1-x_1)u, \quad (16.71)
$$

we can rewrite (16.68) as

$$
= -\frac{ig^3\mu^{\frac{\epsilon}{2}}}{(4\pi)^3} \int_0^1 dx_1 \int_0^1 du \,(1-x_1) \left(\frac{2}{\epsilon} - \ln \frac{(-x_1(1-x_1)Q^2)}{4\pi\mu^2} - \gamma\right),
$$
(16.72)

16.4 BPHZ RENORMALIZATION

where in the new variables

$$x_1(1-x_1)Q^2 = x_1(1-x_1)(p+up_1)^2 + u(1-u)p_1^2. \quad (16.73)$$

Equation (16.68) leads to the one loop counter term (the factor of $\frac{1}{2}$ arises from doing the integrals over x_1, x_2)

$$\cdots\!-\!\!\otimes\!\!\cdots = \frac{ig^3\mu^{\frac{\epsilon}{2}}}{2(4\pi)^3}\frac{2}{\epsilon}. \quad (16.74)$$

Let us next calculate the self-energy at two loops. Let us first look at the graphs with non-overlapping divergence of the form shown in Fig. 16.12 (there would also be diagrams with a one loop self-energy and counter term insertion on the other internal line)

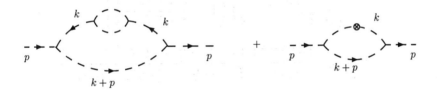

Figure 16.12: Non-overlapping two loop self-energy diagrams in the ϕ^3 theory.

$$= \left(-ig\mu^{\frac{\epsilon}{2}}\right)^2 \int \frac{d^nk}{(2\pi)^n} \frac{i}{k^2} \left(i\Pi^{(1)}(k^2)\right) \frac{i}{k^2} \frac{i}{(k+p)^2}$$

$$+ \left(-ig\mu^{\frac{\epsilon}{2}}\right)^2 \int \frac{d^nk}{(2\pi)^n} \frac{i}{k^2} \frac{(-ig^2k^2)}{12(4\pi)^3} \frac{2}{\epsilon} \frac{i}{k^2} \frac{i}{(k+p)^2}, \quad (16.75)$$

where we have used the form of the one loop counter term in (16.62). Let us look at the two integrals in (16.75) separately. Using the form of the one loop self-energy in (16.59), the first integral in (16.75) has the form

$$I_1 = ig^2\mu^\epsilon \int \frac{d^n k}{(2\pi)^n} \int dx \, \frac{1}{k^2} \frac{ig^2\mu^\epsilon}{2(4\pi)^{\frac{n}{2}}} \frac{\Gamma\left(-1+\frac{\epsilon}{2}\right)}{(-x(1-x)k^2)^{2-\frac{n}{2}}} \frac{1}{k^2}$$

$$\times \frac{1}{(k+p)^2}$$

$$= -\frac{(g^2\mu^\epsilon)^2}{2(4\pi)^{\frac{n}{2}}} \int dx \frac{\Gamma\left(-1+\frac{\epsilon}{2}\right)}{(x(1-x))^{2-\frac{n}{2}}} \int \frac{d^n k}{(2\pi)^n} \frac{(-1)^{\frac{n}{2}-2}}{(k^2)^{4-\frac{n}{2}}(k+p)^2}$$

$$\simeq -\frac{(g^2\mu^\epsilon)^2}{2(4\pi)^{\frac{n}{2}}} \Gamma\left(-1+\frac{\epsilon}{2}\right) \frac{1}{6}\left(1+\frac{5\epsilon}{6}\right)$$

$$\times \int \frac{d^n k}{(2\pi)^n} dy \, \frac{\Gamma\left(5-\frac{n}{2}\right)}{\Gamma\left(4-\frac{n}{2}\right)\Gamma(1)} \frac{(-1)^{\frac{n}{2}-2}(1-y)^{3-\frac{n}{2}}}{((k+yp)^2+y(1-y)p^2)^{5-\frac{n}{2}}}$$

$$= -\frac{(g^2\mu^\epsilon)^2}{12(4\pi)^{\frac{n}{2}}} \Gamma\left(-1+\frac{\epsilon}{2}\right)\left(1+\frac{5\epsilon}{6}\right) \frac{\Gamma\left(5-\frac{n}{2}\right)}{\Gamma\left(4-\frac{n}{2}\right)}$$

$$\times \int dy \, \frac{i(-1)^{5-\frac{n}{2}}}{(4\pi)^{\frac{n}{2}}} \frac{\Gamma\left(5-\frac{n}{2}-\frac{n}{2}\right)}{\Gamma\left(5-\frac{n}{2}\right)} \frac{(-1)^{\frac{n}{2}-2}(1-y)^{3-\frac{n}{2}}}{(-y(1-y)p^2)^{5-n}}$$

$$\simeq \frac{i\left(g^2\mu^\epsilon\right)^2}{12(4\pi)^n} \frac{\Gamma\left(-1+\frac{\epsilon}{2}\right)}{\Gamma\left(4-\frac{n}{2}\right)} \Gamma(5-n)\left(1+\frac{5\epsilon}{6}\right)(-p^2)^{n-5}$$

$$\times \int dy \, \frac{(1-y)^{3-\frac{n}{2}}}{(y(1-y))^{5-n}}$$

$$\simeq -\frac{i(g^2)^2 p^2}{12(4\pi)^6}(1+\epsilon\ln 4\pi)\left(-\frac{2}{\epsilon}+\gamma-1\right)\left(-\frac{1}{\epsilon}+\gamma-1\right)$$

$$\times \left(1+\frac{\epsilon\gamma}{2}\right)\left(1+\frac{5\epsilon}{6}\right)\left(1-\epsilon\ln\frac{(-p^2)}{\mu^2}\right)\frac{1}{6}\left(1+\frac{5\epsilon}{4}\right)$$

$$\simeq -\frac{ig^4 p^2}{72(4\pi)^6}\left(\frac{2}{\epsilon^2}-\frac{3}{\epsilon}(\gamma-1)+\frac{\gamma}{\epsilon}\right)(1+\epsilon\ln 4\pi)\left(1+\frac{5\epsilon}{6}\right)$$

$$\times \left[1-\epsilon\ln\frac{(-p^2)}{\mu^2}+\frac{5\epsilon}{4}\right]+\text{finite}$$

16.4 BPHZ RENORMALIZATION

$$\simeq -\frac{ig^4 p^2}{72(4\pi)^6}\left(1+\frac{5\epsilon}{6}\right)\left[\frac{2}{\epsilon^2}-\frac{2}{\epsilon}\ln\frac{(-p^2)}{4\pi\mu^2}+\frac{5}{2\epsilon}-\frac{1}{\epsilon}(2\gamma-3)\right.$$

$$\left.+\text{ finite}\right]$$

$$= -\frac{ig^4 p^2}{72(4\pi)^6}\left[\frac{2}{\epsilon^2}-\frac{2}{\epsilon}\ln\frac{(-p^2)}{4\pi\mu^2}+\frac{5}{2\epsilon}-\frac{1}{\epsilon}(2\gamma-3)\right.$$

$$\left.+\frac{5}{3\epsilon}+\text{ finite}\right]$$

$$= \frac{ig^4 p^2}{72(4\pi)^6}\left[-\frac{2}{\epsilon^2}-\frac{2}{\epsilon}\left(-\ln\frac{(-p^2)}{4\pi\mu^2}+\frac{43}{12}-\gamma\right)+\text{ finite}\right]. \tag{16.76}$$

Here we have used (16.64) as well as (15.76) and (16.60) in the intermediate steps. (The y integration is related to the appropriate beta function, see (16.85).) We note here that the $\frac{1}{\epsilon}\ln\frac{(-p^2)}{4\pi\mu^2}$ terms in the above expression are potentially dangerous because if such divergent terms are present, they will require non-local counter terms to cancel them.

Next, let us look at the second integral in (16.75) which leads to

$$I_2 = \left(-ig\mu^{\frac{\epsilon}{2}}\right)^2 \int \frac{d^n k}{(2\pi)^n}\frac{i}{k^2}\frac{(-ig^2 k^2)}{12(4\pi)^3}\frac{2}{\epsilon}\frac{i}{k^2}\frac{i}{(k+p)^2}$$

$$= \frac{g^4 \mu^\epsilon}{12(4\pi)^3}\frac{2}{\epsilon}\int \frac{d^n k}{(2\pi)^n}\frac{1}{k^2(k+p)^2}$$

$$= \frac{g^4 \mu^\epsilon}{12(4\pi)^3}\frac{2}{\epsilon}\int \frac{d^n k}{(2\pi)^n}\,dx\,\frac{1}{((k+xp)^2+x(1-x)p^2)^2}$$

$$= \frac{g^4 \mu^\epsilon}{12(4\pi)^3}\frac{2}{\epsilon}\int dx\,\frac{i\pi^{\frac{n}{2}}}{(2\pi)^n}\frac{\Gamma\left(2-\frac{n}{2}\right)}{(-x(1-x)p^2)^{2-\frac{n}{2}}}$$

$$\simeq -\frac{g^4 p^2}{12(4\pi)^3}\frac{2}{\epsilon}\frac{i\pi^{\frac{n}{2}}}{(2\pi)^n}\Gamma\left(-1+\frac{\epsilon}{2}\right)$$

$$\times \left(1-\frac{\epsilon}{2}\ln\frac{(-p^2)}{\mu^2}\right)\int dx\,x(1-x)\left(1-\frac{\epsilon}{2}\ln x(1-x)\right)$$

$$\simeq -\frac{g^4 p^2}{12(4\pi)^3}\frac{2}{\epsilon}\frac{i}{(4\pi)^3}\left(-\frac{2}{\epsilon}+(\gamma-1)-\ln 4\pi\right)$$

$$\times \frac{1}{6}\left(1-\frac{\epsilon}{2}\ln\frac{(-p^2)}{\mu^2}+\frac{5\epsilon}{6}\right)$$

$$\simeq -\frac{ig^4 p^2}{72(4\pi)^6}\frac{2}{\epsilon}\left[-\frac{2}{\epsilon}+\ln\frac{(-p^2)}{4\pi\mu^2}-\frac{5}{3}+(\gamma-1)+O(\epsilon)\right]$$

$$= \frac{ig^4 p^2}{72(4\pi)^6}\left[\frac{4}{\epsilon^2}-\frac{2}{\epsilon}\left(-\ln\frac{(-p^2)}{4\pi\mu^2}-\frac{8}{3}+\gamma\right)+\text{finite}\right]. \quad (16.77)$$

As a result, adding (16.76) and (16.77), the sum of the two graphs in (16.75) takes the form

$$I_1+I_2 = \frac{ig^4 p^2}{72(4\pi)^6}\left[-\frac{2}{\epsilon^2}-\frac{2}{\epsilon}\left(-\ln\frac{(-p^2)}{4\pi\mu^2}+\frac{43}{12}-\gamma\right)\right.$$

$$\left.+\frac{4}{\epsilon^2}-\frac{2}{\epsilon}\left(\ln\frac{(-p^2)}{4\pi\mu^2}-\frac{8}{3}+\gamma\right)+\text{finite}\right]$$

$$= \frac{ig^4 p^2}{72(4\pi)^6}\left[\frac{2}{\epsilon^2}-\frac{11}{6\epsilon}+\text{finite}\right]. \quad (16.78)$$

We see that the potentially dangerous non-local divergences have cancelled in the non-overlapping divergent graphs and the remaining divergences can be cancelled by local counter terms.

Let us next look at the overlapping divergent graph in the self energy at two loops which, upon using the form of the one loop vertex function in (16.67) (with x_2 expressed in terms of u as in (16.71)), takes the form

16.4 BPHZ RENORMALIZATION

$$
\begin{aligned}
&= \frac{(-ig\mu^{\frac{\epsilon}{2}})}{2} \int \frac{d^n\ell}{(2\pi)^n} dx_1 du \, \frac{i}{\ell^2} \frac{i}{(\ell+p)^2} \left(-\frac{ig^3\mu^{\frac{3\epsilon}{2}}}{(4\pi)^{\frac{n}{2}}}\right) \\
&\quad \times \frac{(1-x_1)\Gamma\left(3-\frac{n}{2}\right)}{\left(-x_1(1-x_1)Q^2\right)^{3-\frac{n}{2}}} \\
&= \frac{g^4\mu^{2\epsilon}(-1)^{\frac{n}{2}-3}}{2(4\pi)^{\frac{n}{2}}} \Gamma\left(3-\frac{n}{2}\right) \int \frac{d^n\ell}{(2\pi)^n} dx_1 du \, \frac{x_1^{\frac{n}{2}-3}(1-x_1)^{\frac{n}{2}-2}}{\ell^2(\ell+p)^2(Q^2)^{3-\frac{n}{2}}},
\end{aligned}
\tag{16.79}
$$

where Q^2 is defined in (16.73) (with appropriate momenta). Using (16.64) to combine denominators, we can write (16.79) as

$$
\begin{aligned}
&= \frac{g^4\mu^{2\epsilon}(-1)^{\frac{n}{2}-3}}{2(4\pi)^{\frac{n}{2}}} \Gamma\left(3-\frac{n}{2}\right) \int \frac{d^n\ell}{(2\pi)^n} dx_1 du \, x_1^{\frac{n}{2}-3}(1-x_1)^{\frac{n}{2}-2} \\
&\quad \times \frac{\Gamma\left(5-\frac{n}{2}\right)}{\Gamma\left(3-\frac{n}{2}\right)} \int dy_1 dy_2 dy_3 \, \frac{y_1^{2-\frac{n}{2}} \delta(1-y_1-y_2-y_3)}{(D^2)^{5-\frac{n}{2}}} \\
&= \frac{g^4\mu^{2\epsilon}(-1)^{\frac{n}{2}-3}}{2(4\pi)^{\frac{n}{2}}} \Gamma\left(5-\frac{n}{2}\right) \int dx_1 du \, x_1^{\frac{n}{2}-3}(1-x_1)^{\frac{n}{2}-2} \\
&\quad \times \int \frac{d^n\ell}{(2\pi)^n} \int dy_1 dy_2 dy_3 \, \frac{y_1^{2-\frac{n}{2}} \delta(1-y_1-y_2-y_3)}{(D^2)^{5-\frac{n}{2}}},
\end{aligned}
\tag{16.80}
$$

where we have identified

$$
\begin{aligned}
D^2 &= y_1 Q^2 + y_2(\ell+p)^2 + y_3\ell^2 \\
&= y_1\left((\ell+up)^2 + \frac{u(1-u)}{x_1}p^2\right) + y_2(\ell+p)^2 + (1-y_1-y_2)\ell^2 \\
&= (\ell + (uy_1+y_2)p)^2 \\
&\quad + \left(y_1(1-y_1)u^2 + y_2(1-y_2) - 2uy_1y_2 + \frac{y_1u(1-u)}{x_1}\right)p^2.
\end{aligned}
\tag{16.81}
$$

The divergence of the two loop integral in (16.80) would appear naively to be contained in the ℓ integral which can only have a one loop divergence structure (note that the singular multiplicative gamma function coming from the one loop vertex has been cancelled by those coming from the Feynman combination formula). However, this is not right and, in fact, the divergence of the subdiagram (one loop vertex) has been transformed to the Feynman parametric integrals. This is a special feature of overlapping divergent graphs and to appreciate this, let us evaluate the integral in (16.80) in some detail. First we note that we can shift the variable of integration $\ell \to \ell - (uy_1 + y_2)p$. Next, we can do the integration over y_3 using the delta function in (16.80). Finally, we can redefine the variable y_2 (as we have already done in (16.71)) as

$$y_2 = (1 - y_1)v, \tag{16.82}$$

so that the (shifted) denominator in (16.81) has the form

$$D^2 = \ell^2 + ((1-y_1)(y_1(u-v)^2 + v(1-v)) + \frac{y_1 u(1-u)}{x_1})p^2$$
$$= \ell^2 + d(x_1, u, y_1, v)p^2. \tag{16.83}$$

Substituting this into (16.80) and carrying out the ℓ integral we obtain

$$= \frac{g^4 \mu^{2\epsilon}(-1)^{\frac{n}{2}-3}}{2(4\pi)^{\frac{n}{2}}} \Gamma\left(5 - \frac{n}{2}\right) \int dx_1 du\, x_1^{\frac{n}{2}-3}(1-x_1)^{\frac{n}{2}-2}$$

$$\times \int dy_1 dv\, (1-y_1) y_1^{2-\frac{n}{2}} \frac{i(-1)^{5-\frac{n}{2}}}{(4\pi)^{\frac{n}{2}} \Gamma\left(5-\frac{n}{2}\right)} \frac{\Gamma(5-n)}{(-dp^2)^{5-n}}$$

$$= -\frac{ig^4 p^2}{2(4\pi)^6}\left(\frac{-p^2}{4\pi\mu^2}\right)^{-\epsilon} \Gamma(-1+\epsilon)$$

$$\times \int dx_1 du\, dy_1 dv\, x_1^{2-\frac{n}{2}}(1-x_1)^{\frac{n}{2}-2} y_1^{2-\frac{n}{2}}(1-y_1)(x_1 d)^{1-\epsilon}; \tag{16.84}$$

where d is defined in (16.83). The one loop pole structure is manifest in the overall multiplicative gamma function. To see the divergence

16.4 BPHZ RENORMALIZATION

structure from the parametric integration, let us look at only the leading order term in $(x_1 d)^{1-\epsilon}$. Using the standard results

$$\int_0^1 dt\, t^{\alpha-1}(1-t)^{\beta-1} = \frac{\Gamma(\alpha)\Gamma(\beta)}{\Gamma(\alpha+\beta)} = B(\alpha,\beta),$$

$$\int_0^1 du\, u(1-u) = \frac{1}{6},$$

$$\iint_0^1 du dv\, (u-v)^2 = \frac{1}{6}, \qquad (16.85)$$

the leading term in the parametric integration in (16.84) gives

$$\int dx_1 du dy_1 dv\, x_1^{2-\frac{n}{2}}(1-x_1)^{\frac{n}{2}-2} y_1^{2-\frac{n}{2}}(1-y_1)(x_1 d)$$

$$= \int dx_1 dy_1 du dv\, x_1^{2-\frac{n}{2}}(1-x_1)^{\frac{n}{2}-2} y_1^{2-\frac{n}{2}}(1-y_1)$$

$$\times \left(x_1(1-y_1)(y_1(u-v)^2 + v(1-v)) + y_1 u(1-u)\right)$$

$$= \frac{1}{6}\int dx_1 dy_1\, x_1^{2-\frac{n}{2}}(1-x_1)^{\frac{n}{2}-2} y_1^{2-\frac{n}{2}}(1-y_1)$$

$$\times \left(x_1(y_1+1)(1-y_1) + y_1\right)$$

$$= \frac{1}{6}\left(\frac{\Gamma(2)}{\Gamma(4)} + \frac{\Gamma(2)\Gamma(\frac{\epsilon}{2})}{\Gamma(3)} + \frac{\Gamma(2)\Gamma(\frac{\epsilon}{2})}{\Gamma(3)}\right)$$

$$\simeq \frac{1}{6}\left(\frac{2}{\epsilon} - \gamma + \frac{1}{6}\right). \qquad (16.86)$$

This demonstrates how the subdivergence is hidden in the parametric integration. The parametric integration in (16.84) can be evaluated exactly using Gegenbauer polynomials and the only modification from the leading order result in (16.86) is that the finite constant inside the parenthesis changes from $-\gamma + \frac{1}{6}$ to 4. Thus, substituting this result into (16.84) gives the value of the overlapping self-energy diagram to be

$$= -\frac{ig^4p^2}{2(4\pi)^6}\left(-\frac{1}{\epsilon}+\gamma-1\right)\left(1-\epsilon\ln\frac{(-p^2)}{4\pi\mu^2}\right)\frac{1}{6}\left(\frac{2}{\epsilon}+4+O(\epsilon)\right)$$

$$= \frac{ig^4p^2}{12(4\pi)^6}\left(\frac{2}{\epsilon^2}+\frac{2}{\epsilon}\left(-\ln\frac{(-p^2)}{4\pi\mu^2}-\gamma+3\right)+\text{finite}\right). \quad (16.87)$$

We see again that the dangerous nonlocal divergent terms of the form $\frac{1}{\epsilon}\ln\frac{(-p^2)}{4\pi\mu^2}$ are present in this diagram.

On the other hand, with the one loop counter term for the three point function (see (16.74)), we also have a contribution from the graph

$$= \frac{(-ig\mu^{\frac{\epsilon}{2}})}{2}\int\frac{d^nk}{(2\pi)^n}\frac{i}{k^2}\frac{ig^3\mu^{\frac{\epsilon}{2}}}{2(4\pi)^3}\frac{2}{\epsilon}\frac{i}{(k+p)^2}$$

$$= -\frac{g^4\mu^\epsilon}{4(4\pi)^3}\frac{2}{\epsilon}\int\frac{d^nk}{(2\pi)^n}dx\frac{1}{((k+xp)^2+x(1-x)p^2)^2}$$

$$= -\frac{g^4\mu^\epsilon}{4(4\pi)^3}\frac{2}{\epsilon}\int dx\frac{i}{(4\pi)^{\frac{n}{2}}}\frac{(-1)^2}{\Gamma(2)}\frac{\Gamma\left(2-\frac{n}{2}\right)}{(-x(1-x)p^2)^{2-\frac{n}{2}}}$$

$$= -\frac{ig^4\mu^\epsilon(-p^2)^{\frac{n}{2}-2}}{4(4\pi)^3(4\pi)^{\frac{n}{2}}}\frac{2}{\epsilon}\Gamma\left(-1+\frac{\epsilon}{2}\right)\int dx\,(x(1-x))^{\frac{n}{2}-2}$$

$$= \frac{ig^4p^2}{4(4\pi)^6}\frac{2}{\epsilon}\left(\frac{(-p^2)}{4\pi\mu^2}\right)^{-\frac{\epsilon}{2}}\Gamma\left(-1+\frac{\epsilon}{2}\right)\frac{\left(\Gamma(\frac{n}{2}-1)\right)^2}{\Gamma(n-2)}$$

$$\simeq \frac{ig^4p^2}{4(4\pi)^6}\frac{2}{\epsilon}\left(1-\frac{\epsilon}{2}\ln\frac{(-p^2)}{4\pi\mu^2}\right)\left(-\frac{2}{\epsilon}+(\gamma-1)\right)\frac{1}{6}\left(1+\frac{5\epsilon}{6}\right)$$

$$= \frac{ig^4p^2}{24(4\pi)^6}\left(-\frac{4}{\epsilon^2}+\frac{2}{\epsilon}\left(\ln\frac{(-p^2)}{4\pi\mu^2}+\gamma-\frac{8}{3}\right)\right)+\text{finite}. \quad (16.88)$$

16.4 BPHZ RENORMALIZATION

Here we have used (16.85) in the intermediate step. Similarly, the graph with the counter term at the left vertex also contributes an equal amount

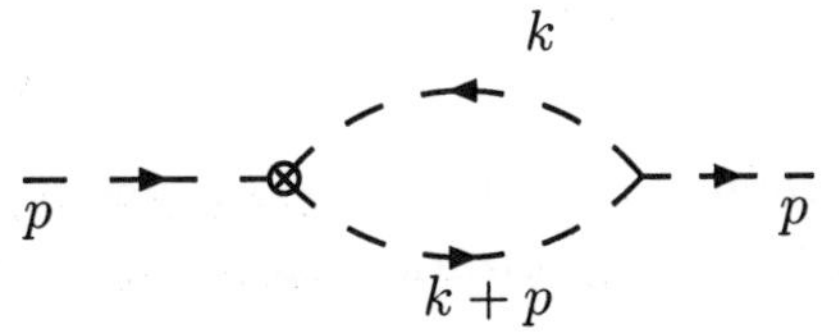

$$= \frac{ig^4 p^2}{24(4\pi)^6}\left(-\frac{4}{\epsilon^2}+\frac{2}{\epsilon}\left(\ln\frac{(-p^2)}{4\pi\mu^2}+\gamma-\frac{8}{3}\right)\right)+\text{finite}, \quad (16.89)$$

so that the sum of the three graphs in Fig. 16.13 gives

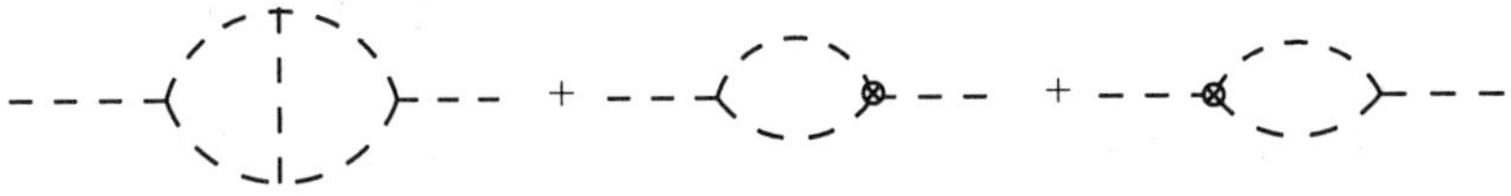

Figure 16.13: Overlapping two loop self-energy diagrams in the ϕ^3 theory.

$$= \frac{ig^4 p^2}{12(4\pi)^6}\left[\frac{2}{\epsilon^2}+\frac{2}{\epsilon}\left(-\ln\frac{(-p^2)}{4\pi\mu^2}-\gamma+3\right)\right.$$
$$\left.-\frac{4}{\epsilon^2}+\frac{2}{\epsilon}\left(\ln\frac{(-p^2)}{4\pi\mu^2}+\gamma-\frac{8}{3}\right)+\text{finite}\right]$$
$$= \frac{ig^4 p^2}{12(4\pi)^6}\left(-\frac{2}{\epsilon^2}+\frac{2}{3\epsilon}+\text{finite}\right). \quad (16.90)$$

We see that all the potentially dangerous nonlocal divergent terms have disappeared from the self-energy and the remaining divergences can be removed by adding a local two loop counter term.

BPHZ is a renormalization procedure where divergences are subtracted in the graph itself so that it works with any regularization.

Consider an arbitrary Feynman graph G. It may contain one particle irreducible subgraphs that are superficially divergent. An 1PI subgraph γ is called proper if it is different from the graph G itself and it is called a renormalization part if

$$D(\gamma) \geq 0, \tag{16.91}$$

namely, if it is superficially divergent. A couple of examples of proper subgraphs which correspond to renormalization parts are shown in Fig. 16.14.

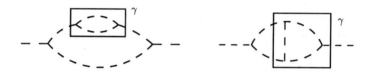

Figure 16.14: Examples of proper renormalization parts in the two loop self-energy graph in the ϕ^3 theory.

Given a Feynman graph G let us draw boxes around all renormalization parts (including the graph G if it is superficially divergent) in all possible ways. Thus, for example, for the three loop self energy graph in the ϕ^3 theory, Fig. 16.15 shows examples of different ways of drawing boxes around the renormalization parts. Two renormalization parts γ_1 and γ_2 are called disjoint if

$$\gamma_1 \cap \gamma_2 = 0, \tag{16.92}$$

while γ_1 and γ_2 are called nested if one is contained inside the other, namely, if

$$\gamma_1 \subset \gamma_2, \quad \text{or} \quad \gamma_2 \subset \gamma_1. \tag{16.93}$$

Finally, two renormalization parts γ_1 and γ_2 are called overlapping if they share lines and vertices, namely, if none of the following holds,

16.4 BPHZ RENORMALIZATION

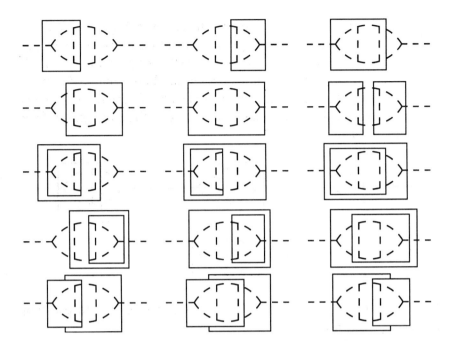

Figure 16.15: Examples of forests in the three loop self-energy graph in the ϕ^3 theory.

$$\gamma_1 \cap \gamma_2 = 0, \quad \gamma_1 \subset \gamma_2, \quad \gamma_2 \subset \gamma_1. \tag{16.94}$$

A set of such laying down of boxes around renormalization parts is called a forest. A forest is called empty if there is no box around any subdiagram. A forest is called normal if there is no box around the complete graph G. On the other hand, a forest is called full if there is a box around G.

Let $\mathcal{D}$ denote the set of laying down of boxes which contain only disjoint renormalization parts. Furthermore, let us add to the set $\mathcal{D}$ the graph with no boxes. Thus, for the two loop self-energy graph in the ϕ^3 theory, the complete set of laying down of boxes (forests) containing disjoint renormalization parts is shown in Fig. 16.16. Similarly, let $\mathcal{N}$ denote the set of laying down of boxes (forests) containing only nested renormalization parts. Fig. 16.17 shows the nested forests in the two loop self-energy graph in the ϕ^3 theory.

Furthermore, we define $\mathcal{O}$ to denote the set of laying down of boxes (forests) containing overlapping renormalization parts, for example, shown in Fig. 16.18 for the two loop self-energy graph in the ϕ^3 theory. The reason why we draw boxes around renormalization parts is because these sub-diagrams are superficially divergent and to make the full diagram finite, we must subtract out the divergences from each of these sub-diagrams.

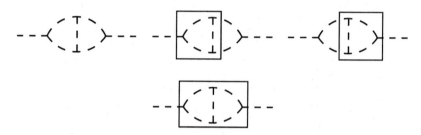

Figure 16.16: The complete set of disjoint forests for the two loop self-energy graph in the ϕ^3 theory.

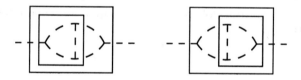

Figure 16.17: The complete set of nested forests for the two loop self-energy graph in the ϕ^3 theory.

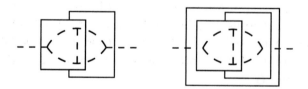

Figure 16.18: The complete set of overlapping forests for the two loop self-energy graph in the ϕ^3 theory.

In the conventional BPHZ method, we introduce an operator t^γ (the Taylor operator or the Taylor expansion operator) for any graph γ such that acting on the integrand of the graph, it Taylor expands

16.4 BPHZ RENORMALIZATION

it in the external momenta (external to the graph) up to terms of order $D(\gamma)$. Thus, (in the self-energy diagram the linear term in the Taylor expansion has been set to zero because of anti-symmetry)

$$\gamma: \quad \underset{p}{\longrightarrow}\!\!\bigcirc\!\!\underset{p}{\longrightarrow} \quad = \int \frac{d^n k}{(2\pi)^n} I(k,p),$$

$$D(\gamma) = 6 - 4 = 2,$$

$$t^\gamma I(k,p) = I(k,0) + \frac{1}{2!} p^\mu p^\nu \left.\frac{\partial^2}{\partial p^\mu \partial p^\nu} I(k,p)\right|_{p=0}.$$

$$\gamma: \quad \underset{p}{\longrightarrow}\!\!\!\bigtriangleup\!\!\!\underset{}{\longrightarrow} \quad = \int \frac{d^n k}{(2\pi)^n} I(k,p,p_1),$$

$$D(\gamma) = 6 - 6 = 0,$$

$$t^\gamma I(k,p,p_1) = I(k,0,0). \tag{16.95}$$

In other words, acting on a graph, the Taylor operator simply separates out the potentially divergent terms in the integrand so that the finite part of the graph can be identified with the integrand

$$(1 - t^\gamma) I_G = \overline{I}_G = \text{ finite.} \tag{16.96}$$

Note that, by definition,

$$t^\gamma I_G = 0, \quad \text{if} \quad D(\gamma) < 0. \tag{16.97}$$

This is the conventional t^γ operator in the BPHZ prescription and it is clear that using this operation, we are simply able to throw away potentially divergent parts inside the integral. Therefore, it never uses any particular regularization and is, in fact, compatible with any regularization. In particular, in the case of dimensional regularization, a graph (integral) at n-loops, has the general form

$$F_G = \sum_{m=-n}^{\infty} a_m \epsilon^m. \tag{16.98}$$

In such a case, the t^γ operation is defined to be (it separates out the pole terms)

$$t^G F_G = \sum_{m=-n}^{-1} a_m \epsilon^m, \tag{16.99}$$

so that the finite part of the graph can be identified with

$$\left(1 - t^G\right) F_G = \sum_{m=0}^{\infty} a_m \epsilon^m = \overline{F}_G = \text{ finite}. \tag{16.100}$$

Once again, this makes it clear that for a superficially convergent graph (with no pole term)

$$t^G F_G = 0. \tag{16.101}$$

Given a renormalization part γ in a Feynman diagram, if the integral has the form

$$I_\gamma = \sum_{m=-n}^{\infty} a_m \epsilon^m, \tag{16.102}$$

then as we have seen the Taylor operation is defined such that

$$\begin{aligned} t^\gamma I_\gamma &= \sum_{m=-n}^{-1} a_m \epsilon^m = \text{ divergent parts}, \\ \left(1 - t^\gamma\right) I_\gamma &= \sum_{m=0}^{\infty} a_m \epsilon^m = \text{ finite parts}. \end{aligned} \tag{16.103}$$

16.4 BPHZ RENORMALIZATION

Thus, the effect of the term $(-t^\gamma I_\gamma)$ can be thought of as that of adding counter terms which subtract the divergence. However, since in this method we do not explicitly use counter terms, this method is correspondingly more general.

Let us consider an arbitrary Feynman diagram G with proper renormalization parts $\gamma_1, \gamma_2, \ldots, \gamma_s$ as subdiagrams. Then, let us define

$$\overline{R}_G I_G = \prod_{i=1}^{s} (1 - t^{\gamma_i}) I_G. \tag{16.104}$$

This, of course, makes all the subdiagrams (proper renormalization parts) in G finite. However, the graph G itself may be superficially divergent in which case, one needs a final subtraction to render the diagram finite. Thus, one defines

$$R_G I_G = \left(1 - t^G\right) \overline{R}_G I_G = \left(1 - t^G\right) \prod_{i=1}^{s} (1 - t^{\gamma_i}) I_G. \tag{16.105}$$

We note that if G is not superficially divergent, then, by definition

$$t^G \overline{R}_G I_G = 0, \tag{16.106}$$

and

$$R_G I_G = \overline{R}_G I_G, \tag{16.107}$$

but this is not true otherwise.

In any case, from the definition of $R_G I_G$ in (16.105), we see that all the superficially divergent subdiagrams as well as the full diagram have been made finite by the subtractions (counter terms) so that the superficial degree of divergence of every subdiagram as well as the full diagram is negative and, by Weinberg's theorem, the graph would now be finite. Thus, graph by graph this procedure can be applied and every graph can be made finite. The only understanding here is that for nested diagrams, the subtraction must be carried out inside out. For disjoint or overlapping divergences, the order of the subtraction is irrelevant.

Although this method of making a Feynman diagram finite is absolutely correct, this does not address the issue of locality of counter terms since we are subtracting out overlapping divergences which, as we have seen, can lead to non-local counter terms. The proof of renormalization by local counter terms, of course, requires that the overlapping divergences be taken care of by lower order counter terms. This is seen through Bogoliubov's R-operator which defines

$$\begin{aligned} R_G I_G &= \left(1 - t^G\right) \overline{R}_G I_G \\ &= \left(1 - t^G\right) \sum_{\Phi_i} \prod_{\gamma_k \in \Phi_i} \left(-t^{\gamma_k} \overline{R}_{\gamma_k}\right) I_G, \end{aligned} \qquad (16.108)$$

where Φ_i denotes the complete set of disjoint proper renormalization parts (subgraphs) of G (Φ_i may be empty and when Φ_i is empty there is no Taylor operation. Its effect is to give I_G itself). Here $\overline{R}_{\gamma_k}$ is the operator (16.104) associated with the superficially divergent renormalization part γ_k. That this is true will be seen shortly through examples. Equation (16.108) is a recursion relation which does not involve nested or overlapping divergences and where at every order the operator $\overline{R}_G$ is determined by the lower order operators $\overline{R}_{\gamma_k}$. This is the BPH formula.

Zimmermann's solution of the BPH recursion relation (16.108) is obtained as follows. Instead of the set of disjoint proper renormalization parts (subgraphs) of a Feynman graph G, let us look at the complete set of subgraphs that include only disjoint and nested proper renormalization parts. Thus, for example, Fig. 16.19 defines all the disjoint and nested proper renormalization parts of the three loop self-energy diagram. With this, we can now write

$$\begin{aligned} R_G I_G &= \left(1 - t^G\right) \overline{R}_G I_G \\ &= \left(1 - t^G\right) \sum_{\Phi_i} \prod_{\gamma_k \in \Phi_i} \left(-t^{\gamma_k}\right) I_G, \end{aligned} \qquad (16.109)$$

where Φ_i is the complete set of proper renormalization parts that are non-overlapping. The BPH recursion relation (16.108) as well as

16.4 BPHZ RENORMALIZATION

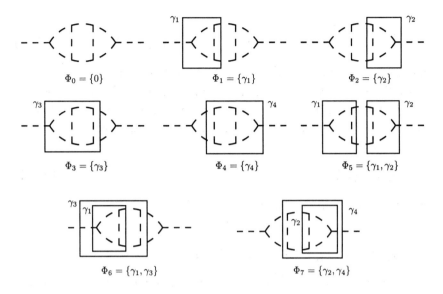

Figure 16.19: The complete set of disjoint and nested proper renormalization parts for the three loop self-energy graph in the ϕ^3 theory.

Zimmermann's solution (16.109) seem quite different from the original relation (16.105). However, let us next show through examples that they are all, in fact, equivalent.

Before we proceed to see this, let us look at some simple examples just to get acquainted with the notions of the BPH procedure. The set of disjoint proper renormalization parts for the two loop vertex correction is shown in Fig. 16.20.

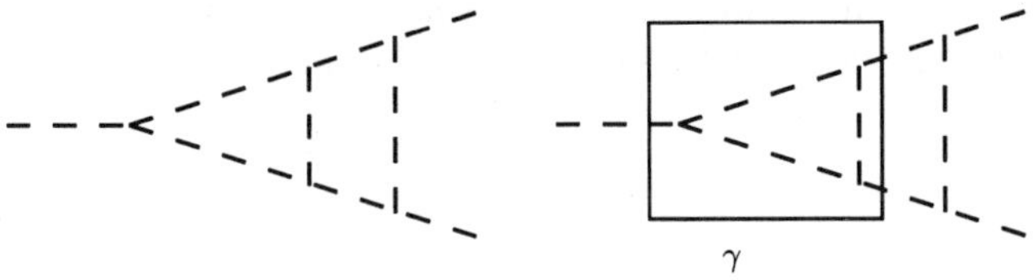

Figure 16.20: The complete set of disjoint proper renormalization parts of the two loop 3 point function graph in the ϕ^3 theory.

Therefore, from the BPH formula (16.108), we have

$$\overline{R}_G I_G = I_G + I_{G/\gamma}\left(-t^\gamma \overline{R}_\gamma I_\gamma\right) = I_G + I_{G/\gamma}\left(-t^\gamma I_\gamma\right), \qquad (16.110)$$

where $I_{G/\gamma}$ is the part of the diagram which does not contain the subdiagram I_γ. We note that, by definition, $\overline{R}_\gamma I_\gamma = I_\gamma$ because there is no superficially divergent proper subdiagram of γ, in fact, there is no subdiagram in this case. From the relation

$$I_G = I_{G/\gamma} I_\gamma, \qquad (16.111)$$

we can write (16.110) also as

$$\begin{aligned}\overline{R}_G I_G &= I_G + (-t^\gamma I_G) = (1 - t^\gamma) I_G, \\ R_G I_G &= \left(1 - t^G\right) \overline{R}_G I_G = \left(1 - t^G\right)(1 - t^\gamma) I_G. \end{aligned} \qquad (16.112)$$

This is exactly the original formula (16.105) and, in fact, this is how we are supposed to renormalize the theory.

Let us next look at the two loop self-energy graphs shown in Fig. 16.21 containing only proper disjoint renormalization parts. Here we have overlapping divergences and following BPH (see (16.108)) we can write

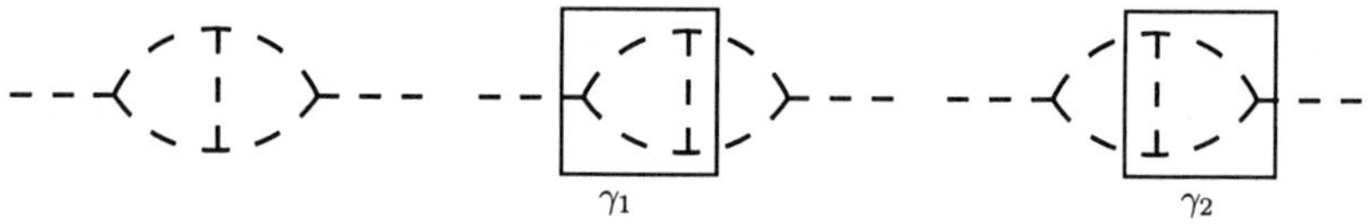

Figure 16.21: The complete set of disjoint proper renormalization parts of the two loop self-energy graph in the ϕ^3 theory.

$$\begin{aligned}\overline{R}_G I_G &= I_G + I_{G/\gamma_1}\left(-t^{\gamma_1} \overline{R}_{\gamma_1} I_{\gamma_1}\right) + I_{G/\gamma_2}\left(-t^{\gamma_2} \overline{R}_{\gamma_2} I_{\gamma_2}\right) \\ &= I_G + I_{G/\gamma_1}\left(-t^{\gamma_1} I_{\gamma_1}\right) + I_{G/\gamma_2}\left(-t^{\gamma_2} I_{\gamma_2}\right) \\ &= I_G + (-t^{\gamma_1} I_G) + (-t^{\gamma_2} I_G) \\ &= (1 - t^{\gamma_1} - t^{\gamma_2}) I_G, \end{aligned} \qquad (16.113)$$

16.4 BPHZ RENORMALIZATION

so that

$$R_G I_G = \left(1 - t^G\right)\left(1 - t^{\gamma_1} - t^{\gamma_2}\right) I_G. \tag{16.114}$$

This does not exactly coincide with the original formula (16.105)

$$R_G I_G = \left(1 - t^G\right)\left(1 - t^{\gamma_1}\right)\left(1 - t^{\gamma_2}\right) I_G. \tag{16.115}$$

The difference between the two is given by

$$\left(1 - t^G\right) t^{\gamma_1} t^{\gamma_2} I_G, \tag{16.116}$$

which we recognize to correspond to the counter term associated with an overlapping divergence if it did not vanish. It is such terms that complicate the proof of renormalizability.

Let us recall some of the calculations we had done earlier. At one loop we had found that (see (16.74), t^γ which is supposed to isolate pole terms in ϵ, sets the factor of $\mu^{\frac{\epsilon}{2}} = 1$ in the graph at one loop)

$$\cdots\!\otimes\!\!<\!\!\cdots \;=\; \frac{ig^3}{2(4\pi)^3}\frac{2}{\epsilon} = \left(-t^{\gamma_2} I_{\gamma_2}\right), \tag{16.117}$$

so that

$$t^{\gamma_1} t^{\gamma_2} \;-\!\!-\!\!\langle\;\stackrel{\top}{{}_{\perp}}\;\rangle\!\!-\!\!-\; =\; -t^{\gamma_1} \;-\!\!-\!\!\langle\;\;\rangle\!\otimes\!-$$

$$= -\frac{ig^3}{2(4\pi)^3}\frac{2}{\epsilon} t^{\gamma_1}\frac{1}{2}\left(-ig\mu^{\frac{\epsilon}{2}}\right)\int \frac{d^n k}{(2\pi)^n}\frac{(i)^2}{k^2(k+p)^2}$$

$$= \frac{g^4}{4(4\pi)^3}\frac{2}{\epsilon}\frac{i}{(4\pi)^3}\frac{(-p^2)}{6}\left(-\frac{2}{\epsilon}\right)$$

$$= \frac{ig^4 p^2}{12(4\pi)^6}\frac{2}{\epsilon^2}, \tag{16.118}$$

where we have used the fact that $n = 6 - \epsilon$ and that t^{γ_1} acts on the integral as well as on $\mu^{\frac{\epsilon}{2}}$ to isolate only pole terms. If we recall our calculation for the two-loop overlapping divergence (see(16.87)), then, the $\frac{1}{\epsilon^2}$ term precisely has the same coefficient and it follows, therefore, that

$$\left(1 - t^G\right) t^{\gamma_1} t^{\gamma_2} I_G = 0. \tag{16.119}$$

Physically, this is the statement that we do not need subtractions associated with overlapping divergences.

In general, there is a theorem that says that if γ_1 and γ_2 denote two proper renormalization parts of G that are overlapping, then we can always find a renormalization part γ_{12} which contains both γ_1 and γ_2 and for which

$$\left(1 - t^{\gamma_{12}}\right) t^{\gamma_1} t^{\gamma_2} I_G = 0. \tag{16.120}$$

This result is very important, physically because it says that we do not need to subtract overlapping divergences, but more importantly, we note that using this, we can now write

$$R_G I_G = \left(1 - t^G\right) \overline{R}_G I_G = \left(1 - t^G\right) \prod_{i=1}^{s} \left(1 - t^{\gamma_i}\right) I_G, \tag{16.121}$$

involving subtractions of only non-overlapping proper renormalization parts, namely, γ_i's are only disjoint or nested proper renormalization parts. The product of the factors in (16.121) can also be written as

$$\prod_i \left(1 - t^{\gamma_i}\right) \equiv \sum_{\Phi_i} \prod_{\gamma_k \in \Phi_i} \left(-t^{\gamma_k}\right), \tag{16.122}$$

where Φ_i's define forests containing only disjoint or nested proper renormalization parts. This makes the connection between (16.105) and Zimmermann's solution (16.109).

To make contact with the formula of BPH, it is best to study an example. Let us recall that the three loop diagram has eight

16.4 BPHZ RENORMALIZATION

forests consisting of disjoint and nested proper renormalization parts which are shown in Fig. 16.19. Thus, writing out explicitly the Zimmermann solution we obtain

$$
\begin{aligned}
R_G I_G &= \left(1 - t^G\right) \left[I_G + I_{G/\gamma_1}\left(-t^{\gamma_1} I_{\gamma_1}\right) + I_{G/\gamma_2}\left(-t^{\gamma_2} I_{\gamma_2}\right)\right. \\
&\quad + I_{G/\gamma_3}\left(-t^{\gamma_3} I_{\gamma_3}\right) + I_{G/\gamma_4}\left(-t^{\gamma_4} I_{\gamma_4}\right) \\
&\quad + I_{G/\{\gamma_1,\gamma_2\}}\left(-t^{\gamma_1} I_{\gamma_1}\right)\left(-t^{\gamma_2} I_{\gamma_2}\right) \\
&\quad + I_{G/\gamma_3}\left(-t^{\gamma_3}\left(I_{\gamma_3/\gamma_1}\left(-t^{\gamma_1} I_{\gamma_1}\right)\right)\right) \\
&\quad \left. + I_{G/\gamma_4}\left(-t^{\gamma_4}\left(I_{\gamma_4/\gamma_2}\left(-t^{\gamma_2} I_{\gamma_2}\right)\right)\right)\right]. \quad (16.123)
\end{aligned}
$$

Let us next recall that since γ_1 and γ_2 do not contain any proper renormalization parts, we can write

$$
\begin{aligned}
\overline{R}_{\gamma_1} I_{\gamma_1} &= I_{\gamma_1}, \\
\overline{R}_{\gamma_2} I_{\gamma_2} &= I_{\gamma_2}. \quad (16.124)
\end{aligned}
$$

Using this, we can rewrite (16.123) as

$$
\begin{aligned}
R_G I_G &= (1 - t^G) \left[I_G + I_{G/\gamma_1}(-t^{\gamma_1}\overline{R}_{\gamma_1} I_{\gamma_1}) + I_{G/\gamma_2}(-t^{\gamma_2}\overline{R}_{\gamma_2} I_{\gamma_2})\right. \\
&\quad + I_{G/\gamma_3}\left((-t^{\gamma_3})\left(I_{\gamma_3} + I_{\gamma_3/\gamma_1}\left(-t^{\gamma_1}\overline{R}_{\gamma_1} I_{\gamma_1}\right)\right)\right) \\
&\quad + I_{G/\{\gamma_1,\gamma_2\}}\left(-t^{\gamma_1}\overline{R}_{\gamma_1} I_{\gamma_1}\right)\left(-t^{\gamma_2}\overline{R}_{\gamma_2} I_{\gamma_2}\right) \\
&\quad \left. + I_{G/\gamma_4}\left((-t^{\gamma_4})\left(I_{\gamma_4} + I_{\gamma_4/\gamma_2}\left(-t^{\gamma_2}\overline{R}_{\gamma_2} I_{\gamma_2}\right)\right)\right)\right] \\
&= \left(1 - t^G\right) \left[I_G + I_{G/\gamma_1}\left(-t^{\gamma_1}\overline{R}_{\gamma_1} I_{\gamma_1}\right) + I_{G/\gamma_2}\left(-t^{\gamma_2}\overline{R}_{\gamma_2} I_{\gamma_2}\right)\right. \\
&\quad + I_{G/\gamma_3}(-t^{\gamma_3})\overline{R}_{\gamma_3} I_{\gamma_3} + I_{G/\{\gamma_1,\gamma_2\}}(-t^{\gamma_1}\overline{R}_{\gamma_1} I_{\gamma_1})(-t^{\gamma_2}\overline{R}_{\gamma_2} I_{\gamma_2}) \\
&\quad \left. + I_{G/\gamma_4}(-t^{\gamma_4})\overline{R}_{\gamma_4} I_{\gamma_4}\right] \\
&= \left(1 - t^G\right)\left(1 - t^{\gamma_1}\overline{R}_{\gamma_1} - t^{\gamma_2}\overline{R}_{\gamma_2} - t^{\gamma_3}\overline{R}_{\gamma_3} - t^{\gamma_4}\overline{R}_{\gamma_4}\right. \\
&\quad \left. + (t^{\gamma_1}\overline{R}_{\gamma_1})(t^{\gamma_2}\overline{R}_{\gamma_2})\right) I_G \\
&= \left(1 - t^G\right) \sum_{\Phi_i} \prod_{\gamma_k \in \Phi_i} \left(-t^{\gamma_k}\overline{R}_{\gamma_k}\right) I_G, \quad (16.125)
\end{aligned}
$$

where Φ_i is the complete set of proper renormalization parts consisting of only disjoint graphs. This is, of course, the BPH formula (16.108). This analysis, therefore, shows how the BPH formula leads to Zimmermann's solution and how, through Weinberg's theorem, it is equivalent to making any Feynman graph finite by local subtractions (counter terms).

So far, we have talked about theories where the index of divergence of interactions in four dimensions (recall that $D = 4 - B - \frac{3}{2}F + \sum_i n_i \delta_i$, see (16.20) and (16.21))

$$\delta_i = d_i + b_i + \frac{3}{2}f_i - 4, \tag{16.126}$$

vanishes so that the superficial degree of divergence of any graph is completely determined by the number of external lines in the Feynman diagram. In this case, as we have seen the theory can be renormalized by adding a finite number of local counter terms which simply redefine the fields and the parameters of the original theory. Such theories are known as renormalizable theories. On the other hand if

$$\delta_i > 0, \tag{16.127}$$

then, it is clear that the superficial degree of divergence of a graph would increase with increasing number of interaction vertices. We would have divergence structures with more and more numbers of external lines. Of course, in this case, we can also remove the divergences by adding local counter terms. However, the number of such counter terms we need will be infinite and we cannot simply absorb them into a redefinition of the finite number of existing parameters and fields of the theory. Such theories are called non-renormalizable theories. In contrast, if the index of divergence were negative, namely,

$$\delta_i < 0, \tag{16.128}$$

not only would we have a finite number of types of graphs that can be divergent, but more importantly only a finite number of graphs (and that too only at low orders in perturbation theory since increasing the number of interaction vertices reduces the superficial degree of divergence) will be divergent. Such theories are conventionally known as super-renormalizable theories.

16.5 Renormalization of gauge theories

The analysis in this chapter demonstrates renormalizability of standard field theories involving scalar and fermion fields. However, some subtleties arise in the case of gauge theories. Let us discuss this briefly with the example of the pure non-Abelian Yang-Mills theory. The total Lagrangian density with the gauge fixing and the ghost Lagrangian densities has the form (see (13.12))

$$\begin{aligned}
\mathcal{L} &= -\frac{1}{4}F_{\mu\nu}^a F^{\mu\nu\,a} + \frac{\xi}{2}F^a F^a + (\partial_\mu F^a)A^{\mu,a} + \partial^\mu \bar{c}^a D_\mu c^a \\
&= \frac{1}{2}\partial_\mu A_\nu^a (\partial^\mu A^{\nu,a} - \partial^\nu A^{\mu,a}) + g f^{abc} A_\mu^b A_\nu^c \partial^\mu A^{\nu,a} \\
&\quad - \frac{g^2}{4} f^{abc} f^{apq} A_\mu^b A_\nu^c A^{\mu,p} A^{\nu,q} + \frac{\xi}{2} F^a F^a + (\partial^\mu F^a) A_{\mu,a} \\
&\quad + \partial_\mu \bar{c}^a \partial^\mu c^a - g f^{abc} (\partial^\mu \bar{c}^a) A_\mu^b c^c.
\end{aligned} \qquad (16.129)$$

Naive power counting shows that the only graphs that are superficially divergent are the two point, the three point and the four point amplitudes involving gauge fields as well as the two point function for the ghosts and the three point ghost interaction vertex function. Let us note here (as we have seen explicitly in the case of QED) that the true degree of divergence in a gauge amplitude is generally lower than its superficial degree of divergence because of gauge invariance. Like QED, here also we can check that only the transverse part of the gauge self-energy diverges (we have seen this explicitly in the calculation of the photon self-energy) so that the longitudinal part is not renormalized (namely, we do not need counter terms for the gauge fixing terms). As a result, we can add counter terms to write

$$\begin{aligned}
\mathcal{L} &= -\frac{(1+A)}{2}\partial_\mu A_\nu^a (\partial^\mu A^{\nu,a} - \partial^\nu A^{\mu,a}) + (1+B) g f^{abc} A_\mu^b A_\nu^c \partial^\mu A^{\nu,a} \\
&\quad - \frac{g^2}{4}(1+C) f^{abc} f^{apq} A_\mu^b A_\nu^c A^{\mu,p} A^{\nu,q} + \frac{\xi}{2} F^a F^a + (\partial^\mu F^a) A_{\mu,a}
\end{aligned}$$

$$+(1+D)\partial_\mu \bar{c}^a \partial^\mu c^a - (1+E)gf^{abc}(\partial^\mu \bar{c}^a)A_\mu^b c^c$$
$$= -\frac{Z_3}{2}\partial_\mu A_\nu^a (\partial^\mu A^{\nu,a} - \partial^\nu A^{\mu,a}) + Z_1 g f^{abc} A_\mu^b A_\nu^c \partial^\mu A^{\nu,a}$$
$$- Z_4 \frac{g^2}{4} f^{abc} f^{apq} A_\mu^b A_\nu^c A^{\mu,p} A^{\nu,q} + \frac{\xi}{2} F^a F^a + (\partial^\mu F^a) A_{\mu,a}$$
$$+ \tilde{Z}_3 \partial_\mu \bar{c}^a \partial^\mu c^a - \tilde{Z}_1 g f^{abc} \partial^\mu \bar{c}^a A_\mu^b c^c, \tag{16.130}$$

where we have identified

$$Z_3 = 1+A, \quad Z_1 = 1+B, \quad Z_4 = 1+C,$$
$$\tilde{Z}_3 = 1+D, \quad \tilde{Z}_1 = 1+E. \tag{16.131}$$

It is worth emphasizing here again that we have not added any counter term for the gauge fixing terms because, as we have seen explicitly in the calculation in QED, the longitudinal part of the photon self-energy is not renormalized by quantum corrections. This is, in fact, a general feature of gauge theories following from the BRST invariance of the theory (see (13.91)).

With these counter terms, of course, all the divergences can be taken care of. But the main question is whether renormalization would preserve the gauge invariance – actually the BRST invariance – present in the original theory. Namely, if we redefine the fields and parameters as

$$A_\mu^{a(u)} = Z_3^{\frac{1}{2}} A_\mu^a,$$
$$c^{a(u)} = \tilde{Z}_3^{\frac{1}{2}} c^a,$$
$$\bar{c}^{a(u)} = \tilde{Z}_3^{\frac{1}{2}} \bar{c}^a,$$
$$g^{(u)} = Z_1 Z_3^{-\frac{3}{2}} g,$$
$$g_1^{(u)2} = Z_4 Z_3^{-2} g^2,$$
$$g_2^{(u)} = \tilde{Z}_1 Z_3^{-\frac{1}{2}} \tilde{Z}_3^{-1} g,$$
$$F^{a(u)} = Z_3^{-\frac{1}{2}} F^a,$$

16.5 RENORMALIZATION OF GAUGE THEORIES

$$\xi^{(u)} = Z_3 \xi, \tag{16.132}$$

naively we would expect that the BRST invariance (or even the gauge invariance in the gauge unfixed theory) would require the different (bare) unrenormalized couplings to be the same, namely,

$$g^{(u)} = g_1^{(u)} = g_2^{(u)}. \tag{16.133}$$

This would, in turn require that the corresponding counter terms have to be related. But the real question is how can the renormalization process guarantee this.

The Slavnov-Taylor identities following from the BRST invariance (see, for example, (13.91)) lead to relations between various amplitudes and we can show from these that they lead to relations among renormalization constants of the form (analogous to the relation $Z_1 = Z_2$ in QED following from the Ward-Takahashi identity)

$$\frac{Z_1}{Z_3} = \frac{\tilde{Z}_1}{\tilde{Z}_3} = \frac{Z_4}{Z_1}. \tag{16.134}$$

Therefore, if we choose a regularization scheme which respects the BRST symmetry, the counterterms will satisfy (16.134) and remove all the divergences from the theory. In this case, therefore, we will have

$$\begin{aligned}
\left(g^{(u)}\right)^2 &= Z_1^2 Z_3^{-3} g^2, \\
\left(g_1^{(u)}\right)^2 &= Z_4 Z_3^{-2} g^2 = Z_1^2 Z_3^{-3} g^2, \\
\left(g_2^{(u)}\right)^2 &= \tilde{Z}_1^2 Z_3^{-1} \tilde{Z}_3^{-2} g^2 = Z_1^2 Z_3^{-3} g^2,
\end{aligned} \tag{16.135}$$

which would lead to

$$\left(g^{(u)}\right)^2 = \left(g_1^{(u)}\right)^2 = \left(g_2^{(u)}\right)^2. \tag{16.136}$$

This is, of course, the same relation as in (16.133) and with this identification we can write (16.130) as

$$\mathcal{L} = -\frac{1}{4}F_{\mu\nu}^{a(u)}F^{\mu\nu\,a(u)} + \frac{\xi^{(u)}}{2}F^{a(u)}F^{a(u)} + (\partial_\mu F^{a(u)})A^{\mu,a(u)}$$
$$+\partial^\mu \overline{c}^{a(u)} D_\mu c^{a(u)}, \tag{16.137}$$

where the covariant derivative is defined with $g^{(u)}$. The BRST invariance of this Lagrangian density is now manifest (with unrenormalized fields and parameters).

As we have seen, dimensional regularization is one of the regularization schemes (and it is quite simple) which respects gauge invariance and BRST symmetry and, consequently, in studying non-Abelian gauge theories it is natural to discuss renormalizability using dimensional regularization. Let us note here that the BRST symmetry is quite fundamental in the study of gauge theories (as we have tried to emphasize in chapter **13**). It is essential in identifying the physical states of the theory, in establishing unitarity as well as in showing the gauge independence of the physical matrix elements. It is for these reasons that in the case of gauge theories, the choice of a regularization which preserves this symmetry is quite crucial. Otherwise, the regularization scheme may introduce unwanted spurious effects into the theory.

16.6 Anomalous Ward identity

Let us next consider a non-Abelian gauge theory (belonging to the group $SU(n)$) interacting with massless fermions (quarks) described by the Lagrangian density (see (12.46))

$$\mathcal{L} = -\frac{1}{4}F_{\mu\nu}^a F^{\mu\nu,\,a} + i\overline{\psi}_k \gamma^\mu D_\mu \psi_k, \tag{16.138}$$

where $a = 1, 2, \cdots, n^2 - 1$ and $k = 1, 2, \cdots, n$. We note that this theory, in addition to having the infinitesimal local gauge invariance

$$\begin{aligned}
\delta A_\mu^a &= D_\mu \epsilon^a(x), \\
\delta \psi_k &= -i\epsilon^a(x)\left(T^a \psi\right)_k, \\
\delta \overline{\psi}_k &= i\epsilon^a(x)\left(\overline{\psi}T^a\right)_k,
\end{aligned} \tag{16.139}$$

16.6 ANOMALOUS WARD IDENTITY

is also invariant under the global transformation

$$\begin{aligned}
\delta \psi_k &= -i\lambda \gamma_5 \psi_k, \\
\delta \overline{\psi}_k &= -i\lambda \overline{\psi}_k \gamma_5.
\end{aligned} \qquad (16.140)$$

Here λ is the space time independent real infinitesimal parameter of transformation and (16.140) describes the chiral symmetry transformations under which the massless theory is invariant. As is the case with continuous symmetries, the invariance of the theory leads to a conserved current. In this theory, therefore, there are two conserved currents,

$$\begin{aligned}
J^{\mu,a} &= \overline{\psi}_j \gamma^\mu (T^a)^{jk} \psi_k, \qquad D_\mu J^{\mu,a} = 0, \\
J_5^\mu &= \overline{\psi}_k \gamma_5 \gamma^\mu \psi_k, \qquad \partial_\mu J_5^\mu = 0.
\end{aligned} \qquad (16.141)$$

We note here that the vector current in (16.141) transforms covariantly under a gauge transformation and, therefore, is covariantly conserved. We can write down in a straightforward manner the Ward identities associated with these conserved currents which are simply expressions of the conservation of currents.

Let us note that, if the fermions are massive, i.e., if there is a term in the Lagrangian density of the form

$$\mathcal{L}_m = -m\overline{\psi}_k \psi_k, \qquad (16.142)$$

then this would not be invariant under the chiral symmetry transformations (16.140). Nonetheless, we can still define the axial vector current as in (16.141) which, however, will not be conserved, rather it would satisfy the equation

$$\partial_\mu J_5^\mu = -\delta \mathcal{L}_m = -2im\overline{\psi}_k \gamma_5 \psi_k, \qquad (16.143)$$

where the variation of the Lagrangian density in (16.143) is without the parameter of chiral transformation. We can also derive identities from this relation that would relate various matrix elements of the theory (also known as broken axial vector Ward identities). However, when we calculate matrix elements regularized in a gauge invariant way, we find that the axial vector Ward identities are violated. When classical identities involving a current do not hold quantum mechanically, we say that the current is anomalous or that the conservation law has anomalies. There are various ways to see and understand this phenomenon in quantum field theories. Let us discuss this in the simple model of two dimensional massless QED known as the Schwinger model.

The Schwinger model is described by the $1 + 1$ dimensional Lagrangian density

$$\mathcal{L} = -\frac{1}{4} F_{\mu\nu} F^{\mu\nu} + i \overline{\psi} \gamma^\mu \left(\partial_\mu + i e A_\mu \right) \psi, \quad \mu, \nu = 0, 1. \quad (16.144)$$

This model can be exactly solved and has been studied from various points of view. Let us note that in $1 + 1$ dimensions, a massless photon does not have any true dynamical degrees of freedom since there cannot be any transverse polarization. Secondly, the electromagnetic coupling (electric charge) in the Schwinger model carries dimensions unlike the four dimensional QED where the electric charge is dimensionless. This is easily seen from the fact that in $1 + 1$ dimensions the Lagrangian density has canonical dimension 2 and, therefore, we have

$$[A_\mu] = 0, \quad [\psi] = [\overline{\psi}] = \frac{1}{2}, \quad [e] = 1. \quad (16.145)$$

Furthermore, by naive power counting we see that only the photon two point function (among photon amplitudes) is superficially divergent. On the other hand, as we have mentioned earlier, gauge invariance reduces the superficial degree of divergence and if treated in a gauge invariant manner even this amplitude is finite.

Let us note that in two dimensions we can choose the two Dirac matrices as

$$\gamma^0 = (\gamma^0)^\dagger = \sigma_1 = \begin{pmatrix} 0 & 1 \\ 1 & 0 \end{pmatrix},$$

16.6 ANOMALOUS WARD IDENTITY

$$\gamma^1 = -(\gamma^1)^\dagger = -i\sigma_2 = \begin{pmatrix} 0 & -1 \\ 1 & 0 \end{pmatrix}, \quad (16.146)$$

so that we have

$$\gamma_5 = \gamma_5^\dagger = \gamma^0 \gamma^1 = \sigma_3 = \begin{pmatrix} 1 & 0 \\ 0 & -1 \end{pmatrix}. \quad (16.147)$$

Furthermore, note that in addition to the diagonal metric tensor $\eta^{\mu\nu} = (+,-)$, in two dimensions we have a second rank anti-symmetric tensor (Levi-Civita tensor) given by

$$\epsilon^{\mu\nu} = -\epsilon^{\nu\mu}, \quad \text{with} \quad \epsilon^{01} = 1. \quad (16.148)$$

There are various simple identities in $1+1$ dimensions involving the Dirac matrices, but the ones that are directly of interest to us are

$$\gamma_5 \gamma^\mu = \epsilon^{\mu\nu} \gamma_\nu, \quad \gamma^\mu \gamma^\nu = \eta^{\mu\nu} + \gamma_5 \epsilon^{\mu\nu}. \quad (16.149)$$

As a result of these identities, in $1+1$ dimensions the vector and the axial vector currents are related as

$$J_5^\mu = \overline{\psi} \gamma_5 \gamma^\mu \psi = \epsilon^{\mu\nu} \overline{\psi} \gamma_\nu \psi = \epsilon^{\mu\nu} J_\nu, \quad (16.150)$$

and at the classical (tree) level both these currents are conserved (recall that the fermions are massless in the Schwinger model and the gauge group in the present case is $U(1)$)

$$\partial_\mu J^\mu = 0, \quad \partial_\mu J_5^\mu = 0. \quad (16.151)$$

Let us now calculate the one loop amplitude involving a photon and an axial vector current as shown in Fig. 16.22 and this is given by

$$I_5^{\mu\nu}(p) = -e^2 \int \frac{d^2k}{(2\pi)^2} \frac{\text{Tr } \gamma_5 \gamma^\mu \not{k} \gamma^\nu (\not{k}+\not{p})}{k^2 (k+p)^2}. \quad (16.152)$$

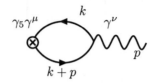

Figure 16.22: The amplitude with an axial vector current and a photon.

Here we have used the propagators for the fermions following from (15.36) in the limit $m = 0$. Furthermore, we note that the amplitude in (16.152) corresponds to the Fourier transform of the matrix element

$$\langle 0|T\left(J_5^\mu(x)A^\nu(0)\right)|0\rangle. \tag{16.153}$$

We can evaluate the integral in (16.152) using dimensional regularization which we know would preserve gauge invariance. Thus, generalizing the integral to $n = 2 - \epsilon$ dimensions would lead to

$$\begin{aligned}
I_5^{\mu\nu}(p) &= -\int \frac{d^n k}{(2\pi)^n} \frac{k_\lambda (k+p)_\rho}{k^2 (k+p)^2} \\
&= -\int \frac{d^n k}{(2\pi)^n} dx \frac{k_\lambda (k+p)_\rho}{\left((k+xp)^2 - x(1-x)p^2\right)^2} \\
&= -\int dx \frac{d^n k}{(2\pi)^n} \frac{k_\lambda k_\rho - x(1-x)p_\lambda p_\rho}{(k^2 - x(1-x)p^2)^2} \\
&= -\int dx \frac{i\pi^{\frac{n}{2}}}{(2\pi)^n} \Bigg[-\frac{\Gamma(1-\frac{n}{2})}{\Gamma(2)} \frac{1}{2} \frac{\eta_{\lambda\rho}}{(-x(1-x)p^2)^{1-\frac{n}{2}}} \\
&\quad - \frac{\Gamma(2-\frac{n}{2})}{\Gamma(2)} \frac{x(1-x)p_\lambda p_\rho}{(-x(1-x)p^2)^{2-\frac{n}{2}}} \Bigg]. \tag{16.154}
\end{aligned}$$

where we have taken out the multiplicative factor $e^2 \text{Tr}(\gamma_5 \gamma^\mu \gamma^\lambda \gamma^\nu \gamma^\rho)$ (to be restored shortly) and we have used the basic formulae of dimensional regularization in (15.71) and (15.73). Furthermore, since

16.6 ANOMALOUS WARD IDENTITY

γ_5 is a two dimensional quantity we have not yet evaluated the trace over the Dirac matrices. We note from (16.154) that the second term inside the bracket is finite and hence for this term the trace over the Dirac matrices can be carried out in two dimensions (because any difference in the trace from extending to n diemnsions would be proportional to ϵ and hence would vanish in the limit $\epsilon \to 0$). On the other hand, the first term is divergent and we have to be careful in evaluating the trace. However, we note that the trace in the first term has the form

$$\begin{aligned} \text{Tr}\,(\gamma_5\gamma^\mu\gamma^\lambda\gamma^\nu\gamma_\rho)\eta_{\lambda\rho} &= \text{Tr}\,(\gamma_5\gamma^\mu\gamma^\lambda\gamma^\nu\gamma_\lambda) \\ &= (2-n)\text{Tr}\,(\gamma_5\gamma^\mu\gamma^\nu), \end{aligned} \quad (16.155)$$

where we have used the gamma matrix identity in n dimension in (15.86). It is clear now that the factor $(2-n) = \epsilon$ makes even the first term finite and, in fact, using this relation, (16.154) takes the form (we now restore the multiplicative factor)

$$I_5^{\mu\nu}(p) = -\frac{ie^2}{(4\pi)^{\frac{n}{2}}}\int dx\bigg[-\frac{1}{2}\frac{\text{Tr}(\gamma_5\gamma^\mu\gamma^\nu)\epsilon\Gamma(\frac{\epsilon}{2})}{(-x(1-x)p^2)^{\frac{\epsilon}{2}}} \\ -\frac{\text{Tr}(\gamma_5\gamma^\mu\gamma^\lambda\gamma^\nu\gamma^\rho)x(1-x)p_\lambda p_\rho}{(-x(1-x)p^2)^{1+\frac{\epsilon}{2}}}\bigg]. \quad (16.156)$$

Since each of the terms is now completely finite, the trace over the Dirac matrices can be carried out in two dimensions using the identities (16.149) and the final result is obtained to be (the x integration is trivial)

$$I_5^{\mu\nu}(p) = -\frac{ie^2}{2\pi}\left(\eta^{\lambda\rho}\epsilon^{\nu\mu} + \eta^{\mu\lambda}\epsilon^{\nu\rho} + \eta^{\nu\rho}\epsilon^{\mu\lambda}\right)\frac{p_\lambda p_\rho}{p^2}. \quad (16.157)$$

The result in (16.157) is quite interesting for it leads to

$$\begin{aligned} p_\mu I_5^{\mu\nu}(p) &= -\frac{ie^2}{2\pi}\left(\eta^{\lambda\rho}\epsilon^{\nu\mu} + \eta^{\mu\lambda}\epsilon^{\nu\rho} + \eta^{\nu\rho}\epsilon^{\mu\lambda}\right)\frac{p_\mu p_\lambda p_\rho}{p^2} \\ &= -\frac{ie^2}{\pi}\,\epsilon^{\nu\mu}p_\mu. \end{aligned} \quad (16.158)$$

Recalling that this amplitude is the Fourier transform of the matrix element in (16.153), this leads to the operator relation

$$\partial_\mu J_5^\mu(x) = \frac{e^2}{\pi}\, \epsilon_{\mu\nu}\partial^\mu A^\nu = \frac{e^2}{2\pi}\, \epsilon_{\mu\nu}F^{\mu\nu}. \tag{16.159}$$

This shows that although the axial vector current is divergence free (conserved) at the tree level, quantum corrections have made the conservation law anomalous and the anomaly in the divergence of the current is given by (16.159). The origin of this result can be traced to the fact that the dimensional regularization which we have used in our calculation violates chiral symmetry although it preserves gauge invariance. In fact, it is possible to show that there is no regularization which will simultaneously preserve gauge invariance and chiral invariance. As a result, one of the two currents (or a linear combination of them) would always become anomalous because of quantum corrections. On the other hand, since gauge invariance is so vital to the physical theory, we always choose a regularization scheme which would preserve gauge invariance in the quantum theory so that the chiral current becomes anomalous. In some gauge theories, the particle multiplets are carefully chosen such that the anomalous contributions coming from various multiplets cancel completely making the gauge current anomaly free. This is very important in building models of particle interactions.

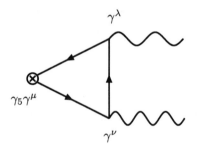

Figure 16.23: The triangle diagram in four dimensional QED contributing to the anomaly.

Although our discussion here has been within the context of the two dimensional QED (Schwinger model), the same behavior is obtained in any dimension. For example, in the four dimensional QED a graph of the form shown in Fig. 16.23 leads to an anomaly in the chiral current. An explicit evaluation of this diagram (that we do not go into) using dimensional regularization leads to an anomaly in the axial vector current of the form (remember that electrons are massive unlike the fermions in the Schwinger model)

$$\begin{aligned}\partial_\mu J_5^\mu &= -2im\overline{\psi}\gamma_5\psi + \frac{e^2}{16\pi^2}\epsilon^{\mu\nu\lambda\rho}F_{\mu\nu}F_{\lambda\rho} \\ &= -2im\overline{\psi}\gamma_5\psi + \frac{e^2}{8\pi^2}F_{\mu\nu}\tilde{F}^{\mu\nu},\end{aligned} \qquad (16.160)$$

where the dual field strength tensor is defined as

$$\tilde{F}_{\mu\nu} = \frac{1}{2}\epsilon_{\mu\nu\lambda\rho}F^{\lambda\rho}, \qquad (16.161)$$

and the second term in (16.160) corresponds to the anomaly in four dimensional QED. We note here that, in this case, it is possible to define a modified axial vector current as

$$\tilde{J}_5^\mu = J_5^\mu - \frac{e^2}{8\pi^2}\epsilon^{\mu\nu\lambda\rho}A_\nu F_{\lambda\rho}, \qquad (16.162)$$

which would satisfy a nonanomalous divergence. However, as is clear from its structure, such a current is not gauge invariant (although the associated charge is gauge invariant). We note from (16.159) that a similar statement can be made in the case of the Schwinger model as well.

As we have discussed in the last chapter, anomalies have observable effect. For example, the correct pion life time can be calculated only if the chiral anomaly is taken into account. Furthermore, if we have a theory where there are both vector and axial vector gauge couplings of the fermions, the model cannot be renormalized unless the anomaly is cancelled. In grand unified theories where we treat both quarks and leptons as massless, they have precisely this kind of coupling. Therefore, a consistent grand unified theory cannot be constructed unless we can cancel out the chiral anomaly and this leads to the study of groups and representations which are anomaly free. In string theories similar anomaly free considerations fix the gauge group uniquely to be $SO(32)$ or $E(8) \times E(8)$. However, these topics are beyond the scope of these lectures.

16.7 References

1. F. J. Dyson, Physical Review **75**, 486 (1949); **75**, 1736 (1949).

2. J. C. Ward, Proceedings of Royal Society **A64**, 54 (1951).

3. A. Salam, Physical Review **82**, 217 (1951); **84**, 426 (1951).

4. S. Weinberg, Physical Review **118**, 838 (1960).

5. N. N. Bogoliubov and O. S. Parasiuk, Acta Mathematica **97**, 227 (1957).

6. J. Schwinger, Physical Review **128**, 2425 (1962).

7. K. Hepp, Communications in Mathematical Physics **2**, 301 (1966).

8. S. Adler, Physical Review **177**, 2426 (1969).

9. J. Bell and R. Jackiw, Nuovo Cimento **60A**, 47 (1969).

10. W. Bardeen, Physical Review **184**, 1848 (1969).

11. A. S. Wightman in "Renormalization Theory", Proceedings of the International School of Mathematical Physics "Ettore Majorana", Erice (1976), ed G. Velo and A. S. Wightman.

12. W. E. Caswell and A. D. Kennedy, Physical Review **25**, 392 (1982).

13. S. Coleman, *Aspects of symmetry*, Cambridge University Press (Cambridge, 1985).

14. T. Muta, *Foundations of Quantum Chromodynamics*, World Scientific, Singapore (1987).

Chapter 17
Renormalization group and equation

17.1 Gell-Mann-Low equation

In the early 1950s, Gell-Mann and Low were interested in studying how the electric force and the potential behave at extremely short distances. We have seen earlier at the end of section **8.10** that the Yukawa potential can be related to the Born amplitude for the scattering of two electrons exchanging a (pseudo) scalar meson. In the same manner, the Coulomb potential can also be obtained from the Born amplitude for the scattering of two electrons exchanging a photon, namely,

$$\longrightarrow V(r) = \frac{\alpha_p}{r}, \qquad (17.1)$$

where in CGS units $\alpha_p^2 = \frac{e_p^2}{\hbar c} \simeq \frac{1}{137}$ and α_p, e_p denote the fine structure constant and the physical electric charge of the fermion respectively. (Physical electric charge is determined from the scattering of on-shell electrons.) It is clear from (17.1) that at the lowest order the potential is independent of the mass of the fermion and it scales with distance as $\frac{1}{r}$. We can, of course, calculate quantum corrections to this potential and if we add the one loop correction to the potential shown in Fig. 17.1, then the potential is obtained to have the form

$$V(r) = \frac{\alpha_p}{r} \left[1 + \frac{2\alpha_p}{3\pi} \ln \frac{1}{r m_e} + \cdots \right]. \qquad (17.2)$$

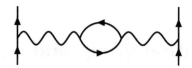

Figure 17.1: The one loop correction to the electromagnetic potential.

It is clear that the one loop corrected potential has a very different character from the lowest order potential. For example, we see from (17.2) that the potential no longer scales as $\frac{1}{r}$ and furthermore, it depends on the mass of the electron in such a way that we cannot take the limit $m_e \to 0$. In fact, in this limit, the potential diverges which is surprising because there is no infrared divergence in the theory and it is not clear what gives rise to such a behavior of the potential.

To understand the origin of this behavior, we note that the one loop potential calculated with a cut off Λ leads to the form (together with the lowest order term)

$$V(r) = \frac{\alpha}{r}\left[1 - \frac{2\alpha}{3\pi}\ln\sqrt{\frac{1+r^2\Lambda^2}{1+r^2m_e^2}} + \cdots\right], \tag{17.3}$$

where α denotes the renormalized fine structure constant of the theory and we note from this result that the potential is well behaved in the limit $m_e \to 0$. Furthermore, $V(r) \to \frac{1}{r}$ as $r \to 0$ (at extremely short distances) as naive scaling would imply. When r is large, namely $r \gg \frac{1}{m_e} \gg \frac{1}{\Lambda}$, it also behaves as

$$V(r) \to \frac{\alpha}{r}\left(1 - \frac{2\alpha}{3\pi}\ln\frac{\Lambda}{m_e}\right) = \frac{\alpha_p}{r}, \tag{17.4}$$

where we have identified

$$\alpha_p = \alpha\left(1 - \frac{2\alpha}{3\pi}\ln\frac{\Lambda}{m_e}\right). \tag{17.5}$$

This relation can be inverted to express the renormalized parameter in terms of the physical parameter as

17.1 Gell-Mann–Low equation

$$\alpha = \alpha_p \left(1 + \frac{2\alpha_p}{3\pi} \ln \frac{\Lambda}{m_e}\right) + \cdots. \tag{17.6}$$

Expressing the potential (17.3) in terms of this (physical) parameter, in the limit $\Lambda \to \infty, rm_e \ll 1$ we obtain

$$\begin{aligned}
V(r) &= \frac{\alpha_p \left(1 + \frac{2\alpha_p}{3\pi} \ln \frac{\Lambda}{m_e}\right)}{r} \left[1 - \frac{2\alpha_p}{3\pi} \ln \sqrt{\frac{1 + r^2\Lambda^2}{1 + r^2 m_e^2}} + \cdots \right] \\
&= \frac{\alpha_p}{r} \left[1 + \frac{2\alpha_p}{3\pi}\left(\ln \frac{\Lambda}{m_e} - \ln \sqrt{\frac{1 + r^2\Lambda^2}{1 + r^2 m_e^2}}\right) + \cdots \right] \\
&= \frac{\alpha_p}{r} \left[1 + \frac{2\alpha_p}{3\pi} \ln \left(\frac{\Lambda}{\sqrt{1 + r^2\Lambda^2}} \frac{\sqrt{1 + r^2 m_e^2}}{m_e}\right) + \cdots \right] \\
&\simeq \frac{\alpha_p}{r} \left[1 + \frac{2\alpha_p}{3\pi} \ln \frac{1}{rm_e} + \cdots \right]. \tag{17.7}
\end{aligned}$$

This expression coincides with (17.2) and also shows that the physical fine structure constant can be identified with

$$\alpha_p = rV(r)\Big|_{r=\frac{1}{m_e}}. \tag{17.8}$$

Furthermore, this derivation clarifies that the singularity in the potential at $m_e \to 0$ arises purely because of the renormalization of charge at large distances. We also see that the naive scaling of the potential breaks down and mass dependence comes in due to renormalization effects.

Let us next investigate what would happen if we define the electric charge at some arbitrary (intermediate) distance R and not as in (17.8). For example, if we define

$$\alpha_R = RV(R), \tag{17.9}$$

then, on dimensional grounds we see that the potential can be expressed as

$$V(r) = \frac{1}{r} F\left(\alpha_R, \frac{r}{R}, Rm_e\right), \tag{17.10}$$

where F is a dimensionless function (function of the independent dimensionless variables $\alpha_R, \frac{r}{R}, Rm_e$). (We note that $rm_e = Rm_e \times \frac{r}{R}$ so that it is not an independent quantity.) This function should not be singular as $m_e \to 0$ which implies that for $r, R \ll \frac{1}{m_e}$, the dependence of the potential on m_e can be neglected, which can be seen from (17.3) as follows. We note that

$$\alpha_R = RV(R) = \alpha \left[1 - \frac{2\alpha}{3\pi} \ln \sqrt{\frac{1+R^2\Lambda^2}{1+R^2 m_e^2}} + \cdots \right]. \tag{17.11}$$

Inverting this relation we obtain

$$\alpha = \alpha_R \left[1 + \frac{2\alpha_R}{3\pi} \ln \sqrt{\frac{1+R^2\Lambda^2}{1+R^2 m_e^2}} + \cdots \right]. \tag{17.12}$$

Putting this back into the potential (17.3) leads, in the limit $\Lambda \to \infty$ and $rm_e, Rm_e \ll 1$ (the second of these limits corresponds to $m_e \to 0$), to

$$\begin{aligned} V(r) &= \frac{\alpha_R}{r} \left(1 + \frac{2\alpha_R}{3\pi} \ln \sqrt{\frac{1+R^2\Lambda^2}{1+R^2 m_e^2}} + \cdots \right) \\ &\quad \times \left(1 - \frac{2\alpha_R}{3\pi} \ln \sqrt{\frac{1+r^2\Lambda^2}{1+r^2 m_e^2}} + \cdots \right) \\ &= \frac{\alpha_R}{r} \left(1 + \frac{2\alpha_R}{3\pi} \ln \sqrt{\frac{1+R^2\Lambda^2}{1+r^2\Lambda^2}} \sqrt{\frac{1+r^2 m_e^2}{1+R^2 m_e^2}} + \cdots \right) \\ &\simeq \frac{\alpha_R}{r} \left(1 + \frac{2\alpha_R}{3\pi} \ln \frac{R}{r} + \cdots \right), \end{aligned} \tag{17.13}$$

which is well behaved as we have already mentioned.

Let us also note that if we define the electric charge (fine structure constant) as

17.1 Gell-Mann-Low equation

$$\alpha_r = rV(r), \tag{17.14}$$

then we see from (17.12) that we can identify

$$\alpha = \alpha_R \left[1 + \frac{2\alpha_R}{3\pi} \ln \sqrt{\frac{1+R^2\Lambda^2}{1+R^2 m_e^2}} + \cdots \right]$$

$$= \alpha_r \left[1 + \frac{2\alpha_r}{3\pi} \ln \sqrt{\frac{1+r^2\Lambda^2}{1+r^2 m_e^2}} + \cdots \right], \tag{17.15}$$

which, in the limit $\Lambda \to \infty$ and $r, R \ll \frac{1}{m_e}$, leads to the relation

$$\alpha_r = \alpha_R \left(1 + \frac{2\alpha_R}{3\pi} \left(\ln \sqrt{\frac{1+R^2\Lambda^2}{1+R^2 m_e^2}} - \ln \sqrt{\frac{1+r^2\Lambda^2}{1+r^2 m_e^2}}\right) + \cdots \right)$$

$$= \alpha_R \left(1 + \frac{2\alpha_R}{3\pi} \ln \frac{R}{r} + \cdots \right), \tag{17.16}$$

and this is compatible with (17.13) (see also the definition (17.14))

$$V(r) = \frac{\alpha_r}{r}. \tag{17.17}$$

From this analysis we see that in the limit of $m_e \to 0$ we can, in general, write (see (17.10))

$$V(r) = \frac{1}{r} F\left(\alpha_R, \frac{r}{R}\right) = \frac{\alpha_r}{r},$$

$$\alpha_r = F\left(\alpha_R, \frac{r}{R}\right). \tag{17.18}$$

This shows that the charge (fine structure constant) changes (runs) with distance and from the second of the equations in (17.18) (see also (17.16)) we see that the equation governing this change is given by

$$r\frac{d\alpha_r}{dr} = -\beta(\alpha_r), \tag{17.19}$$

where

$$\beta(\alpha_r) = -\left.\frac{\partial F(\alpha_r, x)}{\partial x}\right|_{x=1}, \tag{17.20}$$

and this is known as the Gell-Mann-Low equation. We see from this discussion that through proper renormalization the dependence of the potential on the electron mass m_e disappears. Gell-Mann-Low equation basically relates the electric charge at one distance (scale) to another (namely, the electric charge at small distances can be obtained from its value at large distances and *vice versa* through the Gell-Mann-Low equation) and viewed in this manner, we note that α_p and m_e arise only as initial conditions in this evolution, namely,

$$\alpha_{r=\frac{1}{m_e}} = \alpha_p. \tag{17.21}$$

We see explicitly from (17.16) that, in the present case,

$$r\frac{d\alpha_r}{dr} = -\frac{2\alpha_R^2}{3\pi} = -\frac{2\alpha_r^2}{3\pi}. \tag{17.22}$$

There are two possibilities for the solution of the equation (17.18)

$$\alpha_r = F\left(\alpha_R, \frac{r}{R}\right). \tag{17.23}$$

1. As $R \to 0$, α_R does not exist and consequently, the bare charge becomes infinity. In this case, the theory develops ghosts and other problems.

2. As $R \to 0$, α_R does have a limit which is nonzero and is independent of the value $\frac{1}{137}$ which arises only as an initial condition. In this case, for example, if we let $r, R \to 0$ with $\frac{r}{R} = x$ fixed, then (17.18) leads to ($\alpha_0 = \alpha_{r=0}$)

$$\alpha_0 = F(\alpha_0, x). \tag{17.24}$$

This is known as a fixed point of the evolution equation (17.19) (also where the beta function vanishes). In QED, for example, as $r, R \to 0$, $\alpha_0 \not\to 0$ which is obvious from (17.16) (remember that $\frac{r}{R}$ is fixed as the limit is taken) and, therefore, it corresponds to the second possibility.

It is important to note that in a quantum field theory scale invariance is broken by the masses of particles. However, the effects of masses are negligible at high energies if we renormalize the theory in an appropriate way. The only remaining breaking of scale invariance is due to renormalization itself. We can keep track of this by using a running coupling constant. We will discuss these ideas in more detail in section **17.5**.

17.2 Renormalization group

We see from our discussion of QED in the last section that there is an arbitrariness in defining the renormalized parameters of the theory. There are two possible sources for this arbitrariness. First, the renormalization prescription makes a difference in the sense that how much of the finite part we subtract out along with the divergent parts changes the renormalized parameters of the theory. For example, in dimensional regularization, the MS scheme (minimal subtraction) corresponds to subtracting out only the pole parts (in ϵ) of an amplitude. On the other hand, in the $\overline{\text{MS}}$ scheme we not only subtract out the pole parts (in ϵ) of an amplitude, but some constants as well, such as the Euler's constant etc. The renormalized parameters, therefore, are different in the two renormalization schemes even though the regularization is the same (dimensional regularization). Even within a given regularization, a particular renormalization scheme may be preferrable if it makes the higher order contributions in perturbation theory smaller. The second source of arbitrariness in the renormalized parameters arises from the fact that any renormalization prescription introduces a mass scale (or a length scale). Even in dimensional regularization, we saw that the coupling constants are defined with an arbitrary mass scale μ. This introduces an additional arbitrariness into the renormalized parameters depending on the arbitrariness in the mass scale used. (In the example of QED that we discussed in the last section, we defined the charge at some length scale which illustrates the source of arbitrariness of the second kind.)

Let us consider a renormalizable theory, say the ϕ^4 theory in 4-dimensions and let us assume that (in this chapter we will use the terms unrenormalized fields and unrenormalized parameters denoted by a subscript "u" for the bare fields and bare parameters, see the discussion after (16.32))

$$\begin{aligned} \phi_u &= Z_3^{\frac{1}{2}} \phi, \\ m_u^2 &= Z_m Z_3^{-1} m^2, \\ \lambda_u &= Z_1 Z_3^{-2} \lambda, \end{aligned} \qquad (17.25)$$

define a set of renormalized parameters. On the other hand, a different renormalization scale prescription would lead to

$$\begin{aligned} \phi_u &= Z_3'^{\frac{1}{2}} \phi', \\ m_u^2 &= Z_m' Z_3'^{-1} m'^2, \\ \lambda_u &= Z_1' Z_3'^{-2} \lambda', \end{aligned} \qquad (17.26)$$

where we are denoting $Z_3' = Z_3(\mu')$ and so on. Namely, we are using the same regularization but different μ scales and the assumption here is that the unrenormalized field and the unrenormalized parameters are independent of the renormalization prescription.

Calculations with either of these sets of parameters will lead to finite quantities. If we are calculating physical quantities such as the scattering amplitudes, they should not only be finite, they must also have the same value independent of which set of parameters are used to calculate them. Namely, we must have (S and S' denote representative S-matrix elements calculated in the two prescriptions)

$$S'(p, \lambda', m', \mu') = S(p, \lambda, m, \mu), \qquad (17.27)$$

so that if we expand the S matrix elements in a perturbation series

17.2 RENORMALIZATION GROUP

$$S(p, \lambda, m, \mu) = \sum_n a_n \lambda^n,$$
$$S'(p, \lambda', m', \mu') = \sum_n a'_n \lambda'^n, \qquad (17.28)$$

the coefficients of expansion in each of the two series will be renormalization prescription dependent in such a way that the two series would have exactly the same value. If we can calculate the exact series, we can check this. However, in perturbation theory, we can only calculate up to a given order, say the nth order. In such a case, all we can say and check is that

$$S'(p, \lambda', m', \mu') - S(p, \lambda, m, \mu) = O\left(\lambda^{n+1}\right). \qquad (17.29)$$

In a theory, such as QED, where $\alpha \simeq \frac{1}{137}$, the higher order corrections become negligible very quickly. On the other hand, in a theory like QCD (where the value of the coupling constant is not small), the residual terms can be non-negligible.

Let us next see that this arbitrariness in the renormalized parameters is due to a finite renormalization. We can invert the relations in (17.25) and (17.26) so that we can write the renormalized fields in the two prescriptions as

$$\phi = Z_3^{-\frac{1}{2}}(\lambda, \mu)\phi_u, \quad \phi' = Z_3^{-\frac{1}{2}}(\lambda', \mu')\phi_u, \qquad (17.30)$$

which leads to a relation between the two

$$\phi' = z(\mu', \mu)\phi. \qquad (17.31)$$

Here we have identified

$$z(\mu', \mu) = \left(\frac{Z_3(\mu')}{Z_3(\mu)}\right)^{-\frac{1}{2}}. \qquad (17.32)$$

Similarly, from (17.25) and (17.26) we can also write

$$m^2 = Z_m^{-1}(\mu)Z_3(\mu)m_u^2, \quad m'^2 = Z_m^{-1}(\mu')Z_3(\mu')m_u^2, \qquad (17.33)$$

which implies that

$$m'^2 = z_m(\mu', \mu) m^2, \qquad (17.34)$$

with

$$z_m(\mu', \mu) = \left(\frac{Z_m(\mu')}{Z_m(\mu)}\right)^{-1} \left(\frac{Z_3(\mu')}{Z_3(\mu)}\right). \qquad (17.35)$$

Finally, (17.25) and (17.26) imply

$$\lambda = Z_1^{-1} Z_3^2 \lambda_u, \quad \lambda' = Z_1'^{-1} Z_3'^2 \lambda_u, \qquad (17.36)$$

which leads to the relation

$$\lambda' = z_\lambda \lambda, \qquad (17.37)$$

with

$$z_\lambda(\mu', \mu) = \left(\frac{Z_1(\mu')}{Z_1(\mu)}\right)^{-1} \left(\frac{Z_3(\mu')}{Z_3(\mu)}\right)^2. \qquad (17.38)$$

Since (Z_3, Z_m, Z_1) and (Z_3', Z_m', Z_1') differ only in finite parts (the divergent parts are the same since the "μ" dependence comes only in the finite parts), it follows that (z, z_m, z_λ) are completely finite. Thus, parameters corresponding to two distinct renormalization prescriptions are related by a finite renormalization (which can be thought of as a finite transformation much like the symmetry transformations we have talked about in earlier chapters).

Let us next show that these finite renormalizations (finite transformations) define a group. In fact, from the definitions in (17.32), (17.35) and (17.38) we note that

17.2 RENORMALIZATION GROUP

$$z\left(\mu'',\mu'\right)z\left(\mu',\mu\right) = \left(\frac{Z_3(\mu'')}{Z_3(\mu')}\right)^{-\frac{1}{2}}\left(\frac{Z_3(\mu')}{Z_3(\mu)}\right)^{-\frac{1}{2}}$$

$$= \left(\frac{Z_3(\mu'')}{Z_3(\mu)}\right)^{-\frac{1}{2}} = z\left(\mu'',\mu\right),$$

$$z_m\left(\mu'',\mu'\right)z_m\left(\mu',\mu\right) = \left(\frac{Z_m(\mu'')}{Z_m(\mu')}\right)^{-1}\left(\frac{Z_3(\mu'')}{Z_3(\mu')}\right)$$

$$\times \left(\frac{Z_m(\mu')}{Z_m(\mu)}\right)^{-1}\left(\frac{Z_3(\mu')}{Z_3(\mu)}\right)$$

$$= \left(\frac{Z_m(\mu'')}{Z_m(\mu)}\right)^{-1}\left(\frac{Z_3(\mu'')}{Z_3(\mu)}\right) = z_m\left(\mu'',\mu\right),$$

$$z_\lambda\left(\mu'',\mu'\right)z_\lambda\left(\mu',\mu\right) = \left(\frac{Z_1(\mu'')}{Z_1(\mu')}\right)^{-1}\frac{Z_3^2(\mu'')}{Z_3^2(\mu')}$$

$$\times \left(\frac{Z_1(\mu')}{Z_1(\mu)}\right)^{-1}\frac{Z_3^2(\mu')}{Z_3^2(\mu)}$$

$$= \left(\frac{Z_1(\mu'')}{Z_1(\mu)}\right)^{-1}\frac{Z_3^2(\mu'')}{Z_3^2(\mu)} = z_\lambda\left(\mu'',\mu\right). \quad (17.39)$$

Namely, two successive finite renormalizations lead to an equivalent single finite renormalization. These transformations further satisfy

$$\begin{aligned} z(\mu,\mu) &= 1 = z_m(\mu,\mu) = z_\lambda(\mu,\mu), \\ z(\mu,\mu') &= z^{-1}(\mu',\mu), \\ z_m(\mu,\mu') &= z_m^{-1}(\mu',\mu), \\ z_\lambda(\mu,\mu') &= z_\lambda^{-1}(\mu',\mu). \end{aligned} \quad (17.40)$$

The set of finite renormalizations, therefore, define an Abelian group known as the renormalization group and all physical quantities must be invariant under the renormalization group transformations. Namely, physical quantities cannot depend on the arbitrary scale introduced to carry out perturbation calculations in a theory.

17.3 Renormalization group equation

To understand the consequences of the renormalization group of transformations described in the last section, let us next look at the Green's functions of the scalar field theory. The connected renormalized Green's functions of the theory are defined as

$$\begin{aligned} G^{(n)}(x_1,\ldots,x_n) &= \langle T(\phi(x_1)\ldots\phi(x_n))\rangle_c \\ &= Z_3^{-\frac{n}{2}} \langle T(\phi_u(x_1)\ldots\phi_u(x_n))\rangle_c \\ &= Z_3^{-\frac{n}{2}} G_u^{(n)}(x_1\ldots x_n). \end{aligned} \quad (17.41)$$

Such a relation between the renormalized and the unrenormalized Green's functions of the theory holds true in momentum space as well, namely,

$$G^{(n)}(p_1,\ldots,p_n) = Z_3^{-\frac{n}{2}} G_u^{(n)}(p_1,p_2,\ldots p_n), \quad (17.42)$$

where only $(n-1)$ of the external momenta are independent (because of overall energy-momentum conservation). The 1PI vertex functions are obtained from the connected Green's functions by removing the external legs or the external propagators (two point function scales as Z_3) so that we have

$$\Gamma^{(n)}(p_1,\ldots,p_n) = Z_3^{\frac{n}{2}} \Gamma_u^{(n)}(p_1,\ldots,p_n). \quad (17.43)$$

Written out explicitly, this has the form

$$\Gamma^{(n)}(p, m(\mu), \lambda(\mu), \mu) = Z_3^{\frac{n}{2}}(\mu) \Gamma_u^{(n)}(p, m_u, \lambda_u), \quad (17.44)$$

which implies that the renormalized 1PI functions in two renormalization prescriptions are related as

$$\begin{aligned} \Gamma^{(n)}(p, m(\mu'), \lambda(\mu'), \mu') &= Z_3^{\frac{n}{2}}(\mu') \Gamma_u^{(n)}(p, m_u, \lambda_u) \\ &= z^{\frac{n}{2}}(\mu', \mu) \Gamma^{(n)}(p, m(\mu), \lambda(\mu), \mu), \end{aligned} \quad (17.45)$$

17.3 RENORMALIZATION GROUP EQUATION

which follows from the definition of z in (17.32). Here we have used p to denote collectively all the independent external momenta of the 1PI vertex function. Thus, we see that a change in the renormalization scale $\mu \to \mu'$ gives rise to a finite multiplicative renormalization or a finite group of transformations acting on the vertex functions of the theory. As we have noted this defines the Abelian group of transformations known as the renormalization group.

In dimensional regularization, the couplings are defined with an arbitrary mass scale. Let us define the renormalized quartic coupling in the scalar field theory as

$$\mu^\epsilon \lambda, \tag{17.46}$$

while the unrenormalized coupling can be defined as (in discussions in earlier chapters we had used $\mu_0 = 1$)

$$\mu_0^\epsilon \lambda_u. \tag{17.47}$$

Both λ_u and λ are dimensionless and are related as (compare with (16.39))

$$\mu^\epsilon \lambda = Z_1^{-1}(\mu) Z_3^2(\mu) \mu_0^\epsilon \lambda_u = \mu_0^\epsilon \, Z_\lambda^{-1}(\mu) \lambda_u,$$
$$\text{or,} \quad \lambda_u = \left(\frac{\mu}{\mu_0}\right)^\epsilon \lambda Z_\lambda(\mu), \tag{17.48}$$

where we have identified $Z_\lambda = Z_1 Z_3^{-2}$. Here μ_0 is assumed to be fixed and μ is the variable renormalization scale. Since λ_u is μ-independent (independent of the renormalization prescription), it follows that (see also (17.19))

$$\mu \frac{d\lambda}{d\mu} = \beta(\lambda), \tag{17.49}$$

where we see from (17.48) that

$$\beta(\lambda) = -\left(\epsilon + \frac{\mu}{Z_\lambda} \frac{dZ_\lambda}{d\mu}\right) \lambda, \tag{17.50}$$

and the derivative on the right-hand side should be thought of as $\frac{\partial Z}{\partial \mu}$ with λ_u, m_u held fixed. Similarly, from

$$m^2 = Z_m^{-1}(\mu) Z_3(\mu) m_u^2 = \overline{Z}_m^{-1}(\mu) m_u^2, \tag{17.51}$$

we obtain

$$\mu \frac{dm}{d\mu} = -m\gamma_m, \tag{17.52}$$

where

$$\gamma_m = \frac{\mu}{2\overline{Z}_m} \frac{d\overline{Z}_m}{d\mu}. \tag{17.53}$$

Both β and γ_m are finite functions of μ as $\epsilon \to 0$ since the divergences in the numerator and denominator cancel out. They are, in general, functions of λ, m, μ and, in fact, their dependences are of the form (the μ dependence comes in only through the coupling)

$$\beta = \beta(\lambda), \quad \gamma_m = \gamma_m(\lambda), \tag{17.54}$$

in the MS ($\overline{\text{MS}}$) scheme. This is because in these schemes the renormalization constants can be expressed in perturbative series of the forms

$$Z_\lambda = 1 + \frac{a}{\epsilon} \lambda^2 + \left(\frac{b}{\epsilon^2} + \frac{c}{\epsilon}\right) \lambda^4 + \cdots,$$

$$\overline{Z}_m = 1 + \frac{q}{\epsilon} \lambda^2 + \left(\frac{s}{\epsilon^2} + \frac{t}{\epsilon}\right) \lambda^4 + \cdots, \tag{17.55}$$

where the constants $a, b, c, q, s, t, \ldots$ are independent of μ because the pole terms (in ϵ) are independent of μ as there are no overlapping divergences in a renormalizable theory. By dimensional reasoning then they cannot depend on m either. We note that this is a consequence of the fact that dimensional regularization is a mass independent regularization. As a result, the equation for the coupling constant (17.49) decouples from the mass equation (17.52) and can be solved independently which we will analyze in the next section.

17.3 RENORMALIZATION GROUP EQUATION

Let us next look at the 1PI amplitudes (17.44)

$$\Gamma_u^{(n)}(p, m_u, \lambda_u) = Z_3^{-\frac{n}{2}}(\mu)\, \Gamma^{(n)}(p, m(\mu), \lambda(\mu), \mu), \qquad (17.56)$$

which leads to the relation (since the unrenormalized quantities are independent of the renormalization prescription)

$$\mu \frac{d\Gamma_u^{(n)}}{d\mu}(p, m_u, \lambda_u) = 0,$$

or, $\quad \mu \dfrac{d}{d\mu}\left[Z_3^{-\frac{n}{2}}(\mu)\, \Gamma^{(n)}(p, m(\mu), \lambda(\mu), \mu) \right] = 0,$

or, $\quad \left(\mu \dfrac{\partial}{\partial \mu} + \mu \dfrac{\partial \lambda(\mu)}{\partial \mu} \dfrac{\partial}{\partial \lambda(\mu)} + \mu \dfrac{\partial m(\mu)}{\partial \mu} \dfrac{\partial}{\partial m} - n\gamma \right) \Gamma^{(n)} = 0,$

or, $\quad \left(\mu \dfrac{\partial}{\partial \mu} + \beta(\lambda) \dfrac{\partial}{\partial \lambda} - m\gamma_m \dfrac{\partial}{\partial m} - n\gamma \right) \Gamma^{(n)} = 0, \qquad (17.57)$

where we have defined (the derivatives are taken with λ_u, m_u fixed, see also (17.49) and (17.52))

$$\begin{aligned}
\beta(\lambda) &= \mu \frac{\partial \lambda}{\partial \mu}, \\
\gamma_m &= -\frac{\mu}{m} \frac{\partial m}{\partial \mu}, \\
\gamma &= \frac{\mu}{2Z_3} \frac{\partial Z_3}{\partial \mu}.
\end{aligned} \qquad (17.58)$$

We can also argue in the same way (as we have done for β and γ_m) that γ is also finite and is a function of λ alone (in the MS/$\overline{\text{MS}}$ scheme), namely,

$$\gamma = \gamma(\lambda). \qquad (17.59)$$

Equation (17.57) is known as the renormalization group equation for the 1PI vertex functions and expresses how amplitudes change when the renormalization scale is changed.

17.4 Solving the renormalization group equation

As we have seen, the renormalization group equation (17.57) describes how renormalized amplitudes (1PI vertex functions) change as the renormalization scale is changed

$$\left(\mu\frac{\partial}{\partial\mu} + \beta(\lambda)\frac{\partial}{\partial\lambda} - m\gamma_m\frac{\partial}{\partial m} - n\gamma\right)\Gamma^{(n)}(p_i, m, \lambda, \mu) = 0, \tag{17.60}$$

where p_i denotes the independent external momenta and as we have seen in (17.58) (the derivatives are taken with λ_u, m_u fixed)

$$\begin{aligned}
\beta(\lambda) &= \mu\frac{\partial\lambda}{\partial\mu}, \\
\gamma_m &= -\frac{\mu}{m}\frac{\partial m}{\partial\mu}, \\
\gamma &= \frac{\mu}{2Z_3}\frac{\partial Z_3}{\partial\mu}.
\end{aligned} \tag{17.61}$$

For simplicity, let us set $m = 0$ in which case, the equation takes the form

$$\left(\mu\frac{\partial}{\partial\mu} + \beta(\lambda)\frac{\partial}{\partial\lambda} - n\gamma\right)\Gamma^{(n)}(p_i, \lambda, \mu) = 0. \tag{17.62}$$

Let us note from our earlier discussions (see (16.24)) that the n-point 1PI amplitude $\Gamma^{(n)}$ in the ϕ^4 theory in four dimensions has the canonical dimension (the 4 simply represents the fact that the vertex is defined without the energy-momentum conserving delta function)

$$\left[\Gamma^{(n)}\right] = 4 - n. \tag{17.63}$$

Consequently, if we define the Mandelstam variable associated with the n-point function as

$$s = p_1^2 + p_2^2 + \cdots + p_n^2, \tag{17.64}$$

17.4 SOLVING THE RENORMALIZATION GROUP EQUATION

then, scaling all momenta by a factor of $\sqrt{s}$, we can write

$$\Gamma^{(n)}(p_i, \lambda, \mu) = s^{\frac{4-n}{2}} f^{(n)}\left(\frac{s}{\mu^2}, \lambda, \frac{p_i \cdot p_j}{s}\right), \quad (17.65)$$

where we have used the fact that the n-point vertex function of a scalar field theory is Lorentz invariant and, consequently, $f^{(n)}$ is a dimensionless function of independent dimensionless variables which are Lorentz invariant.

Let us relate the renormalization scale μ to a new dimensionless variable t as (this has the advantage that for fixed s, a change in t can be thought of as changing μ or for a fixed μ, it can be thought of as scaling the momentum)

$$\mu = \sqrt{s} e^t, \quad \text{or,} \quad t = -\frac{1}{2} \ln \frac{s}{\mu^2}, \quad (17.66)$$

so that for a fixed s, changing the renormalization scale merely corresponds to changing t. It follows now that

$$\mu \frac{\partial}{\partial \mu} = \mu \frac{\partial t}{\partial \mu} \frac{\partial}{\partial t} = \frac{\partial}{\partial t}, \quad (17.67)$$

(where from (17.66) we have used $\frac{\partial t}{\partial \mu} = -\frac{1}{2} \frac{\partial}{\partial \mu} \ln \frac{s}{\mu^2} = \frac{1}{\mu}$) so that the renormalization group equation (17.62) can be written as

$$\left(\frac{\partial}{\partial t} + \beta(\lambda) \frac{\partial}{\partial \lambda} - n\gamma\right) \Gamma^{(n)} = 0, \quad (17.68)$$

and

$$\Gamma^{(n)} = s^{\frac{4-n}{2}} f^{(n)}\left(t, \lambda, \frac{p_i \cdot p_j}{s}\right). \quad (17.69)$$

Solving the renormalization group equation (17.68) now becomes straightforward from the observation that a linear first order differential equation of the form

$$\frac{\partial \rho(x,t)}{\partial t} + v(x) \frac{\partial \rho(x,t)}{\partial x} = L(x) \rho(x,t), \quad (17.70)$$

is of hydrodynamic type and is best solved by the method of characteristics. The characteristics, in this case, are defined by the equation

$$\frac{\mathrm{d}\bar{x}(x,t)}{\mathrm{d}t} = v(\bar{x}), \tag{17.71}$$

with the boundary condition

$$\bar{x}(x,0) = x. \tag{17.72}$$

Using (17.71), we note that the differential equation (17.70) can be written as

$$\frac{\mathrm{d}\rho(\bar{x})}{\mathrm{d}t} = L(\bar{x})\rho(\bar{x}), \tag{17.73}$$

which can be integrated to give

$$\rho\left(\bar{x}(x,t)\right) = \rho\left(\bar{x}(x,t=0)\right) e^{\int_0^t \mathrm{d}t' L(\bar{x}(x,t'))}. \tag{17.74}$$

Translating t by $-t$, we then obtain

$$\rho\left(\bar{x}(x,0)\right) = \rho\left(\bar{x}(x,-t)\right) e^{\int_{-t}^0 \mathrm{d}t' L(\bar{x}(x,t'))},$$

or, $\quad \rho(x) = \rho\left(\bar{x}(x,-t)\right) e^{\int_{-t}^0 \mathrm{d}t' L(\bar{x}(x,t'))}, \tag{17.75}$

where we have used (17.72). Therefore, we see that solving the equation (17.70) clearly depends on determining its characteristics $\bar{x}(x,t)$.

We can, of course, directly identify the quantities in the renormalization group equation (17.68) with the hydrodynamic problem (17.70) as

$$\begin{aligned}
t &\leftrightarrow t, \\
\lambda &\leftrightarrow x, \\
\beta(\lambda) &\leftrightarrow v(x), \\
\Gamma^{(n)} &\leftrightarrow \rho, \\
n\gamma &\leftrightarrow L,
\end{aligned} \tag{17.76}$$

17.4 SOLVING THE RENORMALIZATION GROUP EQUATION

so that the solution of the renormalization group equation (17.68) requires solving the characteristic equation

$$\frac{d\overline{\lambda}}{dt} = \beta(\overline{\lambda}), \tag{17.77}$$

and once we know the solution to this equation, we can obtain

$$\Gamma^{(n)} = s^{\frac{4-n}{2}} f^{(n)}(\overline{\lambda}(\lambda, -t), \frac{p_i \cdot p_j}{s}) e^{n \int_{-t}^{0} dt' \gamma(\overline{\lambda}(\lambda, t'))}. \tag{17.78}$$

Note that if we scale all momenta (keeping μ fixed) as

$$p_i \to \alpha p_i, \qquad s \to \alpha^2 s, \tag{17.79}$$

so that (remember that $-t = \frac{1}{2} \ln \frac{s}{\mu^2}$)

$$t \to t - \ln \alpha, \tag{17.80}$$

then, for large α (large momentum), the leading behavior of the amplitude is given by

$$\begin{aligned}
\Gamma^{(n)} &\to (\alpha^2 s)^{\frac{4-n}{2}} f^{(n)}(\overline{\lambda}(\lambda, -t + \ln \alpha), \frac{p_i \cdot p_j}{s}) \\
&\quad \times e^{n \int_{-t+\ln\alpha}^{\ln\alpha} dt' \gamma(\overline{\lambda}(\lambda, t' - \ln \alpha))} \\
&\to \alpha^{4-n} (\ln \alpha)^a (\cdots) \\
&= (\alpha)^{D(\Gamma^{(n)})} (\ln \alpha)^a (\cdots),
\end{aligned} \tag{17.81}$$

where $D(\Gamma^{(n)})$ represents the superficial degree of divergence of the n-point function and 'a' is a constant which can be determined from the explicit form of the amplitude. This is known as Weinberg's theorem (the same theorem which also made its appearance in renormalization theory).

Let us next note that if λ_* is a fixed point of the coupling constant evolution equation (17.77), namely, if

$$\beta(\lambda_*) = 0, \tag{17.82}$$

then, at the fixed point,

$$\overline{\lambda} = \lambda_* = \text{constant}. \tag{17.83}$$

Near the fixed point, the n-point amplitude (17.78) behaves as

$$\begin{aligned}\Gamma^{(n)} &= s^{\frac{4-n}{2}} f^{(n)}\left(\lambda_*, \frac{p_i \cdot p_j}{s}\right) e^{nt\gamma(\lambda_*)} \\ &= s^{\frac{4-n}{2}} e^{-\frac{n}{2}\gamma(\lambda_*) \ln \frac{s}{\mu^2}} f^{(n)}\left(\lambda_*, \frac{p_i \cdot p_j}{s}\right) \\ &= s^{\frac{4-n}{2}} \left(\frac{s}{\mu^2}\right)^{-\frac{n}{2}\gamma(\lambda_*)} f^{(n)}\left(\lambda_*, \frac{p_i \cdot p_j}{s}\right).\end{aligned} \tag{17.84}$$

As a result, near the fixed point of the renormalization group equation, the n-point amplitude will scale, under a scaling $p_i \to \alpha p_i$ (17.79), as

$$\Gamma^{(n)} \simeq \alpha^{4-n(1+\gamma(\lambda_*))} \Gamma^{(n)}. \tag{17.85}$$

We note that this is almost like the naive canonical scaling behavior of the amplitude except for an extra anomalous scale dimension $\gamma(\lambda_*)$ for the scalar fields (recall (16.24)). If the fixed point of the beta function happens to be at the origin, namely, if

$$\lambda_* = 0, \tag{17.86}$$

then

$$\gamma(\lambda_*) = 0, \tag{17.87}$$

which follows from the fact that in perturbation theory $\gamma(\lambda)$ is a power series in λ. In such a case the amplitudes will have almost naive scaling behavior at very high energies ("almost" because $f^{(n)}$ also depends on t through the explicit dependence on $\frac{s}{\mu^2}$ which we have not displayed).

17.4 SOLVING THE RENORMALIZATION GROUP EQUATION

It is clear, therefore, that the solution of the coupling constant equation is quite crucial in understanding the behavior of the renormalized amplitudes at high energy as well as under a change of the renormalization scale. We note that

$$\frac{d\overline{\lambda}}{dt} = \beta\left(\overline{\lambda}\right), \tag{17.88}$$

implies that (recall that $\overline{\lambda}(\lambda, t = 0) = \lambda$)

$$\int_\lambda^{\overline{\lambda}} \frac{d\lambda'}{\beta(\lambda')} = \int_0^t dt' = t. \tag{17.89}$$

Before solving this equation explicitly, let us analyze the qualitative behavior of the solution. In perturbation theory the beta function has the form of a power series in the coupling constant

$$\beta(\lambda) = \beta_0 \lambda^n + O\left(\lambda^{n+1}\right). \tag{17.90}$$

Therefore, the first thing that we notice is that

$$\lambda = 0, \tag{17.91}$$

is a fixed point of the beta function.

If there are no other fixed points of $\beta(\lambda)$ and it is positive (namely if $\beta_0 > 0$ and λ restricted to the positive half line), this leads to the behavior of the β function as shown in Fig. 17.2. In this case, the function $\frac{d\overline{\lambda}}{dt}$ is monotonically increasing for positive t and we say that the origin in the coupling constant space is an infrared fixed point. Qualitatively, the behavior of the coupling constant is given by

$$\overline{\lambda}(t) \to \begin{cases} \infty, & t \to \infty, \\ 0, & t \to -\infty. \end{cases} \tag{17.92}$$

On the other hand, if $\beta(\lambda)$ is negative ($\beta < 0$, and $\lambda > 0$), the behavior of $\beta(\lambda)$ is shown in Fig. 17.3 and in this case, qualitatively we have

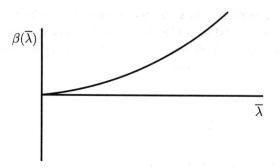

Figure 17.2: The graph showing that origin in the coupling constant space is an infrared fixed point of the β function.

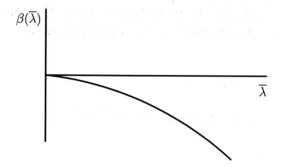

Figure 17.3: The graph showing that the origin in the coupling constant space is an ultraviolet fixed point of the β function.

$$\overline{\lambda}(t) \to \begin{cases} 0, & t \to \infty, \\ \infty, & t \to -\infty, \end{cases} \tag{17.93}$$

and we say that the origin in the coupling constant space is an ultraviolet fixed point.

Let us next analyze the behavior of the solutions in some detail. Let us assume that the beta function has a nontrivial fixed point at λ_* on the positive axis, then expanding around this fixed point, we have two distinct cases to consider (when $\lambda_* = 0$, we recover the two cases we have discussed earlier), namely,

17.4 Solving the Renormalization Group Equation

1. $\beta(\lambda) = \beta_0 (\lambda - \lambda_*) + O\left((\lambda - \lambda_*)^2\right),$ (17.94)

2. $\beta(\lambda) = \beta_0 (\lambda - \lambda_*)^r + O\left((\lambda - \lambda_*)^{r+1}\right), \quad r > 1.$ (17.95)

In the first case (17.89) leads to

$$t = \int_\lambda^{\bar{\lambda}} \frac{d\lambda'}{\beta(\lambda')} = \int_\lambda^{\bar{\lambda}} \frac{d\lambda'}{\beta_0 (\lambda' - \lambda_*)} = \frac{1}{\beta_0} \ln \frac{\bar{\lambda} - \lambda_*}{\lambda - \lambda_*},$$

or, $\quad \bar{\lambda} - \lambda_* = (\lambda - \lambda_*) e^{\beta_0 t},$ (17.96)

which implies that

1. if $\beta_0 < 0$, then as

$$t \to \infty, \quad \bar{\lambda} \to \lambda_*. \tag{17.97}$$

2. if $\beta_0 > 0$, then as

$$t \to -\infty, \quad \bar{\lambda} \to \lambda_*. \tag{17.98}$$

The limits $t \to \infty$ ($t \to -\infty$) are known respectively as the ultraviolet (infrared) limits and fixed points which arise in these limits are correspondingly known as ultraviolet (infrared) fixed points (note from (17.66) that when $t \to \infty, \mu \to \infty$ and when $t \to -\infty, \mu \to 0$ for fixed s). In the diagram Fig. 17.4 (which shows the behavior of a general β function), the arrows indicate the direction in which $\bar{\lambda}$ moves as $t \to \infty$ (if we start near the fixed point).

In the second case (see (17.95)), (17.89) leads to

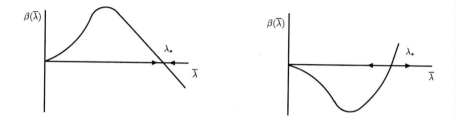

Figure 17.4: Ultraviolet and Infrared fixed points of the β function. The graph on the left shows an ultraviolet fixed point at λ_* while the one on the right corresponds to a nontrivial infrared fixed point.

$$t = \int_\lambda^{\bar{\lambda}} \frac{d\lambda'}{\beta(\lambda')} = \int_\lambda^{\bar{\lambda}} \frac{d\lambda'}{\beta_0 (\lambda' - \lambda_*)^r}$$

$$= -\frac{1}{(r-1)\beta_0} \left[\frac{1}{(\bar{\lambda} - \lambda_*)^{r-1}} - \frac{1}{(\lambda - \lambda_*)^{r-1}} \right],$$

or, $$\frac{1}{(\bar{\lambda} - \lambda_*)^{r-1}} = \frac{1}{(\lambda - \lambda_*)^{r-1}} - (r-1)\beta_0 t$$

$$= \frac{1 - (r-1)\beta_0 (\lambda - \lambda_*)^{r-1} t}{(\lambda - \lambda_*)^{r-1}},$$

or, $$(\bar{\lambda} - \lambda_*)^{r-1} = \frac{(\lambda - \lambda_*)^{r-1}}{1 - (r-1)\beta_0 (\lambda - \lambda_*)^{r-1} t}. \qquad (17.99)$$

Both in QED and QCD, the beta function has the form (namely, the series starts at the cubic order in the coupling)

$$\beta(g) = \beta_0 g^3 + O\left(g^5\right), \qquad (17.100)$$

where β_0 is a constant calculable in perturbation theory. The difference between the two theories is that for QED $\beta_0 > 0$ (see, for example, (17.22)), while for QCD $\beta_0 < 0$. Thus, in both cases, $g = 0$ is

17.4 SOLVING THE RENORMALIZATION GROUP EQUATION

the only fixed point – for QED it is the infrared fixed point while for QCD it is the ultraviolet fixed point. So (setting $g_* = 0$, identifying the coupling with electric charge $g = e$ and recalling $\beta_0 > 0$), for QED we have from (17.99)

$$\bar{e}^2(t) = \frac{e^2}{1 - 2\beta_0 e^2 t} = \frac{1}{\frac{1}{e^2} - 2\beta_0 t}. \tag{17.101}$$

As t increases, we see that the coupling grows and, in particular, diverges when

$$t \to \frac{1}{2\beta_0 e^2}, \tag{17.102}$$

and perturbation theory breaks down. Such a pole in the evolution of the coupling constant at a particular momentum scale is called the Landau pole.

For QCD, on the other hand, ($g_* = 0$, $\beta_0 < 0$) we have from (17.99)

$$\bar{g}^2(t) = \frac{g^2}{1 + 2|\beta_0|g^2 t} = \frac{1}{\frac{1}{g^2} + 2|\beta_0|t}. \tag{17.103}$$

As $t \to \infty$, the coupling becomes weaker and weaker and ultimately vanishes. At infinite energy, therefore, the theory behaves like a free theory and such theories are called asymptotically free theories. In such theories, as we have seen, at large momenta almost naive scaling sets in (see discussion after (17.87)).

Let us now illustrate how the β function is calculated in the context of the ϕ^4 theory in four dimensions which we have studied in some detail. Let us recall that with the MS (minimal subtraction) scheme at one loop we have (see (17.46)-(17.48) as well as (16.38))

$$\lambda_u = \left(\frac{\mu}{\mu_0}\right)^\epsilon \lambda \left(1 + \frac{3\lambda}{16\pi^2 \epsilon}\right). \tag{17.104}$$

Since the unrenormalized (bare) coupling does not depend on the renormalization scale, it follows that

$$\mu \frac{\partial \lambda_u}{\partial \mu} = 0,$$

or, $\left(\dfrac{\mu}{\mu_0}\right)^\epsilon \left(\epsilon\lambda\left(1 + \dfrac{3\lambda}{16\pi^2\epsilon}\right) + \mu\dfrac{\partial \lambda}{\partial \mu}\left(1 + \dfrac{3\lambda}{8\pi^2\epsilon}\right)\right) = 0.$ (17.105)

This leads to

$$\begin{aligned}
\mu \frac{\partial \lambda}{\partial \mu} &= \lim_{\epsilon \to 0} -\epsilon\lambda \left(1 + \frac{3\lambda}{16\pi^2\epsilon}\right)\left(1 + \frac{3\lambda}{8\pi^2\epsilon}\right)^{-1} \\
&= \lim_{\epsilon \to 0} -\epsilon\lambda \left(1 + \frac{3\lambda}{16\pi^2\epsilon}\right)\left(1 - \frac{3\lambda}{8\pi^2\epsilon} + O(\lambda^2)\right) \\
&= \lim_{\epsilon \to 0} -\epsilon\lambda \left(1 + \frac{3\lambda}{16\pi^2\epsilon} - \frac{3\lambda}{8\pi^2\epsilon} + O(\lambda^2)\right) \\
&= \frac{3\lambda^2}{16\pi^2},
\end{aligned}$$ (17.106)

which is finite as discussed earlier. Therefore, at one loop in the ϕ^4 theory in four dimensions we obtain

$$\beta(\lambda) = \mu\frac{\partial \lambda}{\partial \mu} = \frac{3\lambda^2}{16\pi^2},$$ (17.107)

so that we have

$$\int_{\lambda(\overline{\mu})}^{\lambda(\mu)} \frac{d\lambda}{\lambda^2} = \frac{3}{16\pi^2} \int_{\overline{\mu}}^{\mu} \frac{d\mu}{\mu} = \frac{3}{16\pi^2} \ln\frac{\mu}{\overline{\mu}},$$

or, $\quad -\dfrac{1}{\lambda(\mu)} + \dfrac{1}{\lambda(\overline{\mu})} = \dfrac{3}{16\pi^2} \ln\dfrac{\mu}{\overline{\mu}},$

or, $\quad \dfrac{1}{\lambda(\mu)} = \dfrac{1}{\lambda(\overline{\mu})} - \dfrac{3}{16\pi^2} \ln\dfrac{\mu}{\overline{\mu}},$

or, $\quad \lambda(\mu) = \dfrac{1}{\frac{1}{\lambda(\overline{\mu})} - \frac{3}{16\pi^2} \ln\frac{\mu}{\overline{\mu}}}.$ (17.108)

This is precisely the same behavior as we find in QED (see (17.101)). (In fact, only theories involving non-Abelian gauge interactions may be asymptotically free, all other theories have the same character as QED.)

Finally, we note that unlike the general renormalized amplitudes, physical quantities of the theory denoted by P (such as the physical masses of particles etc) satisfy the renormalization group equation

$$\left(\mu\frac{\partial}{\partial\mu} + \beta(\lambda)\frac{\partial}{\partial\lambda}\right) P = 0, \tag{17.109}$$

which simply emphasizes the fact that they are invariant under a change of the renormalization scale.

17.5 Callan-Symanzik equation

Let us consider the ϕ^4 theory in four dimensions described by the Lagrangian density

$$\mathcal{L} = \mathcal{L}_{\text{inv}} + \mathcal{L}_m, \tag{17.110}$$

where we have identified

$$\mathcal{L}_{\text{inv}} = \frac{1}{2}\partial_\mu\phi\partial^\mu\phi - \frac{\lambda}{4!}\phi^4,$$

$$\mathcal{L}_m = -\frac{m^2}{2}\phi^2. \tag{17.111}$$

The rationale for dividing the Lagrangian density in this way will become clear shortly. We note that under a scale transformation the coordinates transform as

$$x^\mu \to e^\alpha x^\mu, \tag{17.112}$$

where α denotes the finite global parameter of scale transformation. For an infinitesimal scale transformation ($\alpha = \epsilon = $ infinitesimal), therefore, we have

$$\delta x^\mu = \epsilon x^\mu. \tag{17.113}$$

Under the scale transformation (17.112), the scalar field transforms as

$$\phi(x) \rightarrow e^{\alpha d}\phi(e^{\alpha}x), \tag{17.114}$$

which has the infinitesimal form

$$\phi(x) \rightarrow (1+\epsilon d)\phi(x+\epsilon x),$$
$$\text{or,} \quad \delta\phi = \epsilon\left(x^{\mu}\partial_{\mu}+d\right)\phi(x). \tag{17.115}$$

Here d represents the scale dimension of the field which coincides with the canonical dimension at the tree level. As we have discussed earlier, for a scalar field in four dimensions $d=1$ at the tree level, but we have also seen that the scaling dimension of a field is different in different dimensions.

Under an infinitesimal scale transformation, we note from (17.111) that (here we will use $d=1$ in (17.115) for the scalar theory in four dimensions)

$$\begin{aligned}
\delta\mathcal{L}_{\text{inv}} &= \partial_{\mu}\phi\partial^{\mu}\delta\phi - \frac{\lambda}{3!}\phi^{3}\delta\phi \\
&= \partial_{\mu}\phi\partial^{\mu}\left(\epsilon\left(1+x^{\nu}\partial_{\nu}\right)\phi\right) - \frac{\lambda}{3!}\phi^{3}\epsilon\left(1+x^{\nu}\partial_{\nu}\right)\phi \\
&= \epsilon\left[\partial_{\mu}\phi\partial^{\mu}\phi - \frac{\lambda}{3!}\phi^{4} + \partial^{\mu}\phi\partial_{\mu}\phi\right.\\
&\qquad \left.+x^{\nu}\partial_{\mu}\phi\partial^{\mu}\partial_{\nu}\phi - \frac{\lambda}{3!}x^{\nu}\phi^{3}\partial_{\nu}\phi\right] \\
&= \epsilon\left[4\left(\frac{1}{2}\partial_{\mu}\phi\partial^{\mu}\phi - \frac{\lambda}{4!}\phi^{4}\right) + x^{\nu}\partial_{\nu}\left(\frac{1}{2}\partial_{\mu}\phi\partial^{\mu}\phi - \frac{\lambda}{4!}\phi^{4}\right)\right] \\
&= \epsilon\left(4\mathcal{L}_{\text{inv}} + x^{\mu}\partial_{\mu}\mathcal{L}_{\text{inv}}\right) \\
&= \epsilon\partial_{\mu}\left(x^{\mu}\mathcal{L}_{\text{inv}}\right). \tag{17.116}
\end{aligned}$$

Therefore, the corresponding action would be invariant under the infinitesimal scale transformations (17.115) (which is the reason for the subscript in the Lagrangian density). The Nöther current for

17.5 CALLAN-SYMANZIK EQUATION

scale transformations can be constructed as (see (6.13) and $T^{\mu\nu}_{\text{imp}}$ corresponds to the improved energy-momentum tensor of the theory)

$$S^\mu = x_\nu T^{\mu\nu}_{\text{imp}}, \tag{17.117}$$

and the conservation of this current leads to

$$\partial_\mu S^\mu = T^\mu{}_{\text{imp}\,\mu} + x_\nu \partial_\mu T^{\mu\nu}_{\text{imp}} = T^\mu{}_{\text{imp}\,\mu} = 0, \tag{17.118}$$

where we have used the fact that the improved energy-momentum tensor is divergence free. Namely, the conservation of the scale current requires that the energy-momentum tensor be traceless. (When $m = 0$, this can be checked explicitly to hold using the equations of motion.) For completeness, we note here that the improved energy-momentum tensor differs from the conventional $T^{\mu\nu}$ by total divergence terms and has the explicit form

$$T^{\mu\nu}_{\text{imp}} = T^{\mu\nu} + \frac{1}{6}(\eta^{\mu\nu}\Box - \partial^\mu\partial^\nu)\phi^2. \tag{17.119}$$

This improved tensor is important in studying quantum field theories in a curved (gravitational) background.

On the other hand, under the infinitesimal scale transformation in (17.115), the mass term in the Lagrangian density (17.111) behaves as

$$\begin{aligned}
\delta\mathcal{L}_m &= -m^2\phi\delta\phi \\
&= -m^2\phi\epsilon\left(1 + x^\mu\partial_\mu\right)\phi \\
&= -\epsilon\left(m^2\phi^2 + m^2 x^\mu\phi\partial_\mu\phi\right) = -\epsilon\left(m^2\phi^2 + \frac{m^2}{2}x^\mu\partial_\mu(\phi^2)\right) \\
&= -\epsilon\left(m^2\phi^2 + \partial_\mu\left(\frac{m^2}{2}x^\mu\phi^2\right) - 2m^2\phi^2\right) \\
&= -\epsilon\partial_\mu\left(\frac{m^2}{2}x^\mu\phi^2\right) + \epsilon m^2\phi^2 \\
&= \epsilon\left(\partial_\mu(x^\mu\mathcal{L}_m) + m^2\phi^2\right). \tag{17.120}
\end{aligned}$$

The divergence term vanishes upon integration in the action. However, there is a second term that does not vanish. This shows that the mass term breaks the invariance under scale transformations (which is the reason for the particular decomposition of the Lagrangian density in (17.111)). In fact, any term in the Lagrangian density with a dimensional coupling will break scale invariance. As a result, the scale current will not be conserved in the full theory, rather the divergence of the scale current will satisfy a relation of the type (this can be checked easily from the definition (17.119) using the equations of motion)

$$\partial_\mu S^\mu = T^\mu_{\text{imp}\,\mu} = \Delta, \quad \Delta = m^2 \phi^2. \tag{17.121}$$

This is analogous to the case of fermion masses breaking the chiral invariance of the theory leading to an identity (in the conservation equation for the axial current) of the form (see (16.143))

$$\partial_\mu J_5^\mu = -2im\overline{\psi}\gamma_5\psi. \tag{17.122}$$

Correspondingly, the Ward identity for the scale current, in this case, would have the form

$$\partial_\mu \langle T\left(S^\mu(x)\phi(x_1)\ldots\phi(x_n)\right)\rangle = \langle T\left(\Delta(x)\phi(x_1)\ldots\phi(x_n)\right)\rangle$$
$$+ \delta\left(x^0 - x_1^0\right) \langle T\left([S^0(x), \phi(x_1)]\ldots\phi(x_n)\right)\rangle + \cdots$$
$$+ \cdots + \delta\left(x^0 - x_n^0\right) \langle T\left(\phi(x_1)\ldots[S^0, \phi(x_n)]\right)\rangle. \tag{17.123}$$

When integrated over x the left-hand side in (17.123) vanishes and the integrated Ward identity takes the form

$$i\langle T\left(\delta\phi(x_1)\cdots\phi(x_n)\right)\rangle + \cdots + i\langle T\left(\phi(x_1)\cdots\delta\phi(x_n)\right)\rangle$$
$$= -\int d^4x \,\langle T\left(\Delta(x)\phi(x_1)\cdots\phi(x_n)\right)\rangle, \tag{17.124}$$

where we have used the fact that the integral of $S^0(x)$ over all space corresponds to the charge associated with the scale current and generates infinitesimal scale transformations of the scalar fields through

17.5 CALLAN-SYMANZIK EQUATION

commutators (see, for example, (13.37)). If $G^{(n)}(p_i)$ and $G^{(n)}_\Delta(p,p_i)$ denote the connected Green's functions respectively for n scalar fields and n scalar fields as well as a Δ composite operator, then using the definitions (we are not writing explicitly a subscript "c" to denote it is a connected Green's function, but this is to be understood)

$$(2\pi)^4 \delta^4\left(\sum_i p_i\right) G^{(n)}(p_i) = \int \prod_{i=1}^n d^4x_i\, e^{ip_i\cdot x_i}\, \langle T(\prod_i \phi(x_i))\rangle,$$

$$(2\pi)^4 \delta^4\left(\sum_i p_i\right) G^{(n)}_\Delta(0,p_i) = \int d^4x \prod_{i=1}^n d^4x_i\, e^{ip_i\cdot x_i}$$
$$\times \langle T(\Delta(x) \prod_i \phi(x_i))\rangle, \quad (17.125)$$

as well as (17.115), we can write (17.124) in the momentum space to have the form

$$\left(n(d-4) + 4 - \sum_{i=1}^{n-1} p_i^\mu \frac{\partial}{\partial p_i^\mu}\right) G^{(n)}(p_i) = i G^{(n)}_\Delta(0,p_i). \quad (17.126)$$

Note here that here the factor "4" arises from the overall energy-momentum conserving delta function (in simplifying the action of the differential operator on the delta function, we have used identities such as $x\partial_x \delta(x) = -\delta(x)$) and the factor $(n(d-4)+4)$ denotes the naive canonical dimension of the n-point Green's function (with d denoting the dimension of the scalar field).

We are, however, interested in the 1PI vertex functions. Let $\Gamma^{(n)}(p_i)$ denote the 1PI vertex function with n scalar fields and $\Gamma^{(n)}_\Delta(p,p_i)$ denote the 1PI vertex with a Δ insertion (which is defined in (17.121) and we are suppressing the dependence on other variables for simplicity). Then, we can obtain from (17.126) the identity satisfied by the n-point 1PI vertex functions which takes the form

$$\left(\sum_{i=1}^{n-1} p_i^\mu \frac{\partial}{\partial p_i^\mu} + nd - 4\right) \Gamma^{(n)}(p_i) = -i\Gamma^{(n)}_\Delta(0,p_i). \quad (17.127)$$

This relation is quite easy to understand. In fact, the 1PI vertex functions are obtained from the connected Green's functions by removing external lines (propagators). The canonical dimension of the propagator, in this case (with d denoting the dimension of the scalar field), is given by $(2d-4)$ so that the naive canonical dimension of the n-point 1PI vertex function is obtained to be $(n(d-4)+4-n(2d-4) = 4-nd$ which is reflected in (17.127). Since at the tree level (where $d=1$) the n-point function has a canonical dimension of $(4-n)$ we can write

$$\Gamma^{(n)}(p_i,\lambda) = m^{4-n} f^{(n)}\left(\frac{p_i}{m},\lambda\right), \tag{17.128}$$

where $f^{(n)}$ is a dimensionless function (of dimensionless variables). This leads to

$$m\frac{\partial \Gamma^{(n)}(p_i,\lambda)}{\partial m} = \left(4 - n - \sum_i p_i^\mu \frac{\partial}{\partial p_i^\mu}\right) \Gamma^{(n)}(p_i,\lambda), \tag{17.129}$$

which leads to

$$\sum_i p_i^\mu \frac{\partial \Gamma^{(n)}}{\partial p_i^\mu} = \left(4 - n - m\frac{\partial}{\partial m}\right) \Gamma^{(n)}. \tag{17.130}$$

Using this in (17.127) we obtain

$$\left(\sum_i p_i^\mu \frac{\partial}{\partial p_i^\mu} + nd - 4\right) \Gamma^{(n)}$$

$$= \left(4 - n - m\frac{\partial}{\partial m} + nd - 4\right) \Gamma^{(n)}$$

$$= -\left(m\frac{\partial}{\partial m} + n(1-d)\right) \Gamma^{(n)}$$

$$= -i\Gamma_\Delta^{(n)}(0,p_i,\lambda), \tag{17.131}$$

which can also be written as

17.5 CALLAN-SYMANZIK EQUATION

$$\left(m\frac{\partial}{\partial m} + n(1-d)\right)\Gamma^{(n)}(p_i,\lambda) = i\Gamma_\Delta^{(n)}(0,p_i,\lambda). \tag{17.132}$$

In the deep Euclidean region where $s \to \infty$ (recall (17.65) and note that a vertex with a Δ insertion (17.121) has two extra scalar fields)

$$\begin{aligned}\Gamma^{(n)} &\to s^{\frac{4-n}{2}}, \\ \Gamma_\Delta^{(n)} &\to s^{\frac{2-n}{2}},\end{aligned} \tag{17.133}$$

so that the leading behavior of the relation (17.132) has the form

$$\left(m\frac{\partial}{\partial m} + n(1-d)\right)\Gamma^{(n)} = 0. \tag{17.134}$$

We note that if $d = 1$ as in the tree level, this relation would imply $m\frac{\partial}{\partial m}\Gamma^{(n)} = 0$. However, as we have seen through explicitly calculations, this is not true for the renormalized amplitudes since they depend on mass through logarithm terms (see, for example, (15.10) or (15.19)). Therefore, there must be some violation of the naive scale Ward identity (17.132).

To understand this question better, let us note that we can relate the renormalized vertex functions to the unrenormalized ones as

$$\begin{aligned}\Gamma^{(n)}(p_i) &= Z_3^{\frac{n}{2}}\Gamma_u^{(n)}(p_i), \\ \Gamma_\Delta^{(n)}(p,p_i) &= ZZ_3^{\frac{n}{2}}\Gamma_{\Delta_u}^{(n)}(p,p_i), \\ m_u\frac{\partial \Gamma_u^{(n)}}{\partial m_u} &= i\Gamma_{\Delta_u}^{(n)}(p,p_i),\end{aligned} \tag{17.135}$$

where Z represents the renormalization of the composite operator (17.121) (related to mass renormalization). Thus, we have

$$i\Gamma_\Delta^{(n)}(0,p_i) = iZZ_3^{\frac{n}{2}}\Gamma_{\Delta u}^{(n)}(0,p_i) = ZZ_3^{\frac{n}{2}}m_u\frac{\partial \Gamma_u^{(n)}}{\partial m_u}$$

$$= ZZ_3^{\frac{n}{2}}m_u\frac{\partial}{\partial m_u}\left(Z_3^{-\frac{n}{2}}\Gamma^{(n)}\right)$$

$$= Zm_u\frac{\partial \Gamma^{(n)}}{\partial m_u} - \frac{nZm_u}{2Z_3}\frac{\partial Z_3}{\partial m_u}\Gamma^{(n)}$$

$$= Zm_u\left[\left(\frac{\partial m}{\partial m_u}\frac{\partial}{\partial m} + \frac{\partial \lambda}{\partial m_u}\frac{\partial}{\partial \lambda}\right) - \frac{n}{2}\left(\frac{\partial \ln Z_3}{\partial m_u}\right)\right]\Gamma^{(n)},$$

(17.136)

which can be written as

$$\left(m\frac{\partial}{\partial m} + \beta(\lambda)\frac{\partial}{\partial \lambda} - n\gamma\right)\Gamma^{(n)} = i\Gamma_\Delta^{(n)},$$
(17.137)

with

$$\beta(\lambda) = Zm_u\frac{\partial \lambda}{\partial m_u},$$

$$m = Zm_u\frac{\partial m}{\partial m_u},$$

$$\gamma = \frac{Zm_u}{2}\frac{\partial \ln Z_3}{\partial m_u}.$$
(17.138)

When $\beta(\lambda) = 0$, (17.137) reduces to the earlier equation (17.132) with the identification

$$\gamma = -(1-d), \quad \text{or,} \quad d = 1 + \gamma.$$
(17.139)

However, in general, this shows that there is a violation of the naive scale Ward identity in a renormalized theory. This equation is known as the Callan-Symanzik equation. It simply reflects the fact that there is an anomaly in the scale Ward identity due to renormalization. Renormalization introduces a mass scale and thereby violates scale invariance. It is impossible to find any regularization which will respect scale invariance which is the essence of the Callan-Symanzik equation.

17.6 References

1. M. Gell-Mann and F. Low, Physical Review **95**, 1300 (1954).

2. E. C. G. Stuckelberg and A. Peterman, Helvetica Physica Acta **26**, 499 (1953).

3. N. N. Bogoliubov and D. V. Shirkov, *Introduction to the theory of Quantized Fields*, Nauka, Moscow (1984).

4. C. G. Callan, Physical Review **D2**, 1541 (1970).

5. K. Symanzik, Communications in Mathematical Physics **18**, 227 (1970).

6. C. G. Callan, S. Coleman and R. Jackiw, Annals of Physics **59**, 42 (1970).

7. S. Weinberg in *Asymptotic realms of physics: essays in honor of Francis E. Low*, eds. A. H. Guth, K. Huang and R. L. Jaffe, MIT Press (Cambridge, 1983).

8. S. Coleman, *Aspects of symmetry*, Cambridge University Press (Cambridge, 1985).

9. D. J. Gross in *Methods in field theory*, eds. R. Balian and J. Zinn-Justin, North Holland Publishing (Amsterdam, 1981).

Index

R_ξ gauge, 594
S-matrix, 229, 230
$V - A$ theory, 597
$\overline{MS}$ scheme, 739
ϕ^4 theory, 219
$\mathcal{CPT}$ theorem, 479
   Electric charge, 480
   Equality of life times, 480
   Equality of masses, 479
't Hooft gauge, 594
't Hooft trick, 521, 536
"In" and "out" states, 231
1PI vertex function, 567

Abelian group, 130
Adiabatic hypothesis, 229
Adjoint representation, 500
Adjoint spinor, 36
Anomalous current, 726
Anomalous magnetic moment, 111
Anomalous scale dimension, 752
Anomalous Ward identity, 724
Anomaly, 726
Anti-BRST transformation, 549
Anti-commutation relation, 285
Anti-commutator, 20
Asymptotically free theory, 757
Axial gauge, 517, 578
Axial vector current, 725

Bare field, 684

Bare parameter, 684
Baryon number, 299
Bhabha scattering, 353
Bianchi identity, 533
Bjorken-Drell metric, 5
Bogoliubov's R operator, 714
Born amplitude, 733
Born approximation, 313, 324
BRST charge, 553
BRST invariance, 545
BRST symmetry, 545
   Nöther current, 552
BRST transformation
   Nilpotency, 546, 548, 556

Callan-Symanzik equation, 759
Canonical dimension, 220, 669
Canonical Poisson bracket, 167, 168
Casimir of representation, 499
Casimir operator, 138, 147, 151
Charge, 214
Charge conjugation, 267, 436
   Dirac bilinears, 446
   Dirac field, 442
   Eigenstates, 453
   Spin zero field, 437
   Weak current, 457
   Weyl fermions, 449
Charge operator, 263, 266
Charge renormalization, 662

Chiral anomaly, 666
Chiral transformation, 725
Chirality, 99, 308
    Negative, 100
    Positive, 100
Cini-Touschek transformation, 116
Classical field, 566
Clifford algebra, 20
Cogradient, 7
Compton scattering, 351
Conjugate momentum, 167, 168
Constrained system, 379
Contraction, 244
Contragradient, 7
Coulomb gauge, 360, 517
Coulomb potential, 733
Counter term, 681, 683
Covariant derivative, 282
Covariant quantization, 550
Current, 214

D'Alembertian, 8
Detailed balance, 474
Dirac bracket, 384, 389
Dirac equation, 19
    Completeness relation, 84
    Continuity equation, 44
    Covariance, 72
    Helicity, 92
    Hole theory, 47
    Non-relativistic limit, 105
    Normalization of wave function, 34
    Plane wave solution, 27
    Projection operators, 84
    Properties, 65
    Solution, 27
    Spin, 40
Dirac field
    Left-handed, 308
    Massless, 308
        Quantization, 308
    Right-handed, 308
Dirac field theory, 285
    Charge operator, 297
    Covariant anti-commutation relation, 303
    Field decomposition, 292
    Green's function, 300
    Normal ordered product, 305
    Quantization, 286
    Time ordered product, 305
Dirac matrices, 23, 49
    Majorana representation, 57
    Pauli-Dirac representation, 23, 57
    Properties, 49
    Weyl representation, 57
Dirac quantization, 379
    Canonical Hamiltonian, 385
    Dirac field theory, 401
    Maxwell field theory, 407
    Particle on a sphere, 390
    Primary Hamiltonian, 386
    Relativistic particle, 395
Direct sum, 137
Discrete symmetry, 415
Dispersion relation, 246
    Subtracted, 252

Effective action, 567
Ehrenfest theorem, 122
Einstein relation, 8
Electric dipole interaction, 111
Electric field, 328
    Non-Abelian, 503
Electron in magnetic field, 107
Energy eigenfunction, 11

INDEX 771

Energy eigenstates
    Physical meaning, 190
Energy-momentum operator, 218
Euclidean metric, 3
Euclidean space, 2
Euler's constant, 626
Euler-Lagrange equation, 165
External line, 319
External line factor, 320

Faddeev-Popov determinant, 518
Faddeev-Popov ghost, 522
Fermi coupling, 597
Fermi theory, 596
Fermi's Golden rule, 475
Fermi-Dirac statistics, 285
Fermion number, 299
Feynman diagram, 318
Feynman gauge, 575
Feynman Green's function, 268
Feynman parameterization, 580, 632
Feynman parametric integral, 704
Feynman propagator, 319
Feynman rule, 321
Feynman rules
    QED, 349
Feynman-Fermi gauge, 363
Field strength tensor, 329
First class constraint, 387
Fixed point, 739, 751
Fock space, 193
Foldy-Wouthuysen
    transformation, 111
Forest, 688
Four vector, 3
Four vector potential, 328
Four velocity, 395
Functional differentiation, 513

Functional integration, 514
Furry's theorem, 456, 679
Future light-cone, 7

Gauge fixing action, 521
Gauge fixing condition, 517
Gauge fixing parameter, 521
Gauge independence, 564
Gauge invariance, 330
Gauge orbit, 516
Gauge transformation, 283, 330
Gell-Mann-Low equation, 733
Generating functional, 566
    Green's function, 513
Ghost action, 522
Ghost fields, 522
Ghost number, 550, 556
Ghost scaling charge, 553
Ghost scaling symmetry, 550
Goldstone field, 278
Goldstone theorem, 270, 280, 593
Grassmann variable, 286, 381
Green's function, 194
    Advanced, 198
    Feynman, 201
    Negative energy, 205
    Positive energy, 205
    Retarded, 200
Gribov ambiguity, 511, 520
Gupta-Bleuler quantization, 373
Gyro-magnetic ratio, 110

Haar measure, 517
Handedness, 99
Heisenberg picture, 223
Helicity, 101, 308
Higgs field, 593
Higgs mechanism, 600
Higgs phenomenon, 583, 589, 593
Higgs potential, 605

Hilbert space, 13

Index of divergence, 674
Index of representation, 498
Induced representation, 154
Infrared fixed point, 755
Interaction picture, 223
Interaction vertex, 320
Internal symmetry, 263
Internal symmetry group, 488
    Generator, 488
    Structure constant, 489
Internal symmetry
    transformation, 212
Invariant length, 6
Invariant scalar product, 5

Jacobi identity, 129, 496

Kalb-Ramond field, 531
Klein paradox, 14
Klein-Gordon equation, 10
    Continuity equation, 12
    Energy eigenvalue, 11
    Negative energy, 11
    Plane wave solution, 11
Klein-Gordon field theory
    Complex, 257
    Covariant commutation
        relations, 205
    Creation and annihilation operators, 175, 183
    Energy eigenstates, 186
    Field decomposition, 171
    Free, 161
    Hamiltonian, 169, 184
    Lagrangian density, 163
    Quantization, 167, 171
    Self-interacting, 211
    Vacuum, 187

Kramers-Kronig relation, 251

Lagrange bracket, 389
Lagrange multiplier, 386
Landau gauge, 548, 575
Landau pole, 757
Large component, 106
Left derivative, 287, 289
Left-handed, 100
left-handed particle, 97
Legendre transformation, 167
Lepton number, 299
Levi-Civita tensor, 9, 50, 126, 727
Lie algebra, 488
Lie derivative, 213
Lie group, 488
    Semi-simple, 497
    Simple, 497
Light-cone gauge, 578
Light-like, 6
Little group, 154, 157, 158
Loop expansion, 621
Lorentz algebra, 55
Lorentz boost, 66
Lorentz factor, 66
Lorentz group
    Representations, 135
Lorentz transformation, 4, 65, 130
    Generator, 80
    Orthochronous, 71
    Proper, 71
    Transformation of bilinears, 82
Lorentz/Landau gauge, 517

Möller scattering, 352
Magnetic field, 328
    Non-Abelian, 503
Magnetic moment, 110
Majorana fermion, 449

Index

Massless Dirac particle, 94
Maxwell field theory, 327
    Canonical quantization, 330
    Covariant quantization, 360
    Field decomposition, 335
    Hamiltonian, 335
Maxwell's equations, 327
Microscopic causality, 208
Minimal coupling, 14, 108, 281
Minkowski metric, 4
Minkowski space, 4
MS scheme, 739

Nöther current
    Scale transformation, 761
Nöther's theorem, 211
Nambu-Goldstone boson, 589
Nambu-Goldstone
        phenomenon, 589
Nielsen ghost, 539
Nielsen identity, 571
Non-Abelian gauge theory, 485
Non-Abelian group, 130, 132
Non-minimal coupling, 111
non-relativistic, 1
Non-relativistic expansion, 112
Non-renormalizable theory, 720
Nonlinear sigma model, 390
Normal ordered product, 233
Normal ordering, 184
Notations, 2

Optical theorem, 251
Overlapping divergence, 679, 702

Pairing, 235
Parity, 415, 597
    Dirac bilinears, 434
    Dirac field, 429
    Photon field, 428

Quantum mechanics, 417
    Spin zero field, 424
Parity violation, 98
Past light-cone, 7
Path integral, 565
Pauli coupling, 111
Pauli exclusion principle, 285
Pauli matrices, 22
Pauli's fundamental theorem, 23, 75
Pauli-Dirac representation, 142
Pauli-Lubanski operator, 147
Phase transformation
    Global, 264
    Local, 283
Photon helicity, 341
Photon pair production, 350
Physical state condition, 551
Plane wave solution, 192
Poincaré algebra, 134
Poincaré group
    Massive representation, 151
    Massless representation, 155
    Unitary representation, 147
Poincaré transformation, 133
Polarization vector, 336, 368
Positronium decay, 432, 448
Primary constraint, 385
Proca equation, 584
Proca Lagrangian density, 584
Proper time, 627

QED, 639
    Fermion self-energy, 639, 657
    Photon self-energy, 642, 659
    Vertex correction, 662
Quantization of tensor field, 530
Quantum chromodynamics, 495
Quantum electrodynamics, 347

Reciprocity relation, 476
Regularization, 619
  ζ function, 666
  Cut-off, 623
  Dimensional, 647
  Higher derivative, 666
  Lattice, 666
  Pauli-Villars, 638
  Point splitting, 666
Relativistic equation, 1
Renormalizable theory, 720
Renormalization, 222, 669
  BPHZ, 680, 692
  Gauge theories, 721
  History, 679
  Multiplicative, 680
Renormalization group, 733, 739
Renormalization group equation, 733, 744
  Solving, 748
Renormalization part, 688
  Disjoint, 708
  Nested, 708
  Overlapping, 708
Retarded function, 249
Right-handed, 100
right-handed particle, 97
Rotation, 65, 125

Scalar product, 3
Scale dimension, 760
Scale transformation, 759
  Ward identity, 762
Scattering matrix, 230
Schrödinger equation, 1
Schrödinger picture, 224
Schwinger function, 201
Schwinger model, 726
Schwinger-Dyson equation, 690

Second class constraint, 387
Second quantization, 171
Secondary constraint, 387
Self-energy, 691
Semi-detailed balance, 477
Semi-direct sum, 135
Similarity transformation, 140
Skeleton graph, 677
Slash notation, 28
Slavnov-Taylor identity, 565
Small component, 106, 112
Space-like, 6
Space-time symmetry transformation, 212
Space-time translation, 215
Spectral function, 248
Spectral representation, 246
Spontaneous symmetry breaking, 270
Stückelberg formalism, 583
Standard model, 308, 583, 596
  Electric charge, 614
  Field content, 599
  Higgs field, 600
  Lagrangian density, 601
  Spontaneous symmetry breaking, 605
  Yukawa interaction, 604
Step function, 242
Stress tensor, 217
Super-renormalizable theory, 721
Superficial degree of divergence, 669
Surface term, 165
Symmetry algebra, 125

T-matrix, 233
Tachyon, 273
Taylor expansion operator, 710

Temporal gauge, 332, 517, 578
Time evolution operator, 223, 226
Time ordered product, 241
Time ordering, 228
Time reversal, 458
    Chirality, 473
    Consequences, 473
    Dirac bilinears, 471
    Dirac field, 467
    Electric dipole moment, 477
    Helicity, 473
    spin zero field, 464
Time-like, 6
Timereversal
    Photon field, 464
Translation, 129
Transverse delta function, 334
Transverse projection operator, 362

Ultrarelativistic expansion, 117
Ultrarelativistic limit, 116
Ultraviolet fixed point, 755
Unitarity, 561
Unitary gauge, 586
Universal covering group, 138
Universality of charge, 687
Unrenormalized field, 684
Unrenormalized parameter, 684

Vacuum expectation value, 188, 591
Velocity operator, 118
Vertex function, 567

Ward-Takahashi identity, 355, 565, 666
Weak current, 614
Weak equality, 385
Weak hypercharge, 597

Weak isospin group, 597
Weak mixing angle, 609
Weinberg angle, 609, 615
Weinberg's theorem, 690
Weinberg-Salam-Glashow theory, 596
Weyl equation, 95
Weyl field, 308
Weyl representation, 142
Wick's theorem, 233, 241

Yang-Mills theory, 485
    Canonical quantization, 502
    Covariant derivative, 490
        Adjoint representation, 501
    Feynman rules, 572
    Field strength tensor, 494
    Ghost free gauge, 578
    Path integral quantization, 512
Yukawa coupling, 604
Yukawa interaction, 312
Yukawa potential, 313, 325
Yukawa theory, 631
    Fermion self-energy, 633
    scalar self-energy, 635

Zitterbewegung, 117